PROCEEDINGS
of the 10th ANNIVERSARY
HTS WORKSHOP
on Physics, Materials
and Applications

Effort sponsored by the Air Force Office of Scientific Research, Air Force Material Command, USAF, under grant number F49620-96-1-0048. The U.S. Government is authorized to reproduce and distribute reprints for Governmental purposes notwithstanding any copyright notation thereon.

The views and conclusions contained herein are those of the authors and should not be interpreted as necessarily representing the official policies or endorsements, either expressed or implied, of the Air Force Office of Scientific Research or the U.S. Government.

This material is based upon work supported by the National Science Foundation under Grant No. DMR-9622660.

Any opinions, findings, and conclusions or recommendations expressed in this material are those of the author(s) and do not necessarily reflect the views of the National Science Foundation.

This material is based on work partially supported by the State of Texas through the Texas Center for Superconductivity at the University of Houston.

PROCEEDINGS of the 10th ANNIVERSARY HTS WORKSHOP
on Physics, Materials and Applications

Edited by

B. Batlogg
Lucent Technologies, USA

C. W. Chu
Texas Center for Superconductivity
University of Houston, USA

W. K. Chu
Texas Center for Superconductivity
University of Houston, USA

D. U. Gubser
Naval Research Laboratory, USA

K. A. Müller
IBM Zürich, Switzerland

March 12-16, 1996
Doubletree Hotel at Allen Center
Houston, Texas, USA

World Scientific
Singapore • New Jersey • London • Hong Kong

Published by
World Scientific Publishing Co. Pte. Ltd.
P O Box 128, Farrer Road, Singapore 912805
USA office: Suite 1B, 1060 Main Street, River Edge, NJ 07661
UK office: 57 Shelton Street, Covent Garden, London WC2H 9HE

British Library Cataloguing-in-Publication Data
A catalogue record for this book is available from the British Library.

Proceedings of the 10th Anniversary HTS Workshop on Physics, Materials and Applications

Copyright © 1996 by World Scientific Publishing Co. Pte. Ltd.

All rights reserved. This book, or parts thereof, may not be reproduced in any form or by any means, electronic or mechanical, including photocopying, recording or any information storage and retrieval system now known or to be invented, without written permission from the Publisher.

For photocopying of material in this volume, please pay a copying fee through the Copyright Clearance Center, Inc., 222 Rosewood Drive, Danvers, MA 01923, USA. In this case permission to photocopy is not required from the publisher.

ISBN 981-02-2715-9

Printed in Singapore.

10th ANNIVERSARY HTS WORKSHOP on PHYSICS, MATERIALS and APPLICATIONS

Co-Chairs: C. W. Chu & K. A. Müller

Organizing Committees

International Advisory Committee

P. W. Anderson (Princeton U), J. G. Bednorz (IBM Zürich), P. Chaudhari (IBM Yorktown Heights), M. L. Cohen (UC Berkeley), R. C. Dynes (UC San Diego), T. H. Geballe (Stanford U), K. Kitazawa (U Tokyo), M. V. Klein (U Illinois, Urbana-Champaign), B. Raveau (U Caen), J. R. Schrieffer (NHMFL, Florida State U), S. Tanaka (ISTEC), M.-K. Wu (National Tsing Hua U), Z. X. Zhao (Chinese Academy of Sciences)

Program Committee

Co-Chairs: B. Batlogg (Lucent Technologies), W. K. Chu (TCSUH), D. U. Gubser (NRL)

U. Balachandran (Argonne), B. Batlogg (Lucent Technologies), M. Cardona (Max Planck Inst.), W. K. Chu (TCSUH), T. Claeson (Chalmers U of Technology), G. W. Crabtree (Argonne), J. Daley (DOE), F. de la Cruz (Centro Atómico Bariloche), G. Deutscher (Tel Aviv U), Ø. Fischer (U Geneva), H. Freyhardt (U Göttingen), H. Fukuyama (U Tokyo), J. B. Goodenough (UT Austin), P. M. Grant (EPRI), R. L. Greene (U Maryland), D. U. Gubser (NRL), R. A. Hawsey (Oak Ridge), H. C. Ku (National Tsing Hua U), D. C. Larbalestier (U Wisconsin, Madison), A. Lauder (DuPont), W.-Y. Liang (U Cambridge), K. K. Likharev (SUNY Stony Brook), A. P. Malozemoff (ASC), D. E. Peterson (Los Alamos), R. Ralston (MIT Lincoln Lab), C. N. R. Rao (Indian Institute of Science), T. M. Rice (ETH Hönggerberg), J. M. Rowell, T. R. Schneider (EPRI), D. T. Shaw (SUNY Buffalo), R. Sokolowski (IGC), C. S. Ting (TCSUH), M. Tachiki (Tohoku U, Sendai), H. Weinstock (AFOSR), H. H. Wickman (NSF), S. A. Wolf (NRL)

Local Committee

Co-Chairs: A. L. de Lozanne (UT Austin), D. G. Naugle (Texas A&M), K. Salama (TCSUH)
Workshop Secretary: S. W. Butler

W. K. Chu (TCSUH), A. L. de Lozanne (UT Austin), P.-H. Hor (TCSUH), A. Ignatiev (TCSUH), A. J. Jacobson (TCSUH), D. G. Naugle (Texas A&M U), K. Salama (TCSUH), C. S. Ting (TCSUH), J. C. Wolfe (TCSUH), Q. Xiong (U Arkansas, Fayetteville)

Workshop Sponsors

Air Force Office of Scientific Research

Apple Computers

Department of Energy

Electric Power Research Institute

Global Services

Houston Chronicle

National Science Foundation

Texas Center for Superconductivity
at the University of Houston

University of Houston

Workshop Exhibits

Adtech Nepth, Inc.

Argonne National Laboratory

American Superconductor Corporation

Department of Energy

Intermagnetics General Corporation

Naval Research Laboratory

Oxford Instruments

Pirelli

Superconductive Components, Inc.

TCSUH High Pressure & Low
Temperature Physics Laboratory

TCSUH HTS Manufacturing Division

TCSUH Levitation Laboratory

PREFACE

The *10th Anniversary HTS Workshop on Physics, Materials and Applications* held in Houston, Texas, USA, March 12–16, 1996, was an outgrowth of a conversation between Alex Müller and myself in late April 1995 at the annual meeting of the National Academy of Sciences in Washington, D.C. We felt that 1996 would be an appropriate time to commemorate the 10th anniversary of the discovery of high temperature superconductivity (HTSy),* to review the present status of the field of HTSy, and to assess the future of HTSy science and technology. The ensuing enthusiastic response from colleagues in the field and the generous support from sponsors made the 1996 Houston Workshop a success. Four-hundred participants from 27 countries contributed to the discussions of results and exchanges of ideas.

The seminal discovery of the first cuprate high temperature superconductor (HTSr) by Georg Bednorz and Alex Müller in 1986 heralded the era of HTSy. Many of the participants in this workshop were also attendees of the special session of the March 1987 Meeting of the American Physical Society in New York City, which was proposed by me and dubbed "The Woodstock of Physics" by the late Mike Schluter, to witness the unfolding of one of the most exciting developments in modern physics on March 16, 1987. The presence of the original five panel speakers from the 1987 "Woodstock of Physics" at the commemoration of the 10th anniversary of HTSy in Houston gave new meaning to the expression, "old soldiers never die."

Over the last 10 years, unprecedentedly rapid progress has been made in all areas of HTSy science and technology, from materials through physics to applications. For instance, more than 75 cuprate HTSrs were discovered with transition temperatures up to 134 K in $HgBa_2Ca_2Cu_3O_{8+\delta}$ under ambient pressure and to 164 K under high pressure, making it possible to achieve the superconducting state by using passive cooling in space or everyday air-conditioner technology. Many anomalous normal state properties have been detected in these HTSrs, providing important insights in unraveling of the mystery of HTSy. Numerous theoretical models have subsequently been proposed to account for the observations. Various techniques have also been developed to process the high temperature superconducting (HTSg) materials into usable forms, e.g. wire and film. Numerous HTSg prototype devices have been fabricated and tested successfully, and some of them are already being considered for system integration.

HTSy technology can touch every aspect of our life where electricity is used. The technology of HTSy has therefore been predicted to be a critical technology which will have profound impacts on the present and future generations. History has shown repeatedly that it takes time and effort to bridge the gap between discoveries and applications, and that the understanding of the science of discoveries expedites the process. I believe that vision and determination of HTSy scientists coupled with patience and support from the public will bring us to the HTSy wonderland in the not too distant future.

---

* To avoid confusion between high temperature superconductivity, high temperature superconducting, and high temperature superconductors, which have all been abbreviated HTS in the past, I would like to propose the adoption of HTSy, HTSg, and HTSrs, respectively, in the future, as I have used in this Preface.

The Proceedings of this wonderful Workshop stand as a testimony to the impressive achievements made by all of us in the field in the last 10 years. They also pave the way for an optimistic decade to come in HTSy science and technology.

Thanks are due to the Honorable Tom DeLay, Majority Whip and Member of Congress, District 22, for welcoming the participants and sharing his commitment to a strong scientific and technological future; and to Dr. Neal Lane, Director of the National Science Foundation, who urged us all to take a proactive role in making our science accessible to the public; and to my colleague, Richard Smalley, who, in his banquet speech, showed all present that, however different our fields might be, we all share a strong commitment to the present and future strength of materials research.

I would also like to offer my thanks to my co-chair, Prof. Dr. Alex Müller, for sharing his support, insight, and vision for this project; to the members of the International Advisory Committee for making valuable suggestions of topics and speakers; to the Program Committee Chairs, Bertram Batlogg, Wei-Kan Chu, and Donald U. Gubser, for planning a coherent program; to the Local Committee chairs, Alex L. de Lozanne, Donald G. Naugle, and Kamel Salama, for attending to the details of the program; and to the Workshop Secretary, Susan W. Butler, and the Workshop Finance Coordinator, Don Waterman, and their staffs for their tireless and expert efforts to sail through the Workshop smoothly. Particularly outstanding among them is the contribution of Troy Christensen, who designed this Proceedings and was responsible for the collection and indexing of its papers.

Thanks are also due to the Air Force Office of Scientific Research, the Department of Energy, the Electric Power Research Institute, the National Science Foundation, the Texas Center for Superconductivity at the University of Houston, Apple Computers, Global Services, the Houston Chronicle, and to the University of Houston for their generous sponsorship of this event.

C. W. Chu
Houston, Texas
October 1996

CONTENTS

Workshop Organizing Committees   v
Workshop Sponsors and Exhibits   vi
Preface   vii

OVERVIEWS

The Development of Superconductivity Research in Oxides   3
K. Alex Müller

Superconductivity Above 90 K and Beyond   17
C. W. Chu

Superconductors and Semiconductors   32
Shoji Tanaka

Critical Current Limiting Mechanisms in Polycrystalline High Temperature Superconductor, Conductor-Like Forms   41
David C. Larbalestier

Bulk Applications of HTS Wire   47
A. P. Malozemoff

Superconducting Electronics — The Next Decade   52
John M. Rowell

Applications of HTS to Electronics — A DoD Perspective   58
Stuart A. Wolf and Francis W. Patten

The Symmetry of the Order Parameter in the Cuprate Superconductors   63
J. F. Annett, N. D. Goldenfeld and A. J. Leggett

Correlations Between T_c and n_s/m^* in High-T_c Cuprates   68
Y. J. Uemura

Raman Spectroscopy: Phonons   72
Manuel Cardona, Xingjiang Zhou and Thomas Strach

I. MATERIALS

HBCCO

Fabrication of $HgBa_2Ca_2Cu_3O_{8-\delta}$ Tape   81
R. L. Meng, B. R. Hickey, Y. Q. Wang, L. Gao, Y. Y. Sun, Y. Y. Xue and C. W. Chu

Strong Flux Pinning, Anisotropy and Microstructure of $(Hg,Re)M_2Ca_{n-1}Cu_nO_y$ (M = Ba,Sr)   85
Jun-ichi Shimoyama, Koichi Kitazawa and Kohji Kishio

Fabrication of High Quality Hg-1212 and Hg-1223 Thin Films 89
J. Z. Wu, S. H. Yun, W. N. Kang, B. W. Kang, A. Gapud, S. C. Tidrow and D. Eckart

Phase Stability and Defects of $HgBa_2Ca_2Cu_3O_{8+\delta}$ 93
Y. Y. Xue, R. L. Meng, Q. M. Lin, B. Hickey, Y. Y. Sun and C. W. Chu

Single Crystals of $HgBa_2Ca_{n-1}Cu_nO_{2n+2+\delta}$ Compounds: Growth at 10 kbar Gas Pressure and Properties 97
J. Karpinski, H. Schwer, R. Molinski, K. Conder, G. I. Meijer, A. Wisniewski, R. Szymczak, M. Baran and Ch. Rossel

Synthesis Conditions and T_c of $Hg_{1-x}Ba_2Ca_{n-1}Cu_nO_{2n+2+\delta}$ with $n \leq 6$ 99
Q. M. Lin, Y. Y. Sun, Y. Y. Xue and C. W. Chu

New HTS and Related Compounds

Atomic-Layer Synthesis of Artificial High-T_c Superconductors and Their Use in Tunnel Junctions 103
Ivan Bozovic and J. N. Eckstein

Properties of Li-Doped La_2CuO_4 107
J. L. Sarrao, D. P. Young, *Z. Fisk*, P. C. Hammel, Y. Yoshinari and J. D. Thompson

Enhancement of T_c for $(Cu_{0.5}C_{0.5})Ba_2Ca_2Cu_3O_{8.5+\delta}$ from 67 to 120 K by Heat Treatments 113
Catherine Chaillout and *Massimo Marezio*

Generation of High T_c Superconducting Oxycarbonates: From the Ceramics to the Thin Films 117
B. Raveau, C. Michel and M. Hervieu

Superconducting Phases in the Sr-Cu-O System 123
Z. L. Du, Y. Cao, Y. Y. Sun, L. Gao, Y. Y. Xue and C. W. Chu

Effect of Pressure on La_2O_3-CuO Subsolidus Phase Equilibria 125
Joel Geny, James K. Meen and Don Elthon

Anomalous Pr Ordering and Filamentary Superconductivity for the Pr2212 Cuprates 127
H. C. Ku, Y. Y. Hsu, S. R. Lin, D. Y. Hsu, Y. M. Wan, V. B. You and S. R. Sheen

Superconductivity in $Sr_2YRu_{1-x}Cu_xO_6$ System 129
D. C. Ling, S. R. Sheen, C. Y. Tai, J. L. Tseng, M. K. Wu, T. Y. Chen and F. Z. Chen

Melting Relations in Bi_2O_3-CuO-Cu_2O 131
James K. Meen and Don Elthon

Growth of Epitaxial $LaAlO_3$ and CeO_2 Films Using Sol-Gel Precursors 134
Shara S. Shoup, Mariappan Paranthaman, David B. Beach and E. D. Specht

RBCO

Strong Flux Pinning in RE123 Grown by Oxygen-Controlled Melt Process 139
N. Chikumoto and M. Murakami

Biaxially Textured YBaCuO Thick Films on Technical Substrates — 143
H. C. Freyhardt, J. Dzick, K. Heinemann, J. Hoffmann, J. Wiesmann, F. Garcia-Moreno, S. Sievers and A. Usoskin

Thick Film YBCO Growth by Photo-Assisted MOCVD: J_c Enhancement by Ion Irradiation — 147
Alex Ignatiev, Qun Zhong, Pen Chu Chou, Xin Zhang, Jia Rui Liu and Wei Kan Chu

Anisotropy as the Determining Factor of Irreversibility Field and its Control by Chemical Doping — 151
K. Kitazawa, J. Shimoyama, H. Ikuta and K. Kishio

A Novel Approach to High Rate Melt-Texturing in 123 Superconductors — 155
K. Salama, A. S. Parikh, P. Putman and L. Woolf

New Process to Control Critical Currents of $Nd_{1+x}Ba_{2-x}Cu_3O_{7-\delta}$ — 159
Y. Shiohara, M. Nakamura, Y. Yamada, T. Hirayama and Y. Ikuhara

Improvement of Flux Pinning in High Temperature Superconductors by Artificial Defects — 163
Harald W. Weber

Fabrication of Highly Textured Superconducting Thin Films on Polycrystalline Substrates Using Ion Beam Assisted Deposition — 167
Q. Xiong, S. Afonso, F. T. Chan, K. Y. Chen, G. J. Salamo, G. Florence, J. Cooksey, S. Scott, S. Ang, W. D. Brown and L. Schaper

Superconductivity and Structural Aspects of $Y_{1-x}Ca_xBa_2Cu_3O_{7-\delta}$ with Variable Oxygen Content — 171
V. P. S. Awana, *J. Albino Aguiar*, S. K. Malik, W. B. Yelon and A. V. Narlikar

Fractal Grain Boundaries in Composite $YBa_2Cu_3O_{7-\delta}/Y_2O_3$ Resulting from a Competition Between Growing Grains: Experiments and Simulations — 173
N. Vandewalle, *M. Ausloos*, N. Mineur and R. Cloots

Twin Structures in Large Grains of $YBa_2Cu_3O_{7-\delta}$ as Affected by the Dispersion and Volume Fractions of Y_2BaCuO_5 — 175
Manoj Chopra, *Siu-Wai Chan*, V. S. Boyko, R. L. Meng and C. W. Chu

Crystal Engineering of Chemically Stabilized, Cation Substituted $YBa_2Cu_3O_{7-\delta}$ Structures — 177
J. T. McDevitt, J. P. Zhou, C. E. Jones and J. Talvacchio

TEM Study of Low-Angle Grain Boundaries in Polycrystalline YBCO — 179
M. Mironova, G. Du, S. Sathyamurthy, I. Rusakova and K. Salama

Molecular Level Control of the Surface Properties of High-T_c Superconductor Materials — 181
J. T. McDevitt, R.-K. Lo, *J. E. Ritchie*, J. Zhao, C. A. Mirkin, F. Xu and K. Chen

Microstructural Changes of YBCO Induced by Lanthanide Doping — 183
I. Rusakova, R. L. Meng, P. Gautier-Picard and C. W. Chu

Control of 211 Particle Distribution and J_c Property in Melt-Textured YBCO Superconductor — 185
Y. Shiohara and A. Endo

BSCCO and TBCCO

Processing and Properties of Ag-Clad BSCCO Superconductors   **189**
R. Jammy, A. N. Iyer, M. Chudzik, U. Balachandran and P. Haldar

Local Texture, Current Flow, and Superconductive Transport Properties of Tl1223 Deposits
on Practical Substrates   **193**
D. K. Christen, E. D. Specht, A. Goyal, Q. He, M. Paranthaman, C. E. Klabunde, R. Feenstra,
F. A. List, D. M. Kroeger, J. E. Tkaczyk, J. A. Deluca, Z. F. Ren, C. A. Wang and J. H. Wang

Flux Pinning by Ti Doping in $Bi_2Sr_2Ca_1Cu_2O_{8+x}$ Single Crystals   **197**
T. W. Li, P. H. Kes and H. W. Zandbergen

Growth of Superconducting Epitaxial $Tl_2Ba_2CuO_{6+\delta}$ Thin Films with Tetragonal Lattice
and Continuously Adjustable Critical Temperature   **199**
Z. F. Ren, C. A. Wang and J. H. Wang

Role of Constituents on Composite BPSCCO Tapes   **201**
S. Salib, M. Mironova and C. Vipulanandan

Transport Critical Currents of Bi-2212 Tapes Prepared by Sequential Electrolytic Deposition   **203**
F. Legendre, L. Schmirgeld-Mignot, P. Regnier, H. Safar, J. Y. Coulter and M. Maley

Monolithic Terminations for Multifilamentary BSCCO Wires   **205**
Y. K. Tao, C. H. Kao and M. K. Wu

Partial-Melt Processing of Bulk MgO-Whisker Reinforced $(Bi,Pb)_2Sr_2Ca_2Cu_3O_{10-x}$ Superconductor   **207**
J. S. Schön and S. S. Wang (presented by M. S. Wong)

II. EXPERIMENTS

Symmetry

Josephson Tunneling Between a Conventional Superconductor and $YBa_2Cu_3O_{7-\delta}$ **213**
R. C. Dynes, A. G. Sun, A. S. Katz, A. Truscott, S. I. Woods, D. A. Gajewski, S. H. Han, M. B. Maple,
B. W. Veal, C. Gu, J. Clarke, R. Kleiner, R. Summer, E. Dantsker, B. Chen and K. Char

Determination of the Pairing State of High-T_c Superconductors Through Measurements
of the Transverse Magnetic Moment   **219**
J. Buan, Branko P. Stojkovic, A. Bhattacharya, I. Zutic, Nathan Israeloff, A. M. Goldman,
D. Grupp, C. C. Huang and Oriol T. Valls

Microwave Measurements of the Penetration Depth in High T_c Single Crystals   **223**
W. N. Hardy, S. Kamal, Ruixing Liang, D. A. Bonn, C. C. Homes, D. N. Basov and T. Timusk

Specific Heat of High Temperature Superconductors in High Magnetic Fields   **228**
A. Junod, M. Roulin, B. Revaz, J.-Y. Genoud, G. Triscone and A. Mirmelstein

Magnetic Flux Distribution in Grain and Twin Boundaries in YBCO   **232**
K. A. Moler, J. R. Kirtley, J. Mannhart, H. Hilgenkamp, B. Mayer, Ch. Gerber, Ruixing Liang,
Douglas A. Bonn and Walter N. Hardy

Single Grain Boundary Josephson Junctions – New Insights from Basic Experiments 236
J. Mannhart, H. Hilgenkamp, B. Mayer, Ch. Gerber, J. Kirtley, K. A. Moler and M. Sigrist

Anomalous Energy Gap in the Normal and Superconducting State of Underdoped
$Bi_2Sr_2Ca_{1-x}Dy_xCu_2O_{8+\delta}$ Thin Film and Single Crystals 240
Z.-X. Shen, J. M. Harris and A. G. Loeser

Intrinsic SIS Josephson Junction in Bi-2212 and Symmetry of Cooper Pair 244
K. Tanabe, Y. Hidaka, S. Karimoto and M. Suzuki

Half-Integer Flux Quantum Effect and Pairing Symmetry in Cuprate Superconductors 248
C. C. Tsuei, J. R. Kirtley, J. Z. Sun, A. Gupta, Z. F. Ren, J. H. Wang, K. A. Moler and C. C. Chi

Paramagnetic Meissner Effect from Spin Polarization 252
A. Kallio, M. Rytivaara and K. Honkala

Use of Tricrystal Microbridges to Probe the Pairing State Symmetries of Cuprate Superconductors 254
J. Lin, J. H. Miller, Jr., Z. G. Zou and Q. Xiong

Transport

Normal-State Resistivity of Superconducting $La_{2-x}Sr_xCuO_4$ in the Zero-Temperature Limit 259
Yoichi Ando, G. S. Boebinger, A. Passner, R. J. Cava, T. Kimura, J. Shimoyama and K. Kishio

High Pressure Study of the T_c and Thermopower of Hg-Based Cuprates 263
F. Chen, X. D. Qiu, Y. Cao, L. Gao, Y. Y. Xue, C. W. Chu and Q. Xiong

Puzzling Behavior of the Electrothermal Conductivity of High-T_c Superconductors 267
J. A. Clayhold, Y. Y. Xue, C. W. Chu, J. N. Eckstein and I. Bozovic

Normal State Transport Properties of Tl-2201 Single Crystals 271
A. M. Hermann, W. Kiehl and R. Tello

Pressure Effect on the Superconducting and Normal-State Properties for the YBCO/PBCO Superlattice 276
J. G. Lin, M. L. Lin, H. C. Yang, C. Y. Huang and Z. J. Huang

Anomalous Resistivity in the Mixed State of $CeRu_2$ 280
J. Herrmann, N. R. Dilley, S. H. Han and *M. B. Maple*

Temperature Dependence of the Upper Critical Magnetic Field in BiSrCuO and NdCeCuO 284
M. S. Osofsky, R. J. Soulen, Jr., S. A. Wolf, J. M. Broto, H. Rakato, J. C. Ousset, G. Coffe, S. Askenazy, P. Oswald, P. Pari, I. Bozovic, J. N. Eckstein, G. F. Virshup, J. Cohn, X. Jiang, S. Mao and R. Greene

Nonlinear Microwave Switching Response of BSCCO Single Crystals 288
T. Jacobs, Balam A. Willemsen, *S. Sridhar*, Qiang Li, G. D. Gu and N. Koshizuka

Thermoelectric Power of High-T_c Cuprates 292
J. L. Tallon

Scattering Time: A Unique Property of High-T_c Cuprates 296
I. Terasaki, Y. Sato and S. Tajima

Spin Gap Effects on the Charge Dynamics of High T_c Superconductors — 300
S. Uchida, K. Takenaka and K. Tamasaku

Thermal Conductivity of High-Temperature Superconductors in a Magnetic Field — 304
Ctirad Uher

Thermopower of the Cuprates Under High Pressure — 310
J.-S. Zhou and J. B. Goodenough

Behavior of ScN and ScS Contacts Under Microwave Irradiation — 314
A. B. Agafonov, D. A. Dikin, V. M. Dmitriev and A. L. Solovjov

Temperature Dependence of Radiofrequency Absorption and Critical Current Density in High-T_c YBaCuO Thin Samples — 316
N. T. Cherpak, G. V. Golubnichaya, E. V. Izhyk, A. Ya. Kirichenko, I. G. Maximchuk and A. V. Velichko (presented by *A. B. Agafonov*)

Normal State Magnetoresistance and Hall Effect in LaSrCuO Thin Films — 318
E. F. Balakirev, I. E. Trofimov, S. Guha and P. Lindenfeld

Ground State of Superconducting $La_{2-x}Sr_xCuO_4$ in 61-Tesla Magnetic Fields — 320
G. S. Boebinger, Yoichi Ando, A. Passner, K. Tamasaku, N. Ichikawa, S. Uchida, M. Okuya, T. Kimura, J. Shimoyama and K. Kishio

The Effect of Sr Impurity Disorder on the Magnetic and Transport Properties of $La_{2-x}Sr_xCuO_4$, $0.02 \leq x \leq 0.05$ — 322
R. J. Gooding, N. M. Salem, R. J. Birgeneau and F. C. Chou

The Normal State Transport Properties in Pure, Pr- and Ca-Doped $RBa_2Cu_3O_{7-\delta}$ Systems — 324
Weiyan Guan

Anisotropy of Thermal Conductivity of YBCO and Selectively Doped YBCO Single Crystals — 326
P. F. Henning, G. Cao and J. E. Crow

Thermal Conductivity of High-T_c Superconductors — 328
M. Houssa and M. Ausloos

Angular Dependence of the c-Axis Normal State Magnetoresistence in $Tl_2Ba_2CuO_6$ — 330
N. E. Hussey, J. R. Cooper, I. R. Fisher, A. P. Mackenzie and J. M. Wheatley

Thermoelectric Power of Superconducting Alloys YNi_2B_2C and $LuNi_2B_2C$ — 332
J. H. Lee, Y. S. Ha, Y. S. Song, *Y. W. Park* and Y. S. Choi

Anomalous Dielectric Effect of $La_{2-x}Sr_xCuO_4$ Film at $x = 1/4^n$ — 335
M. Sugahara, X.-Y. Han, H.-F. Lu, S.-B. Wu, N. Haneji, H. Kaneda and N. Yoshikawa

Investigation of the Microwave Power Handling Capabilities of High-T_c Superconducting Thin Films — 337
J. Wosik, L. M. Xie, D. Li, P. Gierlowski, J. H. Miller, Jr. and S. A. Long

Spectroscopy

Thermodynamic Evidence on the Superconducting and Normal State Energy Gaps in $La_{2-x}Sr_xCuO_4$ — 341
J. W. Loram, K. A. Mirza, J. R. Cooper, N. Athanassopoulou and W. Y. Liang

Neutron Scattering Measurements on $YBa_2Cu_3O_{7-\delta}$ — 345
H. A. Mook, P. Dai, F. Dogan, K. Salama, G. Aeppli and M. E. Mostoller

High-Accuracy Specific-Heat Study on $YBa_2Cu_3O_7$ and $Bi_2Sr_2CaCu_2O_{8.2}$ Around T_c in External Magnetic Fields — 349
A. Schilling, O. Jeandupeux, C. Waelti, H. R. Ott and A. van Otterlo

Phonon Anomaly in High Temperature Superconducting $YBa_2Cu_3O_{7-\delta}$ Crystals — 353
R. P. Sharma, Z. Zhang, J. R. Liu, R. Chu, T. Venkatesan and W. K. Chu

Electrodynamical Properties of High T_c Superconductors Studied with Polarized Angle Resolved Infrared Spectroscopy — 357
D. van der Marel, J. Schützmann, H. S. Somal and J. W. van der Eb

High-Pressure Raman Study of the Mercury-Based Superconductors and the Related Compounds — 361
In-Sang Yang, Hye-Gyong Lee, Nam H. Hur and Sung-Ik Lee

Ion Channeling Studies in YBCO Thin Film at Low Temperature — 365
Xingtian Cui, Z. Zhang, Quark Chen, J. R. Liu and W. K. Chu

Electronic Raman Scattering of $YBa_2Cu_4O_8$ at High Pressure — 367
T. Zhou, K. Syassen, M. Cardona, J. Karpinski and E. Kaldis

Raman Study of $HgBa_2Ca_{n-1}Cu_nO_{2n+2+\delta}$ (n = 1, 2, 3, 4 and 5) Superconductors — 369
Xingjiang Zhou, M. Cardona, C. W. Chu, Q. M. Lin, S. M. Loureiro and M. Marezio

Magnetic

Magnetic and Structural Properties and Phase Diagrams of $Sr_2CuO_2Cl_2$ and Lightly-Doped $La_{2-x}Sr_xCuO_{4+\delta}$ — 373
D. C. Johnston, F. Borsa, J. H. Cho, L. L. Miller, B. J. Suh, D. R. Torgeson, P. Carretta, M. Corti, A. Lascialfari, R. J. Gooding, N. M. Salem, K. J. E. Vos and F. C. Chou

Anisotropy Properties of High-T_c Microcrystals by a Miniaturized Torquemeter — 377
C. Rossel, P. Bauer, D. Zech, J. Hofer, M. Willemin, H. Keller and J. Karpinski

Electronic Structure and Magnetic Properties of $RENi_2B_2C$ (RE = Pr, Nd, Sm, Gd, Tb, Dy, Ho, Tm, Er) — 381
Z. Zeng, Diana Guenzburger, *E. M. Baggio-Saitovitch* and D. E. Ellis

The Magnetic Properties of the Quarternary Intermetallics RNiBC (R = Er, Ho, Tb, Gd, Y) — 383
Julio C. T. Mondragón, *E. M. Baggio-Saitovitch* and M. El Massalami

Mössbauer Studies on Oxyanions Substituted Related Y-Ba-Cu-O System — 385
Angel Bustamente Dominguez, R. B. Scorzelli and *E. Baggio-Saitovitch*

Magnetic Properties of Some High-Temperature Superconductors — 387
C. Y. Huang, J. G. Lin, P. H. Hor, R. L. Meng and C. W. Chu

Mössbauer Studies of $RE_{1.85}Sr_{0.15}CuO_4$ T''-Phase — 389
Ada López, M. A. C. de Melo, D. Sánchez, I. Souza Azevedo, E. Baggio-Saitovitch and F. J. Litterst

Observation of a Pair-Breaking Field at the Ni Site in Non-Superconducting $RENi_2B_2C$ — 391
D. R. Sánchez, M. B. Fontes, S. L. Bud'ko and E. Baggio-Saitovitch

Flux Dynamics

Comparative Study of Vortex Correlation in Twinned and Untwinned $YBa_2Cu_3O_{7-\delta}$ Single Crystals — 395
F. de la Cruz, D. López, E. F. Righi and G. Nieva

First- and Second-Order Vortex-Lattice Phase Transitions in $Bi_2Sr_2CaCu_2O_8$ — 399
B. Khaykovich, T. W. Li, M. Konczykowski, D. Majer, E. Zeldov and P. H. Kes

On the Thickness Dependence of Irreversibility Line in $YBa_2Cu_3O_{7-\delta}$ Thin Films — 403
Pablo Menezes and J. Albino Aguiar

Magneto-Optic Imaging of Melt-Textured $YBa_2Cu_3O_{6+x}$ Bicrystals — 405
Michael B. Field, Anatoly Polyanskii, Alex Pashitski, David C. Larbalestier, Apurva Parikh and Kamel Salama

Vortex Phase Transition in $Bi_2Sr_2CaCu_2O_y$ Single Crystal — 407
H. Ikuta, S. Watauchi, J. Shimoyama, K. Kitazawa and K. Kishio

Three-Dimensional Vortex Fluctuation in $HgBa_2Ca_{0.86}Sr_{0.14}Cu_2O_{6-\delta}$ — 409
Mun-Seog Kim, Sung-Ik Lee, Seong-Cho Yu and Nam H. Hur

Discontinuous Onset of the c-Axis Vortex Correlation at the Melting Transition in $YBa_2Cu_3O_{7-\delta}$ — 411
D. López, E. F. Righi, G. Nieva, F. de la Cruz, W. K. Kwok, J. A. Fendrich, G. W. Crabtree and L. Paulius

Resistive Transitions of HTS Under Magnetic Fields: Influence of Fluctuations and Viscous Vortex Motion — 413
E. Silva, R. Marcon, R. Fastampa, M. Giura, V. Boffa and S. Sarti

Meissner Holes in Remagnetized Superconductors — 415
V. K. Vlasko-Vlasov, U. Welp, G. W. Crabtree, D. Gunter, V. I. Nikitenko, V. Kabanov and L. Paulius

Measurement of the Total Transverse Force on Moving Vortices in YBCO Films — 417
X.-M. Zhu, E. Brändström, B. Sundqvist and G. Bäckström

Occurrence

Phase Separation and Staging Behavior in $La_2CuO_{4-\delta}$ — 421
R. J. Birgeneau, F. C. Chou, Y. Endoh, M. A. Kastner, Y. S. Lee, G. Shirane, J. M. Tranquada, B. O. Wells and K. Yamada

X-Ray Search for CDW in Single Crystal $YBa_2Cu_3O_{7-\delta}$ — 425
P. Wochner, E. Isaacs, S. C. Moss, P. Zschack, J. Giapintzakis and D. M. Ginsburg

Superconducting Properties of Nb Thin Films — 429
Ana Luíza V. S. Rolim, J. C. C. de Albuquerque, E. F. da Silva, Jr., J. M. Ferreira and J. Albino Aguiar

Pressure Effects on T_c of $HgBa_2Ca_{n-1}Cu_nO_{2n+2+\delta}$ with $n \geq 4$ *Y. Cao, X. D. Qiu, Q. M. Lin, Y. Y. Xue and C. W. Chu*	431
Low Temperature Scanning Tunneling Microscopy and Spectroscopy of the CuO Chains in $YBa_2Cu_3O_{7-x}$ *D. J. Derro, T. Koyano, Hal Edwards, A. Barr, J. T. Markert and A. L. de Lozanne*	433
Zn and Ni Impurities in 123 Materials *D. Goldschmidt and Y. Eckstein*	435
Physical Properties of Infinite-Layer and T"-Phase Copper Oxides *J. T. Markert, B. C. Dunn, A. V. Elliott, K. Mochizuki and R. Tian*	437
X-Ray Single Crystal Structure Analysis of $HgBa_2Ca_{n-1}Cu_nO_{2n+2+\delta}$ Compounds *H. Schwer, J. Karpinski, K. Conder, R. Molinski, G. I. Meijer and C. Rossel*	439
In-Plane Ordering in Phase-Separated and Staged Single Crystal $La_2CuO_{4+\delta}$ *X. Xiong, P. Wochner and S. C. Moss*	441

III. THEORY

Mechanisms of Superconductivity

Antiferromagnetic Real-Space Scenario for the Cuprates *Elbio Dagotto*	447
Charge Inhomogeneity and High Temperature Superconductivity *V. J. Emery and S. A. Kivelson*	451
Superconductivity Mediated by Screened Coulomb Field *J. D. Fan and Y. M. Malozovsky*	455
Inside HT_c Superconductors: An Electronic Structure View *A. J. Freeman and D. L. Novikov*	459
Pair-Breaking and the Upper Limit of T_c in the Cuprates *Vladimir Z. Kresin, Stuart A. Wolf and Yu N. Ovchinikov*	463
What Does d-Wave Symmetry Tell Us About the Pairing Mechanism? *K. Levin, D. Z. Liu and Jiri Maly*	467
Spin Fluctuations, Magnetotransport and $d_{x^2-y^2}$ Pairing in the Cuprate Superconductors *David Pines*	471
Pairing Mechanism in the Two-Dimensional Hubbard Model *D. J. Scalapino*	477
A Possible Primary Role of the Oxygen Polarizability in High Temperature Superconductivity *M. Weger, M. Peter and L. P. Pitaevskii*	481
Absence of Exchange Scattering by Cuprate-Plane Impurities in High-Temperature Superconductors *Howard A. Blackstead and John D. Dow*	485

Hall-Effect Scaling and Chemical Equilibrium in Normal States of High-T_c Superconductors **487**
A. Kallio and K. Honkala

Pairing Instability and Anomalous Responses in an Interacting Fermi Gas **489**
Y. M. Malozovsky and J. D. Fan

Inter-Band Theory of Superconductors: Resolution of Observed s and d-Wave Tunneling with Isotropic s-Wave Pairing **491**
Jamil Tahir-Kheli

The Connections of the Experimental Results of Universal Stress Experiments and of Thermal Expansion Measurements and the Mechanisms of Microscopic Dynamics Process on CuO_2 Planes **493**
Dawei Zhou

Strongly Correlated Aspects

Quasiparticles in a Strongly Interacting Regime **497**
A. Ferraz

Hole Spectrum in the Three Band Model **501**
L. P. Gor'kov and P. Kumar

Gauge Theory of the Normal State Properties of High-T_c Cuprates **505**
Naoto Nagaosa

Novel Cuprate Materials **509**
T. M. Rice and B. Normand

Phase String, Superconductivity, and Spin Dynamics in the t-J Model **513**
Z. Y. Weng, D. N. Sheng and C. S. Ting

Spin-Susceptibility of Strong Correlated Bands in Fast Fluctuating Regime **517**
M. V. Eremin, S. G. Solovjanov, S. V. Varlamov and I. M. Eremin (presented by A. Bill)

Short-Range Antiferromagnetic Correlations and the Photoemission Spectrum **519**
N. Bulut

Magnetic Excitations in High-T_c Cuprates: The 41 meV Problem **521**
Hiroshi Kohno, Bruce Normand and Hidetoshi Fukuyama

Anomalous Charge Excitation Spectra in the t-J Model **523**
T. K. Lee, R. Eder, Y. C. Chen, H. Q. Lin, Y. Ohta and C. T. Shih

Exact Diagonalization Study of the Single-Hole t-J Model on a 32-Site Lattice **525**
P. W. Leung and R. J. Gooding

Entropy of 2D Strongly Correlated Electrons **527**
W. O. Putikka

Spin Diffusion as a Test for Spin-Charge Separation **529**
Qimiao Si

c-Axis Electronic Structure of Copper Oxides 531
J. M. Wheatley and *J. R. Cooper*

Order Parameter Symmetry

Impurity Scattering and Localization in d-Wave Superconductors 535
M. Franz, *C. Kallin* and *A. J. Berlinsky*

Aspects of d-Wave Superconductivity in High T_c Cuprates 539
K. Maki, *Y. Sun* and *H. Won*

s+d Pairing Symmetry and Vortex Structures in YBCO 543
J. H. Xu, *Y. Ren* and *C. S. Ting*

Boundary Effects and the Order Parameter Symmetry of High-T_c Superconductors 547
Safi R. Bahcall

Anisotropic Pairing Caused by Unscreened Long Range Interactions 549
V. Hizhnyakov, *A. Bill* and *E. Sigmund*

Zero-Bias Conductance Peak as a Result of Midgap Interface States – A Model Study 551
Chia-Ren Hu

Ginzburg-Landau Equations for a d-Wave Superconductor with Nonmagnetic Impurities 553
W. Xu, *Y. Ren* and *C. S. Ting*

Flux Dynamics

Pancake Vortices in High-Temperature Superconducting Thin Films 557
John R. Clem, *Maamar Benkraouda* and *Thomas Pe*

New Aspects of Vortex Dynamics 561
Anne van Otterlo, *Vadim Geshkenbein* and *Gianni Blatter*

Numerical Studies on the Vortex Motion in High-T_c Superconductors 565
Z. D. Wang

Hall Anomaly in the Mixed State of Type II Superconductors 569
P. Ao

Nonlinear Dynamics in the Mixed State of High Temperature Superconductors 571
M. W. Coffey

Vortex Dynamics in Superfluids: Cyclotron Type Motion 573
E. Demircan, *P. Ao* and *Q. Niu*

Numerical Study of Washboard Effect in High T_c Superconductors 575
Z. D. Wang and *K. M. Ho*

Time-Window Extension for Magnetic Relaxation from Magnetic Hysteresis Loop Measurements 577
Qianghua Wang, Xixian Yao, Z. D. Wang and *Jian-Xin Zhu*

Transport and Spectroscopy

Quasilocalized States as an Explanation of Some Properties of Cuprates 581
M. Cyrot

Neutron Scattering and Gap Anisotropy in High-T_c Superconductors 585
A. Bill, V. Hizhnyakov and E. Sigmund

Impurity Vertex Correction in the NMR Coherence Peak of Conventional Superconductors 587
Han-Yong Choi

Current Instabilities in Reentrant Superconductors 589
David M. Frenkel and Jeffrey A. Clayhold

IV. APPLICATIONS

Large Current

Progress and Issues in HTS Power Cables 595
A. Bolza, P. Metra and M. M. Rahman

Imaging of Vortices in Superconductors with a Magnetic Force Microscope 599
Chun-Che Chen, Qingyou Lu, Caiwen Yuan, Alex de Lozanne, James N. Eckstein and Marco Tortonese

Superconducting Magnetic Bearing and its Applications in Flywheel Kinetic Energy Storage 602
Z. Xia, K. B. Ma, R. Cooley, P. Fowler and W. K. Chu

Superconducting Homopolar Motor 606
Donald U. Gubser

High Temperature Superconducting Fault Current Limiter 610
Eddie M. Leung

Electronic Eyes Based on Dye/Superconductor Assemblies 613
J. T. McDevitt, D. C. Jurbergs, S. M. Savoy, S. J. Eames and J. Zhao

Progress of HTS Bismuth-Based Tape Applications 617
Ken-ichi Sato

Development of High T_c Superconducting Wires for Applications at 20 K 621
T. Haugan, F. Wong, J. Ye, S. Patel, D. T. Shaw and L. Motowidlo

Very High Trapped Fields: Cracking, Creep, and Pinning Centers 625
R. Weinstein, J. Liu, Y. Ren, R.-P. Sawh, D. Parks, C. Foster and V. Obot

Magnetic Levitation Transportation System by Top-Seeded Melt-Textured YBCO Superconductor 629
In-Gann Chen, Jen-Chou Hsu and Gwo Jamn

High-T_c Ceramic Superconductors for Rotating Electrical Machines: From Fabrication to Application 631
A. G. Mamalis, I. Kotsis, I. Vajda, A. Szalay and G. Pantazopoulos

Properties of Jointed BPSCCO Composites ... 633
G. Yang and C. Vipulanandan

Strain Tolerance of Superconducting Properties and Cryogenic Mechanical Behavior of Bulk MgO-Whisker-Reinforced HTS BPSCCO Composite ... 635
G. Z. Zhang, M. S. Wong and S. S. Wang

Small Current

Magnetocardiography in a Magnetically Noisy Environment Using High-T_c SQUIDs ... 639
J. H. Miller, Jr., N. Tralshawala, J. R. Claycomb, J. H. Xu and K. Nesteruk

Near-Term Commercialization of HTS Technology at Conductus ... 643
Randy W. Simon

Mutual High-Frequency Interaction of High T_c Josephson Junctions ... 647
M. Darula, G. Kunkel and S. Beuven

Directly Coupled dc-SQUIDs of YBCO Step-Edge Junctions Fabricated by a Chemical Etching Process ... 649
Junho Gohng, C. Y. Dosquet, J. P. Hong, E.-H. Lee and J.-W. Lee

Noise Reduction Techniques for Operating High-T_c SQUIDs in an Unshielded Environment ... 651
N. Tralshawala, J. R. Claycomb, J. H. Xu and J. H. Miller, Jr.

Enhanced-Resolution Magnetic Resonance Imaging Using High-T_c Superconducting rf Receiver Coils ... 653
J. Wosik, K. Nesteruk, L. M. Xie, X. P. Zhang, P. Gierlowski, C. Jiao and J. H. Miller, Jr.

Interface Roughness Effect on Differential Conductance of High-T_c Superconductor Junctions ... 655
Jian-Xin Zhu, Z. D. Wang, D. Y. Xing and Z. C. Dong

Workshop Program ... 657

Workshop Participants ... 673

Workshop Photos ... 693

Author Index ... 709

OVERVIEWS

OVERVIEWS

THE DEVELOPMENT OF SUPERCONDUCTIVITY RESEARCH IN OXIDES

K. Alex MÜLLER

IBM Research Division, Zurich Research Laboratory, 8803 Rüschlikon and University of Zurich, Physics Department, 8057 Zurich, Switzerland

Starting with the first observation of superconductivity in an oxide, namely $SrTiO_3$, the history of its development is traced. Basically, and consecutively, three kinds of oxide superconductors have been found: compounds with normal transition-metal-based conduction bands, oxides with cations exhibiting charge disproportionation, and finally the cuprates with large Coulomb on-site repulsion, U. The doped La_2CuO_4 was the first oxide discovered in this new class of materials. The discussion will then lead over to a characterization of the high-T_c materials, with regard to their main physical properties.

1 The Initial Two Decades of Superconductivity in Oxides

The first oxide in which superconductivity was found is reduced $SrTiO_3$ as reported by Schooley, Hosler and Cohen[1] in 1964. A $T_c \simeq 0.25$ K was observed with only $n = 3 \times 10^{19}$ electron carriers per cubic centimeter, i.e. at a carrier concentration three orders of magnitude below that of a normal metallic conductor, see Fig. 1. This phenomenon was clearly outside the accepted BCS

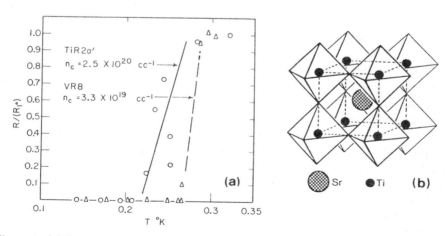

Figure 1: (a) Ratio of sample resistance at temperature T to that at 1 K, as function of temperature. From Ref. 1. (b) Perovskite lattice structure of $SrTiO_3$. From Ref. 5.

picture of electron coupling by *shielded* phonons. Only recently — 17 years later — was the underlying process understood: because the carrier concentration is so low, the plasma edge is *below* the highest optical phonon branch in $SrTiO_3$. This phonon branch is therefore unshielded and the electron-phonon coupling parameter λ sufficiently large to induce superconductivity. Upon Nb doping of $SrTiO_3$, the electron concentration and the plasma edge increases, the edge passes this phonon branch which becomes shielded, and superconductivity vanishes at $n \simeq 10^{21}$ cc^{-1} and $T_c \simeq 1.3$ K.[2] Nine months after the $SrTiO_3$ discovery, Matthias' group reported superconductivity in the sodium tungsten bronze Na_xWO_3 with $x \simeq 0.3$, $n \simeq 10^{22}$ cc^{-1} and a $T_c = 0.57$ K.[3] So it was clear that the phenomenon also occurred in another oxide, but with rather "normal" electron concentrations. Interestingly, in the Na_xWO_3 paper no reference to the over half a year older report on $SrTiO_3$ is given.

More recently the $SrTiO_3$ crystal showed another interesting property. Band calculations and experimental findings had been controversial until Mattheiss' calculations[4] yielded two conduction bands 20 meV apart at the center of the Brillouin zone ($\vec{k} = 0$), the lower one having "starfish" shape. Thus, if he was correct, by increasing the doping one expected that after the first the second band would begin to fill. Consequently, one- and two-band superconductivity in the *same* material was expected and indeed observed by tunneling with In contacts in Rüschlikon,[6] see Fig. 2. The progression of the observed two gaps is shown in Fig. 3 as a function of doping at helium temperature. One sees that the second gap appears at 32 meV, i.e. at a somewhat higher energy than the one calculated,[4] but basically supporting it. The inset depicts the evolution of the two gaps in sample 8 as function of temperature T. It is very clear from it that the two gaps have the same transition temperature T_c. This is of relevance with respect to the controversy regarding the symmetry of the superconducting wave function in the copper oxides, namely, whether it is of "s" *or* rather "d" character.[7] After having considered some of the best tunneling data, this author has recently suggested that also in certain cuprates two gaps with "s" *and* "d" character *and the same* T_c may be present.[8]

In 1965, superconductivity was also reported in TiO and NbO at temperatures of 0.68 and 1.25 K, respectively.[9] However, these results did not meet with great interest because in NbN, with the same NaCl structure, a T_c of 16 K had already been observed back in 1941.[10] After the superconducting bronzes had been found, the next substantial step forward in the oxides was the observation of a transition in the lithium titanium spinel $LiTi_2O_4$ with $T_c \simeq 11$ K by Johnston *et al.* in 1973 at San Diego,[11] Fig. 4. This represented already a quite respectable transition temperature. However, research in the spinel compound was not so intensive, probably because no single crystals became

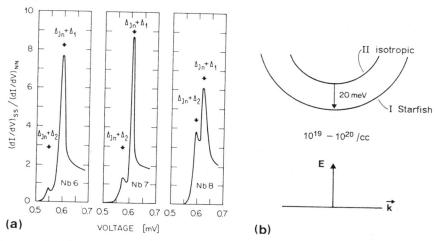

Figure 2: (a) Normalized tunneling conductance of Nb-doped $SrTiO_3$–In junctions in the range of two-band superconductivity ($\mu_F > \mu_c \simeq 32$ μeV) measured at $T = 100$ mK. Only the portion of the voltage scale near the sum of the gaps of $SrTiO_3$ and In is shown. The latter could be determined within ± 2 μeV for a given sample and had a mean value of $\Delta_{In} = 535$ μeV. From Ref. 6. (b) Schematic of the conduction band structure.

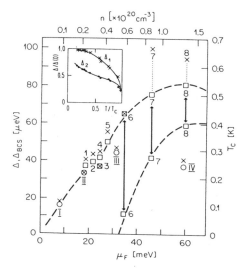

Figure 3: Superconducting order parameters Δ_1 and Δ_2 (inferred from peaks in the measured tunneling conductance of n-type $SrTiO_3$–In junctions at $T \simeq 0.2\, T_c$: open circles, reduced samples; open squares, Nb doped. The inset shows the temperature dependence of both order parameters measured on sample Nb 8. From Ref. 6.

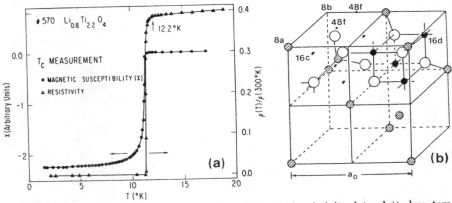

Figure 4: (a) Low-temperature magnetic susceptibility and resistivity data plotted vs. temperature for a sample prepared by sintering $Li_2Ti_2O_5$ and Ti_2O_3 at 800°C for one day. Reprinted from Ref. 11. Copyright 1973, with kind permission from Elsevier Science Ltd. (b) Illustration of the cubic spinel $Fd3m$, from Ref. 12.

available, despite the fact that a T_c of 13.7 K was reached in the mixed phase $Li_{1+x}Ti_{2-y}O_4$.[13]

Transition temperatures of comparable magnitude were reported only two years later in the perovskite $BaPb_{1-x}Bi_xO_3$ system by Sleight et al.,[14] Fig. 5. Remarkably this quaternary compound did not contain transition metals. Subsequent studies on hot-pressed ceramics,[15] see Fig. 6, and single crystals[16] allowed, in conjunction with expert band calculations,[17] the characterization of its properties. The highest T_c was near 11.2 K with $x \simeq 0.25$ and a carrier concentration of 2.4×10^{21} cc^{-1}; correlations lengths were 60–70 Å, and the Ginzburg–Landau parameter $\chi \simeq 70$–80. Therefore, according to the BCS theory, a large electron–phonon coupling was present. The transition temperature was found to be proportional to the *intensity* of a phonon band at 100 cm^{-1},[18] probably indicative of a substantial Bi(Pb)–O *anharmonicity* in σ-bonding. Thus one could expect to find still higher T_c's in other metallic oxides if the electron–phonon interactions and the carrier densities, $n(E_F)$, at the Fermi level could be further enhanced. We were not aware at the time that $n(E_F)$ can be increased by going from three- to quasi-two-dimensional lattices, owing to the possible occurrence of a Kohn anomaly at E_F.[19] More recently, by doping the charge-disproportionated insulator $BaBiO_3$ with K$^+$ on Ba^{2+} sites, an increase of T_c to near 30 K was realized,[20] Fig. 7. This is so far the oxide superconductor with the highest T_c that does not contain transition-metal ions. Its isotope effect is substantial, namely $\beta = 0.22$ to 0.4, i.e. nearly fully developed,[21] indicating that dynamic atomic motion is responsible for the

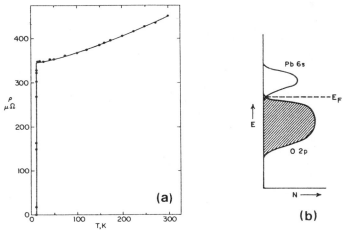

Figure 5: (a) Electrical resistivity vs. temperature for a crystal of $BaPb_{0.8}Bi_{0.2}O_3$. (b) Schematic energy level diagram for $BaPbO_3$. Reprinted from Ref. 14. Copyright 1975, with kind permission from Elsevier Science Ltd.

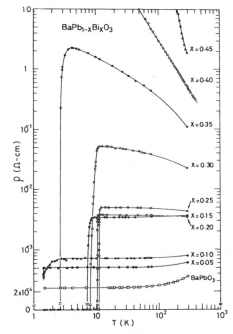

Figure 6: Temperature dependence of the electrical resistivity of $BaPb_{1-x}Bi_xO_3$ with various Bi content. From Ref. 15, © by Springer-Verlag 1980.

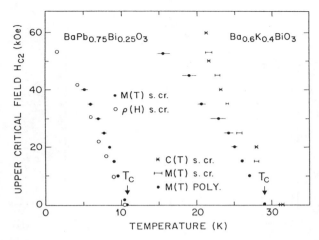

Figure 7: Upper critical field of $BaPb_{0.75}Bi_{0.25}O_3$ and $Ba_{0.6}K_{0.4}BiO_3$. From Ref. 20a. Copyright 1989 by International Business Machines Corporation; reprinted with permission.

coupling. The third class of oxide superconductors was found in 1986 with the Ba^{2+}-doped La_2CuO_4.[22]

2 The Concept and Discovery of Superconductivity in Ba–La–Cu Oxide Ceramics

Strong electron–phonon interactions can occur in oxides, owing to polaron formation as well as mixed-valence states. This can go beyond the standard BCS theory. A phase diagram with a superconducting to bipolaronic insulator transition was proposed early by Chakraverty.[23] A mechanism for polaron formation is the Jahn–Teller (JT) effect as studied by Höck et al. in a linear chain model.[24] From it, one expects heavy polaron masses if the JT stabilization energy becomes comparable to or larger than the bandwidth of the degenerate orbitals, and thus localization. Intermediate polarons are expected if the JT energy is not too large compared to the bandwidth. We recall that the JT theorem states the following: A nonlinear molecule or a defect in a crystal lattice exhibiting an electron degeneracy will spontaneously distort in lowering its symmetry, thereby removing its degeneracy. Isolated Fe^{4+}, Mn^{3+}, Ni^{3+} and Cu^{2+} in an octahedral oxygen environment show strong JT effects because their incompletely occupied e_g orbitals, transforming as $3z^2 - r^2$ and $x^2 - y^2$, point towards the negatively charged oxygen ligands[25] (Fig. 8). Although $SrFe^{4+}O_3$ is a distorted perovskite insulator, $LaNiO_3$ is a JT undistorted metal in which

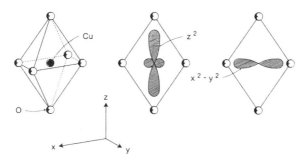

Figure 8: The partially filled $3d$ wave functions of the mixed-valence states. Positions of copper and oxygen atoms are shown at left. From Ref. 26, Copyright 1987 by the American Association for the Advancement of Science.

the transfer energy b_π of the e_g electrons of the Ni^{3+} is large enough to quench the JT distortion.[27] On the other hand, $LaCuO_3$ is a metal containing only the non-JT Cu^{3+}. Therefore, it was decided to investigate and "engineer" nickel- and copper-containing oxides, with reduced bandwidth $\simeq b_\pi$, partially containing Ni^{3+} or Cu^{2+} states. The JT polaron proposed in 1983[24] was one envisaged to lead to a heavy mass of the particle; an intermediate, mobile polaron suited for superconductivity was not considered. However, in the ferromagnetic conductor $La_{1-x}Ca_xMnO_3$, a giant oxygen isotope effect has been discovered most recently and ascribed to the presence of intermediate JT polarons due to the Mn^{3+} ions present in the oxide.[28] This important and new evidence for the existence of intermediate JT polarons is quite in favor of the original concept.[22]

In Rüschlikon, there was a tradition of more than two decades of research in insulating oxides that undergo structural and ferroelectric transitions, which was a strong motivation to pursue the program. Furthermore, in 1979 the present author had started to work in the field of granular superconductors in which small Al grains are surrounded by amorphous Al_2O_3.[29] In these systems T_c's had been reported to be as high as 5 K, compared to pure Al with a T_c of 1.1 K. In our laboratory, the search for superconductivity was initiated together with J.G. Bednorz in mid-summer of 1983. Our efforts first concentrated on Ni^{3+}-containing perovskites, such as mixed crystals of $LaNiO_3$ and $LaAlO_3$. In these unpublished efforts, the metallic behavior of the various synthesized double and triple oxides was measured, and at low temperatures they exhibited localization upon cooling. This indicated the possible existence of JT polarons, however, without any sings of superconductivity. In Fig. 9, results of these efforts are reproduced, see Ref. 30. In late summer of 1985, the efforts were shifted to copper-containing compounds, such as $LaCuO_3$. Because Cu^{3+} has

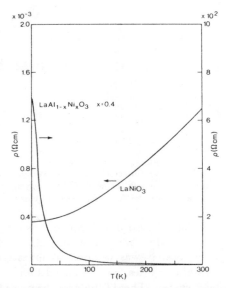

Figure 9: Temperature dependence of the resistivity for metallic $LaNiO_3$ and $LaAl_{1-x}Ni_xO_3$, where substitution of Ni^{3+} by Al^{3+} leads to insulating behavior for $x = 0.4$. From Ref. 30, © The Nobel Foundation 1988.

two electrons in the e_g subshell, the latter is half-filled. Thus, its ground state is not degenerate. It was clear that an oxide with mixed Cu^{2+}/Cu^{3+} or Cu^{3+}/Cu^{4+} valence had to be tried.

At this stage, Georg became aware of a paper by Michel, Er-Rakho and Raveau[31] on the mixed perovskite $BaLa_4Cu_5O_{13.4}$, exactly meeting the requirements of mixed valence. The French authors had shown that this mixed oxide, a metal at room temperature and above, contained Cu^{2+} and Cu^{3+}. Thus, we tried to reproduce it, at the same time continuously varying the Cu^{2+}/Cu^{3+} ratio by changing the Ba concentration in $Ba_xLa_{5-x}Cu_5O_{5(3-y)}$, and we looked for superconductivity. A representative and concise account of the discovery of superconductivity in $Ba_xLa_{5-x}Cu_5O_{5(3-y)}$ and the relevant superconducting phase present appeared in the September 4, 1987, issue of Science[26] and, in more detail, in the first of the two Nobel lectures in 1987,[30] and would exceed the scope of this contribution.

In Fig. 10 original data are shown as they were submitted for publication in April 1986:[22] Upon cooling the resistivity first decreases near linearly, then shows a minimum and increases towards localization as in the case of $LaAl_{0.6}Ni_{0.4}O_3$ in Fig. 9. Then, upon further cooling, a sharp drop in $\rho(T)$ to very low values occurs that for higher currents is partially suppressed (upper

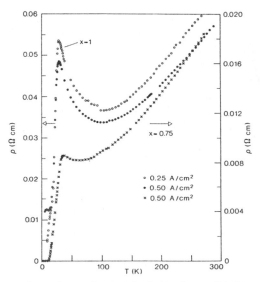

Figure 10: Temperature dependence of resistivity in $Ba_xLa_{5-x}Cu_5O_{5(3-y)}$ for samples with $x = 1$ (upper curves, left scale) and $x = 0.75$ (lower curve, right scale) (x nominal). The first two cases also show the influence of the current density. From Ref. 22, © by Springer-Verlag 1986.

curves, left scale). The maximum onset could be shifted to 35 K by sample preparation. It was interpreted as a percolative onset to superconductivity. X-ray analysis revealed that the system consisted of three phases: CuO_2, the $Ba_xLa_{5-x}Cu_5O_{5(3-y)}$ originally wanted, and a K_2NiF_4 phase containing perovskite layers. A detailed powder X-ray analysis combined with susceptibility measurements showed that the phase becoming superconducting was indeed the layer-like oxide La_2CuO_4,[32] which Michel and Raveau had characterized by X-rays earlier.[33] The La_2CuO_4 crystal contains a *checkerboard-type CuO_2 layer that so far is present in all high-T_c superconductors*. Furthermore, the doping with Ba^{2+} — or another two-valent ion — on the La site was crucial for inducing a mixed-valent *hole conductivity, found also in all other copper-oxide highest-T_c superconductors*.

The first to confirm our discovery were researchers from those laboratories that had investigated the oxide superconductor $BaPb_xBi_{1-x}O_3$ until 1986. This meant that their expertise in oxide superconductivity and equipment were still in place, and ready. The earliest confirmation of the existence of high-T_c superconductivity in the Ba–La–Cu–O came from Tanaka's group at Tokyo University.[34] Their onsets in $\rho(T)$ and diamagnetic crossovers were in the same temperature range as those for the samples investigated in Rüschlikon. Moreover, their independent structure analysis also showed good agreement.[35] The Japanese confirmation stimulated research work in the United States. Both Chu's and Batlogg's groups, as well as that a Bellcore, not only confirmed, but had by the end of the year 1986 surpassed the Rüschlikon results in two ways. At AT&T, they started directly with Sr^{2+} substitution for lanthanum.[36] Their expertise in oxide ceramics allowed AT&T to obtain sharper onsets, with full superconductivity reached a few degrees below onset, and up to 30% Meissner effect at low temperatures, thus proving the presence of three-dimensional superconductivity. Tarascon et al. at Bellcore achieved a transition width of only 2 K with T_c very near 40 K.[37] At Houston, first the Rüschlikon results were confirmed,[38] and then resistivity measurements under hydrostatic pressure revealed onsets up to 52 K.[39] Therefore, Chu et al. foresaw still higher T_c's for the future,[39] which they did indeed find,[40,41] and are reviewed in his contribution in these proceedings. At the Academy of Science in China, a long tradition of research in oxide ceramics exists. The scientists there also optimized the barium–strontium replacement of lanthanum and had reached a T_c of 48 K by the end of 1986.[42]

3 Three Classes of Oxide Superconductors

Characteristic of this last class of oxide superconductors is their layered structure containing perovskite Cu–O planes. These planes can be stacked singly, doubly, triply or quadruply with either Cu–O chains, planes or single-double planes of other oxides (La, Bi, Tl, etc.) between them. More than two dozens of such compounds have been synthesized, with a confirmed maximum of 133 K for $HgBa_2Ca_2Cu_3O_{8+\delta}$[43] and of 163 K at high pressure.[44] The compounds with the highest T_c's are all *hole* superconductors despite the fact that electron ones have also been found.[45] The coherence length ξ is highly anisotropic, 10 to 30 Å along the planes and 1 to 7 Å perpendicular to them. The Ginzburg–Landau parameter χ is of the order of 100 in plane and perpendicular to it over 1000. This is relevant for applications concerning both critical currents and magnetic fields. The latter become very high in the megagauss range, because $H_{c2} = \phi_2/\pi\xi^2$. The flux pinning observed is atypical. Their carrier concentration is of the order of a few times $10^{21} cc^{-1}$,[30] substantially lower than that of the earlier high-T_c superconductors, which were all Nb intermetallic alloys or compounds. So far, no other class of high-T_c superconductors has been found — which makes one wonder whether the copper oxides are unique.

In order to make progress, an understanding of the microscopic mechanism is quite relevant. This is reflected in the still large research effort world-wide.

Of course, development for use in industry is the other motor. Industrial applications can now progress empirically and do so successfully, based on the experimental findings and expert materials research. These have so far advanced the field, and allowed certain microscopic mechanisms to be excluded: on the conservative side those using *harmonic* electron–phonon interactions, on the exotic side the original resonance valence bond and the fractional quantum state theories.[46] Central to all attempts is the insight that a sizeable *atomic* coulomb repulsion U is operative at the copper ions in order to exclude hole carriers on the same site. Thus the ratio U/t, where t stand for the transfer integral, is substantial compared to the ratio Δ/t, where Δ is the energy needed to transfer Cu d-charge onto the neighboring oxygen p orbitals. This is opposite to the process in d–d oxide superconductors with low Hubbard U discovered first.[47] Figure 11 is a modified version of a figure that appeared in the reference cited, and shows the accepted picture. One notes that on both sides of the diagonal, $U \simeq \Delta$, there is a large area in which the transition-metal oxides are insulators. This is true for most oxides containing 3d metals from V to Ni. The d–d metals are mainly titanates, tungstates or molybdates, but there are exceptions.

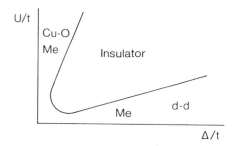

Figure 11: Modified metal–insulator diagram of the reduced Hubbard electron correlation energy, U, vs. reduced charge transfer energy Δ. t: transfer matrix element ($\propto W$, the bandwidth). Adapted from Ref. 47.

Therefore, one can categorize three classes of known oxide superconductors as follows: The first ranges from $SrTiO_3$ to $LiTi_2O_4$ and T_c's from 0.2 to 13 K, with $\Delta > U$. In the second class are the "valence skippers," bismuthates with T_c's from 13 to 32 K. The third, with $U \gg \Delta$, comprises *layered* compounds with T_c's from 35 K as in La_2CuO_4 to well over 100 K. Regarding their transition temperatures, we therefore note that *each of the three categories has its own transition temperatures adjacent to those of next one*: the first up to 13 K, valence skippers up to 32 K, and cuprates above them up to 165 K so far (Fig. 12). Is there perhaps a fourth category?

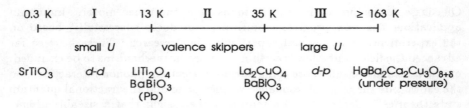

Figure 12: The development of superconductivity in oxides. U: on-site d-Coulomb repulsion.

The author thanks Ch. Bolliger for her help in editing and finalizing this paper, the topic of which has been addressed under the same title in the form of a long abstract without figures elsewhere.[48] Some of the context that led to the discovery of the copper oxide superconductors and the early confirmation[26] has also been included.

References

1. J.F. Schooley, W.R. Hosler and M.L. Cohen, *Phys. Rev. Lett.* **12**, 474 (1964).
2. A. Baratoff, G. Binnig, J.G. Bednorz, F. Gervais and J.L. Servoin, in: *Superconductivity in d- and f-Band Metals*, Proc. IV Conf. on Superconductivity in d- and f-Band Metals, eds. W. Buckel and W. Weber (Kernforschungszentrum Karlsruhe GmbH, Karlsruhe, W. Germany, 1982), p. 419.
3. Ch.J. Raab, A.R. Sweedler, M.A. Jensen, S. Broadston and B.T. Matthias, *Phys. Rev. Lett.* **13**, 746 (1964).
4. L.F. Mattheiss, PRB **6**, 4740 (1972).
5. K.A. Müller, *Helv. Phys. Acta* **31**, 173 (1958).
6. G. Binnig, A. Baratoff, H.E. Hoenig and J.G. Bednorz, *Phys. Rev. Lett.* **45**, 1352 (1980).
7. At this conference, the entire Session C is devoted to this important topic.
8. K.A. Müller, *Nature* **377**, 133 (1995).
9. J.K. Hulm, C.K. Jones, R. Mazelsky, R.C. Miller, R.A. Heim and J.W. Gibson, in: *Low Temperature Physics*, LT-9 (Plenum, New York, 1965). Part A, p. 600.
10. G. Aschermann, E. Friedrich, E. Justi and J. Kramer, *Phys. Z.* **42**, 349 (1941). A historical NbN sample from January 15, 1943, with $T_c = 16.1$ K is now in the author's possession.

11. D.C. Johnston, H. Prakash, W.H. Zachariasen and R. Viswanathan, *Mater. Res. Bull.* **8**, 777 (1973).
12. M.R. Harrison, P.P. Edwards and J.B. Goodenough, *Philos. Mag. B* **52**, 679 (1985).
13. D.C. Johnston, *J. Low Temp. Phys.* **25**, 145 (1976).
14. A.W. Sleight, J.L. Gillson and P.E. Bierstedt, *Solid State Commun.* **17**, 27 (1975).
15. T.D. Thanh, A. Koma and S. Tanaka, *Appl. Phys.* **22**, 205 (1980).
16. B. Batlogg, *Physica* **126 B**, 275 (1984).
17. L.F. Mattheiss and D.R. Hamann, *Phys. Rev. B* **26**, 2682 (1982).
18. S. Sugai, S. Uchida, K. Kitazawa, S. Tanaka and A. Katsui, *Phys. Rev. Lett.* **55**, 426 (1985).
19. J.E. Hirsch and D.J. Scalapino, *Phys. Rev. Lett.* **56**, 2732 (1986).
20. (a) B. Batlogg, R.J. Cava, L.F. Schneemeyer and G.P. Espinosa, *IBM J. Res. Develop.* **33**, 208 (1989); (b) L.F. Mattheiss, E.M. Gyorgy and D.W. Johnson Jr., *Phys. Rev. B* **37**, 3745 (1988).
21. D.G. Hinks, D.R. Richards, B. Dabrowski, D.T. Marx and A.W. Mitchell, *Nature* **335**, 419 (1988); S. Kondoh, M. Sera, Y. Ando and M. Sato, *Physica C* **157**, 469 (1989).
22. J.G. Bednorz and K.A. Müller, *Z. Phys. B* **64**, 189 (1986).
23. B.K. Chakraverty, *J. Phys. (Paris)* **40**, L99 (1979), *ibid.* **42**, 1351 (1981).
24. K.-H. Höck, H. Nickisch and H. Thomas, *Helv. Phys. Acta* **56**, 236 (1983).
25. R. Englman, *The Jahn-Teller Effect in Molecules and Crystals* (Wiley-Interscience, London, 1972).
26. K.A. Müller and J.G. Bednorz, *Science* **237**, 1133 (1987).
27. J.B. Goodenough and M. Longo, *Magnetic and Other Properties of Oxides and Related Compounds*, Vol. 4, Part a, of Crystal and Solid State Physics, Group 3 of Landolt-Boernstein Numerical Data and Functional Relationships in Science and Technology, New Series, eds. K.W. Hellwege and A.M. Hellwege (Springer-Verlag, New York, 1970), p. 262, Fig. 73.
28. Guong-Meng Zhao, K. Conder, H. Keller and K.A. Müller, submitted to *Nature*, March 1996.
29. K.A. Müller, M. Pommerantz, C.M. Knoedler and D. Abraham, *Phys. Rev. Lett.* **45**, 832 (1980).
30. J.G. Bednorz and K.A. Müller, 1987 Nobel Lectures in Physics: "Perovskite-Type Oxides — The New Approach to High-T_c Superconductivity," in: *Les Prix Nobel en 1987* (The Nobel Foundation, Stockholm, 1988), pp. 59–98. Reprinted in *Angew. Chem.* **100**, 757 (1988).

31. C. Michel, L. Er-Rhako and B. Raveau, *Mater. Res. Bull.* **20**, 667 (1985).
32. J.G. Bednorz, M. Takashige and K.A. Müller, *Europhys. Lett.* **3**, 379 (1978).
33. C. Michel and B. Raveau, *Rev. Cim. Miner.* **21**, 407 (1984).
34. S. Uchida, H. Takagi, K. Kitazawa and S. Tanaka, *Jpn. J. Appl. Phys.* **26**, L151 (1987).
35. H. Takagi, S. Uchida, K. Kitazawa and S. Tanaka, *Jpn. J. Appl. Phys.* **26**, L123 (1987).
36. R.J. Cava, R.B. van Dover, B. Batlogg and E.A. Rietman, *Phys. Rev. Lett.* **58**, 408 (1987).
37. J.M. Tarascon, L.H. Greene, W.R. McKinnon, G.W. Hull and T.H. Geballe, *Science* **235**, 1373 (1987).
38. C.W. Chu, P.H. Hor, R.L. Meng, L. Gao, Z.J. Huang and Y.Q. Wang, *Phys. Rev. Lett.* **58**, 405 (1987).
39. C.W. Chu, P.H. Hor, R.L. Meng, L. Gao and Z.J. Huang, *Science* **235**, 567 (1987).
40. M.K. Wu, J.R. Ashburn, C.J. Torng, P.H. Hor, R.L. Meng, L. Gao, Z.J. Huang, Y.Q. Wang and C.W. Chu, *Phys. Rev. Lett.* **58**, 908 (1987); P.H. Hor, L. Gao, R.L. Meng, Z.J. Huang, Y.Q. Wang, K. Forster, J. Vassilious, C.W. Chu, M.K. Wu, J.R. Ashburn and C.J. Torng, *ibid.*, p. 911.
41. J.M. Tarascon, L.H. Greene, W.R. McKinnon and G.W. Hull, *Phys. Rev. B* **35**, 7115 (1985).
42. Z.X. Zhao *et al.*, Institute of Physics, Beijing. *Academia Acta Sinica* (1987).
43. A. Schilling, M. Cantoni, J.D. Goa and H.R. Ott, *Nature* **363**, 56 (1993). (1994)
44. C.W. Chu, L. Gao, F. Chen, Z.J. Huang, R.L. Meng and Y.Y. Xue, *Nature* **363**, 323 (1993).
45. Y. Tokura, H. Takagi and S. Uchida, *Nature* **337**, 345 (1989).
46. See Sessions E and G in this issue.
47. J. Zaanen, G.A. Sawatzky and J.W. Allen, *Phys. Rev. Lett.* **55**, 418 (1985).
48. K.A. Müller, in *Advances in Superconductivity II*, Proc. 2nd Int'l Symp. on Superconductivity (ISS'89), eds. T. Ishiguro and K. Kamimura (Springer-Verlag, Tokyo, 1990), p. 3.

SUPERCONDUCTIVITY ABOVE 90 K AND BEYOND

C. W. CHU

Department of Physics and Texas Center for Superconductivity, University of Houston
Houston, Texas 77204-5932, USA

ABSTRACT

The discovery of high temperature superconductivity, first in 1986 at 35 K in $La_{2-x}Ba_xCuO_4$ and later in 1987 at 93 K in $YBa_2Cu_3O_7$, has been considered one of the most exciting developments in physics in the final decades of this century, with profound technological implications. In this 10th anniversary celebration, I shall begin by describing briefly two approaches adopted in the long and tortuous search for intermetallic superconductors with a higher transition temperature (T_c) that have had a significant bearing on our later work on the non-intermetallic superconductors. I shall then recall crucial steps we took in 1986, after the seminal observation made by Bednorz and Müller, which led to the exciting discovery in 1987 by the combined effort of our group in Houston and Wu's group in Huntsville. Efforts to raise the T_c over the last 10 years will be summarized, and the future prospects for T_c's above the current record of 134 K at ambient and 164 K at high pressure will be contemplated.

1. Introduction

It has always been an exhilarating experience to be a practitioner in the field of superconductivity, where intellectual challenge and technological promise coexist. This has been especially true over the past ten years, following the discovery of high temperature superconductivity (HTSy) [1,2]. HTSy is a rapidly evolving field, where records are shattered soon after their establishment by new discoveries, and models are made obsolete shortly after their proposition by new insights. The rapid pace of progress made in the field has been unprecedented. Voluminous exciting results in all areas of HTSy science and technology have been achieved. For instance, the transition temperature (T_c) has been quadrupled, several crucial aspects of HTSy have been understood, various models have been proposed, numerous material processing techniques have been developed, and a wide variety of prototype devices have been or are being constructed and tested.

It was said by Emerson that "there is no history; there is only biography." This is especially true when the events are recounted by a person who, himself, has been heavily involved and the line between history and autobiography can become blurred. To minimize possible injustices that may be done to researchers in the field, I plan to restrict myself to one small but important area of the HTSy research, *i.e.* the search for materials with higher T_c. Some of these events has been described elsewhere [3]. In this presentation, I summarize and discuss selected events that occurred: before 1986 that sowed the seeds in our group for later HTSy-development; in 1986, that were crucial to our discovery of 93 K-HTSy soon after the discovery of the 35 K high temperature superconductor (HTSr) [1]; in 1987 when the liquid nitrogen boiling temperature of 77 K barrier was finally conquered [2]; and from 1988 to 1996, during which time advancements [4-11] were achieved continuously in raising the T_c to the current records of 134 K at ambient [9] and 164 K at high pressures [10,11]. The prospect for higher T_C's in the future will also be contemplated.

2. Before 1986

In spite of the enormous impact the search for novel HTSr has had on condensed matter physics research, the road to compounds with a higher T_C before 1986 was tortuous and slow, as shown in Figure 1. Until 1986, the record T_C remained at 23 K found in intermetallic Nb3Ge-films in 1972 [12], representing only a 19 K° increase since the discovery of the phenomenon in 1911. All superconductors with such a relatively high T_C then were intermetallic compounds. During this period of time, there were two general approaches adopted in the search for compounds with a T_C higher than the existing record: the BCS approach and the enlightened empirical approach.

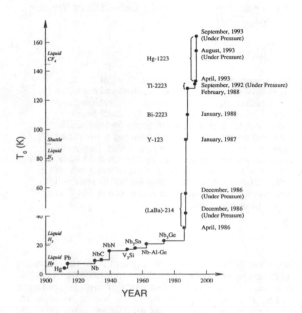

Figure 1. The evolution of T_C with time.

Although neither of the two approaches was able to lift the T_C above 23 K, studies by various groups including our own prior to 1986, give us the proper perspective on the high T_C problem, including: the possibility of a $T_C > 23$ K; the realization of the importance of optimization rather than maximization of parameters in raising T_C; the appreciation of the significance of instabilities and dimensionality to superconductivity at higher temperatures; skills in material synthesis and characterization; and knowledge of physics and chemistry of superconductors in general and of oxides in particular.

2.1. The BCS Approach

According to the BCS theory [13], T_C is given as

$$T_C = 1.14\,\Theta_D \exp[-1/N(E_F)V]$$

where Θ_D is the Debye temperature, $N(E_F)$ the electron-density of states at the Fermi surface (E_F), and V the electron-phonon interaction. In a simplistic sense, T_c is expected to increase with an increase in one or more of the three parameters Θ_D, $N(E_F)$, and V.

Many experiments on various compounds were made and analyzed in the ensuing years, based on the BCS relation. For instance, a large V was sought in unstable compounds [14], a higher N in low-dimensional compounds [15], and a greater characteristic energy in lieu of the conventional $k\Theta_D$ in one- or two-dimensional material systems [15]. During these studies, it was realized that the three parameters in the BCS T_c-relation are not each independent of the other. Unusually large N and/or V can trigger structural and/or electronic instabilities, which in some extreme cases, give rise to a structural collapse or an electronic ordering (such as charge-density waves, spin-density waves, *etc.*), before a really high T_c is achieved. By examining the competition between instabilities and superconductivity using the high pressure technique, we found [14] that the incipient instabilities associated with large N and/or V are not an obstacle to higher T_c. To avoid catastrophic instabilities, novel superconducting mechanisms such as interfacial superconductivity [16] were also proposed and explored in unusual material systems, including oxides near their metal/insulator boundaries [17].

2.2. The Enlightened Empirical Approach

The most successful example of the enlightened empirical approach before 1986 was the so-called "Matthias Empirical Rule" proposed in 1953 [18]. The rule correlates T_c with the ratio of valence electrons to number of atoms (e/a). According to the rule, the maximum T_c of a compound system usually occurs at e/a ~ 4.75 and ~ 6.4. Hundreds of intermetallic compounds and alloys were made with a T_c up to the 1972-record of 23 K in Nb3Ge. While this correlation works well for crystalline superconductors containing transition metal elements, it fails for non-transition metal compounds and alloys, or for amorphous superconductors.

3. 1986

The work on La-Ba-Cu-O (LBCO) with a T_c ~ 35 K in 1986 by Bednorz and Müller [1] ushered in the era of HTSy. We read their paper in early November 1986 with immense excitement. We fabricated multiphased samples of LBCO by the solid-state reaction technique that we acquired in the study of $(Ba_{1-x}Pb_x)BiO_3$ and reproduced their observation immediately. It seemed that the greatest experimental challenge at the time was to find a proper thermometer to measure in a temperature region alien to most superconductivity researchers. Even more exciting, in late November we detected a large resistance (R) drop by a factor of 80 at ~ 75 K in a multiphased sample of LBCO [19] (Fig. 2), suggesting the possibility of an even higher T_c. More R-drops were detected later above 60 K. Unfortunately, sample-aging problems prevented us from carrying out the magnetic experiment to provide more definitive evidence for a superconducting transition at such high temperatures. However, after reviewing all possible non-superconducting

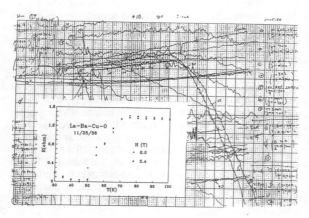

Figure 2. The R(T)-curve of a multiphase LBCO sample obtained on November 25, 1986 (Ref. 3).

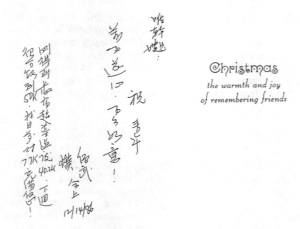

Figure 3. The Christmas card dated December 14, expressing our confidence of T_c's above 77 K. Translation: "Just obtained highest T_c at 40.2 K, next week very likely will reach 50 K. Presently, I am full of confidence of 77 K."

transitions in various compounds, I stated that "I am now full of confidence about the 77 K (for superconductivity)," in a Christmas card to Wei-Kan Chu dated December 14, 1986 (Fig. 3).

With LBCO-samples, even though multiphased, on hand, we examined the pressure effect on their superconducting transition in an attempt to determine if the nature of the superconductivity in this compound was similar to that of the superconductivity in the low temperature superconductors (LTSrs). We found its T_c to increase at a rate almost 10 times that observed in LTSrs without a structural transition to 40.2 K at 1.3 GPa [19], a temperature previously predicted to be impossible. This observation suggested to us that the superconductivity in LBCO was different from that in LTSrs, and that T_c may not have a ceiling, a faith that played an important role in our later work.

We presented our first LBCO-results at the Materials Research Society Meeting in Boston in early December, where I invited my former student Maw-Kuen Wu then of the University of Alabama at Huntsville to join our study. Koichi Kitazawa of Tokyo told me at the Meeting that they had identified the superconducting phase responsible in LBCO to be layered $La_{2-x}Ba_xCuO_4$ (La-214), which consists of one (CuO_2)-layer per unit formula. Unfortunately, the frequency of detecting a R-drop above 60 K started to plummet as our samples were made more La-214 pure. This suggested that superconductivity above 60 K might exist in compounds structurally different from La-214. The pressure data implied that reducing the interatomic distance might favor a higher T_c. Replacement of Ba by Sr and La by the non-magnetic Y, Yb, and Lu was therefore contemplated. Urged by the University, a patent was drafted in the first week of January and filed on January 12, 1987 with the US Patent Office. By replacing Sr for Ba in La-214, Maw-Kuen and students did manage to raise the T_c to 42 K at ambient [19].

4. 1987

After destroying two of our three Pt-crucibles in an attempt to grow La-214 single crystals for a systematic study, we shifted our efforts to the R-drops in multiphased LBCO samples at high temperatures. While waiting for the Y, Yb and Lu we had ordered to arrive, we deliberately prepared LBCO samples with a composition distribution across them and heated them under different conditions. On January 12, 1987, a diamagnetic shift signaling the onset of superconductivity was unambiguously detected at ~ 90 K for the first time in one of the samples (Fig. 4).

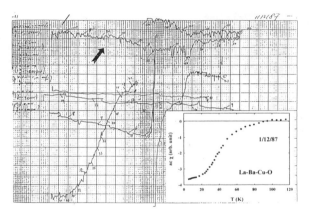

Figure 4. $\chi(T)$ for a multiphased LBCO-sample determined on January 12, 1987 (Ref. 3). Inset: Measurements after subtraction of the background.

However, the insulating exterior of the sample prevented us from doing the resistive measurements. Since a large diamagnetic signal (~ 26% at 4 K) is a more definitive indicator than a R-drop, there was no doubt in our mind that superconductivity above 77 K had finally been observed. The X-ray diffraction pattern of the sample taken on January 12, 1987 (Fig. 5) did display a structure different from La-214; but was indexed as $LaBa_2Cu_3O_7$ (La-123) for the major phase only after the structure of $YBa_2Cu_3O_7$ (Y-123) was determined in early March. Unfortunately, the diamagnetic signal in the sample disappeared the next day. The only questions that remained then were how to stabilize and isolate the high temperature superconducting (HTSg) phase. After careful consideration, we decided to publish the LBCO data so that others better equipped could stabilize the 90 K superconducting phase, provided that we stated clearly the conditions under which the observation was made.

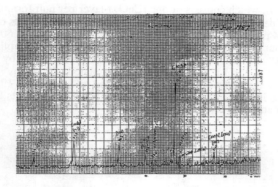

Figure 5. X-ray diffraction pattern for the LBCO-sample taken on January 12, 1987 and indexed later in March to be La-123 for the major phase (Ref. 3).

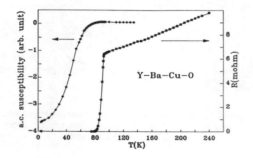

Figure 6. R(T) and χ(T) for a YBCO sample measured on January 30, 1987 (Ref. 2).

As I started to draft a paper on these results, Maw-Kuen called on January 29, 1987 and informed my former student Peiherng Hor and me that he and his students had just observed an R-drop to zero above 77 K. We were all ecstatic. Next day, Maw-Kuen and his student Jim Ashburn flew to Houston with their sample of Y-Ba-Cu-O (YBCO) for a definitive magnetic check. As shown in Figure 6, the R-drop between 80 and 93 K observed at Huntsville was reproduced and a diamagnetic signal indicative of a superconducting transition below 91 K was detected. Using the newly arrived Y, several superconducting samples were made and tested on the same day at Houston. The long sought stable and reproducible superconductivity above the 77 K-temperature barrier was finally established. Although we did not know how to determine the structure of the superconducting phase, the X-ray pattern of the sample and the pressure effect on its T_c [20] led us to conclude that the 90 K-superconducting phase must be structurally different from the 30 K La-214 phase.

The YBCO-results easily preempted my desire to write the paper on the unstable 90 K LBCO results. With the exciting YBCO results and the thought invested in the stillborn LBCO-paper, it did not take more than one evening for me to draft two papers on YBCO [2,20] which the LBCO-data was briefly included. After being reviewed by Mau-Kuen, Peiherng, and Ruling Meng, the papers were sent by express mail on February 5, 1987 to *Physical Review Letters* for publication. I was notified by Myron Strongin, editor of *Physical Review Letters*, of the acceptance of the papers on February 11.

I still vividly remember the extraordinary emotion I felt when I wrote the second sentence of the paper [2], "To obtain a superconductor reaching beyond the technological and psychological temperature barrier 77 K, the liquid nitrogen boiling point, will be one of the greatest triumphs of scientific endeavor of this kind," knowing that we had achieved it. In spite of all my confidence, deep down I still felt that, no matter how remote the possibility, it might be too good to be true. There were momentary scares after the submission of the papers that my career in superconductivity could end abruptly if the 90 K superconductivity we reported were to be proven untrue. More than once, I asked my colleagues, "Can there be phenomena other than superconductivity that are able to account for our observations? Please think and think hard!" The emotional burden at the time was enormous and was relieved only after news from other labs that they had reproduced the observation [21].

The structure of the superconducting phase was soon determined to be $YBa_2Cu_3O_7$ (Y-123) in collaboration with Bob Hazen and Dave Mao at the Geophysical Lab [22]. Other labs had also cracked the structure code at about the same time [23]. It is a layered structure with the stacking sequence of $(Y)(BaO)(CuO_2)(CuO)(CuO_2)(BaO)$, with two (CuO_2)-layers separated by a (CuO)-linear chain layer per unit cell.

Once the structure was determined, we set out to determine the role of Y in 90 K HTSy by partial replacement of magnetic rare-earth for Y. We found that, even with a large fraction of Y replaced by Gd and Eu, no T_c-depression was detected, suggesting that Y is electronically isolated from the superconducting carrier system and serves mostly as a stabilizer in the compound. A whole new series of $RBa_2Cu_3O_{7-\delta}$ (R-123) with R = Y, La, Nd, Sm, Eu, Gd, Ho, Er and Lu, with a $T_c \sim 90$ K was quickly synthesized [24] in our first trial. The series was synthesized in a reduced atmosphere which we used to significantly shorten the synthesis time and thus to give us a large advantage over other groups in the early days of HTSy. Our results prompted us to propose to attain higher T_c's by increasing the number of CuO_2-layers per unit cell [24]. Some of the R-123 were independently discovered at about the same time by other groups [25].

After submitting the paper on R-123, I departed for the special HTSy session of the 1987 March APS Meeting held on March 16 in New York. The special session was triggered by a "Dear Colleague" rejection letter from APS in the second week of December 1986, because our abstract on LBCO was half a line over the limit. By this time, a higher T_c was attained by pressure and R-drops at temperatures above 75 K were observed. I strongly felt that an exciting physics story was on the verge of unfolding. It did not take me long to convince the chairman of the Condensed Matter Physics Council of APS, Neil Ashcroft, and the other council members to agree to add a Special Session to the Meeting. I was asked to organize the special session. Alex Müller of IBM Zürich, Shiji Tanaka of the University of Tokyo, Paul Chu of the University of Houston, Zhong-Xian Zhao of the Physics Institute of the Chinese Academy of Sciences at Beijing, and Bertram Batlogg of AT&T Bell were invited to join a panel to address their work, in the same chronological order in which their respective work on La-214 appeared in journals. The dramatic events that took place within less than two months prior to the Meeting turned the Special Session on HTSy into the "Woodstock of Physics," a moniker coined by the late Mike Schlüter. I could not have dreamed that any such event would have happened when I first proposed it.

5. 1988-96

The years between 1988 and 1996 represent a period of solid advancement and many cuprate superconductors were discovered. They may be considered to belong to or be derivable from the layered compound systems of R_2CuO_4 [R-214 (1T), with R = La; or R-214(1T'), with R = Nd, Sm, Eu and Gd], $RBa_2Cu_3O_7$ (R-123, where R= rare earths except Ce, Pr, and Tb), $Bi_2Sr_2Ca_{n-1}Cu_nO_{2n+4}$ [Bi-22(n-1)n, with n = 1, 2, 3, ...], $Tl_2Ba_2Ca_{n-1}Cu_nO_{2n+4}$ [Tl-22(n-1)n, with n = 1, 2, 3, ...], $HgBa_2Ca_{n-1}Cu_nO_{2n+2+\delta}$ [Hg-12(n-1)n, with n = 1, 2, 3, ...], $CuBa_2Ca_{n-1}Cu_nO_{2n+2+\delta}$ [Cu-12(n-1)n, with n = 1, 2, 3, ...]; and $A_{1-x}B_xCuO_2$ (with A = alkaline earth, B = alkaline earth or vacancy), and Sr_2CuO_4. They have also set new T_c-records and contributed to debates on the occurrence of HTSy. In this section I shall briefly comment on these systems except the first two, that have already been discussed.

5.1. Bi-22(n-1)n

As 1987 came to a close, T_c stagnated at 93 K. Some noted that the accumulated manhours devoted to HTSy in 1987 probably exceeded all those devoted to low temperature superconductivity over the preceding 75 years, over-zealously concluding that $T_c > 93$ K could only be found in non-cuprate materials, if it existed at all. Their premature predictions were shattered by Maeda *et al.* who discovered [4] superconductivity above 100 K in the Bi-Sr-Ca-Cu-O (BSCCO) system in January 1988.

In the summer of 1987, Michel *et al.* reported [26] an important observation of superconductivity in Bi-Sr-Cu-O at ~ 8 K, which turned out to be the n = 1 member of the Bi-22(n-1)n system. Independently, in an attempt to expand the HTSg material base, Maeda *et al.* decided to replace the trivalent rare-earth element in R-123 by elements from the V-b group in the Periodic Table, such as Bi and Sb, which are trivalent and have ionic radii similar to the rare earths. They succeeded in detecting superconductivity above 105 K in multiphase samples of BSCCO [4]. The crystal structures of three members of the homologous series Bi-12(n-1)n with n = 1, 2, and 3 were soon determined [27], showing layered stacking sequence of $(BiO)_2(BaO)(CuO_2)(Ca)(CuO_2) ... (Ca)(CuO_2)(BaO)$ with n (CuO_2)-layers separated by n-1 (Ca)-layers, for Bi-22(n-1)n. The new record $T_c = 110$ K was attributed to the n = 3 members; and 22 and 80 K to members for n = 1 and 2, respectively. T_c clearly increased with n, as was predicted by us earlier [20]. Unfortunately, T_c was later shown to decrease with n > 3. It is interesting to note that it took Hazen *et al.* less than 48 hours to determine the Bi-1212 structure after receiving the samples from us. The acceleration of HTSy-research was clearly evident.

5.2. Tl-22(n-1)n

Following the similar rationale in forming R-123 [2,20], Sheng and Hermann started to substitute the trivalent nonmagnetic Tl for R by the end of 1987. After overcoming some problems associated with sample synthesis due to the low melting point and high volatility of Tl_2O_3, they detected superconductivity above 90 K in a multiphased sample with a nominal composition of TlBa$_2$Cu$_3$O$_x$ in November 1987 [28]. By partially replacing the Ba by Ca, they discovered [5] a T$_c$ ~ 120 K in the multiphase sample of Tl-Ba-Ca-Cu-O (TBCCO) in February 1987. The structures of members of the homologous series were soon determined afterward to be rather similar to Bi-2223 but with no modulation in the (TlO)$_2$-double layer [29]. T$_c$ = 90, 110, and 125 K were assigned to the n = 1, 2, and 3 members, respectively. In September 1992, the record T$_c$ = 125 of Tl-2223 was further enhanced to 131 K by the application of pressure [30].

5.3. Hg-12(n-1)n

As early as 1991, attempts were made to substitute the linearly-coordinated Hg^{+2} for the Cu in the (CuO)-chain layer in R-123. Later, Putilin *et al.* [8] synthesized HgBa$_2$CuO$_{4+\delta}$, the n = 1 member of Hg-12(n-1)n, and found a T$_c$ = 94 K, the highest T$_c$ detected in a single-layered cuprate. It has been shown that T$_c$ increases with n at least up to 3 or 4. Schilling *et al.* [9] succeeded in raising the T$_c$ to 133 K in a multiphase sample of Hg-12(n-1)n with n = 2 and 3. Later studies demonstrated that n = 1, 2, or 3 members of Hg-12(n-1)n possess T$_c$ = 97, 127, and 134 K, when optimally doped. The crystal structure of Hg-12(n-1)n displays the stacking sequence of (HgO$_\delta$)(BaO)(CuO$_2$)(Ca)(CuO$_2$) ... (Ca)(CuO$_2$)(BaO), with n (CuO$_2$)-layers separated by n-1 (Ca)-layers. This layered structure is similar to TlBa$_2$Ca$_{n-1}$Cu$_n$O$_{2n+\delta}$, except Hg-12(n-1)n exhibits a large vacancy concentration in the (HgO$_\delta$)-layer for oxygen to occupy. Due largely to the local structure of the (HgO$_\delta$)-layers, unusually large T$_c$ enhancement by pressure was achieved, pushing T$_c$ first to ~ 154 K at ~ 16 GPa [10], and then to 164 K at ~ 30 GPa [11], setting new T$_c$-records (Fig. 7) in a temperature region attainable in the Space Shuttle on the side opposite to the sun or by air-conditioner technology. Hg-12(n-1)n serves as an excellent system to probe the physics of HTSy due to the unusually large doping-range associated with the large void concentration in the (HgO$_\delta$)-layer for oxygen.

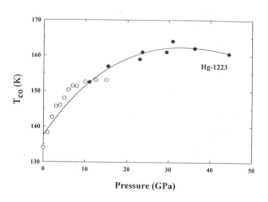

Figure 7. T$_c$(P) for Hg-1223. (Refs. 10,11)

5.4 $CuBa_2Ca_{n-1}Cu_nO_{2n+2+\delta}$

It has been observed that the T_c of the optimally doped homologous series $A_mX_2Ca_{n-1}Cu_nO_{2n+m+2+\delta}$, where m = 1 or 2; X = Ba or Sr; and n = 1, 2, 3, ... , increases as A changes progressively from the Group VB element Bi, through the Group IIIB element Tl, to the Group IIB element Hg. It may, therefore, be possible to raise the T_c by changing A further to the Group IB element Au, Ag, or Cu. Unfortunately, our attempt to synthesize $A_mX_2Ca_{n-1}Cu_nO_{2n+m+2+\delta}$ with A = Au or Ag failed due to the chemical inertness of Au and Ag. We have, therefore, tried Cu for A and discovered a 124 K superconducting phase in the system [31]. This becomes the highest T_c of cuprate without the toxic element, such as Tl or Hg.

In earlier studies, $CuBa_2Ca_{n-1}Cu_nO_{2n+2+\delta}$ with n = 3 and 4 was formed under high pressure with a T_c of 60 and 117 K in the n = 3 and 4 members, respectively [32]. $Cu_2Ba_2Cu_2Cu_3O_{9+\delta}$ was also synthesized under high pressure to show a $T_c \sim 110$ K [33]. All $Cu_mBa_2Ca_{n-1}Cu_nO_{2n+m+\delta}$ made were layered cuprates with the Cu in the (CuO_δ)-layer partially replaced by C due to contamination from the high pressure graphite furnace used. The reported T_c-values are lower than the extrapolated value from T_c's of cuprates with A = Bi, Tl, and Hg. By varying the synthesis conditions, such as lower pressure (~ 5.5 GPa) and temperature (~ 870 °C) than those previously reported, we succeeded [31] in synthesizing Cu-Ba-Ca-Cu-O samples with or without C-contamination with a bulk superconducting transition up to 124 K, as shown in Figure 8. The T_c of this 124 K-phase degraded with age to ~ 80 K as the sample was kept in the desiccator for 2 months and 6 months, successively, as shown in the same Figure. These different T_c's displayed drastically different pressure effects. The structure of the 124 K-phase has yet to be determined. However, Marezio et al. suggested in this Workshop [34] that the 124 K superconductivity might be associated with optimally-doped Cu-1223, although the maximum T_c of Cu-1223 achieved so far is 120 K [34].

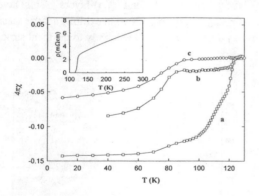

Figure 8. $\chi(T)$ and $\rho(T)$ for Cu-Ba-Ca-Cu-O at different ages: (a) < 3 days, (b) 2 weeks, and (c) 6 months.

5.5. R-214(1T')

In the beginning of 1989, Tokura et al., discovered [6] that the charge carriers in $Nd_{2-x}Ce_xCuO_4$ with a $T_c \sim 24$ K are electron-like, in contrast to almost all others HTSg cuprates. The compounds have a 1T" structure distinctly different from but closely related to the 1T-structure of $La_{2-x}Ba_xCuO_4$. While the 1T-phase has two neighboring apical oxygen ions to each Cu in the (CuO_2)-layer, the 1T'-phase does not. The subtle structural difference between 1T and 1T'-phases, which lead to the subtle difference in charge carrier characters in the two phases stems from the stress induced by the different ionic radii of R in the compounds. By varying the doping (*i.e.* Ce) as well as R (*i.e.* R = Gd, Eu, Sm, Nd and Pr), an electron-hole symmetry in the induction of HTSy in cuprates by doping appeared to exist [35], in consistence with some models [36] but not others [37]. Unfortunately, the highest T_c of the 1T'-phase does not exist above 30 K and many of the unusual normal state properties characteristic of the hole-doped cuprates with a very high T_c, *e.g.* > 90 K, are absent from the 1T'-compounds. Whether the 1T'-compounds form a distinct group of their own, more similar to the conventional LTSrs than the hole-doped high T_c cuprates, remains unknown. A resolution of this issue may provide critical insights to the understanding of HTSy.

5.6. $A_{1-x}B_xCuO_2$

It was pointed out earlier that T_c appeared to increase with n, the number of CuO_2-layers per unit formula. The homologous series of layered $Ca_{n-1}Cu_nO_{2n-2}$ with large n, which form the center part of Bi-22(n-1)n or Tl-22(n-1)n had become an interesting candidate for the study for quite a while.

In 1988, Siegrist *et al.* [38] succeeded in stabilizing $(Ca_{0.85}Sr_{0.15})CuO_2$, the n = ∞ member of $Ca_{n-1}Cu_nO_{2n-2}$ known as the infinite layered compound. Unfortunately, it was not superconducting. Three years later, Smith *et al.* [39] synthesized electron-doped $Sr_{1-y}Nd_yCuO_2$ under 2.4 GPa and found it superconducting with a $T_c \sim 40$ K. Later, Takano *et al.* [40] detected superconductivity up to 110 K in $(Sr_{1-x}Ca_x)_{0.9}CuO_2$ prepared under 6 GPa. The nominal composition suggested hole-doped for the sample. Great excitements followed the observation, since one-model [41] seemed to suggest that hole-doped infinite layered or the 1T'-cuprate should have the highest T_c. The ensuing microstructure study revealed that the $(Sr_{1-x}Ca_x)_{0.9}CuO_2$ samples were full of defects and the exact superconducting phase is yet to be determined.

Stoichiometric $SrCuO_2$ samples were made by us recently in bulk and thin film forms and found to be insulating. This is not surprising since the Cu-valence in the compound is exactly +2. Recently, an alternative was proposed for the observed superconductivity in the defective $SrCuO_2$ to be $CuSr_2Sr_2Cu_3O_{8+\delta}$ or other homologous members of $CuSr_2Sr_{n-1}Cu_nO_{2n+2+\delta}$ [42]. A perfect structure of $SrCuO_2$ does not provide layer-block to act as a charge reservoir to carry out the so-called modulation doping such as the CuO in R-123, (HgO_δ) in Hg-12(n-1)n, $(TlO)_2$ in Tl-22(n-1)n, or $(BiO)_2$ in Bi-22(n-1)n. We propose a possible way to make $ACuO_2$ superconducting and retain its structural integrity through partial replacement of the alkaline earth A by atomically small alkaline elements such as Li, Na, K, *etc.*

5.7. Sr_2CuO_4

Superconductivity up to 90's K has been reported in layered Sr_2CuO_4 synthesized under high pressures [43]. Sr_2CuO_4 so prepared exhibits a similar structure to that of R-214 (1T). Unfortunately, according to the Cu-valence effect on T_c, Sr_2CuO_4 is not expected to be superconducting. Recently, oxygen vacancies were also found [44] in the CuO_2-layers of $Sr_2CuO_{4-\delta}$, degrading the CuO_2-layer integrity which has been considered to be detrimental to HTSy. Furthermore, the superconducting volume faction in samples studied appears to be small compared with the volume-fraction of $Sr_2CuO_{4-\delta}$ present in these samples. A serious question concerning the identification of the reported superconductivity with the Sr_2CuO_4 phase has arisen.

We have recently succeeded in preparing samples with various volume fractions of Sr_2CuO_4 and $Sr_3Cu_2O_5$, under different conditions up to 10 GPa and 1200 °C. Some pure samples of Sr_2CuO_4 were obtained and found to be insulating and not superconducting. Superconductivity at ~ 90's K was detected in samples with the presence of the $Sr_3Cu_2O_5$ and the superconducting volume fraction increases as the $Sr_3Cu_2O_{5+\delta}$-content increases [45]. This demonstrates that the superconductivity at 90's K reported in the Sr_2CuO_4 is due to the impurity, such as $Sr_3Cu_2O_{5+\delta}$, present in the samples. Attempts to induce superconductivity in $Sr_2CuO_{4-\delta}$ by reduction have not yet been successful, perhaps, due to the O-defects in the CuO_2-layer previously reported.

VI. Prospects for a Higher T_c

There appears to be a lack of consensus as to the occurrence of HTSy at such high temperatures in such an unusual class of materials, not to mention the anomalous properties of the cuprate HTSrs in their normal state. The discovery of superconductors with a higher T_c's will no doubt pose new challenges to HTSy science and make applications more practical. Thermodynamically, the higher the T_c is, the more efficient operating a HTSg device becomes. Some of the target T_c's for applications can be determined by the coolants to be used, *e.g.* liquid nitrogen up to 77 K, passive cooling on the Space Shuttle in the space environment up to ~ 100 K, Freon (CF_4) up to 148 K, dry ice up to 198 K, and room temperature up to ~ 300 K. At present, there exists neither experimental nor theoretical reasons for T_c's not to reach these target temperatures.

It is my view that a T_c above the present record of 134 K at ambient or 164 K under pressure may still be found in the layered cuprate family by increasing the number of CuO_2-layers (n) per unit cell followed by proper doping. It has been shown that the T_c of a specific layered cuprate system increases with n up to 3 or 4 and that maximum T_c for all cuprates takes place at a carrier concentration (p) of ~ 0.16 holes per Cu-ion. The T_c drop for n > 3 or 4 has been attributed [46] to the possible depletion of p in the interior layers in a unit cell. We recently found [47] that p of cuprates with T_c > 3 or 4 indeed were underdoped with a p smaller than the optimal value of ~ 0.16 for maximum T_c to occur. Our attempt to enhance p by charging the sample with oxygen at high pressure was found to result in the decomposition of the n > 3 members of the homologous series to the n ≤ 3 ones. This may be understood in terms of the structural instability associated with large Coulomb repulsion that arises from the build up of excess charge as n increases, making doping the CuO_2-layers and doping them uniformly in all layers increasingly difficult. The suggestion is consistent with a small but noticeable decrease of the optimal p with increasing n [47]. We believe to properly dope cuprates with large n's to their optimal p by developing steps to overcome the doping instability described above without degrading the CuO_2-layer integrity will enable us to achieve a T_c higher than what we have presently. Experiments are under way to test this conjecture.

Suggestions that cuprate is necessary for HTSy have been made but not yet proven unequivocally. Therefore compounds other than cuprates should not be ignored, provided that catastrophic instabilities such as structural collapse, magnetic ordering, formation of charge density waves or spin density waves, *etc.* are avoided. Novel materials with possible novel mechanisms are promising candidates for higher T_c's.

Many reports of a sharp resistivity drop (but not to zero) or a diamagnetic shift (but small and always superimposed on a large paramagnetic background) have appeared over the last ten years [48]. Unfortunately, none could satisfy the four criteria I set in 1987 for the existence of superconductivity, *i.e.* in addition to zero resistivity and Meissner effect, the compound has to be stable (enough for definitive diagnosis) and the observation has to be reproducible (from sample to sample and from lab to lab). Therefore the reported observations can at best be called Unidentified Superconducting Objects (USO's). The most recent reports in 1994-95 of possible superconductivity above 200 K in the Hg-Ba-Ca-Cu-O, Bi-Sr-Ca-Cu-O, and Y-Ba-Cu-Se systems still belong to the same USO category. Some of them seem to be attributable to experimental artifacts or misinterpretation of data. However, the persistent sighting of USO's in multiphase oxides, many by reputable laboratories, in a similar temperature range over the last ten years, is too tantalizing to ignore although too fleeting to confirm. Could this be due to the ever shortening of the coherence length and the ever increasing of the compound stability for compounds with increasing T_c that make it more difficult to satisfy the four criteria for superconductivity to exist, using conventional diagnostic tools, or could this be caused by the very sensitive dependence of the electrical properties of many oxides on the environment and thus not by a superconducting transition? A definitive answer has yet to be determined.

VII. Acknowledgement

I feel extremely blessed to be in the right place at the right time so that I could witness one of the most exciting developments in physics and play a role in it. For that, I am forever grateful to many dedicated, hard working, and able colleagues of mine which include Jeff Bechtold, Laurence Beauvais, Daniel Campbell, Feng Chen, Wei-Kan Chu, Jeff Clayhold, Ken Forster, Li Gao, Jason Gibson, Bob Hazen, Peiherng Hor, Chao-Yuan Huang, Zhijun Huang, Allan Jacobson, Theresa Lambert, Qiu-Ming Liu, Jeff Lynn, Dave Mao, Ruling Meng, Diego Ramirez, Irina Rusakova, Chin-Sen Ting, Maw-Kuen Wu, Quan Xiong, Ya-Qi Wang and Yu-Yi Xue. Last but not least, I am also extremely thankful to my mentor, the late Bernd Matthias, who was a strict Edisonian with a deep skepticism about theories. His style of doing physics and his taste in selecting problems have influenced my work, for better or for worse, in condensed matter physics in general and superconductivity in particular. Some of my students, past and present, and I have therefore benefitted greatly by paying attention to materials which are considered to be too mundane by many traditional physicists and by not being intimidated by theoretical predictions which are treated as Sacred Writ by many experimentalists.

The work at Houston is supported in part by NSF, DoE, EPRI, the State of Texas, and the T. L. L. Temple Foundation.

References

1. J. G. Bednorz and K. A. Müller, Z. Phys. B 64, 189 (1986).
2. M. K. Wu et al., Phys. Rev. Lett. 58, 908 (1987).
3. C. W. Chu, Proc. Natl. Acad. Sci. USA 84, 468 (1987); and C. W. Chu, to appear in *Proceedings of the International Conference on the History of Original Ideas and Basic Discoveries in Particle Physics*, Erice, Sicily, Italy, July 29 - August 4, 1994 (Plenum, New York, 1996).
4. H. Maeda et al., Jpn. J. Appl. Phys. 27, L209 (1988).
5. Z. Z. Sheng and A. M. Hermann, Nature 332, 138 (1988).
6. Y. Tokura et al., Nature 337, 345 (1989).
7. M. Azuma et al., Nature 356, 775 (1992).
8. S. N. Putilin et al., Nature 362, 226 (1992).
9. A. Schilling et al., Nature 363, 56 (1993).
10. C. W. Chu et al., Nature 363, 323 (1993).
11. L. Gao et al., Phys. Rev. B 50, 4260 (1994).
12. J. R. Gavala et al., J. Appl. Phys. 46, 3009 (1974); and L. R. Testardi et al., Solid State Comm. 15, 1 (1974).
13. J. Bardeen, L. N. Cooper and J. R. Schrieffer, Phys. Rev. 106, 162 (1957).
14. C. W.Chu, *High Pressure and Low Temperature Physics*, ed. C. W. Chu and J. A. Woolam (Plenum, New York, 1978), p. 359; C. W. Chu and M. K. Wu, *High Pressure Science and Technology*, ed. C. Homan, R. K. Ma Crone and E. Whalley (North Holland, Amsterdam, 1983), p. 3; C. W. Chu, P. H. Hor and Z. X. Zhao, Physica B 131, 439 (1986); and references therein.
15. W. A. Little, Phys. Rev. A 134, 1416 (1964); V. L. Ginzburg, Phys. Lett. 13, 101 (1964); and F. R. Gamble et al., J. Chem. Phys. 55, 3525 (1971).
16. D. Allender, J. Bray and J. Bardeen, Phys. Rev. B 7, 1020 (1973).
17. D. C. Johnston et al., Matr. Res. Bull. 8, 77 (1973); C. W. Chu et al., *Solid State Physcis Under Pressure*, ed. S. Minimura (KTK Scientific, Tokyo, 1985), p. 223.
18. B. T. Matthias, Phys. Rev. 92, 874 (1953).
19. C. W. Chu et al., Phys. Rev. Lett. 58, 405 (1987).
20. P. H. Hor et al., Phys. Rev. Lett. 58, 911 (1987).
21. Z. X. Zhao et al., K. X. Tongbao 32, 522 (1987); J. M. Tarascon et al., Phys. Rev. B 35, 7119 (1987); S. Hikami et al., Jpn. J. Appl. Phys. 26, L314 (1987); H. Takagi et al., Jpn. J. Appl. Phys. 26, L320 (1987); R. J. Cava et al., Phys. Rev. Lett. 58, 1676 (1987); L. C. Bourne et al., Phys. Lett. 120, 494 (1987).
22. R. M. Hazen et al., Phys. Rev. B 35, 7238 (1987).
23. P. M. Grant et al., Phys. Rev. 35, 7224 (1987); T. Siegrist et al., Phys. Rev. B 35, 7137 (1987); Y. LePage et al., Phys. Rev. B 35, 7245 (1987); S. B. Qadri et al., Phys. Rev. B 35, 7235 (1987); and K. Samba et al., Jpn. J. Appl. Phys. 26, L429 (1987).
24. P. H. Hor et al., Phys. Rev. Lett. 58, 189 (1987).
25. A. R. Moodenbaugh et al., Phys. Rev. Lett. 58, 1995 (1987); Z. Fisk et al., Solid State Comm. 62, 743 (1987); and D. W. Murphy et al., Phys. Rev. Lett. 58, 1888 (1987).
26. C. Michel et al., Z. Phys. B 68, 21 (1987).
27. R. M. Hazen, et al., Phys. Rev. Lett. 60, 1174 (1988); M. K. Subramanian et al., Science 239, 1015 (1988).
28. Z. Z. Sheng and A. M. Hermann, Nature 332, 55 (1988).
29. R. M. Hazen et al., Phys. Rev. Lett. 60, 1657 (1988); C. C. Torardi et al., Phys. Rev. B 38, 225 (1988); and S. S. P. Parkin et al., Phys. Rev. Lett. 60, 2539 (1988).

30. D. D. Berkeley et al., Phys. Rev. B 47, 5524 (1993).
31. L. Gao et al., Mod. Phys. Lett. B 9, 1397 (1995).
32. H. Ihara et al., Jpn. J. Appl. Phys. 33, L300 (1994); X. J. Wu et al., Physica C 223, 243 (1994).
33. T. Kawashima et al., Physica C 227, 95 (1994).
34. M. Marezio et al., this Workshop (1996).
35. Y. Y. Xue et al., Physica C 165, 357 (1990); and references therein.
36. P. W. Anderson and Z. Zhou, Phys. Rev. Lett. 60, 2557 (1988).
37. V. Emery, Phys. Rev. Lett. 58, 2794 (1988).
38. T. Siegrist et al., Nature 334, 231 (1988).
39. M. O. Smith et al., Nature 351, 519 (1991).
40. M. Takano et al., Physica C 176, 441 (1991).
41. S. Maekawa, Y. Ohta and T. Tohyama, Physica C 185, 168 (1991).
42. S. Tao and H.-U. Nissen, Phys. Rev. B 51, 8638 (1995).
43. Z. Hiroi et al., Nature 364, 315 (1993); and P. D. Han et al., Physica C 228, 129 (1994).
44. Y. Shimakawa et al., Physica C 228, 73 (1994).
45. Z. L. Du et al., this Workshop (1996).
46. M. Distasio et al., Phys. Rev. Lett. 64, 2827 (1990).
47. F. Chen et al., to appear in Phys. Rev. B (1996).
48. See, for example, B. G. Levi, Physics Today, February 1994, p. 17; R. F. Service, Science 265, 2015 (1994).

Superconductors and Semiconductors

Shoji TANAKA

Superconductivity Research Laboratory, ISTEC

1-10-13 Shinonome, Koto-ku Tokyo,135 Japan

Great success of semiconductor technology is based on the enormous efforts on the material science of semiconductors in the past fifty years. The quality of material has reached the limit of material control. In the science of high-Tc superconductors, the material science on those materials has just begun recently. In order to realize the possible future applications, it seems to be necessary to introduce the semiconductor technology in this field as much as possible. To do so, the quality of materials must be close to that of semiconductors. In this talk, the similarities and the differences between both technologies are discussed, and the recent progress on the research and developments in high-Tc superconductors are shown. Single crystal growth, high quality thin film syntheses and surface and boundary problems are also discussed.

1. Introduction

Since the discovery of high-Tc Superconductors in 1986, the research and developments have made great progress. In the past ten years, we have learned that the oxide high-Tc superconductors have quite different characters from metallic superconductors in crystal structures, conduction properties and son on. The comparisons between two kinds of superconducting materials are shown in Table 1.
As indicated in this table, we recognize that the two dimensionality, where the superconductivity occurs in very near region of CuO_2-planes in the crystal, and the short coherence length is the most important factors in taking into account of the future development of science and technology of materials.

Furthermore, since the carrier concentrations in high-Tc superconductors are much lower than those of metallic superconductors by more than one order of magnitude, the superconductivity in oxides depends on the carrier concentration significantly.

Table.1 Superconductivity in Metals and HTSC

	Metals	Oxides
Dimensionality	Three Dim.	Two Dim.
Crystallinity	not sensitive	very sensitive
Coherence Length	Long (>100Å)	very short (<30Å)
Boundary Effect	small	very large

These facts indicate that the artificial controllability is essentially important for both pure science and application technologies of high-Tc superconductors. And those requirements rather resemble to that of semiconductors than to metallic superconductors.

There are still big differences between oxide superconductors and semiconductors. The peculiar transport properties in normal state of oxide superconductors have not been explained thoroughly and now it is considered that it comes from the strongly correlated nature of the electronic system including electronic spins. Furthermore, we are not sure whether or not the electronic band theory can be applied to these systems.

However, in the sense of material control, we can find many similarities between oxide superconductors and semiconductors ; single crystal growth, thin film preparations, boundary and surface problems and so on.

In this paper, the recent progress on the material technology of high-Tc superconductors are reported in comparison with those of semiconductors.

2. Single Crystal Growth

In order to reach reliable scientific results, it is necessary to obtain pure single crystals and further, as was mentioned in the preceding section, it is important for future practical applications. The high-Tc superconductors have complicated crystal structures composed of more than 4 elements, and basically they are incongruent melting materials. Thus special techniques are necessary to obtain large and pure single crystals of high-Tc oxide superconductors. The growth of large single crystals, especially of 1-2-3 materials, has been succeeded by using crystal pulling technique in our laboratory as shown in Fig.,1.[1] And now we can obtain pure single crystals of YBCO larger than $15 \times 15 \times 15 mm^3$. The crystallinity of these crystals are investigated by minimum yields of RBS and FWHM of X-ray rocking curves and it was proved that those crystals of YBCO is better in perfections than commercially available MgO single crystals and slightly inferior to $SrTiO_3$ single crystals. It is rather surprising that the single crystals of YBCO reach such perfection nevertheless of such complicated structure, even though GaAs single crystals have higher degree of perfection.

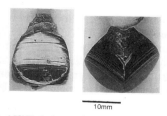

Fig.1 Y123 single crystal grown along the c-axis

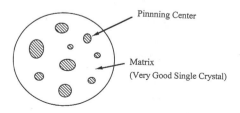

Fig.2 Image of "Pseudo Single Crystal"

The as-grown crystals of the 1-2-3 material have tetragonal structure and it changes to orthoromibic superconducting phase by annealing in O_2-atmosphere, and twinned structure comes out. But we succeeded in obtaining fairly large size of twin-free single crystals, size of which reached $2 \times 2 \times 2 mm^3$, by applying uniaxial pressing during the annealing process. The twin-free crystal of YBCO has high critical temperature of 93K and the transition width at the critical temperature is less than 0.1K. Single crystals of PrBCO, SmBCO and NdBCO were also grown by the pulling method and it was found that the oxygen partial pressure of the atmosphere is very crucial in obtaining high quality single crystals. It must be also noted that the floating zone method is also useful for obtaining single crystals of BSCCO (2212) and NdBCO.

3. Imperfections (pinning centers)

The largest difference between oxide superconductors and semiconductors is the problem of imperfections. In the case of semiconductors, carriers are supplied from donors and/or acceptors. In oxide superconductors, on the other hand, carriers are supplied from so-called charge reservoirs outside of CuO_2-plane. Furthermore, the main characteristics of semiconductors are carrier transport properties and emission and absorption of photons due to inter-band transition or to inter-impurity level transition. In the case of superconductors, the behavior of magnetic vortex is one of the most important characteristics, which determines the superconductivity characteristics. Thus in oxide superconductors, most of impurities such as Zn, Al, Mg and so on deteriorate the superconducting characteristics ; that is the critical temperature decreases. The most important imperfections in superconductor are pinning centers, which pin magnetic vortex.

By pinning effect, the critical current of superconductor increases. At present, it is considered that many kinds of imperfections can be pinning centers ; domain boundaries, dislocations, stacking faults and so on. However, it becomes clear that some kinds of the lattice defects having macroscopic size, from $100 Å$ to several thousands $Å$, survive as pinning center at liquid N_2 temperature. This was confirmed in the YBCO bulk material containing finely dispersed particles of the $Y_2Ba_1Cu_1O_5$ phase. The traces of ion bombardments in YBCO also increase the critical current[2]

On the other hand, those materials, having high critical currents showed rather good superconducting characteristics ; the critical temperature is almost the same as that of pure single crystals and the super-normal transition is also fairly sharp. These facts lead us to the new concept of pseudo-single crystals. The image of the pseudo-single crystal is shown in Fig., 2. It is expected that the large pinning force comes from the boundary between the pinning center and the matrix, and the sharper the boundary, the pining force larger. Probably this sharp change is crystal structure at the boundary (and also sharp change in order parameter) may be the origin of strong pinning force.[3]

In the field of applications, the magnetic field dependence of the pinning force is essentially important. Usually, the critical current decreases with increasing

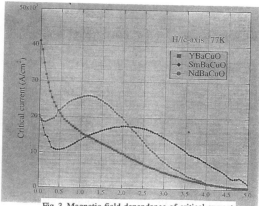

Fig. 3 Magnetic field dependence of critical current for different superconducting material

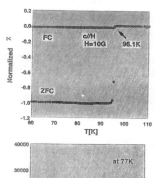

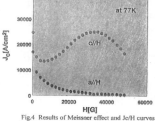

Fig.4 Results of Meissner effect and Jc/H curves measurements of NdBCO single crystal. After quenching as-grown crystal, followed the heat treatment (600-350°C for 200h, 340°C for 200h in O_2)

magnetic fields and this prevent the wide range of applications of superconductivity.

In the 1993, Yoo et al [4] in our laboratory found that bulk of $Nd_1Ba_2Cu_3O_7$ and $Sm_1Ba_2Cu_3O_7$, which were made by the OCMG (Oxygen Controlled Melt Growth) method, showed a prominent peak in magnetic field dependencies of critical currents at 77K as shown in Fig.3, and the peak current much higher than the Jc of the YBCO bulk in the high magnetic field region.

And further, they succeeded in increasing critical current in the lower magnetic field region (0∼2T) by introducing $Nd_4B_2Cu_2O_8$ particles. By those processes, it becomes possible to control the critical current in a wide magnetic field range (0∼5T) at liquid N_2 temperature. They also found that the irreversible field, which is the measure of pinning force of pinning centers, reached beyond 10T at 77K.

Even in single crystals of $Nd_1B_2Cu_3O_7$ grown by the pulling method, Shiohara et al [5] found the strong peak effect near 4T after suitable annealing process as shown in Fig.4. It is the most striking fact that very strong pinning centers are introduced even in single crystals simpl by the thermal annealing process, even though the structure of pinning centers is still under investigation at present.

4. Problems in Films and Boundaries

Problems in films and boundaries are very important for future developments both in semiconductor electronics and superconducting electronics. In the case of semiconductor electronics, size of transistors in memory chips is becoming smaller and smaller and now reaches less than 1 micrometer. And size of capacitors joints with transistors is also becoming smaller. In order to maintain the capacity of the capacitor almost constant, the thickness of the SiO_2 layer on the surface of Si is becoming thiner and thiner and now reaches less than $50 Å$. Further it has been tried that the SiO_2 layer is replaced by some ferroelectric materials to increase the capacity. In these cases, the perfection of very thin films is vitally important to prevent even very small leak currents through films.

The multi-layer system consisted of several different kinds of thin layers is also important in semiconductor electronics, especially in semiconductor lasers consisted of several films of III-V compound. In this case, boundaries between the layers dominate operation characteristics of devices.

In the case of oxide superconductors, the problems in films and boundaries are similar to the case of semiconductors. The film preparation is the first problem to be solved. There are several methods in preparing oxide superconductor films, including Sputtering method, Laser ablation method, Molecular beam method, MOCVD method and Liquid Phase Epitaxy (LPE) method. Selection of substrate materials is also critical.

Recently, liquid sources were introduced to the MOCVD method instead of solid sources, and high quality thin films of YBCO can be prepared steadily as shown in Fig.5.[6] As to the LPE method, we already reported that the thick film ($10 \sim 50 \mu$ m) of single crystalline YBCO is grown on MgO substrates.

Both of these methods are performed under nearly thermal equilibrium conditions and this may be the reason of obtaining high quality films. This fact is in good correspondence to the fact that in semiconductor technology only MOCVD method and LPE method survived in obtaining high quality thin layers of compound semiconductors.

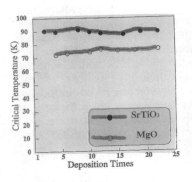

Fig. 5 Reproducibility of critical temperature measured by the dc four-probe method

Fig.6 a-YBCO film grown on a-YBCO single crystal by MOCVD

Recently, it becomes possible to grow films of the 1-2-3 material on the surface of the single crystal of 1-2-3 materials. We can call this as homoepitaxial growth. This success will lead to the growth of the multi-layer system with the highest quality. Fig.6 is an example of a TEM photograph of homoepitaxial growth of an a-axis oriented YBCO single crystal.[7] In this figure, the boundary is hardly recognized.

On the other hand, the lattice parameters of YBCO is slightly different from PrBCO, and consequently a beautiful array of misfit dislocations can be observed as is shown in Fig.7 [8] This kind of misfit dislocations are frequently observed in semiconductor devices like semiconductor lasers.

The lattice misfit between NdBCO and PrBCO is rather small, and so we can make the multi-layer system using both materials. In Fig.8, very thin films of NdBCO and PrBCO were deposited alternately, but no boundary can be observed.[9] Furthermore, only one pattern of 1-2-3 structure, which differs slightly from either material, is observed in X-ray diffraction. Therefore, it can be said this is a so-called strained-super-lattice, which is used in semiconductors electronics.

As mentioned above, the thin film technology especially in 1-2-3 materials is approaching the semiconductor technology day by day, and this will give us a bright future of superconducting electronics.

5. Surface

As to surfaces of oxide superconductors, we know very little at present. Atomic structures and also electronic structures of these materials must be very complicated and sometimes unstable. Accordingly, it is necessary to study quite a lot of things on treatments of surfaces of single crystals. Even though the surface problem must become very important for future applications, only the facts are written in this paper due to page limitation.

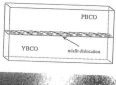

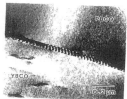

Fig.7 TEM image of the misfit dislocation at the interface of heteroepitaxially grown Y123/Pr123

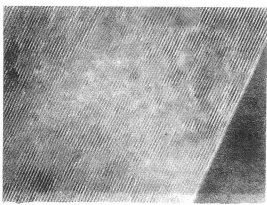

Fig.8 Multi-layer system of PrBCO and NdBCO

If we assume the possibility of application of S-N-S Josephson junctions, the roughness of the surface of oxide superconductors must be hopefully less than 10 Å. In single crystal of YBCO, it was found that the roughness of the polished surface measured by AFM is nearly 10 Å as shown in Fig.9. This result indicates the possibility of making a very thin insulating layer on the superconductor surface, thickness of which is as small as a coherence length. Detailed investigation on this surface is now in progress.

Very recently, it was found that the surfaces of NdBCO films deposited by the laser ablation method and also by the sputtering method are very flat and very stable even in the ordinary atmosphere.[10]

The roughness of the film is nearly 10 Å and arrays of surface atoms (probably Cu ions) can be observed by AFM. Later it was also found that similar arrays were also observed on the surface of NdBCO single crystals. Reasons of such peculiar phenomena are not clear yet, but it is somewhat indicative that the surface problems of these materials are very interesting and important, though they are still mysterious at present.

6. Some topics in applications

Here some applications, which have been made recently in our laboratory, are introduced.

(a) High sensitive mixer antenna in 100GHz region.
The millimeter-wave technology is now developing very rapidly. It is applied to a car-radar system and other sensing equipments. In such a high frequency range, it is hardly possible to operate even by the advanced semiconductor technology. We developed a new type of the high sensitivity mixer antenna, which is a combination of wide-band antenna and Josephson junctions, and this can be operated up to THz region.[11]

Whole picture of this mixer antenna is shown in Fig 10. Series of Josephson junctions, which are made by means of the focused ion beam method, are put at the center of wide-band antenna of thin YBCO film. Because of its strong non-linearity in I-V characteristics at 0-Bias, higher harmonics of local frequency are easily created. In our case, 99GHz, 10th harmonics of local 9.9 GHz, is generated and this is mixed with transmitted 100GHz wave, and the intermediate frequency of 1GHz comes out of the antenna. This intermediate frequency wave is easily detected and

amplified by ordinary semiconductor devices. This antenna is very easy to make and is very simple for high frequency operation. It is proved that this type of antenna has actually very high sensitivity.

(b) Magnetic levitation by YBCO bulk superconductors

YBCO bulk materials, made by the MPMG method, have been improved recently in critical currents and in strength of the trapped magnetic field at 77K. In

Fig.11, a new levitation equipment is shown. The superconducting part (below) consists of about 200 pieces of YBCO bulk, and the permanent magnet part (above) is improved recently to extend the magnetic field to a longer distance than before. As shown in this picture, total weight (above) is about 200kg, and still there is a 2.5cm gap between the two parts. Thus, we expect that this equipment can levitate more than 300kg in weight, and there must be very wide fields of applications, for instance, electricity storage system, transportation system and so on.

7. Conclusions

Great progress has been recently made in the material technologies in high-Tc superconductors. Especially, crystal growth technology and high quality thin film preparations are approaching the advanced semiconductor technology rapidly. The creation of very strong pinning centers in high magnetic fields and appearance of very flat and stable surfaces in NdBCO and SmBCO seem to be indicate of a bright future of applications of high-Tc superconductors.

Such progress in material technologies must excite the material science of these materials in a very near future.

Fig.9 Surface roughness of the polished YBCO single crystal measured by AFM

Fig.10 Mixer Antenna of 100GHz

Fig.11 Example of Magnetic Levitation

References

1. Y.Namikawa, M. Egami, and Y. Shiohara, J.J.Metals 59, 1053 (1995)
2. H.Fujimoto, M. Murakami, S.Gotho, N.Koshizuka, and S.Tanaka, Adv.Supercond. 2, 285 (1990)
3. M.Murakami, Melt Processed high Temperature Superconductors, World Scientific (1991)
4. S.I.Yoo, N. Sakai, T. Higuchi, and M.Murakami, appl. Phys.Lett.,65,633 (1994)
5. M.Nakamura, Y.Yamada, T. Hirabayashi, Y. Ikuhara, Y. Shiohara and S.Tanaka, Physica C, Physica C 259, 295 (1996)
6. Y.Yoshida, Y. Ito, and I. Hirabayashi, Submitted for publication in Appl.Phys.Lett.
7. J.G.Wen, T.Zama, T. Morishita, and N.Koshizuka, to be submitted
8. M.Tagami and Y.Shiohara, to be submitted
9. G.A.Alvarez, J.G.Wen, F. Wang and Y.enomoto: submitted for publication in Appl. Phys. Lett.
10. M. Badaye, Wu Ting, K. Fukushima, N. Koshizuka, T. Morishita and S.Tanaka. Appl.Phys.Lett.67(15), 2155 (1995)
11. K.Suzuki, M.Ban, S.Tokunaga, M.Ohtsuka and Y. Enomoto :submitted for publication in Appl. Phys. Lett.to IE^3

CRITICAL CURRENT LIMITING MECHANISMS IN POLYCRYSTALLINE HIGH TEMPERATURE SUPERCONDUCTOR, CONDUCTOR-LIKE FORMS

DAVID C. LARBALESTIER

Applied Superconductivity Center, University of Wisconsin-Madison,
1500 Engineering Drive, Madison, WI, 53706 USA

ABSTRACT

To develop the highest critical current density in conductor-like forms of HTS requires a knowledge of the mechanisms controlling the critical current density. These mechanisms are discussed for the case of the most important present HTS conductor material, BSCCO-2223. A hierarchy of current limiting mechanisms ranging from flux pinning to cracking operates, reducing the overall J_c orders of magnitude below the local flux pinning J_c values. Examples of the current limiting mechanisms are drawn from experiments with BSCCO-2223 tapes and from experiments with BSCCO-2212 bicrystals. An important conclusion is that multiple limits operate simultaneously. Increasing the connectivity of superconducting filaments is a particularly effective way to raise the overall J_c.

1. Introduction

One of the most critical underlying conditions for practical applications of HTS is the need to produce conductors with high critical current density (J_c) under realistic service conditions. Perhaps because the transition temperature, T_c, is a well defined critical parameter of the superconducting state, it is often assumed that J_c is a similarly well defined parameter. The complex vortex dynamics and materials science of HTS make this not true. This paper briefly describes the hierarchies of current limiting mechanisms operating in conductor forms, and the way in which they reduce the overall critical current density below the local limits by about two orders of magnitude.

2. Definitions of the critical current density

From the application point of view, the crucial definition of critical current density is: Given a certain cross-sectional area of superconductor A, what is the maximum current I that can be passed along the conductor? In the rest of this paper we define J_c in terms of this critical current I_c, defined at a dissipation of 1 µV/cm, divided by A. We then pose the question: What mechanisms limit J_c? To make the discussion relevant to practical materials, we concentrate on $(Bi,Pb)_2Sr_2Ca_2Cu_3O_x$ (BSCCO-2223). Since no good single crystals have yet been characterized, one must deduce the intrinsic properties from polycrystalline samples or from quasi-single-crystal thin films, supplemented by studies of $Bi_2Sr_2CaCu_2O_x$ (BSCCO-2212). The background to our discussion is that electrotechnology requires an overall conductor J of 10^4-10^5 A/cm^2 to be economic.

Present BSCCO conductors attain ~10^4 A/cm^2 at 77K and 10^5 A/cm^2 at 4 K. J_c is improving rapidly[1] as current limiting mechanisms become better understood.

2.1. The depairing critical current density J_d

The fundamental limit is the depairing current density, J_d, whose magnitude is given approximately by $J_d = H_c/\lambda$, where H_c is the critical field (0.3 - 1 T) and λ the penetration depth[2]. For BSCCO-2223, this gives $J_d = 1\text{-}3 \times 10^8$ A/cm^2 in the zero temperature limit, reduced by about an order of magnitude at 77 K. Thus J_d is quite high enough to provide no fundamental limit to economic operation.

2.2. The intragranular or flux pinning critical current density J_{fp}

A fundamental issue for any HTS compound is its anisotropy since this determines the range of reduced field (H/Hc2) over which flux pinning can be effective. Unfortunately BSCCO-2223 is strongly anisotropic, leading to a strong suppression of the range of practical use, the irreversibility field H*(77 K) lying below 1 T. Below H* we would expect that current densities of 0.01-0.1 J_d would be feasible, thus leading to J_{fp} ~10^6-10^7 A/cm^2 (at 0 K with no flux creep), derated by about an order of magnitude at 77 K. To determine J_{fp} experimentally, epitaxial BSCCO-2223 films grown on single crystal substrates are the best choice. Two studies[3,4] report J_c(77K, 0T) greater than 10^6 A/cm^2 (T_c 105K and 96-99K) and one[5] greater than 10^5 A/cm^2 (T_c 92-97K), thus making J_{fp} ~0.1 J_d at 77 K, indicating very strong flux pinning, in spite of T_c being lower than the 105-110 K found in bulk 2223 samples. These high J_{fp} values are 10-100 times the threshold

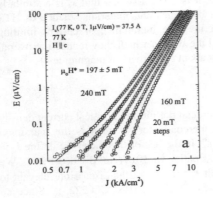
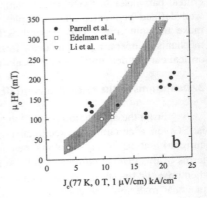

Figure 1: a.) extended electric field, critical current density curves for a representative BSCCO-2223 tape near the irreversibility field H*, showing the change of sign in the curvature of E-J at H*, and b.) the variation of H* with processing condition (Edelman et al.[6]) and with post processing treatment (Parrell et al.[7]).

required for economic operation.

Polycrystalline conductor forms have much smaller J_c, reproducible values falling into the range 10-30 kA/cm^2, only 1-3% of the best thin film values. One rationalization is that polycrystals contain many weak links and other barriers to current flow. To deduce what part of J_c is determined by flux pinning and what part by the dilution of the active cross-section by weak links we can study the irreversibility field H*, as in Fig. 1a. There is interesting detail to the way that H* changes with the processing condition in BSCCO-2223 tapes, as is shown by two recent experiments of Edelman et al.[6] and Parrell et al.[7] in Fig. 1b. In contrast to the very high J_c films cited above[4,5] for which H* was <100 mT at 77K, H* for these much *lower* J_c tapes is 2 to 3 times *higher* (200-300 mT). This anti-correlation is a strong hint that the large difference in *magnitude* of J_c of tapes and films below H* is not determined by flux pinning. H* in tapes undergoes systematic changes with the degree of reaction[6] and of the effective equilibration temperature of the 2223 phase[7], perhaps reflecting the influence of varying cation stoichiometry on flux pinning. Fig. 1 also shows that there is a marked tendency within tapes for increasing J_c (= I_c/A) to be correlated to increased H*. We interpret these data as showing that J_{fp} is sensitive to formation and processing conditions but that the magnitude of J_c is mainly determined by the percolative connectivity of the BSCCO filaments.

2.3. The intergranular or grain boundary critical current density J_b

The way that GBs impede current flow in BSCCO tapes has been much debated[8,9], though there is little single GB data from bicrystals (and this is all on BSCCO-2212)[10-13] that leads to definitive conclusions. Some recent bicrystal

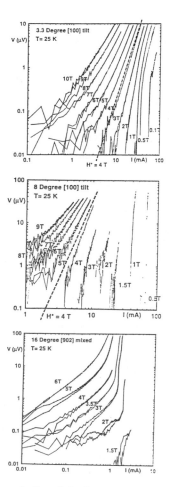

Figure 2: Extended voltage-current characteristics for BSCCO-2212 bicrystals showing the progressive change occurring for "Railway Switch" types[9] of grain boundaries. The 3 and 8° [100] tilt bicrystals have characteristics like those in Fig. 1 while the basal-plane faced grain boundary of the 16° [902] bicrystal is qualitatively different.

experiments of Wang et al.[14] that are more conclusive are shown in Fig. 2. A key issue is which GBs can be in the current path without degrading J_c. The V-I curves in Fig. 2 show that low angle "Railway Switches" are acceptable provided that the GBs are not basal-plane faced. We interpret the low V linear resistance across the 16° [902] GB as being the signature of c-axis current flow, which

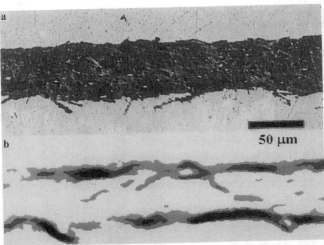

Figure 3: Upper image shows the BSCCO-2223 core surrounded by its Ag sheath. Lower image shows the local current densities reconstructed from magneto-optical studies of the tape. The current flows preferentially in larger grains found near the Ag sheath. Gray scale J_c values are from darkest to lightest: >3.6, 2.6-3.6, 1.6-2.6, <1.6 x 10^5 A/cm^2 at 44 mT at 10 K.

is both more dissipative and qualitatively different (e.g. it does not have a definable H*) from the ab plane flow seen in the 3 and 8° [100] GB. To get the characteristics seen in Fig. 1a appears to require connections that are of the type in Fig. 2a and 2b, that is tilt type GB connections.

2.4. Percolation current density J_p of a polycrystalline tape

Grain boundaries are only one barrier to current flow through the polycrystalline network. Additional, more macroscopic

Figure 4: Magneto-optical images of exposed tapes of BSCCO-2223 after a third heat treatment. The left tape was pressed, while the right was rolled. Lighter regions indicate preferential flux penetration, which occurs preferentially transverse to the tape axis for the rolled tapes and parallel to the tape axis for the pressed sample. Many of the lines of penetration are cracks produced in the deformation step that cannot be healed in the subsequent heat treatment. Both tapes were heat treated identically. The cracks introduced by rolling are more deleterious to transport J_c since they run transverse to the tape axis while pressing cracks run parallel to the tape axis.

barriers are unconverted non-superconducting phases, cracks, and section irregularities due to filament sausaging. All of these defects reduce the cross-section carrying transport current. Thus it is easy to see that the magnitude of J_c defined by I_c/A can be much less than the flux pinning limit. Because it seems likely that less than 10% of typical tapes are carrying current effectively, it becomes important to analyze the possible role of all defects that appear in tapes. Recently, magneto-optical imaging has been applied to this task[15-16], as shown in Fig. 3. Data such as this shows directly that the percolative path through the network is the prime variable controlling the magnitude of J_c, not J_d, J_{fp}, or J_b. This insight is vital in the drive to strongly enhance J_c.

Table 1: Summary of current limiting mechanisms.

CURRENT DENSITY TYPE	MAGNITUDES	COMMENTS
Depairing (J_d)	$10^8 - 10^7$ A/cm^2	Determined by basic physics
Flux pinning (J_{fp}) Pinning centers, Irreversibility field	Can achieve about 10% of J_d: 10^7 A/cm^2 at 4 K- 10^6 A/cm^2 at 77 K	So far little explicit success in raising J_{fp} except through radiation experiments. C-axis coupling controls vortex dimensionality and limits H* to << H_{c2}. *Evaluate through H* measurement.*
Grain boundary (J_b)	$10^7 - 10^2$ A/cm^2 Very strong dependence on misorientation	Dislocation core size controls J_b at low angles. Basal plane faced GBs may be bad because c-axis current flow required. *Evaluate with bicrystal measurements.*
Percolation (J_p) current density is the residual that can contribute to long range transport. It is still << J_{fp} and J_b and is additionally reduced by the effects of macroscopic barriers introduced by 2nd phase, cracks, sausaging of the Ag interface etc.	~10^5 A/cm^2 (4K)- few 10^4 A/cm^2 (77K) in good cases but varies strongly from place to place and wire to wire.	Large scale disruption to grain alignment and local chemistry by 2nd phase, amplified by cracking due to lack of bonding or differential thermal contraction. *Evaluate all current limiting mechanisms and devise intelligent processing strategies to defeat principal ones.*

2.5. The role of cracks in controlling the percolation current density J_p of tapes

An important hypothesis is that cracks are a major barrier to current flow in BSCCO tapes. To test this hypothesis explicitly is not easy but a recent experiment by Parrell et al.[17] used magneto-optical imaging to directly demonstrate the important role that cracks can play. Partial results are shown in Fig. 4, where rolled and pressed samples (otherwise identically treated) after the third and final heat treatment are contrasted. The pressed sample exhibits strong flux exclusion and a high J_c (20 kA/cm^2), while the rolled sample is very much more non-uniform and has only some 20% of the J_c in the rolled tape. This contrast between rolled and pressed tape is common. One reason that cracks are hard to avoid at final stages is that the liquid that eases their healing is used up in reacting to form the 2223 phase. This area of process control[18] seems to be a very fertile one in raising J_c.

3. Summary

There is no fundamental limit to developing economically attractive values of J_c in BSCCO-2223 tapes. Even at 77K, the flux pinning current densities are greater than 10^6 A/cm^2. However, real conductor forms have J_c values that are about 5-10% of this. Great strides in J_c are coming from carefully identifying those defects that limit the connectivity of the polycrystal. These points are summarized in Table 1.

4. Acknowledgments

I am very grateful to my many colleagues in Madison who have contributed to many aspects of the studies summarized here. I am particularly grateful to H. Edelman, J. A. Parrell, A. Pashitski, A. Polyanskii and Jyh-Lih Wang who have provided me with copies of as yet unpublished results for this paper. Frequent discussions with S. Babcock and E. Hellstrom are also acknowledged, as are many with colleagues in the Wire Development Group, particularly G. N. Riley Jr. and S.E. Dorris. The principal funding for the BSCCO-2223 work comes from ARPA and from EPRI while the bicrystal work was funded by the NSF-MRG Program DMR-9214707.

6. References

1. A. P. Malozemoff, Paper B.1 this symposium
2. Qiang Li, M. Suenaga, T. Hikata, K. Sato. *Phys. Rev* **B46** (1992) 5957.
3. Y. Hakuraku and Z. Mori, *J. Appl. Phys.* **73** (1992) 309
4. P. Wagner, U. Frey, F. Hillmer, and H. Adrian, Phys. Rev. B51 (1995) 1206.
5. H. Yamasaki et al., *IEEE Trans. on Supercond.* **3** (1993) 1536.
6. J. A. Parrell, G. N. Riley, D. C. Larbalestier, work in progress.
7. H. Edelman, J. A. Parrell, D. C. Larbalestier, subm. to *J. Appl. Phys.* (1996)
8. L. N. Bulaevskii, J. R. Clem, L. I. Glazman, and A. P. Malozemoff, *Phys. Rev.* **B45** (1992) 2545.
9. B. Hensel, J.-C. Grivel, A. Jeremie, A. Perin, A. Pollini, and R. Flukiger, *Physica C* **205** (1993) 329.
10. N. Tomita, Y. Takahashi, M. Mori, and Y. Ishida, *Jpn. J. Appl. Phys.* **31** (1992) L942.
11. J. L. Wang, X.Y. Cai, R.J. Kelley, M.D. Vaudin, S.E. Babcock, and D.C. Larbalestier, *Physica C* **230** (1994) 189.
12. J. L. Wang, X.Y. Cai, R.J. Kelley, M.D. Vaudin, S.E. Babcock, and D.C. Larbalestier, *J. Mater. Res.* to appear April 1996.
13. Qiang Li and M. Suenaga, Materials Res. Soc. Fall 1995 Meeting.
14. J. L. Wang et al., Materials Res. Soc. Fall 1995 Meeting.
15. A. Pashitski, A, Polyanskii, A. Gurevich, J. A. Parrell, D. C. Larbalestier, *Physica C* **246** (1995) 133.
16. U. Welp, D. O. Gunter, G. Crabtree, J. S. Luo, V. A. Maroni, W. L. Carter, V. K. Vlasko-Vlasov, V. I. Nikitenko, *Appl. Phys. Lett.* **66** (1995) 1270.
17. J. A. Parrell, A. Polyanskii, A. Pashitski, D. C. Larbalestier, *Super. Sci. and Tech.* to appear May 1996.
18. J. A. Parrell, S. E. Dorris, D. C. Larbalestier, *J. Mater. Res.* **11** (1996) 555.

BULK APPLICATIONS OF HTS WIRE

A. P. MALOZEMOFF

American Superconductor Corporation
Two Technology Drive, Westborough MA 01581

ABSTRACT

A retrospective on the development of HTS wire for bulk applications reveals remarkable breakthroughs but also the linear progress of BSCCO-2223 rolled multifilament composite wire over time, attaining 44,000 A/cm^2 (77 K, 0 T) recently, with significant opportunities for the future.

A 200 hp superconducting ac synchronous motor[1], a 50 m superconducting machine-stranded cable conductor for power transmission carrying 1800 A at 77 K[2], a two foot diameter superconducting current limiter coil with 3 mH inductance[3], a 3.4 T (4.2 K) superconducting magnet for magnetic separation[4], and a 160 hp superconducting homopolar motor[5] are the latest in a series of significant new applications prototypes based on high temperature superconducting (HTS) wire and coils from American Superconductor Corporation (ASC). Robust cryogenic current leads based on HTS wire have been introduced last year as a first ASC commercial product trademarked Cryosaver™. Other reports in this session describe these major achievements from an applications perspective[1-5]; so here I concentrate on the HTS wire which underlies this progress. At the tenth anniversary of the discovery of HTS materials, it is fitting to ask, what can we learn for the future from the last ten years of HTS wire development?

First, where is the technology headed? Addressing the huge range of commercial opportunities in electric power and magnet technology requires attractive cost-performance, characterized by a $/kA-m figure of merit in the range of 10, comparable to that of the low temperature superconductor Nb$_3$Sn. Furthermore, engineering critical current density (J_e) performance (across the full cross-section of the wire) must exceed 10-20,000 A/cm^2 for most applications. Given typical fill-factors of 30% in sheathed multifilament wire, this implies the need for current density (J_c) in the core of 30-60,000 A/cm^2, all at operating conditions and over long lengths. Because J_c simultaneously determines electrical performance and enters into the $/kA-m figure of merit, J_c-enhancement has been the predominant focus of wire development worldwide. But many other wire characteristics are also important: mechanical properties, dimensional tolerances, sheath resistivity and filament dimensions which affect ac loss characteristics, etc.

Work on long-length flexible HTS wires started in earnest with the discovery of YBa$_2$Cu$_3$O$_7$ (YBCO), as first evidenced by the multiple patents filed in spring 1987, including for example Yurek and Vander Sande's concept of an HTS/noble-metal composite plus their metallic precursor process[6]. However, a basic challenge to long-length flexible YBCO wire emerged from the work of Dimos et al.[7]. Growing epitaxial YBCO films on bicrystal substrates, they demonstrated that grain boundaries with less than 10° misorientation angle carried currents at bulk levels, but higher angles caused a precipitous drop in critical current. J_c(B) of these so-called weak links is further suppressed in magnetic field B.

This was bad news for long-length HTS wire. No known manufacturing process can produce single-crystal wires in the required lengths and volumes at commercially competitive costs. In contrast, costeffecive manufacturing typically involves polycrystalline structures with a random array of grain boundary misorientation angles and so unacceptable weak-link behavior. Indeed all the early long-wire efforts, such as Pacific Superconductor's electrophoretic deposition of YBCO on a 1 km wire-substrate[8], foundered on this problem: the wire could not carry enough current.

In 1989, an unexpected discovery dramatically changed this picture. Tenbrink et al.[9], melt-processing $Bi_2Sr_2CaCu_2O_8$ (BSCCO-2212) in a silver sheath, achieved significant c-axis texture (though the ab-plane was still random), and they found an apparently non-weak-linked or strongly coupled $J_c(B)$. Most exciting, $J_c(B)$ at 4.2 K extended as a plateau beyond 20 T, well above conventional NbTi and even Nb_3Sn superconductors.

This breakthrough launched a major research effort around the world. Both melt-dipped[10] and sheathed wire[11] processes have given 4.2 K high-field performance of order 200,000 A/cm^2, making BSCCO-2212 a leading contender for high field insert magnets for applications such as high-field NMR. Soon, Sato et al. at Sumitomo Electric (SEI) extended the sheathed-wire composite concept to BSCCO-2223[12], finding a high-field plateau exceeding 150,000 A/cm^2 at 4.2 K, and superior performance to BSCCO-2212 at temperatures above about 20 K. Because of the broader range of commercial opportunities at the higher temperature, American Superconductor Corporation (ASC) has also focused its program on BSCCO-2223, pushing performance to levels[13] comparable to those of Sumitomo and establishing pilot manufacturing with throughput exceeding 10000 m of wire per month and up to 1km piece lengths. High yields to rigorous specs [14] have established the viability of the powder-in-tube BSCCO-2223 process for large scale manufacturing and have made possible the recent prototypes[1-5], each of which require 1000's of meters of wire. J_c performance up to 44,000 A/cm^2 at 77 K, 0 T in rolled multifilaments[13], along with fill factors of 30%, imply the potential for J_e over 13000 A/cm^2, well into the commercial range discussed earlier. For example, $J_c(B)$ studies[15] show that such performance translates to almost double this level at 20 K and 3 T, which is expected to facilitate commercial level HTS motors[1]. ASC is actively working to put these technology elements together into a future generation of long-length wire.

In spite of all this progress, further J_c/J_e improvement is necessary in the future to provide more cost-performance margin. Prospects for such progress are controversial. Table 1 shows short-length J_c records (at 77 K, 0 T) over time, reported by different organizations throughout the world. Typically, these results are on monofilament samples prepared by a pressing rather than rolling technique. Since the 66,000 A/cm^2 of Yamada et al. at Toshiba[16] and 69,000 A/cm^2 of Li et al. at NKT[17] in 1992, no new records have been set. Furthermore, rolled multifilamentary samples, rather than pressed monos, are required to meet mechanical requirements and to facilitate a continuous manufacturing process. Yet J_c performance of rolled multis has lagged that of pressed monos over time, leading some to question the long-term prospects for the BSCCO-2223 HTS wire technology.

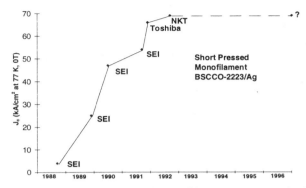

Fig. 1. J_c vs. time for short length pressed monofilamentary composite BSCCO-2223.

The view at ASC is very different. First, we note the many recent studies demonstrating significantly higher performance at least locally in these wires. In so-called "microslice" experiments, slices cut out from the intial monofilament have local current densities up to 76,000 A/cm^2 (77 K, 0 T)[18]. Magneto-optic investigations reveal local performance greater than 80,000 A/cm^2 in regions near the silver interface[19]. And careful measurement of BSCCO surface layers adhering to silver peeled away from the bulk gives up to 110,000 A/cm^2[20]. Headroom for yet higher levels is demonstrated by BSCCO-2223 thin film work which, even in non-epitaxial ab-random microstructures, reached levels of over 1,000,000 A/cm^2[21]. Secondly, we consider the saturating J_c progress in monos to be not unexpected. Given the known improvement in texturing and phase conversion near silver interfaces, the typically 50 μm-thick cores of monos are clearly non-optimal in contrast to the finer cores and distributed silver matrix of multifilaments.

Figure 2 shows J_c progress over time at ASC in short rolled multifilament BSCCO-2223 silver-sheathed wires. The remarkably linear behavior starts in late 1990 when the first multifilaments were manufactured, and culminates last summer at 44,000 A/cm^2, a new world record for this class of wire. As also emphasized by the Sumitomo group[12], such results approach the best achieved in monos either at ASC or Sumitomo. At ASC we expect to stay on this line, implying that soon we will actually exceed the mono records, overturning the old paradigm of mono J_c superiority.

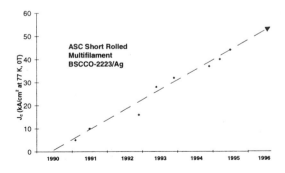

Fig. 2. J_c vs. time for short length rolled multifilamentary composite BSCCO-2223

In the semiconductor world, Gordon Moore of Intel proposed in 1965[22] a now well-known empirical law for DRAM progress over time: performance, expressed through the number of transistors per die, doubles every 18 months, a behavior which has held true for over 30 years to levels of integration inconceivable when the law was first proposed. Obviously, the determinants of this behavior are incredibly complex, involving not only the full complexity of the planar semiconductor fabrication process with its hundreds of steps, but also factors like ongoing technical creativity and the economic willingness to invest huge sums in development and in the construction of semiconductor fabs.

Of course, it remains to be seen if Fig. 2 will become the Moore's law of BSCCO HTS wire. But there are many parallels. Here too, the technology is incredibly complex, involving hundreds of process steps: powder preparation and packing, detailed deformation schedules, and a complex heat treatment interspersed with additional deformation. All these elements are closely linked such that change in any one of these parameters requires reoptimization of many others. Making progress in this kind of multidimensional process space requires both sophisticated statistical optimization techniques and also new insights into the underlying materials science. Because of the complexity of the chemistry of BSCCO-2223 formation (more than 10 active phases), the unusual deformation characteristics of the BSCCO precursor powder and the microstructural complexity, these insights have come slowly. But they have come and will continue to come. An example is the relatively recent discovery of the role of two different BSCCO precursor phases, one with Pb and one without[23]. The complexity explains in part why progress is so smooth; no single breakthrough by itself can easily transform such a technology. Also parallel to the semiconductor story is the creativity of the technical teams working around the world in this technology and the increasing willingness of corporations and governments to invest substantial sums in the required technology development, as the huge economic impact of HTS wire technology becomes clearer.

The linear rather than exponential progress in Fig. 2 may at first seem disappointing. But the increasing cost-performance advantage is enough to predict a massive impact on the mature electric power and magnet industries, where margins have been low. As indicated earlier, performance is already near the desired range, and cost will certainly drop significantly as wire throughput increases. A doubling of performance projected from Figure 2 over the next five years, over and above the impressive levels achieved so far, should coincide with the commercialization timeline for many electric power applications such as power transmission, high power motors, transformers and so forth.

While BSCCO wire moves rapidly towards commercialization, a major development has occurred recently which promises to bring the story of HTS wire full circle. This is the comeback of YBCO, made possible by the development of biaxial texturing, that is, texturing of not just the c-axis but of the ab-axes as well, which reduces large angle grain boundaries. While there is no space to go into details here, two techniques haven been pioneered by Fujikura[24] and Toshiba[25], and developed further at Los Alamos[26] and Oak Ridge[27] National Laboratories. Known as IBAD and Deformation Texturing (DeTex) or RABiTS^{TM}[27] respectively, they biaxially texture a substrate tape; then an YBCO thick film is deposited epitaxially, forming a coated conductor. Los Alamos has reported J_c >1 MA/cm^2 (77 K, 0 T) over a 1.2 μm thick film[26].

If the substrate can be kept to no more than 25 μm thick, a 1 MA/cm^2 1.2 μm film would give a J_e of 50,000 A/cm^2! In addition, the superior $J_c(B)$ performance of YBCO as compared to BSCCO opens up the opportunity for higher field magnets at 77 K. YBCO's advantage is its lower structural anisotropy, leading to stronger superconducting coupling between the CuO planar blocks, which in turn leads to reduced rates of thermal depinning of vortices and improved current density, particularly in field.

But translating the early research results on biaxially textured YBCO into a full-scale manufacturing process will be a major challenge. There is no precedent for a thin-film deposition process of a material as complex as YBCO on a scale as vast as would be required for a bulk wire industry. A decade time scale is likely for the evolution of this technology; in the meantime, BSCCO sheathed wire is the vehicle for prototypes and early commercialization.

In summary, the ten-year history of HTS wire development has, and will continue to have, some surprising twists and turns, but particularly the BSCCO composite wire technology is making inexorable progress towards commercialization. The author thanks many colleagues at ASC for their comments on this manuscript and determined efforts to push this remarkable technology forward.

References

1. D. Driscoll et al., this conference..
2. A. Bolza, P. Metra and M. M. Rahman, this conference
3. E. Leung et al., this conference.
4. J. Willis, Los Alamos National Laboratory, private communication.
5. D. Gubser, Journal of Electronic Materials, to be published.
6. G. Yurek and J. Vander Sande, U. S. Pat. 4,826,808; 5,189,009; 5,204,318, filed Mar. 27, 1987.
7. D. Dimos, P. Chaudhari, J. Mannhart and F. K. LeGoues, Phys. Rev. Lett. 61, **219** (1988).
8. L. Woolf et al., Appl. Phys. Lett. **59**, 534 (1991).
9. K. Heine, J. Tenbrink and M. Thoener, Appl. Phys. Lett. **55**, 2441 (1989).
10. J. Kase et al., IEEE Trans. Magn. **27**, 1254 (1991).
11. M. Okada et al., Jpn. J. Appl. Phys. **34**, L981 (1995).
12. K. Sato et al., IEEE Trans. Magn. **27**, 1231 (1991).
13. Q. Li et al., Proc. CEC/ICMC Conference, Columbus OH, July 17-22, 1995.
14. J. Scudiere et al., Proc. CEC/ICMC Conference, Columbus OH, July 17-22, 1995.
15. S. Fleshler et al., to be published.
16. Y. Yamada et al., Proc. ISS '92, Kobe, Japan, Nov. 16-19 (1992).
17. Q. Li, K. Brodersen, H. A. Hjuler and T. Freltoft, Physica C **217**, 360 (1993).
18. D. C. Larbalestier et al., Physica C **221**, 299 (1993).
19. A. E. Pashitski et al., Physica C **246**, 133 (1995).
20. M. Lelovic, P. Krishnaraj, N. G. Eror and U. Balachandran, Physica C **242**, 246 (1995).
21. H. Itozaki et al., Sumitomo Technical Review **29**, 40 (1990).
22. G. Moore, *Electronics*, 1965
23. J. C. Grivel, A. Jeremie, B. Hensel and R. Flukiger, Supercond. Sci. Technol. **6**, 725 (1993).
24. Y. Iijima, N. Tanabe, O. Kohno and Y. Ikeno, Appl. Phys. Lett. **60**, 769 (1992).
25. H. Yoshino et al., in Advances in Superconductivity VI, ed. T. Fujita (Springer, Tokyo 1994).
26. X.-D. Wu et al., Appl. Phys. Lett. 65, 1961 (1994); S. Foltyn, MRS Spring Mtg., San Francisco, 1995.
27. A. Goyal, MRS Spring Mtg., San Francisco, April 1995; RABiTS is an ORNL Trademark

SUPERCONDUCTING ELECTRONICS -- THE NEXT DECADE

JOHN M. ROWELL
John Rowell Inc., 102 Exeter Drive, Berkeley Heights,
N.J., 07922, USA

The first decade of high temperature superconductivity can be described as the decade of materials science, in that great progress was made in understanding the physical properties of useful forms of the materials, say single crystals, films and wires. To justify the investment of the first decade, the second decade of the oxide superconductors must be the decade of markets. To gain some understanding of the implications of market success for superconducting electronics, I assume success by 2006 in all major markets and attempt to predict the advances that must be made in the technology. I conclude that the barriers to market success are no longer always limitations of the superconducting technology itself, but more often in what might be termed supporting technologies, such as refrigeration, substrates and test instrumentation. Success in 2006 also has broad applications beyond superconductivity, for example in the acceptance of cooled electronic systems.

1. Introduction

In preparing this talk for the Anniversary meeting in Houston, I was tempted to use as a title, "Superconducting Electronics, the Second Decade". However, that seemed unfair to all those, including myself, who built the foundations of superconducting electronics using simple materials such as tin, lead, niobium and aluminum. In comparison to the earlier three decades from 1956 to 1986, that from 1986 to 1996 has not been one of major advances in the technology of superconducting electronics, but rather one of major progress in the understanding and control of the properties of the HTS materials and their utilization in applications that were, except in one case (NMR coils), identified for LTS materials.

If the past decade can be described as the decade of materials science of the oxide superconductors, then the next decade of 1996 to 2006 must surely be the decade of markets. If superconducting electronics remains predominantly an R&D field in 2006, with limited commercial revenues, that will not be an acceptable outcome for funding agencies, shareholders, venture capitalists, or for many of those working in the field.

Given the time taken to transform a research idea, discovery or invention into a technology and then into manufactured products, it is highly likely that the major products of 2006 have already been identified as potential applications. New ideas should be expected, but it is unlikely that they will be translated into products with reasonable volumes within the decade. This implies that if we can define today's products, or those being seriously discussed today, then we can predict the technology required for 2006, at least in broad outline. That is the purpose of this talk, to attempt to describe the technology of 2006, and hence the technology and research needs of the superconducting electronics industry over the next decade.

2. The Products of 2006

Following the argument made above, the products of 2006 are already largely defined within the strategic business plans of the superconducting electronics companies. Within the US (I will confine my discussion largely to the US), these are BTi, Conductus, Dupont, Hypres, ISC, Neocera, Northrup Grumman, Quantum Design, SCT, STI, and TRW. The HTS products being discussed by these companies are, in order of market appearance, magnetic sensor systems, NMR probes, communications systems using passive components, MRI coils, and finally high speed digital instrumentation, switching and computing systems. The first HTS electronics product, "Mr SQUID", was sold by Conductus in 1991, followed by the iMAG system in 1994, while an NMR probe, co-developed with Varian, was just introduced in early 1996. A number of companies, lead by ISC, have demonstrated prototype filter systems in cellular base stations, and it is expected that some of these will be sold later in 1996. The development of HTS digital systems is expected to take some time, but LTS systems are likely to appear earlier, as I will discuss below.

At the system level, the enabling technology is refrigeration. It is clear that potential customers are, in almost all cases, much more concerned about the cost, reliability, size and performance of the cryocooler, than they are interested in the small enabling superconducting component buried somewhere within the cooled package.

At the component level, the superconducting devices that will make the products possible are SQUIDS, NMR and MRI RF coils, RF and microwave filters, and digital circuits. The critical enabling technology for all these components, and hence for all superconducting electronics products, is the ability to grow and process superconducting HTS films into devices with useful performance.

The markets and industry of 2006 will be built on this mix of components, perhaps with a few additional surprises. However, based on these components the variety of product applications that have been suggested and discussed for LTS and HTS electronics is quite large. I doubt that anyone would confidently predict which specific application will dominate the market in 2006. Some examples are as follows.

SQUID based products include research instruments; medical systems for the study of the brain, heart, fetal heart, stomach and nerves; geophysics instruments for the location of natural resources by surface and borehole mapping, for the detection of toxic waste and ordnance, and for earthquake prediction; and non-destructive evaluation of cracks and corrosion in aging aircraft, nuclear power plants, bridges and other metal infrastructure. Progress towards these applications has been quite rapid, as demonstrated by the over 3 orders of magnitude improvement in sensor performance achieved since 1988. The most complex HTSQUID system demonstrated to date is a 32 channel cardiology system built by the Superconducting Sensor Laboratory in Japan.

The use of coils to increase the signal to noise in NMR and MRI systems was the only electronics application of superconductivity that was not extensively discussed for LTS materials. The present product is a probe that replaces the standard NMR probe and is inserted into the base of the magnet of the NMR system, with the sample being inserted from the top. The tuned RF coil, made of YBCO, is cooled to below 30K by a recirculating flow of cold helium gas from a refrigerator, which is located remotely to avoid any disturbance of the highly uniform magnetic field of the spectrometer. Other products that are expected to follow within the next few years are more complex NMR probes with multiple coils, MRI coils, especially for less expensive MRI machines working at lower magnetic fields, and MR microscopes, which will allow rapid imaging of small animals and biopsy samples, and perhaps some types of in situ production monitoring.

It is the wireless communications market that has created the greatest investor interest in the superconducting electronics companies. Again, there are a number of possible applications being considered, both in the base station receivers of today's cellular systems operating near 880 MHz and in the receivers and transmitters of the emerging PCS market at frequencies close to 1.8 GHz. Even in some of the earliest prototypes, it is important to note that the advantages of hybrid systems are being realized, in that the cold package includes both the HTS filter and a low noise semiconductor amplifier. Interestingly, two types of HTS technology have been developed, one using thick YBCO films on the interior of rather conventional cavity filters, the other using a single layer of highly oriented HTS film on an epitaxial substrate with a ground plane, which is also an HTS film. The latter technology has the advantage of using all the thin film processing methods developed for semiconductors, with their implied cost reductions as volumes increase. Also, the small size of the thin film filters will allow multiple filters, say six or more, to be cooled by one cryocooler.

The digital superconducting applications of 2006 are expected to include voltage standards, high speed research and test instruments, communications switching systems, high speed optical/electrical interface components, and perhaps digital signal processors. Whether these systems will be built from LTS or HTS technology is an open question that I will discuss again later.

3. The State of Today's Technology

It is sensible to ask whether today's refrigeration and superconducting technologies are ready and adequate for the "decade of markets". I believe that the answer depends on the application under discussion.

For some products, today's cryocoolers are adequate, for others they are too expensive and not yet reliable enough.

The single layer HTS film technology is adequate for a variety of products now. Or at least, some films and devices have been made which demonstrate acceptable performance! These include single layer SQUID magnetometers, NMR and MRI coils and communications filters.

The HTS multilayer film technology is not yet adequate for SQUID magnetometers and gradiometers (although good progress has been reported recently), for multi-chip modules, or for digital circuits. However, the LTS digital technology is adequate for such products, in that fully integrated magnetometers and gradiometers have been sold for some time and 1 volt voltage standard chips are commercial, while the 10 volt chip under development contains 20,208 junctions. Prototype digital integrated Josephson junction circuits have been demonstrated to operate fully up to about 1000 gates.

4. The "Wildest Dreams Test"

To identify what needs to be done to improve and advance the relevant technologies over the next decade, I found it instructive to apply the "widest dreams test". This question I learned from Eric Nussbaum, a friend from Bellcore, who on hearing a proposal for research from one of his scientists or engineers, would ask, "If you succeed beyond your wildest dreams, then what?" He would expect to be told of real impact, either on science or technology.

In the spirit of this question, let me *assume* that *every* application discussed above is a success in the marketplace of 2006. (Please note that I am not *predicting* this will happen!). The interesting exercise is then to identify what success implies, to infer what advances and progress will be needed in both refrigerator and HTS technologies, and in manufacturing capabilities.

This exercise has two benefits. First, if (as is likely) every application is not a success, at least we have planned the technology development more than adequately! Second, capable technologies and performance are required for *any* market growth, so only the manufacturing volume in 2006 is in question.

5. Implications of success

Imagining this complete success in 2006, I concluded that the implications are broad and numerous, some examples being the following.

5.1. Acceptance of cryocoolers

The most important implication is that there will be widespread acceptance of refrigerators, including those at 60K, by a number of industries. This acceptance will be as complete as by the silicon chip manufacturers, who use cryopumps routinely without even knowing what 12 Kelvin means. The performance of all electrical devices improves at low temperatures. Cooling of different devices might result in one or more of these advantages; higher speed, lower loss, lower noise, increased signal to noise ratio, improved sensitivity, and decreased power dissipation. Hence, if cryocoolers are as routine in 2006 as fans are today, even if not as cheap, many devices and systems will be cooled below 300K. A further implication is that hybrid systems will be commonplace. We have already noted the first example of an HTS filter with a cooled low noise amplifier. There will be renewed interest in cold CMOS, and semiconductor technologies will be optimized for low temperature operation.

5.2. Thin Film Manufacturing Process

Market success implies that volume production methods for reproducible and uniform films, say from 5 to 15 cm diameter, will be in place with high yields and at acceptable costs. Epitaxial substrates, without the twinning exhibited by lanthanum aluminate, will have to be available in the required numbers and at lower cost than today. The most suitable film deposition methods will be selected as manufacturing processes on the basis of yield and cost, as all current growth methods seem to produce good films occasionally. While lithographic and processing tools will be adopted from the semiconductor industry, this can not be true for testing tools. Hence new instruments, which determine the important properties and performance of large area films, will have to be developed. These instruments will need to give answers in a time that is short compared to that taken to process and test a completed device.

5.3. Passivation, Packaging and Operating Lifetime

As the industry is already discussing the need for products to have a 5 year lifetime, it seems reasonable to assume that by 2006 it will be necessary for the thin film device, its package, the vacuum of the dewar package, and the cryocooler itself to all have an expected lifetime of 10 years.

5.4. The US Superconducting Electronics Companies

The HTS companies, which in the past have been an interactive part of the R&D community, will, given success, behave as typical US businesses. This seems to imply, at least for many US corporations, a declining support of R&D and decreasing contact with the knowledge base of the universities and national laboratories. The R&D community will find that ensuring their advances have impact on the SC companies will become increasingly difficult.

6. An Estimate of Film and Device Manufacturing Volumes in 2006

If we define market success in 2006, we can also estimate the volume of films and devices which will have to be manufactured, and identify which products will be the "drivers" of manufacturing volumes. This will raise the question of whether today's deposition processes will be extendible to those volumes.

6.1. Magnetic Sensor Products.

Of the SQUID applications mentioned earlier, the medical applications will require the manufacture of the largest number of sensors each year. I define success in this market as follows. All the 100 research hospitals of the US would have a 256 channel MEG system. Each of the 5,500 hospitals would have a 128 channel MCG system for arrhythmia diagnosis. Each of the 500,000 doctors (I will continue to use US numbers) would have a replacement for the ECG, which would probably use about 7 sensors, including those for ambient field suppression.

If I assume market adoption of these products is spread over 10 years, then 400,000 sensors, each 1 cm square, must be made, assuming 100% yield. I will address the question of yield later. Although larger substrates will be used, it is convenient to quote volumes in terms of 5 cm diameter wafers. The 400,000 sensors would mean 33,000 5 cm diameter wafers would have to be made and processed.

Other implications of this market success are that the currently used bicrystal SQUID junctions will have to be replaced by a more manufacturable junction process, a multilayer film technology for gradiometers must be developed, and unshielded operation in unfriendly magnetically noisy environments will be essential.

6.2. NMR/MRI Products.

Success in the NMR and MRI markets does not imply such a large volume of films. There are 3,000 existing NMR systems in the US, and 400 new systems per year. Assuming these numbers double by

2006, and that success is defined as one HTS probe per new system with retrofits spread over 5 years, then the market is 2000 probes per year, or 2000 5 cm diameter wafers per year (again at 100% yield).

At present there are 3000 MRI units in the US, so I estimate 8,000 by 2006 and define success as 2 coils per unit, adopted over 5 years. This means 3200 films of various diameters, probably from 5 to 15 cm, but for convenience I use 5 cm only.

6.3. Wireless Communications Products.

The communications markets have the potential to create the greatest demand for films. There are currently 20,000 cellular base stations in the US, with 50,000 estimated in 2006. Success can be defined as 10% of these stations being equipped with an HTS system each year, each system having 6 filters, for a total of 30,000 5 cm wafers per year. The PCS market has almost no base stations at present, but 70,000 are projected by 2000. There are no projections to 2006, so I doubled the number to 140,000 and defined success as 10,000 systems per year, each with 12 HTS filters, which would require 60,000 5 cm wafers per year.

6.4. Digital Instrumentation, Switching and Computing Products.

Predictions of needs for superconducting digital electronics are more difficult to make, in that the relative maturity of two technologies in 2006 needs to be estimated. The question is whether LTS digital technology, which is relatively mature compared to HTS, combined with a 4K refrigerator, for which the technology is immature, is more likely in the commercial products of 2006 than HTS digital technology (immature) used with 50K coolers (mature). It is my understanding that Boreas, a small US company, is now funded to develop by the end of 1996 a compact 4K cryocooler which will be self contained in a relay rack sized unit about 60 cm high, and which will operate on less than 15 amps at 110 volts from a standard US wall outlet. While space applications of digital SC technology might still require more efficient refrigeration, and hence HTS electronics, many customers would find this 4K cooler to be quite acceptable, if it is reliable and cost effective. For the purposes of the present estimates of film volumes, I will assume that digital applications will not be a major factor in producing demand for HTS films. However, HTS digital electronics is a major factor in the technology strategy, as it demands the development of a multilayer film and device technology.

6.5. Film Volumes in 2006

While success in digital and the other markets mentioned earlier will add somewhat to the total number of wafers, success in medical SQUID applications, NMR/MRI and wireless communications implies, for the US alone, about 125,000 5 cm wafers per year. This number will be reduced as larger area wafers become routinely available. If I assume a yield of 50%, the number increases to 250,000 wafers which must be made, processed and tested in 2006. As most of the communications wafers will have film on both sides, the number of depositions is about 430,000. Within the accuracy of these estimates, and including the other applications, I can use 500,000 depositions as a useful number.

7. Film Manufacture, Substrates and Refrigerators.

It is natural to ask whether the deposition methods being used at present can be extended to half a million films per year. As an example, consider a coevaporation system of the type developed at the Technical University of Munich at Garching. It has a 25 cm diameter heater which can produce nine 5 cm wafers per run, with two runs per day being easily possible. So one such system can make 6,000 single sided films per year. Roughly eighty systems of even the current type could satisfy the demand in 2006, and it is straightforward to make the systems larger. So a path to the required film volume, using this or other deposition methods, seems predictable.

The projections for substrate availability are not so optimistic. The number of lanthanum aluminate substrates made and sold in 1995 was between 3000 and 4000. Success in 2006 will demand 250,000, an increase in production of close to 100 times. A further challenge is that lanthanum aluminate is far from an ideal substrate, so it is probably true to say that the ideal substrate for the SC electronics industry has not yet been manufactured in *any* volume.

However, these volumes are small compared to the silicon industry. In 1995, 2.645 billion (10^9) square inches of silicon substrates were manufactured worldwide. Although these were of larger areas, this is equivalent to almost one billion 5 cm wafers, a volume 4000 times greater than projected for SC. So, if the demand does indeed increase for lanthanum aluminate or for an improved substrate, the silicon industry is an example that such volumes are quite possible. (It is amusing to use these wafer areas to estimate market revenues. Although it is difficult to get good estimates, it is believed that the 2.6 billion square inches of silicon substrates enable a total market of about one trillion dollars, or about $385 of enabled product per square inch. This would imply a superconducting electronics market of close to $300 million enabled by the 750,000 square inches of substrates for the SC industry. It is also likely, for early products, that the cost of the system level product made possible by each wafer will be higher for superconductors than for semiconductors, which would increase this estimate of the SC market size.)

There is both good and bad news regarding the impact of success on the cryocooler industry. The total number of systems described for 2006 is 5560 for Squids, 5200 for NMR/MRI, and 15,000 for communications, for a total of about 26,000. The production of cryocoolers of all types in 1995 was about 30,000. So the good news is that the industry would seem quite capable of manufacturing the volume needed for SC electronics. The bad news is that the increase in demand is probably not large enough to drive down the price to levels required for some of the SC applications. However, the demand from the power SC industry should also be considered, and the biggest factor by far could be the widespread use of cooled semiconductor electronic systems.

8. Implications of Success for the Research Community

While the implications of success for the superconductor industry have been the primary topic of this talk, there are obvious implications for the research community that are worth lengthy discussion. However, space constraints allow only a brief summary.

The need for high yield of all HTS films and devices requires a much better understanding of the relationships between their growth, microstructure, physical properties *and relevant device performance* than presently available. As one route to optimization of film and device performance, much more work needs to be done on doping of the 123 compounds. Far too much current research deals only with YBCO, even though substitutions of other elements can be made on every site in the structure. Improved substrates must be developed. New test instruments need to be invented to rapidly screen the *device relevant* properties of films and the performance of devices. The reliability of components must be improved by passivation and by doping of the films. I know of only one such program at present in the US, at the University of Texas at Austin. A large fraction of the research which presently deals only with single layer films should instead be focused on issues of growth on non-planar substrates, e.g. over steps and into vias. Questions of optimum oxygen doping at interfaces, over steps and in the lower levels of multilayer structures must be investigated. And finally, as implied earlier, as the superconducting electronics companies become typical US businesses, they will do less applied research and more product development. Thus the community must find ways to do more HTS applied research, which is neglected in US universities, and explore ways to make that research of value to the growing SC electronics industry.

9. Summary

My "wildest dreams success" exercise has not identified any insuperable barriers to the growth of the SC electronics industry through the next decade. Many of the challenges are in the supporting infrastructure of the industry, which has not yet received the attention given to the superconducting technology itself.

APPLICATIONS OF HTS TO ELECTRONICS-A DOD PERSPECTIVE

STUART A. WOLF
Naval Research Laboratory,
Washington, D.C. 20375-5000
and
DARPA/Defense Sciences Office
Arlington VA, 22203

and

FRANCIS W. PATTEN
DARPA/Defense Sciences Office
Arlington VA, 22203

ABSTRACT

Over the last decade there have emerged many electronic applications of HTS technology which have important military as well as commercial implications. Examples of these include space communications, terrestrial wireless communications, radar, surveillance, countermeasures, medical applications and cryo-electronics. I will review many of the efforts in this area that have been supported by the DoD which has invested over $350M on this technology. In addition I will discuss the efforts to develop cryocoolers to support the HTS technology.

1. Introduction

In this review article, I will briefly describe some of the electronic applications of HTS technology that have been supported by the DoD since 1987. Most of these applications have important military potential as well as commercial implications. These efforts that I will describe have had substantial support from DARPA (formerly ARPA). The areas in which these applications fall are space electronics, communications, radar, countermeasures, medical applications and cryo-electronics. In addition, I will describe several efforts to provide reliable, affordable cryocoolers which are an enabling technology for any superconductivity application.

2. The High Temperature Superconductivity Space Experiment (HTSSE)

The HTSSE program was one of the first and largest programs that was initiated to advance HTS technology. This program started the end of calendar 1988 and was aimed at demonstrating the viability of superconducting devices in a space environment. The program was envisioned to have three phases. Phase I was the demonstration of single components. Phase II, the demonstration of small systems and subsystems and Phase III would be an operational satellite. As Phase I evolved it became clear that the devices that were be maturing fastest were passive microwave components and they quickly became the focus of this phase of the program. In fact it became clear that the focusing of HTSSE on microwave devices was instrumental in the rapid development of high quality films and the techniques for patterning them that have enabled this area to approach commercialization. Industry was galvanized and there were more than 20 contributors to this first effort which successfully built a payload that consisted of 15 microwave devices that included resonators, filters, a bulk cavity resonator a patch antenna, and a coupler that was fully tested and space qualified. The payload was fully integrated onto a launch vehicle. The devices were to be measured in space with a special NRL built and space qualified network analyzer and the results were to be transmitted back to NRL. Unfortunately, the satellite never achieved orbit and this opportunity was lost. The HTSSE II payload has recently been built and space qualified. It consists of eight communications subsystems including multiplexers, receivers, filterbanks etc. and is ready to be shipped to Rockwell to be integrated onto the ARGOS I spacecraft. This system will operate like a conventional communication satellite in that signals will be send from the ground, processed on the satellite and returned to the ground. The launch date has suffered several setbacks but is currently slated for mid 1997. We expect that the successful completion of this mission will give spacecraft designers and builders the confidence they need to embark on the next phase which will be a operational satellite.

3. Cryogenic-Radar

One of the most pressing problems faced by the Navy is the threat of cruise missiles launched from a hostile shore. In this littoral environment, the clutter that is associated with various land masses can mask the missile and render it nearly impossible to detect. The very low surface resistance of superconducting materials and the complementary low loss of high quality sapphire have enable the development of a highly stable reference oscillator (STALO) that can lower the phase noise floor of a ship defense radar by a significant amount enough so, that it is projected that a cruise missile can be picked out of the doppler clutter. A STALO with these characteristics will be tested this summer with a state of the art ship defense radar. In addition to the STALO, a switched filterbank will be incorporated at the front end, behind the dish to filter any unwanted signals from overloading the radar. This can be very effective against a jammer attempting to defeat the radar set.

The expected reduction in the noise floor of the radar with a superconducting front end puts an additional burden on the A/D converter that is essential in any radar to digitize the waveforms for processing. Superconducting digital technology offers a unique solution for the very high resolution A/D converters that are being targeted for this application (20 bits at 100 MHz). This will allow the full dynamic range of the radar to be processed.

Recently it was realized that a very low phase noise waveform generator can be built using superconducting technology. This waveform generator will allow the production of many single tones as well as the production of complicated waveforms that will significantly enhance the capabilities of advanced radar sets and push superconductivity into other modalities of radars. This may be one of the most important military applications of superconducting electronic technology.

4. Cryo-Communications

Interference in communications is a significant problem for the military as well as commercial communications. The ability to build very low loss, extremely high performance filters will push superconductivity into many markets. One of the first commercial applications of superconducting filters and cryogenic low noise semiconducting amplifiers will be for cellular and PCS systems. Several superconductivity vendors have demonstrated system performance that significantly exceed the performance of conventional filters. The remaining issues are reliability and cost. The ability to reduce the intereference from competing systems has very significant military implications as well. There are a plethora of military communication systems spanning many decades of frequency space. Often these systems are co-located and they can and do interfere with each other since small non-linearities present in one system cause spillover into a neighboring band causing interference. Superconductivity offers the ability to provide excellent filtering on small, mobile platforms as well as on aircraft and ships. This interference problem exists within all of the military services and often is very severe.

5. Countermeasures

Many of our advanced aircraft suffer from an interoperability problem between the radar and the radar warning receivers. The newest generation of radars hop in frequency so that they avoid being jammed. However the radar warning receivers that are currently in operation do not have the ability to notch out the rapidly hopping radars and are thus overloaded by the radar on the same platform or from a neighbor in the formation. Again, superconductivity provide a very convenient solution to this problem. Very compact and switchable HTS filters can be built that can effectively notch out the frequency hopping radar and enhance the survivability of the aircraft. Simulations show that many more types of missions can be flown if this capability becomes operational. Currently there are programs targeting upgrades for the B1-B and the F-15.

6. Medical Applications

Superconducting pickup coils can reduce the noise inherent in nuclear magnetic resonance systems. When the signal to noise ratio is limited by the instrument rather than the sample than superconductivity can play a role. MRI applied to limbs or breasts are an example of when the signal is limited by the pickup coil rather than the object being measured. For whole body MRI, it is usually the body that is the dominant noise source. Thus, in the former case a superconducting pickup coil can enhance the quality of image or reduce the time necessary to produce an image. For NMR spectroscopy and or pathology superconductivity can also significantly enhance the performance in the same way. Either smaller samples can be measured or they can be scanned more rapidly.

7. Cryo-Electronics

Early in the DARPA (ARPA) HTS program, a study pointed to an opportunity to utilize superconducting interconnects on a multi-chip-module (MCM) to provide an unprecedented level of interconnection between very high performance semiconductor chips. A program was established to develop all of the required infrastructure to build such a superconducting MCM. This program was a true precompetitive consortium involving many of the HTS vendors, universities and other labs. It reached a significant level of maturity, however the missing link was the semiconductor chips to operate at 77 K and the cryo-cooler. The effort punted and now is focused on the low temperature optimization of the semiconductor chips. In this regard a potential breakthrough was uncovered which, if realized would allow the development of a new type of FET which would be highly optimized at cryogenic temperatures and would offer an order of magnitude higher speed at two orders of magnitude lower power. Thus even if the power to cool the chip were included there would still be a significant overall gain. If this all comes to fruition then the opportunities presented by superconducting interconnects will again come to the forefront.

8. Cryocooler Development

The cryocooler development that has been sponsored by DARPA is extremely important for the future of superconducting electronics. Without reliable, affordable cryocoolers it would be impossible for superconducting or cryo-electronic systems to be commercialized and it would make the military applications more difficult in the current military procurement environment where leveraging the commercial marketplace is considered highly desirable if not essential. In this regard we have sponsored a program run through NRL that has contracted with six of the major cooler manufacturers to develop a series of cryocoolers that have greater than a three year mean time to failure, will cost about $1000.00 and are targeted at several applications including the ones that have been mentioned above. This program started about one year ago and several of the vendors will have very reliable coolers. The cost goal may be too ambitious for many of the vendors but significant cost reductions over the current state-of-the-art coolers will certainly be achieved. A new effort is slated to begin next year to further the cryocooler development as well as to expand into the area of thermoelectric cooler development. This thermoelectrics program will initially focus on materials improvements with the goal of finding new compounds or structures that will enable thermoelectrics to overlap with superconductivity.

9. Acknowledgments

I have not specifically named any of the companies involved in these efforts in the text since this is a generic review. However this work would not be possible without the efforts of APD Cryogenics, Com Dev, Conductus, Cryomech, CTI, Dupont, Hughes Aircraft, Lincoln Laboratory, Loral, MMR, NASA-Lewis, NRL, JPL, SAI, SCT, STI, TRW, Westinghouse, Wright Laboratory, and many others.

THE SYMMETRY OF THE ORDER PARAMETER IN THE CUPRATE SUPERCONDUCTORS

J. F. ANNETT
Department of Physics, University of Bristol, Royal Fort, Tyndall Avenue
Bristol BS8 1TL, UK

N. D. GOLDENFELD
Department of Physics, University of Illinois at Urbana-Champaign, 1110 W. Green St.
Urbana, IL 61801, USA

and

A. J. LEGGETT
Department of Physics, University of Illinois at Urbana-Champaign, 1110 W. Green St.
Urbana, IL 61801, USA

ABSTRACT

The symmetry of the Cooper-pair internal wave function (order parameter) under the operations of the relevant crystal group is an important clue to possible microscopic mechanisms of superconductivity in the cuprates. We review the currently available experimental evidence concerning this symmetry, with particular emphasis on the Josephson experiments of the last three years. We particularly stress (a) the relevance of thermodynamic considerations in constraining possible forms of the order parameter (b) the importance, in interpreting the Josephson experiments, in distinguishing those conclusions which rely on symmetry arguments alone from those which require additional assumptions about the form of matrix elements, etc., and (c) the fact that in the analysis of experiments of this type performed on YBCO the complications associated with the orthorhombic anisotropy and bilayer structure of this material are very largely irrelevant. We conclude that while there is no single choice of order parameter which is compatible with all the existing experiments, the choice which is consistent with by far the largest fraction is the so-called $d_{x^2-y^2}$ state.

1. Introduction

In this talk we assume that the cuprate superconductors, like the classical ones, are characterized by an "order parameter" (anomalous expectation value, Cooper pair wave function...) of the form

$$F_{\alpha\beta}(\mathbf{r},t: \mathbf{r'},t') \equiv \langle \psi_\alpha(\mathbf{r}t) \psi_\beta(\mathbf{r'}t') \rangle \qquad (1)$$

and review the current experimental evidence regarding the symmetry of this quantity under the operations of the relevant crystal point group. In view of space limitations we will mainly confine ourselves to stating our conclusions, and refer the reader to our longer paper (ref. (1)), hereafter referred to as AGL, for derivations and qualifications. We make no pretense, in this abbreviated account, to complete citation of the relevant literature.

To keep the discussion manageable we will make from the start a few additional assumptions:
(1) The quantity F defined by eqn. (1) is nonzero for $t = t'$.
(2) The spin structure of F is a singlet.
(3) F is nonzero for (at least some values of) $\mathbf{r}$ and $\mathbf{r}'$ within the same CuO2 plane.
(4) F is invariant under crystal translation of the center of mass coordinate $(\mathbf{r} + \mathbf{r}')/2$, the relative coordinate $\mathbf{r} - \mathbf{r}'$ being held fixed.

With these assumptions it follows that for a single-plane material we may take as the object of our study the Fourier transform with respect to relative coordinate of the expression (1) evaluated for $\alpha = -\beta$, $t = t'$, i.e. the quantity

$$F_{\mathbf{k}} \equiv \langle a_{\mathbf{k}\uparrow} \, a_{-\mathbf{k}\downarrow} \rangle \qquad (2)$$

where $\mathbf{k}$ is a two-dimensional vector lying in the ab-plane, and moreover that $F_{\mathbf{k}}$ must have even parity under inversion.

2. The Allowed States in Tetragonal Symmetry: Thermodynamic Constraints

The irreducible even-parity representations (I.R.'s) of the group C_{4v} of the square are given in the Table. The third and fourth columns indicate the parity under $\pi/2$ rotation and inversion in a crystal axis respectively.

Informal name	Group-theoretic notation	$R_{\pi/2}$	I_{axis}	Representative state
$s+$	A_{1g}	+1	+1	const.
$s-$ ("g")	A_{2g}	+1	−1	$xy(x^2 - y^2)$
d_{x2-y2}	B_{1g}	−1	+1	$x^2 - y^2$
d_{xy}	B_{2g}	−1	−1	xy

There is, of course, no *a priori* constraint that the pair wave function (2) must correspond to one of the forms listed in Table I, i.e. to a single I.R.; it could in principle be a superposition of two or more of them with arbitrary complex coefficients.

However, the free energy, expressed as a function of the complex amplitudes of the different I.R.'s which can be represented in the order parameter F, must be invariant under any operation of the crystal point group, and inspection of the third and fourth columns of the Table then shows that in any term involving three or less (2) I.R.'s only even powers of each can occur. But with any free energy expression of this type it is easy to show that the occurrence in the actually realized F of more than one I.R. inescapably requires not one but at least two phase transitions, which in the absence of pathological degeneracy will occur at different temperatures and show up as (at least) slope discontinuities in experimentally measured quantities such as the penetration depth (as well as in thermal anomalies). To the extent that these are observed not to occur, only single I.R.'s are possible.

3. Bulk Experiments

If the microscopic theory is BCS-like, the energy gap $|\Delta_\mathbf{k}|$ is (roughly) proportional to $|F_\mathbf{k}|$. From the earliest days of research on the cuprates considerable evidence has accumulated that many of the low-temperature properties have a power-law temperature dependence rather than the exponential form predicted by BCS theory and verified in the traditional superconductors; on these experiments, see e.g. ref. (3). If taken at face value, this evidence would suggest that the minimum value of the energy gap is either zero or at least very small (probably less than 1K in YBCO). The most convincing experiments in this class are probably those on the electromagnetic (London) penetration depth, which shows a crossover from linear dependence on temperature, as $T \to 0$, to quadratic with increasing doping; this is precisely the behavior predicted in a BCS-type theory in which the gap is assumed to have nodes (see AGL section 5.1), and it is not easy to think of alternative explanations.

More recently, various experiments have attempted to probe the k-dependence of the energy gap directly. While none of these is completely clear-cut, the least ambiguous are probably the experiments on angularly resolved photoemission spectroscopy (ARPES). All such experiments appear to agree that the value of the energy gap along the 45° axes (ΓX direction) is, at least, considerably smaller than along the crystal axes, and most though not all are consistent with its being zero.

Not all experiments on the bulk cuprates are limited to studying the energy gap; in a BCS-like theory, the expressions for various scattering probabilities, etc., involve also coherence factors which are sensitive to the sign as well as the magnitude of the order parameter. In most such cases the predictions for different pairing states differ quantitatively rather than qualitatively, but a striking exception is Raman scattering in the so-called B_{1g} geometry[4] compared with that for A_{1g} and A_{2g}; for an s^+ (or d_{xy}) state the absorption should vary as ω in all three cases, while for a $d_{x^2-y^2}$ (or s^-) state it should vary as ω^3 for the first case but as ω in the latter two; these predictions do not depend on the detailed form of the energy gap but only on the symmetry, and it is the latter which is verified experimentally.

4. Josephson (Quantum Phase Interference) Experiments

These involve two fundamental principles. Principle A applies to a single junction, and states that two I.R.'s which behave differently under a symmetry operation which leaves both the bulk crystal lattices and the junction itself invariant cannot couple so as to produce a Josephson current. Principle B relates to a complete circuit: It states that under the normally realized conditions, the change in phase of each of the I.R. amplitudes are going around a complete circuit must be equal to $2\pi\Phi/\phi_0$, where Φ is the total flux (external plus self-induced) threading the circuit and ϕ_0 is the usual flux quantum, h/2e.

We may divide existing experiments on quantum phase interference into three types. Type I experiments use a single c-axis-oriented Josephson junction between (say) YBCO and an "old-fashioned" superconductor such as Pb in which the order parameter is reliably known to be of the simple s^+ type. They exploit only principle A: In tetragonal symmetry, the only state which can couple to s^+ is $d_{x^2-y^2}$, and hence observation of a finite critical current (hence finite ΔE) establishes prima facie that the order parameter of YBCO has at least some s+ component. (For the complications associated with orthorhombic symmetry, see section 5).

Type-II experiments involve, again, YBCO and an "old-fashioned" superconductor, but now in a "dc-SQUID" type geometry or something similar, with a closed circuit interrupted by two junctions parallel to the ac- and bc-planes of the YBCO respectively (see fig. 4 of AGL). It is immediately clear that in such a geometry any s^- or d_{xy} component in the YBCO can give no contribution to the coupling energy across the individual junctions and hence will not be detected in the experiment. On the other hand the s-s coupling coefficient is the same for the two junctions (assumed for simplicity to be identical), while the s-d one changes sign (here s is shorthand for s^+ and d for $d_{x^2-y^2}$). Let us consider (despite the arguments of section 3!) an arbitrary superposition of the s^+ and $d_{x^2-y^2}$ states with a relative phase ϕ. Then, minimizing the total energy of the circuit (i.e. the sum of the ΔE's of the two junctions) [5] we find that the total phase accumulated around the circuit (which may be measured by the flux it produces, or by the interference effect on the parallel critical current of the circuit) is given by the expression

$$\Delta\phi_{tot} = \tan^{-1}\left(\frac{-2\gamma\sin\phi}{1-\gamma^2}\right), \quad \gamma \equiv A_{ss}/A_{ds} \tag{8}$$

where A_{ss} is the YBCO s-wave amplitude times an appropriate coupling coefficient, etc., and where to avoid ambiguity we specify that for $\gamma \sin\phi > 0$ we take $\Delta\phi_{tot} = \pi$ for $\gamma \sin\phi \to 0+$ and continue the inverse tangent from there (so that pure s corresponds to $\Delta\phi_{tot} = 0$ and pure d to $\Delta\phi_{tot} = \pi$). Note that for a real s-d mixture this experiment can give only the sign of $|A_{ss}| - |A_{sd}|$.

Type-III experiments involve only different grains of YBCO or some other high-temperature superconductor, which in the general case are oriented at arbitrary angles relative to one another and to the grain boundaries; they invoke both our principles, A and B. It should be emphasized that for the general case the only conclusion which can be drawn from symmetry arguments <u>alone</u> is that observation of a finite $\Delta\phi_{tot}$ excludes a pure (or dominating) nodeless s^+ state. However, for certain particular grain and boundary orientations (fortunately realized, approximately, in some of the existing experiments) one can relate some or all of the coupling constants for the different junctions to one another and hence draw more detailed conclusions: see AGL.[6]

5. Effects of Orthorhombicity and Bilayer Structure

Space allows us to state only our conclusions concerning these complications and we must refer the reader to AGL for their justification: Provided that the order parameter behaves "gyroscopically" across twin boundaries (an assumption for which we believe the experimental evidence to be convincing), and provided we are in the "thermodynamic limit" (an unbiased mixture of many twins), then we believe that, at least as far as macroscopic properties such as the Josephson behavior are concerned, we can simply introduce a "twin-averaged" order parameter which by construction transforms according to C_{4v} and whose analysis then proceeds exactly as above (including the arguments of section 3). Thus heavy twinning in effect simply restores the tetragonal symmetry. (For untwinned samples, or for local probes such as ARPES, orthorhombicity of course still needs to be taken into account, but does not usually make a qualitative difference). As to the bilayer structure of YBCO, the more complicated forms of order parameter permitted by it exacerbate rather than alleviating the problem of reconciling the various experiments.

6. Conclusion

If we are prepared, despite the arguments of section 3, to consider superpositions of more than one I.R., then there is one (and, as far as we can see, only one) uniform choice of order parameter which is consistent with all the experiments, namely a state which is predominantly $d_{x^2-y^2}$ but with a fairly small real admixture of s^+: see AGL section 9. We emphasize that this admixture has nothing to do with the orthorhombicity of YBCO, but is the manifestation of an extra "broken symmetry"; and that it would require thermodynamic behavior for which to put it conservatively, there is at present no positive evidence.

If we regard the last remark as excluding bulk superpositions, and also assume absence of surface T-violation, then there appears to be no choice of order parameter which is consistent with all the experiments. However, the one which is consistent with by far the largest fraction is the pure $d_{x^2-y^2}$ state.

This work was supported by grants NSF-DMR-93-14938 (NG) and NSF-DMR-91-20000 through the Technology Center for Superconductivity (AJL) and by the Office of Naval Research grant ONR-N00014-95-1-3038 (JA). In addition to the colleagues already acknowledged in AGL, we would like to take this opportunity to thank D. J. Scalapino and C. Irwin for helpful discussions.

References

1. J. F. Annett, N. Goldenfeld and A.J. Leggett, in *Phyical Properties of High Temperature Superconductors V*, ed. D. M. Ginsberg, World Scientific, Singapore 1996.
2. Mixing of all four I.R.'s does not invalidate the argument.
3. J. F. Annett, N. D. Goldenfeld and S. R. Renn, in *Physical Properties of High Temperature Superconductors II*, ed. D. M. Ginsberg, World Scientific, Singapore 1990.
4. The notation "B_{1g}" etc. refers (here only!) to the symmetry of the photon pair, not that of the order parameter.
5. For simplicity of presentation we neglect the self-inductance term (for which see AGL).
6. We take this opportunity to note an error in eqn. (7.3) of AGL: the sine in the numerator should be squared.

Correlations between T_c and n_s/m^* in High-T_c Cuprates [*]

Y.J. Uemura

Physics Department, Columbia University, New York, NY 10027, USA

Muon spin relaxation (μSR) measurements of the magnetic field penetration depth λ in cuprate superconductors revealed interesting correlations between T_c and n_s/m^* (superconducting carrier density / effective mass). These correlations and the "pseudo gap" behavior lead us to a picture of high-T_c superconductivity in terms of crossover from Bose-Einstein to BCS condensation.

Muon spin relaxation (μSR) is one of the best existing methods for determining the magnetic field penetration depth λ. Soon after the discovery of the cuprates, we started plotting T_c versus the muon spin relaxation rate $\sigma(T \to 0) \propto 1/(\lambda_{ab})^2 \propto n_s/m^*_{ab}$ (superconducting carrier density / in-plane effective mass) to study the doping dependence of T_c [1]. With further accumulation of data in cuprates [2] and other superconductors [3], we obtained Fig. 1, which demonstrates that T_c increases linearly with increasing n_s/m^* in the underdoped region, with a slope common to most of the cuprates. Other exotic type-II superconductors, such as organic BEDT, doped C_{60}, BKBO, Chevrel phase and heavy-fermion systems also showed high ratios of $T_c/(n_s/m^*)$.

Using values of m^* from quantum oscillation measurements, one can estimate the number of carriers per area of the coherence length squared ξ_{ab}^2 on the conducting plane. Both cuprates and organic BEDT systems have only several pairs per ξ^2, which is a situation between the limit of BCS (more than 10,000 carriers / ξ^2) and Bose Einstein (BE) (1 molecule / ξ^2) condensation, but impressively close to the latter [3].

The horizontal axis of Fig. 1 can be converted to an energy scale [3], as the Fermi temperature T_F is proportional to n_{2d}/m^* in 2-d systems. σ can be combined to the Pauli susceptibility or the Sommerfeld constant to obtain T_F in 3-d systems. With this procedure, we obtain an "effective Fermi temperature" which represents an energy scale for translational motion of superconducting carriers, i.e., the spectral weight of the sharp Drude part condensing into the superfluid below T_c in the terminology of optical conductivity.

In the plot of T_c versus T_F shown in Fig. 2, we find that $T_c/T_F = 0.1 - 0.01$ for these "exotic" superconductors is much higher than $T_c/T_F < 0.001$ for more conventional BCS superconductors. The broken line indicates the BE condensation temperature T_B for a free non-interacting Bose gas with the mass $2m^*$ and density $n_s/2$. Compared to T_B, the T_c of underdoped cuprates and BEDT systems is reduced only be a factor of 4 to 5. This reduction is not surprising in view of overlapping several pairs / ξ^2 in plane, since even the lambda transition at T=2.2 K of ^{4}He is about 50 % reduced from $T_B = 3.2$ K for an ideal Bose gas.

Figure 2 represents an empirical method to classify different superconductors in the limit of BE condensation (in real space with a non-retarded strong interaction) and BCS condensation (in momentum space with a retarded weak interaction). In this sense, the underdoped cuprates are impressively close to BE condensation. Moreover, the linear dependence of T_c versus n_s/m^* can be expected in BE condensation, while BCS theory predicts much less explicit and weaker dependence of T_c on carrier density. These results suggest the importance of considering the physics in terms of a crossover from BE to BCS condensation, as we mentioned in Ref. 3.

Let us now consider a simple model with an effective attractive interaction mediated by an exchange boson having an energy scale $\hbar\omega_B$. We start doping this system with carriers. In the low density limit where $\epsilon_F \ll \hbar\omega_B$, pairs are formed at a temperature corresponding to the binding energy, and then undergo a BE condensation at a much lower temperature. In this low density BE limit, T_c will follow the behavior of T_B. The doping level of $\epsilon_F \sim \hbar\omega_B$ will separate the low density BE-like region and high-density BCS-like region. The pair formation and superconducting condensation will occur at separate temperatures in the former low density region, while at the same temperature in the latter high density region. Assuming that these two regions connect smoothly, we obtain a phase diagram as sketched in Fig. 3(a).

If we identify the "pseudo gap" behavior observed in NMR, neutron, and dc- and ac- conductivity measurements as a signature of a pair formation in the normal state, we can

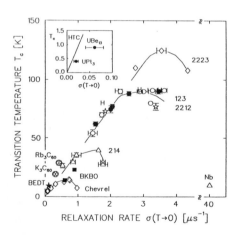
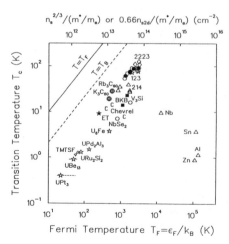

Fig. 1. A plot of the muon spin relaxation rate versus T_c [2,3].

Fig. 2. A plot of the effective Fermi temperature versus T_c [3].

regard Fig. 3(a) as the phase diagram of the high-T_c cuprates, with the pseudo-gap temperature T^* as the pair formation line. The underdoped region corresponds to the low-density side while the overdoped region to the high-density BCS side. The optimum-T_c occurs where these two descriptions cross over. A strong support for this picture comes from the c-axis conductivity in the underdoped region. The insulating behavior below the pseudo-gap temperature T^* results from a diminished interplanar hopping (or tunnelling) probability of 2-e pairs as compared to that of a single fermion. The present author has proposed this crossover picture at several conferences since 1993 [4-6].

The expected behavior of the energy gap is shown in Fig. 3(b). The real BCS gap will connect to the pseudo gap in the BE-region, and then to the (molecular-like) binding energy in the low density limit. This is why the BE condensation is not associated with a gap at T_B. Various phenomena should reflect both BCS and BE characters in the crossover region. Some k-dependence for a pseudo gap, for example, can be expected as a remainder of the BCS situation. In this sense, the "pre-formed" pairs will be substantially extended but not quite a local object in most of the underdoped region.

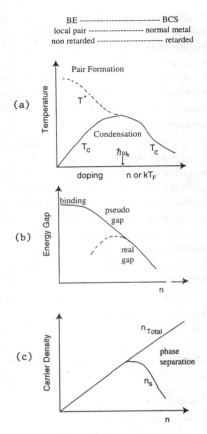

Fig. 3. Conceptual (a), (b), and observed (c) behavior in the high-T_c cuprate superconductors.

Using this picture, we can estimate the energy scale of the exchange boson $\hbar\omega_B$ to be comparable to $T_F \sim 2,000$ K of the cuprates in the optimum T_c (i.e., the crossover) region. This energy scale corresponds well to the mid-infrared reflection (MIR), suggesting that MIR is a manifestation of the pairing interaction unlike the Drude part for the translational motion. This energy scale $\sim 2,000$ K agrees very well with the antiferromagnetic exchange interaction J, thus providing a strong support to the picture that the exchange of spin fluctuations is indeed the microscopic origin of high-T_c superconductivity.

A strong deviation from the naive picture in Fig. 3(a) has been found in the behavior of n_s/m^* in the overdoped Tl2201 systems by μSR [7,8]. There, n_s/m^* *decreases* with increasing carrier doping. The overdoped Tl2201 cuprates are extremely good conductors with mean free paths more than 200-500 Å. Then, this reduction can not be ascribed to the pair-breaking scattering. Instead, the most likely scenario is a microscopic phase separation into superconducting and normal-metallic regions [7]. Figure 3(c) illustrates this situation.

The decrease of n_s/m^* with increasing Zn concentration in our recent μSR measurements on Zn-doped 214 and 123 systems indicates that the superconducting pairs in the area of $\pi\xi_{ab}^2$ around each Zn are localized, while other parts of the planes maintain superconductivity. This "Swiss cheese" like situation is reminiscent of superfluid ^{4}He in porous media, again suggesting fundamental importance of BE condensation.

Emery and Kivelson [9] recently considered low-density superconductors, and presented an argument similar to our picture using the terminology of the amplitude of the order parameter instead of "pre-formed pairs" and long-range phase order instead of BE-like condensation in the underdoped region. With the most-recent photo-emission results of pseudo gaps [10], the "pre-formed pair" can now be considered as a real possibility. Further experimental and theoretical studies will elucidate this crossover and hopefully lead to the understanding of the mechanism of high-T_c superconductivity.

References

* Supported by NSF (DMR-95-10453, 10454) and NEDO (Japan)
[1] Y.J. Uemura et al., Phys. Rev. B38 (1988) 909.
[2] Y.J. Uemura et al., Phys. Rev. Lett. 62 (1989) 2317.
[3] Y.J. Uemura et al., Phys. Rev. Lett. 66 (1991) 2665.
[4] Y.J. Uemura, presented at Euroconference on Superconductivity in Fullerenes, Oxides and Organic Materials, Pisa, Italy, January, 1993: proceeding of this conference was not published.
[5] Y.J. Uemura, proceedings of the Workshop on Polarons and Bipolarons in High-T_c Superconductors and Related Materials, Cambridge, UK, April, 1994, *Polarons and Bipolarons in high-T_c Superconductors and Related Materials*, ed. by E.K.H. Salje, A.S. Alexandrov and W.Y. Liang, Cambridge University Press, 1995, p.p.453.
[6] Y.J. Uemura, lecture given at CCAST Symposium, Beijing, May, 1994, in *High-T_c Superconductivity and the C_{60} Family*, ed. by S. Feng and H.C. Ren, Gordon and Breach, 1995, p.p.113.
[7] Y.J. Uemura et al., Nature 364 (1993) 605.
[8] Ch. Niedermayer et al., Phys. Rev. Lett. 71 (1993) 1764.
[9] V.J. Emery and S.A. Kivelson, Nature 374 (1995) 434. Emery and Kivelson argued that they provided a "re-interpretation" of the results in Refs. 1-3. However, most of the seed ideas leading towards the crossover picture were given in Refs. 1-3, and the crossover picture was obtained in Refs. 5-6 as a natural extension of these seed ideas, without involving "re-interpretation".
[10] See, for example, papers by Z.X. Shen and by J.C. Campuzano at this conference.

RAMAN SPECTROSCOPY: PHONONS

MANUEL CARDONA, XINGJIANG ZHOU, THOMAS STRACH
Max-Planck-Institut für Festkörperforschung
Heisenbergstr. 1, D-70569 Stuttgart, Germany

The contributions of Raman spectroscopy to the basic properties and the characterization of high T_c superconductors are briefly reviewed with particular emphasis on the role of phonons.

1 Introduction

It was realized soon after the discovery of the high T_c superconductors that Raman spectroscopy should be an excellent tool to study their low temperature excitations, some of which play an important role in superconductivity and related phenomena.[1,2] The early measurements were performed on ceramic samples: in the case of the Zurich oxides (214), however, such samples led to spurious data because of "burning" of the material by the laser power.[3] Reliable data for 214's had to await the availability of single crystals.[3,4] The Houston compounds (123) were shown to yield reliable data for phonons even with ceramic samples.[1,2] However, an identification of the observed phonon modes was first made possible through measurements on single crystals.[5] $YBa_2Cu_3O_7$ was shown to exhibit five strong modes of A_g symmetry corresponding to Ba (120 cm^{-1}), CuII (150 cm^{-1}), OII-OIII vibrating in opposite directions in each plane (340 cm^{-1}), OII-OIII vibrating in the same direction (440 cm^{-1}) and the apical oxygens (500 cm^{-1}), all vibrations along the c-axis. The 340 cm^{-1} phonon has B_{1g} symmetry if the crystal is assumed to be tetragonal (D_{4h} point group like $YBa_2Cu_3O_6$) and thus displays very characteristic selection rules, of $d_{x^2-y^2}$ type, which led to its definitive identification.[5] The Raman modes of B_{2g} and B_{3g} symmetry (polarized along z, x and z, y), considerably weaker, were observed and interpreted[6] on the basis of *ab initio* calculations of their frequencies performed within the LDA electronic theory.[7] Particularly striking was the demonstration that the observed absolute scattering efficiencies, and their resonances versus laser frequency, agree with *ab initio* electronic calculations of the Raman polarizabilities.[8]

The B_{1g} (340 cm^{-1}) phonon of 123 played an important role in the investigation of electron-phonon interaction. It undergoes an anomaly at T_c in its frequency [9,10] and linewidth [11] which was early attributed to phonon renormalization by electron-phonon interaction.[10] Similar effects have been recently observed by inelastic neutron scattering spectroscopy.[12] They have been used to estimate the superconducting gaps and the electron-phonon interaction constants.[13,14]

These phonon renormalizations imply the presence of a continuum of electronic excitations which can also be seen in Raman scattering and discriminated from spurious effects by the observed opening of a gap below T_c,[15] no doubt related to the superconducting gap. It has been suggested that polarized Raman measurements at low T give information about gap symmetry and, in the cases of YBaCuO, BiSrCuO, and TlBaCuO, support a $d_{x^2-y^2}$-like gap. This is presently the object of a heated controversy.[16,17,19] A remarkable achievement,

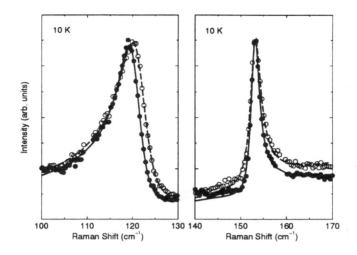

Figure 1: The lowest frequency A_g Raman phonon modes of YBa$_2$Cu$_3$O$_7$ for two different Ba isotopes (● ^{138}Ba and ○ ^{134}Ba). The solid and dashed lines represent fits with Fano lineshapes.

however, is the calculation of the absolute scattering efficiencies of the electronic continuum which agree rather well with experimental observations.[18,19]

Other types of electronic excitations can be observed with Raman spectroscopy. For compounds containing lanthanides (RE) crystal field transitions involving the $4f$ levels of the RE-atom can be observed, either as mixed electron-phonon excitations [20] or as pure electronic excitations, in the materials with the T' structure (e.g. Nd$_2$CuO$_4$).[21]

Another type of electronic excitations accessible to Raman spectroscopy are magnons. Scattering by two magnons, each at the edge of the Brillouin zone, was early observed at ≈ 3000 cm^{-1} in insulating or underdoped cuprates [22] but the high intensity of the observed scattering was a puzzle. Calculations based on the t-J model show that strong scattering is indeed expected.[23]

2 Phonons: Some Recent Results

2.1 Mixing of the Cu and Ba A_g-modes in $(RE)Ba_2Cu_3O_7$

These modes appear at ~ 120 and 150 cm^{-1}. Usually, the lower frequency mode would be attributed to vibrations of the heavier atom (Ba), the higher one to those of the CuII in the CuO$_2$ planes. However, since both frequencies are rather close, mixing of the pure Cu or Ba eigenvectors can occur. Evidence for strong mixing (about 50%!) was provided by the phonon eigenvectors based on an *ab initio* evaluation of the total energy vs. atomic displacement from LDA electronic band structure calculations.[24] A number of experiments suggested, however, nearly unmixed Cu and Ba eigenvectors.[25] This matter was settled by measuring the effect of isotopic substitution on the phonon frequencies: if the eigenvector involves vibrations of only one kind of atom of mass M, its frequency should vary like

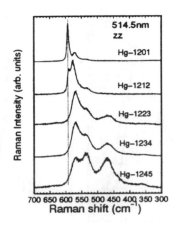

Figure 2: Raman spectra of as-grown (before laser annealing) crystallites of Hg-12$(n-1)n$.

$M^{-1/2}$. If this is not the case, a simple calculation yields the mixture of eigenvectors.[26]

Figure 1 shows the 120 and 150 cm^{-1} modes for YBa$_2$Cu$_3$O$_7$ prepared with the two isotopes ^{138}Ba and ^{134}Ba. The lines represent fits with Fano line shapes (which imply the interaction of theses phonons with an electronic continuum). One can see with the naked eye that the 120 cm^{-1} mode shifts appreciably when the Ba mass is changed, while the 150 mode barely shifts. A quantitative analysis reveals that the mode admixture is no more than 10%. Similar results have been obtained by isotopic substitution of the copper.[25,26] The reason why the *ab initio* calculations led to a much higher degree of mixing is to be found in inaccuracies of the calculations which yield nearly degenerate *uncoupled* modes. A small amount of coupling results then in ~50% mixing. An analysis of the isotopic substitution data yields a much larger splitting of the *uncoupled* modes.[26] The small coupling matrix element found in the calculations does not suffice to induce significant coupling.

2.2 Raman modes related to oxygen in the mercury-based cuprates

The family of mercury cuprates HgBa$_2$Ca$_{n-1}$Cu$_n$O$_{2n+2+\delta}$ (referred to as 12$(n-1)n$) has thus far exhibited the highest critical temperature, in particular under pressure ($T_c \approx$ 160 K). The stoichiometric materials ($\delta = 0$) are insulators, an excess of oxygen (most likely to go to the Hg-planes) introduces holes as charge carriers and leads to superconductivity.

The vibrational Raman spectra of the Hg-12$(n-1)n$ compounds are still controversal, especially in the region where oxygen vibrations occur (300 – 600 cm^{-1}).[27-31] Most available samples are polycrystalline but it is possible to select single crystals for measurements by micro-Raman techniques. The spectra are strongest in zz configuration (A_{1g}) and are shown in Fig. 2 for $n = 1$ to 5. The peak seen for $n = 1$ at 592 cm^{-1} has been attributed to A_g phonons of the apical oxygens. It decreases in intensity and even disappears with increasing n. This fact can be interpreted as induced by increasing oxygen disorder (most likely in the Hg planes): this disorder lifts the k = 0 selection rule and thus replaces the sharp peak at 592 cm^{-1} by broad bands. Three bands are seen in all compounds of Fig. 2 below the 592 cm^{-1} peak (centered at ~570, 540 and 470 cm^{-1}). An assignment of all these

FABRICATION OF HgBa$_2$Ca$_x$

R. L. MENG, B. R. HICKEY, Y. Q. WANG, L. GAO,
C. W. CHU

Texas Center for Superconductivity and Departm.
University Houston, Houston TX 77204-.

ABSTRACT

HgBa$_2$Ca$_2$Cu$_3$O$_{8+\delta}$ (Hg-1223) exhibits superconducting properties superio. other cuprates. Thus, it is a good candidate for making superconducting ta, wires. Unfortunately, the complex chemistry of Hg-1223 makes such fabri. difficult. Therefore, we have carried out a systematic study on the formation chemical stability of Hg-1223, effect of substitution, Hg-substrate reaction an. intergrain couplings. By selecting Ni as the metallic substrate with a thin buffer layer of Cr, we have prepared thick tapes of Hg-1223 with a $T_c \sim 130$ K and a $J_c \sim 2.5 \times 10^4$ A/cm^2 in self-field at 77 K.

1. Introduction

In addition to the highest transition temperature (T_c) of 134 K at ambient pressure [1] and 164 K at 30 GPa [2], the Hg-based compound exhibits a modest flux pinning strength between that of YBCO and the Bi-based cuprates, suggesting a very high critical current density (J_c) at 77 K [3]. Indeed, an epitaxial thin film of Hg-1212 showed a T_c of 128 K and a J_c of 10^6 A/cm^2 at 100 K [4]. This implies that much higher J_c of 10^7 A/cm^2 might be achievable at 77 K. Given its superior superconducting properties, Hg-1223 is therefore a desirable material for superconducting tapes or wires.

However, Hg-1223 is extremely difficult to form due to its complicated chemistry. The significant Hg-loss above 350 °C makes it difficult to align the grains through the high-temperature melt-texture technique commonly used in other HTS materials. In addition, Hg attacks many metals, especially at high temperatures. Therefore, the selection of an appropriate substrate also becomes a challenge. To overcome these problems, we carried out a systematic study on the formation and chemical stability of Hg-1223. We also investigated the effect of cation substitution, and the effect of heat treatment on grain alignment. By selecting Ni as the metallic-substrate with a thin buffer layer of Cr, we have obtained $T_c \sim 130$ K and $J_c \sim 2.5 \times 10^4$ A/cm^2 at 77 K in as-synthesized Hg-1223 tapes.

2. Compound Formation and Chemical Stability

...tion of Hg-1223 depends critically on the Ba-Ca-Cu-O precursor used. We ...posed [5] a procedure for the production of a better precursor that is less ... to CO_2/H_2O and easier to form Hg-1223 by optimizing its phase composition. ...ther discovered that the grinding, mixing, and pressing of precursors can all be ... out in open air instead of inside a glove box, if small amount of Pb-, Mo-, or Re-... was added to the starting material, in agreement with previous reports [6,7]. ...iously, the time required for the synthesis of Hg-1223 was very long, *i.e.* 5 to 65 hr ...], even with these cation substitutions. By partially substituting Hg-halogens for ...O, we shortened the formation time to 1 hr, and increased the grain-size of Hg-1223 ...].

The chemical stability of the Hg-1223 samples with different cation substitutions was examined by measuring their Hg-vapor pressure P_{Hg}. The results are shown in Figure 1. It is evident that P_{Hg} of Hg-1223 at 750 °C is reduced by cation substitution of Mo, Re, and Pb. In other words, the chemical stability at 750 °C increases in the order of Hg-1223, (Hg,Mo)-1223, (Hg,Re)-1223, and (Hg,Pb)1223. However, at higher temperatures, *e.g.* at 850 °C, (Hg,Re)-1223 has the lowest P_{Hg} and thus the highest stability. Therefore, (Hg,Re)-1223 is the most desirable if the grain alignment, which requires a high synthesis temperature, can be achieved. However, for low-temperature process, which is preferred for the suppression of the Hg-substrate reaction, the best choice should be (Hg,Pb)-1223.

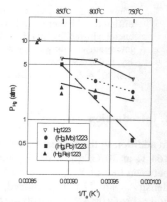

Figure 1. Hg vapor pressure of undoped and doped Hg-1223 between 750 and 850 °C.
* denotes decomposition.

3. Tape Processing and Characterization

The selection of a proper substrate is very important in making tapes. Preferably the substrate of a HTS tape should be metallic, flexible, chemically stable under processing conditions, strongly adhesive to the HTS layer, and inexpensive. Ag, the most commonly used substrate in HTS wire and tapes, reacts with Hg even at low temperatures while stainless steel is easily oxidized at high temperatures. Therefore, Ni was chosen for its low solubility in Hg, *e.g.* 8.5×10^{-3} wt% at 550 °C. Unfortunately, the precursor does not stick well on Ni-substrates after pressing. And prolonged exposure to Hg vapor at ~ 870 °C still causes corrosion on the Ni-surface. Therefore, various buffer layers were tested. We observed that a Cr buffer-layer drastically improves the adhesion of the precursor. It is also known that Cr is ~ 100 times less reactive with Hg than Ni. Hence, we coated both sides of the Ni-substrates with a 500 to 800 Å thick buffer-layer of Cr by sputtering.

The spray process was chosen for the coating of the precursor due to its relatively low cost and its potential for commercial scale-up. To avoid any contamination from the organic binder, isopropyl alcohol was used as the solvent. Cold pressing or cold rolling was applied to improve the density, eliminate voids, and promote the grain alignment of Hg-1223 after processing, which is known to be related to intergranular connection.

A set of quench experiments were carried out to obtain the optimal synthesis temperature. In these experiments, the cold-pressed precursor/substrate assembly was sealed inside a quartz tube together with a solid Hg-source. The tube was first heated up to a elevated temperature T_a in 3 hr and then kept at the temperature for 2 hr before being quenched to room temperature. As shown in Figure 2, $CaHgO_2$ and $BaCuO_x$ are the dominant phases for T_a between 600 °C and 700 °C; Hg-1212 starts to appear around $T_a \sim$ 600 °C and becomes the major phase around $T_a \sim$ 800 °C. Hg-1223 forms only in a rather narrow T_a window of 800 to 870 °C. Above 880 °C, Hg-1223 decomposes. The formation of Hg-1223 appears to be diffusion-rate limited. A synthesis temperature of 840 to 870 °C was chosen based on these measurements.

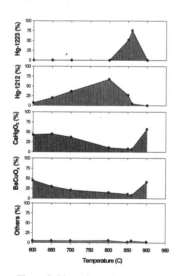

Figure 2. Phase formation of the various compounds at different temperatures.

The effects of the heating and cooling rates were also tested. In a reference schedule, the sealed quartz tube was first heated to the synthesis temperature in 5 hr, maintained at the temperature for 5 hr, and then slowly cooled to room temperature over 4 hr. The scanning electron micrograph of the tape so prepared displays stacks of c-axis oriented Hg-1223 platelets interrupted by grains with other orientations. Unreacted $BaCaO_2$ and $CuCa_2O_3$ can also be found. In an alternate heating schedule, we heated the sample up to the synthesis temperature in 5 hr., maintained it at this temperature for 2 hr. and cooled it to 820 °C in 5 hr., followed by a rapid cooling to room temperature. Bigger Hg-1223 grains were obtained but with greater amount of impurity phases, presumably due to the decomposition of Hg-1223 resulting from a prolonged exposure to temperatures above 820 °C. We have also shortened the initial heating time to 1 hr. while keeping the remainder of the original heating schedule unchanged. Samples so prepared were found to be covered by platelets of c-oriented Hg-1223 grains with a small amount of impurity phases, as shown by the SEM micrographs in Figure 3. Using this last heating schedule, we have obtained highly c-oriented as-synthesized tapes that show a T_c onset at 130 K and reach zero resistance at ~

117 K. A J_c of 2.5×10^4 A/cm^2 at 77 K was observed in self-field over a length of 1 cm, using a pulse technique and a 50 μV/cm criteria [10].

In conclusion, highly c-axis oriented Hg-1223 thick films (~ 10 μm) have been made for the first time on a flexible Ni tape with a thin Cr buffer layer using the controlled vapor/solid reaction. $T_c \sim 130$ K and $J_c \sim 10^4$ A/cm^2 were obtained at 77 K. The morphology and the orientation of the Hg-1223 grains of the tape depend sensitively on the processing parameters in a yet-to-be determined manner. However, it is expected that a higher J_c is achievable once the processing parameters are optimized.

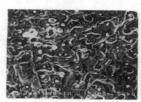 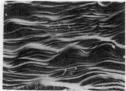

Figure 3. SEM micrographs of the tape made in the fast-heating schedule. Top view (left) and cross sectional view (right).

Acknowledgment

This work is supported in part by EPRI RP-8066-04, NSF Grants No. DMR 95-00625, the State of Texas through the Texas Center for Superconductivity at the University of Houston, and the T. L. L. Temple Foundation.

References

1. A. Schilling, M. Cantoni, J. D. Guo and H. R. Ott, Nature 363, 56 (1993).
2. L. Gao, Y. Y. Xue, F. Chen, Q. Xiong, R. L. Meng, D. Ramirez, C. W. Chu, J. H. Eggert and H. K. Mao, Phys. Rev. B 50, 6260 (1994).
3. Z. J. Huang, Y. Y. Xue, R. L. Meng and C. W. Chu, Phys. Rev. B 49, 4218 (1994).
4. L. Krusin-Elbaum, C. C. Tsuei and A. Gupta, Nature 373, 679 (1995).
5. Q. M. Lin, Z. H. He, Y. Y. Sun, L. Gao, Y. Y. Xue and C. W. Chu, 254, 207 (1995).
6. J. Shimoyama, K. Kishio, S. Hahakura, K. Kitazawa, K. Yamaura, Z. Hiroi and M. Takano, in *Advances in Superconductivity*, vol. 2, edited by K. Yamafuji and T. Morishita (Springer, Tokyo, 1995), pp. 287-290.
7. Y. Y. Xue, Z. J. Huang, X. D. Qiu, L. Beauvais, X. N. Zhang, Y. Y. Sun, R. L. Meng and C. W. Chu, Mod. Phys. Lett. B 7, 1833 (1993).
8. T. Goto, Jpn. J. Appl. Phys. 34, 555 (1995).
9. R. L. Meng, B. R. Hickely, Y. Y. Sun, Y. Cao, C. Kinalidis, J. Meen, Y. Y. Xue and C. W. Chu, to appear in Physica C.
10. R. L. Meng, B. R. Hickey, Y. Q. Wang, Y. Y. Sun, L. Gao, Y. Y. Xue and C. W. Chu, to appear in Appl. Phys. Lett.

STRONG FLUX PINNING, ANISOTROPY AND MICROSTRUCTURE OF $(Hg,Re)M_2Ca_{n-1}Cu_nO_y$ (M=Ba,Sr)

Jun-ichi Shimoyama, Koichi Kitazawa and Kohji Kishio

Department of Superconductivity, University of Tokyo,
7-3-1 Hongo, Bunkyo-ku, Tokyo 113, Japan

ABSTRACT

Effects of Re-doping on the superconducting properties were studied for nearly single-phased samples of Hg-based superconductors, $(Hg,Re)Ba_2Ca_{n-1}Cu_nO_y$ (n=1~4, x=0.02~0.3) and $Hg_{0.75}Re_{0.25}Sr_2Ca_2Cu_3O_y$. The $HgRe_xBa_2Ca_2Cu_3O_y$ sample showed high T_c ~135K and quite high irreversibility fields up to 8T at 86K. This dramatic improvement in the flux pinning is explained by a decrease of the electromagnetic anisotropy, which originated from the introduction of highly conductive local structure, ReO_6, into the Hg(O)-layer as well as the carrier overdoping state.

1. Introduction

The Hg(Ba)-based superconductors, $HgBa_2Ca_{n-1}Cu_nO_y$ (n=1,2,3,4,···; Hg(Ba)1201, Hg(Ba)1212, Hg(Ba)1223, Hg(Ba)1234, ···), have been well known to show record-high T_c's up to 134K[1-3]. However, they are chemically unstable against an ambient air unless they are completely single phased specimen, which are hard to be obtained due to the difficulty in the control of Hg content during the high temperature heat-treatment.
In addition, the irreversibility fields (B_{irr}) for $HgBa_2Ca_{n-1}Cu_nO_y$ at high temperatures were relatively low in spite of their possible advantages of being high T_c's. The reported B_{irr} for $HgBa_2Ca_2Cu_3O_y$ was approximately 2T at 77K[4] which is significantly lower than that of the $YBa_2Cu_3O_y$ polycrystalline samples which has much lower T_c of 92K.
Recently, we have found that the small amount of Re-doping was effective to overcome these problems[5]. For the Re-doped samples, $HgRe_{0.1}Ba_2Ca_{n-1}Cu_nO_y$, any deterioration was not observed after storage for a long period in open air and high irreversibility fields up to 6.5T at 77K was achieved.
In the present paper, we report the synthesis of $HgBa_2Ca_{n-1}Cu_nO_y$ (n=1,2,3,4) having various doping level of Re and dramatic improvement in the flux pinning properties by an increase of Re concentration as well as the additional oxidizing treatment.

2. Experimental

Samples with nominal compositions of $HgRe_xBa_2Ca_{n-1}Cu_nO_y$ [n=1,2,3,4; x=0.02~0.3] were prepared through the two-step reaction technique using high purity powders (≥99.9%) of HgO, $BaCO_3$, $CaCO_3$, CuO and ReO_3. Starting materials except HgO were first mixed, calcined at 905°C for 12h, thoroughly ground, re-calcined at 900~930°C for 12h and quenched to room temperature. The resulting precursors were mixed with HgO and pressed into pellets 10mmϕ × 1~3mm in dimension. All these procedures were carried out in ambient air atmosphere. Each pellet was sealed in an evacuated quartz ampoule, sintered at 670~860°C for 5~55h and finally quenched to room temperature. Some samples were post-annealed at 350~500°C for 20h under $P(O_2)$~2atm.

A $Hg_{0.75}Re_{0.25}Sr_2Ca_2Cu_3O_y$ sample was prepared by the high-pressure synthesis technique, under 6GPa at 1200°C, starting from a powder mixture of HgO and precursor material, $Re_{0.25}Sr_2Ca_2Cu_3O_y$ (nominal composition)[6]. Vacuum annealing at 450°C were carried out on the as-prepared sample.

Crystal structure was analyzed by the powder X-ray diffraction method using Cu-K_α radiation. Magnetic susceptibility and magnetic hysteresis loops were measured by means of a SQUID susceptometer and a vibrating sample magnetometer, respectively. The B_{irr} was determined as the field where the width of magnetic hysteresis fell below 0.02emu/cm^3. Since observed mean grain size of samples in the present study was ~5μm, this criterion corresponds to intragrain J_c of approximately 10^{3}A/cm^2.

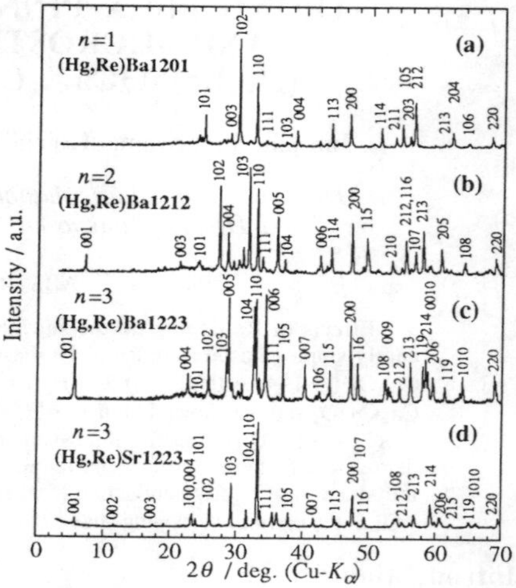

Figure 1.
Powder X-ray diffraction patterns of $HgRe_{0.25}Ba_2Ca_{n-1}Cu_nO_y$ [n=1~3; (a)~(c)] and $Hg_{0.75}Re_{0.25}Sr_2Ca_2Cu_3O_y$ (d) (all nominal compositions).

3. Results and Discussion

3.1. Crystal Structure

Figures 1(a)-(d) show the powder X-ray diffraction patterns of $HgRe_{0.25}Ba_2Ca_{n-1}Cu_nO_y$ (n=1~3) and $HgRe_{0.25}Sr_2Ca_2Cu_3O_y$. Small amounts of impurity phases still remained, however, the aimed Hg12(n-1)n phases formed as the main phase. On the contrary, the Re-free $HgBa_2Ca_{n-1}Cu_nO_y$ (n=1~4) prepared with the same conditions (their XRD patterns are not shown) contained large amounts of impurity phases, such as $CaHgO_2$, $BaCuO_2$, besides none or very small amount of the Hg-based superconducting phases. The c-axis lengths, c_0, of the $HgRe_xBa_2Ca_{n-1}Cu_nO_y$ samples slightly depend on the Re-doping level, x, and were 0.1~0.2Å shorter than the reported values for Re-free compounds[1-3]. The c_0's were expressed as 9.3$_8$+3.1$_6$(n-1)Å and 9.3$_3$+3.1$_8$(n-1)Å for sample series of x=0.1 and 0.25, respectively. The shorter c-axis lengths in the Re-doped samples provide an evidence of Re incorporation in the crystal structure. The c_0 of $Hg_{0.75}Re_{0.25}Sr_2Ca_2Cu_3O_y$ was much shorter, ~15.1Å, reflecting the smaller ionic size of Sr than Ba. Recently, Chmaissem et al. reported that the substitution site of Re in $(Hg,Re)Sr_2Ca_{n-1}Cu_nO_y$ (n=2,3) was the Hg site and each Re ion formed a slightly distorted ReO_6 octahedron[7]. The same structure is supposed to be formed in the present Re-doped $HgBa_2Ca_{n-1}Cu_nO_y$ compounds.

3.2. Superconducting Properties

D.C. susceptibility measurements revealed that the all of the obtained Re-doped $HgBa_2Ca_{n-1}Cu_nO_y$ were superconductors showing large diamagnetism and relatively sharp transition. Figure 2 summarizes the T_c's of the $HgRe_xBa_2Ca_{n-1}Cu_nO_y$ (n=1~4, x=0.02~0.3, as-prepared) samples and those reported for Re-free compounds. The T_c's of n=1 and n=2 samples drastically decreased with increasing x, while those for n=3 were almost constant approximately 135K against x variation. In the case of the n=4 compound, the T_c gradually increases with increasing x. Assuming the Re ions also form ReO_6 octahedra in the crystal structure of $HgRe_xBa_2Ca_{n-1}Cu_nO_y$ samples, each introduces four excess oxygen ions into

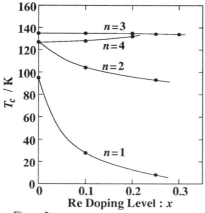

Figure 2. Dependence of T_c on the Re-doping level for HgRe$_x$Ba$_2$Ca$_{n-1}$Cu$_n$O$_y$ (n=1~4; nominal compositions).

the Hg(O)-plane. Since the Re ions are heptavalent at most, the Re-doping essentially results in hole carrier doping to the present system. The observed T_c behavior is well explained by this hole doping mechanism by the Re-doping. For as-prepared samples, we concluded that the carrier doping states were overdoped for n=1 and 2, optimally-doped for the n=3 and lightly underdoped for the n=4. After the oxygen annealing, T_c of the n=3 sample decreased to 127~132K corresponding to its carrier overdoped state.

The T_c of Hg$_{0.75}$Re$_{0.25}$Sr$_2$Ca$_2$Cu$_3$O$_y$, whose T_c was originally ~100K, was optimized by the moderate vacuum annealing up to 123K.

3.3. Performances under Magnetic Fields

Magnetic field dependence of intragrain J_c's at 77K, which are calculated from the magnetic hysteresis using Bean's critical state model, for HgRe$_x$Ba$_2$Ca$_2$Cu$_3$O$_y$ and Hg$_{0.75}$Re$_{0.25}$Sr$_2$Ca$_2$Cu$_3$O$_y$ samples are shown in Fig. 3. For comparison, intragrain J_c -B curves for YBa$_2$Cu$_3$O$_y$ and Re-free HgBa$_2$Ca$_2$Cu$_3$O$_y$[4] are also shown. It was found that the intragrain J_c's of the present Re-doped compounds were higher than those of YBa$_2$Cu$_3$O$_y$ and HgBa$_2$Ca$_2$Cu$_3$O$_y$ and their dependence on the magnetic field was relatively small. Increasing Re-doping level and further oxidizing treatment were apparently effective to enhance the intragrain J_c.

Figure 4 summarizes the irreversibility lines obtained in the present study. The data with error bars were estimated by extrapolation of ΔM vs magnetic field curves. The B_{irr}'s of as-prepared HgRe$_{0.25}$Ba$_2$Ca$_2$Cu$_3$O$_y$ sample was approximately 4.6T at 77K which is almost doubled compared with the reported Re-free compound[4]. Moreover, both the oxygen annealing and the increase of Re doping level lead further enhancement of the B_{irr}. The highest B_{irr} was observed for oxygen annealed HgRe$_{0.25}$Ba$_2$Ca$_2$Cu$_3$O$_y$, at approximately 8T at 86K. This value is believed to be the highest one among all of the superconductors as for the randomly oriented polycrystals. It was already reported for (La,Sr)$_2$CuO$_4$, Bi$_2$Sr$_2$CaCu$_2$O$_y$ and YBa$_2$Cu$_3$O$_y$ that the irreversibility line strongly depends on the doping state and the samples in over-doping state show steeper slope of irreversibility line against temperature corresponding to the decrease in the electrical anisotropy[8]. Therefore, it is reasonable that the oxygen annealed sample shows the steeper irreversibility line. As was discussed before, if the Re substitutes in the Hg-site with forming ReO$_6$ octahedron similarly as in the binary oxide ReO$_3$, which is well known to have extraordinarily high conductivity, it is possible to explain that the drastic improvement in magnetic properties by the Re-doping originates from a decrease in the electrical anisotropy.

The irreversibility line of the vacuum annealed Hg$_{0.75}$Re$_{0.25}$Sr$_2$Ca$_2$Cu$_3$O$_y$, which is believed to be in the optimally-doped state, locates at higher field than that of the as-prepared HgRe$_{0.25}$Ba$_2$Ca$_2$Cu$_3$O$_y$, while it has lower T_c by ~12K. This can be explained in terms of the shorter interlayer distance between CuO$_2$-planes, which is believed to further improve the flux pinning properties[9,10].

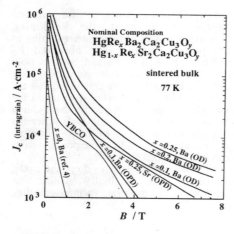
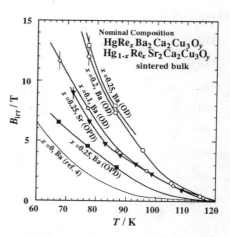

Figure 3.
Intragrain J_c vs. B curves for Re-doped Hg-based superconductors.
[OD=overdoped; OPD=optimally-doped]

Figure 4.
Irreversibility lines of $HgRe_xBa_2Ca_2Cu_3O_y$ and $Hg_{0.75}Re_{0.25}Sr_2Ca_2Cu_3O_y$.
[OD=overdoped; OPD=optimally-doped]

4. Conclusion

The Re-doped Hg-based compounds showed excellent performance in the mixed state. Both an increase of Re-doping level and the additional oxidation are effective to enhance the intragrain J_c' and B_{irr} at high temperatures. Our present study suggests that the Hg-based compounds should be considered again as promising candidate materials for practical high-field applications at high temperatures.

5. Acknowledgment

The authors are very grateful to Mr. K. Yamaura, Profs. Z. Hiroi and M. Takano of Kyoto University for high-pressure synthesis and useful discussions.

6. References

1. S.N. Putilin et al., *Nature(London)* **362** (1993) 226.
2. A. Schilling et al., *Nature(London)* **363** (1993) 56.
3. E. V. Antipov et al., *Physica* **C215** (1993) 1.
4. U. Welp et al., *Physica* **C218** (1993) 373.
5. J. Shimoyama et al., *Adv. Superconductivity* **VII** (1994) 287.
6. K. Yamaura et al., *Physica* **C246** (1995) 351.
7. O. Chmaissem et al, submitted to *Phys. Rev.* **B**.
8. K. Kishio et al., *Physica* **C235-240** (1994) 2775.
9. T. Nabatame et al., Physica **C193** (1992) 390.
10. J. Shimoyama et al., *Physica* **C235-240** (1994) 2795.

FABRICATION OF HIGH QUALITY Hg-1212 AND Hg-1223 THIN FILMS

J.Z. WU, S.H. YUN, W.N. KANG, B.W. KANG and A. GAPUD
Department of Physics & Astronomy, University of Kansas, Lawrence, KS 66045

S.C. TIDROW and D. ECKART
US Army Research Laboratory, Fort Monmouth, NJ 07703-5601

ABSTRACT

A fast temperature ramping Hg-vapor annealing method has been adopted to reduce the formation of $CaHgO_2$ impurity phase and the film/substrate interface chemical reaction in fabrication of Hg-based superconducting thin films. Using this technique, high-quality Hg-based cuprate thin films with zero-resistance T_c above 130 K have been obtained reproducibly.

Superconductivity above 130 K discoverd recently in Hg-based cuprates ($HgBa_2Ca_{n-1}Cu_nO_{2n+2+\delta}$, n =1,2,3,4) has generated much excitement because a higher superconducting transition temperature (T_c) implies a higher device operation temperature and better performance at a give temperature. To make an effective use of these materials in microelectronic device applications, it is important to fabricate high quality Hg-based superconducting thin films.

Limited progress has been achieved [1-3] due to high volatility of the Hg-based compounds which causes two major problems in thin film growth: film/substrate interface chemical reaction and formation of $CaHgO_2$ impurity phase. In the conventional annealing process previously used for Hg-based cuprates (both bulk and thin films), the sample temperature is increased slowly (4-6 hours) to the annealing temperature ($\sim$ 780-860 °C) in order to maintain a phase equilibrium between the precursor and the unreacted $HgO+Ba_2Ca_{n-1}Cu_nO_x$ pellet which are sealed together in a evacuated quartz tube. Since HgO decomposes at around 500 °C, two problems can arise using this slow heating cycle. First, above 500 °C the Hg^{+2} begins to react with Ca to form $CaHgO_2$ which seriously degrades the superconducting properties of the sample. Second, the high temperature processing time (500 °C to annealing temperature) is unnecessarily long which increases the problem of film/substrate interface chemical reaction and interdiffusion. In order to solve these problems, we adopted a fast temperature ramping Hg-vapor annealing process in which the furnace temperature is ramped to the annealing temperature within 1 to 15 minutes and the sample is annealed for 5-30 minutes at the annealing temperature. With an additional follow-up O_2 anneal at 300-400 °C, good-quality Hg-1223 cuprate thin films have been obtained on $LaAlO_3$ substrates. In the article, we reported our experimental results on fabrication and characterization of high quality Hg-1212 and Hg-1223 thin films using the FTRA technique.

The fabrication of Hg-1223 films consists of three steps [4,5]: (1) rf sputtering of non-Hg-containing precursor films ($Ba_2Ca_2Cu_3O_x$); (2) post-annealing of the precursor films at 780-840 °C with an unreacted $HgBa_2Ca_2Cu_3O_x$ pellet and a non-Hg-containing $Ba_2Ca_2Cu_3O_x$ pellet encapsulted in a precleaned and evacuated quartz

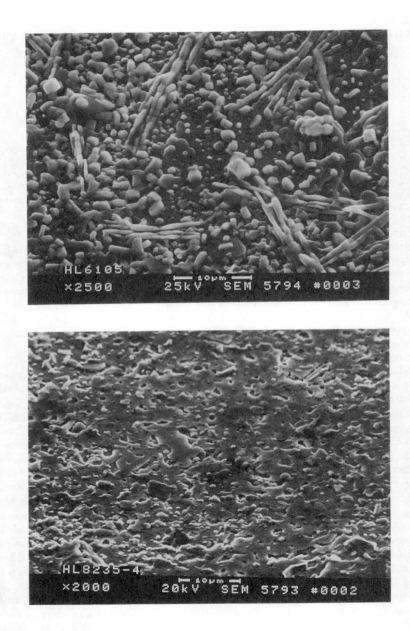

Figure 1. SEM micrographs of Hg-1223 films gorwn on LaAlO$_3$ substrates using in (a) STRA and (b) FTRA processes respectively.

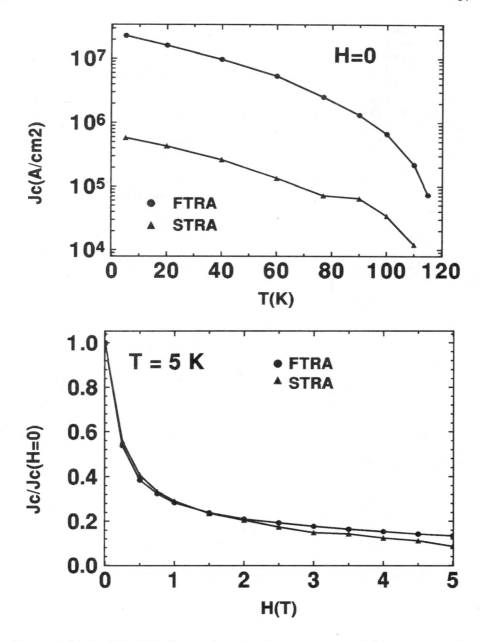

Figure 2. (a) J_c of Hg-1223 films as function of temperature and (b) magnetic field applied parallel to the c-axis.

tube using the FTRA method; and (3) low temperature O_2 annealing of as-made films in flowing O_2 at 400 °C for 24 hours to optimize the oxygen content.

The microstructures of Hg-based cuprate films have been studied carefully using the scanning electron microscope (SEM) and energy dispersive X-ray (EDX) spectrometer. Figs. 1(a) and 1(b) are typical SEM micrographs of Hg-1223 thin films made in the slow temperature ramping Hg-vapor annealing process (STRA) and in a FTRA process. On the STRA sample as shown in Fig. 1(a), besides other impurity phases indentified by EDX, the massive small bright blocks are found to be $CaHgO_2$ phase. As a comparison, samples made in FTRA process show much smoother surface morphology and the $CaHgO_2$ has been significantly suppressed [Fig. 1(b)]. Moreover, a preliminary Auger study indicates that the film/substrate interface quality has also been greatly improved in FTRA samples as the high temperature Hg-vapor processing time is effectively reduced. In particular, Hg-based thin films grown on $LaAlO_3$ substrates using STRA method have poor superconducting properties while those made in FTRA process have equivalent quality to that on the $SrTiO_3$ substrates [5].

The sample crystalline structures are characterized using x-ray diffraction (XRD) using a Cu K_α radiation source. The observation of a series of (00l) peaks indicates that the films are c-axis oriented Hg-1223 [5] and Hg-1212 films [6], respectively. The c-axis lattice constants are estimated to be 12.7 Å and 15.7 Å for Hg-1212 and Hg-1223 phases respectively and are consistent with those reported for bulk samples. The T_c for the film was determined from both magnetic and electrical transport measurements. Zero resistance T_c's for Hg-1223 films are in the range of 128-130.5 K [5] and for Hg-1212 films, are between 120-124 K [6].

The critical current density (J_c) of these films is estimated from the measurement of magnetization at different temperatures and fields using the Bean model. Typical temperature and field dependences of J_c are shown in Figs. 2(a) and 2(b) for samples made in FTRA and STRA processes. It is worthy to notice that the J_c has been significantly improved in FTRA process, especially at higher fields.

This work is supported in part by the University of Kansas GRF fund and NSF EPSCoR fund. J. Wu would like to thank Dr. E. Potenziani for technical help. The authors are very grateful to the Midwest Superconductivity Inc. for support in material, laboratory space, and various experimental facilities.

1. Y.Q. Wang, R.L. Meng, Y.Y. Sun, Z.J. Huang, K. Ross and C.W. Chu, Appl. Phys. Lett. **63**, 3084 (1993).
2. C.C. Tsuei, A. Gupta, G. Trafas, and D. Mitzi, Science **263**, 1259 (1994).
3. S.H. Yun, J.Z. Wu, B.W. Kang, A.N. Ray, A. Gapud, Y. Yang, R. Farr, G.F. Sun, S.H. Yoo, Y. Xin, and W.S. He, Appl. Phys. Lett. **67**, 2866 (1995).
4. S.H. Yun and J.Z. Wu, Appl. Phys. Lett. **68**, 862 (1996).
5. S.H. Yun, J.Z. Wu, S. Tidrow and D. Eckart, Appl. Phys. Lett. **68**, April 29 issue (1996).
6. S.H. Yun, B.W. Kang, W.N. Kang, and J.Z. Wu, preprint.

PHASE STABILITY AND DEFECTS OF $HgBa_2Ca_2Cu_3O_{8+\delta}$

Y.Y. XUE, R.L. MENG, Q.M. LIN, B. HICKEY, Y.Y. SUN AND C.W. CHU
Physics Department and Texas Center for Superconductivity at University of Houston
University of Houston, Houston TX 77204-5932

ABSTRACT

The stability of $HgBa_2Ca_2Cu_3O_{8+\delta}$ (Hg1223) and $HgCaO_2$ was studied in terms of their equilibrium Hg vapor pressure P_{Hg}. It was observed that the partial formation Gibbs-energy of these two phases was very small and nearly the same. Therefore, various defects occur and the phase balance is severely affected by contamination and substitution. Especially, traces of CO_2 and H_2O stabilized $HgCaO_2$ and may have caused the C-substitution for Hg. On the other hand, doping by Re/Pb reduces the P_{Hg} of Hg1223 and stabilizes it. The effects of the substitutions on Hg1223 are discussed.

1. Introduction

The synthesis of $HgBa_2Ca_{n-1}Cu_nO_{2n+2+\delta}$ [Hg12(n-1)n] is still a challenge [1,2,3,4]. It was proposed [5] that the equilibrium Hg vapor pressure P_{Hg} is the main factor affecting the phase stability, which was supported by our investigation of Hg1201 [6]. However, it was argued recently that the competition between Hg1201 and $HgBaO_2$ was actually dominated by the oxygen partial pressure, which shifts the phase equilibria between various Ba-Cu oxides [7]. The phase equilibrium of Hg1223 is even less clear. Therefore, P_{Hg} was measured for $HgCaO_2$ and Hg1223. The partial formation Gibbs-energy was deduced from P_{Hg}. The data show that both the formation entropy and enthalpy of fresh $HgCaO_2$ are roughly equal to those of Hg1223. Hence, direct synthesis of Hg1223 will be marginally possible under ambient pressure. The presence of H_2O and CO_2 decreases the formation entropy of $HgCaO_2$, and makes the competition between Hg1223 and $HgCaO_2$ unfavorable for Hg1223. Partial substitution by Re and Pb for Hg, on the other hand, strongly reduces P_{Hg} of Hg1223 and stabilizes it. The formation energy of Hg1223, however, is very small even with such substitutions, and various defects, *e.g.* C-substitution of Hg, will naturally occur. By adjusting the synthesis conditions, $(Hg_{1-x}C_x)1223$ has been made reproducibly with $x \leq 0.3$. However, our data show that both the oxidation thermodynamics and the superconductivity are almost unaffected by x, which is in contrast to some earlier reports [8,9].

2. Experiments

(Hg,Re)1223 samples were synthesized at ambient pressure by the controlled vapor-pressure method [5]. Undoped Hg1223, (Hg,Pb)1223, (Hg,Mo)1223 and

(Hg,C)1223 were synthesized at 3 GPa [2]. Fresh $HgCaO_2$ samples were synthesized at 700 °C. In addition, several samples, which were made by reacting HgO and aged Ba-Ca-Cu oxide precursors, were used as "contaminated" $HgCaO_2$.

The phase purity of the samples was checked by X-ray diffraction (XDR) using a Rigaku DMAX/BIII diffractometer. XDR indicated that all samples, except the "contaminated" $HgCaO_2$, were nearly single phase with a few percents of impurities. The "contaminated" $HgCaO_2$ samples were a mixture of various Ba-Ca-Cu oxides and $HgCaO_2$. The grain morphology of the sample was observed by a JEOL JSM6400 scanning electron microscope (SEM), and the cation stoichiometry by energy disperse spectrum (EDS) on a LINK energy disperse X-ray spectrometer attached to the SEM.

The equilibrium Hg vapor pressure was measured by the weight loss due to annealing. A sample was sealed inside a quartz tube with a fixed volume ~ 7.5 cm^3 and heated to a temperature T_a for 1 hr before being quenched in liquid nitrogen. The chemical reactions during annealing were checked by examining the samples before and after annealing. It was concluded that only ~ 10% of the sample was decomposed during annealing according to

$$HgBa_2Ca_2Cu_3O_{8+\delta} \leftrightarrow Hg(g) + yO_2 + (Ba_2, Ca_2, Cu_3)O_{8+\delta-2y}$$

is a good approximation for our annealing process, although some evidence suggests that Hg1223 is not a stoichiometric compound. P_{Hg} is calculated from the weight loss ΔM of the samples, assuming that $y \sim 0.5$ and the released Hg vapor behaves like an ideal gas. It should be kept in mind, however, that our measurements were done at an oxygen partial pressure of $P(O_2) = yP_{Hg}$. The P_{Hg} at other oxygen partial pressure may be calculated based on the law of mass action.

The partial formation Gibbs-energy, enthalpy and entropy were calculated as $\Delta G \sim -\frac{1}{2}(1+y)RT_a ln P_{Hg} \sim -\frac{3}{4}RT_a ln P_{Hg}$; $\Delta H \sim \frac{3}{4}R\frac{\partial ln P_{Hg}}{\partial(1/T_a)}$; and $\Delta S \sim -\frac{3}{4}R ln P_{Hg}$ when $T_a \rightarrow \infty$, respectively.

3. Results and Discussions

The difference in P_{Hg} between the undoped Hg1223 and fresh $HgCaO_2$ is less than a factor of two, or a Gibbs-energy difference of ~ 4 kJ/mole, with Hg1223 being the stable phase only between 800 °C and the decomposition temperature (Figure 1). This observation is in agreement with many early reports that Hg1223 can be synthesized at ambient pressure only following strict procedures. The contamination by H_2O and/or CO_2 has been assumed by many groups [1,3,4,5] as the main reason for failures. The contamination may affect the phase equilibrium by either stabilizing $HgCaO_2$ or deducing $-\Delta G$ of Hg1223, which should be distinguishable in terms of their P_{Hg}. Therefore, the P_{Hg} of several reactant pellets [5] from previous failed runs in synthesizing Hg1223 was measured. We interpreted this P_{Hg} as that of "contaminated" $HgCaO_2$, since $HgCaO_2$ is the only Hg-containing phase observed in those samples. The P_{Hg} of the contaminated $HgCaO_2$ displays the same temperature dependency (i.e. ΔH) as that of fresh $HgCaO_2$, although their ΔS's are different (Fig. 1). The P_{Hg} or ΔG of

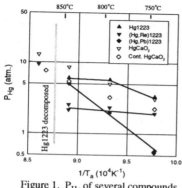
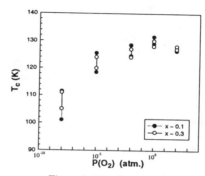

Figure 1. P_{Hg} of several compounds

Figure 2. T_C after annealing

Hg1223 is always larger than that of the contaminated $HgCaO_2$ below the decomposition temperature of Hg1223 (Fig. 1).

It has been widely observed that the partial substitution of Hg by Re/Pb stabilizes Hg1223 [3,4]. The P_{Hg} of $Hg_{0.8}Pb_{0.2}Ba_2Ca_2Cu_3O_{8+\delta}$ and $Hg_{0.8}Re_{0.2}Ba_2Ca_2Cu_3O_{8+\delta}$ was measured (Figure 1). The observed P_{Hg}'s of both (Hg,Re)1223 and (Hg,Pb)1223 were below the P_{Hg}'s of Hg1223 and the contaminated $HgCaO_2$ between 750 and 850 °C, demonstrating a significant doping effect. However, their temperature dependence are rather different with $\Delta H \sim -160$ kJ/mole for (Hg,Pb)1223, but only ~ -33 kJ/mole for (Hg,Re)1223. Compared with $\Delta H \sim -65$ kJ/mole in the undoped Hg1223, it seems that, while (Hg,Pb)1223 is an enthalpy-stabilized compound, Re-doping stabilizes Hg1223 by entropy only, i.e. increasing ΔS from ~ -75 J/(mole·K°) to ~ -35 J/(mole·K°). A reaction entropy is usually dominated by the amount of gas released during the reaction. Therefore, we tentatively suggest that the value of y may change with Re-substitution. In other words, the valences of Re and/or Cu may change significantly during the decomposition of (Hg,Re)1223.

The ΔG of these Hg-based cuprates is comparable to the ΔG of $YBa_2Cu_3O_{7+\delta}$ related to oxygen-loss, and much smaller than the formation Gibbs energy of typical oxides. Even the ΔG of the Tl-based cuprates against Tl-loss will be a few times higher, judged by the Tl-vapor pressure. Therefore, it is only natural that various defects will occur in Hg-1223, similar to the oxygen non-stoichiometry in $YBa_2Cu_3O_{7+\delta}$. A noticeable off-stoichiometry of Hg was indeed observed in quenched Hg1201 samples, and shear planes were identified by TEM as the main defects [6]. Similar Hg non-stoichiometry was also observed in undoped Hg1223 samples synthesized at ambient pressure. These plane-defects seem to affect the maximum T_c significantly [6]. However, the dominant defects of the Hg1223 synthesized under 3 GPa are carbon contamination. We observed that the sintering temperatures of the sample and its precursor change the average carbon-content of Hg1223 significantly. Based on these observations and the published thermodynamics data, the synthesis conditions for $(Hg_{1-x}C_x)1223$ with $x \sim 0.3$ was estimated as $T \sim 950$ °C and $P(CO_2) \sim 10^{-8} - 10^{-2}$ atm under 3 GPa. Samples with various C-content up to $x \sim 0.3$ can be reproducibly synthesized.

The effects of the C-contamination on the superconductivity is another controversial issue [8,9]. To explore the topic, the average C-content was measured by gas effusion spectrum [10]. A Hg-deficiency, roughly the same as the measured C-content, was observed by EDS. The lattice constants also similar to those of (Hg,C)1223 reported earlier [8]. Therefore, we use this C-content as the x in $(Hg_{1-x}C_x)1223$. Two samples with x ~ 0.1 and 0.3 were annealed at 350 °C under various oxygen partial pressures. The T_c of the samples was deduced from the field cooled magnetization at 10 G. The onset and the middle point of the superconducting transition are shown in Figure 2 as the top and bottom symbols, respectively. It shows that the T_c of $(Hg_{1-x}C_x)1223$ is not affected by x within our experimental resolution up to x ~ 0.3, which is in disagreement with some early reports [8,9].

4. Summary

The stability of several Hg-containing compounds was measured by determining their equilibrium P_{Hg}. P_{Hg} and the related partial formation Gibbs energy are nearly the same for undoped Hg1223 and fresh $HgCaO_2$. Contamination with H_2O and/or CO_2 reduces the P_{Hg} of $HgCaO_2$, while the partial substitutions of Re and Pb reduces the P_{Hg} of Hg1223 significantly. The small formation energy of Hg1223 causes various structure defects, *e.g.* shear planes and C-contamination. The effects of these defects were measured, and it was found that the C-contamination have a negligible effect on T_c.

ACKNOWLEDGMENTS

This work is supported in part by NSF Grant No DMR 95-00625, EPRI RP-8066-04, DoE Grant No DE-FC-48-95-R810542, the State of Texas through the Texas Center for Superconductivity at University of Houston, the T. L. L. Temple Foundation.

REFERENCES

1. E. V. Antipov *et al.*, Physica C215 (1993) 1.
2. Z. H. He *et al.*, Physica C241 (1995) 211.
3. J. Shimoyama *et al.*, in *Advance in Superconductivity*, Vol. 2, edited by K. Yamafuji and T. Morishita (Springer-Verlag, Tokyo, 1995), p. 289.
4. R.L. Meng *et al.*, accepted by Physica C; Y.Y. Xue *et al.*, Mod. Phys. Lett. B7 (1994) 1833.
5. R.L. Meng *et al.*, Physica C216 (1993) 21.
6. Y.Y. Xue *et al.*, Phyica C255 (1995) 1.
7. V.A. Alyoshin, D.A. Mikhailova and E.V. Antipov, Physica C255, (1995) 173.
8. E.K. Kopnin *et al.*, Physica C 243 (1995) 222; S.M. Loureo *et al.*, Physica C 257 (1996) 117.
9. Y. Shimakawa *et al.*, Phys. Rev. B50 (1994) 16008.
10. A. Hamed *et al.*, to be appeared in Phys. Rev. B.

SINGLE CRYSTALS OF $HgBa_2Ca_{n-1}Cu_nO_{2n+2+\delta}$ COMPOUNDS: GROWTH AT 10 KBAR GAS PRESSURE AND PROPERTIES

J.Karpinski, H.Schwer, [1]R.Molinski, K.Conder, G.I.Meijer
Laboratorium für Festkörperphysik ETH 8093 Zürich
[1]also High Pressure Research Center Polish Academy of Science 01-142 Warsaw
A.Wisniewski, R.Szymczak, M.Baran
Institut of Physics Polish Academy of Science 02-668 Warsaw
Ch.Rossel IBM Research Division Zürich

ABSTRACT

High-pressure gas-phase technique has been used for the growth of Hg-12(n-1)n single crystals for $n=2,3,4,5$ and ∞ of a size up to $1mm^2$. Crystals have been grown from a flux of $BaCuO_2$-CuO or PbO. Many important physical studies like electron tunneling spectroscopy, anisotropy, magnetic and transport measurements, structural investigation, have been performed on our crystals. The T_c depends on conditions of crystal growth and n and varies from 130 to 70K.

At the temperature close to the melting point of $HgBa_2Ca_{n-1}Cu_nO_{2n+2+\delta}$ compounds decomposition partial pressures of volatile components HgO, Hg and O_2 are very high and reach values of several hundreds bar. Therefore crystal growth is very difficult. From the other hand single crystals are of crucial importance for the studies of both crystallographic and anisotropic physical properties. A failure to understand the significance of anisotropy caused a lot experiments to be incorrectly interpreted. Polycrystalline samples with randomly oriented grains were thought to be representative materials.

High pressure of an active gas component is required for a chemical equilibrium between a solid and a gas phase during crystal growth. Synthesis of compounds from $YBa_2Cu_4O_8$ (124), $Y_2Ba_4Cu_7O_{15-x}$ (247) and $HgBa_2Ca_{n-1}Cu_nO_{2n+2+\delta}$ (Hg-12(n-1)n) families can be given as examples. The required pressure of active gas component (O_2, HgO or Hg vapors) can reach values up to 3 kbar at the temperatures close to melting points.

Hg-12(n-1)n compounds melt peritectically, but at ambient pressure they decompose before melting and the volatile components evaporate. We prevent decomposition by an encapsulation of the sample with an argon pressure 10 kbar. In a gas-phase high-pressure chamber partial pressures have been controlled by application of well known amount of a precursor and known volume of a crucible. Application of a $BaCuO_2$-CuO flux alowed growth of single crystals below the peritectic decomposition temperature. The crystals have been grown by slow cooling (6-10°C/h) from maximum temperature 1020-1070°C. The maximum size of crystals is $1\times 1mm^2$. Fig.1 shows the surface of Hg-1223 crystals.

The T_c of as grown crystals varies from 130 to 70K as a function of n, doping and conditions of growth. Fig.2 shows the T_c dependence on n. For Hg-1234 the anisotropy γ of the penetration depth λ has been determined with torque magnetometer (D.Zech et al. Phys.Rev B, accepted). The mean value for temperatures 103-110K is $\gamma=52\pm 1$.

The irreversibility line (IL) for Hg-1223 ($T_c=120K$) single crystal was determined from magnetization measurements (SQUID, magnetic field up to 5T). The irreversibility field H_{irr} was determined from the M(H) hysteresis loops recorded at a given temperature (Fig.3). On a log(H_{irr}) versus log(1-T_{irr}/T_c) plot (Fig.4) the IL shows two slopes. At higher temperatures (100-85K) the IL shows a power law dependence $H_{irr} = \beta(1-T_{irr}/T_c)^\alpha$ with an exponent $\alpha \approx 2.2$. This is close to the value $\alpha = 2$ predicted for conventional melting of three-dimensional vortex lattice. At lower temperatures (25-60 K) more rapid temperature dependence is observed. The IL can be also described by a power law dependence $H_{irr} = \beta(1-T_{irr}/T_c)^\alpha$ but with larger exponent $\alpha \approx 4.1$. This increase of value of an exponent α is an indication of increasing two-dimensionality of the slope of the system at lower temperatures. At temperature of about 74 K a distinct change of the slope of the IL indicates that at higher temperatures pancakes couple into vortices and form three-dimensional (anisotropic) system.

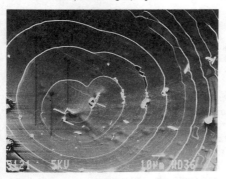

Fig.1. Hg-1223 single crystal with characteristic growth steps.

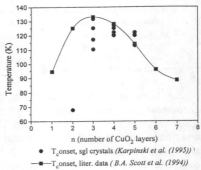

Fig.2. T_c's of Hg-12(n-1)n crystals as a function of n.

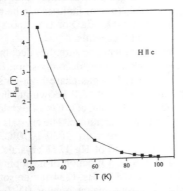

Fig.3. The irreversibility line in single crystal of Hg-1223 (T_c=120K) for H∥c.

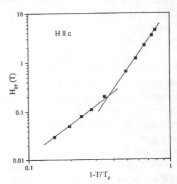

Fig.4. The irreversibility line of Fig.3 on a double logarithmic scale.

SYNTHESIS CONDITIONS AND T_C OF $Hg_{1-x}Ba_2Ca_{n-1}Cu_nO_{2n+2+\delta}$ WITH $n \leq 6$

Q. M. LIN, Y. Y. SUN, Y. Y. XUE and C. W. CHU

Department of Physics and Texas Center for Superconductivity at University of Houston
University of Houston, Houston TX 77204-5932

Intergrowth and phase mixing are persistent problems in the formation of $Hg_{1-x}Ba_2Ca_{n-1}Cu_nO_{2n+2+\delta}$ compounds, especially when $n > 3$. We have therefore investigated the formation of $Hg_{1-x}Ba_2Ca_{n-1}Cu_nO_{2n+2+\delta}$ with different Hg-deficiency (x) and synthesis temperature (T_S) under pressure. We found that as n increases, the compounds form more easily with increasing x and T_S, leading to an increase in carbon contamination from the graphite-furnace. Pure samples have been made for n = 3, 4 and 5, with x = 0.25, ~ 0.25, and 0.35 respectively. The related carbon substitution effect has also been investigated. It is our conclusion that carbon substitution has little effect on T_C's of Hg compounds.

Since the discovery of $HgBa_2CuO_{4+\delta}$, much effort has been devoted to the synthesis of high layer members of $HgBa_2Ca_{n-1}Cu_nO_{2n+2+\delta}$ [Hg-12(n-1)n]. However, intergrowth and phase mixing of several members were persistent problems reported by many groups. By adjusting initial composition and synthesis temperature, we can get nearly single phase samples of Hg-12(n-1)n with n up to 5 as determined by X-ray diffraction pattern.

Samples were synthesized with a piston cylinder high pressure apparatus [1]. BaO, CuO and CaO were mixed according to nominal composition $Ba_2Ca_{n-1}Cu_nO_{2n+1}$ to make the precursor, which was thoroughly ground with an appropriate amount of HgO according to an initial composition $Hg_{1-x}Ba_2Ca_{n-1}Cu_nO_{2n+2+\delta}$. Synthesis temperature (T_S) and Hg-deficiency x were adjusted to obtain the best results.

X-ray diffraction (XRD) measurement was performed with a Rigaku D-Max/BIII powder diffractometer. The phase fraction was estimated by comparing the intensity of the strongest XRD peak of the phase in the sample with that of the corresponding phase in a pure sample. The carbon contamination was measured by an effusion cell equipped with a gas mass spectrometer [2]. T_C was measured magnetically by a SQUID magnetometer.

The synthesis pressure was usually between 20 and 30 kbar. The synthesis time is between 2 and 4 hr. No significant change in the sample quality was observed within these pressure and time ranges. It was found that higher layer cuprates form easily with increasing synthesis temperature. Table 1 gives our T_S's for different values of n.

Table 1. T_S's for different values of n.

n	1	2	3	4	5	6
T_S (°C)	800-860	820-880	840-960	950-1020	1000-1040	1000-1040

Several groups have shown that the optimal Hg-deficiency is 0.25 for Hg-1223 synthesis [1] [3]. Our experimental results show that the optimal Hg-deficiency is nearly the same for Hg-1234. For Hg-1245, however, larger Hg-deficiency x ~ 0.35 has to be used to obtain a single phase sample (Table 2). The Hg-1234 phase appears to be the main impurity for x < 0.35, and Hg-1256 is the main impurity when x > 0.35. When the Hg-deficiency is as large as 0.5, Hg-1256 becomes the majority phase.

Table 2. XRD results for nominal $Hg_{1-x}Ba_2Ca_4Cu_5O_{12+\delta}$ samples with different values of x at the same synthesis condition.

x	0.25	0.30	0.35	0.40	0.50
Hg-1234	~ 20%	~ 10%	~ 0%	~ 0%	~ 0%
Hg-1245	~ 75%	~ 85%	> 95%	~ 80%	~ 25%
Hg-1256	~ 0%	~ 0%	~ 0%	~ 15%	~ 70%
others	~ 5%	~ 5%	< 5%	~ 5%	~ 5%

It has been reported [4] that carbon substitution for Hg is the main cause for the Hg-deficiency in high pressure synthesized Hg samples. The use of the graphite heater could be the source of carbon contamination. To verify the effect of graphite heater, two Hg-1223 samples were made from the same batch of precursors but at 850 °C and 950 °C. Carbon-content measurement shows that the sample synthesized at 850 °C contains ~ 0.15 carbon per unit cell, but ~ 0.30 for the 950 °C synthesized sample. The observation demonstrates that carbon contamination increases with T_S.

The measured T_C's are 126 K, 109 K and 105 K for n = 4, 5 and 6, respectively. To examine the effect of carbon contamination, the two Hg-1223 samples mentioned above were annealed at different oxygen pressures. The maximum T_C ~ 132 K was the same for both samples. Therefore, we conclude that the carbon contamination has little effect on T_C when the carbon content is no more than 0.30 per unit cell.

Acknowledgments

The authors would like to thank Dr. A. Hamed for the carbon content measurement. The work is supported in part by NSF, EPRI, the State of Texas through the Texas Center for Superconductivity at the University of Houston.

References

1. Z. H. He *et al.*, Physica C 241 (1995) 211
2. A. Hamed *et al.*, Am. Phys. Soc. Bull. 40 (1995) 392
3. M. Hirabayashi *et al.*, Physica C 219 (1994) 6
4. E. M. Kopnin *et al.*, Physica C 243 (1994) 222

I. MATERIALS

New HTS and Related Compounds

Atomic-layer synthesis of artificial high-T_c superconductors and their use in tunnel junctions

Ivan Bozovic and J. N. Eckstein

Edward L. Ginzton Research Center, Varian Associates, Inc.
Palo Alto, California 94704-1025, USA

Atomic-layer-by-layer molecular beam epitaxy (ALL-MBE) has been used to engineer novel cuprate superconductors. In particular, we have investigated Bi-Sr-Ca-Cu-O and La-Sr-Ca-Cu-O phases containing multiple (4-10) CuO_2 planes. So far, the highest T_{c0} (75 K) was achieved in Bi-1278 compound epitaxially stabilized by Bi-2201. We have utilized this compound to fabricate the first HTS *tunnel* (SIS) junctions. Replacing the central Ca layer in Bi-1278 by Dy, one creates a bizarre compound which contains, *within a single unit cell*, the bottom superconducting electrode, an insulating barrier layer (less than 1 nm thick), and the top superconducting electrode. These artificial *intra-cell* Josephson junctions exhibit very sharp quasiparticle tunneling I-V characteristics, consistent with tunneling between 2D superconductors with d-wave pairing.

Our technique for deposition of single-crystal thin films of cuprate superconductors and other complex oxides, ALL-MBE (atomic layer-by-layer molecular beam epitaxy), has been described in detail elsewhere[1-3], so here we just summarize its main features. The ultra-high-vacuum chamber contains eight thermal evaporation sources with computer-controlled shutters. Currently, these are used for Bi, Sr, Ca, Cu, Dy, Mn, La, and Ag, which enables synthesis of various superconducting or insulating phases from the Bi-Sr-Ca-Cu-O and La-Sr-Ca-Cu-O families, as well as La-Sr-Ca-Mn-O compounds that show colossal magnetoresistance[4]. Silver is used to deposit electrical contacts *in-situ*. To monitor the atomic fluxes, we use an atomic absorption spectroscopy system, accurate enough to detect changes of less than one percent and fast enough to allow real-time feedback control. By using a pure ozone beam, sufficient oxidation is achieved under high vacuum conditions, which permits *in-situ* monitoring of the surface structure by RHEED. This allows us to follow the sequence of transient chemical states as the surface reactions occur, and thus to find precise extent of overlap of deposition from various sources that corresponds to the path through the phase space which avoids formation of secondary-phase defects.

Scanning tunneling microscopy (STM) and atomic-force microscopy (AFM) studies of our films have shown that surfaces are atomically flat except for occasional steps with the height corresponding to one molecular layer. The terraces are aligned with similar ones which occur in the substrates and which originate from slight misalignment of the polished surface from the ideal (100) orientation. STM, AFM, and transmission electron microscopy (TEM) studies and transport measurements on a variety of superlattices and multilayers containing ultrathin (i.e., one unit cell thick) layers of cuprate superconductors or other complex oxides (titanates, manganites) have demonstrated that virtually atomic perfection of interfaces between such compounds can be achieved[5,6].

Utilizing this technique we have synthesized a number of samples of various metastable compounds in Bi-Sr-Ca-Cu-O system, with the number of CuO_2 planes n in a single molecular layer being larger than 3. For example, in this manner we have synthesized thin films of $Bi_2Sr_2Ca_7Cu_8O_{20+x}$ (2278). Substantial interest in this compound has been aroused by a report[7] that it exhibited superconductivity at 250 K. The highest T_c we measured[8] in single-crystal 2278 thin films was around 60 K. The material was metallic, but its relatively high resistivity, $\rho \approx 600$ $\mu\Omega$cm at T=300K, indicated that it was substantially underdoped, even after vigorous *in-situ* post-annealing in ozone. This was also confirmed by subsequent mutual inductance measurements which showed a very low superfluid density.

For this reason, we have investigated the possibility of modulation doping by omitting entire BiO layers. Since each BiO unit donates one electron to the CuO_2 state manifold, a Bi-1 compound (such as Bi-1278) should have about 1 hole more per formula unit compared to its Bi-2 counterpart (Bi-2278). [This hole count assumes, however, that the oxygen content in the remaining layers stays the same, which indeed need not be true in reality, as discussed below.]

Indeed, Bi-1278 is an even less stable compound, and for this reason single-phase 1278 films are very difficult to grow. However, a very thin layer of 1278 can be epitaxially stabilized if it is sandwiched between layers of stable phases such as 2212 or 2201. We have therefore grown superlattice films consisting of many repetitions of the 1278-1201-$(2201)_m$ sequence, with m=1, 2, or 3. In this case, RHEED showed a very smooth growth, and X-ray diffraction indeed confirmed the desired superlattice structure. As an example, in Fig. 1 we show an X-ray diffraction pattern for a 2201-2201-1278-1201-2201 superlattice. The diffraction peaks can be indexed as (*00l*) with $c = 75.73$Å, close to what was predicted. Notice the pronounced finite-thickness oscillations which indicate that the film was indeed very smooth. In similar superlattice samples, 2201-1278-1201-2201, we measured[6] T_c = 75 K. This result demonstrated the capability of ALL-MBE method to artificially assemble novel high-T_c materials.

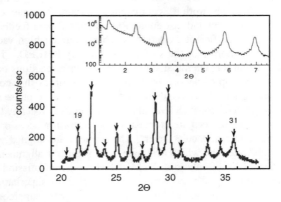

Figure 1. X-ray diffraction pattern of a $[2201\text{-}1278\text{-}1201\text{-}2201\text{-}2201]_n$ superlattice, with n-14. Notice the pronounced finite-thickness oscillations at low angles (the inset).

We have also fabricated Josephson junctions by inserting a single molecular layer of 1278 to act as a barrier layer between the two 2212 electrodes. This, in general, provides *c*-axis critical currents that are too high for practical devices. For the best performance, we substitute Ca^{2+} in the central layer by Dy^{3+}, generating a structure illustrated in Fig. 2.

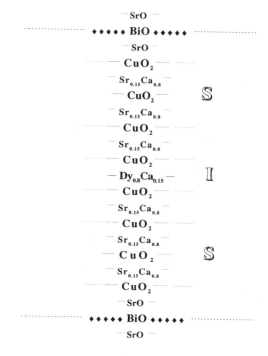

Figure 2. Dy-doped Bi-1278: *intra-cell* SIS tunnel junction

Again, by studying superlattices containing Dy-doped Bi-1278, in similar 2201-1278-1201-2201 sequences, we have found that even for complete one-monolayer Dy substitution this compound, $Bi_2Sr_2DyCa_6Cu_8O_{20+x}$, is still superconducting, with T_c = 45 K. However, according to the penetration length studies from mutual inductance measurements, in this case superconductivity is confined to the outer CuO_2 planes, which are underdoped ($n_s \approx$ 0.3 pairs per formula unit). The central slab is actually insulating, with typical $R_n \approx$ 50 Ω (at low bias) in 30×30 μm² mesa structures, for c-axis transport; this corresponds to $\rho \approx 10^4$ Ωcm (at T=4.2 K).

So, in this way we have fabricated, *within a single unit cell*, the bottom superconducting electrode, an insulating barrier layer (less than 1 nm thick), and the top superconducting electrode. These *intra-cell* Josephson junctions show a high I_cR_N product (10 mV), which indeed makes them very attractive for applications. Furthermore, they exhibit very sharp quasiparticle tunneling I-V characteristics. As the temperature is raised, these gap-like structure moves to lower voltages and gets smeared as expected, see Fig. 3.

Evidence that these devices are tunnel (SIS) junctions is mounting. First, if $\rho = 10^4$ Ωcm, these indeed can not be SNS junctions; the barrier is way too insulating, by 4-5 orders of magnitude. Next, we have also fabricated similar structures, but where the insulating barrier were created by inserting $SrTiO_3$ layers in the middle of the Bi-1278 layer. The advantage here is that it was possible to vary the insulating barrier thickness systematically from one to seven unit cells (4-28 Å). The barrier resistance indeed showed an exponential thickness dependence, clearly indicating tunneling[6].

To the best of our knowledge, these are the first artificial HTS tunnel (SIS) junctions reported.

However, these junctions do not show the ordinary BCS-type SIS behavior. A detailed quantitative analysis, which will be presented elsewhere, actually shows that these I-V curves are consistent with tunneling between 2D superconductors with d-wave pairing.

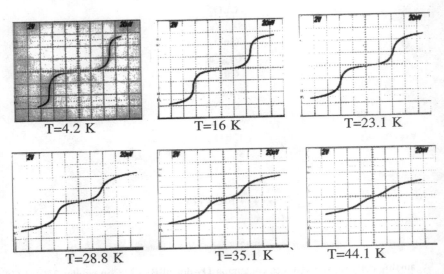

Figure 3. Temperature dependence of the I-V characteristic of an SIS tunnel junction consisting of a single Dy-doped Bi-1278 molecular layer. (Vertical scale: 20 mA/div. Horizontal scale: 20 mV/div.)

References

1. J. N. Eckstein, I. Bozovic, M. E. Klausmeier-Brown, G. F. Virshup, and K. S. Ralls, *MRS Bulletin* **18** (8), 27 (1993).
2. J. N. Eckstein, I. Bozovic, and G. F. Virshup, *MRS Bulletin* **19**, (9), 44 (1994).
3. J. N. Eckstein and I. Bozovic, *Ann. Rev. Mater. Sci.* **25**, 679 (1995).
4. See e.g. S. Jin, M. McCormack, T. Tiefel, and R. Ramesh, *J. Appl. Phys.* **76**, 6929 (1994).
5. I. Bozovic, J. N. Eckstein, M. E. Klausmeier-Brown, and G. F. Virshup, *J. Supercond.* **5**, 19 (1992).
6. I. Bozovic, J. N. Eckstein, G. H. Virshup, A. Chaiken, M. Wall, R. Howell, and M. Fluss, *J. Supercond.* **7**, 187 (1994).
7. M. Laguës, X. M, Xie, H. Tebbji , X. Z. Xu, V. Markert , C. Hattere, C. F. Beuran, C. Deville-Cavelin, *Science* **262**, 1850 (1993).
8. I. Bozovic, J. N. Eckstein, and G. F. Virshup, *Physica C* **235-240,** 178 (1994).

PROPERTIES OF Li-DOPED La_2CuO_4

J.L. Sarrao, D.P. Young, and Z. Fisk
*NHMFL and Dept. of Physics, Florida State University
Tallahassee, FL 32306*

P.C. Hammel, Y. Yoshinari, and J.D. Thompson
*MST Division, Los Alamos National Laboratory
Los Alamos, NM 87545*

ABSTRACT

Li substitutes for Cu in La_2CuO_4 up to the limiting stoichiometry $La_2Cu_{0.5}Li_{0.5}O_4$, which has superstructure order. The effects of this in-plane hole doping on the structural and magnetic properties of La_2CuO_4 are very similar to those due to Sr-doping, except that here the holes appear to be localized. In the Cu NMR of the diamagnetic end-member $La_2Cu_{0.5}Li_{0.5}O_4$, the Cu nuclear spins relax via a magnetic excitation with characteristic energy 1500 K.

1. Introduction

Extensive work exists on the effects of chemical substitutions on the properties of La_2CuO_4. The original work, of course, involved Ba and other alkaline earth substitution for La, leading to superconductivity. Later it was found that interstitial oxygen could introduce hole conductivity into the CuO_2-planes and lead to superconductivity as well. There is a long history of studies of the effects of impurities on T_C in low T_C materials, and this approach was used as well in the cuprates in an attempt to understand both the nature of the superconductivity and, equally importantly, the puzzling nature of the normal state of these materials.

In-plane substitutions for Cu were found to have strong effects on T_C as well as on conductivity. Quite generally, such substitutions were found to decrease T_C rapidly, independent of whether or not the substitutions carried a magnetic moment, just as in heavy Fermion superconductors. Zn substitutions in particular were found to have a marked effect on T_C, 3 atomic percent Zn substitution for Cu being sufficient to completely suppress T_C in $La_{1.85}Sr_{0.15}CuO_4$ [1]. The conclusion from this and similar in-plane substitution experiments was that disturbing the CuO_2-planes was very hostile to the superconductivity.

A particularly interesting in-plane substitution is that of Li for Cu. Li^{+1} has ionic radius essentially the same as that of Cu^{+2}, and brings a hole with it to the plane. This hole appears to be localized, and it is interesting to compare in-plane substituted holes with the mobile ones introduced by out-of-plane substitutions. There has been considerable experimental work reported on this [2,3,4,5]. Our aim has been to understand the apparent loss of Cu magnetic moment due to Li substitution and its relationship to the effect of Li-substitution on the superconducting properties of $La_{1.85}Sr_{0.15}CuO_4$. Although all of the data presented here were taken with sintered specimens, it has been possible to grow single crystals by flux methods out to approximately $Li_{0.1}$. The high volatility of Li_2O makes traveling solvent methods problematic and dictates low sintering temperatures (900 C) for ceramic specimens.

2. Structural Effects

The effect of Li-substitution into pure and Sr-doped La_2CuO_4 is shown in Figure 1. The data shown for the in-plane lattice parameters collapse onto one plot versus hole count: the effect of Sr or Li addition is essentially identical structurally. This is true for both the concentration dependence of the orthorhombic-tetragonal transition temperature, as well as the variation of tetragonal a_0 lattice parameter at room temperature. It appears that the introduced hole has a particular size independent of whether or not the hole is mobile. Perhaps the way to think about the contraction of the in-plane lattice parameter with hole addition is as a relaxation about the hole in the highly negatively charged CuO_2 background. Note here the surprising fact, first reported by Demazeau et al. [7], that substitution can be made all the way to $La_2Cu_{0.5}Li_{0.5}O_4$, which forms the simple ordered superstructure of alternating Cu/Li positions in the planes. Rietveld refinement of neutron powder diffraction data shows the ordering to be 80%.

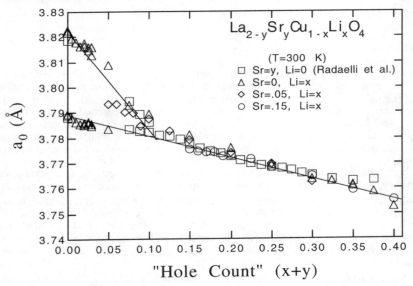

Figure 1. In-plane lattice constant, a_0, of $La_{2-y}Sr_yCu_{1-x}Li_xO_4$ at room temperature. a_0 is a function only of the net hole count (i.e., x+y) and not the individual Sr and Li concentrations. Data for y=0 are after Radaelli et al. [6].

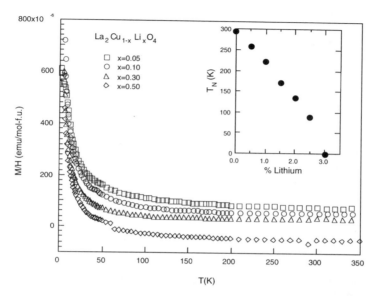

Figure 2. Magnetic susceptibility as a function of temperature for $La_2Cu_{1-x}Li_xO_4$ for various values of x. The inset shows the suppression of the T_{Neel} with Li substitution.

3. Magnetic Effects

Sr and Li have nearly identical effects as well on the magnetic phase diagram of La_2CuO_4: the long range magnetic order of the Cu-spins is lost by x=0.03 [2]. This behavior contrasts strongly with the effect of in-plane substitution of Zn or Mg. For both of these cases, the 3-D magnetic ordering temperature is depressed at nearly but less than the percolation rate, with roughly 30% substitution finally suppressing T_N completely [8]. The development of the temperature dependent magnetic susceptibility with Li-substitution is shown in Figure 2. The remarkable fact evident here is that the end point $La_2Cu_{0.5}Li_{0.5}O_4$ is diamagnetic at room temperature, the small low-temperature Curie tail being, we believe, an extrinsic effect. While the Cu is formally +3 here, the stability of this material, coupled with the NMR data presented below, indicate that this diamagnetism is unlikely to be a property of Cu^{+3}.

4. Electronic Properties

We show two sets of data here. The first is the reduction of T_c of $La_{1.85}Sr_{0.15}CuO_4$ by Li substitution (Figure 3). The effect is considerably slower than that of Zn substitution, and in fact the effect beyond x=0.05 Li trails off quite slowly, while the shielding fraction becomes very much smaller than might be expected. We have made samples to test if mobile holes, say by Sr substitution, can "free up" Li-introduced holes in samples such as

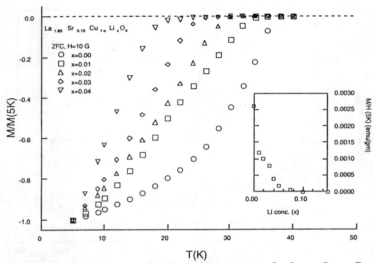

Figure 3. ZFC magnetization as a function of temperature for $La_{1.85}Sr_{0.15}Cu_{1-x}Li_xO_4$ showing the weak suppression of T_c with Li substitution. The inset shows the corresponding and more rapid decrease in shielding fraction.

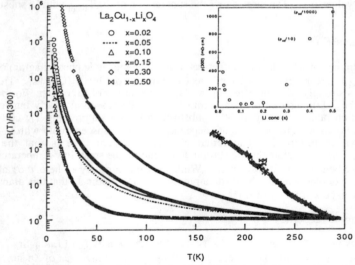

Figure 4. Normalized resistance as a function of temperature for $La_2Cu_{1-x}Li_xO_4$ for various values of x. The inset shows the room temperature resistivity as function of Li concentration. Note that the values for x=0.4 and x=0.5 have been reduced by a factor of 10 and 1000 respectively to bring the values on scale.

$La_{1.9}Sr_{0.1}Cu_{0.95}Li_{0.05}O_4$. This does not appear to work: this is not equivalent to $La_{1.85}Sr_{0.15}CuO_4$.

The room temperature resistivity as a function of Li concentration as well as its temperature dependence is shown in Figure 4. These results confirm the earlier data from Kastner et al. [4]. The resistance of the samples increases substantially with decreasing temperature for all Li concentrations. The $Li_{0.5}$ compound is extremely resistive, $\rho=1k\Omega$-cm at room temperature. We find that neither simple activation nor variable range hopping really fit our data. What is interesting is the variation of resistivity at room temperature with x. We see a drop through x=0.15 of more than an order of magnitude from pure La_2CuO_4. It is apparent, however, that the introduction of holes via Li never leads to mobile holes in contrast to Sr substitution. The recent results of Ando et al. at high magnetic fields for underdoped material come to mind here [9].

5. NMR results

Particularly interesting are the NMR data on Cu in the end member, diamagnetic insulator $La_2Cu_{0.5}Li_{0.5}O_4$ [10]. These are presented in Figure 5. The relaxation times are fitted using a stretched exponential with exponent 0.58. Although the origin of the distribution of T_1 is not known, the distribution is temperature independent. A plot of $\ln(1/T_1T)$ versus $1/T$ gives a good straight line, with slope 1510 K. We know from the ratio of T_1's for the two Cu isotopes ^{63}Cu and ^{65}Cu that the relaxation mechanism is magnetic [10].

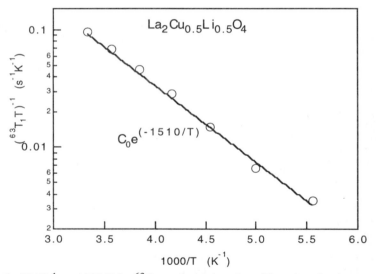

Figure 5. $(T_1T)^{-1}$ vs. 1000/T for ^{63}Cu nuclear relaxation. The relaxation is magnetic and follows an activated form with characteristic temperature 1500 K. After [10].

This result is surprising. The simple interpretation is that Cu possesses a spin here, but that this spin is compensated into a singlet, given the diamagnetism of the compound, with characteristic temperature of the same order of magnitude as the exchange coupling, J, of the pure La_2CuO_4. This is perhaps coincidental, and we see no easy explanation for it.

6. Concluding remarks

The in-plane doping of holes by Li substitution leads to effects which remind one of a localized version of the Zhang-Rice singlet [11]. The effect on superconductivity is substantially smaller than that from the divalent Zn or Mg substitution. The fact that these holes appear to be bound by the Li perhaps is a result of the relative ease of binding in 2D, the difference between Li and Sr being that the Sr is a much weaker perturbation on hole motion since it is not lying in the CuO_2-planes. The diamagnetism of the 0.5 Li-doped material, which the NMR shows to have moment present, suggests that the bound holes can compensate the Cu spins in a manner similar to the mobile ones. Various L-edge x-ray studies have attempted to ascertain whether Cu is really +3 in $La_2Cu_{0.5}Li_{0.5}O_4$, with equivocal results [12,13]. The generally accepted idea, and one with which we agree, is that the low temperature state of this material involves a strong coupling between the Cu^{+2} and a ligand hole, forming a magnetic singlet: the Zhang-Rice insulator.

There is much still to be understood however. Where does the activation energy of 1510 K come from, what is the origin of the stretched exponential, and what causes the deep minimum in room temperature resistivity versus Li concentration?

Acknowledgments

The NHMFL is supported by the NSF and the State of Florida through cooperative agreement #DMR-9016241. Work at LANL was performed under the auspices of the U.S. Dept. of Energy.

References

1. G. Xiao et al., *Phys. Rev. B* **42**, 8752 (1990).
2. A.I. Rykov et al., *Physica C* **247**, 327 (1995).
3. J.P. Attfield and G. Ferey, *J. Solid State Chem.* **80**, 112 (1989)
4. M.A. Kastner et al., *Phys. Rev. B* **37**, 111 (1988)
5. J.L. Sarrao et al. to appear in *Proceedings of the PPHMF-II Conference*, ed. L.P. Gor'kov (World Scientific, Singapore, 1996).
6. P.G. Radaelli et al., *Phys. Rev. B* **49**, 4163 (1994).
7. G. Demazeau et al., *Mat. Res. Bull.* **7**, 913 (1972).
8. S.-W. Cheong et al., *Phys. Rev. B* **44**, 9739 (1991).
9. Y. Ando et al., *Phys. Rev. Lett.* **75**, 4662 (1995).
10. Y. Yoshinari et al., submitted to *Phys. Rev. Lett.* (1996).
11. F.C. Zhang and T.M. Rice, *Phys. Rev. B* **37**, 3759 (1988).
12. N. Merrien et al., *J. Phys. Chem. Solids* **54**, 499 (1993).
13. J.-H. Choy et al., *Phys. Rev. B* **50**, 16631 (1994).

ENHANCEMENT OF T_c FOR $(Cu_{0.5}C_{0.5})Ba_2Ca_2Cu_3O_{8.5+\delta}$ FROM 67 TO 120 K BY HEAT TREATMENTS

CATHERINE CHAILLOUT

Laboratoire de Cristallographie, CNRS-UJF, BP 166, 38042 Grenoble cedex 9, France

and

MASSIMO MAREZIO

Laboratoire de Cristallographie, CNRS-UJF, BP 166, 38042 Grenoble cedex 9, France
and MASPEC-CNR, Via Chiavari 18/A, 43100 Parma, Italy

ABSTRACT

T_c of $(Cu_{0.5}C_{0.5})Ba_2Ca_2Cu_3O_x$ has been increased from 67 K to 120 K by heat treatments in reducing atmosphere. The corresponding variations of the lattice parameters show that the as prepared sample is in an overdoped state. The dT_c/dP of the as prepared sample is negative while that of the 120 K sample is almost pressure independent.

High pressure has played an important role in the enhancement of superconducting transition temperatures of various compounds. For example, in-situ measurements have shown that the T_c of $HgBa_2Ca_2Cu_3O_{8.4}$ can increase from 135 K at ambient pressure to 164 K under 35 GPa. In general, copper mixed oxide superconductors are very susceptible to pressure because they contain large bonding anisotropy. One strategy to increase T_c at ambient pressure would be to simulate through substitutions the structural arrangement responsible for the 164 K transition. Such an experiment, if successful, will be very rewarding.

Not all samples of high T_c superconductors exhibit positive dT_c/dP coefficients. But even when this coefficient is negative, the results can be useful to ultimately increase T_c. In fact, an empirical rule shows that overdoped samples exhibit negative coefficients, whereas underdoped and optimally-doped samples usually exhibit positive ones. In those cases for which the structure is not known in detail, the knowledge of the doping status is extremely important. It can show the right direction in the search for the appropriate heat treatments which will optimize the doping. We will describe herein the enhancement of about 55 K when overdoped samples of $(Cu_{1-y}C_y)Ba_2Ca_2Cu_3O_x$ are heat-treated in a reducing atmosphere. This is a classical example of the above strategy as the structural arrangement of the charge reservoir of this compound is not known in detail, but in-situ high pressure experiments on the as prepared samples showed that the dT_c/dP coefficient is negative.

In 1994 a new superconducting homologous series, $(Cu_{1-y}C_y)Ba_2Ca_{n-1}Cu_nO_x$, was synthesised at high pressure and high temperature[1-3]. The presence of carbon as CO_3 groups was detected by ultra-high resolution electron microscopy[4] and by EELS[5]. Superconducting transition temperatures around 67 K and 117 K were measured for the $n = 3$ and $n = 4$ members, respectively. These values showed that this would have been the first homologous series of copper-mixed oxide superconductors for which the highest T_c did not correspond to the $n = 3$ member.

These compounds were found to be tetragonal with lattice parameters $a = b \approx 3.86$ Å and $c \approx 14.75$ Å for $n = 3$ and ≈ 18.0 Å for $n = 4$. A superstructure, doubling the a and c axes, was also detected by electron diffraction and attributed to the ordering of Cu and C in the basal plane of the structure. By analogy with the other layered cuprate superconductors, the structures of the $(Cu_{1-y}C_y)Ba_2Ca_{n-1}Cu_nO_x$ compounds comprise $CuO_2[(Ca)(CuO_2)]_{n-1}$ blocks intercalated by $(BaO)(Cu_{1-y}C_yO_\delta)(BaO)$ blocks. The δ concentration of the oxygen atoms in the basal plane includes those bonded to the C atoms and the additional oxygen atoms which might be bonded to the Cu cations of the basal plane as in $YBa_2Cu_3O_{6+x}$. But one could also envisage that these extra oxygen atoms are located in the centre of the (3.86x3.86 Å) mesh as in the Hg-based cuprates, in which case they would not be bonded to the Cu cations of the basal plane. These Cu cations, called herein Cu1, would then have a dumbbell coordination. A structural refinement of the $n = 4$ member, based on powder neutron diffraction data and carried out by using the subcell[6], indicated that the extra oxygen atoms were present around the Cu1 cations as in $YBa_2Cu_3O_7$.

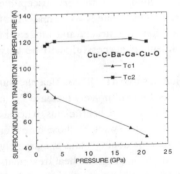

Fig. 1: $T_{c1} \rightarrow (Cu_{0.5}C_{0.5})Ba_2Ca_2Cu_3O_x$, $T_{c2} \rightarrow (Cu_{0.5}C_{0.5})Ba_2Ca_3Cu_4O_x$

Jaime et al.[7] reported that the dT_c/dP coefficient for $(Cu_{1-y}C_y)Ba_2Ca_2Cu_3O_x$ was appreciably negative, while that of the member with $n = 4$ was very close to zero (see Fig. 1). According to the empirical rule that overdoped samples exhibit a decrease of T_c for increasing pressure, these data indicate that the $n = 3$ sample is in the overdoped state while that of the $n = 4$ sample was somewhat close to optimal doping. They motivated Chaillout et al.[8] to study the possibility of changing the oxygen stoichiometry of these samples by reducing heat treatments.

Fig. 2 shows the variation of T_c vs the a parameter for four (Cu,C)-1223 samples, one as prepared and the other three obtained after successive heat treatments under Ar at 400°C for 2, 3, and 4 hrs, respectively.

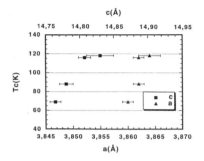

Fig. 2: Tc vs a and c parameters for 4 samples of $(Cu_{0.5}C_{0.5})Ba_2Ca_2Cu_3O_x$, one as prepared (69 K) and the other 3 obtained after successive heat treatments for 2 (88 K), 3 (116 K), and 4 (118 K) hrs, respectively.

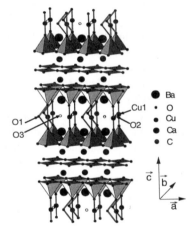

Fig. 3: Model for $(Cu_{0.5}C_{0.5})Ba_2Ca_2Cu_3O_x$. The two possible positions, O2 and O3, for the mobile oxygen atoms in the basal plane are indicated.

As the a parameter increases, T_c increases from 67K for the as prepared sample to 120K for the samples heat treated for 4 hours. Since in these types of compounds the CuO_2 layers are almost flat, the a parameter is closely determined by the in-plane Cu-O distance and in turn by the valence state of the Cu cations which is strictly related to the oxygen stoichiometry. Therefore, heat treatments in reducing atmosphere induce a decrease of the oxygen stoichiometry, a reduction of the valence state of the Cu cations of the CuO_2 layers, an increase of the Cu-O distances and of the a parameter, and finally an increase of T_c toward its optimal value. By analogy with the variation of the c parameter with oxygen stoichiometry in $YBa_2Cu_3O_{6+x}$, the increase of the c parameter with heat treatment duration, shown in Fig. 2, can be associated to the loss of oxygen atoms located in the basal plane of the structure and linked to Cu1.

The average cation positions of the structure of $(Cu_{1-y}C_y)Ba_2Ca_2Cu_3O_{8+\delta}$ have been determined by several techniques and all of them seem to agree to the schematic structural model shown in Fig. 3. However, the detailed arrangement of the $(Cu_{1-y}C_y)O_\delta$ layer is not known and this does not allow one to determine on one side the oxygen concentration δ and on the other the doping mechanism. The δ concentration includes the oxygen atoms linked to the C and those around the Cu cations. The results of the heat treatments may give us a clue about these two important parameters.

A qualitative analysis of the electron diffraction photographs corresponding to $(Cu_{1-y}C_y)Ba_2Ca_2Cu_3O_{8+\delta}$, showing well defined superstructure reflections, indicates that the Cu/C ratio is very close to unity. If we assume that at optimal doping, the valence state of the planar Cu cations is $\approx 2.15+$ and the valence of a CO_3 group is 2-, the concentration of the extra oxygen atoms, besides those linked to the C atoms, is given by

$x \approx 0.25(v(Cu1)-1)$. The valence of Cu1 may vary between 1+ and 2+; correspondingly x may vary between 0 and 0.25. In the former case, the optimal doping corresponds to a chemical formula such as $(Cu_{0.5}C_{0.5})Ba_2Ca_2Cu_3O_{8.5}$, in which the arrangement of the reservoir layer, $(Cu_{0.5}C_{0.5})O_{0.5}$, corresponds to all the Cu1^{1+} in dumbbell coordination. Additional oxygen atoms in the reservoir layer increase the valence of the planar Cu cations, the sample becomes overdoped and a decrease of T_c is observed. The as prepared sample would correspond to a value of x greater than 0 and the reducing heat treatments would remove the extra oxygen atoms located in the reservoir layer. As reported in reference 8, the dT_c/dP coefficient for the samples exhibiting $T_c = 120$ K was found to be closely pressure-independent.

For $x = 0.25$ and all the Cu1 in the 2+ state, additional oxygen atoms in the reservoir layer would necessarily overdope the CuO_2 layers and induce a decrease of T_c. In this case, an underdoped region would exist for $x < 0.25$. However, this has not yet been found experimentally. Moreover, the arrangement corresponding to the optimal doping with $x = 0.25$ is not satisfactory from the crystallochemical point of view, because in this case half of the Cu^{2+} cations would have the dumbbell coordination which has never been observed for Cu cations except when in the 1+ state.

The extra oxygen of the reservoir layer which are not linked to the C atoms can be located either in the (1/2,0,0) position as in $YBa_2Cu_3O_7$ to form squares around the Cu1 cations or in the (1/2,1/2,0) position as in the Hg based cuprates, in which case the Cu1-O bonds are extremely weak or non-existing. These two postions are indicated in Fig. 3.

When the same heat treatments are applied to the $(Cu_{1-y}C_y)Ba_2Ca_3Cu_4O_{10+\delta}$ no variations of lattice parameters and consequently of T_c are observed after the treatments. A qualitative valence calculation for the $n = 4$ member shows that the optimal doping is obtained for x very close to 0.05. This means that the as prepared samples of this phase correspond to this small concentration. A reducing heat treatment could only induce the loss of this amount of oxygen. A further reduction would decompose the compound.

References

1. M.A. Alario-Franco et al., Physica C, **222**, 52 (1994)
2. H. Ihara et al., Jpn. J. Appl. Phys. **33**, L503 (1994)
3. C.Q. Jin et al., Physica C **223**, 238 (1994)
4. T. Kawashima et al., Physica C **224**, 69 (1994)
5. M.A. Alario-Franco et al., Physica C, **231**, 103 (1994)
6. Y. Shimikawa et al., Phys. Rev. B **50**, 16008 (1995)
7. M. Jaime et al., Solid State Comm., **97**, 131 (1996)
8. C. Chaillout et al., Physica C, accepted (1996)

GENERATION OF HIGH T_c SUPERCONDUCTING OXYCARBONATES : FROM THE CERAMICS TO THE THIN FILMS

B. RAVEAU, C. MICHEL and M. HERVIEU
Laboratoire CRISMAT, ISMRA et Université de Caen, URA 1318 associé au CNRS
Bd du Maréchal Juin - 14050 Caen Cedex - France

ABSTRACT

The copper oxycarbonates form a very promising family of high T_c superconductors with closely related structures. The synthesis of these materials is based on the infinite layer (IL) structure considered as a host lattice for the formation of the oxycarbonate $Sr_2CuO_2CO_3$ (S_2CC) and on the combination of the S_2CC structure and with the $Tl_{0.5}Pb_{0.5}Sr_2CuO_5$ ("1201") and $Bi_2Sr_2CuO_6$ ("2201") type structures. The application of intergrowth and shearing mechanisms to these compounds is described, showing the possibility to generate numerous new superconductors.

Besides the cuprates family, that of copper oxycarbonates, discovered more recently, appears as very promising, owing to the fact that these compounds which derive structurally from the cuprates often exhibit a higher critical temperature than the corresponding cuprates. Another interesting feature of these materials deals with their metastable character, which makes that sometimes particular conditions are required for their synthesis. The latter property has opened the route to the thin film chemistry of such materials, suggesting that so far many other metastable superconductors remain to be discovered.

The present review shows the structural filiation that can be proposed between all the oxycarbonates, allowing new directions to be emphasized for the exploration of these superconductors.

If we put aside the chronological aspect of the discovery of these materials, the infinite layer compound ("IL"), $Sr_{0.85}Ca_{0.15}CuO_2$ [1] synthetized rather lately and which was only found to be superconductor by using high-pressure, represents according to us the host matrix for the generation of oxycarbonates. This structure (Fig. 1a) can be considered as a privileged host lattice to accomodate rows of carbonate groups (fig. 1b) leading in fact to the oxycarbonate $Sr_2CuO_2CO_3$ [2].

At this point, the S_2CC structure appears as the parent structure for the generation of new superconductors by applying in a first step intergrowth mechanisms. The first kind of intergrowth mechanism is based on the similarity of the IL and S_2CC structures that exhibit identical stacking of the $[CuO_2]_\infty$ layers, so that intergrowths of these two structures $(IL)_m (S_2CC)_m$ should be expected. None of such phases could be synthesized as bulk materials to date. But, in contrast, the application of laser ablation to the realisation of such metastable framework, in the form of thin films, has been successful, leading to a new series of metastable superconductors with the generic formulation $(CaCuO_2)_m (Ba_2CuO_2CO_3)_n$ [3]. The structure of these oxycarbonates derives from the "123" cuprate by replacing rows of CuO_4 groups by CO_3 groups as shown for the member m=3, n=1 (Fig.2) which consists of quadruple copper layers built up from double $[CuO_2]_\infty$ square planar layers sandwiched between two pyramidal copper layers and interconnected through carbonate layers. Although they are not so far optimized, due to the coexistence of several members, the superconducting properties of these films appears as most promising. One indeed observes zero resistance at 75 K and a T_c (onset) of 110 K, significantly higher that obtained for $YBa_2Cu_3O_7$. There is no doubt that ultra vacuum laser deposition will allow the different members to be isolated, so that the route is opened to superconductivity at 110 K in this series of oxycarbonates.

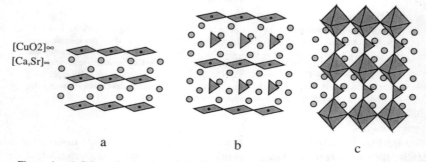

Figure 1 : a) Schematic drawing of the IL-type structure built up from $[Sr, Ca]_\infty$ and $[CuO_2]_\infty$ layers. (b) Triangular carbonate groups can be accomodated between two strontium layers, leading to the S_2CC-type structure (c).

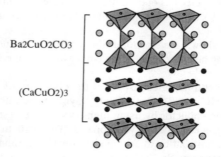

Figure 2 : Structure of the member m = 3, n = 1 of the $(CaCuO_2)_m (Ba_2CuO_2CO_3)_n$ family

The second kind of intergrowth mechanism results from the fact that the rock salt structure forms easily intergrowth with the perovskite. Consequently, the single perovskite layers of the S_2CC structure and of the 1201 or 2201 structures, can play the role of interface to build $[1201]_m [S_2CC]_n$ and $[2201]_m [S_2CC]_n$ intergrowths. This is the case of several oxycarbonates with the formula $Tl_{1-x}M_xSr_4Cu_2O_7CO_3$ [4,5] with M = Pb,

Bi and $Hg_{1-x}M_xSr_4Cu_2O_7CO_3$ with M = Pb, Bi, Mo, Cr, V [6-9]. All these oxides, represented by the prototype $Tl_{0.5}Pb_{0.5}Sr_4Cu_2O_7CO_3$, correspond to the member m = n = 1 of the series. Their structure consists of the intergrowth of double rock salt-type layers $[(Tl_{0.5}Pb_{0.5}O)(SrO)]_\infty$ and single perovskite layers, linked through layers of CO_3 groups. In other words, this structure is a single intergrowth of the S_2CC and of the "1201" $Tl_{0.5}Pb_{0.5}Sr_2CuO_5$ structures. It is remarkable that all these compounds are superconductors, with T_c ranging from 17 K to 76K. Concerning the $[2201]_m$ $[S_2CC]_n$ intergrowth, the investigation of the bismuth oxycarbonates has also been fruitful. Three oxycarbonates with the generic formulation $(Bi_2Sr_2CuO_6)_m$ $(Sr_2CuO_2CO_3)_n$ have been synthetized for m = 1 and n = 1, 2, 3 [10-12]. Their structure corresponds to the intergrowth of a single 2201-type layer (i.e. with the $Bi_2Sr_2CuO_6$ structure) with a multiple S_2CC layer. Again, it is remarkable that the critical temperature of these oxycarbonates is significantly higher than those of the parent structures. One indeed observes a T_c of 30 K, 40 K and 34 K for n = 1, 2 and 3 respectively whereas $Bi_2Sr_2CuO_6$ exhibits a T_c of 22 K [13] and $Sr_2CuO_2CO_3$ does not superconduct.

The second step for the generation of new layered oxycarbonates, deals with the application of shearing mechanisms to the intergrowths that have been generated in the first step of this study. Such shearing phenomena can be applied transversally to the $[CuO_2]_\infty$ layers to three families of intergrowths $[IL]_m$ $[S_2CC]_n$, or $[1201]_m$ $[S_2CC]_n$ or $[2201]_m$ $[S_2CC]_n$. The application of crystallographic shearing (C.S.) to the first family of intergrowths leads to the oxycarbonates (Y, Ca)$_n$(Ba, Sr)$_{2n}$Cu$_{3n-1}$(CO$_3$)O$_{7n-3}$ [14-18] which are in fact derived from the structure of $YBa_2Cu_3O_7$ by an ordered substitution of CO_3 groups for CuO_4 square planar groups.

Two distinct C.S. mechanisms can be applied to the $[1201]_1$ $[S_2CC]_1$ intergrowth along $\bar{c}$, i.e. transversally to the $[CuO_2]$ layers. They correspond to the (100)p (Fig. 3a) and (110)p (Fig. 3b) crystallographic shear planes repered with respect to the perovskite layer of these oxycarbonates. Thus, in both cases, a translation by c/2 is applied between two parts of the "1201" structure either in the (100)p plane (Fig. 3c) or in the (110)p plane (Fig. 3d). It is remarkable that such a crystallographic shearing does not interrupt the copper layers (Fig. 3c - 3d). On the contrary, the C.S. leads to an interruption of the $[TlO]_\infty$ layers and, vice versa, of the layers of CO_3 groups, that form, in that way, mixed layers built up from (CO) and (TlO) ribbons. In fact this shearing mechanism implies a waving of the $[CuO_2]_\infty$ layers, due to the size difference between the thallium and carbon atom that are interleaved between the $[CuO_2]_\infty$ layers. The first class of shear structures, called (100)-collapsed "1201-S_2CC" oxycarbonates are obtained by replacing partly strontium by barium. The shearing phenomenon often appears every four CuO_6 octahedra, as shown for $TlBa_2Sr_2Cu_2O_7CO_3$ [19] and $Tl_{0.8}Hg_{0.2}Ba_2Sr_2Cu_2O_7CO_3$ [20] whose structure consists of waving octahedral copper layers interconnected through almost flat mixed layers $[(CO)_4(TlO)_4]_\infty$ built up of ribbons involving four rows of carbonate groups and thallium alternately. Varying the Ba/Sr ratio allows the shearing periodicity to be varied ; this is the case of the oxycarbonates $TlSr_{4-x}Ba_xCu_2O_7CO_3$ [21,22] for which the shearing may also appear every three CuO_6 octahedra, or even every three and four octahedra alternatively. Similarly a more complicated sequence of shearings has been obtained for $Tl_{0.3}Hg_{0.4}Ba_2Sr_2Cu_2O_7CO_3$ [20] that involves three shearing every four octahedra followed by one shearing every three octahedra. The most important feature deals with the fact that all these phases are superconductors, characterized by a narrow transition and a critical temperature ranging from 50 K to 70 K. This demonstrates, if needed, that the planar character of the $[CuO_2]_\infty$ layers is not necessary for the appearance of superconductivity, in contrast to the statements of several theories.

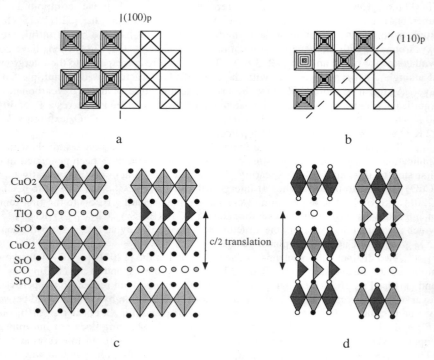

Figure 3 : a) $(100)_p$ and (b) $(110)_p$ crystallographic shear planes viewed along $[001]_p$, shaded and opened octahedra are translated by c/2. Illustration of the C.S. mechanisms viewed along $[100]_p$ (c) and $[110]_p$ (d).

In order to generate the (110)-collapsed "1201-S$_2$CC" structure, one can substitute partly a transition element for thallium. This is the case of the oxycarbonates $Tl_{2/3}Cr_{1/3}Sr_4Cu_2O_7CO_3$ and $Tl_{0.8}V_{0.2}Sr_4Cu_2O_7CO_3$[11] which correspond to a shearing mechanism in the (110) plane every five CuO_6 octahedra. Note that the structure of this phase is also characterized by an ondulation of the $[CuO_2]_\infty$ layers in order to accommodate the interleaved flat mixed layers $[(CO)_5(TlO)_5]_\infty$ that are built up of ribbons involving five rows of carbonate groups and thallium alternatively. The substitution of barium for strontium also allows the (110) C.S. mechanism to be generated, as shown for the oxycarbonate $Hg_{0.5}Pb_{0.5}Sr_{4-x}Ba_xCu_2(CO_3)O_7$ [23] that exhibits a (110) C.S. phenomenon every five and six CuO_6 octahedra alternatively. In a similar way the oxycarbonate $HgBa_2Sr_2Cu_2(CO_3)O_7$ [24] exhibits a (110) C.S. mechanisms but in that case the sequence of the C.S. planes is more complex. Again, like for the (100)-collapsed $[1201]_1[S_2CC]_1$ oxycarbonates, it is remarkable that all the (110)-collapsed $[1201]_1[S_2CC]_1$ oxycarbonates are superconductors with T_c ranging from 50 K to 68 K, in spite of the waving of their $[CuO_2]_\infty$ layers.

The application of a shearing mechanism to the bismuth oxycarbonates $(Bi_2Sr_2CuO_6)$ $(Sr_2CuO_2CO_3)_1$ is also possible. However, in that case the $[CuO_2]_\infty$ layers are interrupted due to the fact that double bismuth layers are facing a single carbonate layer after shearing. Only one collapsed bismuth oxycarbonate, $Bi_{15}Sr_{29}Cu_{12}(CO_3)_7O_{56}$, has been synthesized to date [25]. Although it does not superconduct, this phase is of great interest, since it must be considered as a potential candidate for improving pinning in the superconductor $Bi_2Sr_4Cu_2O_8CO_3$, if introduced as an extended defect in the matrix of the latter.

This overview of the crystal chemistry of oxycarbonates demonstrates that the application of intergrowth and shearing mechanisms to the combination of layered cuprates with the S_2CC oxycarbonate should allow other new to be synthesized in the future.

The mechanisms of superconductivity in theses phases, such as the role of CO_3 groups and of the copper mixed valence are so far not understood. In the same way, the physics of those superconductors should be developed in the future, and particularly the pinning of the vortices is a very important issue, which should be investigated, especially if one takes into consideration the possibility of superconductivity at 110 K in the "BaCaCu" oxycarbonates, that may be very promising for electronic applications.

References

1- T. Siegrist, S.M. Zahurak, D.W. Murphy and R.S. Roth, *Nature* **334** (1988), 231.
2- D.V. Fomichev, A.L. Kharlanov, E.V. Antipov and L.M. Kovba, *Superconductivity* **3** (1990) 126.
3- J.L. Allen, B. Mercey, W. Prellier, J.F. Hamet, M. Hervieu and B. Raveau, *Physica C* **241** (1995) 158.
4- M. Huvé, C. Michel, A. Maignan, M. Hervieu, C. Martin and B. Raveau, *Physica C* **205** (1993) 219.
5- A. Maignan, M. Huvé, C. Michel, M. Hervieu, C. Martin and B. Raveau, *Physica C* **208** (1993) 149.
6- C. Martin, M. Hervieu, M. Huvé, C. Michel, A. Maignan, G. Van Tendeloo and B. Raveau, *Physica C* **222** (1994) 19.
7- D. Pelloquin, M. Hervieu, C. Michel, A. Maignan and B. Raveau, *Physica C* **227** (1994) 215.
8- D. Pelloquin, M. Hervieu, S. Malo, C. Michel, A. Maignan and B. Raveau, *Physica C* **246** (1995) 1.
9- A. Maignan, D. Pelloquin, S. Malo, C. Michel, M. Hervieu and B. Raveau, *Physica C* **249** (1995) 220.

10- D. Pelloquin, M. Caldès, A. Maignan, C. Michel, M. Hervieu and B. Raveau, *Physica C* **208** (1993) 121.
11- D. Pelloquin, A. Maignan, M. Caldès, M. Hervieu, C. Michel and B. Raveau, *Physica C* **212** (1993) 199.
12- D. Pelloquin, M. Hervieu, A. Maignan, C. Michel, M.T. caldès and B. Raveau, *Physica C* **232** (1994) 75.
13- C. Michel, M. Hervieu, M.M. Borel, A. Grandin, F. Deslandes, J. Provost and B. Raveau, *Z. Phys.* **B 68** (1987) 421.
14- Y. Miyazaki, H. Yamane, N. Ohnishi, T. Kajitani, K. Hiraga, Y. Morii, S. Funahashi and T. Hirai, *Physica C* **198** (1992) 7.
15- J. Akimitsu, M. Uehara, M. Ogawa, H. Nakata, K. Tomimoto, Y. Miyazaki, H. Yamane, T. Hirai, K. Kinoshita and Y. Matsui, *Physica C* **201** (1992) 320.
16- A. Maignan, M. Hervieu, C. Michel and B. Raveau, *Physica C* **208** (1993) 116.
17- M. Hervieu, P. Boullay, B. Domengès, A. Maignan and B. Raveau, *J. Solid State Chem.* **105** (1993) 300.
18- B. Domengès, M. Hervieu and B. Raveau, *Physica C* 207 (1993) 65.
19- F. Goutenoire, M. Hervieu, A. Maignan, C. Michel, C. Martin and B. Raveau, *Physica C* **210** (1993) 350.
20- T. Noda, M. Ogawa, J. Akimitsu, M. Kikuchi, E. Ohshima and Y. Syono, *Physica C* **242** (1995) 12.
21- Y. Matsui, M. Ogawa, M. Uehara, H. Nakata and J. Akimitsu, *Physica C* **217** (1993) 287.
22- M. Kikuchi, E. Ohshima, N. Ohnishi, Y. Muraoka, S. Nakajima, E. Aoyagi, M. Ogawa, J. Akimitsu, T. Oku, K. Hiraga and Y. Syono, *Physica C* **219** (1994) 200.
23- M. Huvé, G. Van Tendeloo, M. Hervieu, A. Maignan and B. Raveau, *Physica C* **231** (1994) 15.
24- M. Uehara, S. Sahoda, J. Akimitsu and Y. Matsui, *Physica C* **222** (1994) 27.
25- M. Hervieu, M.T. Caldès, D. Pelloquin, C. Michel, S. Cabrera and B. Raveau, *J. Mater. Chem..* **6** (1996) 175.

SUPERCONDUCTING PHASES IN THE Sr-Cu-O SYSTEM

Z.L. DU, Y. CAO, Y.Y. SUN, L. GAO, Y.Y. XUE AND C.W. CHU
Texas Center for Superconductivity at The University of Houston
University of Houston, TX, 77204-5932

ABSTRACT

We have examined various Sr-Cu-O samples with the superconducting transition temperatures ranging from 70 to 98 K. It was observed that prolonged annealing in a reduced atmosphere merges the different T_c's to ~90 K without changing their Meissner volume fraction. Semi-quantitative XDR analysis suggests that the phase responsible for superconductivity is $Sr_3Cu_2O_5$ instead of Sr_2CuO_4.

Superconductivity in the Sr-Cu-O samples prepared under 4-8 GPa with a transition temperature T_c up to ~100 K has been reported by many groups. $SrCuO_2$ [1], $Sr_2CuO_{3+\delta}$ [2], $Sr_3Cu_2O_{5+\delta}$ [2] and $Sr_4Cu_3O_{7+\delta}$ [3] have been suggested to be the corresponding superconducting phases. Recently, Shaked *et al.* [3] suggested that both $SrCuO_2$ and $Sr_2CuO_{3+\delta}$ are not superconducting based on the measured Cu-O bond length. To clarify this situation, an investigation was carried out.

Our samples of Sr:Cu=2:1 were synthesized at 800 -1100 °C in a multi-anvil press at 6-8 GPa. 3-30% $KClO_4$ by weight was mixed with precursor to offer the needed oxidation condition. The powder mixture was then pressed into pellets, wraped in Au-foil and fired under pressure. The phase composition of the samples was analyzed by XRD. SiO_2 powder, 20% by weight, was added as an internal standard. The relative superconducting volume fraction of the samples was estimated using the field cooling (FC) magnetization at 10 G.

The phase composition of the as-synthesized samples depends on the synthesis condition and the amount of oxidation agent present: $SrCuO_2$ was the major phase for samples synthesized below 1050 °C with large $KClO_3$ addition (15% or more); $Sr_2CuO_{3+\delta}$ was the dominant phase if the samples were made with less than 4% of the $KClO_3$ addition. However, $Sr_3Cu_2O_{5+\delta}$ and $Sr_4Cu_3O_{7+\delta}$ were the main phases for samples synthesized at 1000-1200 °C with a $KClO_3$ addition of 5-10%. The samples synthesized in this condition usually had a large FC diamagnetization and a T_c between 70 to 98 K.

Several as-synthesized samples with superconducting volume fraction 5% or more were annealed at 280 °C under pure N_2 gas. Typical FC magnetizations measured before and after annealing are shown in Figure 1. The FC magnetization at 5 K is almost unaffected by annealing, suggesting that the superconducting phase has not been altered by such low-temperature annealing. However, various initial T_c's merge to ~90 K after the annealing. Therefore, we suggest that the superconductivity observed between 70 and 100 K in our samples is related to one particular phase.

To identify this phase, semi-quantitative XDR analysis was carried out. The total peak-area (I) under the low-angle diffraction peaks characteristic of the phase of interest was calculated and normalized by the peak-area $I_{101}(SiO_2)$ of the (101) line of SiO_2. The ratio $R = I / I_{101}(SiO_2)$ was plotted against the FC magnetization. $Sr_2CuO_{3+\delta}$ and $Sr_3Cu_2O_{5+\delta}$ were the main phases in the samples with the superconducting volume frac-

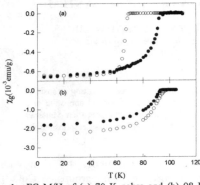
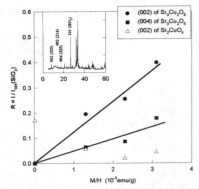

Fig. 1: FC M/H of (a) 70 K pahse and (b) 98 K phase before (open circles) and after (filled cirlces) annealing.

Fig. 2: Normalized line intensity R vs. M/H. Inset: XRD pattern for a typical sample.

tion larger than ~15%. The R of the (002), (004) lines of $Sr_3Cu_2O_{5+\delta}$ and (002) line of $Sr_2CuO_{3+\delta}$ are shown in Figure 2. It shows that the relative intensities of both (002) and (004) lines of $Sr_3Cu_2O_{5+\delta}$ are proportional to M/H within our experimental resolution, but no superconductivity was detected in the $Sr_2CuO_{3+\delta}$ sample with a R ~ 0.17. Therefore, $Sr_3Cu_2O_{5+\delta}$, rather than $Sr_2CuO_{3+\delta}$, was assigned as the superconducting phase in our samples, which had a T_c of 70 - 98 K. However, it should be noted that samples in Ref. 3 were synthesized with a different initial composition of Sr:Cu = 1:1 in which $Sr_3Cu_2O_{5+\delta}$ was only a trace impurity in XDR, although they showed a comparable M/H and T_C. $Sr_4Cu_3O_{7+\delta}$ was suggested by Shaked et al. as the superconducting phase based on the estimated phase composition. It is possible that there are several superconducting phases with similar T_c in the $Sr_{n+1}Cu_nO_{2n+\delta}$, and the relative abundance of these phases in a particular sample is determined by detailed synthesis conditions.

In summary, a superconducting phase in Sr-Cu-O has been identified as $Sr_3Cu_2O_{5+\delta}$ with T_c between 70 and 100 K. On the other hand, the phase of $Sr_2CuO_{3+\delta}$ synthesized at 800 - 1100 ºC, 6-8 GPa over a broad oxidation range is not superconducting down to 4.2 K.

ACKNOWLEDGMENTS

This work is supported in part by NSF Grant No DMR 95-00625, EPRI RP-8066-04, DoE, the State of Texas through the Texas Center for Superconductivity at University of Houston and the T. L. L. Temple Foundation.

References

1. M. Azuma et al., Nature 356 (1992) 775.
2. Z. Hiroi et al., Nature 364 (1993) 315.
3. H. Shaked et al., Phys. Rev B51 (1995) 11784.

EFFECT OF PRESSURE ON La_2O_3-CuO SUBSOLIDUS PHASE EQUILIBRIA

JOEL GENY, JAMES K. MEEN, and DON ELTHON
Department of Chemistry and the Texas Center for Superconductivity,
University of Houston, Houston, TX 77204 U.S.A.

ABSTRACT

The relative stabilities of La_2CuO_4 and $La_2Cu_2O_5$ as a function of pressure are reviewed. The reaction $La_2CuO_4 + CuO = La_2Cu_2O_5$ occurs at lower temperatures at progressively higher pressures so that La_2CuO_4 is progressively destabilized with respect to $La_2Cu_2O_5$ as pressure increases. This negative slope of the pressure-temperature curve is very unusual for reactions involving condensed phases. It requires that either the change in entropy or the change in volume of the reaction given above is negative.

1. Introduction

The system La_2O_3-CuO is an important one in studies of superconductivity because La_2CuO_4 is parental to solid solutions of the form $La_{2-x}AE_xCuO_4$ (AE is an alkaline earth metal), that include high-temperature superconductors. We have investigated phase relations in the La_2O_3-CuO system as a function of pressure up to 30 kbar at 950°C. These data, combined with a knowledge of 1 atm. phase relations[1], allow estimation of subsolidus pressure-temperature stabilities of binary La-Cu oxides.

2. Prior Studies

A 1 atm study of La_2O_3-CuO in air[1] found three intermediate phases: La_2CuO_4, which melts incongruently to La_2O_3 + liquid at 1375°C; $La_2Cu_2O_5$, which forms at 1002°C by reaction between La_2CuO_4 and CuO and melts incongruently to La_2CuO_4 + liquid at 1035°C; $La_8Cu_7O_{19}$, which forms at 1012°C by reaction between La_2CuO_4 and $La_2Cu_2O_5$ and decomposes at 1027°C.

3. Discussion

Figure 1 is a pressure-composition section for La_2O_3-CuO at 950°C based on published results[1,2,3]. At this temperature, the only binary phase stable at low P is La_2CuO_4. $La_2Cu_2O_5$ is stable at 30 kbar[3] and higher P. It is also stable at 1 atm above 1002°C[1]. Thus, $La_2CuO_4+CuO=La_2Cu_2O_5$ moves to progressively lower T as P increases. Such a slope on a P-T diagram is very unusual for reactions that involve no volatile phase.

The change in Gibbs free energy of reactants, δG_r, due to a change in pressure, δP, and in temperature, δT, is given by $\delta G_r = V_r \delta P - S_r \delta T$ where V_r and S_r are the volume and entropy of reactants, respectively. Similarly the change in Gibbs free energy of the products in response to the same change in P and T is $\delta G_p = V_p \delta P - S_p \delta T$ (p specifies products). For equilibrium to be maintained, changes in Gibbs free energy of products and reactants are the same so that $V_r \delta P - S_r \delta T = V_p \delta P - S_p \delta T$ or $(V_p - V_r)\delta P = (S_p - S_r)\delta T$ so that $dP/dT = \Delta S/\Delta V$. For $La_2CuO_4 + CuO = La_2Cu_2O_5$, dP/dT is negative so either ΔS

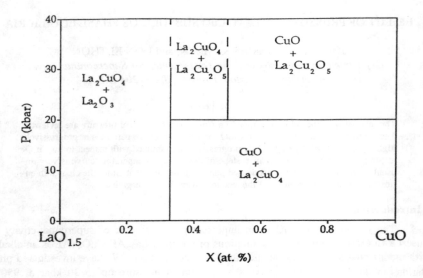

Figure 1. Pressure-composition phase diagram for the system La_2O_3-CuO at 950°C estimated from data at 1 atm, 10 kbar, and 30 kbar. The pressure of the reaction $La_2CuO_4 + CuO = La_2Cu_2O_5$ is only constrained to lie between 10 and 30 kbar and is arbitrarily shown to lie at 20 kbar.

or ΔV of the reaction, but not both, is negative. Molar volumes estimated from cell parameters of La_2CuO_4 and $La_2Cu_2O_5$ (at room temperature) indicate that ΔV of the reaction is ~9 cm^3/mole. If that positive number also applies to the reaction at 1002°C, then the reaction has a negative change in entropy. Of course, the La-Cu oxides may have different coefficients of thermal expansivity so the reaction does have a negative volume change at equilibrium conditions.

The negative slope inferred for $La_2CuO_4+CuO=La_2Cu_2O_5$ indicates that increasing pressure, at least to 30 kbar, destabilizes La_2CuO_4 relative to $La_2Cu_2O_5$. Temperatures of melting of $La_2Cu_2O_5$ and of $CuO+La_2Cu_2O_5$, while unknown at 10-30 kbar, are likely to be higher than at 1 atm. (ΔS and ΔV of melting are almost always positive for melting condensed phases, so $\{dP/dT\}_{melting}$ is positive.) The stability field of $La_2Cu_2O_5$ is expected to increase to both higher and lower temperatures at elevated pressure.

4. Acknowledgments

This work was supported by grants from the State of Texas to the Texas Center for Superconductivity at the University of Houston. Funds to develop the high-pressure facility used here were provided by a grant from the Texas Advanced Research Program.

5. References

1 J. M. S. Skakle and A. R. West, *J. Am. Ceram. Soc.*, 77 (1994) 2199-202
2 J. Geny, J.K Meen, D.Elthon, *submitted to J. Am. Ceram. Soc.*
3 J. Geny, J.K Meen, D.Elthon, *submitted to J. Am. Ceram. Soc.*

ANOMALOUS Pr ORDERING AND FILAMENTARY SUPERCONDUCTIVITY FOR THE Pr2212 CUPRATES

H. C. KU, Y. Y. HSU, S. R. LIN, D. Y. HSU, Y. M. WAN AND Y. B. YOU
Department of Physics, National Tsing Hua University, Hsinchu, Taiwan 300, Republic of China

S. R. SHEEN
Material Sciences Center, National Tsing Hua University, Hsinchu, Taiwan 300, Republic of China

The anomalous Pr antiferromagnetic ordering and filamentary superconductivity for the cuprate Pr2212 system $M_2A_2PrCu_2O_8$ (M = Cu, Bi, Pb; A = Ba, Sr) has been studied through the synthesis of (Pb,Cu)-2212, Bi-2212, and Cu-2212C cuprates. A tetragonal (Pb,Cu)-2212 $(Pb_{0.5}Cu_{0.5})_2(Ba_{0.5}Sr_{0.5})_2PrCu_2O_{8+\delta}$ with $T_N(Pr)$ of 9 K and filamentary superconductivity with T_c of 6 K have been observed. Preliminary studies on the orthorhombic Bi-2212 $Bi_2Sr_2PrCu_2O_{8+\delta}$ single crystals show a superconducting transition around 15 K. For the orthorhombic Cu-2212C $Cu_2Ba_2PrCu_2O_{8-\delta}$ ($PrBa_2Cu_4O_{8-\delta}$), an antiferromagnetic-like transition was observed around 18 K.

The orthorhombic $PrBa_2Cu_3O_7$ compound with an anomalously high Pr antiferromagnetic ordering temperature $T_N(Pr)$ of 17 K, is the only nonsuperconducting member of the $RBa_2Cu_3O_7$ system (R = Y or a rare earth). Since $PrBa_2Cu_3O_7$ compound can be re-categorized as the Cu-1212C-type compound (C stands for chain) and a total replacement of Ca by R between the two CuO_2 layers can be achieved in the orthorhombic/tetragonal two-CuO_2-layered $M_mA_2RCu_2O_y$ m212-type structure (m = 1, 2, 3; M = Cu, Bi, Hg, Tl, Pb and A = Ba, Sr),[1] questions arise whether the anomalous Pr effect observed in the Pr1212-type $MA_2PrCu_2O_7$ system [2] will occur in the Pr2212-type $M_2A_2PrCu_2O_8$ system. Recently, we were able to observe the first Pr anomaly in the tetragonal (Pb,Cu)-2212 $(Pb_{0.5}Cu_{0.5})_2(Ba_{0.5}Sr_{0.5})_2PrCu_2O_{8+\delta}$ compound.[3] In addition to the magnetic Pr anomaly, variations of oxygen stoichiometry in this (Pb,Cu)-2212 compound might partially induce the metal-insulator transition, making possible the appearance of filamentary superconductivity with T_c of 15 K for R = Y and T_c = 6 K for R = Pr.[4] In this paper, we further report the trace of filamentary superconductivity in the Bi-2212 compound $Bi_2Sr_2PrCu_2O_{8+\delta}$ and the trace of anomalous $T_N(Pr)$ in $PrBa_2Cu_4O_{8-\delta}$ which can be re-classified as the Cu-2212C $[Cu]_2Ba_2PrCu_2O_{8-\delta}$ compound.

$Bi_2Sr_2PrCu_2O_{8+\delta}$ crystals were prepared by solid-state reaction in flowing oxygen. The stoichiometric oxides was fired quickly to 1000 °C, held at the temperature for 10 hours, then slowly cooled at 2 °C/h to 900 °C, and at 100 °C/h to room temperature. Thin crystallites of millimeter squares in area and several micron meters in thickness were extracted mechanically from an alumina crucible. Post-annealing in oxygen for 2 days was to ensure the homogeneity of oxygen. $PrBa_2Cu_4O_{8-\delta}$ samples was prepared by the chemical sol-gel method and sintered at 860–875 °C for 8 days in 1 atm of oxygen pressure. The X-ray data shows a dominant phase of the Cu-2212C $PrBa_2Cu_4O_{8-\delta}$ and some traces of $BaCuO_2$, CuO and $PrBaO_3$. Resistivity measurements were performed using the four-probe method with a resistance bridge (Linear Research LR-700) at 16 Hz with an excitation current of 1 mA. All measurements were carried out in a magnetically shielded magnetometer (Quantum Design $MPMS_2$ SQUID) on bulk samples under the zero-field-cooled condition.

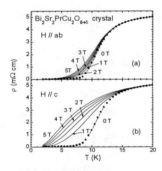
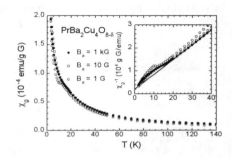

Figure 1. Resistivity curves of $Bi_2Sr_2PrCu_2O_{8+\delta}$ crystal with various applied fields (a) paralleled to the ab plane and (b) paralleled to the c axis.

Figure 2. Magnetic susceptibility of $PrBa_2Cu_4O_{8-\delta}$. The inset shows the inverse susceptibility and the Curie-Weiss fit with negative intercept on 1 kG data points.

As shown in Fig. 1, superconducting transition with T_c(onset) ≈ 15 K and T_c(zero) ≈ 8 K can be seen from the in-plane resistivity ρ_{ab} of the Bi-2212 crystal. To investigate the anisotropy we examine the field dependence of ρ_{ab} with the external fields parallel to the ab-plane and the c-axis while normal to the bias current. The results shown in Fig. 1a and 1b demonstrate the anisotropy. The magnitude of the anisotropy is estimated to be ≈ 100 from the data of $\rho_{ab}(B_a = 100 G // ab)$ and $\rho_{ab}(B_a = 1 T // c)$, which is also confirmed by magnetic susceptibility measurements.

The magnetic (mass) susceptibility $\chi_g(T)$ under various applied magnetic fields are shown in Fig. 2 with inset reveals the corresponding inverse $\chi_g^{-1}(T)$. All $\chi_g^{-1}(T)$ curves suggest Curie-Weiss behavior at temperatures higher than 18 K with an antiferromagnetic-like, negative paramagnetic intercept. Further analysis on the 1 kG data shows that the derivative $d\chi_g(T)/dT$ has a local minimum at 18 K coincides with the temperature around which $\chi_g^{-1}(T)$ deviates from the Curie-Weiss fit. We have observed T_N(Pr) of the (Pb,Cu)-2212 occurs at the local minimum of $d\chi_g(T)/dT$.[3] Therefore, it is plausible to suggest that an antiferromagnetic transition occurs around 18K. Further specific heat studies on this compound are required to confirm the transition.

In conclusion, superconductivity around 15 K is observed in $Bi_2Sr_2PrCu_2O_{8+\delta}$ crystals. Our studies suggest that a proper oxygen post-annealing which increases oxygen stoichiometry parameter δ thus generates more electrical carriers is the key to synthesize a superconducting Pr-based 2212 compound. Our results of anomalous Pr ordering in the Cu-2212C compound, along with earlier reports on the (Pb,Cu)-2212 compound, demonstrate the appearance of Pr-ordering in the Pr2212. The anomalously high T_N(Pr) observed in these systems indicates the importance of the quasi-2D Pr-O-Pr superexchange magnetic coupling mechanism through the strong hybridization between the Pr 4f and the eight O $2p_\pi$ orbitals in the adjacent two-CuO_2 layers.

This work was supported by the NSC of ROC contract NSC85-2112-M007-043 and -044PH.

References

1. H. C. Ku et al., *J. Appl. Phys.* **79**, in press (1996), and reference cited therein.
2. C. H. Chou et al., *Phys. Rev.* B **53**, in press (1996), and reference cited therein.
3. C.L. Yang et al., *Phys. Rev.* B **52**, 10452 (1995).
4. C.L. Yang et al., *Chin. J. Phys* **34**, in press (1996).

SUPERCONDUCTIVITY IN $Sr_2YRu_{1-x}Cu_xO_6$ SYSTEM

D.C. LING[1], S.R. SHEEN[1], C.Y. TAI[1], J.L. TSENG[1], M.K. WU[1], T.Y. CHEN[2], AND F.Z. CHIEN[2]

[1]*Materials Science Center, National Tsing Hua University, Hsinchu, 30043, Taiwan, ROC*
[2]*Department of Physics, Tamkang University, Tamsui, Taiwan, ROC*

A systematic study on the superconducting properties of $Sr_2YRu_{1-x}Cu_xO_6$ system with $x = 0.0$-0.5 has been carried out. With Cu doping, the system undergoes a phase transition from an insulator to a superconductor. In addition, a ferromagnetic order can coexist with superconductivity. The partial substitution of Ru by Cu not only creates movable carriers also induces a ferromagnetic order through the double exchange interaction.

1 Introduction

The common features of the high temperature superconducting cuprates are their parent compounds being antiferromagnetic (AF) insulators and consist of layered structure based on a perovskite template with CuO_2 square planes. This raises a question whether superconductivity can occur by proper doping in an AF insulator compound that contains no CuO_2 planes. The double perovskite compound A_2YRuO_6, where A stands for divalent alkaline earth, is a well known antiferromagnetic insulating system. A recent systematic study on the structural stability and superconductivity of the transition-metal doped $YSr_2(Cu_{3-x}M_x)O_y$ led us to the discovery of superconductivity in $Sr_2Y(Ru_{1-x}Cu_x)O_6$. The superconducting transition (onset) temperature, determined by both resistive and magnetic measurements, of the compound ranges from 85 K to 30K depending on the Cu content and the processing conditions. Detailed structure refinement shows that this Cu-doped system has the same crystal symmetry as its parent compound, and there exists no component that is related to the known cuprate superconductors.

All compounds investigated were prepared by a solid-state reaction method. Stoichiometric starting powders of $SrCO_3$, Y_2O_3, RuO_2, and CuO were thoroughly mixed, then repeated calcined at 1000 °C for several days in air in an Al_2O_3 crucible. Subsequently, the powder was ground and pressed into a pellet, then annealed at 1390°C in O_2 for 2 hours followed by 6 hours oxygen annealing at 1330 °C. The last step was repeated two or three times. Sample structures were characterized by the powder x-ray diffraction technique. A single-phase $Sr_2YRu_{1-x}Cu_xO_6$ with orthorhombic structure was obtained for $x = 0.0$-0.3. Some unidentified peaks were observed in the samples with $x = 0.4$ and 0.5, suggesting that the limiting stoichiometry of the phase is just below $x = 0.3$. The resistivity measurements were made by a standard 4-probe method. The magnetic susceptibility was performed by the SQUID magnetometer.

2 Results and Discussion

Figure 1 shows the typical temperature dependence of the resistivity of $Sr_2YRu_{0.8}Cu_{0.2}O_6$. The resistivity for this sample just above T_c is ~10 Ω-cm, considerably higher than that of the cuprate. The resistivity decreases with increasing Cu doping, suggesting that the concentration of movable holes are related to the Cu doping. The onset of the resistive transition is at 83 K, with a midpoint of 47 K and a zero-resistance state at 22 K. The rather broad might arise from either a suppression of superconductivity by magnetic scattering of Ru ions or an intrinsic microstructure disorder. The magnetic field dependence of ρ(T) is rather complex with the zero-resistance state shifting toward lower temperatures with increasing fields. Using the WHH equation[1], a value of $H_{c2}(0)$ can be estimated to be about 50 T for the system investigated.

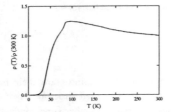

Figure 1. Resistivity of $Sr_2YRu_{0.8}Cu_{0.2}O_6$ as a function of temperature, normalized to the value at 300 K.

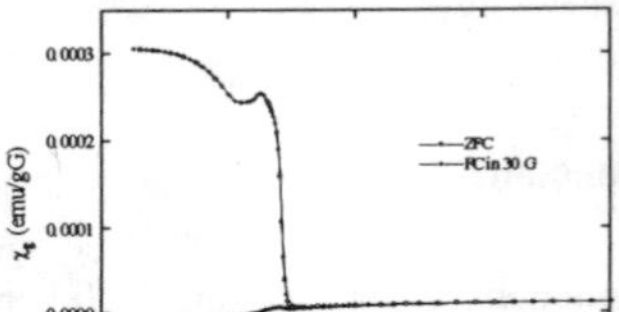

Figure 2. Temperature dependence of magnetic susceptibility of $Sr_2YRu_{0.8}Cu_{0.2}O_6$ sample.

The temperature dependence of ZFC and FC magnetic susceptibility of $Sr_2YRu_{0.8}Cu_{0.2}O_6$ is shown in Fig. 2. The ZFC curve shows a weak diamagnetic response below 30 K. The FC curve exhibits a sharp rise in susceptibility that can be associated with a ferromagnetic order. In addition, there is a kink at 26 K in the FC associated with antiferromagnetic order of the parent compound, indicating that antiferromagnetism (AFM) is not destroyed by doping. The coexistence of FM and AFM can be explained by the double exchange[2] idea which occurs indirectly via spin coupling to itinerant electrons traveling from one ion to the next. It appears that the ferromagnetic ordering temperature and the onset temperature of the diamagnetic response in the ZFC are on the same order of magnitude. The small diamagnetic signals are probably due to a compensation of the significant ferromagnetic response. In any case, a detailed study is required to further understand the striking feature of magnetic properties of this system.

Acknowledgments

This work was supported by the National Science Council of ROC under grant No. NSC84-2212-M-007-005PH.

References

1. N.R. Werthamer, E. Hefland, and P.C. Hohenberg, *Phys. Rev.* **147**, 295 (1966).
2. C. Zener, *Phys. Rev.* **82**, 403 (1951); P.W. Anderson et al., *Phys. Rev.* **100**, 675 (1955).

MELTING RELATIONS IN Bi_2O_3-CuO-Cu_2O

JAMES K. MEEN and DON ELTHON
Department of Chemistry and the Texas Center for Superconductivity,
University of Houston, Houston, TX 77204 U.S.A.

ABSTRACT

Most liquids in the system Bi_2O_3-"CuO" even at $f(O_2)=1$ contain CuO and Cu_2O. The temperature of melting of the assemblage Bi_2CuO_4+CuO decreases as $f(O_2)$ is lowered, consistent with an increase in Cu^+ in the liquid at lower $f(O_2)$ despite the fact that coexisting phases contain only divalent Cu. Solid-liquid phase relations determined at fixed $f(O_2)$ act as if the section is binary even though liquids contain Bi_2O_3, CuO, and Cu_2O. The projected melting point of CuO determined from Bi_2O_3-"CuO" melting relations at $f(O_2)=1$ is less than the 1117°C melting temperature of pure CuO.

1. Introduction

Many studies of melting relations of oxide systems that include copper oxide have been conducted in the last decade in response to the discovery of high-temperature superconductivity in cuprates[1]. Despite widespread recognition that copper potentially exists in both univalent and divalent forms in phases formed in the melting intervals of many of these systems, little attention has been devoted to the influence of the presence of these two states to the phase relations of the systems. This paper investigates melting relations in CuO-Cu_2O-bearing systems with particular attention to the Bi-Cu-O system. The treatment developed here shows that if some copper in liquids in a "binary" such as Bi_2O_3-CuO is univalent, the liquid-solid phase relations are determined by the topology of the Bi_2O_3-CuO-Cu_2O phase diagram even at temperatures well below those of the CuO-Cu_2O buffer at the relevant $f(O_2)$. Tsang *et al.*[2] present analytical evidence that most liquids in the system Bi_2O_3-"CuO" (in pure O_2) contain both univalent and divalent copper.

2. Suppression of the Eutectic Temperature by Decrease in Oxygen Fugacity

Figure 1 shows the temperature of the eutectic at which Bi_2CuO_4 and Bi_2O_3 melt at different $f(O_2)$. There is a regular non-linear decrease in eutectic temperature as $f(O_2)$ decreases. It falls from 784°C at $f(O_2)=1$ to 776° at $f(O_2)=10^{-2}$ but then decreases more rapidly to 747°C at $f(O_2)=6.6 \times 10^{-4}$. At $f(O_2)=1$, CuO melts before forming Cu_2O. The CuO-Cu_2O eutectic is at 1075°C, $f(O_2)=0.43$[3]. The CuO-Cu_2O buffer is at 883°C at $f(O_2)=10^{-2}$ and at 778°C at $f(O_2)=6.6 \times 10^{-4}$. The eutectic temperatures are lower than those for $2CuO=Cu_2O+1/2O_2$ at constant $f(O_2)$ but the difference decreases at lower $f(O_2)$.

The obvious explanation for decrease in eutectic temperature at lower $f(O_2)$ is that liquids formed under more reducing conditions have higher contents of Cu^+. This not only reduces the melting temperature by freezing point depression due to adding another component but also decreases the activity of Bi_2CuO_4 in the liquids by decreasing the concentration of Cu^{2+}. As the temperature of the CuO-Cu_2O reaction approaches the eutectic temperature (near $f(O_2)=10^{-2}$), Cu^+:Cu^{2+} in the liquid increases rapidly, causing a greater depression in the eutectic temperature.

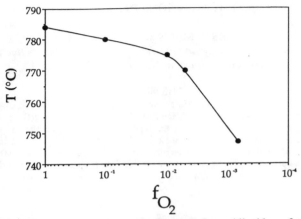

Figure 1. Temperature of the eutectic between Bi_2O_3, Bi_2CuO_4, and liquid as a function of oxygen fugacity. Errors on temperature estimate are ~±5°C; errors on fugacity measurements are <3%.

3. Melting relations in Bi_2O_3-CuO-Cu_2O

Figure 2 represents phase relations in Bi_2O_3-CuO-Cu_2O. In experimental work with fixed $f(O_2)$, liquids must lie on an oxygen isobar. The highest temperatures of melting are along the join CuO-Cu_2O so that liquids in equilibrium with solids have higher $Cu^+:Cu^{2+}$ as they become more copper-rich at a fixed $f(O_2)$.

The liquidus surface slopes gently from CuO (melts at 1117°C) to the CuO-Cu_2O eutectic (1075°C, $f(O_2)$=0.43, 34.5% Cu_2O)[3] and rises rapidly to melting of Cu_2O (1242°C, $f(O_2)$~8×10^{-4}). Temperatures marked on the liquid saturated in Bi_2CuO_4 and Bi_2O_3 are from Figure 1; those marked on the liquid saturated in CuO and Cu_2O are calculated from thermodynamic data[3].

Even the oxygen isobar for $f(O_2)$=1 lies well away from the Bi_2O_3-CuO join particularly at higher Cu contents. The first melting at a given $f(O_2)$ occurs at the temperature at which the isobar crosses $L(Bi_2O_3,Bi_2CuO_4)$. Because the $f(O_2)$ of the system is controlled externally, the liquid cannot remain on $L(Bi_2O_3,Bi_2CuO_4)$ so the system acts like a binary. Increasing the temperature results in movement of the liquid across the ternary but along the relevant isobar. Appropriate bulk compositions move toward $L(Bi_2CuO_4,CuO)$. On reaching this line, the liquid reacts with Bi_2CuO_4 and crystallizes CuO isothermally. The external control of $f(O_2)$ again results in the system behaving as if it were a binary even though the liquid contains three components. At higher temperatures, liquids migrate across the CuO field. Liquids of appropriate composition migrate toward the CuO-Cu_2O join. Even at $f(O_2)$=1, however, the apparent melting temperature of CuO projected from Bi_2O_3-CuO data is <1117°C (but >1075°C). At $f(O_2)$<0.43, the liquid can reach $L(CuO,Cu_2O)$ at <1075°C. An increase in temperature results in reaction of CuO to form Cu_2O, and migration of the liquids toward the CuO-Cu_2O join.

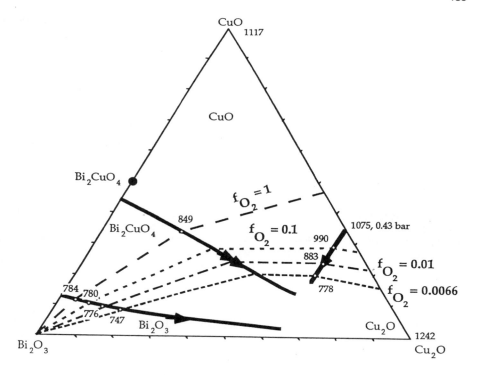

Figure 2. Phase relations in the ternary Bi_2O_3-CuO-Cu_2O. $Cu^+:Cu^{2+}$ values in the liquids are largely conjectural. Heavy lines represent multiply-saturated liquid lines that separate the primary phase fields labelled by phase. The dashed lines indicate the liquid compositions formed at the indicated values of oxygen fugacity. The temperatures of intersection of isobars and multiple saturation lines, where known, are given in degrees Celsius.

4. Acknowledgments

This work was supported by grants from the State of Texas to the Texas Center for Superconductivity at the University of Houston.

5. References

1. J. G. Bednorz and K. A. Müller, *Z. Phys.*, **B64** (1986) 189-93
2. C-F. Tsang, J. K. Meen, D. Elthon, *J. Am. Ceram. Soc.*, **77** (1994) 3119-24
3 N. H. Santander and O. Kubaschewski, *High Temp. High Press.*, **7** (1975) 573-82

GROWTH OF EPITAXIAL LaAlO$_3$ AND CeO$_2$ FILMS USING SOL-GEL PRECURSORS

SHARA S. SHOUP, MARIAPPAN PARANTHAMAN, DAVID B. BEACH, and E.D. SPECHT[§]
Chemical and Analytical Sciences Division, Metals and Ceramics Division[§]
Oak Ridge National Laboratory, P.O. Box 2008, MS 6100
Oak Ridge, TN 37831-6100

LaAlO$_3$ and CeO$_2$ films have been successfully grown using sol-gel precursors. LaAlO$_3$ precursor solution has been prepared from a metal alkoxide route and spun-cast on a SrTiO$_3$ (100) single crystal to yield an epitaxial film following pyrolysis at 800°C in a rapid thermal annealer. A CeO$_2$ precursor solution has been made using both an aqueous and an alkoxide route.

1. Introduction

Sol-gel techniques have emerged as viable non-vacuum methods for the fabrication of long-length conductors. Buffer layers such as LaAlO$_3$ and CeO$_2$ are of interest because critical current densities exceeding 10^6 A/cm^2 at 77K have been repeatedly observed from superconductors grown on these layers.[1] Also, the lattice mismatch between these layers and Y-123 or Tl-1223 is small.

2. Growth of Epitaxial LaAlO$_3$ Films

We have prepared an all alkoxide LaAlO$_3$ precursor solution from which epitaxial films can be grown. Aluminum *tris*-sec-butoxide (3.89 g, 15.8 mmol) was refluxed in 50 ml of 2-methoxyethanol. Approximately 30 mL of sec-butanol/2-methoxyethanol were distilled from the solution which was repeatedly rediluted with 30-50 mL of fresh 2-methoxyethanol followed by distillation. When the boiling point reached that of 2-methoxyethanol (124°C), exchange of sec-butoxide by methoxyethoxide (to help control hydrolysis) was presumed finished. A stoichiometric amount of lanthanum *tris*-isopropoxide (5.03 g, 15.9 mmol) was then added to the refluxing aluminum solution. Approximately 30 mL of isopropanol/2-methoxyethanol were distilled off, and the solution was repeatedly rediluted with 30-50 mL of fresh 2-methoxyethanol and further distilled to insure ligand exchange. The final volume of the light yellow solution was adjusted with 2-methoxyethanol to 32 mL to make a 0.5 M LaAlO$_3$ precursor solution.

The resulting precursor solution was stoichiometrically or partially hydrolyzed using a 1 M H$_2$O in 2-methoxyethanol solution to produce, respectively, a light yellow, transparent gel or a solution suitable for spin-coating. The gel was decomposed and then fired between 500° and 800°C in air for 1 to 12 hours. X-ray diffraction (XRD) studies show that crystallization begins around 600°C. A total of four coatings of the partially hydrolyzed solution were applied to a SrTiO$_3$ (100) single crystal which was pyrolyzed between coatings in O$_2$ in a rapid thermal annealer at 800°C for 2 minutes. The resulting film is approximately 4000 Å thick and exhibits c-axis preferred orientation on the SrTiO$_3$ substrate (Fig. 1). The LaAlO$_3$ peaks are considered to correspond to (001), (002), and (003) planes in a cubic structure. A rocking curve scan of the (003) LaAlO$_3$ reflection (inset of Fig. 1) verifies the good c-axis alignment with the full-width at half-maximum (FWHM) determined to be 0.87°. A φ scan made from the (202) plane indicates that the film had good in-plane texture with a FWHM=1.07°, and a typical x-ray pole figure of the cubic (202) reflections shows a single cube-on-cube epitaxy: SrTiO$_3$(001)||LaAlO$_3$(001)$_{cubic}$ and SrTiO$_3$[100]||LaAlO$_3$[100]$_{cubic}$. The microstructure of the film, as seen by scanning electron microscopy, indicates that the grains are small, surface roughness is low, and the film is continuous.

3. CeO$_2$ Film Growth

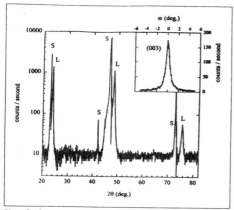

Fig.1 θ-2θ scan of c-axis oriented LaAlO$_3$ (L) on SrTiO$_3$ (100) (S). Inset is rocking curve of LaAlO$_3$ (003). (FWHM=0.87°)

Cerium(III) acetate 1.5H$_2$O (1.38 g, 4.00 mmol) was dissolved in 10 mL of a 9:1 H$_2$O/acetic acid solution to yield a colorless 0.4 M CeO$_2$ aqueous precursor solution. The solvent from a portion of the solution was evaporated, and the remaining residue was fired between 300° and 800°C in air for 4 to 18 hours where oxidation to cerium (IV) occurred. Crystallization begins around 300°C as determined by XRD. The remainder of the solution was used for spin-coating on single crystal SrTiO$_3$. The solution did not always coat well but instead would pile on the ends of the substrate. Only random films could be grown on the SrTiO$_3$. In attempts to achieve epitaxial growth, different combinations of pyrolysis conditions were employed including atmospheres of Ar, O$_2$, or air, times from minutes to hours, and temperatures from 450° to 800 C, but all were unsuccessful. Additives such as tartaric acid, ethylene glycol, or a surfactant were added to the solution to obtain better coating, but little improvement was noticed. The solutions were also unstable after about one week when an unidentified phase would appear in the XRD pattern.

Another CeO$_2$ precursor solution was obtained using a commercial cerium(IV) methoxyethoxide (18-20%) in 2-methoxyethanol dark red solution. Yellow, transparent gels were formed upon complete hydrolysis using the H$_2$O/2-methoxyethanol solution and heat. The product was then pyrolyzed, fired between 200° and 800°C in air for 4 to 18 hours, and found by XRD to begin crystallization at a slightly lower temperature (200°C) than seen for the aqueous solution. A partially hydrolyzed solution was spun-cast on (100) SrTiO$_3$, but like the aqueous route, no epitaxial film could be grown so far and the same unidentified phase was detected by XRD pattern when older solutions, which turned yellow, were used.

Acknowledgments

This research was sponsored by the U.S. D.O.E., Division of Materials Sciences, Office of Basic Energy Sciences and Office of Energy Efficiency and Renewable Energy, Office of Utility Technology-Superconductivity program and managed by Lockheed Martin Energy Research Corporation under contract # DE-AC05-96OR22464. S.S.S. acknowledges support by an appointment to the ORNL Postdoctoral Research Associates Program administered jointly by ORNL and by the Oak Ridge Institute for Science and Education.

References

1. P.C. Chou, Q. Zhong, Q.L. Li, K. Abazajian, A. Ignatiev, C.Y. Wang, E.E. Deal, and J.G. Chen, Physica C **254**, 93 (1995).

I. MATERIALS

RBCO

STRONG FLUX PINNING IN RE123 GROWN BY OXYGEN-CONTROLLED MELT PROCESS

N. CHIKUMOTO and M. MURAKAMI

ISTEC-SRL, 16-25, Shibaura 1-chome, Minato-ku, Tokyo 105, JAPAN.

It has been recently revealed that a reduced oxygen atmosphere during melt processing is critical for the fabrication of good $REBa_2Cu_3O_y$ (RE: Nd, Sm, Eu) superconductors. Nd123 superconductor melt processed in the mixed gas of 0.1 % O_2 in Ar showed a sharp superconducting transition at around 96K, while that fabricated in air showed very broad transition with depressed T_c. In addition, fabricated RE123 superconductors exhibit excellent pinning properties : J_c values in a high field regions exceed those of good quality melt-processed Y123 and the irreversibility field with $H \parallel c$-axis reaches 8 T at 77 K. Such a strong flux pinning is likely to be provided by Nd-rich regions (low T_c regions) distributed in the high-T_c matrix.

1 Introduction

For the realization of bulk application, it is necessary to obtain high critical current density (J_c) by overcoming the weak link problem and introducing effective pinning centers. Since the first report by Jin *et al.* [1], the melt processing has been known as the most successful method to solve the weak link problem in $YBa_2Cu_3O_y$ (Y123). It is well-known that $REBa_2Cu_3O_y$ (RE: rare earth ions) also show a superconducting transition around 90 K. However, it had been long recognized that it was impossible to fabricate the melt-textured RE123 superconductors with light RE ions (Nd, Sm, Eu) possessing both high superconducting transition temperature (T_c) and a sharp superconducting transition, because of the heavy substitution of RE on the Ba site. Recently Yoo *et al.* [2] developed so-called Oxygen-Controlled Melt Growth (OCMG) method which enable us to fabricate the melt-textured light RE123 superconductor with good superconducting properties. In this brief article, some of the aspects of this OCMG method are summarized.

2 Oxygen-Controlled Melt Growth Process

2.1 Preferential Formation of High T_c Phase by Use of OCMG-process

The key point of the OCMG process is the control of oxygen partial pressure, P_{O2}, during the stage of the melt growth. The details of processing conditions are described elsewhere [2,3,4]. Figure 1 shows typical magnetic susceptibility curves of Nd123 samples melt-textured in various P_{O2}. Similar results were obtained for Eu and Sm-systems [5,6]. It is clearly seen that a reduced oxygen atmosphere is necessary for obtaining high T_c and a sharp superconducting transition.

The underlying mechanism for the preferential formation of high-T_c phase in low P_{O2} is the difference in the phase stability of the $RE_{1+x}Ba_{2-x}Cu_3O_y$ solid solution. Figure 2 shows a stability phase diagram for Nd-system [8]. The stability phase boundaries for $Nd_{1+x}Ba_{2-x}Cu_3O_y$ solid solutions diverge with decreasing of P_{O2} and hence the boundary between small x and large x is widely separated at low P_{O2}. When the sample is processed in

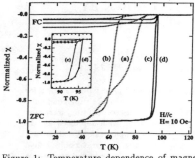
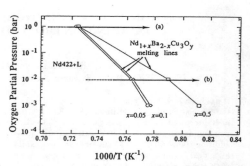

Figure 1: Temperature dependence of magnetic susceptibility for Nd123 melt-processed in (a) O_2, (b) air, (c) 1% O_2 in Ar and (d) 0.1% O_2 in Ar[4].

Figure 2: A stability phase diagram of $log[P_{O2}]$ vs $1/T$ for $Nd_{1+x}Ba_{2-x}Cu_3O_y$. After Yoo et al.[6]

air or in pure O_2 (arrow (a) in Fig.2), all $Nd_{1+x}Ba_{2-x}Cu_3O_y$ solid solutions within solubility limit have an equal chance to grow because of small difference in the phase stability. On the other hand, when the sample is processed in low P_{O2} atmosphere (arrow (b) in Fig.2), high-T_c phase with small x can preferentially form.

2.2 Pinning Properties of OCMG-Processed RE123

In addition to successful fabrication of RE123 with high-T_c and a sharp superconducting transition, the OCMG processed samples exhibit excellent pinning properties [7].

Figure 3 compares the field dependence of J_c at 77 K for a good quality melt-processed Y123 sample and the OCMG-processed Nd123 and Sm123 samples. We can see that the OCMG-processed samples show much higher J_c than that of Y123 in a high field region due to the broad anomalous peak effect. In Fig. 4, temperature dependence of the irreversibility field with $H \parallel c$, determined from the closing of the hysteresis loop, is displayed. Nd123 samples processed in the low P_{O2} show the irreversibility field higher than that of Y123 and it reaches 8 T at 77 K.

Strong flux pinning in OCMG-processed samples may be attributed to Nd-rich regions (low T_c regions) distributed in the high-T_c matrix [7]. Even when RE123 is melt-processed in a reduced oxygen atmosphere, there exist regions with slightly depressed T_c, as recently identified by TEM [10] and STM [11] observations. Such weakly superconducting regions are likely to be driven normal with the applied field and act as additional pinning centers, causing the peak effect in the magnetization hysteresis loop.

2.3 Effect of Oxygen Annealing

Since the pinning characteristic is affected by the value of the electric anisotropy and the type of defects, such as oxygen vacancy, it is important to see the effect of oxygen annealing in the present system. The dependence of T_c on the oxygen annealing temperature (T_a) is shown in Fig.5. T_c monotonously decreases with increasing T_a. It must be noted that the

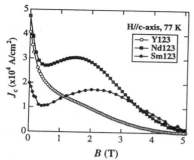

Figure 3: The field dependence of J_c for a good Y123 sample and the OCMG processed RE123 samples (RE:Nd, Sm)[5,8].

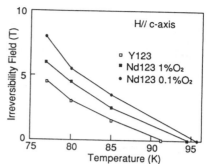

Figure 4: Irreversibility lines of melt-textured Y123 and Nd123 samples[7].

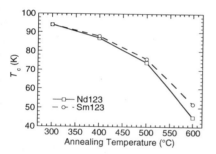

Figure 5: The dependence of T_c on the oxygen annealing temperature in Nd123 and Sm123.

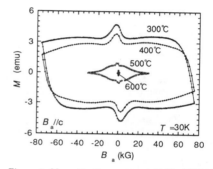

Figure 6: Magnetization curves at 30 K of Nd123 oxygen-annealed at various temperature.

highest T_c could be obtained for $T_a = 300°C$, showing that we should perform the oxygen-annealing at lower temperatures than 400°C which is a typical annealing temperature in Y-system.

Figure 6 shows the magnetization curves at 30 K of Nd123 annealed at various temperatures. The hysteresis width increases drastically with lowering T_a. The irreversibility line was also shifted upward with decreasing T_a. This result shows that low-temperature annealing is necessary not only for achieving high T_c but also for obtaining better pinning properties.

2.4 Second Phase Inclusion

Like the '211' phase in Y-system, '211' ($RE_4Ba_2Cu_2O_y$ ('422') for Nd-system) inclusions are produced in the melt-processed RE123. In the case of MPMG-processed Y123, finely dispersed '211' is responsible for large J_c. On the other hand, the second phase in RE-system have nothing to do with the characteristic pinning behavior of OCMG-processed samples, since the peak effect in the hysteresis loop was found to be almost independent of '211'('422') content and size[12]. However, some improvement in J_c in low field by decreasing the size of the second phase was observed[5,9], indicating that J_c can be further enhanced by controlling the microstructure with the optimization of processing.

2.5 Irradiation Effect

Irradiation is one of the way to introduce pinning centers. It was revealed that irradiation with fast neutrons largely enhanced for an OCMG-processed Nd123 J_c at 77K[13].

Acknowledgments

The authors would like to thank Dr. S.I. Yoo, N. Sakai and all other members in Div. 7 of Superconductivity Research Laboratory for their help and useful discussions in preparing this review. This work was partially supported by NEDO for the R & D of Industrial Science and Technology Program.

References

1. S. Jin, T. Tiefel, R.C. Sherwood, M. E. Davis, R.B. van Dover, G.W.Kammlott, R. A. Fastnacht and H.D. Keith, *Appl. Phys. Lett.* **52**, 2074 (1988)
2. S.I. Yoo, N. Sakai, H. Takaichi, T. Higuchi and M. Murakami, *Appl. Phys. Lett.* **65**, 633 (1994).
3. M. Murakami, S.I. Yoo, T. Higuchi, N. Sakai, J. Weltz, N. Koshizuka and S. Tanaka, *Jpn. J. Appl. Phys.* **33**, L715 (1994).
4. S.I. Yoo, M. Murakami, N. Sakai, T. Higuchi and S. Tanaka, *Jpn. J. Appl. Phys.* **33**, L1000 (1994).
5. N. Sakai, S.I. Yoo, K. Sawada, T. Higuchi, N. Hayashi, F. Frangi and M. Murakami, in *Fourth Euro Ceramics Vol.7*, ed. A. Barone *et al.* (Gruppo Editoriale Faenza Editrice, Italy, 1995) p74.
6. S.I. Yoo and M. Murakami, in *Recent Research Development in Cryogenics*, (1996, India), in print.
7. M. Murakami, S.I. Yoo, T. Higuchi, N. Sakai, M. Watahiki, N. Koshizuka and S. Tanaka, *Physica* C **235-240**, 2781 (1994).
8. S.I. Yoo, N. Sakai, K. Segawa, M. Watahiki and M. Murakami, in *Advances in superconductivity VII*, ed. K. Yamafuji and T. Morishita (Springer-Verlag, Tokyo, 1995) p705.
9. H. Kojo *et al.*, unpublished.
10. T. Egi, J.G. Wen, K. Kuroda, H. Unoki and N. Koshizuka, *Appl. Phys. Lett.* **67**, 2406 (1995).
11. W. Ting, T. Egi, R. Itti, K. Kuroda, N. Koshizuka and S. Tanaka, unpublished.
12. S.I. Yoo, M. Murakami, N. Sakai, T. Ohyama, T. Higuchi, M. Watahiki and M. Takahashi, *J. Elec. Mater.* **24**, 1923 (1995).
13. H.W. Weber, private communication.

BIAXIALLY TEXTURED YBaCuO THICK FILMS ON TECHNICAL SUBSTRATES

H.C. FREYHARDT, J. DZICK, K. HEINEMANN, J. HOFFMANN, J. WIESMANN
Institut fuer Metallphysik: LMM/KL, University of Goettingen
Windausweg 2, D-37073 Goettingen

F. GARCIA-MORENO, S. SIEVERS, A. USOSKIN
Zentrum fuer Funktionswerkstoffe gem. GmbH Goettingen/Clausthal,
Windausweg 2, D-37073 Goettingen

ABSTRACT

High-current applications require homogeneous well textured HTS YBaCuO films on technical substrates in the form of tapes, tubes, planar or curved sheets (metallic: Ni, Hastelloy; nonmetallic: Al_2O_3, YSZ). Ion-beam-assisted deposition was employed to grow biaxially textured YSZ and CeO_2 buffer films, which serve as template for the YBaCuO and as a diffusion barrier. High-quality thick ($\leq 9\mu m$) films were deposited on moving planar and curved substrates by pulsed laser ablation. Critical current densities (0T, 77K) of 0.9×10^6 A/cm^2 were reached on Ni and of 5×10^5 A/cm^2 on YSZ, both with intermediate YSZ buffer layers. The critical superconducting parameters of the YBaCuO films were investigated under tension and compression.

1. Introduction

For high current applications in electrical engineering or power distribution systems, multifilamentary high-temperature-superconducting (HTS) tapes on the basis of Bi-2223 have been developed even in lengths beyond 1km[1] with engineering critical current densities around 10kA/cm^2 in low magnetic fields. Because of the weak-link problem, the powder-in-tube technique, which works for the Bi-based HTS, cannot be employed for YBaCuO. However, YBaCuO films epitaxially grown on surfaces of single crystals yield current densities, J_c, of 10^7 A/cm^2 at 77K[2] and exhibit high irreversibility fields. Films as thick as 9 μm have been deposited[3] on $SrTiO_3$ with J_c between $(3.5-1.5) \times 10^6$ A/cm^2 (77K). It is, therefore, of high interest to further develop this thin-film technology into an industrially relevant processing method, which is versatile and flexible and which allows to produce thick/thin YBaCuO films on large areas, tapes or sheets, and on curved surfaces, e.g. on cylinders and tube conductors. In this contribution we report on the recent success to deposit a biaxially textured yttria-stabilized ZrO_2 (YSZ) layer on large areas and on cylindrical surfaces on which subsequently high-quality YBaCuO film are grown by pulsed laser deposition.

2. Biaxially Textured Buffer Films

As technically relevant substrates, Ni, Hastelloy, YSZ, poly-Al_2O_3 tapes, tubes or sheets were used. To achieve the required surface roughness of 30-60 nm, the Ni and Hastelloy surfaces were mechanically and electro-chemically polished. Commercially polished Al_2O_3 and YSZ tapes were available with a roughness of better than 30 nm.

For the deposition of CeO_2 and YSZ buffer layers, an IBAD[4] (ion-beam-assisted-deposition) technique was employed with two ion sources of the Kaufman type. In general both sources are operated with Ar. At an oxygen partial pressure of 1×10^{-4} mbar, the ablation source 1 and the assisting IBAD source 2 use an accelaration energy of 1500eV and 300eV, respectively. The assisting beam is directed parallel to the <111> axis of the growing YSZ film. In the up-scaled IBAD system (4" sources), with a deposition rate of ≤ 0.1 nm/sec, YSZ layers up to 1.6 µm were deposited.

For CeO_2 reasonably good in-plane alignment was obtained only at reduced oxygen pressures of 10^{-5} mbar. On the other hand, YSZ layers grow with an excellent (100) texture and good in-plane alignment, characterized by the full width half maximum (FWHM) of a φ scan for a (111) reflex. The recently obtained remarkable results can be summarized as follows: (1) One finds a considerable improvement of the in-plane alignment with increasing YSZ-layer thickness. The average FWHM decreases from around 28°-30° for a layer thickness of ≈ 250 nm to 10.8° at 1.6 µm, which is amongst the best values obtained so far (Fig.1). These layers can possess, however, stresses (up to 3 MPa). (2) Starting from lab-size substrates (10 x 10 mm^2) we were able to increase the deposition area for the bi-YSZ films to a maximum of now of 200 x 200 mm^2. An inhomogeneity of the FWHM distribution is observed for large substrate areas, which in principle could be improved by introducing a motion of the substrates. (3) For cylindrical substrates, ideal IBAD conditions can be maintained only within a small section of the circumference.

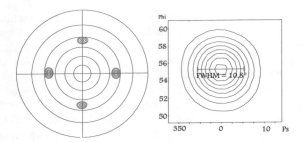

Figure 1: (111)-pole figure of a 1.6 µm-thick bi-YSZ layer on poly-Al_2O_3

Nevertheless, biaxially textured YSZ buffer films could successfully be prepared for the first time on rotating cylindrical surfaces by employing an appropriately positioned deposition window. An efffective deposition rate of 20-40 nm/h for the outer surface of a 12 mm tube was obtained. The in-plane alignment of the bi-YSZ again improves with layer thickness but deteriorates with increasing speed of rotation. So far the best in-plane textures achieved are characterized by a FWHM of ≈ 27°.

3. Pulsed laser deposition of YBaCuO films

A conventional pulsed laser deposition (PLD) technique, which uses an excimer laser (Siemens XP 2020: 308 nm, 2J /pulse) was employed for high-J_c film preparation.[5] This method was further improved to enable YBaCuO PLD on moving substrates in the form of tapes and tubes and extends the conventional process by two principal features: (1) A quasi-equilibrium radiative heating technique is capable of maintaining a constant surface temperature of the growing HTS film to an accuracy of better than 0.5° C (at about 750°C). In addition to an equilibrium heater with rotating substrate holders, a clyindrical heater arrangement, which can also be converted into a channel heater systems, allows the continuous YBaCuO deposition on tapes and tubes as well as the control of the oxygen partial pressure between pulses and thus the oxygenation of the HTS. (2) For a technologically relevant PLD a long-term stability of the ablation process is needed. Unfortunately, even for rotating YBaCuO targets, an increasing surface roughness develops with increasing ablation time. The introduction of a variable azimuth ablation not only leads to a stable plasma plume but also to an almost constant and higher (about a factor of 2) ablation rate.

The ablation parameters found earlier for stationary substrates were applied and transferred to moving tapes and tubes. Several critical superconducting parameters obtained for the deposition on moving substrates are summarized in Table 1. The critical current densities, J_c, of (microstructured) films were measured with dc current pulses (30 µs) and of unstructured films with an ac (100kHz) method, which employs the effect of magnetic shielding and which also gives information on the spatial distribution of J_c values.

Table 1: Superconducting critical parameters for YBaCuO films on technical substrates

Buffer/Substrate	T_c in K	ΔT_c in K	J_c in A/cm^2	
			max. achievable	well reproducible
YSZ tape	88	1.2	1.8×10^5	8×10^4
bi-YSZ/YSZ	90	0.5	5×10^5	2×10^5
bi-YSZ/Al$_2$O$_3$	87	1.5	1×10^5	-
bi-YSZ/Ni	89.5	1	0.9×10^6	4×10^5

Both methods yield similar results. In earlier experiments on single crystalline SrTiO$_3$, J_c values between $(3.5-1.5) \times 10^6$ A/cm^2 (0T, 77K) were found for thick (≤ 9 µm) YBaCuO films[3], which only slightly decreased with film thickness, D. J_c for bi-YSZ/Ni on the other hand remains nearly unchanged for a D of several µm, however starts to decrease for D below 400 nm (a reaction layer forms between YBaCuO and buffer). At 77K (0T) remarkably high maximum achievable critical currents of 0.9×10^6 A/cm^2 are reached, which are only slightly smaller than the 1.5×10^6 A/cm^2 at 75 K of Wu et al.[6]. The deposition of an IBAD buffer enhances, furthermore, the J_c values by a factor of ≈ 2.5 for YSZ tapes. The technologically relevant well reproducible current densities turn out to be at present by a factor of ≈ 2 smaller than the optimum J_c's for the various substrate materials. Deformation experiments of YBaCuO on bi-YSZ/Ni (Figure 2) under tension and compression reveal that T_c remains almost constant up to a strain, ε, of +0.6% in tension and even slightly increases in compression. A tension up to ε $\approx$ 0.5 does not change the critical currrent density and under compression even a maximum in J_c is observed. This behavior is presently investigated in connection with the stresses observed after the buffer deposition and the mismatch stresses between YBaCuO and YSZ. J_c remains reversible for strains up to $\approx 0.5\%$ and $\approx -0.7\%$ in tension and compression, respectively. For larger strains cracks form and deteriorate J_c. YBaCuO on polycrsytalline YSZ ribbons seems to be more susceptible to deformation.

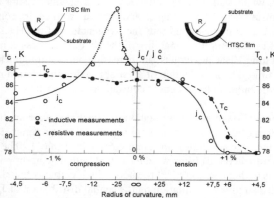

Figure 2: T_c and J_c variation for YBaCuO on bi-YSZ/Ni during bending deformation.

4. Conclusions

The pulsed laser deposition proves to be a highly attractive way for the preparation of high-J_c HTS films on technical substrates. The method is versatile because it allows the YBaCuO deposition on quite differently shaped substrates: tapes, cylindrical tubes and sheets. As a prerequisite, however, highly in-plane textured YSZ buffer layers are necessary, which serve as a template for the YBaCuO and as a diffusion barrier. This became feasible by using an IBAD method for the YSZ deposition onto (large) planar but also curved technical substrates. With an average current density of $\approx 5 \times 10^5$ A/cm^2 (77K,0T), which can be realized now one could expect a 4 µm-thick YBaCuO film to carry an attractive current of ≈ 200 A per 10 mm width of the film. Functional models of resisitive fault-current limiters and (HF) feeder cables are developed together with industry on the basis of the present results. This work was supported by the German BMBF under grant numbers 1.3N6009 and 13N6482 and the Siemens AG.

5. References

1. G. Papst, J. Kellers and A.P. Malozemoff, *Inst. Phys. Conf. Sec.* **148**, 359 (1995)
2. G. Kormann et al. *J. Appl. Phys.* **71**, 3419 (1992)
3. A. Usoskin, H.C. Freyhardt, W. Neuhaus and M. Damaske, *Critical Currents in Superconductors* (Singapore World Scientific, 1994), p.383
4. Y. Iijima, N. Tanabe, O. Kohno and Y. Ikeno, *Appl. Phys. Lett.* **60**, 769 (1992)
5. A. Usoskin, H.C. Freyhardt, et al. *Inst. Phys. Conf. Ser.* **148**, 499 (1990)
6. X.D. Wu, S.R. Foltyn et al. *Appl. Phys. Letters* (1955) in print.

Thick Film YBCO Growth by Photo-Assisted MOCVD: J_c Enhancement by Ion Irradiation

Alex Ignatiev, Qun Zhong, Pen Chu Chou and Xin Zhang
Space Vacuum Epitaxy Center and
Texas Center for Superconductivity
University of Houston
Houston, TX 77204-5507

Jia Rui Liu, and Wei Kan Chu
Texas Center for Superconductivity
University of Houston
Houston, TX 77204-5932

I. Introduction

Currently, tremendous interest is focused on the vapor deposition of thick $YBa_2Cu_3O_{7-\delta}$ (YBCO) films (>1 μm thick) for application to high current superconducting wires and tapes [1-3]. For this purpose, large critical current density (J_c), or ultimately large critical currents (I_c), for the thick films is the major interest. The main reason of choosing YBCO as the target material instead of Bi- or Tl-based cuprate high temperature superconductors is that at 77 K the irreversibility lines of both of these two latter classes of superconductors can be displaced well below their corresponding upper critical fields under even small applied magnetic fields (< 1T), with a resultant drastic decrease in critical current density [4]. YBCO does not have this problem, however, the powder-in-tube technique applied successfully to the Bi and Tl-based is not successful for YBCO materials [2,5]. YBCO can however, be made with very high superconducting performance (T_c > 89K and J_c > 10^6 A/cm^2) in thin film form. Therefore, the application of YBCO to superconducting wires and tapes requires the deposition of YBCO films on flexible substrates as the approach to fabrication.

These films require thicknesses >1 μm with high critical current density ($J_c \sim 10^6$ A/cm^2), and deposition onto flexible metal substrates in short times, i.e. under high growth rates. Most YBCO film physical deposition techniques and chemical vapor deposition techniques suffer from slow growth rates for good quality (high J_c, high T_c) films [6], while non-vapor deposition thick film techniques suffer from poor quality material [7]. Photo-assisted metalorganic chemical vapor deposition (PhAMOCVD), has however, shown the growth of high quality YBCO films under very high growth rates [8]. The PhAMOCVD technique has been systematically optimized by a very effective statistical method, Robust Design [9] to reproducibly yield thin films (~200 nm -500 nm) with J_c = 3 - 5x10^6 A/cm$_2$. In addition, film orientation can be controlled to yield films with either pure c-axis or pure a-axis orientation. [8]. Film growth rates of these orientation controlled films can be as high as 0.7 μm/min, but with high crystalline quality and good superconducting properties (J_c and T_c). However, it has been found as shown in Fig. 1, and elsewhere [10] that as soon as the thickness of YBCO films become thicker than about 1 µm, J_c of these films starts to drop appreciably. Specifically for the 4.5 µm thick film identified in Figure 1, XRD measurement indicates crystalline quality as good as the 0.5 µm thick film noted in Figure 1. Specifically the FWHM of the (005) diffraction peak shows a 0.26° value for the 0.5 µm thick film, and a 0.28° value for the 4.5 µm thick film. In addition, SEM measurement of the surface morphology and the fracture surface cross sections of the two films indicate continuous, dense films.

It is believed, therefore that reduced crystalline quality is not responsible for the decrease of J_c with increasing film thickness. In fact, the thick film looks as good microstructurally as the thin film one. The telltale decrease of the PhAMOCVD-grown thick film J_c with thickness, therefore, could be the result of the film quality being too good. If the pinning centers responsible for high J_c in thin films are only localized to the film interface region, then as the film thickens with the high structural quality imparted by the PhAMOCVD method, the effective J_c of the film decreases due to the lack of pinning center within the film.

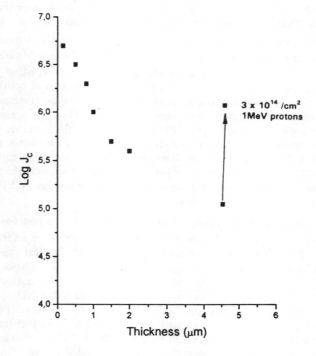

Figure 1. Critical current density J_c as a function of film thickness for YBCO films grown by photo-assisted MOCVD. The 4.5 µm thick film data shows J_c as measured before and after irradiation by 1 MeV protons at a dose of 3×10^{14} /cm^2.

This hypothesis can be tested by the introduction of pinning center into a PhAMOCVD-grown thick YBCO film. It is well know that high energy ion irradiation of YBCO single crystal samples can increase J_c. Such irradiation, however, has done little for YBCO thin films often showing a decrease of J_c due to crystal damage and the reduction of intergrain J_c [11]. This means, however, that if a thick, PhAMOCVD-grown c-axis oriented YBCO film sample of high crystalline quality, i.e., with near single crystalline structure, were appropriately ion-irradiated, an enhancement effect to either magnetic or transport J_c could be very apparent. It was for this consideration that an ion-irradiating technique was adopted for preliminary study of possible J_c enhancement of thick, c-axis oriented YBCO films prepared by PhAMOCVD under high growth rate.

II. Experimental

A thick PhAMOCVD-grown YBCO sample (4.5 μm thick) of near single crystal quality was selected for ion-irradiation. This sample pre-irradiation had, as noted above high crystalline quality, and moderate to low J_c (~1×10^5 A/cm^2). The YBCO film was deposited at a substrate temperature (T_s) range of about 750-800 °C, and was grown at a rate of 0.6 μm/min on a LaAlO$_3$ substrate. The photo-assisted MOCVD apparatus has been previously reported[8]. The sole energy source for the MOCVD reactor was tungsten-halogen quartz lamps (TH lamps) which were used for not only thermal but also photo-stimulation of the chemical and physical processes involved in the MOCVD reaction [12]. J_c for the sample was measured both by 4-point probe transport measurements, and by the magnetic induction technique.

Ion irradiation of the sample was done with a 1 MeV proton beam extracted from a 2 x 1.7 MV Tandem accelerator at the Texas Center of Superconductivity. The total proton dosage of 3×10^{14} protons/cm^2 was selected from extrapolation of previous experiments on single crystal samples with 200 KeV protons [13]. The projected range of a 1 MeV proton in YBCO is ~ 9μ, therefore, the protons penetrated the YBCO film completely, and irradiated the film quite uniformly as a function of depth into the film.

III. Results and Discussion

Figure 1 also shows the J_c results (as measured by magnetic induction) for the 4.5μ YBCO film after 1 MeV proton irradiation. It can be seen in Figure 1 that the J_c of the irradiated thick film increases dramatically from ~ 1×10^5 A/cm^2 to a level of ~3×10^6 A/cm^2 after irradiation. This phenomenal increase in J_c supports the hypothesis that the PhAMOCVD-grown YBCO films were crystallographically too good, and that pinning centers induced by proton irradiation were required to increase the J_c of the films. In addition, this data also supports the concept that pinning centers responsible for high J_c's in thin YBCO films are localized principally at the interface of film and the substrate.

Summary

It has been shown that photo-assisted MOCVD, when applied to the growth of YBCO films, results in films with very high crystalline quality approaching single crystal samples. Such films show reduced J_c with increased thickness principally because the pinning responsible for high J_c in YBCO films is localized to the film substrate interface, and the thicker films have no or minimal pining centers in their bulk. The injection of pinning centers throughout a thick YBCO film by 1 MeV ion irradiation has resulted in a significant increase of the J_c of the film beyond the 1×10^6 A/cm^2 value. It is now clear that high J_c films (not necessarily high crystalline quality films) can be made by PhAMOCVD at very high growth rates when accompanied by additional pinning center generation, e.g. exposed to ion irradiation. Such films can now be applied to flexible metal substrates for utilization as high current superconducting wires and tapes.

Acknowledgments

The authors wish to acknowledge the help of Dr. Pei Hor and Dr. H. S. Feng with the magnetic induction J_c measurements. This work was supported in part by NASA, the Texas Center for Superconductivity, the State of Texas through the Advanced Technology Program, and the R. A. Welch Foundation.

References

1. S.R. Foltyn, S.D. Wu, Q.X. Jia, P. Tiwari, P.N. Arendt, and D.E. Peterson, Proc. MRS Symposium, **388**, (1995).
2. U.S. Department of Energy sponsored "HTS Wire Development Workshop", held at St. Petersburg, Florida (January 31- February 1, 1995).
3. X.D. Wu, S.R. Foltyn, P. Arendt, J. Townsend, C. Adams, I.H. Camplell, P. Tiwari, Y. Corlter, and D.E. Peterson, App. Phys. Lett., **65**, 1961 (1994).
4. S.Y. Ting, "Second-type superconductors", in Ch. 7 of *High-T_c Superconductivity*, ed. by C.R. Zhang, Zhechiang University Press, Zhechiang, China, p. 371 (1992).
5. D.C. Larbalestier and M.P. Maley, MRS Bull., **XVIII** (8), 50 (1993).
6. R. Biggers, M. Norton, I. Maartense, T. Peterson, E. Moser, D. Dempsey, and J. Brown, Proc. MRS Symposium, **388**, 67 (1995).
7. M. Sanda and O. Ishii, J. Appl. Phys., **69**, 6586,(1991).
8. Q. Zhong, P.C. Chou, Q.L. Li, G.S. Taraldsen, and A. Ignatiev, Physica C, **246**, 288 (1995).
9. P.C. Chou, Q. Zhong, Q.L. Li, K. Abazajian, A. Ignatiev, C.Y. Wang, E.E. Deal, and J.G. Chen, Phsica C, **254**, 93 (1995).
10. H. Busch, A. Fink, and A. Müller, J. Appl. Phys. **76**, 2449 (1991).
11. J.R. Liu, J. Kulik, Y.J. Zhao, and W.K. Chu, Nucl. Inst. Meth. Phys. Res., **B80/81**, 1255 (1993).
12. Q. Zhong, Ph.D. Thesis, University of Houston, December 1995.
13. Y.J. Zhao, J.R. Liu, Y.K. Tao, P.H. Hor, and W.K. Chu, Mater. Res. Soc. Symp. Proc., **235**, 577 (1992).

ANISOTROPY AS THE DETERMINING FACTOR OF IRREVERSIBILITY FIELD AND ITS CONTROL BY CHEMICAL DOPING

K. KITAZAWA, J. SHIMOYAMA, H. IKUTA and K. KISHIO
Department of Superconductivity, University of Tokyo
Hongo 7-3-1, Bunkyo-ku, Tokyo 113, Japan

The strong anisotropy has turned out to be the most important factor to determine the behaviors of the high temperature superconductors (HTS) under magnetic fields. This paper describes that the irreversibility field B_{irr}, the convenient index to express the performance of the HTS under magnetic fields, can be estimated as far as the anisotropy factor γ^2 is known irrespective of differences in the specific material system. Consequently, it is highly desirable for their practical application to control the anisotropy factor by chemical means because of their readiness in the actual processing. Substitution of ions for adjusting the carrier concentration and for strengthening the interlayer coupling are discussed.

1 Introduction

It has been discussed that the high temperature superconductivity is induced by electron spin interaction in the two dimensional CuO_2 plane whose energy scale amounts to as high as ~ 1000 K. If it were in a three dimensional medium, the strong interaction would have induced antiferromagnetic insulator and superconductivity would not have been obtained. In this context, the strong anisotropy is the inevitable factor in order to achieve the high T_c. On the other hand, it has been shown that the strong anisotropy brings in the HTS the strong tendency towards quantum and thermodynamical fluctuations [1] which create serious problems in their magnetic properties. It therefore seems that the two important parameters T_c and the performance of the HTS in magnetic fields, typically represented by the irreversibility field B_{irr}, are not compatible to each other but rather in a trade-off relationship. In this paper we report that the relationship between T_c and B_{irr} is not so simple but there are ways to improve the magnetic properties without deteriorating the high T_c.

2 Crystal Structure and Anisotropy

Figure 1 compares the electrical anisotropy obtained on single crystals of typical HTS compounds [2]. The lower curves are for the CuO_2 in-plane resistivity ρ_a or ρ_b and the upper ones are for the out-of-plane resistivity ρ_c. As the carrier density increases by doping; oxygen in cases of Y123 and Bi2212 and Sr in La214, the resistivity decreases monotonically but more drastically in ρ_c. Hence the anisotropy weakens by increasing the carrier density at first rapidly in the under-doped regions and then slowly in the over-doped regions. It should be noted that among the three cited materials Bi2212, La214, Y123, the order in anisotropy γ^2 differs from that in T_c. Hence the larger γ^2 does not necessarily mean the higher T_c.

The anisotropy factor γ^2 ($\equiv (\rho_c/\rho_{ab})$) at $T=T_c+10$ K) varies significantly from a material to another but follows a certain relationship [2] with the inter-layer distance when the value for the optimum composition (the highest T_c composition) is adopted as shown in Fig. 2.

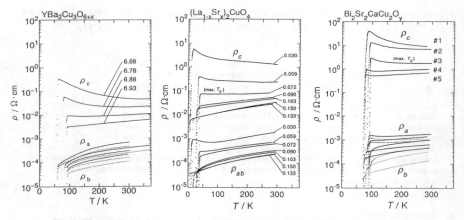

Fig. 1: Electrical resistivity ρ_c, ρ_a, ρ_b, plotted in the same scale, of (a)YBCO, (b)LSCO and (c)Bi2212 single crystals with different chemical compositions and carrier doping states[2].

In case of Y123, the point falls on the line if the spacing between CuO_2 and CuO planes is taken. We might generalize this relationship as

$$\log \gamma^2 = 3.1 \times d + 0.56 \quad (1)$$

where d is the inter-planar distance between CuO_2 planes in nm. It has been found that the anisotropy factor measured in the normal and superconducting states by various methods; ac surface impedance[3], optical conductivity[4] and torque magnetometry[5] as well as dc resistivity[2], agree reasonably well. The fact that the anisotropy factor obtained in the superconducting state as $\gamma_s^2 = (\lambda_c/\lambda_{ab})^2$ retains its value into the normal state across the superconducting transition indicates that the anisotropy is determined by the same mechanism in both of the states.

$$\begin{aligned} \gamma_s^2 &= (\lambda_c/\lambda_{ab})^2 \\ &= m_c/m_{ab} = \rho_c/\rho_{ab} = \gamma_n^2 \end{aligned} \quad (2)$$

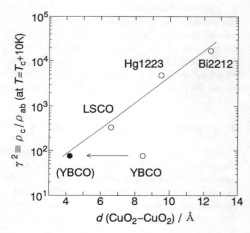

Fig. 2 : Resistivity anisotropy of optimally doped YBCO, La124, Hg1223 and Bi2212 as a function of interlayer distance of CuO_2 blocks. For YBCO(●), the distance between CuO_2 plane and CuO chain plane is considered.

3 Vortex Lattice Melting and Dimensional Crossover

It has been widely accepted that in the conventional superconductors the Abrikosov vortex lattice is formed and persists up to right below the B_{c2} line. However, the strong anisotropy in the HTS reduces the rigidity of the vortex line and makes the 3D vortex lattice unstable. Thus the magnetic phase diagram of HTS becomes significantly complicated compared to the conventional systems; e.g. the Abrikosov lattice is formed much below the expected B_{c2} line.

In Y123 the existence of the first order phase transition was pointed out by the observations of step-like changes both in resistivity [6] and magnetic susceptibility [7]. Also for Bi2212, clear jumps in the magnetization curves were reported [8] and corresponding resistivity drop was also confirmed [9]. This transition has been believed to correspond to the melting of the vortex lattice and a recent study [10] on Bi2212 crystals showed the systematic shift of this melting line with the carrier doping level and associated change in the electrical anisotropy.

In Bi2212 system, dimensional crossover of the vortex lattice has been proposed [11] to take place at a characteristic field, $B_{3D-2D} \sim \Phi_0/\gamma^2 s^2$ where Φ_0 is the magnetic flux quantum and s is the distance between superconducting CuO_2 blocks. Recent μSR measurements [12] clearly showed that this crossover field does shift by a change in the anisotropy factor γ^2.

4 Irreversibility Field and Its Chemical Control

In order to express how high a temperature or magnetic field a superconductor can stand with zero resistivity, the experimental quantity B_{irr} is used as a convenient index. B_{irr} under fields perpendicular to the CuO_2 plane is especially important because this is the weakest configuration for the HTS. Therefore the irreversibility fields and the anisotropy factors have been measured in various single crystals of various carrier densities. By plotting $B_{irr}(H//c$-axis), we have found that B_{irr} is nearly proportional to γ^{-2} at each temperature regardless of change from a material to another or of change in the carrier density of a specific material as shown in Fig. 3 [13].

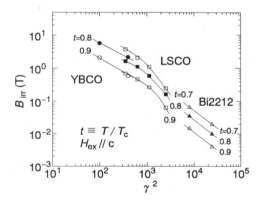

Fig. 3 : Irreversibility field, $B_{irr}(H \parallel c)$ at various reduced temperatures of YBCO, LSCO and Bi2212 single crystals plotted as a function of their electrical anisotropy, $\gamma_n^2 (\equiv \rho_c/\rho_{ab})$.

Therefore we can draw an empirical relationship for the irreversibility field;

$$B_{irr}[\text{T}] \sim 3.34 \times 10^3 \, \gamma^{-2}(1 - T/T_c)^{1.13}$$

in the temperature region $T/T_c \geq 0.7$.

Although there are complicated factors involved in determining the irreversibility field in detail, the fact that the empirical relationship (3) roughly holds on various material systems is significant. This suggests that the practical materials should better be optimized following this line. Also it implies that the vortex structure and its pinning behavior can be dominantly described by the single parameter, anisotropy factor γ^2.

5 Substitution of High Valency Transition Metal Ions

Shimoyama et al. [14] have found that substitution of Re ions for Hg in $HgBa_2Ca_{n-1}Cu_nO_y$ makes the synthesis process of this material systems much easier without deteriorating its T_c and besides that it significantly enhances B_{irr}. They attributed the enhancement of B_{irr} to the large reduction of the CuO_2 interlayer spacing and the subsequent weakening of the anisotropy. They also paid attention on the well-known fact that the binary oxide ReO_3 exhibits the lowest metallic resistivity of all the known oxides. Furthermore the system $Hg_{1-x}Re_xBa_2Ca_{n-1}Cu_nO_y$ seems to form an ordered structure [15] on the Hg-Re plane which may bring about a possibility of creating a novel type of pinning centers. This example clearly shows that the reduction of anisotropy and hence improvement of the HTS in performances under magnetic fields may be made without necessarily involving deterioration in T_c.

References

1. K. Kitazawa, in *"Coherence in High Temperature Superconductors"* eds. A. Revcolevschi and G. Deutscher, (1996) World Scientific Pub.
2. K. Kishio, *ibid.*
3. T. Shibauchi *et al.*, *Phys. Rev. Lett.* **72** 2263 (1994).
4. K. Tamasaku *et al.*, to be published.
5. T. Kimura *et al.*, in *"Advances in Superconductivity VII"*, 539 (1995).
6. H. Safar *et al.*, *Phys. Rev. Lett.* **69** 824 (1992).
7. U. Welp *et al.*, preprint.
8. E. Zeldov *et al.*, *Nature (London)*, **375** 373 (1995).
9. S. Watauchi *et al.*, *Physica* **C259** in press (1995).
10. H. Ikuta *et al.*, in This Proceedings.
11. R. Cubitt *et al.*, *Nature (London)* **365** 407 (1993).
12. C. Bernhard *et al.*, *Phys. Rev.* **B52** R7050 (1995).
13. K. Kishio *et al.*, *Physica* **C235-240** 2775 (1994).
14. J. Shimoyama *et al.*, *"Advances in Superconductivity VII"*, 287 (1995).
15. O. Chmaissem *et al.*, *Phys. Rev. B*, in press.

A NOVEL APPROACH TO HIGH RATE MELT-TEXTURING IN 123 SUPERCONDUCTORS

K. SALAMA, A. S. PARIKH, P. PUTMAN

Texas Center for Superconductivity, University of Houston Houston, TX 77204

and

L. WOOLF

General Atomics San Diego, CA 92121

ABSTRACT

Initial hurdles of processing the Y-123 ($Y_1Ba_2Cu_3O_x$) compound to satisfy high current applications were overcome by the melt texturing process developed in 1988. This process yielded pseudo-single crystals that have transport current densities of 10^5 Amp/cm^2 at 77 K in self field and 10^3 Amps/cm^2 at 77 K and in 30 T. Recently, it was found that certain RE-123 type compounds containing the rare earth element neodymium have a very high rate of recrystallization. This high rate of recrystallization causes the recombination and solidification rates to increase significantly, thus making these compounds highly suitable for the directional solidification process. By processing bars (50 mm × 5 mm × 5 mm) of these compounds through a high temperature gradient, fully textured microstructures were obtained at rates up to 50 mm/hr, and up to 100 mm/hr in case of electrophoretically coated wires of Yb-123. These processing rates are 50-100 times more faster than the texturing rates employed in Y-123. After oxygen annealing, these textured bars exhibited a T_c onset of 93K and a transport J_c, at 77K and zero applied magnetic field, on the order of 5,000 A/cm^2. These results are expected to increase significantly as more progress takes place.

1. Introduction

The melt texturing process developed in 1988[1] demonstrated the large potential of the Y-123 ($Y_1Ba_2Cu_3O_x$) compound for high current applications. This process yielded pseudo-single crystals that had transport current densities of 10^5 Amp/cm^2 at 77 K in self field and 10^3 Amps/cm^2 at 77 K and in 30 T[2,3]. When heated above the peritectic temperature, solid Y-123 decomposes into solid Y-211 and a liquid phase that is rich in Ba and Cu. Slow cooling through the peritectic temperature allows the recombination of Y-211 and the liquid phase to form melt-textured Y-123. In order to obtain a faceted growth front, typical melt-texturing rates for Y-123 are restricted to 1°C/hr. Also, grain boundaries in Y-123 act as weak links because of the compound's very small coherence length. Directional solidification has been used to process long lengths (100 mm) of Y-123 bars free of high angle grain boundaries[4]. This involves the relative motion between the hot zone and the sample, which can be achieved by the movement of the sample in the furnace. The presence of a temperature gradient provides the directionality to the growth process. Y-123 bars processed using this technique are found to have a single domain along the length of the bar and carry current densities on the order of 10^4 A/cm^2 at 77 K and zero applied magnetic field.

Recently, it was found that certain RE-123 type compounds with RE = Nd and Yb have a very high rate of recrystallization[5]. The high rate of recrystallization causes the recombination and solidification rates to increase significantly, thus making these compounds highly suitable for the directional solidification process. We have examined the texturing of these compounds at high rates by processing bars made from the Nd-123 compound and electrophoretically coated Au-wires of Yb-123 using the directional solidification processing method. Results on bulk samples processed at rates as high as 50 mm/hr seem to be very encouraging.

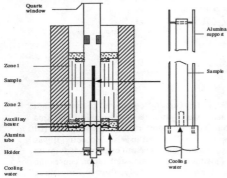 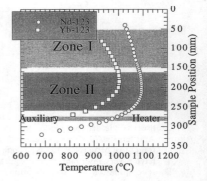

Figure 1. Modified Bridgman furnace for Directional Solidification.

Figure 2. Temperature profile in the modified Bridgman furnace.

2. Experimental

Calcined Nd-123 precursor powder prepared by solid state reaction was obtained from General Atomics, San Diego. The actual composition of the powder consisted of 0.4 moles of Nd_2BaCuO_5 per mole of $Nd_{1.05}Ba_{1.95}Cu_3O_y$ and is hereafter referred to as Nd-123. DTA of this compound in air indicates that the peritectic reaction commences at 1040°C and partial melting is completed near 1090°C. Bars of dimensions 50 mm in length, 10 mm in width and 5 mm in thickness were pressed under a pressure of 10^7 N/m^2 uniaxially using a stainless steel split die. These green compacts were then sintered in a box furnace in air for a duration of 24 hours at 950°C. After sintering, the bars were sectioned longitudinally using a speed-controlled diamond saw into bars of equal dimensions of 50 × 5 × 5 mm. Each of these bars was then subsequently placed on a stainless steel stage that traveled through the hot zone in a modified Bridgman furnace in air for undergoing melt texturing and directional solidification[4].

Electrophoretically coated Au-wires of Yb-123 were also obtained from GA. Wires 20 mm long and 650 μm dia. are processed at 995°C since the peritiectic point for Yb-123 is 970°C. A schematic drawing of the processing set-up is shown in figure 1 and the temperature profile is shown in figure 2. In case of the bars, the top 50 mm is introduced into the hot zone for a short period before it is withdrawn at a constant rate. For wires, the wire is moved at a constant speed from top to bottom in the vertical furnace.

The processing speeds studied in this work were 1, 4, 6, 8, 10, 12, 15, 20, 30, 50, 70, 100 and 200 mm/hr for the bars and 10, 25, 50, 75 and 100 mm/hr for the wires. After texturing, the bars and the wires were oxygen annealed to obtain the high-T_c superconducting phase. The samples were heated to 950°C for 12 hrs in flowing 1% O_2 / 99% Ar atmosphere to allow for the formation of the high-T_c phase[9]. They were then held in flowing 100% O_2 at 600°C, 500°C and 350°C for duration of 12, 24 and 60 hrs respectively[6]. The four point probe method using continuous current was used to measure T_c and the transport critical current density, J_c, in the bars. Silver leads of 0.25 mm diameter were attached using a silver colloidal paste prior to the oxygen annealing.

3. Results and Discussions

Bars processed with the rates of 1, 4, 6, 8, 10, 12, 15, 20 30, and 50 have shown complete texturing. A significant amount of faceted grain and domain growth along with shrinkage was observed in these bars. Bars processed at rates 70 and 100 mm/hr showed shrinkage and some grain growth indicating the partial melting and rapid solidification and growth. The growth front in these bars was intermittently faceted indicating the rapid rate of nucleation and solidification. These bars, however, had small grains with high density indicating liquid phase sintering and recrystallization. Also the microstructure of the bar processed at the rate of 100 mm/hr contained evenly distributed spherical pores indicating that at very high processing rates, gases cannot escape due to rapid solidification and hence get trapped in the bulk. The bar processed at the rate of 200 mm/hr showed shrinkage demonstrating that partial melting had occurred, and the microstructure consisted of very fine crystallites, which indicated that the rate of processing was very high and thus there was insufficient time for solidification and growth. The amount of liquid loss in these bars was negligible.

The microstructures of the bars processed at rates up to 100 mm/hr contained the two phases Nd-123 and Nd-211. The Nd-211 phase externally added is present in the microstructure in the form of spherical shaped particles and the peritectic Nd-211 phase is present in the form of acicular translucent needles that are golden brown in color. These 211 phases are homogeneously distributed throughout the bulk of all the samples, indicating that homogenous melting and solidification has occurred at all these processing rates.

Wires processed with the rates of 10, 25, 50, 75 and 100 mm/hr have shown complete texturing. The microstructure of these processed wires showed faceted grains, however, the grain size was less than 1mm for all processing rates. A primary cause for this could be the very thin nature of the wires which will have extremely high rate of heat transfer. Upon oxygenation, the microstructure indicated the presence of the Yb-123 phase in the cross polarized light. The peritiectic Yb-211 particles had a typical length of 10 μm and diameter of 2-3 μm for all the wires.

After the oxygen annealing heat treatment, the T_c was determined by measuring the ohmic resistance as a function of temperature for a bar and a wire processed at the rate of 50 mm/hr (figure 3). An onset of approximately 96K reaching zero resistance at about 94K was observed in the case of the bar and an onset of 87K was observed in case of the wire. The critical current density, J_c, in self field measured in a small specimen of dimensions 6 mm × 0.6 mm × 0.6 mm cut from the bar processed at the rate of 30 mm/hr was found to be 5,000 A/cm^2 at 77K and 6,700 A/cm^2 at 65K. Although these values are lower than those obtained in textured YBCO, they are high enough to indicate that texturing has occurred even at such a high growth rate.

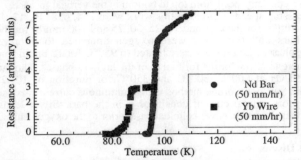

Figure 3. Resistance vs. temperature of bar and wire processed at 50 mm/hr.

4. Conclusion

In this work the possibility of high speed melt texturing of RE-123 type superconductors is successfully demonstrated in rapidly recrystallizing Nd-123 compounds. Bars of these compounds (50 mm × 5 mm × 5 mm) were processed at rates ranging from 1 to 200 mm/hr through a high temperature gradient. Fully melt textured microstructures were obtained at processing rates up to 50 mm/hr, which is more than 50 times faster than that employed in the processing of Y-123. After oxygen annealing, these textured bars showed a T_c onset of 93K and a transport J_c, at 77K and zero applied magnetic field, on the order of 5,000 A/cm^2. These results lend promise to applications using these high rate processing compounds.

This work is supported by the Texas Center for Superconductivity at University of Houston and the State of Texas.

List of References

[1] S. Jin, T. H. Tiefel, R. C. Sherwood, M. E. Davis, R. B. van Dover, G. W. Kammlott, R. A. Fastnacht and H. D. Keith, Appl. Phys. Lett. 52 2074 (1988).

[2] K. Salama, V. Selvamanickam and D. F. Lee, Bulk Materials (Processing and Properties of High-Tc Superconductivity 1) edited by S. Jin (Singapore: World Scientific, 1993) p 155.

[3] J. W. Ekin, K. Salama and V. Selvamanickam, Nature 350 26 (1991).

[4] V. Selvamanickam, C. Partsinevelos, A. V. McGuire and K. Salama, Appl. Phys. Lett. 60 3313 (1992).

[5] L. D. Woolf, S. S. Pak and T. L. Figueroa, Appl. Phys. Lett., 1995 (Submitted).

[6] S. Yoo, M. Murakami, N. Sakai, T. Higuchi and S. Tanaka, Jpn. J. Appl. Phys. 33 L1000 (1994)

NEW PROCESS TO CONTROL CRITICAL CURRENTS OF $Nd_{1+x}Ba_{2-x}Cu_3O_{7-\delta}$

Y. SHIOHARA, M. NAKAMURA, Y. YAMADA

Superconductivity Research Laboratory, ISTEC, 1-10-13 Shinonome, Koto-ku, Tokyo, 135 Japan

T. HIRAYAMA, Y. IKUHARA

Japan Fine Ceramics Center, 2-4-1 Mutsuno Atsuta-Ku, Nagoya, 456 Japan

To investigate whether for $NdBa_2Cu_3O_{7-\delta}$ (Nd123) an anomalous peak effect in the magnetization hysteresis (M-H) loops is an intrinsic property or not, different series of heat treatments were performed in a pure oxygen gas flow using Nd123 single crystal, and critical current density (J_c)-H curves at 77K were determined from M-H loops. J_c-H curve with either H//a-axis or H//c-axis of the obtained crystal heat treated at 340°C for 200h shows no anomalous peak effect. On the other hand, when two stage heat treatment (500°C for 100h and 340°C for 200h) was applied, J_c-H curve at 77K with H//c-axis exhibited an anomalous peak effect. Further annealing (900°C for 100h, quenched, 340°C for 200h) almost resulted in disappearance of the peak effect. These results together with transmission electron microscopy (TEM) observation confirm that a reversible solid state reaction such as spinodal decomposition modifies nano-scale microstructure of the bulk Nd123 crystal. The nano-scale microstructure is considered to be responsible for an anomalous peak effect in J_c-H curve.

1 Introduction

Recently, it has been reported that $Nd_{1+x}Ba_{2-x}Cu_3O_{7-\delta}$ (Nd123) superconductors prepared by the oxygen controlled melt growth (OCMG) method,[1,2] have higher T_c (96K) and higher the critical current density (J_c) at 77K in high magnetic field region than those of the melt processed Y123 with finely distributed Y211 inclusions, because Nd123 samples prepared by the OCMG method exhibited an anomalous peak effect in the J_c-magnetic field (H) curve [in other word, magnetization hysteresis (M-H) loop]. However, this anomalous peak effect for Nd123 in M-H loop has not been clarified yet.

In this work, to investigate whether for Nd123 the anomalous peak effect in M-H loop is an intrinsic property or not, different series of heat treatments were performed using bulk Nd123 single crystals produced separately by the top-seeded solution-growth (TSSG) method. J_c-H curves at 77K were determined from M-H loops measured by a SQUID magnetometer.

2 Experimental Procedure

In this experiments, Nd123 single crystals were used which were grown in Nd_2O_3 crucible under 1% oxygen partial pressure atmosphere ($P(O_2)$=0.01atm) by the TSSG method. Crystal growth is described in detail elsewhere.[3] The composition of grown crystals were analyzed by inductivity coupled plasma atomic emission spectrometry (ICP-AES) and confirmed to be Nd:Ba:Cu=1.01:1.97:3.00. The crystal structure was identified by X-ray structure analysis. This analysis indicated that as-grown single crystals was identified to have the $NdBa_2Cu_3O_6$ tetragonal structure with space group P4/mmm and no significant substitution of Nd ion into Ba site was detected within this experimental accuracy. Noteworthy is that crystal growth temperature were let to be constant, and as-grown crystals were removed from the furnace immediately after crystal growth was finished. The typical removing time was approximately 10 min corresponding to about 100°C/min of cooling rate.

Different seres of heat treatments in a pure oxygen gas flow were performed using the Nd123 single crystals produced separately by the TSSG method. Temperature dependence of dc magnetization under zero filed cool (ZFC) and field cool (FC) conditions and M-H loops with H//a-axis and H//c-axis at 77K were measured by a SQUID magnetometer. The magnetic J_C values in A/cm^2 were determined from M-H loops applying the extended Bean's critical state model.[4]

3 Results and Discussion

The heat treatment at 340°C for 200h for the oxygenation was carried out. Figure 1(a) shows the temperature dependence of the normalized magnetic susceptibility and a steep superconductive transition at around 96.5 K ($\Delta T \approx 1K$). Figure 2(a) exhibits the J_C-H curve with H//a-axis and H//c-axis at 77 K. For this heat treatment, no anomalous peak effect with either H//a-axis or H//c-axis in J_C-H curve was observed and J_C value for this Nd123 single crystal was as small as those for Y123 single crystal.[5] This small J_C value shows that these Nd123 single crystals do not have any strong pinning centers or crystal defects.

On the other hand, the continuous cooling heat treatment from 600°C to 350°C for 200h was performed, and then this sample was applied to the oxygenation heat treatment (340°C for 200h). Figure 1(b) shows the temperature dependence of the normalized magnetic susceptibility and figure 2(b) exhibits the J_C-H curve at 77K. As can be seen from Fig. 2(b), J_C-H curve with only H//c-axis shows an anomalous peak effect at around 40000G (4T), although with H//a-axis exhibits no anomalous peak effect. The anisotropic property in J_C-H curve with H//a-axis and H//c-axis is thought to be caused by anisotropic microstructure in Nd123 crystals. These results indicate that for Nd123 anomalous peak effect in J_C-H curve is considered to be not an intrinsic property but a process dependent. Moreover, it is thought that the solid reaction, through which microstructure in Nd123 crystal causes the anomalous peak effect and anisotropic property in J_C-H curve, occurs at the temperature range between 600°C and 350°C.

In addition, this sample was heated to 900°C, and held for 100h. In this case, immediately after this isothermal holding, samples were quenched in air. Further the oxygenation at 340°C for 200h was also performed. Figure 1(c) shows the temperature dependence of the normalized magnetic susceptibility and figure 2(c) exhibits the J_C-H curve at 77K. As shown in Fig. 2(c), the anomalous peak effect with c//H in the J_C-H curve almost disappeared. This result indicates that this solid reaction is found to be a reversible reaction. Accordingly, the existence of the anomalous peak in the J_C-H curve could be considered to be controlled by heat treatment. At this stage, we tentatively propose that this solid reaction may be a phase separation derived from somewhat spinodal decomposition of the Nd123 unstable solid solution even though no significant amounts of Nd-Ba substitution was detected by X-ray analysis. But the composition of the separated phases could not be identified yet because of very fine structure of separated phases. To confirm the separated phase, further investigations on composition analyses are in progress.

To investigate the solid reaction temperature, two stage heat treatments were conducted. The first stage heat treatment is to investigate the solid reaction temperature and the second stage one is to oxygenate. In this experiment, the first stage temperature was changed as a parameter (600, 500 and 400°C for 100h) and the second stage temperature was fixed to be 340°C for 200h. In this case, immediately after the first isothermal holding, samples were quenched in air. When the first stage was only let to be 500°C, an anomalous peak effect with only H//c-axis was observed similar to Fig. 2(b). Therefore it is found that the solid reaction, through which microstructure causes the anomalous peak effect and anisotropic property in J_C-H curve, occurred at around 500°C. But temperature below 400°C is so low that this solid reaction could not take place but oxygenation only could occur.[6]

Finally, to investigate microstructures, specimens of the Nd123 single crystal were observed by transmission electron microscopy (TEM).[6] The TEM image of the Nd123 crystal which has shown an anomalous peak effect in J_C-H curve, observed from (001) direction, exhibits the modulated structure (tweed like image) similar to that reported for the typical spinodal decomposition.[7] The electron diffraction patterns show that this Nd123 crystal has orthorhombic structure and high order reflections have the spot extended along the (110) direction. This crystal also has shown the twin structures similar to Y123 crystal in orthorhombic phase. On the contrary, the Nd123 crystal which has shown no anomalous peak effect J_C-H curve, was found to posses no modulated structure but twin structure was observed. Therefore in the Nd123 crystal which has shown an anomalous peak effect J_C-H curve, the twin structure and the modulated structure coexist and we assume that the nano-scale microstructure (the modulated structure) causes the anomalous peak effect and anisotropic property in J_C-H curve. But this nano-scale fine structure imposes difficulty in identifying the composition of the separate phases. These TEM observation also supports that the spinodal decomposition mechanism might be responsible for the phase separation leading to creation of nano-scale microstructure .

Acknowledgments

This work was supported by New Energy and Industrial Technology Development Organization for the R&D of Industrial Science and Technology Frontier Program.

References

1. M. Murakami, S. I. Yoo, T. Higuchi, N. Sakai, J. Weltz, N. Koshizuka and S. Tanaka, Jpn. J. Appl. Phys. **33**, L715(1994).
2. S. I. Yoo, N. Sakai, H. Takaichi, T. Higuchi and M. Murakami, Appl. Phys. Lett. **65**, 633(1994).
3. M. Nakamura H. Kutami and Y. Shiohara, Physica C, in print.
4. E. M. Gyorgy, R. B. van Dover, K. A. Jackson, L. F. Schneemeyer and J. V. Waszcazk, Appl. Phys. Lett. **55**, 283(1989).
5. Y. Yamada H. Kutami, Y. Shiohara and N. Koshizuka, in: Adv. Superconductivity VII, eds. K. Yamafuji and T. Morishita (Springer, Tokyo,1994) p.133.
6. M. Nakamura, Y. Yamada, T. Hirayama, Y. Ikuhara, Y. Shiohara and S. Tanaka, Phisica C, in print.
7. M. Doi and T. Miyazaki, Philos. Mag. B, **10**, 305 (1993).

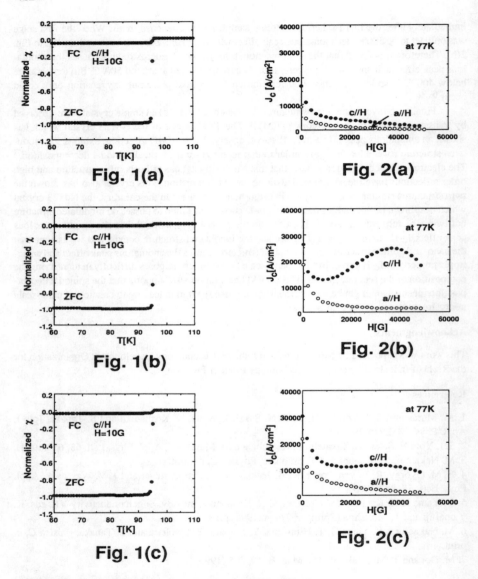

Figure 1: Temperature dependence of the normalized dc magnetic susceptibiliry.

Figure 2: Jc-H curve at 77K determined from M-H loop applying the extended Bean's critical state model

(a) 340°C for 200h. (b) 600°C to 350°C for 200h and 340°C for 200h. (c) 600°C to 350°C for 200h and 340°C for 200h. In addition, heated up 900°C for 100h, quenched in air, 340°C for 200h.

IMPROVEMENT OF FLUX PINNING IN HIGH TEMPERATURE SUPERCONDUCTORS BY ARTIFICIAL DEFECTS [1]

HARALD W. WEBER
*Atominstitut der Österreichischen Universitäten, Schüttelstraße 115,
A-1020 Wien, Austria*

Many attempts to improve flux pinning in high temperature superconductors have been reported on, but only a few were found to be effective in all families of the cuprate superconductors. This paper is intended to review the progress achieved by introducing artificial defects by fast neutron irradiation into single crystals, textured bulk materials as well as tapes of Y-, Bi- and Tl-based cuprates. The beneficial effects of these defects are examplarily demonstrated, in particular with regard to enhancements of the critical current densities and the shifts of the irreversibility lines, which are especially remarkable at higher fields and at temperatures of practical interest.

1 Introduction

The most crucial parameter for successful applications of high temperature superconductors is the critical current density J_c and its temperature and field dependence. Enormous efforts have been made so far to understand the principles of flux pinning and to improve the current carrying capability by any feasible means. However, apart from the introduction of normal conducting precipitates into bulk Y-123 [1] and the recent demonstration of pinning by arrays of stacking faults and associated dislocations [2], "metallurgical" methods were of limited success. On the contrary, radiation-induced defects were shown to improve J_c early on, but a more detailed understanding of their flux pinning effects emerged only gradually from systematic work on single crystals. Protons with energies of 10 MeV, e.g., enhance the critical current densities, but leave the irreversibility lines unchanged [3]. This is due to the nature of the proton-induced defects, which are point defects, similar to those present in the materials prior to irradiation. Fast neutrons on the other hand, both enhance J_c [4] *and* shift the irreversibility line to higher fields and temperatures as was demonstrated on Bi-2212 single crystals for the first time [5]. Details of the defect production and on their nature will be summarized in section 2. More recently, high energy heavy ion irradiation was employed [6], but the penetration of these particles is limited to a few 10 µm and the defects ("columnar tracks") are all basically aligned parallel to the incident beam. Hence, their flux pinning action is very different for "three-dimensional" or for "two-dimensional" systems with pancake vortices [7]. Both of these "drawbacks" can be avoided by bombarding suitable materials with very high energy protons (e.g. 800 MeV), which induce fission of certain nuclei, e.g. ^{209}Bi in Bi-2212 [8] or Bi-2223 [9], and thus create fission tracks. These defects are again amorphous with diameters of $\approx$ 10 nm and lengths of 20-40 µm, but their main advantage is their random orientation, which ensures flux pinning for any orientation of the magnetic field.

In summary, fast neutrons, heavy ions and fission all lead to the formation of extended defects in the superconductors, which modify the pinning-active defect structure in such a way that "strong" pinning conditions prevail, because the sizes of the defects are comparable to the diameters of the flux line cores or pancakes, especially at higher temperatures. In the following, neutron irradiation effects will be discussed in more detail.

---

[1] Supported in part by the Austrian Science Foundation and by the European Union in the framework of the Brite Euram program.

2 The Defects

Neutron irradiation experiments are usually done in the core region of fission reactors, where the entire spectrum of neutron energies (from 10^{-8} to ≈ 20 MeV) interacts with matter [10]. Detailed calculations [11] showed, that the fast neutron spectrum, i.e. neutrons with E>0.1 MeV, leads to cross sections between 4 and 8 barns for the production of recoil atoms with energies in excess of 1 keV. Therefore, these primary knock-on atoms have sufficient energy for the creation of collision cascades. Extensive TEM work [11] [12] has led to the following results for Y-123 single crystals. The displacement cascades are formed in a direct process and surrounded by an inwardly directed strain field of approximately the same size as the cascade. A statistical analysis of the defect size distribution on several hundred cascades showed that their average size is ≈ 2.5 nm and that the mean strain field diameter is ≈ 6 nm. A high resolution TEM picture of such a cascade is shown in Figure 1 [13]. In most cases, the cascades are found to be amorphous and of almost spherical shape, but in some cases the cascade's interior was re-crystallized Y-123 [12], but tilted with respect to the matrix orientation and presumably fully oxygen depleted. Annealing experiments established a high defect stability, i.e. no loss of cascades was detected up to 400 °C. Furthermore, the density of defects scales with neutron fluence, e.g. 5×10^{22} cascades per m^3 are created by 1×10^{22} fast neutrons (E>0.1 MeV) per m^2. In addition, due to the random impact positions of the neutrons and in view of their enormous mean free path (≈ 5 cm) the cascade distribution within the superconductor is completely statistical.

Of course, the lower energy part of the reactor neutron spectrum will also interact with the superconductor. In this case point defects and / or point defect clusters of varying size are produced, but their diameters will in all cases be below 1 nm. Oxygen *displacements* (not loss!) play a particular role, since they tend to reduce T_c. However, suitable annealing treatments (up to 400 °C) can lead to a recovery of T_c without loss of the pinning-effective cascades (cf. above).

3 Irreversibility Lines and Critical Current Densities

The clearest signature of changes in the pinning mechanisms is provided by the location of the irreversibility line, above which all of the defects have lost their flux pinning capability. Many

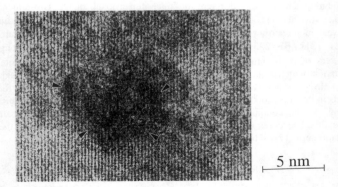

Figure 1: High resolution TEM picture of a collision cascade in Y-123.

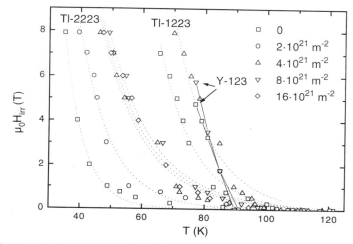

Figure 2: Irreversibility lines of Tl-2223, Tl-1223 and Y-123 single crystals (H∥c) prior to and following sequential neutron irradiation.

experiments on Y-, Bi- and Tl-based single crystals [14] [15] [16] have confirmed the general trend that pronounced shifts of the irreversibility lines result from the introduction of collision cascades. A summary of results in shown in Figure 2. Clearly, the largest effects are observed in those materials, which are entirely characterized by decoupled pancake vortices, such as, e.g., Tl-2223. In this case, we also note the first slight degrading effect of displaced oxygen at the highest neutron fluence (1.6×10^{22} m^{-2}), where the transition temperature has dropped by ≈ 8 K.

Similar experiments on melt-textured Y-123, have led to qualitatively similar, but much smaller effects [17]. This can be understood in terms of the pre-irradiation defect structure. Because of the "strong pinning" characteristics provided by the Y-211 precipitates (or some related defects) and in view of the rather "3D"-character of Y-123, especially at higher temperatures, only minor improvements of the overall pinning potentials can be expected.

With regard to the critical current densities, the following patterns have emerged. In single crystals, we generally find enhancements of J_c by factors of ≈ 10 at low temperature (e.g. 5 K), which increase to factors of ≈ 100 at higher temperature (e.g. at 77 K in Y-123). Of course, any quotation of enhancement factors does not make sense in those vast parts of the H,T-plane, where the pre-irradiation J_c's were zero. For melt-textured Y-123, the improvements of J_c are again smaller, but still reach factors of 10-30 at 77 K. An example of this kind is depicted in Figure 3, where results on low fluence irradiation (2×10^{21} m^{-2}, E>0.1 MeV) of the newly developed Nd-123 compound [18] are shown [19]. The low-field critical current density reaches $\approx 2 \times 10^9$ Am^{-2} and is still above 10^8 Am^{-2} at 6 T.

Whereas all of the results presented so far were based on magnetic measurements, direct transport experiments were made on tapes and thin films as well. Neutron irradiation of Y-123 thin films proved unsuccessful, since the pre-irradiation defect structures seem to be fully optimized (which is actually demonstrated by their record high J_c's) and additional defects immediately start to degrade them. On the contrary, moderate enhancements of J_c were found recently in Tl-based thin films following neutron and 230 MeV Au irradiations [20]. However, from STM work and based

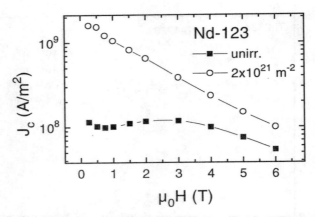

Figure 3: Critical current density versus applied field for melt-processed Nd-123 (H∥c) at 77 K.

on their pre-irradiation magnetic properties, these films were clearly not yet optimized in their composition and defect structures.

As a final remark, I wish to comment on transport measurements on neutron irradiated Bi-2223 tapes [21] [22] carried out recently. In contrast to expectations, according to which the essential parameter determining J_c would be an improved alignment of grain colonies, the results show clear enhancements of J_c by radiation-induced pinning *within* the grains, once the low-field weak-link dominated regime is surpassed. Enhancement factors between 2 and 4 were found at 77 K for all orientations of the field with respect to the tape surface after low fluence irradiation. However, because of the high level of radioactivity induced in the Ag-tube, this kind of experiment is somewhat tedious, but certainly deserves further attention.

References

[1] M. Murakami *et al.*, *Supercond. Sci. Technol.* **4**, 43 (1991)
[2] P.X. Zhang, H.W. Weber, and L. Zhou, *Supercond. Sci. Technol.* **8**, 701 (1995)
[3] L. Civale *et al.*, *Phys. Rev. Lett.* **65**, 1164 (1990)
[4] A. Umezawa *et al.*, *Phys. Rev. B* **36**, 7151 (1987)
[5] W. Kritscha *et al.*, *Europhys. Lett.* **12**, 179 (1990)
[6] L. Civale *et al.*, *Phys. Rev. Lett.* **67**, 648 (1991)
[7] J.R. Clem, *Phys. Rev. B* **43**, 7837 (1991)
[8] L. Krusin-Elbaum *et al.*, *Appl. Phys. Lett.* **64**, 3331 (1994)
[9] H.Safar *et al.*, *Appl. Phys. Lett.* **67**, 130 (1995)
[10] H.W. Weber *et al.*, *J. Nucl. Mat.* **137**, 236 (1986)
[11] M.C. Frischherz *et al.*, *Physica* C **232**, 309 (1994)
[12] M.C. Frischherz *et al.*, *Phil. Mag.* A **67**, 130 (1995)
[13] O. Eibl and H.W. Weber, *unpublished*
[14] F.M. Sauerzopf *et al.*, *Phys. Rev. B* **51**, 6002 (1995)
[15] W. Kritscha *et al.*, *Physica* C **179**, 59 (1991)
[16] G. Brandstätter *et al.*, *submitted to Phys. Rev. B*
[17] C. Czurda *et al.*, *Adv. Cryog. Eng.* **40**, 1015 (1994)
[18] S.I. Yoo *et al.*, *Appl. Phys. Lett.* **65**, 633 (1994)
[19] C. Kern, H.W. Weber, and M.Murakami, *work in progress*
[20] G. Samadi Hosseinali *et al.*, *submitted to Physica* C
[21] Q.Y. Hu *et al.*, *Appl. Phys. Lett.* **65**, 3008 (1994)
[22] G.W.Schulz, H.W. Weber, and H.W. Neumüller, *work in progress*

FABRICATION OF HIGHLY TEXTURED SUPERCONDUCTING THIN FILMS ON POLYCRYSTALLINE SUBSTRATES USING ION BEAM ASSISTED DEPOSITION

[1]Q. Xiong, [1]S. Afonso, [1]F.T. Chan, [1]K.Y. Chen, [1]G.J. Salamo, [2]G. Florence, J. Cooksey, S. Scott, [2]S. Ang, [2]W.D. Brown and [2]L. Schaper,
[1]Physics Department, [2]Electrical Engineering Department / High Density Electronics Center, University of Arkansas, Fayetteville, AR 72701

ABSTRACT

$Tl_2Ba_2CaCu_2O_y$(Tl2212) thin films on ceramic Al_2O_3 substrates with J_c(77K) of about 10^5 A/cm^2 and high quality $YBCO/YSZ/SiO_2/YSZ/YBCO/LaAlO_3$ mutilayers with J_c(77K) of about 6×10^5 A/cm^2 in the top YBCO layer have been successful deposited for the first time. These Mirror-like, highly c-axis oriented films were grown on highly textured YSZ buffer layers, which were deposited through Ion Beam-Assisted Laser Ablation. The zero resistance temperature is 95-108K for the Tl2212 films, and 85-90K for the multilayer YBCO films. The results suggest that using cheap non-single crystal substrates to fabricate good HTS films is possible.

I. Introduction

Since the discovery of high temperature superconductors(HTS), many possible applications of HTS thin films have been design and demonstrated, from very useful fault current limiters to the electronic interconnects on multichip module (MCM) substrates. All of these applications would benefit from the use of HTS thin films. One of the major problems for the application of HTS's is that HTS's can carry only a limited amount of current without resistance. This problem is related to their two dimensional layered structure. According to earlier studies[1-2], if the layers do not line up properly, the critical current density will decrease dramatically in the misaligned region. One way to overcome this problem is to grow micron-thin layers of the material on well organized substrates, epitaxally. The process has the effect of lining up the superconducting layers more accurately. HTS thin films grown on single crystal substrates of $LaAlO_3$ or $SrTiO_3$ have good lattice match between the HTS and substrate, and have critical current densities of about $\sim10^6$ A/cm^2, which is large enough for most HTS thin film applications. While this effort is impressive, it is far from useful, since the films so developed are much too expensive because the single crystal substrates are very expensive and available only in relatively small sizes. Moreover, for some electronic applications such as HTS multi-chip modules(MCM's), there are at least two HTS thin films that are separated by thick dielectric layers. Useful dielectric layers such as SiO_2 have noncrystalline structures. For this reason, HTS thin film layers cannot be grown on the dielectric layer with good alignment by conventional methods. Recent studies[3-7] have shown that highly biaxially aligned YSZ layers can be grown on non-crystalline substrates, which are cheaper and easy to get in any size needed by Ion beam-assisted laser deposition(IBAD). Using this method, high quality $YBa_2Cu_3O_y$ (YBCO) films have been deposited on biaxially aligned YSZ layers grown on non-crystalline substrates with J_c's of up to $8\times10^5 A/cm^2$ at 75K.

Here, we report the results from Tl2212 films fabricated on polycrystalline Al_2O_3 substrates with an IBAD deposited YSZ buffer layer and $YBCO/YSZ/SiO_2/YSZ/YBCO/LaAlO_3$ mutilayers.

II Experimental

Fine polished Al_2O_3 was used for the substrates of the Tl2212 films. Ion-beam assisted pulsed laser deposition was used to prepare highly biaxial aligned YSZ buffer layers at room temperature. While YSZ was deposited by pulse laser deposition, an Ion-beam (argon ion source) bombarded the substrates at an angle of 55° from the substrate normal. The laser used here was an ArF excimer laser (wavelength = 193 nm, shot frequency = 5 Hz). The energy density of the laser beam, which was focused on the YSZ target, was set to about 2 - 3 J/cm^2. The ion beam energy was typically set at about 250 eV, and the beam current density was about 120 $\mu A/cm^2$. The oxygen partial pressure was maintained at 3×10^{-4} Torr, and with the ion beam the total pressure in the vacuum chamber was about 8×10^{-4} Torr during the deposition of the YSZ layers. The substrate temperature increased to about 100 °C due to the assisting ion beam. The growth rate was about 0.5 Å/S. The thickness of the YSZ buffer layers in our experiment was about 5000 Å. For the YBCO multilayer structure, 2000 Å thick YBCO layers were deposited by pulsed laser ablation at ~ 750 °C. A 4-5 micron thick amorphous SiO_2 layer was deposited as a dielectric layer at room temperature by rf sputtering. For the Tl2212 films, 2000~3000Å $Ba_2CaCu_2O_y$ precursor films were deposited on the YSZ-buffer layer at a temperature of 400°C using pulsed laser ablation. The films were then treated in a thallination process at 810°C.

The structures of deposited films were characterized by x-ray diffraction. T_c and J_c of HTS layers were measured by both transport method using the standard four-lead technique and the magnetization method using a Quantum Design Magnetometer.

III. Results and Discussion

All YSZ films deposited on the polycrystalline Al_2O_3 substrates for Tl2212 or on first YBCO layer, or on amorphous SiO_2 layer were (001) oriented with the <001> axis normal to the substrate plane. A typical X-ray θ-2θ diffraction pattern for a YSZ film deposited on a polycrystalline Al_2O_3 substrate is shown in Fig.1. Observation of only the YSZ(002) peak in the

X-ray θ-2θ diffraction pattern indicates that the YSZ was highly (001) oriented. The φ scan width of the {111} peaks is about 25° (FWHM).

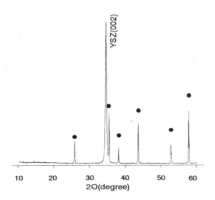

Fig. 1. X-ray diffraction θ-2θ scan of YSZ on Al_2O_3 (● Al_2O_3 peak)

Typical resistance versus temperature data for Tl2212 is shown in Fig.2. The zero resistance temperature T_c is 105.6K which is comparable with the results from the high quality films grown on $LaAlO_3$ or $SrTiO_3$ substrates. The critical current densities of these films are about 8×10^4 A/cm^2 by transport method. Fig.3 shows the typical resistance as a function of temperature for the top YBCO layer in the multilayer structures. The zero resistance temperature for this YBCO layer is about 86K with a transition width of about 2 K. The Messier effect M(T) for the same mutilayer sample is shown in the inset of Fig.3. As in the inset of Fig.3, the onset transition of the M(T) curve is at about 89 K, a second transition (indicated by the arrow) in the inset of Fig.3 is consistent with the T_c of the top YBCO layer measured by the transport method (Fig.3). In order to verify this, T_c was remeasured again after the top layer of YBCO was etched

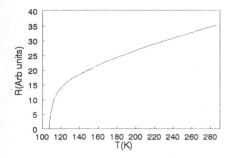

Fig. 2. Resistance vs. Temperature for the $Tl_2Ba_2CaCu_2O_8$ on YSZ/Al_2O_3.

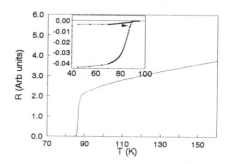

Fig. 3. Resistance vs temperature for the top YBCO layer in the $YBCO/YSZ/SiO_2/YSZ/YBCO/LaAlO_3$ multilayer. Inset: Magnetization M as a function of temperature for the entire multilayer.

away by using a dilute EDTA solution, and the onset transition was found to be at about 89 K without the second transition. The critical current density, J_c of the samples were measured by the magnetization method and calculated using Bean's model before and after the top layer of YBCO was etched away. The J_c of the bottom YBCO layer is about 7.2×10^5 A/cm^2 at 77 K, and of the top YBCO layer is about 6×10^5 A/cm^2 at 77 K, For comparison, we deposited both YBCO and Tl2212 on ceramic Al_2O_3 substrates. Even though the T_c of these films were about the same, J_c is $< 10^4$ A/cm^2. Which is about two orders of magnitude lower than that of the film on the biaxially aligned YSZ layer

IV. SUMMARY

Tl2212 films on ceramic Al_2O_3 substrates have been grown for the first time and good quality YBCO multilayer structures have been prepared using laser ablation and ion beam assisted rf sputtering. Our results demonstrate that good quality HTS films can be grown on cheap nonsingle crystal substrates though the Ion Beam-assisted deposition technique.

ACKNOWLEDGMENTS

This work is supported in part by NSF DMR 9318946, DARPA (Contract No. MDA 972-90-J-1001) through Texas Center for Superconductivity at the University of Houston.

REFRENCE

1. D. Dimos, P. Chaudhari, J. Mannhart, and F. K.LeGoues, Phys. Rev. Lett.61(1988)219.
   D. Dimos, P. Chaudhari, J. Mannhart, Phys. Rev. B 41(1990) 4038.
2. "Effect of Pressure on the Critical Current Density of YBa2Cu3O7-d Thin Films," Q. Xiong, Y. Y. Xue, Y. Y. Sun, P. H. Hor, C. W. Chu, M. F. Davis, J. C. Wolfe, S. C.Deshmukh and D. J. Economou, Physica C 205, 307 (1993).
3. R. P. Reade, P. Berdahl, R. E. Russo, and S. M. Garrison, Appl. Phys. Lett. **61**, 2231 (1992).
4. 2. Y. Iijima, N. Tanabe, O. Kohno, and Y. Ikeno, Appl. Phys. Lett. **60**, 769 (1992).
5. X. D. Wu, S. R. Foltyn, P. Arendt, J. Townsend, C. Adams, I. H. Campbell, P. Tiwari, Y. Coulter, and D. E. Peterson, Appl. Phys. Lett. **65**, 1961 (1994).
6. F. Yang, E. Narumi, S. Patel, and D. T. Shaw, Physica, C **244**, 299 (1995).
7. X. D. Wu, S. R. Foltyn, P. N. Arendt, W. R. Blumenthal, I. H. Campbell, J. D. Cotton, J. Y. Coulter, W. L. Hults, M. P. Maley, H. F. Safar, and J. L. Smith, Appl. Phys. Lett. **67**, 2397(1995).

SUPERCONDUCTIVITY AND STRUCTURAL ASPECTS OF $Y_{1-x}Ca_xBa_2Cu_3O_{7-\delta}$ WITH VARIABLE OXYGEN CONTENT

V.P.S. AWANA, J. ALBINO AGUIAR,
DEPARTAMENTO DE FÍSICA, UFPE 50670-901 RECIFE-PE, BRASIL
S.K. MALIK,
TATA INSTITUTE OF FUNDAMENTAL RESEARCH, BOMBAY, 400005, INDIA
W.B. YELON
UNIVERSITY OF MISSOURIE RESEARCH REACTOR, COLUMBIA, MO 65211, USA
and
A.V. NARLIKAR,
NATIONAL PHYYSICA LABORATORY, NEW DELHI, 110012, INDIA.

We have mapped nearly full phase diagram of $Y_{1-x}Ca_xBa_2Cu_3O_{7-\delta}$ system with both the different values of x and δ. While this substitution decreases the T_c of the $YBa_2Cu_3O_{7-\delta}$ system having δ = 0.0, the same enhances the T_c of the oxygen defficient Y:123 systems i.e for the δ > 0.0 values. Also the same decreases the oxygen content in the former situation and leaves the overall oxygen of pristine sample nearly unaltered in the later situations. Our x-ray and neutron diffraction results indicate that, while an increase in the T_c with this substitution is always accompanied with an increased c-parameter, the nearly unchanged c-parameter always warrents a decrease in the T_c. Further the variations of Cu(2)-O(2) bond distances indicate that this substitution dopes directly the conduction holes into the Cu(2)-O(2) planes for the oxygen defficient systems, but not in the case of fully oxygenated systems.

1. INTRODUCTION

Plenty of work has already been reported on the Y^{3+}/Ca^{2+} substitution in Y:123 system[1-3]. Most of the researchers agree that this substitution decreases the T_c of fully oxygenated Y:123 system, i.e from T_c = 90K[1,3], while the same increases the T_c of oxygen defficient system[2]. Unfortounately a more simplistic picture, i.e the introduction of mobile holes due to Y^{3+}/Ca^{2+} substitution has been applied by a number of workers to explain the results in both situations, in terms of the overdoping and optimum doping of the parent system. In this paper we present the results of x-ray and neutron diffraction with superconductivity and oxygen content values to prove that besides the injection of mobile holes due to Y^{3+}/Ca^{2+} substitution, this dopant gives rise to interesting crystallographic changes.

2. EXPERIMENTAL

The bulk samples of $Y_{1-x}Ca_xBa_2Cu_3O_{7-\delta}$ with different x and δ values were synthesized by the solid state reaction route[1]. The final sintering temperatures of set I were observed to be 10^0C and 20^0C higher respetively for x = 0.20 and 0.30 samples. For the various δ values, the as prepared set I, samples were further annealed in Argon atmosphere at 400^0C and 550^0C respectively for 12 hours, named as set II[2] and set III. Details of x-ray and neutron diffraction measurements along with the superconductivity and oxygen content are given else where[1,2,4].

3. RESULTS AND DISCUSSION

The quantitative data obtained on all the samples of set I, II and III is summarised in Table.1. It is clear from the Table that the T_c decreases for set I, while the same increases for set II and III, with the progressive substitution of Ca^{2+} at Y^{3+} site. The ionic size of Ca^{2+} matches well with that of Y^{3+} in 6-fold co-ordination, while the same is higher than, in the 8-fold co-ordination[1,2]. This implies that in set I samples the 6-fold co-ordinated Ca^{2+} substitutes 8-fold co-ordinated Y^{3+}, creating necessarily the vacancies in the adjacent Cu(2)-O(2) planes, and hence decreases the T_c. Further the increased Cu(2)-O(2) distances with substitution in set I descardes the possiblity of increase in p-type carriers in these samples[5]. As far as set II and III are concerned the Ca^{2+} stays in the same co-ordination as that of Y^{3+} and hence warrents an increase in the c-parameter, which is also dependent on x and heat treatments. Also in these cases the decrease in Cu(2)-O(2) distances indicates an increase in p-type carriers, and hence an increased T_c with this substituion for the samples of set II and III[5]. As far as the Buckling angle Cu(2)-O(2)-Cu(2) is concerned it increases in all the cases, i.e the Cu(2)-O(2) planes become more flat.

TABLE1.
Quantitative structural and superconducting data for $Y_{1-x}Ca_xBa_2Cu_3O_{7-\delta}$ system.

x	Lattice Parameter			T_c (K)	Cu(2)-O(2) (A)	Cu(2)-O(2)-Cu(2) (degree)	Oxygen content
	a(A)	b(A)	c(A)				
Set I.							
0.0	3.821	3.885	11.684	90	1.929	164.24	6.99
0.05	3.827	3.879	11.682	85	-----	-----	-----
0.10	3.828	3.876	11.683	80	1.930	164.86	6.96
0.15	3.822	3.878	11.684	70	-----	-----	-----
0.20	3.832	3.873	11.684	60	1.931	165.64	6.94
Set II.							
0.0	3.838	3.889	11.728	57	1.938	163.97	6.76
0.05	3.838	3.883	11.739	70	1.936	164.70	6.76
0.10	3.839	3.878	11.746	75	1.934	166.10	6.72
0.15	3.847	3.878	11.741	76	1.941	164.41	6.60
Set III.							
0.0	3.863	*	11.783	NSC	-----	-----	
0.10	3.864	*	11.788	NSC	-----	-----	
0.20	3.867	*	11.793	NSC	-----	-----	
0.30	3.866	*	11.799	43	-----	-----	6.25@

*:b=a ---Neutron diffraction data not available, NSC: Non superconducting: @iodometry result

4. ACKNOWLEDGEMENT

One of the authors V.P.S. Awana thanks the Brazilian Science Agency (CNPq) for an award of a visiting post Doctoral fellowship. Part of this work was supported by MURR and by US Department of Energy grant # DE-FG02-90ER45427.

5. REFERENCES

[1] V.P.S. Awana and A.V. Narlikar, Phys. Rev. B, 49, 6353 (1994).
[2] V.P.S. Awana, Ashwin Tulapurkar and A.V. Narlikar, Phys. Rev.B, 50, 594 (1994).
[3] H. Yakabe, M. Kosuge, Y. Shiohara and N. Koshizuka, Jap. J. Appl. Phys. 34, 4754 (1995).
[4] V.P.S. Awana, S.K. Malik and W.B. Yelon, Submitted for Publicaion.
[5] M.H. Whangbo and C.C. Torardi, Science, 249, 1143 (1990).

FRACTAL GRAIN BOUNDARIES IN COMPOSITE $YBa_2Cu_3O_{7-\delta}$ / Y_2O_3 RESULTING FROM A COMPETITION BETWEEN GROWING GRAINS: EXPERIMENTS AND SIMULATIONS

N.Vandewalle(*), M.Ausloos(*), N.Mineur(#) and R.Cloots(#)
S.U.P.R.A.S, (*) Institut de Physique B5, (#) Institut de Chimie B6,
Université de Liège, B-4000 Liège, Belgium

Fractal grain boundaries are found in $YBa_2Cu_3O_7$ melt-textured compounds grown in presence of some Y_2O_3 oxide additions. The nature of this "no-scale" geometry in such a compound is found to result from the competition between different grains growing from one single "nucleation center" in the melt. The grain boundary fractality was investigated for several samples. The fractal dimension D_f of the boundaries is found to be close to 4/3 and this result is reproducible. Fractal scaling is valid from 5μm to 500μm, i.e. ranging from the 211 particle size to the grain size. This value of D_f is predicted by a simple multicomponent Eden growth model. This work opens new ways of investigations for the synthesis and the patterning of superconducting and non-superconducting ceramics.

The liquid phase processing methods performed on YBaCuO bulk materials above the peritectic decomposition of the 123 phase, lead to an increase of <u>intragrain</u> critical current density. This is due to Y_2BaCuO_5 (211) precipitates within the 123 superconducting grains[1]. It is also known that the addition of the 211 phase in excess in Y-123 oxide further improves the <u>intergrain</u> critical current density[2]. Moreover, due to the fact that the peritectic reaction does not go to completion, the excess liquid phase in the system offers the possibility to consume this excess liquid (i) by forming additional YBaCuO phase, and (ii) by avoiding the liquid phase at the grain boundaries, whence "cleaning" them very much. Moreover, different characteristic lengths must correspond to the temperature dependent vortex core sizes for better pinning, whence 123-211 but also 123-123 boundaries should be examined.

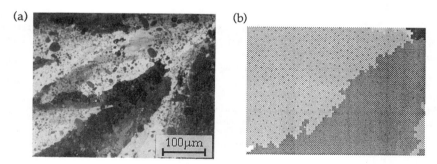

Fig.1 (a) Micrography of an $YBa_2Cu_3O_{7-\delta}$ set of grains in the vicinity of a Y_2O_3 seed (upper right hand corner); (b) the corresponding q-state Eden model simulation.

Recent works on 123-211 composite systems show that their physical properties strongly depend on the microstructure. Sandiumenge et al. discussed them from microstructural investigations[3]. Studies of the <u>intergranular</u> regions of polycrystalline samples are thus of importance in order to understand their electrical as well as their magnetic properties. Thus, examination of the grain interfaces might lead to understand how to enhance the field dependence of the critical current densities.

It is also known that the grain boundaries (and the 211 size) can be controlled by introducing PtO_2, CeO_2, Y_2O_3,...[4]. We investigated the role of such Y_2O_3 seeds. We have found out that the growth of the YBCO grains can start from such a seed and lead to specific microstrutral patterns(Fig. 1a). We have investigated the structure of such patterns, and observed that they correspond to multi-grain growth competition features. Image analysis of the planar grain boundary structure cross section (Fig. 1a) on different scales allowed us to observe that in the investigated scale range, i.e. between 5 μm and 500 μm, the grain boundary pattern can be represented by one fractal dimensionality D_f= 1.44±0.08 close to 4/3, therefore by a value D_f=2.44±0.08≈7/3 in the bulk.

Following our investigations on the growth of YBCO systems[5,6] in the light of the kinetic Eden growth models, we propose a spherulitic growth model for the above grain forming feautres. Starting from a central seed on a square lattice we let the Eden growth to be generalized to the case of a q-number of degrees of freedom, as was done in the q=2 Ising model case[7]. In so doing the number of possible different domains which can be formed is q. They might be disconnected[8]. The lower left hand corner of a simulation for q=8 is shown in Fig. 1b and can be compared to the optical micrograph of Fig. 1a

However the most surprising finding is that the computer analysis of the simulation data for such a model leads to the fractal dimension D_f=7/3 in very good agreement with the observation. Further work has shown that this fractal dimension is independent of q and of the lattice symmetry. Therefore we conclude that a simple physical and numerically easy to implement model can represent such growth features. They seem to be independent of the basic chemical kinetics, though more work should be done to investigate the influence of other kinetic parameters. These findings should be now related to physical measurements. We recall that the $J_c(H)$ behavior can be understood in terms of field percolation mechanism through grain boundaries[2]. Thus, it should be interesting to relate the power law exponents of $J_c(H)$ to the conductivity percolation dimension (4/3) which is quite close to the above grain boundary fractal dimensionality.

Acknowledgements

Part of this work has been financially supported through the ARC 94-99/174 contract of the Ministery of Higher Education and Scientific Research through the University of Liège Research Council.

References

1. M. Murakami et al., *Jpn. J. Appl. Phys.* **28**, (1989) L1125.
2. R. Cloots et al., *Z. Phys. B*, in press (1996).
3. F. Sandiumenge et al., *Phys. Rev. B* **50**, (1994) 7032.
4. C.-J. Kim et al., *Materials Lett.* **19**, (1994) 185.
5. N. Vandewalle et al., *J. Mater. Res.* **10**, (1995) 268; *Phil. Mag. A* **72**, (1995) 727.
6. R. Cloots et al., *Appl. Phys. Lett.* **65**, (1994) 3386.
7. M. Ausloos et al., *Europhys. Lett.* **24**, (1993) 629.
8. N. Vandewalle and M. Ausloos, *Phys. Rev. E* **52**, (1995) 3447.

TWIN STRUCTURE IN LARGE GRAINS OF YBa$_2$Cu$_3$O$_{7-\delta}$ AS AFFECTED BY THE DISPERSION AND VOLUME FRACTIONS OF Y$_2$BaCuO$_5$.

Manoj Chopra[†], Siu-Wai.Chan[†], V.S.Boyko[†].
[†]*Henry Krumb School of Mines, Columbia University NY,NY -10027.*
R.L.Meng[‡] and C.W.Chu[‡]
Texas Center of Superconducting Research, Houston. T.X[‡]

Quantitative Microscopy study[1] has been conducted on large grained YBa$_2$Cu$_3$O$_{7-\delta}$ (Y123) samples with different volume fractions of Y$_2$BaCuO$_5$ (211). Samples containing 0.5wt% platinum facilitate the dispersion of the 211 particles. Both the 211 particle size and distribution are analyzed. Transmission electron microscopy (TEM) shows the following: 1) Average twin spacing (Tw) increases linearly with the square root of the local interparticle spacing (S$_{211}$) when S$_{211}$ is larger than 4 times the diameter of adjacent 211 particles. The twin boundary energy is evaluated to be 28mJ/m^2 for samples containing no Pt and 11.4 mJ/m^2 for those with Pt. 2) For sub micron 211 particles the twin spacing decreases as one approaches the Y123/211 interface. Furthermore, this interfacial twin spacing is dependant on the effective radius (R) of the 211 particle. Twin spacing measurements from various samples are related to the 211 particle shape and distribution.

1. Variation of Twin spacing with interparticle spacing

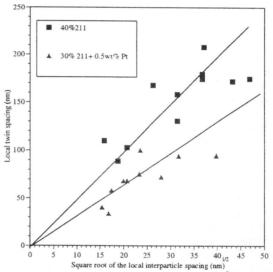

Twinning in Y123 has attracted considerable attention due to the possible flux pinning behavior exhibited by it. Here the variation of the twin structure as a function of 211 size, shape and distribution is examined. The local twin spacing corresponding to the local 211 interparticle spacing are shown in Fig.1 for 40 volume% 211 and 30%211 + 0.5wt% Pt respectively. The mean local twin spacing increases linear;y with the square root of the local interparticle spacing for the two samples. Based on a double pile up dislocation model[2] the twin spacing is related to the local interparticle spacing as Tw = $\sqrt{\dfrac{4\pi\gamma\ G}{\mu\varepsilon^2}}$ where G is the grain size in a polycyrstalline sample. However, in the present case the constraint size is the spacing of the 211 particles (S$_{211}$) instead of G. Here Tw is the mean local twin spacing, γ is the twin boundary energy, S$_{211}$ is the shortest local distance in the Y123 matrix between two 211/Y123 interfaces, μ is the shear modulus for Y123, ε is a measure of the twinning shear strain and is given by (b/a -1), where `b' and `a' refer to the lattice parameters of the basal plane of Y123. The best fit lines for our data (see Fig.1). give slopes that yield γ based on the above equation. From the slopes of these lines, the mean twin boundary energy is ≈ 28.0 ± 0.6 mJ/m^2 for the sample with no Pt and 11.4± 0.6 mJ/m^2 for the Pt doped sample. Studies done on a sample containing Pt reveal no changes in the orthorhombicity. This implies that the shear strain associated with the t→o transformation is the same for both cases. The ratio of the twin energies between the Pt doped and undoped case can be expressed as $\gamma_{tw}^{No\ pt} = 2.5 \gamma_{tw}^{pt}$.

2. Dependence of the twin spacing near the Y123/211 interface

Microscopic examinations of regions around the 211/Y123 interface reveal a non uniform twin spacing. This non uniformity in twin spacing occurs near small 211 particles (< 1μm). Quantitatively the twin spacing around the Y123/211 interface is found to increase with R. Here R is the effective radius (i.e. the sum of the local radius of curvature of the Y123/211 and the distance from the 211/Y123 interface of the closest 211 particle). Twins closer to the 211/Y123 interface have a smaller twin spacing than the average local twin spacing. As shown in Fig.2 the log - log plots of the local twin spacing around the particle under consideration and the effective radius R for different samples. Around each particle, a Tw ∝ R^n relationship is observed where `n' lies between 2 and 8.

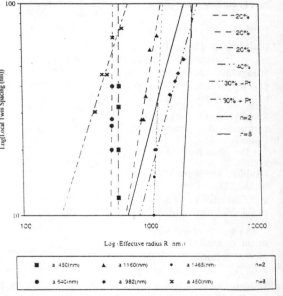

3. Conclusions

The twin spacing is found to be affected by the 211 interparticle spacing, twin boundary energy and Pt addition that lowers γ_{tw}. Furthermore submicron 211 lead to finer twin spacings due to a local increase in strain energy density around small 211 particles.

4. References

1. M.Chopra,Siu-Wai Chan,R.L,Meng and C.W.Chu. *Journal of .Mater.Res in press.*
2. T.E.Mitchell and J.P.Hirth *Acta.Metall.mater*,**39**,1711.(1991).

CRYSTAL ENGINEERING OF CHEMICALLY STABILIZED, CATION SUBSTITUTED $YBa_2Cu_3O_{7-\delta}$ STRUCTURES

J. T. McDEVITT*, J.P. ZHOU, C. E. JONES and J. TALVACCHIO‡
Department of Chemistry and Biochemistry, University of Texas at Austin, Austin, TX 78712
‡Northrop Grumman STC, Pittsburgh, PA 15235

ABSTRACT

To produce stable forms of $YBa_2Cu_3O_{7-\delta}$ superconductors, a series of cation substituted reactions have been completed. It is found that the corrosion resistance increases with increasing substitution level in systems of $Y_{1-y}Ca_yBa_{2-y}La_yCu_3O_{7-\delta}$. Interestingly, the composition of $Y_{0.6}Ca_{0.4}Ba_{1.6}La_{0.4}Cu_3O_{7-\delta}$ (T_c =80 K) is found to be at least 100 times more stable than the parent compound, $YBa_2Cu_3O_{7-\delta}$. Similar stable cuprate systems with transition temperatures above 85K have been prepared suggesting that the surface reactivity, processability and superconducting properties can be tailored with the use of the appropriate cation composition.

Recently, researchers worldwide have made considerable progress in the fabrication of high-T_c superconductor products such as tapes, wires and thin film devices. Unfortunately, only in a few cases have efforts been successful to date to commercialize high-T_c products. Much of the slow progress in this area can be traced to the poor materials properties, high chemical reactivity and poor oxygen stability exhibited by these systems. To improve the chemical stability, we have doubly doped $YBa_2Cu_3O_{7-\delta}$, forming $Y_{1-x}Ca_xBa_{2-y}La_yCu_3O_{7-\delta}$. These substitutions disrupt the integrity of the $CuO_{1-\delta}$ chains while keeping the total oxidation state of the Cu(2)-O(2)/O(3) arrays nearly constant. In this study, the degradation behavior of $YBa_2Cu_3O_{7-\delta}$ (YBCO) and substituted compound, $Y_{0.6}Ca_{0.4}Ba_{1.6}La_{0.4}Cu_3O_{7-\delta}$ (TX-YBCO), in water vapor environments are established for bulk and thin film structures.

A dramatic demonstration of the increased stability afforded with the cation substitution is shown in Figure 1. Here bulk samples of YBCO and TX-YBCO following their exposure to aerated water at room temperature is provided. For YBCO, two days water exposure causes the formation of a significant amount of $BaCO_3$ crystals which decorate the entire surface of the ceramic sample. This behavior is reminiscent of a sample that has decomposed to a large extent. On the other hand, the TX-YBCO sample treated for 30 days in a similar manner shows no visible signs of corrosion. A greater than 100 fold increase in stability is thus noted for the modified superconductor compound.

Thin films of YBCO and TX-YBCO have also been evaluated to explore their processability and durability. Lifetime measurements of the various samples were performed using our previous described resistivity measurement.[1,2] Accordingly, film samples were placed in a water vapor chamber equilibrated at constant temperature with controlled humidity. From such measurements, increases in the film resistivity are utilized to judge the rate at which damage to the cuprate structure occurs. Typically, the films decompose initially at a modest rate with little increase in resistivity over time. However, once the degradation process creates a sufficient number of corrosion defects to produce a high resistive barrier in the film, a marked increase in resistance is noted. Because sampling current, contact method and measurement intervals influence to some extent the measured lifetime, identical parameters were utilized to evaluate the various samples described herein.

It is clear that temperature and humidity are two important parameters that influence the degradation rate of YBCO. For example, YBCO samples studied at a constant temperature of 23°C while varying the humidity from 60% to >98% exhibited a lifetime decrease from >10 days for the former to <24 hours for the latter. However, while maintaining the relative humidity at ≥98%, and increasing the temperature from 23°C to 75°C, the YBCO lifetime decreased from >10 days to ~3 hours. In comparison, the TX-YBCO films showed no signs of degradation for >10 days under the severe conditions of 75°C water vapor and ≥98% humidity. From these studies and numerous other trials not included here, it is clear that major factors which influence the degradation rate are the high-T_c structure (i.e. YBCO vs. TX-YBCO), the humidity content and the temperature.[3]

 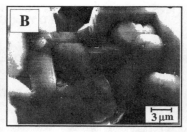

Figure 1: Scanning electron micrographs depicting the surface of a) $YBa_2Cu_3O_{7-\delta}$ exposed to a room temperature water solution for 2 days and b) $Y_{0.6}Ca_{0.4}Ba_{1.6}La_{0.4}Cu_3O_{7-\delta}$ exposed to a room temperature water solution for 30 days. The $YBa_2Cu_3O_{7-\delta}$ surface is decorated with corrosion products while the cation substituted compound shows little sign of degradation.

Considerations of the structure/corrosion reactivity relationships for the YBCO and TX-YBCO materials can provide essential mechanistic information. The parent compound contains $Cu(1)$-$O(1)$ chains as well as ordered oxygen vacancies (V_O) at the $O(5)$ sites. Previous reports have suggested that in the early stages of corrosion, water molecules enter the open channels along the $O(5)$ sites and the oxide ions ultimately dock at the $O(5)$ sites.[4] Proton transfer from water to the oxide lattice components leads to the formation of two moles of OH^- which occupy both vacancy site and the preexisting occupied oxide site. Internal charge transfer reactions within the modified lattice leads to the evolution of molecular oxygen. Following these initial steps, a series of reactions ensue ultimately leading to the bulk decomposition of the lattice.

Interestingly, the TX-YBCO is expected to be involved in many of the same decomposition steps. However, two important differences exist for the modified compound. First, internal stresses introduced by bond-length mismatch in the intergrowth structures appear to be an important factor that dictates the decomposition kinetics of the YBCO superconductor. Our calculations completed using the bond valence sum theory suggest that the substitution of La for Ba leads to a reduction of the internal strain energy from 1.79% to 0.35%. This reduction in strain may lower the driving force for corrosion. Second, the TX-YBCO material possesses a tetragonal structure in which occupancy at the $O(1)$ and $O(5)$ sites are equivalent. The disordering of these sites may influences the relative electrochemical potentials for the $O(1)$ and $O(5)$ sites, thereby altering the driving force for the charge transfer reaction. Without sufficient internal strain acting as a driving force for charge transfer, the substituted phase may remain resistant to corrosion. Moreover, changes in the structure may alter the rates of water diffusion into and oxygen diffusion out of the high-T_c lattice.

In summary, the preparation of the chemically robust cuprate bulk and film samples has been accomplished. Detailed studies of the corrosion mechanism for these samples suggest that changes in the lattice stress and strain features as well as minor changes in the oxygen ordering properties are responsible for the enhanced stability. Importantly, oxygen mobility within the modified structures is suppressed also making these systems quite attractive from a processing perspective.

Acknowledgements This work was supported at the University of Texas by ONR and NSF and at Northrop Grumman by AFOSR.

References

1. J.-P. Zhou; S.M. Savoy; R.-K. Lo; J. Zhao; M. Arendt; Y.T. Zhu; A. Manthiram; J.T. McDevitt *Appl. Phys. Lett.* **66**, 2900-2902 (1995).
2. J.-P. Zhou; S.M. Savoy; J. Zhao; D.R. Rilet; Y.T. Zhu; A. Manthiram; R.-K. Lo; D. Borich; J.T. McDevitt *J. Am. Chem. Soc.* **116**, 9389-9390 (1994).
3. J.-P. Zhou; R.-K. Lo; S.M. Savoy; M. Arendt; J. Armstrong; D.-Y. Yang; J. Talvacchio; J.T. McDevitt *Physica C* **submitted**, (1996).
4. J.G. Thompson; B.G. Hyde; R.L. Withers; J.S. Anderson; J.D. Fitzgerald; J. Bitmead; M.S. Paterson; A.M. Stewart *Mat. Res. Bull.* **22**, 1715 (1987).

TEM STUDY OF LOW-ANGLE GRAIN BOUNDARIES IN POLYCRYSTALLINE YBCO

M. MIRONOVA, G. DU, S. SATHYAMURTHY, I. RUSAKOVA AND K. SALAMA
Texas Center for Superconductivity, University of Houston, 4800 Calhoun, Houston, Texas 77204-4793, USA

The low-angle grain boundaries and platelet boundaries in polycrystalline melt-textured YBCO fabricated by Liquid Phase Removal Method have been studied by TEM. Platelet boundaries are found to terminated in the bulk and are accompanied by parallel low-angle grain boundaries. Both types of boundaries have similar misorientation characteristics: the misorientation axis, misorientation angles within the range of 0.2-5° containing tilt and twist components. Platelet and corresponding low-angle grain boundaries are found to maintain the horizontal alignment of the bulk as a whole to within ± 6°. These results are used to explain the possibility for high-angle grain boundaries to carry large currents.

1. Introduction

Grain boundaries are extremely important for the ability of High Temperature Superconductors to carry high currents. In regular melt-textured YBCO, the bulk consists of one large grain that contains different low-angle grain boundaries [1]. The YBCO superconductors fabricated by Liquid Phase Removal Method (LPRM), however, contain large number of grains separated by high angle grain boundaries [2]. In these polycrystalline melt-textured YBCO, grain boundaries are found to have current densities up to 30,000 A/cm^2 at 77 K in self field [2]. To understand the current carrying capability of these polycrystalline YBCO, its intra- and intergrain microstructures are investigated. In this paper, we report the results of TEM analysis of intragrain microstructure of polycrystalline melt-textured YBCO.

2. Experiment

Platelet and low-angle grain boundaries were studied in melt-textured samples grown by LPRM [3]. For TEM examinations, the specimens were polished down to 100 μm, dimpled and ion-milled at 4 kV. TEM studies were conducted using JEOL 2000FX electron microscope at 200 kV. The chemical composition was determined by EDS analysis. Misorientation of platelet boundaries and low-angle grain boundaries was characterized by rotation axes and rotation angles of the tilt and twist components. Rotation angles and axes were determined from the rotation matrix calculations using beam and beam stopper directions as two fixed directions [3]. The tilt and twist components of the grain boundary misorientation were determined as described in ref. [4]. The study included 45 platelet and low-angle grain boundaries, with 24 of them between adjacent grains. The density of platelet and grain boundaries was estimated as the average line length per unit area of 5-10 films covering ~3 000 μm^2.

3. Results and Discussion.

The grains formed in polycrystalline YBCO consist of fine platelets of 0.2-2 μm thick with boundaries of 2-30 nm. Platelet boundaries are observed to be perpendicular to the *c*-axis, taper inside the grain [5] and terminate with low-angle grain boundaries. Such boundaries are also found to be parallel to the platelet boundaries (Fig. 1) and could have been formed during the platelet growth [5] instead of the platelet boundaries. Platelet boundaries contain thin liquid phase layer rich in Ba and Cu, often are only partially filled and also may be cracked.

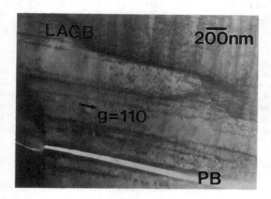

Fig. 1. Platelet boundary (PB) and parallel low-angle grain boundary (LAGB) in polycrystalline YBCO with misorientation angles 1.2 and 0.9°, respectively. Misorientation axis for both boundaries is [110].

Low-angle grain boundaries are observed to be clean, their chemical compositions are the same as that of the bulk and the inner boundary structure consists of dislocation arrays. These grain boundaries lie in the ab-plane. The total density of platelet and grain boundaries is $\sim 10^8$ cm^{-2}. The observed misorientation angles for these boundaries are within the range of 0.2-5° and contain tilt and twist components. Platelet boundaries, however, have higher misorientation angles than the corresponding low-angle grain boundaries. Many of the rotation axes are of [hk0] type.

Analyses of large intragrain areas containing adjacent boundaries revealed that the twist direction of platelet and low-angle grain boundaries alternates, resulting in maintaining of the lateral alignment of the bulk. However, the misorientation between two separated points may reach ±6°. In polycrystalline YBCO, platelets and low-angle grain boundaries are found to be dense, cover the whole grain and run into high-angle grain boundaries in a systematic manner. This means that alternating misorientation of platelet boundaries and low-angle grain boundaries may affect the misorientation of high-angle grain boundaries and can be used to explain the possibility of high-angle grain boundaries to carry large currents.

4. References

1. K.Salama, V.Selvamanickam, L.Gao and K.Sun, *Appl. Phys. Lett.* **54**, (1989), 2352.
2. A.S.Parikh, B.Meyer and K.Salama, *Supercond. Sci. Technol.,* **7**, 455 (1994).
3. Y.Zhu, H.Zhang, H.Wang and M.Suenada, *J. Mater. Res.* **6**, 2507 (1991).
4. L.M.Utevsky, *Electron Microscopy Diffraction in Metal Science*, Moscow, Metallurgy, 583 (1973) (*in Russian*).
5. K.B.Alexander, A.Goyal, D.M.Kroeger, V.Selvamanickam and K.Salama, *Phys. Rev. B* **45** (1992), 5622.

MOLECULAR LEVEL CONTROL OF THE SURFACE PROPERTIES OF HIGH–T_C SUPERCONDUCTOR MATERIALS

J. T. McDEVITT, R. -K. LO, J. E. RITCHIE, AND J. ZHAO
Department of Chemistry and Biochemistry, The University of Texas at Austin, Austin, TX 78712

and

C. A. MIRKIN, F. XU, AND K. CHEN
Department of Chemistry, Northwestern University, Evanston, IL 60208

ABSTRACT

It has been established that molecules containing alkylamine and arylamine functionalities bind tenaciously to the cuprate superconductor interfaces. The derivatization of bulk ceramic and thin film samples of $YBa_2Cu_3O_7$ using amine-tagged hydrocarbon and fluorocarbon molecules that are capable of spontaneous adsorption onto the cuprate materials is described. The formation and performance characteristics of corrosion protection barriers formed from these self-assembled monolayers is discussed.

One of the major stumbling blocks that has plagued the practical utilization and fundamental studies of the high-T_c superconductors has been the poor interfacial properties exhibited by these systems. Development of reliable methods to control the interfacial properties of cuprate systems is particularly important now as the initial high-T_C superconductor products approach the marketplace. The organization of molecular reagents into monolayer films on the surfaces of solid–state materials provides a convenient and rational approach towards controlling the surface and interfacial properties of these materials. The best characterized and most extensively studied systems have involved linear alkanethiol reagents adsorbed onto gold and other noble metal substrates. The use of long chain hydrocarbons and crystalline substrates yields monolayers which are densely packed, oriented, and highly ordered. A recent survey of the surface coordination chemistry of $YBa_2Cu_3O_7$ has led to the identification of reliable methods for preparing persistent monolayers on the surface of a cuprate superconductor.[1,2]

Herein, carefully chosen adsorbate molecules, such as linear alkylamines and fluorinated alkylamines, are used to tailor the corrosion resistance and surface adhesion properties of $YBa_2Cu_3O_7$, Figure 1a. The most striking demonstration of the utility of the monolayer methodology for modification of the cuprate superconductor interfacial properties is shown in Figure 1b and 1c. The illustration provides a dramatic comparison of the ability of a single monolayer formed from $CF_3(CF_2)_7(CH_2)_2NH_2$ to exclude water from the surface of a high-T_C ceramic pellet. Here, the uncoated $YBa_2Cu_3O_7$ pellet degrades rapidly over a period of one day upon soaking in an aerated water solution, as evidenced by the formation of $BaCO_3$. This impurity phase collects on the surface of the superconductor, and its presence is visualized readily by scanning electron microscopy (i.e. the white crystals in Figure 1). On the other hand, $YBa_2Cu_3O_7$ modified with a monolayer of $CF_3(CF_2)_7(CH_2)_2NH_2$ shows little signs of corrosion under identical conditions. This re-

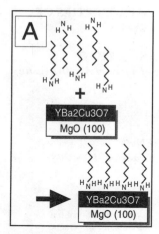  

Figure 1— (A) Schematic of the spontaneous adsorption of alkylamines on $YBa_2Cu_3O_7$.

(B) Bulk polycrystalline sample of $YBa_2Cu_3O_7$ after 1 day water exposure.

(C) Bulk polycrystalline sample of $YBa_2Cu_3O_7$, coated with fluorinated alkyl amine reagent, after one day water exposure.

sult supports the notion that densely packed adsorbate layers capable of excluding even small molecules such as water.

Moreover, in recent studies we have also shown that fluorocarbon amine layers can be exploited to alter the surface adhesion properties of $YBa_2Cu_3O_7$ thin film structures. In the absence of treatment of $YBa_2Cu_3O_7$ with the fluoroamine reagent, Teflon AF, a fluorocarbon polymer, adheres poorly to the high-T_c surface. However, with a monolayer of $CF_3(CF_2)_7(CH_2)_2NH_2$, the polymer is found to adhere more strongly. Thus, the combination of monolayer adhesive and polymer layers serves to create an excellent packaging method for the long term protection of vulnerable $YBa_2Cu_3O_7$ thin film structures. These results demonstrate that corrosion resistance, adhesion and wetting properties of the high-T_c material can be tailored through judicious choice of adsorbate molecule. Thus, through appropriate choice of adsorbate molecule, superconductor surfaces can now be altered in a controlled manner so as to suit a variety of applications. These new methods provide for the first time a procedure through which interfacial properties of superconductors can be controlled at the molecular level with an easily exploited method. These new developments will likely influence in a positive manner the processing of high-T_c structures.

Acknowledgements

This work was supported by NSF, ONR, and AFOSR.

References

(1) K. Chen; C.A. Mirkin; R.-K. Lo; J. Zhao; J.T. McDevitt *J. Am. Chem. Soc.* **117**, 6374-6375 *(1995)*.
(2) J.T. McDevitt; C.A. Mirkin; R.-K. Lo; K. Chen; J.-P. Zhou; F. Xu; S.G. Haupt; J. Zhao; D.C. Jurbergs *Chem. Mater.* (1996) *accepted.*

MICROSTRUCTURAL CHANGES OF YBCO INDUCED BY LANTHANIDE DOPING

I. RUSAKOVA, R. L. MENG, P. GAUTIER-PICARD, and C. W. CHU
Texas Center for Superconductivity and Department of Physics at the University of Houston, Houston, TX 77204-5932, USA

Melt-textured YBCO reveales an increase of J_c after partial substitution of Y by the lanthanides Sm, Eu, Gd, Ho, Dy, respectively. This effect might be due to the random distribution of the magnetic atoms which act as effective pinning centers as well as to some changes of twins and dislocation structure that were determined by TEM analysis.

1. Introduction

One of the many approaches that have been proposed to improve the flux pinning characteristics of YBCO is partial substitution of Y by other rare earth ions. Previous studies have determined that such substitution leads to enchanced J_c [1-3]. Various origins have been proposed for this effect: variation in the number and quality of grain boundaries [2], size and distribution of Y-211 particles [2], atomic radius of substitutes [3], etc. However, the actual mechanism of the observed enchanced pinning remains unclear. In this work we study the structure of YBCO after partial substitution of Y by Dy, Ho, Eu, Gd, and Sm.

2. Experiment

The doped samples were prepared from pure Y-123 and Y-211 powder by a solid state reaction method. The appropriate compositions were mixed, pressed into disks, and melt textured. Due to the different peritectic reaction temperatures of different rare-earth partial-substitutes, the melt-texturing temperature profile was slightly modified to optimize the processing conditions. The electrical connectivity was determined by a magnetic-field profiling technique in which the sample was field cooled to 77 K and the distribution of the field at the surface of the sample after the removal of the external field was mapped out with a Hall probe. The chemical composition was checked by EDS and the concentration of lanthanides was determined as ~1.8 at.%. The microstructure of the samples was studied by scanning electron microscopy (SEM) and transmission electron microscopy (TEM). Samples for TEM were prepared by ion-milling with 4keV Ar ions using a liquid nitrogen cooled sample holder. The TEM study was performed using a JEOL 2000 FX operating at 200 kV.

3. Results and discussion

In our study we find that partial substitution of Y by rare earth ions such as Sm, Eu, Gd, Ho, and Dy, respectively, results in increasing values of J_c. Such increase of J_c

may correlate with the increasing magnetic moment of substitutional ions and it might be due to the random distribution of the magnetic atoms which act as effective pinning centers.

The TEM study reveales changes in the twin structure as result of doping by all types of lanthanides (Fig. 1). First, the d-spacing between twins decreases from d≈100 nm to d≈60 nm in doped YBCO, and second, after doping twin boundaries have more imperfections (steps) which are actually twinning dislocations. Areas exibiting intersections of two systems of twins are more frequently observed in the Dy-doped sample. These structural changes can increase vortex pinning in the doped YBCO.

Fig.1. Twin structure of YBCO after doping with Sm. Typical changes of twins after doping with lantanides such as smaller d-spacing and more imperfections on the twin boundaries are observed.

The dislocation structure of YBCO samples doped by Eu, Gd and Sm is similar to the dislocation structure in undoped YBCO specimens. Single randomly oriented dislocations and dislocation tangles believed to contribute to flux pinning were observed in all of them. In addition, arrays of long parallel dislocations were observed in Dy and Ho doped samples. Such dislocations could be responsible for enhancing pinning in the Dy and Ho doped materials.

Density and size of Y-211 particles in doped YBCO are about the same as in undoped YBCO if the processing conditions are optimized by modification of the melt-texturing temperature profile.

4. References

1. R.L. Meng and C.W. Chu, Growth and possible size- limitation of large single-grain $YBa_2Cu_3O_7$, 124-th TMS Annual Meeting, February 12-16, 1995, Las Vegas, Nevada, p.203.
2. Y. Feng, L. Zhou, P. Zhang, P. Ji, X. Wu, C. Luo, and M. Xing, Effect of Ho addition on the properties of Y-system superconductors, Journal of Superconductivity, 5, No.2, 95, 1992.
3. M.H. Fang, G.J. Hu, M. Cai, Z.K. Jiao, Q.R. Zhang, Flux pinning effect of small size defects in $YBa_2Cu_3O_{7-d}$, Solid State Commun., 89, 93, 1994.

CONTROL OF 211 PARTICLE DISTRIBUTION AND JC PROPERTY IN MELT-TEXTURED YBCO SUPERCONDUCTOR

Y. SHIOHARA, A. ENDO
Superconductivity Research Laboratory, ISTEC
1-10-13, Shinonome, Koto-ku, Tokyo 135, Japan

Macrosegregation of Y_2BaCuO_5 (211) particle distribution in Pt-added $YBa_2Cu_3O_{7-\delta}$ (123) crystals grown by an isothermal solidification method was investigated. It was found that entrapped 211 distribution in 123 crystals depended on the growth direction and the growth rate (R) as a function of undercooling (ΔT). These phenomena could be qualitatively explained by the prevalent inclusion trapping theories. From SQUID measurements, the Jc values were shown to vary with the 211 distribution even though the nominal composition was constant. These results indicated that the Jc properties of the 123 crystals were affected by the trapping phenomena of 211 particles as well as the amount of 211 addition in the nominal composition.

1 Introduction

Melt growth of the $YBa_2Cu_3O_{7-\delta}$ (123) superconductor is generally used to yield high critical current density (Jc). Because it is effective to introduce Y_2BaCuO_5 (211) inclusions, which are considered to act as strong flux pinning sites, dispersed into solidified 123 grains formed through the peritectic reaction between the high temperature phase (211) and a liquid phase. Recently, the significant macrosegregation of 211 particles in 123 crystals has been found to occur during the melt textured growth, especially in the case of finely dispersed 211 particles caused by Pt-addition[1,2]. These phenomena are indispensable subjects from a viewpoint of improving Jc as well as understanding solidification mechanism. In this study, we investigated the macrosegregation phenomena of 211 particles in Pt-added YBCO crystals grown by an isothermal solidification processing combined with seeded growth. Furthermore, the effects of the microsegragetion on Jc properties were studied.

2 Experimental

The sample pellets (20 mm diameter and 7-10 mm height) were prepared using the constant nominal composition of $YBa_2Cu_3O_{7-\delta}$ + 0.4 Y_2BaCuO_5 + 0.5wt%Pt. The 123 crystals were grown from a cleaved Sm123 seed crystal at two different undercoolings ($\Delta T=10, 19K$). The ΔT, which is a driving force for solidification, was defined as the difference between the peritectic temperature ($T_p=1010°C$) of the 123 phase and growing temperature. The macrosegregation and the size distribution of 211 particles were investigated by optical microscopy. In addition, the samples cut out of the 123 crystals grown at each undercooling for a- or c-direction growth were measured by SQUID magnetometry after annealing at 600°C for 12h and at 400°C for 300h in oxygen atmosphere.

3 Results and Discussion

The 123 crystals epitaxially grown from the seeds consisted of a- and c-direction growth regions. Therefore, four kinds of samples were obtained from $\Delta T=10K$ and 19K crystals; (A)$\Delta T=10K$, a-direction growth, (B)$\Delta T=10$, c-direction growth, (C)$\Delta T=19K$, a-direction growth, (D)$\Delta T=19K$, c-direction growth. From microstructure observations, it was clear that the amount of 211 particles in

A region was larger than that in B region while no such a significant difference was observed in C and D regions. The amount of entrapped 211 in C and D is larger than that in A and B. From the quantitative analysis of 211 particles in the four regions, the 211 particles with diameter smaller than 0.5μm were rarely observed in the growth region of ΔT=10K (A and B). Also it is noticed that the total volume fraction of 211 particles in the ΔT=19K regions (C and D) are almost as same as the calculated one using the lever rule, which is expected from the nominal composition. On the other hand, the total volume fraction in ΔT=10K region (A and B) is much smaller than the calculated one. This result suggests that the excess 211 particles could not enter the 123 grain completely.

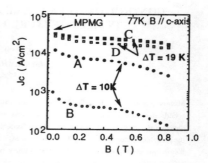

Fig.1 Jc-B properties for the samples grown at diffrent ΔT for each growth direction.

Figure 1 shows the Jc-B properties of the four samples at 77K. The Jc values are qualitatively consistent with the 211 distribution as a function of ΔT or growth direction. These results indicate that Jc is not only uniquely determined by the nominal composition but is affected by process condition such as R(ΔT) or growth direction.

These macrosegregation phenomena of 211 particles in 123 crystals could be qualitatively explained by pushing/trapping theories[3] on foreign particle behavior at an advancing solid/liquid interface during solidification. According to these theories, a particle, which is smaller than the critical size (r*), is pushed out by a solid at the critical growth rate (R*). The R* value is roughly determined by R* and interfacial energy ($\Delta\sigma_0$) as following;

$$R^* \propto \frac{\Delta\sigma_0}{\eta \cdot r^{*n}} \qquad (1)$$

where η is melt viscosity and n is 1-2. Eq. (1) indicates that smaller 211 particles, which can be entrapped at ΔT=19K (i.e large R), are pushed at ΔT=10K(i.e small R) and its trapping phenomenon is dependent of growth direction (i.e. $\Delta\sigma_0$). Further study for quantitative discussion on 211 trapping is in progress, considering the characteristic features in 123/211 system such as reactive inclusion (211 particle) for solid (123 phase), facetted growth (123 crystal) and large volume fraction of solid (211) in liquid, which rarely exist in the cases of ordinary trapping/pushing phenomena.

Acknowledgments

This work was supported by New Energy and Industrial Technology Development Organization for the R&D Industrial Science and Technology Frontier Program.

References

1. M. J. Cima et al., in Extended Abs. Inter. Workshop on Supercond., Maui, Hawaii (1995), **55**.
2. A. Endo, H. S. Chauhan, and Y. Shiohara, in print
3. for example, D. R. Uhlmann et al., J. Appl. Phys. **35**, 2986 (1964).

I. MATERIALS

BSCCO and TBCCO

PROCESSING AND PROPERTIES OF Ag-CLAD BSCCO SUPERCONDUCTORS

R. JAMMY, A. N. IYER, M. CHUDZIK, and U. BALACHANDRAN
Energy Technology Division, Argonne National Laboratory
Argonne, IL 60439

P. HALDAR
Intermagnetics General Corporation
Latham, NY 12110

ABSTRACT

Long lengths of mono- and multifilament Ag-clad BSCCO conductors with critical current densities of >10^4 A/cm^2 at 77 K were fabricated by the powder-in-tube technique. High-T_c magnets were assembled by stacking pancake coils fabricated from long tapes and then tested as a function of applied magnetic field at various temperatures. A magnet that contained ≈ 2400 m of high-T_c conductor generated a field of 3.2 T at 4.2 K. In-situ tensile and bending characteristics of the Ag-clad conductors have been studied. Multilayer Ag/superconductor composites have been fabricated with a novel chemical etching technique. Preliminary results with multilayer tapes show that continuous Ag reinforcement of the BSCCO core improves strain tolerance of the tapes so they can carry 90% of their initial I_c at 1% bend strain despite a higher superconductor/Ag ratio than that of unreinforced tapes.

1. Introduction

The powder-in-tube (PIT) process for the Bi-Sr-Ca-Cu-O (BSCCO) system of superconductors, with silver (Ag) as a sheathing material, is a promising technique for fabricating long-length conductors from high-transition-temperature superconducting (HTS) materials. Although significant progress has been made in the past few years in fabricating PIT tapes and enhancing their critical current density (J_c), the remaining critical issues impede the successful commercialization of HTSs. For many applications, as in electromechanical machinery, continuous lengths of tape, on the order of 1 km, with uniform electrical properties are necessary.[1-5] The electrical properties of these conductors under strain is another important engineering issue.

2. Conductor Fabrication

2.1 Fabrication of long-length conductors

Powder-in-tube fabrication of Ag-clad BSCCO conductors by our group is described in ref.4. High transport current properties in short samples have been achieved by a combination of uniaxial pressing and heat treatment. Critical current values above 40 A were typically attained at 77 K, the highest of which was 51 A.

Implementing a carefully designed two-step rolling and heat-treatment procedure, we have fabricated mono- and multifilament conductors several hundred meters in length.[4,6] A consistent J_c of 1.2 x 10^4 A/cm^2 has also been achieved in 850-m-long, and recently in 1260-m-long, 37-filament tapes at 77 K in self fields, as shown in fig.1. Short lengths from a recent 100-m-long 37-filament tape carried critical currents as high as 50 A at 77 K. The transport properties of both short and long mono- and multifilament tapes are shown in Table 1 with corresponding superconductor/Ag (sheath) ratios (superconductor fill factor).

Table 1. Summary of transport current properties of short and long mono- and multifilament Ag-clad BSCCO conductors.

Conductor	Length (m)	I_c (A)	Core J_c (A/cm^2)	Overall J_c (A/cm^2)	Fill Factor (%)
Monofilament					
Short Pressed	0.03	51	45,000	9,000	20
Short Rolled	0.03	51	29,000	7,800	27
Long Length	70	23	15,000	3,500	24
Long Length	114	20	12,000	3,200	27
Multifilament					
Long Length	20	42	21,000	6,800	32
Long Length	90	35	17,500	5,600	32
Long Length	850	16	10,500	2,500	24
Long Length	1,260	18	12,000	3,500	30

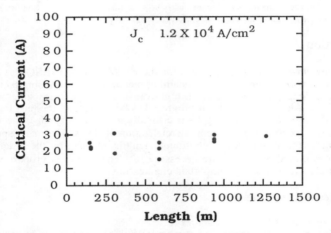

Fig. 1. Critical current vs. length of 1260-m-long 37-filament conductor at 77 K

2.2 Fabrication of prototype magnets and transformers

A test magnet fabricated by stacking 20 pancake coils, with a total conductor length of 2400 m, generated a record-high field of ≈3.2 T at 4.2 K in zero applied field. Another test magnet, generated a field of 1 T at 4.2 K and 0.6 T at 27 K, in a background field of 20 T. Total length of the conductor in this magnet was 770 m. The racetrack-wound solenoid was used to develop a 0.25 kVA high-T_c superconducting transformer that has an iron core and can operate at liquid nitrogen temperature. The primary end of the transformer was made of ≈85 m of high-T_c conductor, and the secondary end was 31 m long. Turns in the primary and secondary winding were 140 and 40, respectively.

3. Strain tolerance of Ag-clad BSCCO conductors

3.1. Mono- and multifilament conductors

Ag-clad BSCCO tapes are susceptible to damage due to axial, bending, and magnetic stresses that come into play during fabrication and operation of devices made from such tapes. Cracks induced in the superconducting core because of these stresses could lead to degradation of the superconducting properties. Therefore, the importance of understanding the strain tolerance of these tapes as a function of their transport properties cannot be overstated.[2-4]

Axial strain tolerance of mono- and 61-filament conductors (reported in Ref. 5) was evaluated by subjecting the tapes to an in-situ tensile test. We obtained bending characteristics of the conductors in-situ, with a custom-designed test fixture.[4] The irreversible strain limit (ε_{irr}) is defined as that strain beyond which a decrease in I_c is irreversible. Bending tests conducted on both mono- and 61-filament conductors at 77 K and zero applied field revealed that multifilament conductors exhibit better strain tolerance than monocore tapes.

3.2. Multilayer conductors

Strain tolerance of Ag-clad BSCCO tapes can be improved by using Ag addition, alternative sheath materials, or multifilament conductors.[4] Singh et al. showed that the strain tolerance of tapes can be improved without much loss in J_c by adding Ag to the superconductor powder.[6] In addition, as reported by several researchers,[4] the interfacial region between the BSCCO core and the Ag sheath with well-aligned BSCCO grains is believed to be responsible for high J_c. Our work on the influence of Ag showed that the strain tolerance of the tapes increased with increasing Ag (sheath) content. Thus, Ag addition should not only enhance ε_{irr}, but the increased Ag/BSCCO interfacial area should also increase the J_c.

Cracks initiated on the outer side (tensile side) of a bent tape eventually propagate through the thickness of the core to the inner side (compressive side) and impede the flow of current. Because ceramics withstand much higher compressive strains than tensile strains, if Ag reinforcement is included in the core to isolate the cracks in the tensile side of the bent tape, current can still flow unimpeded through the part that is under compression. However, it is critical to control the geometry of the additional Ag in the core because in powder or particulate form, the Ag may not provide the same reinforcement provided by a continuous filament. On the other hand, a continuous filament may produce a break in the superconducting core by fusing with the sheath.

Employing an etching technique that was developed to join Ag-clad BSCCO tapes,[4] we have fabricated tapes with a multilayer structure that incorporate Ag foil or wire in the core. The Ag sheath from one side of a pair of tapes was etched while the other side was kept intact. The exposed cores were then aligned and joined with either a single filament of 25-μm-thick Ag foil or 25-μm-thick Ag wires sandwiched between them.[4] The joined tapes were pressed and subjected to the same thermomechanical treatment that was used for regular tapes. Preliminary observations of such tapes indicate enhanced strain tolerance, as shown in fig. 2. A multilayer tape with a fill factor of 40% exhibited higher strain tolerance than did monofilament conductors with fill factors of 23, 30, and 38%; and the multilayer tape retained 90% of its initial I_c at a bend strain of 1%. Work is in progress to understand the influence of the geometry and processing of the sandwiched Ag layer on the electrical and mechanical properties of multilayer tapes.

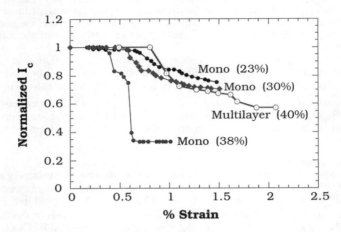

Fig. 2. Strain tolerance of Ag-clad BSCCO mono- and multilayer (foil) tapes with various superconductor fill-factors (in parentheses)

4. Summary

Long-length mono- and multifilament Ag-clad BSCCO-2223 tapes with a consistent J_c of 12000 A/cm^2 at 77 K have been fabricated. High-T_c magnets and a prototype 0.25 kVA transformer have been constructed from such long tapes. In-situ strain tolerance tests were conducted with a custom-made test fixture. Multilayer Ag/BSCCO composite tapes were fabricated by sandwiching Ag foil/wires between the cores of two tapes. Continuous reinforcement provided by the sandwiched Ag appears to enhance the strain tolerance of such tapes in spite of increased superconductor/Ag ratios.

5. Acknowledgments

Work at ANL and part of the work at IGC is supported by the U.S. Department of Energy (DOE), Energy Efficiency and Renewable Energy, as part of a DOE program to develop electric power technology, under Contract W-31-109-Eng-38.

6. References

1. K. Sato et al., *IEEE Trans. Mag.* **27**, 1231 (1991).
2. R. Flukiger et al., *Appl. Supercond.* **1**, 709 (1993).
3. J. Tenbrink et al., *IEEE Trans. Mag.* **27**, 1239 (1991).
4. R. Jammy et al., *Proc. TMS annual meeting*, 1996 (in press)
5. U. Balachandran et al., *JOM* **46**, 23 (1994).
6. J. P. Singh et al., *J. Mater. Res.* **8**, 2458 (1993).

LOCAL TEXTURE, CURRENT FLOW, AND SUPERCONDUCTIVE TRANSPORT PROPERTIES OF Tl1223 DEPOSITS ON PRACTICAL SUBSTRATES

D. K. CHRISTEN, E. D. SPECHT, A. GOYAL, Q. HE, M. PARANTHAMAN,
C. E. KLABUNDE, R. FEENSTRA, F.A. LIST, AND D. M. KROEGER,
Oak Ridge National Laboratory, Oak Ridge, TN 37831, USA

J. E. TKACZYK AND J. A. DELUCA,
General Electric Corporation R&D, Schenectady, NY 12301

Z. F. REN, C. A. WANG, AND J. H. WANG,
SUNY Buffalo, Buffalo, NY 14260

ABSTRACT

Quantitative investigations of the crystal grain orientations and electrical transport properties of high-temperature superconducting (HTS) $TlBa_2Ca_2Cu_3O_{8+x}$ (Tl1223) deposits on polycrystalline substrates show that current flow comprises percolative networks of strongly-coupled material. Superconductive transport properties on different samples, on the same samples at different widths, and on samples with artificially-induced strong flux pinning defects confirm the nature of current flow, and suggest that these materials may be useful as a new class of HTS conductors.

1. Introduction

Among the HTS material classes having three adjacent copper oxide planes, $TlBa_2Ca_2Cu_3O_{8+x}$ (Tl1223) appears most promising as a candidate for current conductors because of its high transition temperature (120 K), ease of formation, and capabilities to operate at temperatures above 40 K in substantial magnetic fields due to a relatively reduced electronic anisotropy. Although attempts to develop these materials as powder-in-tube tapes have been unsuccessful so far, initial results obtained on short segments of Tl1223 conductors, formed as deposits on appropriate substrate materials, have been encouraging.[1] For example, it has been demonstrated that thick films can be synthesized by nitrate-spray pyrolysis of $Ba_2Ca_2Cu_3O_x$ precursors on polycrystalline YSZ, followed by Tl-vapor-phase reaction in a two-zone furnace. These deposits can yield Tl1223 with critical current densities that exhibit strong field dependences similar to those of high-J_c epitaxial thin films, but with an overall reduced magnitude. The thick films have c-axis texture, with $(00l)$ x-ray rocking curve FWHM of less than 2°, but show no long-range in-plane texture using large area x-ray diffraction. However, microstructural grain-orientation mapping using electron beam backscattering patterns (EBSP) and x-ray microdiffraction has revealed the presence of local in-plane grain alignment extending over grain colonies of dimensions 100 µm to 1 mm.[2] These observations have suggested that current transport occurs through a network of percolative paths across adjacent colony boundaries. These paths select the low-angle grain boundaries of the overlapping grain orientation distributions of neighboring colonies.

This type of conduction mechanism should lead to transport properties that are characterized by parallel paths of weak-linked and well-coupled material. In substantial magnetic fields, the weak-linked component is quenched, and the properties take on field dependencies characteristic of strong material occupying an overall reduced volume fraction. In the following we describe some recent experiments in support of this view. The results provide motivation for further development of Tl1223 thick deposits for conductor applications.

2. Experiment and Results

The film precursor deposition and post-annealing conditions have been described previously.[1] Briefly, precursors were deposited on polished polycrystalline YSZ substrates by aerosol spray pyrolysis of nitrate solutions with cation content $Ba_2Ca_2Cu_3Ag_{0.37}$. Typical thicknesses are 3 µm. Tl1223 is formed by reaction of the precursor films with oxygen/Tl_2O vapor in a two-zone furnace, with the sample zone maintained at 860°C and the Tl_2O_3 source zone at ~740 °C. Reaction times are typically ~1 h, although the reaction occurs very rapidly, within 20 min. Electrical transport measurements were made on short samples, approximately 15 mm long and 8 mm wide. Electrical contact to the Tl1223 surface was provided by sputtered gold pads. For studies of sample dimension effects, the samples were photolithographically patterned to 3 mm long bridges of widths down to ~80 µm. Measurements of the critical current density were conducted in fields to 8 Tesla, and J_c was defined at a criterion of 1 µV/cm.

Figure 1 illustrates the dependence of the transport J_c on magnetic fields oriented parallel to the c axis at 77 K. Here data are shown for different Tl1223/YSZ samples, for one sample that has been patterned to progressively narrower widths, and for two fully epitaxial thin films grown on a single crystal $LaAlO_3$ substrates.[3] The logarithmic scale illustrates the common dependence of J_c on magnetic field, i.e. the data in field are similar aside from a multiplicative factor, even among different samples and despite large differences in absolute values. This observation is consistent with current flow occurring in a well-coupled component that occupies a reduced geometrical fraction of the sample. By regarding the epitaxial films with $J_c(H=0) > 1$ MA/cm² as fully-connected benchmarks, one may infer that only 5–10% of the Tl1223/YSZ material is active in a magnetic field, and for that component the $J_c(H)$ characteristics are nearly identical among all samples.

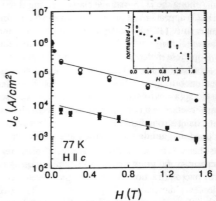

Figure 1. The field dependence of the critical current density at 77 K, for different Tl1223 samples and for a sample of different widths. The epitaxial films, deposited on single crystal $LaAlO_3$, are ~1 µm thick and represents a fully-connected benchmarks. The inset shows the same data, with J_c scaled by appropriate multiplicative factors.

Further confirmation of the strongly-coupled, percolative-path picture of current flow is provided by the effects of artificially induced flux pinning defects. In previous work, we found that Tl1223/YSZ samples irradiated parallel to the c axis with energetic heavy ions showed substantial increases in J_c and the irreversibility fields.[4] In this case, it is reasonable that the effects arise from the columnar-defect-induced flux pinning in the well-coupled component, and that this component is responsible for virtually all the loss-free current flow in substantial magnetic fields (we know of no mechanism whereby damage will raise the tunneling currents through weak-linked grain boundaries).

An as additional test of this current-path picture, we investigated the dependence of the apparent J_c on lateral sample dimensions. This effect arises because well-coupled current path options over long lengths are statistically restricted if sample widths become comparable to or smaller than the colony size. A quantitative

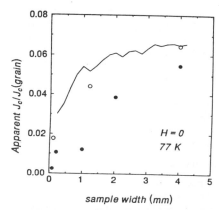

Figure 2. The dependence of the apparent J_c on the width of two Tl1223/YSZ samples. The solid curve is a numerical calculation based on a limiting path model of current flow across basal plane tilt boundaries. The calculation assumes a fully-connected J_c=1 MA/cm^2, consistent with the observed values of epitaxial films.

analysis of this phenomenon has been conducted through both experiment and numerical modeling. In the former case, the transport $J_c(H)$ of two Tl1223/YSZ samples were measured as the samples were patterned to progressively narrower bridge widths, ranging from 8 mm to 75 µm. The numerical modeling was based on a limiting-current-path analysis applied to the actual colony orientation distribution of one of the samples, measured by x-ray microdiffraction using a 100 µm diameter beam.[5] Current transfer between adjacent colonies was calculated from the observed dependence of J_c on grain misorientation as determined in the epitaxial bicrystal film experiments of Nabatame, *et al.*[6] The model describes the effect as essentially a two-dimensional phenomenon, where c-axis currents are assumed to be negligible, in contrast to the view of the "brick-wall" model. Figure 2 illustrates the ratio of observed J_c to intra-grain J_c for different sample widths, compared to the numerical predictions at zero applied field. The measurements show qualitative agreement, but fall somewhat below the numerical predictions. This effect might be explained by the relatively large size of the x-ray beam (comparable to some grain colony dimensions), which may overestimate the intra-colony J_c values within small colonies having a relatively large distribution of grain orientations.

In addition to the decline of J_c at narrow widths, its overall field dependence and the accompanying current-voltage characteristics become qualitatively different. The latter is demonstrated in Fig. 3, where the field dependence at 77 K of electric field E on current density J is compared for examples in the two width regimes: w=4 mm > colony size, and w=0.2 mm ≤ colony size. For the case w=4 mm, the $E(J)$ relations are qualitatively similar to those of the epitaxial film (negative curvature and power law at low fields, evolving toward the resistive TAFF regime at high fields). For the case w=0.2 mm, at low and moderate fields the curves exhibit a constant linear *differential* resistance as dissipation sets in at currents just above the (low) J_c values. At high fields, the behavior again follows that of intra-grain dissipation. Both these observations are consistent with a picture of parallel conductive channels, one being a well-coupled component and the other being a grain-boundary limited, weak-link channel. The differences in the overall character of Figs. 3(a) and 3(b) are simply determined by the relative fraction of each component. In Fig. 3(b), the dashed line represents a dominant weak component having a volume resistivity of $\sim 8 \times 10^{-9}$ $\Omega \cdot$cm.

3. Summary and Conclusions

We have shown that the transport properties of Tl12223 thick deposits are consistent with the observed grain orientations of the as-formed material on smooth, poly-crystalline YSZ surfaces (recently, these observations has been extended to Tl1223 deposits on polycrystalline silver). The growth characteristics, which produces c-axis perpendicular deposits, with large colonies of grains having low-angle in-plane grain boundaries, are necessary for well-coupled current paths. Studies of the dependence of $J_c(H)$ and of the voltage-current characteristics on lateral sample dimensions uphold a picture of percolative, well-coupled current paths through a fraction of the overall sample volume. Although existing J_c values are not high enough to provide practical conductor current levels, large improvements could occur through improved processing for enhanced colony formation, and provide motivation for the further development of Tl1223 thick deposits for conductor applications.

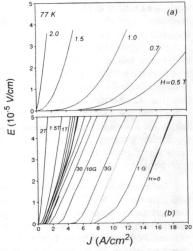

Figure 3. Dependence of electric field on current density for two sample width regimes. (a) w=4 mm > grain colony size; (b) w=0.2 mm ≤ colony size.

Research co-sponsored by the DOE Division of Materials Sciences, and by the DOE Office of Advanced Utility Concepts, Superconductivity Program for Electric Power Systems, both under Contract No. DE-AC05-84OR21400 with Lockheed Martin Energy Systems, Inc.

4. References

1. J. E. Tkaczyk, *et al.*, *Appl. Phys. Lett.* **61** (1992) 610; J. A. DeLuca, *et al.*, *Physica C* **205** (1993) 21; Q. He, *et al.*, *Appl. Phys. Lett.* **67** (1995) 294.
2. D. M. Kroeger, *et al.*, *Appl. Phys. Lett.* **64** (1994) 106; A. Goyal, *et al.*, *J. Electronic Mater.* **23** (1994) 1191; J. E. Tkaczyk, *et al.*, *J. Mater. Res.* **10** (1995) 2003.
3. C. A. Wang, *et al.*, *Physica C* **245** (1995) 171; Z. F. Ren, *et al.*, *Physica C* **258** (1996) 129.
4. J. E. Tkazcyk, *et al.*, *Appl. Phys. Lett.* **62** (1993) 3031.
5. E. D. Specht, *et al.*, *Physica C* **242** (1995) 164.
6. T. Nabatame, *et al.*, *Appl. Phys. Lett.* **65** (1994) 776.

FLUX PINNING BY TI DOPING IN $Bi_2Sr_2Ca_1Cu_2O_{8+x}$ SINGLE CRYSTALS

T.W. LI, P.H. KES.
Kamerlingh Onnes Laboratory, Leiden University; 2300RA Leiden, The Netherlands.

H.W. ZANDBERGEN
Delft University of Technology, 2628AL Delft, The Netherlands.

We demonstrate for the first time that Ti has been partially substituted in $Bi_2Sr_2Ca_1Cu_2O_{8+x}$ (Bi2212) single crystals by the traveling solvent floating zone method in concentrations of about 1 and 2at%. From high resolution electron microscopy (HREM) we discovered that a high density of planar defects is created at ac-plane by the Ti doping. These defects form effective pinning centers when the field is applied along the c–direction. In addition to dramatically changing the irreversible magnetic properties, it is also shown to significantly affect the reversible magnetic properties.

Introduction

One of the most promising high-T_c superconductor compounds from the point of view of applications is Bi2212. Unfortunately, the layered structure, high anisotropy and short superconducting coherence length enormously enhance the role of thermal fluctuations of the vortex lines, which result in a noticeable change in the mixed state and limitation of a variety of applications [1]. In this paper, we show that partial substitution of Ti in Bi2212 single crystal can create a high density of planar defects extended along the ac-plane and they form effective pinning centers when the field is applied along the c-direction. They are not only considerably modifies the irreversible proprieties in the mixed state, but also has a lager effect on superconducting fluctuation.

Experiment and Results

Single crystals were grown by a traveling solvent floating zone method from the feed rod with a nominal composition of Bi:Sr:Ca:Cu:Ti = 2.05:1.95:1.0:(2-x):x, x = 0.04 and 0.1. Cation composition of the grown crystals was determined by Electron Probe Micro-analysis to be 2.01:1.92:1:1.97:0.017 and 1.99:1.93:1:1.95:0.038, respectively The microstructure defects were studied by HREM. Measurements of the magnetization were performed using a commercial magnetometer (Quantum Design MPMS-5S),.with a small scan length of 3.0 cm. All measurements presented below have been corrected by subtracting the magnetization measured at 120K.

a). Defects Structure: The HREM images of these crystals reveal defected areas containing Ti and regions with the structure of pure Bi-2212 in which no Ti could be probed. These defects are parallel to the ac–plane. [2].

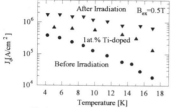

Fig.1 Comparision of critical current densities.

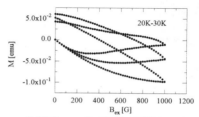

Fig.2 M-H curves of 1at.% Ti-doped Bi2212 crystal at various temperature 20K, 25K and 30K in a field parallel to the c axis.

b). Irreversible magnetic property: We compare in Fig.1 the current densities obtained from SQUID magnetization measurements on our Ti-doped sample with data obtained by D. Prost et al.[3] on as-grown Bi2212 and on Bi2212 with columnar defects created by heavy ion irradiation.. It is seen that the current densities in our Ti-doped sample are appreciably larger than in as-grown Bi2212, but not as large as in the irradiated sample.

Fig.2 shows the M-H curves of 1at.% Ti doped Bi2212 single crystal at various temperatures. In contrast to pure Bi2212 crystal, these curves do not exhibit any second peak anomalous behavior in magnetization hysteresis loops.

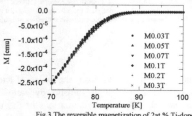
Fig.3 The reversible magnetization of 2at.% Ti-doped Bi2212 crystal vs temperature at different fields.

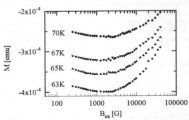
Fig.4 The reversible magnetization of 2at.% Ti-doped Bi2212 crystal vs magnetic field at different temperatures.

c). Reversible magnetic propriety: The measurements of magnetization as a function of temperature and field for 2at.% Ti doped Bi2212 crystal in different constant applied fields and constant temperatures are shown in Fig.3 and Fig.4, respectively. In Fig.3, at the low field (B<0.3T) the "crossing point" in the critical fluctuation regime is suppressed. Fig.4 shows that at low temperature, $-M$ first increases with increasing applied field, until it reaches a maximum at field about 0.3T. Above 0.3T $-M$ decreases proportionally to $\ln B$. The maximum in $-M$ at low temperature was also observed in pure Bi2212 with columnar defects.[4], which is related to the matching fields B_Φ. The maximum in $-M$ in the low field London regime has been explained by considering the decrease of the mixed state free energy due to the localization of vortices on columnar.

d).Thermal Stabilization: Annealing the crystals in air up to 850°C these defects show to be not changed, which were confirmed by HREM and magnetization measurements. On the other hand, the columnar defects created by heavy ion irradiation will be disappeared when the annealing temperature above 450°C - 500°C in air.

Conclusion

The improvements of irreversible and reversible behavior with Ti-substitution is clearly shown, which is promising for application.

This work is supported by the Netherlands Foundation FOM (ALMOS)..

Reference

[1]. G.Blatter, et al., Rev. Mod. Phys.Vol.**66**, 1125(1994)
[2] T.W.Li, H.W. Zandbergen, P.H. Kes, to be published..
[3.] D.Prost et al., Phy.Rev.B **47**, 3457(1993).
[4].C.J.van der Beek, M.Konczykowski, T.W.Li, P.H.Kes, , submitted to Phys.Rev.Lett..

GROWTH OF SUPERCONDUCTING EPITAXIAL $Tl_2Ba_2CuO_{6+\delta}$ THIN FILMS WITH TETRAGONAL LATTICE AND CONTINUOUSLY ADJUSTABLE CRITICAL TEMPERATURE

Z.F. REN, C.A. WANG, and J.H. WANG

Superconductive Materials Laboratory, New York State Institute on Superconductivity and Department of Chemistry, SUNY at Buffalo, Buffalo, NY 14260-3000

High quality epitaxial $Tl_2Ba_2CuO_{6+\delta}$ thin films on single crystal, bicrystal, and tricrystal $SrTiO_3$ substrates have been synthesized by RF magnetron sputtering followed by a two-step post-deposition annealing process. Films with critical temperatures ranging from 11 to 82 K have been obtained by annealing the films in flowing argon without noticeable change in the tetragonal structure. X-ray diffraction data show that the film is c-axis oriented epitaxially grown on the substrate.

1. Introduction

Among all the superconducting systems, $Tl_2Ba_2CuO_{6+\delta}$ (Tl-2201) is of special interest for the following reasons: 1) tunable critical temperatures (T_c) from 0 to 82 K without noticeable change in the tetragonal structure [1,2], 2) structural simplicity: tetragonal symmetry with only one CuO_2 plane in each unit cell and no CuO chain [3], which minimized the complications caused by structural anisotropy and interaction between CuO_2 planes. Therefore, $Tl_2Ba_2CuO_{6+\delta}$ can be used as a model system for basic studies of high T_c superconductivity, for example, the determination of pairing symmetry [4], and deserves further attention, especially in the form of thin film. In this paper, we report the successful synthesis and characterization of epitaxial Tl-2201 films.

2. Experimental

The 0.3 - 0.6 μm thick superconducting Tl-2201 films were made by RF magnetron sputtering and two-step post-deposition annealing. The sputtering source was prepared by pressing and sintering an intimate mixture of $Tl_2O_3 + 2BaO_2 + CuO$ preheated twice with one in-between regrinding. The precursor films were deposited on (100) $SrTiO_3$ substrates by RF magnetron sputtering at room temperature. The as-deposited films were amorphous and non-superconducting. They must be annealed to form the epitaxial film. The annealed films are always shiny and appear to be uniform with T_c of 11 - 12 K due to too much oxygen in the lattice. They must be re-annealed in flowing argon to remove the excess oxygen so that T_c can be continuously raised all the way to 83 K. Below the optimal oxygen concentration, the T_c decreases again.

3. Results and Discussion

The typical x-ray 2θ diffraction pattern, as shown in Fig. 1, shows that phase-pure Tl-2201 film is obtained, with the c-axis aligned along the direction normal to (100) of the substrate. The rocking curve, measured at (0010) of Tl-2201, yielded a full-width-at-half-maximum (FWHM) value of 0.265°, indicating excellent c-axis alignment as shown in the inset. All these XRD data suggest that the Tl-2201 films are characterized by high-quality c-axis oriented film growth.

The in-plane alignment of a Tl-2201 film on a $SrTiO_3$ substrate, measured by φ-scan of the (105) of Tl-2201 film, illustrated in Fig. 2a, with four 90°-shift sharp peaks, indicates a good in-plane alignment of the Tl-2201 film on the $SrTiO_3$ substrate. The φ-scan of the (111) of $SrTiO_3$ substrate is also measured, as shown in Fig. 2b, to determine the epitaxial relationship between the Tl-2201 film and the $SrTiO_3$ substrate. Considering the 45°-shift between the projections of (105) and (111), an epitaxy of the [100] of Tl-2201 film over the [100] of $SrTiO_3$ substrate was deduced from Figs. 2a and 2b.

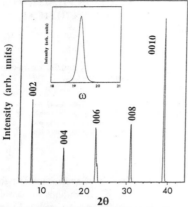

Fig.1 X-ray diffraction pattern of Tl-2201 thin film grown on (100) SrTiO₃ single crystal substrate. The rocking curve measured at (0010) is shown in the inset.

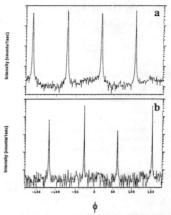

Fig.2 X-ray φ scan of the (105) of Tl-2201 film (a) and the (111) of SrTiO₃ substrate (b).

The temperature dependence of resistance measurements showed that the as-made films were superconductors with a T_c of only 11 - 12 K due to too much oxygen in the lattice (over-doped). After annealing in Ar atmosphere, the shiny films went through stages of over-, optimal-, and under-doping respectively. The optimal T_c was achieved at an annealing time depending on both annealing temperature and film thickness. A typical annealing time dependence of T_c for a film with 0.5 μm thick at the annealing temperature of 400°C was shown in Fig. 3. It is worth noting that the under-doping was observed for the first time.

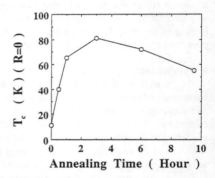

Fig.3 Annealing time dependence of T_c for films annealed at 400°C.

4. Conclusion

High quality epitaxial Tl-2201 thin films were successfully grown. T_c values ranging from 0 to 83 K depending on the doping status were obtained without noticeable change in the tetragonal structure. Under-doping was obtained for the first time in Tl-2201 superconductor.

Acknowledgment

This work was supported by N. Y. State Energy Research and Development Authority.

References

1. Y. Kubo, Y. Shimakawa, T. Manako, and H. Igarashi, *Phys. Rev.* B **43**, 7875 (1991).
2. C.A. Wang, Z.F. Ren, J.H. Wang, D.K. Petrov, M.J. Naughton, W.Y. Yu, and A. Petrou, *Physica* C, (1996).
3. R.M. Hazen, Physical Properties of High Temperature Superconductors II, p121 (by D.M. Ginsburg).
4. C.C. Tsuei, J.R. Kirtley, M Rupp, J.Z. Sun, A. Gupta, M.B. Ketchen, C.A. Wang, Z.F. Ren, J.H. Wang, and M. Bhushan, *Science* **271**, 329 (1996).

ROLE OF CONSTITUENTS ON COMPOSITE BPSCCO TAPES

S. SALIB, M. MIRONOVA and C. VIPULANANDAN
Materials Engineering Laboratory
Texas Center for Superconductivity at the University of Houston
Houston, TX 77204-4791

ABSTRACT

The irreversible strain of BPSCCO tapes was improved by incorporating metal powder in the ceramic core of the tapes. Silver powder (15-30 vol%) was mixed with the BPSCCO powder and packed into silver tubes to fabricate tapes. After reinforcement the average irreversible strain increased from 0.19% to over 0.40%. Silver flakes and silver fibers were also studied as reinforcements. Due to the processing conditions and mismatch in the coefficient of thermal expansion of the BPSCCO core and the metal reinforcements, residual stresses developed in the core. XRD was used to evaluate these residual stresses in the metal sheath. The maximum residual stress occurred after one sintering. All the residual stresses were tensile. A TEM study was conducted on the tapes to evaluate the dislocation densities in the monolithic and composite tapes. Finally a constitutive model was used to investigate the role of each component on the mechanical properties of the tape. The mechanical properties of the ceramic core were the most important parameters influencing the overall behavior of the tapes.

1. Introduction

In order for BPSCCO tapes to be used in commercial applications, their irreversible strain of approximately 0.2% must be further improved. Several methods are being investigated to improve the irreversible strain in the tapes. Some of the methods include changing the sheath material[1] or using multifilament tapes[2] instead of monocore tapes. In this study methods to improve the ceramic core and the failure mechanism for the tapes were investigated.

2. Experimental

A commercially available BPSCCO powder was used in the study. As per manufacturers data the nominal composition of the powder was $Bi_{1.7}Pb_{0.3}Sr_2Ca_2Cu_3O_x$. Processed powder was packed using a uniaxial pressure of 50 MPa in a Ag tube with an outside diameter (O.D.) of 6.35 mm and an inside diameter (I.D.) of 4.24 mm. After the powder was packed the Ag tube was plugged and then drawn and rolled to a final overall width of approximately 2 mm and a thickness of approximately 0.3 mm. After cold working the tapes were cut in small pieces about 6 cm long and sintered at 830°C for 70 hours in 7% oxygen. Then they were pressed uniaxially using a pressure of 2 GPa and sintered for a second time at 830°C for 70 hours in 7% oxygen. For the composite tapes, 15-30 vol% silver powder (99.9% pure, particle size 2-3.5 μm) was mixed with the BPSCCO powder before it was packed into the tube. The fiber composite was fabricated in such a way that the final product had aligned fibers.

In the electromechanical test, tapes were subjected to direct tension in liquid nitrogen with the critical current measured simultaneously. The specimens tested were about 6 cm long and were sintered twice and pressed once. XRD and TEM studies were used to evaluate the residual stresses and the dislocation densities respectively.

3. Results and Discussion

Silver powder improved the irreversible strain to an average of 0.47% from 0.19% (for monolithic). The reason was because the silver powder transformed into strips during the cold working process and reinforced the core uniformly. The ceramic core fractured in multiple regions during the tensile testing.

There are mainly two sources for residual stresses in the BPSCCO tapes. The first source is due to the cold working of the tape, namely drawing and rolling, and the second is due to the thermal stresses which develop in the tape while processing, due to the difference in thermal expansion coefficients of silver and BPSCCO. The maximum residual stress occurred after the first sintering cycle, it was approximately 0.8 GPa. Dislocation density in the monolithic tapes was found to be one order of magnitude higher near the tape's edges ($\approx 10^{10}$ cm^{-2}) compared to that in the middle of the same samples ($\approx 10^9$ cm^{-2}). This is due to the fact that the edges during cold working are exposed to higher stresses than the middle. This difference between the middle and the edges, is believed to be the cause for the superior current carrying ability of the edges over the middle[3]. The model used revealed that the mechanical properties of the core had the most influence on the irreversible strain of the tape, more than the mechanical properties of the sheath.

4. Conclusions

The silver powder improved the irreversible strain of the BPSCCO tapes to over 0.40%. The tapes exhibited multiple fracture during tensile testing. The residual stresses were maximum after the first sintering cycle and the dislocations were maximum in the edges of the tapes. The model showed the importance of the mechanical properties of the ceramic core.

5. Acknowledgments

This work was supported by the Texas Center for Superconductivity at the University of Houston under Prime Grant MDA 972-88-G-0002 from the Defense Advanced Research Agency and the State of Texas.

6. References

1. J. W. Ekin, S. L. Bray, N. F. Bergren and J. Tenbrink, Proceedings of the 9th U.S./Japan Workshop, High Field Materials, Kyoto, Japan, 1995.
2. K. Sato, T. Hikata, H. Mukai, M. Ueyama, N. Shibuta, T. Kato, T. Masuda, M. Nagata, K. Iwata and T. Mitsui, IEEE Transactions on Magnetics, Vol. 27, No. 2, March 1991, pp. 1231-1238.
3. D. C. Larbalestier, Y. Feng, X. Y. Cai, H. Edelman, E. E. Hellstrom, Y. H. High, J. A. Parrell, Y. S. Sung and A. Umezawa, Presented at the 7th IWCC, Alpbach, Austria, 24-27 January, 1994.

TRANSPORT CRITICAL CURRENTS OF Bi-2212 TAPES PREPARED BY SEQUENTIAL ELECTROLYTIC DEPOSITION

F. LEGENDRE, L.SCHMIRGELD-MIGNOT, P. REGNIER
Section de Recherches de Métallurgie Physique, DECM, C.E.A Saclay
F91191 Gif Sur Yvette CEDEX, FRANCE

H. SAFAR, J.Y. COULTER, M. MALEY
Superconductivity Technology Center, Los Alamos National Laboratory
Mail Stop K763, Los Alamos NM 87545, U.S.A

The transport critical currents of Bi-2212 tapes prepared by oxidation of metallic precursors have been measured as a function of temperature, between 4 and 77K, and as a function of magnetic field, for fields up to 9T applied perpendicular and parallel to the ab planes. The tapes have been prepared by a technique, developed with the aim of manufacturing long lengths, and based on the sequential electrolytic deposition of Bi, Cu, Sr, and Ca onto polycrystalline silver ribbons followed by thermal treatments. Optimisation of the deposition sequence and of the thermal treatments have lead to highly textured specimens having transport Jc values of 35 000 A/cm^2 at 77K, and of 2.6 x 10^5 A/cm^2 at 4 K, both under self field, after only 4 hours total processing time (starting from salts of the metallic constituents and not from already synthesized superconductor powders). Values of the order of 4 x 10^4 A/cm^2 remain under magnetic fields of the order of 10T at 4 K.

The technique of tape preparation consists in the sequential electrolytic deposition of Bi, Sr, Ca and Cu onto 50 μm thick polycrystalline silver tapes followed by short thermal treatments in the interval 800-870°C [1,2], and has been optimised according to results from microstructural and electrical characterisations. In its present state the whole process takes only 4 hours. Transport critical current measurements have been performed with the samples immersed in liquid He (T=4K), liquid neon (T=26.2K) liquid nitrogen (T=75K) and pumped liquid nitrogen (T=64K) using the standard 1μV/cm criterion. The magnetic field was applied perpendicular to the ab planes; only in Fig.2 comparative results with H//ab are presented.

Our Jc values are highly reproducible; they are among the best values reported in the literature for Bi-2212 tapes: between 30000 and 35000 A/cm^2 at 77K, and of 2.6 x 10^5 A/cm^2 at 4 K, both under self field. The behaviour under applied magnetic fields is also typical of high quality tapes (Fig. 1 & 2); even for H//c of the order of 10T, Jc's greater than 4 x 10^4 A/cm^2 persist at 4 K. The microstructural characterisation shows that there are good prospects for further progress, as the presence of large sized secondary phases (Fig. 3) implies that only a thin layer of the tape is at work for current transport. The remarkably similar behaviour under magnetic fields (Fig. 4) of the Jc for two tapes, one having the normal 1 hour and the other a 16 hours final treatment points to the fact that the reaction time is long enough and that improvements are not to be expected from prolonging processing times.

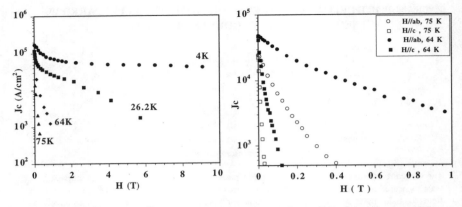

Fig.1.- Transport Jc (A/cm^2) as a function of H //c at different temperatures

Fig.2.- Comparison of the behaviour of the transport Jc for H // c and for H // ab at 75K (open symbols) and 64K (filled symbols).

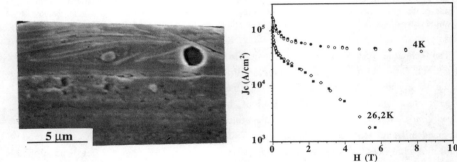

Fig. 3.- SEM micrograph of the cross-section of a sample with Jc = 30 000 A/cm^2 (77K) showing how the presence of secondary phase precipitates disturbs the texture of the central layers of the tape

Fig.4.- Jc as a function of H //c for two slices of the same sample processed with different final annealing times; filled symbols 1h, open symbols 16h

Acknowledgements

The financial support from Electricité de France (EDF) is gratefully acknowledged

References

1. Ph. Gendre, *Thesis, Univ Paris VI*, April 1994
2. Ph. Gendre, L. Schmirgeld, P. Régnier, F. Legendre, S. Sénoussi and A. Marquet, *Appl. Supercond.* **2**, 7 (1995)

MONOLITHIC TERMINATIONS FOR MULTIFILAMENTARY BSCCO WIRES

Y. K. TAO
Teco Electric & Machinery Co. Ltd., Taiwan, R. O. C.

C. H. Kao and M. K. Wu
Materials Science Center, National Tsing Hua University, Taiwan, R. O. C.

ABSTRACT

In this work, monolithic terminations at the ends of multifilamentary Ag-sheathed BSCCO wires have been prepared. Preliminary results show that the transition from the multifilamentary section to the monolithic termination is structurally smooth, and the critical current in the transition zone is basically the same as in the wires.

For some superconductivity applications, it is necessary to have superconducting joints between superconducting wires in order to form persistent current loops. Following the discovery of Bi-based high temperature superconductors, there has been a substantial progress in the fabrication of superconducting Ag-sheathed BSCCO wires by the powder in tube process. Meter long samples exhibiting critical current densities of the order of $10^4 A/cm^2$ at 77K and self-fields have been reported [1]. By using a large area to form the connection after removal of the Ag sheath, single core Ag-sheathed BSCCO wires have been jointed with no degradation in the critical current [2]. However, the technique developed is not quite successful in jointing multifilamentary BSCCO wires. A 30-50% decrease in critical currents measured over the joints was reported [2]. The differences in junction performances reflect the difficult in forming good superconducting joints between multifilamentary BSCCO wires. In this study, we investigated one possible solution to this problem and that is to prepare monolithic terminations at the ends of the multifilamentary BSCCO wires [3].

Powder of BSCCO precursor was prepared from stoichiometric Bi(Pb): Sr: Ca: Cu = 2: 2: 2: 3 oxides and carbonates by solid state reaction. X-ray diffraction pattern indicates that Bi-2212 is the major phase in the BSCCO precursor. Single core Ag-sheathed BSCCO wires were prepared by the powder in tube process. After the wires were drawn to 1-1.5 mm in diameter, 4-8 single core BSCCO wires were filled into a 6 mm OD 4 mm ID Ag tube. The multifilamentary rod was swaged to a diameter ~3.5 mm, then a short piece of the rod was cut off to get a straight end surface. On the cut end of the rod, a 2 cm deep, 2.4 mm diameter hole was drilled. BSCCO powder was filled into the hole to about 1 cm in length and slightly packed by a pestle. The hole was swaged sealed. After an intermediate annealing at 800°C for 10 minutes to releases the stress built up due to the cold deformation, the rod was swaged to ~2 mm diameter, then further drawn to ~1 mm diameter, and finally rolled to a 0.3 mm thick tape. The tapes were treated at 835°C for 50 hours. The critical current was determined by the four probe technique using 1 micro volt/ cm standard. To check the junction morphology, the Ag sheath was removed chemically by $NH_4OH + H_2O_2$ solution and the BSCCO was examined by scanning electron microscopy.

The steps to make the monolithic termination are the same as for the powder in tube process to make the mono-filament Ag-sheathed BSCCO wires. The drilling process makes the end of the wire a tube. The following deformation process produces the densification and preferred orientation necessary for a high critical current. The SEM micrograph (figure 1) shows that the structure at the transition changes smoothly from well-separated BSCCO filaments to a bundle of filaments with barely distinguishable boundaries, and finally to a single BSCCO core. The critical currents measured at 77K and self field at the multifilamentary section and over the transition zone (figure 2) are basically the same. Further work to characterize the properties of the monolithic terminations and the use of the monolithic termination to form the connection between two multifilamentary BSCCO wires are in process.

Acknowledgements
This work is funded by the National Science Council of R. O. C. under grant No. NSC84-2212-M-007-005PH and Teco Electric & Machinery Co. Ltd.

Reference
1. For example, R. S. Sokolowski, V. Selvamanickam, D. W. Hazelton, L. R. Motowidlo, M. S. Walker, and P. Halder, *JOM* 47, no. 8 (1995) 61.
2. J. Fujikami, K. Sato, N. Shibuta, H. Mukai, and T. Kato, *United States Patent* P/N 5, 358,929.
3. S. J. Waldman and M. O. Hoenig, *Advances in Cryogenic Engineering* 26 (1979) 608.

Figure 1: SEM micrograph of an 8 filament BSCCO tape with a monolithic termination.

Figure 2: Critical currents at 77K and self field measured at the multifilament section (filled circles) and over the transition zone (open squares).

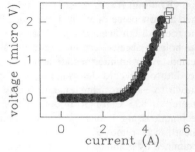

PARTIAL-MELT PROCESSING OF BULK MgO-WHISKER REINFORCED $(Bi,Pb)_2Sr_2Ca_2Cu_3O_{10-x}$ SUPERCONDUCTOR

J.S. SCHÖN and S.S. WANG
Texas Center for Superconductivity and Department of Mechanical Engineering
University of Houston, Houston, TX 77204, U.S.A.

For potential applications, weak mechanical properties associated with monolithic HTS materials need to be improved to sustain long-term thermomechanical loading, while retaining satisfactory superconducting properties. In our laboratory, a comprehensive study of processing the BPSCCO HTS material with high-strength/high-modulus MgO whiskers has been conducted. The $(MgO)_w$/BPSCCO composites has been shown to possess a promising combination of superconducting and mechanical properties suitable for bulk HTS applications. In this paper, a partial-melt processing method is developed to improve further the performance of the $(MgO)_w$/BPSCCO composite. Relationships among the bulk processing parameters, associated microstructures and grain texturing are summarized.

1. Introduction

Monolithic high-temperature superconductors (HTS) are recognized to have weak mechanical properties. In order to sustain thermomechanical stresses and deformations resulting from coupling/interactions among thermal, electrical and magnetic fields, the HTS materials need to be strengthened and capable of withstanding long-term thermomechanical loading, while retaining satisfactory superconducting properties. In previous studies[1-4], a MgO-whisker reinforced HTS BPSCCO composite has been successfully developed by a solid-state processing method.[1] The fabricated $(MgO)_w$/BPSCCO composite has been shown to possess excellent, combined superconducting and mechanical properties.[2-4] In this study, a new partial-melting processing method is developed to further improve overall composite properties of the $(MgO)_w$/BPSCCO HTS composite.

2. Experimental Procedure

2.1 Constituent Materials

The HTS powder used in this study was commercial available (SSC, Woodinville, WA) with 99.9% purity and a normal chemical composition of $Bi_{1.84}Pb_{0.34}Sr_{2.03}Ca_{1.90}Cu_{3.06}O_y$. The as-received powder was in an intermediate precursor form, dominated by the BSCCO-2212 phase. The MgO whiskers used in this study were provided by a supplier. The whiskers were approximately 50-600 μm in length and 0.5-5 μm in diameter.

2.2 Partial-Melting Processing of $(MgO)_w$/BPSCCO Composite

Partial-melting behavior of the monolithic BPSCCO phase was first investigated by using Differential Thermal Analysis (DTA). A broad endothermic peak was observed to begin just below 860°C, indicating the onset of liquid phase formation. Based on this result, an optimal partial-melting process was developed, and the processing variables used are shown in Fig. 1.

3. Results

The $(MgO)_w$/BPSCCO composite processed by the partial-melting method was found to achieve an excellent grain alignment. As shown in Fig. 2, the plate-like BPSCCO grains and the MgO whiskers were well aligned normal to the hot-pressing direction. In addition, high phase purity and matrix density were obtained by a selected, repeated hot-pressing and

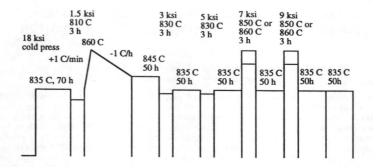

Fig. 1 Processing variables of the partial-melting process developed for $(MgO)_w$/BPSCCO composite.

annealing process. The favorable microstructure led to excellent bulk superconducting properties, as shown in Fig. 3.

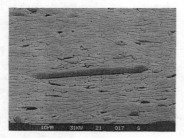

Fig. 2 SEM microstructure of $(MgO)_w$/BPSCCO fabricated by partial-melting method.

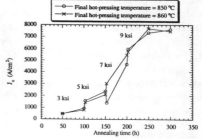

Fig. 3 J_C values of $(MgO)_w$/BPSCCO fabricated by partial-melting method.

4. Summary

(1) The partial-melting process developed in this study enhances superconducting properties of the $(MgO)_w$/BPSCCO composite, by obtaining high phase purity and density, good grain alignment and possibly additional pinning centers.

(2) The presence of the MgO whiskers is not observed to affect processing variables and parameters of the partial-melting method, due to good thermodynamic and thermomechanical compatibility with the BPSCCO HTS phase.

(3) Overall composite properties of the $(MgO)_w$/BPSCCO are correlated well with processing variables and microstructure.

(4) Thermomechanical deformations in the BPSCCO phase at high temperatures and pressures may be beneficial to superconducting properties.

Reference

1. Y.S. Yuan, M.S. Wong and S.S. Wang, *J. Mater. Res.* **11** (1996) 8-17.
2. Y.S. Yuan, M.S. Wong and S.S. Wang, *J. Mater. Res.* **11** (1996) 18-27.
3. Y.S. Yuan, M.S. Wong and S.S. Wang, *Physica C*, **250** (1995) 247-55.
4. Y.S. Yuan, M.S. Wong and S.S. Wang, *J. Mater. Res.* (1996) in press.

II. EXPERIMENTS

II. EXPERIMENTS

Symmetry

JOSEPHSON TUNNELING BETWEEN A CONVENTIONAL SUPERCONDUCTOR AND $YBa_2Cu_3O_{7-\delta}$

R.C. DYNES, A.G. SUN, A.S. KATZ, A. TRUSCOTT, S.I. WOODS, D.A. GAJEWSKI, S.H. HAN AND M.B. MAPLE

University of California, San Diego

B.W. VEAL AND C. GU

Argonne National Lab

J. CLARKE, R. KLEINER, R. SUMMER, E. DANTSKER AND B. CHEN

University of California, Berkeley

K.Char

Conductus, Inc.

ABSTRACT

We report two experiments which are designed to extend our earlier studies on Josephson tunneling between a conventional superconductor and $YBa_2Cu_3O_{7-\delta}$. The first is a study of tunneling into the a-b direction to probe the in plane coupling of the order parameter. The second is a study of the a-c Josephson effect through measurements of the microwave induced Shapiro steps in c-axis junctions. Both experiments are consistent with earlier results which conclude that while we cannot exclude a $d_{x^2-y^2}$ component to the order parameter, there is substantial 1st order coupling to a conventional s-wave superconductor.

I. Introduction

Josephson measurements are sensitive to the phase difference between two superconductors in such a way that has motivated a variety of attempts to determine whether the symmetry of the superconducting order parameter of $YBa_2Cu_3O_{7-\delta}$ (YBCO) can be revealed[1]. In one of those measurements we have attempted to probe via tunneling[2] (as opposed to weak links or SNS junctions) the overlap of the wavefunctions of a conventional superconductor (Pb, Sn, etc.) with the high T_c superconductor YBCO. For tunneling strictly in the c-axis direction, it is expected that if the symmetry of YBCO is purely $d_{x^2-y^2}$, there should be no Josephson current as the overlap between s and

$d_{x^2-y^2}$ is zero. A Josephson current was indeed observed[2] and while not as large as the theoretical value for simple s-wave overlap, the currents (or more correctly the product I_cR, I_c = critical Josephson current, R = normal state junction resistance) were sizable and suggest a strong s component in the order parameter of YBCO.

We have extended these measurements on several fronts. Here we report two of those investigations. The first is an extension to tunneling into the a-b direction[3] and the second is a quantitative study of the ac Josephson effect in c-axis junctions[4]. As a result of our original studies and these continuing investigations, we continue to conclude that while we cannot and do not exclude the possibility of a component of the order parameter with $d_{x^2-y^2}$ symmetry in YBCO, these measurements indicate a substantial component which couples to a conventional s symmetry superconductor.

II. Tunneling Into the a-b planes

A substantial part of the motivation for these studies[3] of tunneling into the a-b direction is to investigate possible differences from our previous observations of c-axis tunneling[2]. The single crystals investigated were grown at Argonne National Lab and had typical edge thicknesses of .3-.5mm. The a-b junctions were fabricated using techniques similar to those used for c-axis junctions[2]. A schematic of the geometry of our junctions is illustrated in Figure 1.

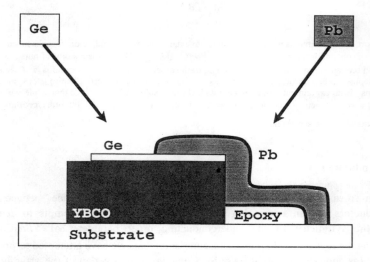

Figure 1: A schematic of the geometry used for tunneling into the a-b direction of single crystals of YBCO.

To date, we have fabricated 10 such junctions and they show low leakage, good quasiparticle tunnel characteristics and a Josephson current unless the junction resistances are too high. From a cursory glance the characteristics look quite similar to our c-axis results but upon more careful examination, the $I_c(B)$ patterns do not show the same simple Fraumhofer patterns we see in the c-axis junctions. A wide variety of behavior is observed from junction to junction and we show the $I_c(B)$ pattern for one of those in Figure 2. A detailed examination shows fine structure around zero field and a pattern which neither shows the classic $\left(Sin \pi\phi/\phi_o / \pi\phi/\phi_o \right)$ nor is really symmetric in B. We interpret this fine structure in terms of a non-uniform current distribution, the model for which is shown in the insert of Figure 2.

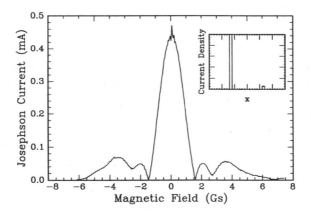

Figure 2: $I_c(B)$ for one of the a-b junctions studied. Note the fine structure around zero field which implies a non-uniform current distribution and "squid-like" behavior. Careful investigation shows this curve is not symmetric in B.

We find flux trapping problems in these a-b junctions are responsible for the lack of real symmetry in B and our efforts to eliminate the flux trapping have not been successful. We believe that flux is very easily trapped in the a-b direction and these junctions are sensitive to that trapping. This trapped flux can produce behavior which mimics some of the expected for π junctions[1].

In spite of these difficulties, we can conclude the following from these studies:

1) The amount of Josephson coupling for a-b tunneling into a conventional superconductor is comparable to that from our earlier c-axis measurements. This is measured by the I_cR product determinations and these are of the same order of magnitude for a-b tunneling as for c-axis tunneling.

2) From $I_c(B)$ studies we can obtain an estimate of the penetration depth for λ_c and obtain a value on the order of 1 micron.

3) Trapped flux in the plane is a serious problem and we see its effects in these tunnel junctions.

III. ac Josephson Effect

We have measured the microwave response of c-axis YBCO/Pb Josephson tunnel junctions. Part of the original motivation for this study was the notion that our original c-axis Josephson tunneling might be due to higher order tunneling processes[5]. Weak links and SNS junctions can easily have higher order terms on the order of ~10%. While conventional wisdom argues against this in tunnel junctions, an investigation of the Shapiro steps results in a very simple answer to this question. Straightforward considerations conclude that higher order processes will result in constant voltage Shapiro steps at voltages $nhf/4e$ where n is an integer, f is the frequency, as opposed to the conventional case $nhf/2e$. Thus a second order effect is predicted to lead to microwave induced steps at one half the voltage interval of a first order effect.

Experiments have been performed both at UC Berkeley and UC San Diego on junctions fabricated on untwinned single crystals, twinned single crystals and thin films of YBCO. The results in all three cases are similar and "typical results" are shown in Figure 3. Here we show the critical current as well as the height of the first and second Shapiro steps vs. applied microwave power. The frequency is $f = 4.56$ GHz and T = 4.2K. The solid lines are Bessel functions J_0, J_1 and J_2. Within resolution of a few percent, we see no harmonics expected for higher order processes and so conclude again that the Josephson tunneling is a result of first order coupling between Pb and YBCO. While this experiment again does not exclude a $d_{x^2-y^2}$ component, it gives strong evidence for a component which couples in first order, to a conventional superconductor.

IV. Summary

We report results of two experiments to further investigate the properties of our Josephson tunnel junctions between YBCO and a conventional superconductor (Pb). We have successfully fabricated tunnel junctions on the "sides" of single crystals and have studied the characteristics for tunneling into the a-b direction. These studies show similar I_cR products as our c-axis studies but point to difficulties of trapped flux in the a-b planes. In studies of the ac Josephson effect we have concluded that the c-axis Josephson coupling is first order and that the s component of the order parameter is sizable. While neither experiment excludes a $d_{x^2-y^2}$ component, they both are confirmation of our earlier conclusion that there is a substantial component of the order parameter in YBCO which couples to a conventional s symmetry superconductor.

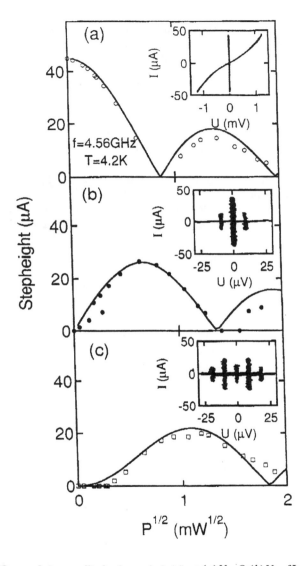

Figure 3: Measured step amplitudes (open circles) I_n at $(a) V=0, (b) V=f\Phi_0, (c) V=2f\Phi_0$ for sample 1 at 4.2K and 11.5 Ghz. Solid lines in (a) - (c) are computed for first-order tunneling $(\alpha=0)$, fitted at first maximum of I_1, dotted lines are for $s\pm d$ with $\alpha=0.2$, and dashed lines are for $s\pm d$ with $\alpha=0.2$. (d) shows amplitude of steps at $V=\pm\frac{1}{2}f\Phi_0$ with the same notation and $\alpha=0.2$. Arrow indicates value of $V_m/f\Phi_0$ at which these steps have their first maximum.

This work has been supported by AFOSR Grant No. F4962092 J007, the NSF Grant No. DMR 91-13631, the DOE, Basic Energy Sciences Contract Numbers DE AC03-76SF00098 and W-31-109-ENG-38 and DEFG0386ER45230 and the NSF-STC Contract NO. DMR 91-20000.

References:

[1] D.J. Van Harlingen, Rev. Mod. Phys. **67** (2), 515 (1995).
[2] A.G. Sun, D.A. Gajewski, M.B. Maple and R.C. Dynes, Phys. Rev. Lett. **72**, 2267 (1994).
[3] A.G. Sun, A. Truscott, A.S. Katz, R.C. Dynes, B.W. Veal and C. Gu, On the direction of tunneling in Pb/I/YBCO tunnel junctions, submitted to Phys. Rev. B.
[4] R. Kleiner, A.S. Katz, A.G. Sun, R. Summer, D.A. Gajewski, S.H. Kan, S.I. Woods, E. Danker, B. Chen, K. Char, M.B. Maple, R.C. Dynes and J. Clarke, Phys. Rev. Lett., in press.
[5] Y. Tanaka, Phys. Rev. Lett. **72**, 3871 (1994).

DETERMINATION OF THE PAIRING STATE OF HIGH-T_C SUPERCONDUCTORS THROUGH MEASUREMENTS OF THE TRANSVERSE MAGNETIC MOMENT

J. BUAN, BRANKO P. STOJKOVIC, A. BHATTACHARYA, I. ZUTIC, NATHAN ISRAELOFF, A. M. GOLDMAN, D. GRUPP, C. C. HUANG, and ORIOL T. VALLS

School of Physics and Astronomy, University of Minnesota
116 Church St. SE, Minneapolis
MN 55455, USA

The transverse magnetic moment of a well-characterized single crystal of LuBaCuO of has been studied in the nonlinear Meissner regime. The expected signature of d-wave pairing, a Fourier component of the transverse moment with angular period $\pi/2$ exhibited a dependence on magnetic field which disagreed with the predicted functional form. The results are consistent with isotropic pairing, but do not rule out nodeless anisotropic pairing.

1. Introduction

An issue related to the mechanism underlying high-temperature superconductivity is the identification of the symmetry of the pairing state. This has attracted considerable interest in the past decade because knowledge of this symmetry may serve to narrow the choice of potential mechanisms. The presence of d-wave, rather than s-wave symmetry would argue for antiferromagnetic spin fluctuations mediating the electron-electron interaction rather than phonons. In d-wave pairing the superconducting gap would have nodes resulting in quasiparticle excitations down to zero temperature. These should be detectable in a number of experiments, and indeed the results of several studies, some of which have been reported at this meeting, have been interpreted in such a manner. A d-wave pairing state would also affect the phase of the superconducting order parameter. A number of Josephson junction experiments in several geometries have been interpreted in this context. On the other hand, there are additional results, including this one, which do not support the d-wave picture.

Here we report on the angular-dependent transverse magnetization in the nonlinear Meissner regime. The samples which were studied have been measured extensively using other techniques. Those results will be reported elsewhere.[1]

2. Angular Dependence of the Transverse Magnetic Moment in the Nonlinear Meissner Regime

The supercurrent response to a magnetic field will deviate from linearity at low temperatures when the supercurrent velocity becomes comparable to a critical velocity $v_c = \Delta(\mathbf{k})/v_F$, where v_F is the Fermi velocity and $\Delta(\mathbf{k})$ is the energy gap. When this critical velocity is reached there will be a quasiparticle current persisting to zero temperature, which reduces the supercurrent response. An earlier experiment probing this nonlinear Meissner effect in terms of the field-dependent penetration depth in the Meissner state was carried out on twinned YBCO crystals. The results were interpreted with the framework of an s-wave order parameter[2]. A more recent penetration depth experiment in the nonlinear regime[3] has been interpreted in terms of d-wave, but the parameter values required seem unreasonable.

An angular dependence of the energy gap in k-space will result in an angle-dependent supercurrent in the a-b plane and a transverse magnetic moment.[4] If the order parameter exhibits d-wave symmetry, the transverse magnetic moment will have four-fold symmetry with respect to rotation about the c-axis, while it will vanish in the isotropic case. The calculation can be generalized to the cases of anisotropic s-wave or $s+id$-wave pairing in which there can be four-fold symmetric responses, but with reduced amplitude.[5] The nonlinear regime is defined by a crossover temperature or a crossover field. For the particular $LuBa_2Cu_3O_{7-d}$ crystal the crossover field has been estimated as 10G at a temperature of a few K. The crossover field and temperature are approximately linearly related to each other.

The transverse magnetic moment was measured in the nonlinear regime at fixed magnetic field as a function of temperature, and at fixed temperature as a function of magnetic field using a modified superconducting susceptometer.[6] The crystal was zero-field cooled to the temperature at which the data was to be taken. Then, the field was ramped to the starting value (for the runs at fixed temperature this was 50 G). Magnetization data was then acquired as a function of angle in the basal plane with the crystal rotated about its c-axis. After each measurement, the sample was rotated by 1.44 degrees. A total of 501 data points were taken in every data set, resulting in two full rotations of the sample. After this, the field was increased to the next value, or the temperature was changed with the field being fixed.

Transverse magnetic moment data, at 2 K consisted of five runs starting at 50 G, ending at 250 G, with the field incremented by 50 G for each data set. To search for a signal with a $\pi/2$ periodicity, a Fast Fourier Transform (FFT) analysis was undertaken. The coefficients corresponding to periodicities of π and 2π, due to demagnetization effects and trapped flux, respectively, were found to be much larger than any of the others. The $\pi/2$ component was found in a continuum with other higher order components, suggesting that it arises in about the same manner as the others, i.e., possibly as a harmonic of the dominant 2π and π components.

To confirm this and to further analyze the results, we note that the trapped flux and demagnetization factor signals and their harmonics are proportional to the field H, while the putative signal from the nonlinear Meissner effect is quadratic in H at low temperatures.[4,5] To extract the nonlinear part the $\pi/2$ component, therefore, we fit our experimental result for this component to the sum of a linear and a quadratic term in H.

As expected, the dominant signal is linear. The remaining signal (i.e., the π/2 signal in absolute value, minus the linear part in the field) is an upper bound to any experimental signal corresponding to nonlinear effects, and it is plotted as the symbols in Fig. 1. It is

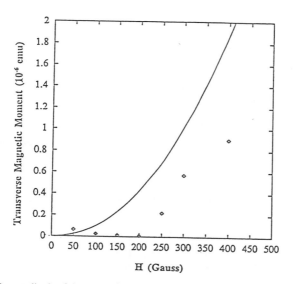

Fig. 1 Fourier amplitude of the magnetic moment at angular period π/2, as a function of magnetic field at a temperature of 2K. The solid line is a numerical evaluation based on the theory described in Ref. 5.

an upper bound because it is derived from the absolute value of the Fourier component: there is an assumption in this analysis that the phase of the signal is fully consistent with theory.

Theoretical values displayed in Fig. 1 were computed using material parameters appropriate to $YBa_2Cu_3O_{7-d}$,[7] using very conservative assumptions. The value of the penetration depth was taken to be equal to λ_{ab}, without including *any* corrections from the much larger λ_c corresponding to return currents in the c-direction, which would have led to much larger estimates.[6] Thus, even though the results in Fig. 1 are upper bounds and the theoretical estimates lower bounds, one still finds that the experimental signal is too small to arise from nonlinear effects. Finally, we find that the phase, found with the real and imaginary parts of the FFT analysis, of the signal at π/2 is consistent with the signal itself being generated as a higher harmonic of the diamagnetic response.

3. Summary

In summary, the angular dependence of the in-plane, off-axis magnetization of the single crystal of $LuBa_2Cu_3O_{7-d}$ was measured. The amplitude of the Fourier component with angular period $\pi/2$ is found to be smaller than expectations for a *d-wave* pairing state. Furthermore, the field and temperature dependencies, as well as the phase, of the signal are consistent with it being generated as a higher harmonic of signals at angular periods π and 2π. As a consequence, the conclusion is that no nodes are being observed in the superconducting energy gap. The results are consistent with either isotropic pairing, nodeless *s-wave* anisotropic pairing, or $s + id$ pairing in which a lower limit on the magnitude of the *s-wave* component of the gap is at least 10% of the maximum gap.[5] The next generation of measurements of the type described above will include several innovations designed to both reduce the background signal from demagnetizing effects and reduce noise associated with position measurement errors. First disk-shape single crystal samples are being used. These are prepared by machining rectangular single crystals using an excimer laser. Second, a positive angular position readout of the crystal orientation is being employed to eliminate noise associated with positioning errors. Finally a theoretical analysis taking into account detailed shape effects is being used to analyze the data.

4. Acknowledgments

This work was supported by the Air Force Office of Scientific Research under Grants AF/F49620-93-1-0076 and AF/F49620-96-1-0043

5. References

1. J. Buan, *et al.*, submitted to Physical Review B.
2. S. Sridhar, Dong-Ho Wu, and W. L. Kennedy, Phys. Rev. Lett. **63**, 1873 (1989).
3. A. Maeda, Y. Iino, T. Hanaguri, N. Motohira, K. Kishio, and T. Fukase, Phys. Rev. Lett. **74**, 1202 (1995).
4. S. K. Yip and J. A. Sauls, Phys. Rev. Lett. **69**, 2264 (1992); D. Xu, S. K. Yip, and J. A. Sauls, Phys. Rev. B **51**, 16233 (1995).
5. B. Stojkovic and O. T. Valls, Phys. Rev. B**51**, 6049 (1995).
6. J. Buan, B. Stojkovic, N. E. Israeloff, C. C.Huang, A. M. Goldman, O. T. Valls, J-Z. Liu, and R. Shelton, Phys. Rev. Lett. **72**, 2632 (1994).
7. D. R. Harshman and A. P. Millis, Phys. Rev. B **45**, 11631 (1993); A. Kapitulnik, M. R. Beasley, C. Castellani, and C. DiCastro, Phys. Rev. B**37**, 537 (1988); C. Strombos, doctoral dissertation, Ecole Normale Superieure, Paris, 1993, unpublished.

Microwave Measurements of the Penetration Depth in High T_c Single Crystals

W.N. Hardy, S. Kamal, Ruixing Liang, D.A. Bonn
Dept. of Physics and Astronomy, Univ. of British Columbia
Vancouver, B.C. V6T 1Z1, Canada

C.C. Homes, D.N. Basov, T. Timusk
Dept. of Physics and Astronomy, McMaster Univ.
Hamilton, ON L8S 4M1, Canada

This paper summarizes the penetration depth part of our efforts to determine the intrinsic electrodynamics of YBCO for the a, b, and c directions, and as a function of oxygen content. The temperature dependence of $\lambda_i(T)$, i=a,b,c, is obtained by precision microwave cavity perturbation measurements on untwinned single crystals; the zero temperature values, $\lambda_i(0)$, are obtained by far infrared reflectivity methods. For oxygen concentrations x = 6.60, 6.95, and 6.99, $1/\lambda_i^2(T)$, which is proportional to the superfluid density, shows a strong linear component at low temperatures for the a and b directions, whereas the c direction shows a much flatter dependence, even for overdoped samples (x=6.99). Near T_c strong fluctuation effects are observed for all 3 directions. In the a and b directions the results fit easily into the d-wave picture of superconductivity for a two-dimensional system. The significance of the unusual c-axis results will be discussed.

1 Introduction

Measurements of low frequency electrodynamics have played an important role in helping to sort out the nature of the high T_c superconducting state. This is due in part to difficulties encountered with many of the traditional tools, but also to the relative simplicity of local electrodynamics, which are ensured by the short coherence lengths and long penetration depths of the high T_c materials. On the other hand, many of the low temperature properties are quite sensitive to impurities, lattice defects etc; only in the last year or so has a consensus begun to emerge with regard to the intrinsic electrodynamics. Our group at UBC has concentrated on measuremnt of the microwave properties of $YBa_2Cu_3O_x$ single crystals, with a range of oxygen content and impurity substitutions such as Ni, Zn, and Ca. In this paper we collect our past results for the penetration depth tensor for the pure system, and present new results for the slightly overdoped system ($O_{6.99}$).

2 Experimental

The crystals were grown by a flux method in yttrium stabilized zirconia crucibles.[1] From a wide variety of measurements (c_p at low and high temperatures, ρ, photoemission, muon spin resonance, microwave loss etc.) we conclude they have a very high degree of crystallinity, with sharply defined superconducting transitions. Samples used for microwave and far infrared work had thicknesses from 7 to 30 μm, were cleaved to rectangular shapes of dimensions $\sim 1mm \times 1mm$, and then mechanically detwinned.

Values of $\Delta\lambda(T) = \lambda(T) - \lambda(1.3K)$ were obtained from the frequency change of a superconducting loop-gap resonator ($f_0 \simeq 900$ MHz) as the sample temperature is changed.[2] The sample is mounted on a thin sapphire plate in such a way that its temperature could be varied from 1.2 K to 120 K with no measureable perturbation on the temperature of the measuring cavity. Typically the resolution achieved was better than 0.2 Å in $\Delta\lambda$. In the thicker samples, the data was corrected for thermal expansion of the YBCO itself. For the data presented here, the microwave field H_1 was always applied parallel to the ab plane, in either the a or b direction. Demagnetizing effects were essentially negligible, given the thin platelet shape of the crystals.

For the field in the b direction the frequency shift is given by $\Delta f = K[a\Delta\lambda_a + c\Delta\lambda_c]$ where λ_a is the penetration depth for currents running in the a-direction, a is the width in the a-direction etc. (this expression is valid for $\lambda_a \ll c$ and $\lambda_c \ll a$). Generally the contribution from $\Delta\lambda_c$ is quite small. However, by cleaving the crystal into a number of long plates (in this case along b), the contribution of $\Delta\lambda_c$ can be multiplied up by the number of plates. In this way $\Delta\lambda_c(T)$ can be extracted from measurements of $\Delta f(T)$ before and after cleaving. This is a robust procedure with an important check: a measurement of $\Delta f(T)$ with H_1 *perpendicular* to the cleaves (but still in the ab plane); in this case $\Delta f(T)$ should, and is observed to be, the same before and after the sample is cleaved. This agreement ensures we have not lost any of the sample in the cleaving process and also that no measureable damage to the sample has occurred.

3   Results for $\Delta\lambda(T)$ and $\lambda^2(0)/\lambda^2(T)$

Figure 1 shows the low temperature region of $\Delta\lambda(T)$ in $O_{6.60}$, $O_{6.95}$, and $O_{6.99}$ crystals, for all three crystallographic directions. It is clear that $\Delta\lambda_a$ and $\Delta\lambda_b$ have a strong, nearly linear component to the temperature dependence for underdoped, optimally doped, and slightly overdoped crystals.

Figure 2 shows $\lambda^2(0)/\lambda^2(T)$ obtained by extrapolating $\Delta\lambda(T)$ to T=0 and using the IR determined values for $\lambda(0)$ [see Table 1].[3,4] Exceptions are the values for the overdoped crystal where the IR measurements have not yet been done for the a and b directions. Following the scenario developed by Tallon et al,[5] based on μSR measurements of ceramic samples, we have assumed $\lambda_a(0)$ to be essentially the same as for the optimally doped sample, and then used the value of λ_a/λ_b given by Tallon et al for $O_{6.99}$. Future adjustments of these numbers are likely to be minor, and will not change the overall picture. The value of $\lambda_c(0)$ for the $O_{6.99}$ crystal came from our own microwave measurements. In the low temperature region of Figure 2 the linear temperature dependence in the a and b directions is again clear. Also, the relatively flatter temperature dependence of $\lambda_c^2(0)/\lambda_c^2(T)$ is very obvious, especially for the overdoped sample. Finally, Figure 3 shows $\lambda^2(0)/\lambda^2(T)$ for the a and c directions plotted as a function of T/T_c so that the temperature dependencies for the three doping levels can be more easily compared.

For the optimally doped ($O_{6.95}$) sample these results can be compared to our earlier results in the literature for twinned and untwinned samples which showed strong linear terms for λ_{plane}.[2,6,7] Whereas in the past there had been many reports of T^2 or exponentially activated behaviour for λ_{ab}, recent measurements on high quality single crystal[8] and thin film samples[9] now have confirmed the existence and magnitude of the linear term in $\lambda^2(0)/\lambda^2(T)$. There have also been other measurements on the 2-layer compounds Tl2212[10]

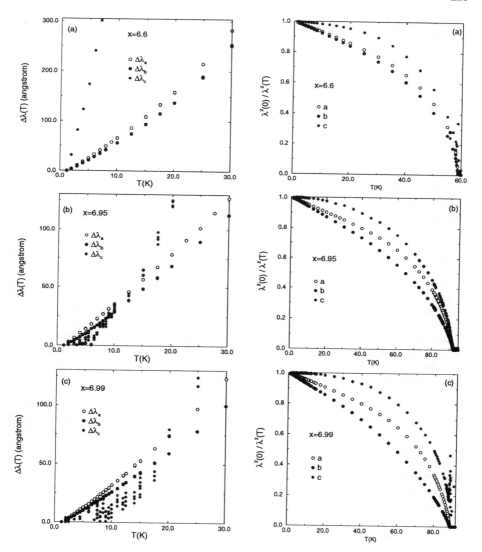

Figure 1: $\Delta\lambda(T) = \lambda(T) - \lambda(1.3K)$ versus T for the a, b, and c directions.

Figure 2: $\lambda^2(0)/\lambda^2(T)$ versus T for the a, b, and c directions. See table 1 for values of $\lambda(0)$ used.

Table 1: Values of $\lambda(0)$ used to generate $\lambda^2(0)/\lambda^2(T)$ from measurements of $\Delta\lambda(T)$. The values with asterisks (*) are not measured, but are instead estimated as described in the text. The value denoted with a dagger (†) was measured directly at microwave frequencies, rather than using a value derived from IR measurements.

x	T_c [K]	$\lambda_a(0)$ [Å]	$\lambda_b(0)$ [Å]	$\lambda_c(0)$ [Å]
6.60	59	2100	1600	65,000
6.95	93.2	1600	1030	11,000
6.99	89	1600*	800*	11,000†

and Bi2212 [11] which show behaviour very similar to that of $YBa_2Cu_3O_{7-\delta}$ crystals. Most recently, Lee et al have made a remarkable set of measurements on Bi2212 which show that the electrodynamics in the ab direction of this compound are very similar to YBCO, both for the penetration depth *and* the surface resistance. [12]

Mao et al [8] have reported results for $\lambda_c(T)$ in $YBa_2Cu_3O_{7-\delta}$ which are in serious disagreement with the UBC results. In particular Mao et al found $\lambda_c^2(0)/\lambda_c^2(T)$ to fall twice as fast as $\lambda_{ab}^2(0)/\lambda_{ab}^2(T)$ away from T=0. We believe our result for the much *flatter* variation is the correct one, given the robustness of the method with its internal checks. Jacobs et al [11] report $\Delta\lambda_c(T)$ results for Bi2212 which show a linear term, in contrast to our observations for YBCO. It seems counterintuitive that the most three dimensional of the cuprates has $\lambda_c(T)$ so different from $\lambda_{ab}(T)$. It is possible that YBCO, with its chains, has different c-axis electrodynamics in the superconducting state than other cuprates. In particular, well formed chains may act as a "filter" whereby the presence of 45° nodes on the planes does not affect the c-axis transport ie. the one dimensional chains have essentially a single gap and only interact with pairs having $\vec{k}$ along $\hat{b}$ which is also a fully gapped direction. [14]

We have previously reported [13] that for a wide region (~10 K) below T_c in optimally doped YBCO, $1/\lambda^2(T)$, which is proportional to the superfluid density, does not have mean field behaviour, but varies as $[(T-T_c)/T_c]^{2/3}$. The exponent $\frac{2}{3}$ is close to the exponent ν expected to govern the superfluid density in the 3D XY model, the critical behaviour expected for a 3D superfluid with a 2 component order parameter. Our more recent results on untwinned crystals ($O_{6.60}$ and $O_{6.95}$) and the latest results ($O_{6.99}$) explicitly show that non-mean field critical behaviour is observed for all three components of λ, as would be required for 3D fluctuations. We note however, that for $O_{6.60}$ the data needs to be improved and that for $O_{6.99}$ the critical region is narrower, of order 3 K.

4 Discussion

Measurements of λ_a, λ_b, and λ_c for underdoped, optimally doped, and slightly overdoped YBCO have revealed two surprising results, in addition to the initial surprise of the strong linear term. First, $\lambda_a^2(0)/\lambda_a^2(T)$ shows almost *no* dependence on doping, when plotted versus T/T_c. Second, $\lambda_c^2(0)/\lambda_c^2(T)$ is *much* flatter than its in-plane counterparts and also shows

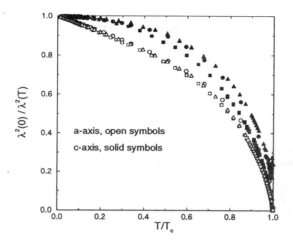

Figure 3: Dependence of $\lambda^2(0)/\lambda^2(T)$ for the a and c directions, plotted versus T/T_c. Circles: $O_{6.60}$; squares: $O_{6.95}$; triangles: $O_{6.99}$

relatively little dependence on doping, even though $\lambda_c(0)$ for $O_{6.60}$ is more than 5 times as large as that for $O_{6.95}$. The linear variation of the in-plane components of $1/\lambda^2$ at low T fits easily into the d-wave scenario for the superconducting groundstate, and its presence in both the a and b directions and its relative insensitivity to the oxygen doping level seem to rule out alternative explanations based on normal layers, phase fluctuations etc. However, the apparent insensitivity of $\lambda_c^2(0)/\lambda_c^2(T)$ to doping seems an interesting puzzle, even for d-wave theories. The relatively weak temperature dependence of this quantity at low temperatures needs to be explained, although this may turn out to be a special property of materials with 1D chains.

References

1. Ruixing Liang et al, Physica C **195**, 51 (1992).
2. W.N. Hardy et al, Phys. Rev. Lett. **70**, 3999 (1993).
3. C.C. Homes et al, Physics C **254**, 265 (1995).
4. D.N. Basov et al, Phys. Rev. Lett. **74**, 598 (1995).
5. J.L. Tallon et al, Phys. Rev. Lett. **74**, 1008 (1995).
6. Kuan Zhang et al, Phys. Rev. **73**, 2484 (1994).
7. D.A. Bonn et al, J. Phys. Chem. Solids **56**, 1941 (1995).
8. Mao et al, Phys. Rev. B **51**, 3316 (1995).
9. L.A. deVaulchier et al, Europhys. Lett. **33**, 153 (1996).
10. Zhengxiang Ma, Ph.D. Thesis, Stanford University, 1995.
11. T. Jacobs et al, Phys. Rev. Lett. **75**, 4516 (1995).
12. Shih-Fu Lee et al, preprint.
13. S. Kamal et al, Phys. Rev. B **73**, 1845 (1994).
14. The authors are indebted to D.J. Scalapino for pointing out this possibility.

SPECIFIC HEAT OF HIGH TEMPERATURE SUPERCONDUCTORS IN HIGH MAGNETIC FIELDS

A. JUNOD, M. ROULIN, B. REVAZ, J.-Y GENOUD, G. TRISCONE and A. MIRMELSTEIN
*Department of Physics of Condensed Matter, University of Geneva,
24 quai E.-Ansermet, CH-1211 Geneva 4, Switzerland*

ABSTRACT

The mixed-state specific heat $C(T,H)$ of numerous type-II superconductors was investigated in fields up to ≥ 14 teslas. High-T_c cuprates are compared with classic superconductors with small coherence length and/or large anisotropy. The behaviour of the quantities $\partial(C/T)/\partial H$, $\partial(C/T)/\partial T$ and $M(H,T)$ at the superconducting transition of HTS is classified as 3D (e.g. Y-123) or 2D (e.g. Bi-2212), depending on the critical exponents. No contribution from d-wave vortex states is observed in the entropy.

1. Introduction

Progress acomplished during the last few years in the precision of specific heat experiments ($\approx 0.01\%$), in the available magnetic fields (≥ 14 T), and in the quality of the samples allowed a more detailed study of the mixed state specific heat of type-II superconductors. Some "anomalous" properties of HTS are already present in classic superconductors. Most results summarized here are detailed and illustrated in Ref. 1.

2. Specific heat of classic type-II superconductors in high fields

$Nb_{77}Zr_{23}$ ($T_c = 10.8$ K) can be considered as a reference type-II classic superconductor. The field causes a shift of the transition toward lower temperatures, and an increase of the electronic specific heat in the superconducting state. The bulk measurement of $T_{c2}(H)$ by closely spaced specific heat experiments from 0 to $\mu_0 H_{c2}(0) = 7.9$ T shows a perfect WHH [2] dependence. The systematic broadening of the initially sharp transition, which simulates the effect of fluctuations, results from local variations of the mean free path in the dirty limit. The mixed-state specific heat $C_m(H)/T \equiv [C(H) - C(0)]/T$ is given by $\approx 1.3 \gamma H/H_{c2}(0)$, in close agreement with theory[3], almost up to $T_{c2}(H)$. This contribution is understood as the specific heat of the non-superconducting electrons in the vortex cores.

$PbMo_6S_8$ ($T_c = 13.9$ K) is an almost isotropic superconductor with a coherence length $\xi \cong 25$ Å that is only a factor of two larger than those of typical HTS. Its mixed-state specific heat is anomalous: rather than being constant and equal to $\approx \gamma H/H_{c2}(0)$, $C_m(H)/T$ tends to zero almost linearly in T below $T_c/2$. This indicates that the electron states localized in the vortex cores do not contribute, owing to the relatively large energy ($\propto \Delta/k_F\xi$) of the first level. The power law dependence of $C_m(T)$ can be explained either by a recent (s-wave) theory[4] stressing the importance of gapless points at the Fermi surface in high fields, or phenomenologically by the London model (see section 3).

2H-NbSe$_2$ (T$_c$=7.1 K) is a classic layered superconductor with an anisotropy that is about half that of optimally doped YBa$_2$Cu$_3$O$_7$ (Y-123). Its specific heat is interesting in relation with the announcement[5] that the contribution of isolated d-wave vortex states C$_m$(T,H)$\propto$TH$^{1/2}$ can be observed in Y-123: the mixed-state specific heat of NbSe$_2$ follows approximately the same TH$^{1/2}$ law in spite of the fact that it is a s-wave superconductor. Therefore a TH$^{1/2}$ law does not generally imply d-wave pairing.

The smallness of the coherence length of PbMo$_6$S$_8$ and the anisotropy of NbSe$_2$ are surpassed by those of HTS. It is then no wonder that the same anomalies persist in HTS. The specific heat of isolated vortex cores is not observed, and the mixed-state specific heat is found to be non-linear in H. A satisfactory phenomenological description can be given by the London model, which considers only magnetic interactions between vortices.

3. Specific heat in the London model

$$M(T,H) = -\frac{\Phi_0}{32\pi^2\lambda^2}\ln\frac{\eta H_{c2}}{eH}, \quad H_{c1} \ll H \ll H_{c2}, \quad \frac{\eta}{e} \approx 0.4 \quad (1)$$

is the London magnetization in the mixed state[3]. Assuming an empirical dependence for the penetration depth given by $\lambda^{-2}(T)=\lambda_0^{-2}[1-(T/T_c)^n]$, $2\leq n\leq 4$, and neglecting the relatively slow variation of $\ln\xi^{-2}(T)$ with respect to that of $\lambda^{-2}(T)$ (the magnetization is well described by $\xi(T)\approx$constant outside of the fluctuation domain[6]), one obtains:

$$S(T,H) - S(T,0) = \int_0^H \frac{\partial M(T,H')}{\partial T}dH' \cong \frac{n\Phi_0}{32\pi^2\lambda_0^2 T_c^n}T^{n-1}H\ln\frac{\eta H_{c2}}{H} = \int_0^T \frac{C_m(T',H)}{T'}dT \quad (2)$$

This result implies C$_m$(T,H)$\propto$T^{n-1}HlnH at low and intermediate temperatures.

4. Specific heat of HTS in high fields at low temperature

The d-wave term announced[5] in Y-123 is based on a fit of the lattice, magnetic and electron specific heat. Can such a term be observed without having to rely on a fitting procedure? A first possibility consists in studying the anisotropic component C$_{total}$(T,H//c)−C$_{total}$(T,H$\perp$c), in which isotropic background terms cancel out. In our measurements of Y-123, the anisotropic component, essentially C$_m$(T,H//c), is found to be non-linear in T, decaying to zero within experimental error in the vicinity of 10 K. Furthermore we do not observe the expected increase $\propto$H$^{1/2}$ at low temperatures (1-2 K) and high fields (8-14 T) where the data should be most significant because the Schottky and lattice contributions are small. The anisotropic component may originate from undamped vortex modes. A second possibility consists in measuring the mixed-state entropy S$_m$(T,H//c), which is obtained directly by integration of C$_{total}$(T,H//c)−C$_{total}$(T,H=0) from $\approx$100 K downwards. The third law of thermodynamics allows an extrapolation down to S$_m$=0 at T=0. We compared the result with (a) the entropy of the s-wave *isolated* vortex cores, S$_m\approx\gamma$TH/H$_{c2}$(0), (b) d-wave isolated vortices[7],

$S_m \approx 1.6\gamma T[H/H_{c2}(0)]^{\frac{1}{2}}$, and (c) magnetically interacting vortices in the London model, $S_m = f(\lambda_0) T^{n-1} H \ln(\eta H_{c2}/H)$. This test was conducted for two Y-123 crystals and one ceramics in fields from 0.25 to 14 teslas. In all cases we found that scenario (a) does not correctly describe the non-linear field dependence of S_m; d-wave scenario (b) overestimates the entropy by a factor of typically 5-10 at $T_c/2$; London model (c) describes the data quantitatively, with $n \approx 2.3$ to 3, $\mu_0 H_{c2} \approx 175$ to 250 T and $\lambda(0) \approx 1700$ Å.

The electronic specific heat of $Bi_2Sr_2CaCu_2O_8$ (Bi-2212) is smaller, and such a procedure is less conclusive. But nature provides another route: unlike Y-123, Bi-2212 is reversible over an extended temperature range, and the quantity $\partial S/\partial T$ can be measured by its thermodynamic equivalent $\partial M/\partial H$. We can again compared the measurements with scenarios (a), (b) and (c). The logarithmic field dependence of $\partial M/\partial H$ favours model (c), rejects scenario (a), and may be compatible with (b) at high fields only.

In conclusion, the mixed-state entropy of Y-123 and Bi-2212 in the mid-T_c range is rather incompatible with the contribution of isolated d-wave vortex states. The field and temperature dependence of the macroscopic quantities M and S_m can be quantitatively described by magnetic interactions between vortices and a suitable power-law dependence for $\lambda^{-2}(T)$, which may result e.g. from strong-coupling effects.

5. Specific heat of HTS in high fields near T_c

We measured systematically the magnetization up to 5.5 teslas and the specific heat up to 14 teslas of numerous ceramics[1] (Y-123, Bi-2212, Bi-2223, Tl-2201, Tl-1223, CLBCO-123 i.e. $(Ca_{0.4}La_{0.6})_1(Ba_{0.68}La_{0.32})_2Cu_3O_{7+x}$, etc.), large Y-123 and Bi-2212 single crystals obtained by TSFZM, and small Y-123 crystals grown in inert $BaZrO_3$ crucibles. On the whole, HTS can be classified as "3D" (Y-123-like) and "2D" (Bi-2212-like) on the basis of the shape of their superconducting transition and its variation with the field.

The transition of the "2D" Bi-2212 family most probably reflects intrinsic HTS, i.e. superconducting planes with vanishing interlayer coupling. It is a third-order transition characterized by a discontinuous change of the slope of the specific heat at T_c. At the same temperature (within experimental accuracy), the magnetization M(T,H) becomes independent of the field over up to four decades of field. This crossing point follows from the general scaling property for the singular part of the free energy[8,9],

$$-F_{\text{singular}} = \xi^{-d} \varphi(H\xi^2/\Phi_0), \quad \xi(t \equiv T/T_c - 1) = \xi_0 |t|^{-\nu} \quad (3)$$

without any reference to vortex fluctuations or to the LLL approximation[10]: $M = -\partial F/\partial H$ does not depend on the field at t=0 in dimension d=2, whatever the value of ν. The critical temperature T_c does not appear to be a function of the field. The differences C(T,H)−C(T,0) or $\partial C/\partial H$ are particularly instructive in this respect: they remain centered on $T_c(0)$. The magnetization does not obey convincingly the scaling properties predicted by the 2D-LLL model[10], but rather those obtained from Eq. (3) with $\nu=3/2$, i.e. $M(T,H) = f_1(t/H^{1/3})$ where f_1 is a universal function. It follows that $\partial(C/T)/\partial H = H^{-2/3} f_2(t/H^{1/3})$, where f_2 is another universal function. The hyperscaling relation $\alpha = 2 - \nu d$ finally implies $\alpha = -1$, i.e. the zero field specific heat "diverges" as $C_{el} \propto \text{constant}-|t|$ at T_c. These scaling laws are experimentally verified. The scaling function $f_2(x)$ is close to a lorentzian.

The transition of the Y-123 family ("3D") is modified with respect to that of Bi-2212 by the presence of interplanar coupling. The shape of the zero-field specific heat anomaly can be reconstructed from the elementary lorentzian function f_2 found for Bi-2212 by using two modifications: (1) a smaller critical exponent, $v=2/3$ rather than 3/2, and (2) a non-constant $T_{c2}(H)$ (note that pushing to the limit $v=0$ and $T_{c2}(H)=T_c\cdot(1-H/H_{c2})^{\frac{1}{2}}$ would lead to the mean-field, two-fluid model[1]). The zero-field transition reconstructed by integrating $\partial(C/T)/\partial H$ from H_{c2} to 0 appears to contain both a 3D mean-field component ("jump") and a quasi logarithmic 3D-XY divergence, consistently with the hyperscaling relation $\alpha=2-vd\approx 0$. Practically, almost perfect fits of the total experimental specific heat at T_c in zero field (at the 0.02% accuracy level, outside of $T_c\pm 0.5$ K) can be obtained by including a few Einstein modes for the lattice, and both a mean-field step and a logarithmic divergence for the electron specific heat. The magnetization no longer has a true crossing point, but follows the scaling law $M=H^{\frac{1}{2}}f_3(t/H^{3/4})$. The measured specific heat in a field obeys the scaling laws $\partial(C/T)/\partial H=H^{-1}f_4(t/H^{3/4})$ and $\partial(C_{el})/\partial H=H^{-3/4}f_5(t/H^{3/4})$. These exponents and amplitudes are consistent with Eq. (3). Gaussian fluctuations and the 3D-LLL approximation in its original form[10] are less satisfactory.

The crossover from the "2D" to the "3D" families is illustrated by the CLBCO system, a "123" structure with disordered chains. The underdoped compound with $T_c=64$ K is "2D" whereas the optimally doped compound with $T_c\cong 80$ K is "3D"[11].

Acknowledgments

Discussions with Prof. J. Muller are gratefully acknowledged. This work was supported by the Fonds National Suisse de la Recherche Scientifique.

References

1. A. Junod, in *Studies of High-Temperature Superconductors Vol. 18*, ed. A.V. Narlikar (Nova Science Publishers, New York, in press).
2. N.R. Werthamer, E. Helfand and P.C. Hohenberg, Phys. Rev. **147**, 295 (1966).
3. P.-G. de Gennes, *Superconductivity of Metals and Alloys*, trad. P.A. Pincus (Benjamin, New York, 1966); C. Caroli *et al*, Phys. Lett. **9**, 307 (1964).
4. S. Dukan and Z. Tesanovic, Phys. Rev. B **49**, 13017 (1994).
5. K.A. Moler *et al*, Phys. Rev. Lett. **73**, 2744 (1994).
6. J.H. Cho, Z. Hao and D.C. Johnston, Phys. Rev. B **46**, 8679 (1992).
7. G.E. Volovik, JETP Lett. **58**, 469 (1993); K. Maki, private communication.
8. M.B. Salamon *et al*, Phys. Rev. B **38**, 885 (1988).
9. D.S. Fisher, M.P.A. Fisher and D.A. Huse, Phys. Rev. B **43**, 130 (1991).
10. Z. Tesanovic *et al*, Phys. Rev. Lett. **69**, 3563 (1992).
11. M. Roulin and A. Junod, Physica C in press.

MAGNETIC FLUX DISTRIBUTION IN GRAIN AND TWIN BOUNDARIES IN YBCO

K.A. MOLER*, J.R. KIRTLEY
IBM T.J. Watson Research Center, Yorktown Heights, NY 10598, USA
Present address: Department of Physics, Princeton University, Princeton, NJ 08544, USA

J. MANNHART, H. HILGENKAMP, B. MAYER, Ch. GERBER
IBM Research Division, Zurich Research Laboratory, 8803 Rüschlikon, Switzerland

RUIXING LIANG, DOUGLAS A. BONN, WALTER N. HARDY
Department of Physics, University of British Columbia, Vancouver, BC V6T 1Z1, Canada

The observation of fractional vortices (isolated vortices with total flux which is not an integral or half-integral multiple of $h/2e$) would be evidence for an unconventional order parameter which breaks time reversal symmetry. In this paper we consider scanning SQUID microscope images of several different geometries of $YBa_2Cu_3O_{7-\delta}$ (YBCO) grain boundaries and twin boundaries in thin films and single crystals, and conclude that these experiments can be explained by a d-wave order parameter without time reversal symmetry breaking.

In conventional superconductors, magnetic flux is quantized in integer multiples of $\Phi_0 = h/2e$ as a result of the requirement that the order parameter be single valued. It is now well established that in special geometries of both $YBa_2Cu_3O_{7-\delta}$[1,2] and $Tl_2Ba_2CuO_{6+\delta}$,[3] magnetic flux is quantized in half-integer fluxoids ($\Phi = (n+1/2)\Phi_0$, where n is an integer), consistent with a sign change in the order parameter resulting from d-wave symmetry.[4] It has been pointed out that observation of fractional vortices (different from $n\Phi_0$ or $(n+1/2)\Phi_0$) in SQUID's would be an indication of a time reversal symmetry breaking order parameter.[5] With a high degree of accuracy, only integer or half-integer fluxoids have been observed in SQUID's of several different geometries.[4,6] It has also been proposed that a local time reversal symmetry breaking may appear at interfaces such as grain boundaries,[7] twin boundaries,[8] or heterojunctions.[9] In this paper we will discuss several relevant experiments, in which various geometries of $YBa_2Cu_3O_{7-\delta}$ grain boundaries and twin boundaries in thin films and single crystals have been imaged using scanning SQUID microscopy.[10]

Localised quantities of flux much smaller than Φ_0 were observed in enclosed geometry grain boundaries of YBCO,[11] as shown in Figure 1. The triangles in Figure 1a represent photolithographically defined 45 degree grain boundaries produced using seed and buffer layers to control the in-plane epitaxy of the deposited YBCO film.[12] Modelling showed that the apparently isolated regions of flux contained much less than Φ_0, but that the total flux summed over a particular triangle contained integer multiples of Φ_0 within experimental error. While the observed flux distribution was consistent with isolated fractional vortices, it was not possible to determine unambiguously that the vortices were isolated because the Josephson penetration depth, the SQUID pickup loop diameter, and the length of the grain boundaries were the same order of magnitude. These results were taken as evidence for local time reversal symmetry breaking in the grain boundaries of YBCO.[7]

An alternate explanation for these observations is that these grain boundaries are locally

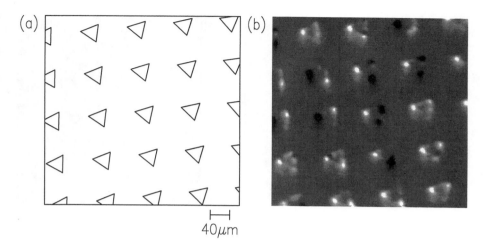

Figure 1: (a) Photolithographically produced pattern of triangular enclosed grain boundaries. (b) Scanning SQUID microscope image of magnetic flux trapped in grain boundaries. Data from reference 11.

rough. With a d-wave order parameter, this roughness results in a complicated array of 0 and π-junctions along the boundary. The grain boundaries spontaneously magnetize with flux that can have values ranging from 0 to $\Phi_0/2$, depending on the local value of $2\pi L I_c/\Phi_0$, where L is the effective inductance of a ring and I_c is the local critical current.[13,3]

This second model was tested by studying macroscopically straight grain boundaries produced by epitaxial growth on bicrystalline $SrTiO_3$ substrates. Figure 2 is a scanning SQUID microscope image of a YBCO asymmetric 45° [001] tilt boundary, cooled in a field less than 1mG, and imaged at 4.2K.[14] In such a boundary, for a d-wave superconductor, a lobe of the order parameter is normal to the boundary on one side, while a node is normal on the other. Faceting of the grain boundary produces slight variations of the grain boundary angle, rocking the local normal from the positive lobe to the negative lobe, varying the local critical current from positive to negative, producing a series of 0 and π-junctions and spontaneous magnetization. Experiments with 18.4°/18.4°, 22.5°/22.5°, and 45°/32° grain boundaries did not show this effect.

Fractional vortices have also been speculated to exist as a result of a local time reversal symmetry breaking state at twin boundaries.[8] Figure 3 shows a scanning SQUID microscope image of a sparsely twinned single YBCO crystal, cooled in a field of about 10 mG. Such images show that vortices trap preferentially at the twin boundaries, but have very uniform total flux. Figure 4 is a histogram of the maximum field intensities for vortices trapped in the central $512\mu m \times 512\mu m$ area of Figure 3. The peak near 0.5 Φ_0 threading through the SQUID sensor loop corresponds to a bulk vortex with Φ_0 total flux. Fractional vortices, whose existence would confirm the speculations of Sigrist et al.,[8] were not observed.

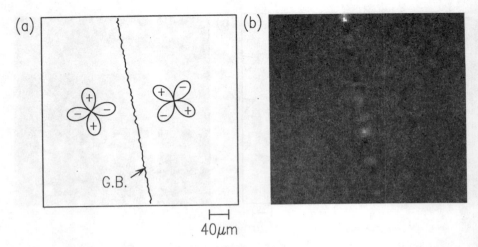

Figure 2: (a) Schematic of an asymmetric 45° [001] bicrystal grain boundary, showing a tetragonal d-wave order parameter locked to the crystal axes. (b) Scanning SQUID microscope image of apparently spontaneous magnetic flux distribution.

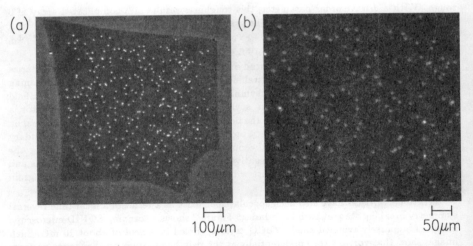

Figure 3: (a) Scanning SQUID microscope image of Abrikosov vortices trapped in sparsely twinned single crystal of YBCO. (b) High-resolution image of the central region of the crystal.

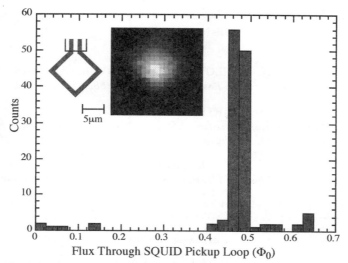

Figure 4: Histograms of observed maximum flux peaks from Figure 3b. The peak at about $0.5\Phi_0$ corresponds to Φ_0 flux in the vortex. Insets: sketch of the pickup loop, and a single vortex from Figure 3b.

In conclusion we find that the apparent existence of fluxoids with values different from $N\Phi_0$ or $(N+1/2)\Phi_0$ can be understood in terms of d-wave symmetry and rough grain boundaries, without the need to invoke time reversal symmetry breaking.

We thank M. Ketchen, M. Bhushan, and A. Ellis for their contributions to the development of the scanning SQUID microscope, and M. Sigrist, C.C. Tsuei, and R.B. Laughlin for stimulating discussions.

References

1. D.A. Wollman *et al, Phys. Rev. Lett.* **71**, 2134 (1993).
2. C.C. Tsuei *et al, Phys. Rev. Lett.* **73**, 593 (1994).
3. C.C. Tsuei *et al, Science* **272**, 329 (1996).
4. J.R. Kirtley *et al, Nature* **373**, 225 (1995).
5. M.R. Beasley, D. Lew, and R.B. Laughlin *Phys. Rev.* B **49**, 12330 (1994).
6. A. Mathai *et al, Phys. Rev. Lett.* **74**, 4523 (1995).
7. M. Sigrist, D.B. Bailey, R.B. Laughlin, *Phys. Rev. Lett.* **74**, 3249 (1995).
8. M. Sigrist, K. Kuboki, P.A. Lee, A.J. Millis, T.M. Rice, to appear in *Phys. Rev. B*.
9. S. Bahcall, *preprint*.
10. J.R. Kirtley *et al. Appl. Phys. Lett.* **66**, 1138 (1995).
11. J.R. Kirtley *et al. Phys. Rev.* B **51**, 12057 (1995).
12. K. Char *et al, Appl. Phys. Lett.* **59**, 733 (1991).
13. Manfred Sigrist and T.M. Rice, *J. Phys. Soc. Japan* **61**, 4283 (1992).
14. J. Mannhart *et al.* submitted to *Phys. Rev. Lett.*.

Single Grain Boundary Josephson Junctions—New Insights from Basic Experiments

J. MANNHART, H. HILGENKAMP, B. MAYER, Ch. GERBER
IBM Research Division, Zurich Research Laboratory, 8803 Rüschlikon, Switzerland

J. KIRTLEY, K.A. MOLER*
IBM T. J. Watson Research Center, P.O. Box 218, Yorktown Heights, NY 10598
*Present address: Dept. of Physics, Princeton University, Princeton, NJ 08536

M. SIGRIST
Theoretische Physik, ETH-Hönggerberg, 8093 Zürich, Switzerland

We report highly anomalous critical current vs. magnetic field dependencies of grain boundaries in $YBa_2Cu_3O_{7-x}$. Direct imaging with scanning SQUID microscopy provides evidence of magnetic flux generated by single grain boundaries. Conventional Josephson junction models cannot explain these effects if a superconducting order parameter with a pure s-wave symmetry is assumed. The results have significant implications for our understanding of the properties of grain boundaries in high-T_c superconductors and for their applications.

Single grain boundaries are widely used as high-quality Josephson junctions in a variety of device applications and for experiments to investigate fundamental properties of the high-T_c cuprates, such as the symmetry of their superconducting order parameter.[1-3] In contrast to the textbook behavior of the commonly used 24° grain boundary, characteristic features of large-angle grain boundaries are highly anomalous and cannot be explained by conventional Josephson junction properties and by an order parameter with s-wave symmetry. If a $d_{x^2-y^2}$ wave component of the order parameter and faceting of the grain boundary are taken into consideration, these unusual properties can be accounted for remarkably well, however. Based on these considerations, it was expected that grain boundaries will spontaneously generate magnetic flux.[4,5] We report evidence of this self-generated flux obtained by scanning *SQUID* microscopy (SSM).

The samples used for the experiments were c-axis-oriented films of $YBa_2Cu_3O_{7-x}$ grown by standard laser deposition on bicrystalline $SrTiO_3$ substrates to a thickness of 20–150 nm. A grain boundary we investigated extensively is the asymmetric 45° [001] tilt boundary, which is otherwise fabricated by biepitaxy. An atomic force microscopy (AFM) image of the surface of such a sample is presented in Fig. 1. This micrograph reveals the meandering of the grain boundary line. Typically, facets and continuously curved sections induced by the growth islands of the film are found on all length scales <100 nm in agreement with published reports (see e.g. Ref. 6).

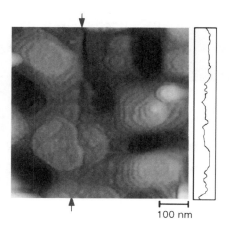

Figure 1: AFM image of the surface of an $\simeq$150-nm-thick $YBa_2Cu_3O_{7-x}$ film with an asymmetric 45° [001] tilt boundary. The location of the grain boundary is indicated by the arrows. The meandering path of the boundary has been replotted at the right of the figure.

100 nm

The asymmetric 45° boundaries display an anomalous dependence of the critical current I_c on a magnetic field H_a applied in the boundary plane.[4, 7-9] An example of such an $I_c(H_a)$ curve is shown in Fig. 2. These characteristics do not change significantly with temperature or oxygen doping.[9] A striking feature is the small critical current in zero field $I_c(0)$, which is significantly exceeded at fields $H_a^* \neq 0$. Such behavior is only consistent with standard junction theories if self-field effects or trapped flux quanta are present, which can both be ruled out in the present case.[9]

These characteristics can be remarkably well explained if a $d_{x^2-y^2}$ symmetry component of the order parameter (see e.g. Ref. 10) and the observed grain boundary faceting is taken into account.[5, 8, 9, 11] In this case, for some facets, the local current crosses the boundary in the opposite direction of I_c. These facets are called π facets because for them the grain misorientation causes a

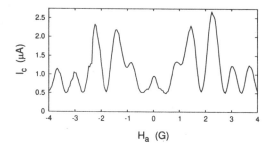

Figure 2: Critical current vs. applied magnetic field for an asymmetric 45° boundary in a 15-nm-thick $YBa_2Cu_3O_{7-x}$ film at 4.2 K, junction width: 16 μm, from Ref. 9).

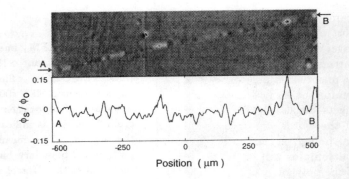

Figure 3: SSM image of a 400×400 μm^2 area including an asymmetric 45° [001] tilt $YBa_2Cu_3O_{7-x}$ bicrystal grain boundary taken at 4.2 K without an applied magnetic field after cooling the sample in a magnetically shielded environment (<1 mG). The arrows indicate the boundary. The bottom shows a profile of the data along the boundary measured in ϕ_0 penetrating the pickup loop.

shift of the phase difference of the order parameters by π.

For zero current bias, highly interesting behavior has been predicted for the grain boundaries because the phase shift of the π facets is expected to induce a Josephson current, which will generate a disordered pattern of unquantized magnetic flux at the boundary.[4, 5] To clarify whether this is indeed the case, we imaged the magnetic fields on the sample surface by scanning the pickup loop of a high-resolution SSM a few microns above the sample.[12] Figure 3 shows an SSM image of a 45° grain boundary. Several bulk vortex pairs are visible, as well as spontaneously generated flux in the boundary. This flux is randomly distributed and also changes its sign randomly, with fine-scale variations limited only by the spatial resolution (~4 μm) of the SSM. The characteristic features of this flux are consistent with expectations based on the $d_{x^2-y^2}$ wave symmetry component of the superconductor and faceting.[13]

These results have several significant implications, both for the understanding of the properties of grain boundaries in high-T_c superconductors and for their applications.

- Experimental proof has been presented for the presence of π facets at grain boundaries in $YBa_2Cu_3O_{7-x}$ films. The π facets are the most prominent for asymmetric 45° boundaries, independent of oxygen concentration and temperature. This provides further evidence of a $d_{x^2-y^2}$ symmetry component of the order parameter in $YBa_2Cu_3O_{7-x}$.

- Transport properties measured in practical grain boundary experiments reflect the averaged behavior of inhomogeneous junctions and are no direct measure of intrinsic grain boundary properties. The dependence of J_c on grain boundary misorientation, the $I(V)$ characteristic and the I_cR_n product are not known for the individual facets.
- The effects discussed provide a missing link to explain the contradicting results of the symmetry experiments performed on tricrystal rings[2] and on biepitaxial polygons.[1] Differences of I_c for the various hexagon sites in the biepitaxial samples will be suppressed.[14] If they are taken into account, the results of the polygon experiment become consistent with a $d_{x^2-y^2}$ wave order parameter.[13-15]
- The magnetic flux is a potential source of noise for grain boundary junctions and for SQUIDs. Asymmetric 45° boundaries are the worst in this respect.
- Of the three orders of magnitude drop of J_c for an increase of the grain boundary angle from 0° to 45°, about one order of magnitude may be attributed directly to a $d_{x^2-y^2}$ wave symmetry component and to faceting.[5]

We thank J.G. Bednorz, R. Berger, M. Bushan, A. Ellis, D. Ertas, P. Guéret, M. Ketchen, A.J. Millis, K.A. Müller, and T.M. Rice for useful discussions and their valuable support. This work was supported by the ESPRIT project 8132 "WELITTD-HTS", the Bundesamt für Bildung und Wissenschaft, Bern, the Consortium for Superconducting Electronics (Adv. Res. Projects Agency Contract No. MDA972-90-C-0021), and the Swiss Nationalfonds (PROFIL fellowship).

1. P. Chaudhari and S.Y. Lin, *Phys. Rev. Lett.* **72**, 1084 (1994).
2. C.C. Tsuei *et al.*, *Phys. Rev. Lett.* **73**, 593 (1994).
3. J.H. Miller *et al.*, *Phys. Rev. Lett.* **74**, 2347 (1995).
4. N.G. Chew *et al.*, *Appl. Phys. Lett.* **60**, 1516 (1992).
5. H. Hilgenkamp *et al.*, to be published, *Phys. Rev.* B.
6. J.A. Alarco *et al.*, Ultramicroscopy **51**, 239 (1993).
7. Z.G. Ivanov *et al.*, in *Proc. Beijing Int'l Conf. on High-Temperature Superconductivity*, ed. Z. Gan *et al.* (World Scientific, Singapore, 1992).
8. C.A. Copetti *et al.* Physica C **253**, 63 (1995).
9. J. Mannhart *et al.*, to be published in *Z. Phys.* B.
10. K.A. Müller, Nature **377**, 133 (1995).
11. M. Sigrist and T.M. Rice, J. Phys. Soc. Jap. **61**, 4283 (1992).
12. J.R. Kirtley *et al.*, *Appl. Phys. Lett.* **66**, 1138 (1995).
13. J. Mannhart *et al.*, submitted to *Phys. Rev. Lett.*
14. A.J. Millis, *Phys. Rev.* B **49**, 15408 (1994).
15. J.R. Kirtley *et al.*, *Phys. Rev.* B **51**, 12057 (1995).

Anomalous Energy Gap in the Normal and Superconducting State of Underdoped $Bi_2Sr_2Ca_{1-x}Dy_xCu_2O_{8+\delta}$ thin film and single crystals

Z.-X. Shen, J.M. Harris and A.G. Loeser
Department of Applied Physics and Stanford Synchrotron Radiation Laboratory
Stanford University, Stanford, CA 94305

We summarize our photoemission data from underdoped $Bi_2Sr_2Ca_{1-x}Dy_xCu_2O_{8+\delta}$ that revealed anomalous properties. First, there is an excitation gap in the normal state. The gap has very similar magnitude and momentum dependence as the superconducting gap, which is consistent with a $d_{x^2-y^2}$ order parameter. Second, the gap disappears at a temperature range much higher than T_c. Finally, the superconducting gap size in the underdoped regime does not decrease with T_c reduction. These results can not be reconciled within the mean-field BCS-Eliashberg theory.
Key Words: photoemission, high-temperature superconductivity, d-wave pairing

I. *Introduction*: This paper summarize results from various underdoped, but still superconducting single crystalline $Bi_2Sr_2Ca_{1-x}Dy_xCu_2O_{8+\delta}$ (Bi2212) thin film and crystal samples. [1-3] Due to space limitations, we refer the readers to our other publications for details of the experiments.[1-3]

The solid line in Fig.1 depicts the Fermi surface (FS) of slightly overdoped Bi2212 that is commonly observed by several groups.[4-7] The FS (dark line) can be described as a large hole pocket centered at the (π, π) point. In a rigid band picture, if one underdopes this material by removing holes from the sample, the FS should look something like the dashed line. Our experiments, on the other hand, found only a section of the FS, indicated by the gray arc in Fig.1. In other positions along the dashed FS, the spectral features move to higher binding energy. We interpret this behavior as a consequence of opening an energy gap in the normal state.

II. *Experimental Results* Without a theory for these superconductors, it is difficult to characterize the gap. Two experimental quantities can possibly be used to quantify the size of the gap. The first is the peak of the spectral feature and the second is the energy position of the leading edge. Since photoemission spectra can be approximated as the spectral function cut off by the Fermi function, we can consider two limiting cases. If the feature is well below the Fermi level with most of the spectral weight unaffected by the Fermi function, the centroid of the spectrum is a reasonable indication of the band's kinetic energy. On the other hand, if the feature is very close to the Fermi level, then the leading edge position of the spectra is a reasonable quantity to characterize the gap. The leading edge method has been used to characterize the superconducting gap from photoemission data,[8] yielding a gap size that is very consistent with results from other techniques. This leads us to use the leading edge method in the rest of the paper because the features are close to the Fermi level in k-space locations selected. It is important to keep in mind that this is a specific convention used to discuss the data. However, it is very important to stress here that both methods yield qualitatively the same conclusion for the existence of the gap.[1]

II.1 *The momentum dependence of the gap* Fig.2 shows the normal and superconducting spectra from an underdoped single crystalline film sample ($T_c \sim 78K$) as a function of momentum along the FS.[3] These spectra were recorded on the underlying FS which is determined by a series of cuts (see arrows in Fig.1). Only the spectra at the underlying FS are shown in Fig.2. We can clearly see that the gap is a very anisotropic function of crystal momentum. At $(\pi, 0.2\pi)$, the leading edge of both the normal and superconducting state spectra are about 20-25 meV below E_F, revealing a gap of that magnitude. At $(0.4\pi, 0.4\pi)$, on the other hand, the leading edges reach the Fermi energy both above and below T_c, reflecting the much smaller or zero energy gap along this direction. The data clearly indicates that the normal and superconducting gap have very similar magnitudes, and almost identical momentum dependence. This suggests that the normal and superconducting gap have the same origin. Furthermore, the momentum dependence of the gap is similar to what one expects from a $d_{x^2-y^2}$ order parameter: the gap has a minimum along the $(0, 0)$-(π, π) line and gets bigger as one goes away from that line. Impurity scattering may cause a region of near zero gap about the $(0, 0)$-(π, π) line, explaining why we see a section of FS .[9] We will come back to this point later.

II.2 *The temperature dependence of the gap* One should be very cautious when extending the leading-edge method to a temperature dependence study because we don't fully understand the photoemission lineshape. However, our data still allows us to draw some conclusions, with larger uncertainty than the results discussed earlier. Fig.3 presents the leading edge positions for an underdoped sample (T_c~84K) at the FS crossing points along the $(0, 0)$ to (π, π) (empty circles) and $(\pi, 0)$ to (π, π) (filled circles) directions, respectively.[2] At low temperature, the two directions differ by 20-25 meV, reflecting the excitation gap of that order of magnitude. The gap is also reflected in the fact that the leading edge at the FS along the $(\pi, 0)$ to (π, π) is below the Fermi level. This gap closes or becomes very small near 200±25 K, well above the superconducting transition temperature.

II.3. *The doping dependence of the gap* The inset of Fig.4 shows the superconducting gap, using the gap maximum (along $(\pi, 0)$ to (π, π)) minus the gap minimum (along $(0, 0)$ to (π, π)) in k, as a function of doping in the underdoped regime.[3] The quantity plotted here may have a slight advantage of canceling out part of the instrumentation broadening effect. Here the doping δ (hole concentration per planar Cu) is obtained using the empirical relation $T_c/T_{c,max} = 1 - 82.6(\delta - 0.16)^2$ [10, 11]. The predicted BCS weak-coupling d-wave result,[12] $\Delta_{SC} = 2.14kT_c$ (dashed line), shows no resemblance to the observed doping dependence. In spite of substantial error bars, it is clear that the familiar BCS mean-field result $\Delta_{SC} \propto T_c$ is violated. This indicates that the Δ_{SC} and T_c represent two distinct energy scales, in striking contrast to the mean-field picture. While the trend in the data is clear, one should be cautious about the quantitative agreement between the measured gap size and the weak coupling calculation near optimal doping. The gap size used here depends on the specific leading edge convention we used.

III. *Discussion and Conclusion* The general observation of an excitation gap in the normal state of underdoped samples is consistent with many other experiments, including NMR, transport, specific heat, and c-axis optical conductivity. These results are discussed by others at the conference. The emerging picture from our complete data set summarized here can be naturally explained by theoretical models in which pairing occurs above T_c in these underdoped, but still superconducting samples (T_c~40-84K).

The idea of pairing above T_c took many forms in the literature over the years.[13-23] It has recently been advocated by Emery and Kivelson on a phenomenological ground which is independent of a microscopic model of HTSC.[13] This idea was also proposed by Doniach and Inui based on a generalized version of the Ginzburg-Landau theory.[14] These models are consistent with the experimental finding by Uemura and his collaborators that the T_c is proportional to superfluid density in the underdoped regime.[15] Fig. 4 reproduces the phase diagram of reference 13. There are two energy scales in the underdoped regime. The first describes the mean-field temperature (T_{MF}) below which the quasi-particles form pairs, while the second describes the temperature (T_θ) below which phase coherence of the pairs occurs. The key difference between this picture and that of the mean-field BCS-Eliashberg theory is that T_{MF} and T_θ are very different, and the lower of the two determines T_c. In the mean-field theory, T_{MF} determines T_c since phase coherence is assumed in the theory, which does not consider fluctuations.

In the underdoped regime, T_θ is expected to be low because the phase stiffness depends on the superfluid density, which is low in underdoped samples. T_c, as determined by T_θ, is lower than T_{MF}, with the result that pairing occurs above T_c. Because photoemission is not a phase sensitive measurement, it detects the normal state gap which reflects the binding energy of the electron pairs. Given that the pairing above and below T_c has the same origin, it is natural that the magnitude and the momentum dependence of the gap are very similar. In this scenario, the temperature range of 200±25 K in our T_c = 84K samples may be the mean-field temperature below which the normal state pairs start to form. The superconducting gap is controlled by T_{MF}; it has nothing to do to T_c. This explains our observation that the gap does not decrease in proportion with T_c reduction. The fact that we see the normal state gap only in the underdoped regime is also consistent with this phase diagram. Hence, all aspects of our data in the underdoped regime can be naturally explained by this phase diagram.

A possible microscopic justification for the phase diagram as well as d_{x2-y2} symmetry of the pairs can be found in various extensions of Anderson's original resonating valence bond (RVB) idea.[16-21] In this picture, spin-charge separation is assumed to occur in the normal state with low energy physics characterized by collective modes of spin and charge excitations called spinons and holons. These microscopic theories predict a phase diagram that is similar to Fig.4 [17, 18] and hence can also naturally explain all aspects of our data. In particular, these theories predict that the spinons start to pair with d_{x2-y2} symmetry at a higher temperature than Tc in the underdoped regime. The holons Bose condense at T_c, resulting in a superconductor with d_{x2-y2} order parameter. This explains very nicely the d-wave like normal state as well as superconducting gap seen in our experiment. We note here that our result discussed here are from underdoped, but still superconducting (Tc > 40K) samples. The physics in overdoped regime and the more underdoped regime as well as the relation to the insulator remains to be investigated.

In conclusion, ARPES data from underdoped Bi2212 reveal evidence for a novel transition into the superconducting state that is very different from the mean-field picture which describes well the superconducting transition in conventional metals.

Acknowledgment We acknowledge D.S. Marshall, D.S. Dessau, C.-H. Park, A.Y. Matsuura, W.E. Spicer, J.N. Eckstein, I. Bozovic, P. Fournier and A. Kapitulnik for collaborations on experiments summarized here. We acknowledge Steve Kivelson for helpful suggestions. Our experiments were performed at SSRL, which is operated by the DOE Office of Basic Energy Sciences, Division of Chemical Sciences. The Office's Division of Material Science provided support for this research. The Stanford work was also supported by ONR grant N00014-95-1-0760 and NSF grant DMR-9311566. The research at Varian was supported in part by NRL contract N00014-93-C-2055 and ONR contract N00014-94-C-2011.

References:
1. D.S. Marshall et al., Phys. Rev. Lett., in press
2. A.G. Loeser et al., Physica C, 263, 208, 213 (1996); Preprint, Submitted to Science; unpublished.
3. J.M. Harris et al., preprint
4. D.S. Dessau et al.; Phys. Rev. Lett. 71, 2781 (1993))
5. P. Aebi et al, Phys. Rev. Lett. 72, 2757-60 (1994)
6. Ding et al.; Phys. Rev. Lett. 76, 1533 (1996)
7. Here we concentrate on the Fermi surface centered at (π,π) only, which is sufficient for the discussion of this paper.
8. Z.-X. Shen et al., Phys. Rev. Lett. 70, 1553 (1993)
9. M.R. Norman et al., Phys. Rev. B 52, 15107 (1995)
10. H. Won and K. Maki, Phys. Rev. B 49, 1397 (1994)
11. M.R. Presland et al., Physica C 176, 95 (1991)
12. W.A. Groen, D.M. de Leeuw, and L.F. Feiner, Physica C 165, 55 (1990)
13. V.J. Emery and S.A. Kivelson, Nature 374, 434 (1995)
14. S. Doniach and M. Inui, Phys. Rev. B 41, 6668 (1990)
15. Y.J. Uemura et al., Phys. Rev. Lett. 66, 2665 (1991)
16. N. Trivedi and M. Randeria, Phys. Rev. Lett. 75, 312 (1995)
17. P.W. Anderson, Science 235, 1196 (1987)
18. G. Kotliar and J. Liu, Phys. Rev. B 38, 5142 (88);
19. T. Tanamoto, K. Kohno, and H. Fukuyama, J. Phys. Soc. Japan 61, 1886 (1992)

20. G. Baskaran, Z. Zou and P.W. Anderson; Solid St. Comm. 63, 973 (1987).
21. A.E. Ruckenstein and P.J. Hirschfeld and J. Appel, Phys. Rev. B36, 857 (1987).
22. Xiao-Gang Wen and P.A. Lee, Phys. Rev. Lett. 76, 503 (1996)
23. B.L. Altschuler, L.B. Ioffe and A.J. Millis, Phys. Rev. B 53, 415 (1996)

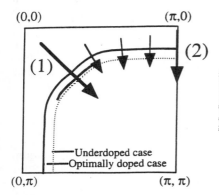

Fig.1 Two-dimensional Brillouin zone and the Fermi surfaces. All arrows indicate the cuts used for data in Fig.2.

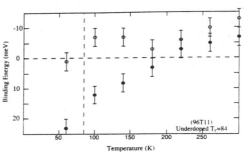

Fig.3. The temperature dependence of the leading edge for spectra taken at FS crossings along the $(\pi, 0)$ to (π, π) line (solid) and along the $(0, 0)$ to (π, π) line (empty).

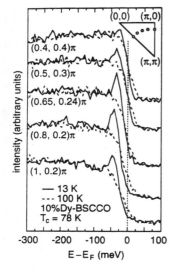

Fig.2. ARPES spectra near the Fermi energy for an underdoped single crystal thin film ($T_c \sim 78K$) in the superconducting and normal states.

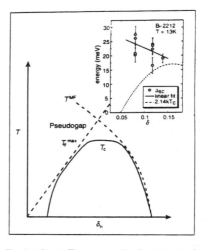

Fig.4. Inset: The superconducting state gap Δ_{SC} measured at 13 K on $Bi_2Sr_2Ca_{1-x}Dy_xCu_2O_{8+\delta}$ plotted vs. doping δ, with δ inferred from $T_c/T_{c,\,max}$ (see text). Δ_{SC} is the gap maximum minus the gap minimum in k-space. Main panel: Phenomenological phase diagram after reference 13.

INTRINSIC SIS JOSEPHSON JUNCTION IN Bi-2212 AND SYMMETRY OF COOPER PAIR

K. Tanabe, Y. Hidaka, S. Karimoto, M. Suzuki

NTT Interdisciplinary Research Laboratories, 162 Shirakata, Tokai, Ibaraki 319-11, Japan

By fabricating intrinsic junction stacks with a controlled number of CuO_2 bilayers, we have made the first observation of both pair and clear quasiparticle tunneling characteristics for underdoped Bi-2212. It is found that these intrinsic junction stacks behave as an SIS junction array with a sufficiently insulating barrier and nearly temperature-independent junction resistance R_N below T_c. Numerical calculation indicates that the observed quasiparticle tunneling characteristics are consistent with the d-wave order parameter in the CuO_2 bilayer.

1 Introduction

In spite of enormous efforts dedicated so far, superconductor-insulator-superconductor (SIS) junctions based on cuprate superconductors have not been fabricated with success. This is due to the very short coherence length of these materials and the difficulty in obtaining monolayer-level perfection at the superconductor-barrier interface. On the other hand, the highly anisotropic structural and physical properties of high-T_c cuprates have strongly suggested that a stack of SIS josephson junctions is naturally formed along the c-axis of these materials. The existence of intrinsic SIS junctions has been confirmed by the recent direct observation of dc and ac Josephson effects in small pieces of $Bi_2Sr_2CaCu_2O_{8+\delta}$ (Bi-2212) single crystals.[1] These intrinsic SIS junction stacks are of particular interest not only from the viewpoint of device applications but also as a direct probe into the magnitude and symmetry of the order parameter. Moreover, their tunneling characteristics can also provide an insight into another current issue, c-axis coherency. However, since even cleaved thin Bi-2212 single crystals contain too many (typically 1000) junctions in series, there has been much difficulty in observing their *intrinsic* tunneling characteristics.

Recently, we have successfully observed both pair and clear quasiparticle tunneling characteristics in intrinsic junction stacks with much less number of CuO_2 bilayers fabricated on underdoped Bi-2212 single crystals.[2] In this paper, we will review our recent study on Bi-2212 intrinsic junction stacks.

2 Fabrication and Properties of Intrinsic Junction Stacks

Intrinsic junction stacks were fabricated using standard photolithographic and ion milling techniques on Bi-2212 single crystals grown by a self-flux method. The details of the fabrication procedures were described elsewhere.[2] The schematic cross-sectional view of the stack is shown in Fig. 1(a). The lateral size of the stack was typically $20 \times 20 \mu m^2$. The junction thickness d and the number of included CuO_2 bilayers N were varied typically between 60 nm and 300 nm and between 40 and 200, respectively, by adjusting the milling time. The temperature dependence of ρ_c measured with a three-probe configuration for a 300 nm thick junction stack is also shown in Fig. 1(a). A negative temperature dependence above T_c and

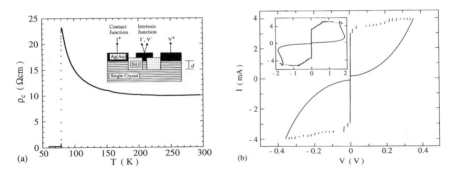

Figure 1: (a) Schematic cross-sectional view and $\rho_c - T$ curve, (b) $I - V$ characteristics at 11 K of a 300 nm thick Bi-2212 intrinsic junction stack.

a ρ_c value at room temperature of approximately 10 Ωcm clearly indicate that the crystals are in the underdoped regime. The contact resistance was as small as 0.1 - 1 Ω and thus has no significant influence on measurements of the current-voltage $(I - V)$ characteristics.

Figure 1(b) shows the $I - V$ curve for a 300 nm thick junction stack. For a bias current just above the critical current I_c, the junction stack exhibits a periodic voltage jump with a period V_j of approximately 20 mV and the switching between multiple quasiparticle branches. These features are typically observed in a series-array or stack of many SIS junctions with a substantial distribution of I_c. For a larger bias current, even larger hysteresis is observed. At a characteristic voltage V_c of approximately 1.8 V, each SIS junction contained in the stack is considered to be switched to the voltage state. This value is much smaller than the product of V_j and the number of SIS junctions in the stack $N = 200$.

This significant suppression of V_c which suggests the reduction in the gap size has already been reported in a junction stack of similar or larger thickness.[1] We have shown that this reduction in the gap size can be explained primarily by nonequilibrium superconductivity induced by quasiparticle injection.[2] Since the effective quasiparticle recombination time is dominated by the phonon escape time in a stacked junction array, the reduction in V_c which reflects the degree of nonequilibrium effect is expected to be approximately proportional to the thickness d of the junction stack.

Figure 2(a) shows the $I - V$ curve for a 60 nm thick junction stack. Approximately 40 voltage jumps with a V_j of 20 mV can be observed, explicitly indicating that SIS junctions are stacked with a period of 1.5 nm and thus each CuO_2 bilayer acts as an S electrode. The V_c value of 0.8 V coincides with $V_j N$, indicating that the effect of nonequilibrium superconductivity is negligible for this current range. Figure 2(b) shows the $I - V$ curve extended to a higher voltage range for the same junction stack. A clear gap-like structure is observed at a voltage V_g of 1.35 V which is much larger than V_c. At higher voltages, a linear part of the quasiparticle tunneling curve is visible, which gives the the normal tunneling resistance for each junction R_N of approximately 2.2 Ω. It was found that all the quasiparticle $I - V$ curves at different temperatures coalesce into this linear line, implying

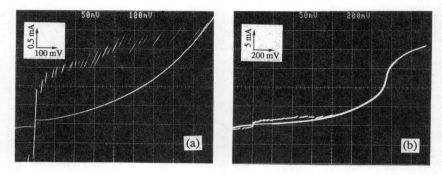

Figure 2: $I-V$ characteristics of a 60 nm thick junction stack at 4.2 K.

that R_N is nearly temperature independent.

The observed $I-V$ curve has two noticeable features quite different from those expected for SIS junctions based on conventional s-wave superconductors. These are the substantial subgap conductance and the reduced I_cR_N product of 4 - 5 mV for each junction which is less than 1/5 of the BCS value $\pi\Delta/2e$. The explanation based on extrinsic origin for these features may be tunneling via nonideal tunnel barrier with substantial localized states or a semiconducting nature. Underdoped Bi-2212 and YBa$_2$Cu$_3$O$_{7-\delta}$ exhibit a substantial increase in ρ_c below typically 200 K which is attributed to, for example, the incoherent hopping via impurity scattering,[3] phonon-assisted hopping, or resonant tunneling.[4]

In contrast, the observed tunneling resistance R_N is nearly constant below T_c, as mentioned above. The nearly temperature independent R_N in Bi-2212 intrinsic junction stack is also suggested by the temperature dependence of I_c which roughly follows the Ambegaokar-Baratoff relation[5] as shown in Fig. 3(a). These facts as well as the McCumber parameter larger than 500 for this junction stack indicate that the SrO-BiO-BiO-SrO block acts as a sufficiently insulating tunnel barrier and also strongly suggest that the increase in ρ_c has an origin different from those associated with the imperfection in this block.

The intrinsic subgap conductance which gradually rises for $V > 0$ implies the existence of low-lying quasiparticle excitations. In the inset of Fig. 3(b), the derivative conductance dI/dV is plotted against V^2, indicating that dI/dV is proportional to V^2. This particular dependence is expected in a two-dimensional (2D) d-wave superconductor. Therefore, we have numerically calculated $I-V$ curves based on the 2D pure d-wave order parameter,

$$\Delta(\theta) = \Delta_0 cos(2\theta) \; ; \quad (\theta = tan^{-1}(k_y/k_x)). \tag{1}$$

The tunneling current along the c-axis is calculated using the normalized quasiparticle density of states,

$$N(E) = \text{Re} \int_0^{2\pi} \frac{1}{2\pi} \left[\frac{E}{\sqrt{E^2 - \Delta^2(\theta)}} \right] d\theta. \tag{2}$$

The dotted line in Fig. 3(b) is the fit to the observed curve at 4.2 K. Although the fit

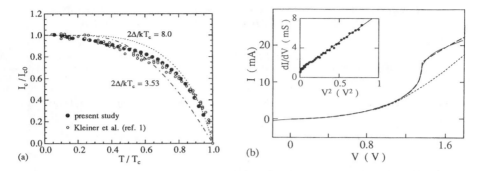

Figure 3: (a) Temperature dependence of I_c and (b) quasiparticle tunneling characteristics at 4.2 K of a 60 nm thick junction stack. The dotted and broken lines are theoretical curves calculated assuming a d-wave order parameter with $2\Delta_0 = 50$ meV. The inset shows the observed conductance for smaller voltages.

is pretty good for $V < 1.2$ V, a significant deviation is seen near the gap-like structure. This is likely due to the effect of the nonequilibrium superconductivity, since much higher quasiparticle current is injected in this region. In order to take this effect into account, we further replaced Δ_0 in Eq. (1) by $\Delta(J_{qp}) = \Delta_0(1 - pJ_{qp}^2)^2$. Here, $J_{qp} = I_{qp}/I_0$ is the quasiparticle current normalized by $I_0 = 2\Delta_0/eR_N$ and the variable parameter p represents the degree of nonequilibrium effect. The calculated broken line fits the experimental curve over the whole V range measured, indicating that the quasiparticle $I - V$ characteristics of the present intrinsic junction stack are consistent with the d-wave order parameter. It should be noted that the deduced $2\Delta_0$ value of 50 meV agrees with those reported in STM measurements. The slight deviations from Ambegaokar-Baratoff relation are observed both at very low temperatures and at temperatures near T_c in Fig. 3(a). These deviations appear to be qualitatively consistent with the d-wave order parameter with a $2\Delta_0$ much larger than the BCS value.

In summary, both pair and clear quasiparticle tunneling characteristics have been observed in intrinsic junction stacks with a reduced number of CuO_2 bilayer fabricated on underdoped Bi-2212 single crystals. We have shown that the observed tunneling characteristics are consistent with a sufficiently insulating SrO-BiO-BiO-SrO barrier for tunneling along the c-axis and the d-wave order parameter with a $2\Delta_0$ of approximately 50 meV in each CuO_2 bilayer.

References

1. R. Kleiner *et al*, *Phys. Rev. Lett.* **68**, 2394 (1992); *Phys. Rev. B* **49**, 1327 (1994).
2. K. Tanabe *et al*, *Phys. Rev. B* **53**, (in press) (1996).
3. A. G. Rojo and K. Levin, *Phys. Rev. B* **48**, 16861 (1993).
4. A. A. Abrikosov, *Phys. Rev. B* **52**, 7026 (1995).
5. V. Ambegaokar and A. Baratoff, *Phys. Rev. Lett.* **10**, 486 (1963).

HALF-INTEGER FLUX QUANTUM EFFECT AND PAIRING SYMMETRY IN CUPRATE SUPERCONDUCTORS

C.C. TSUEI, J.R. KIRTLEY, J.Z. SUN, A. GUPTA
IBM T.J. Watson Research Center, Yorktown Heights, NY 10598, USA

Z.F. REN, J.H. WANG
Superconducting Materials Laboratory, State University of New York, Buffalo, NY 11794, USA

K.A. MOLER
Department of Physics, Princeton University, Princeton, NJ 08544

C.C. CHI
Materials Science Center and Department of Physics, National Tsing-Hua University, Hsin-Chu, Taiwan, R.O.C.

We have used a scanning SQUID microscope to directly observe the spontaneously generated half-integer flux quantum effect in controlled orientation tricrystal cuprate superconducting systems. These experiments show that the order parameter in YBCO and Tl2201 has lobes and nodes consistent with the $d_{x^2-y^2}$ pair state.

After ten years, the mechanism for high-T_c superconductivity remains elusive. A definitive determination of pairing symmetry in copper-oxide superconductors may help to unravel this mystery. To unambiguously demonstrate the existence of unconventional pairing symmetry in copper-oxide superconductors, a phase-sensitive symmetry measurement is required. Wollman et al.[1] performed the first such phase-sensitive experiments using YBCO-Pb dc-SQUID interferometry as suggested by Sigrist and Rice.[2] The tricrystal scanning SQUID imaging technique used in our symmetry tests has the advantages over other phase-sensitive measurements that it is non-invasive, provides an in-situ diagnostic of sample homogeneity and flux trapping, and measures the magnetic flux quantum state of the tricrystal samples in their ground state with no externally applied transport current.

The details of our experiments have been described elsewhere.[3,4,5,6,7] In essence our experiment combines two macroscopic quantum coherence effects, pair tunneling and flux quantization, to decipher the microscopic phase. The sign of the supercurrent $I_c^{ij}(\theta_i, \theta_j)$ across a grain boundary junction as a function of grain boundary misorientation is well-defined for a given pairing symmetry. For example, the supercurrent for a junction made of d-wave cuprate superconductors can be expressed by the following Sigrist-Rice clean limit (perfectly smooth interfaces) formula:[2]

$$I_s^{ij} = (A^{ij} \cos 2\theta_i \cos 2\theta_j) \sin \Delta\phi_{ij}, \qquad (1)$$

where A^{ij} is a constant characteristic of the junction interface ij and angles θ_i and θ_j. The corresponding equation in the dirty limit (maximum allowable roughness) is:

$$I_s^{ij} = (B^{ij} cos2(\theta_i + \theta j)) sin\Delta\phi_{ij}, \qquad (2)$$

where B^{ij} is a constant characteristic of the junction interface.

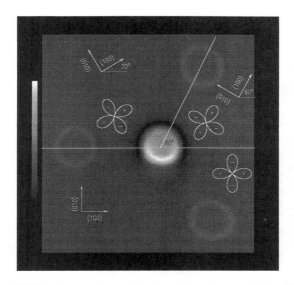

Figure 1: Scanning SQUID microscope image of a tricrystal experiment with four YBCO rings at 4.2K, cooled in nominal zero magnetic field. Solid lines are superposed on the magnetic field image to indicate the positions of the grain boundaries, and polar plots to show the orientation of assumed d-wave symmetry order parameters aligned with the YBCO crystalline axes.

In principle, one can measure I_c^{ij} as a function of θ_i and θ_j to probe the symmetry of the pair wavefunction. In practice, such experiments are plagued by the fact that $I_c^{ij}(\theta_i, \theta_j)$ depends sensitively on the junction characteristics as well as the angular dependences of the pairing interaction. Realistic samples are probably represented by some interpolation between Eqs. (1) and (2). Furthermore, a measurement of the supercurrent of a single junction yields only the magnitude of the pair tunneling current $| I_i^{ij} |$, not its sign: It is not a phase-sensitive measurement. To remedy this situation, we have designed a series of superconducting cuprate rings with three deliberately oriented grains to probe the pairing symmetry. Using the presence or absence of the half-integer flux quantum effect in such tricrystal systems, we can obtain the microscopic phase information of the pair wavefunction, that is, the sign of the supercurrent I_i^{ij}, without the need to measure its magnitude. A superconducting ring with weak links, with (in the absence of a supercurrent) an odd number of sign changes to the normal component of the order parameter (a π-ring) should show a spontaneous magnetization of one-half flux quantum ($\pm\Phi_0/2$) provided that $(I_c)_{min}L \gg \Phi_0$, where $(I_c)_{min}$ is the critical current of the weakest junction in the ring, and L is the self-inductance of the ring.[2,3]

To measure the magnetic flux quantum state of these rings, a high-resolution scanning SQUID microscope (SSM)[8] was used to directly image the magnetic flux threading these rings at 4.2K and near zero ambient magnetic field. Fig. 1 shows a SSM image of four YBCO rings deposited and patterned on a tricrystal (100) $SrTiO_3$ substrate with a d-wave π-ring geometrical configuration based on Eqs. (1) and (2). The fact that only spontaneous magnetization of a half magnetic flux quantum has been directly observed in the 3-junction

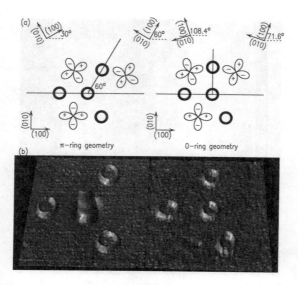

Figure 2: (a) A schematic of tricrystal substrate configurations: on the left is a π-ring geometry and on the right is a 0-ring configuration, both are base on d-wave pairing symmetry (Eqs. (1) and (2)). (b) A perspective view of SSM images of two YBCO tricrystal ring samples, one with a π-ring geometry (left) and the second with a 0-ring geometry (right).

ring, but not in the control rings, has lent strong support for d-wave pairing in YBCO.

By varying the tricrystal geometry and the cuprate systems, we have carried out a number of tricrystal experiments to probe the symmetry of the pair state in high-T_c superconductors and to elucidate the nature of the half-integer flux quantum effect. The main results of our tricrystal experiments are summarized as follows:

1) We have directly observed the half-integer flux quantum effect in YBCO tricrystal disks, blanket films, as well as rings. This demonstrates that our observations do not depend on the macroscopic sample configuration. The blanket geometry provides a direct measurement of the Josephson penetration depth, and a technique for symmetry tests that does not require photolithographic patterning.

2) The absence of the half-integer flux quantum effect in our recent g-wave only tricrystal experiments with YBCO and Tl2201 systems has ruled out an even-parity state with an energy gap $\Delta(k)$ varying as $(\cos k_x + \cos k_y)$.

3) Three separate tricrystal geometries (see e.g. Fig. 2), argue against any symmetry independent mechanisms such as spin-flip scattering and electron correlation effects at the grain boundaries as the cause of the observed π-phase shift.

4) Previous to our work, all phase sensitive measurements were done in YBCO. We have reproduced our tricrystal experimental results in ring and blanket geometries in the Tl2201 system. This indicates that these results do not depend on the special properties of YBCO, such as two CuO_2 planes per unit cell, CuO chains between the planes, orthorhombic symmetry, or the existence of twins. Fig. 3 shows a half-flux quantum Josephson vortex trapped at the tricrystal point of a blanket coverage of Tl2201 on a tricrystal with the

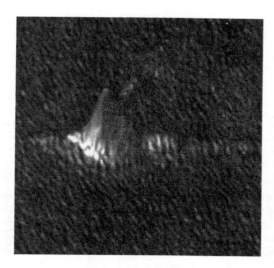

Figure 3: A perspective view of the SSM image of a Tl2201 blanket film deposited on a π-ring (100) $SrTiO_3$ tricrystal substrate (see Fig. 2a). The spontaneously generated Josephson vortex spreads along all three grain boundaries emanating from the tricrystal point.

geometry of Fig. 2a at 4.2K, cooled in nominally zero field.

5) Our technique can be used, with relative ease, to study the effect of doping and/or impurity scattering on pairing symmetry. For example, to test the universality of the $d_{x^2-y^2}$ state and to find out if there is a doping (band-structure) induced transition from d-wave to s-wave or other pair states, we are repeating our tricrystal experiments with the Tl2201 system ($Tl_2Ba_2CuO_{6+\delta}$) as a function of the oxygen content δ. Preliminary results show that a Tl2201 superconductor with a reduced T_c of 66K still has d-wave pairing symmetry.

In summary, we have used a scanning SQUID microscope to make the first direct observation of the half-integer flux quantum effect in controlled orientation tricrystal cuprate rings, disks, and blanket films. Our results show that both YBCO and Tl2201 have gaps with nodes and lobes consistent with $d_{x^2-y^2}$ symmetry.

We thank M. Ketchen, M. Bhushan, and A. Ellis for their contributions to the development of the scanning SQUID microscope.

References

1. D.A. Wollman *et al*, *Phys. Rev. Lett.* **71**, 2134 (1993).
2. Manfred Sigrist and T.M. Rice, *J. Phys. Soc. Japan* **61**, 4283 (1992).
3. C.C. Tsuei *et al*, *Phys. Rev. Lett.* **73**, 593 (1994).
4. J.R. Kirtley *et al*, *Nature* **373**, 225 (1995).
5. C.C. Tsuei *et al.* *J. Phys. Chem Solids* **56**, 1787 (1995).
6. J.R. Kirtley *et al*, *Phys. Rev. Lett.* **76**, 1336 (1996).
7. C.C. Tsuei *et al*, *Science* **272**, 329 (1996).
8. J.R. Kirtley *et al.* *Appl. Phys. Lett.* **66**, 1138 (1995).

PARAMAGNETIC MEISSNER EFFECT FROM SPIN POLARIZATION

A.Kallio, M. Rytivaara and K.Honkala
Department of Physical Sciences, Theoretical Physics, University of Oulu, Linnanmaa, FIN-90570 Oulu, Finland

Abstract

We propose that recent experiments for paramagnetic Meissner effect can be understood in terms of pairing fermion (quasiparticle) spin polarization. With suitable surface topology the polarized spins cannot be screened by supercurrents inside a surface layer. The same theory explains also the double transition in UPt_3 and the observed breakdown of time reversal invariance associated with polarized spins.

The Meissner effect has been corner stone in proofing that a sample is superconducting. One of surprising findings in the high T_c superconductors has been the appearance of paramagnetic Meissner effect (PME) first detected experimentally for cuprates by Svendlidh et al[1] and subsequently by many other groups also in single crystals[2] and most recently even for Nb-disks[3], which are thought to be conventional BCS superconductors. The theoretical explanations for powder sample vary from abnormal order parameter such as d-wave to the so called π-rings[4,5] of Josephson junctions. Since the PME occurs also in single crystals and even for BCS superconductors these explanations seem somewhat doubtful and one must look for other alternative explanations.

From calculations of one component Coulomb Fermi system[6] it is known that at some rather low density range the spin polarized state has lower energy than the unpolarized state. The critical r_s values for 3D are $r_s \approx 60 - 80$. These estimates are valid for jellium background and the critical densities are sensitive to disorder for which only few estimates exists. Since disorder causes additional localization the critical density scales can be different. The following features will most likely remain within the realistic background of chemical lattice: In zero external magnetic field the polarized state has lower energy below certain fairly small density of pairing fermions but the energy difference is small. Clearly the spin polarized state breaks the time reversal invariance. The minimum features needed from a high-T_c model to explain PME is existence of pairs ($\approx$ bosons) and their decay products pairing fermions ($\approx$ quasiparticles) below T_c in quasi chemical equilibrium. These features come out from the *spectator fermion superfluid* – model (SFS)[7] which assumes existence of two active fermion bands, one with light effective mass (spectator band) and the other with heavier mass (pairing fermion band). Within the SFS-model described above the following scenario is possible: In small external field H the pairing fermion spins may get polarized inside a surface layer at temperatures $T_1 \lesssim T_c$ since the mean number of pairing fermions gets small below T_c. They may be polarized also in the bulk but the associated field is screened by the supercurrents. Consider a plane sample with area A. The field B_s produced by the polarized spins in the volume of thickness $\lambda(T)$ within the surface of the sample is proportional to λ, the penetration depth, hence $B_s = \alpha A \lambda(T) = \beta \lambda(T)$. The magnetic field from deeper in polarized spins are screened by the supercurrents. The corresponding susceptibility is

$$\chi = \frac{B_s}{4\pi H} - \frac{B_M}{4\pi} = \frac{\beta \lambda(T)}{4\pi H} + \chi_0. \tag{1}$$

Here B_M or χ_0 are due to Meissner currents associated with the screening of the polarized spins deeper in. For full Meissner fraction one would have $\chi_0 = -1/4\pi$. If $H > H_{c1}$ the sample develops

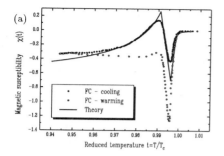 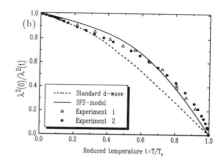

Figure 1: (a)Theoretical predictions from Eq.(1) compared with data from ref.2. (b) SFS penetration depth from ref.7 compared with d-wave prediction from ref.9.

vortices whose energetics dominates over the polarization energy difference and one obtains conventional Meissner effect. This simple theory by Eq.(1) explains the main features observed in single crystals: (i) The paramagnetic term is inversely proportional to external field H. (ii) The maximum value of χ is at T_1 slightly below T_c. The comparison with experiment is given in Figs.1.

Our proposal has further consequences. The spins deep inside may also be polarized and their polarization can be seen by muons. The muons should see local non–zero field B_s inside the sample not visible from outside. In UPt_3 such a field has been observed with associated breakdown of time reversal invariance below the lower specific heat peak T_c^- which we associate with temperature T_1. Further fact is that PME has been observed also in UPt_3 for very small fields. Our proposal here is in accordance with a new heavy fermion model [8] where the double peak in the case of UPt_3 is associated with the change in spin polarized state below T_c^- to unpolarized state both remaining superconducting below upper peak T_c^+. The breakdown of time reversal invariance would therefore not be associated with the superfluid order parameter but is due to the polarized spins deeper in. In the case of YBCO the time reversal invariance of superfluid order parameter has been "shown" in SQUID – experiments by Mathai et al. [10].

1. P. Svedlindh et al Physica C **162-164** (1989) 1365.
2. S. Riedling et al Phys. Rev. **B 49** (1994) 13283.
3. D.J. Thompson et al Phys. Rev. Lett. **75** (1995) 529.
4. W. Braunisch et al Phys. Rev. **B 48** (1993) 4030.
5. F. V. Kusmartsev Phys. Rev. Lett. **69** (1992) 2268.
6. D.M. Ceperly, and B.J. Alder Phys. Rev. Lett. **45** (1980) 566.
7. A. Kallio, V. Apaja, X. Xiong, and S. P öykkö Physica C **219** (1994) 340.
8. A. Kallio, S. Pöykkö, and V. Apaja Proc. of the Fourth International Conference and Exhibition: World Congress on Superconductivity, Eds. K. Krishen and C. Burnham (NASA Conf. Publ., 1994), 3290.
9. R.A. Klemm, and S. Liu Phys. Rev. Lett. **74** (1995) 2343.
10. A. Mathai et al Phys. Rev. Lett. **74** (1995) 4523. ö

USE OF TRICRYSTAL MICROBRIDGES TO PROBE THE PAIRING STATE SYMMETRIES OF CUPRATE SUPERCONDUCTORS

J. LIN, J. H. MILLER, JR., Z. G. ZOU

Department of Physics and Texas Center for Superconductivity, University of Houston
4800 Calhoun Road, Houston, TX 77204-5932 USA

and

Q. XIONG

Department of Physics, University of Arkansas, Fayetteville, AR 72701 USA

ABSTRACT

We have carried out field-modulated critical current measurements of nominally pure, doped, and ion-irradiated YBCO tricrystal microbridges in order to probe superconducting pairing state symmetry. Our results in the short junction limit are consistent with orthorhombically-distorted d- wave pairing symmetry in nominally pure YBCO films. Ion-irradiated samples, however, exhibit more ideal $d_{x^2-y^2}$-like behavior.

Introduction

An accurate determination of the pairing state symmetries of cuprate superconductors over their entire range of carrier concentrations and critical temperatures is important to help elucidate the mechanism responsible for high-temperature superconductivity. Phase-sensitive experiments that directly probe the pairing state symmetries of YBCO and Tl-2201[1] are consistent with $d_{x^2-y^2}$-symmetric pairing, with little or no imaginary s-wave component. However, some experiments on Pb/YBCO tunnel junctions[2] indicate the existence of an s-wave component, suggesting the possibility of a *real* mixture of large d-wave and small s-wave components in YBCO.

Experiment

Our technique of probing superconducting pairing symmetry[3] is to measure the field-modulated critical current of a cuprate thin film microbridge on a $SrTiO_3$ tricrystal substrate, as illustrated in Fig. 1. The critical current is limited by the two shortest of the three grain boundaries crossing the microbridge. This device is analogous to an s-d corner junction, since one of the two short boundaries acts as a π-junction, while the other acts as a 0-junction, if the pairing symmetry is d-wave. Our measurements provide additional information besides the phase shift, and are capable of detecting either a complex or a real mixture of s- and d-wave components of the order parameter.

Two central peaks, of approximately equal height, are observed in the field-modulated critical currents of frustrated nominally pure YBCO tricrystal junctions operating in the short junction limit, consistent with predominantly d-wave pairing symmetry. However, the critical current is nonvanishing at zero field, consistent with a real mixture of large d-wave and small s-wave components. Such a pairing state is equivalent to an orthorhombically distorted d-wave order parameter.

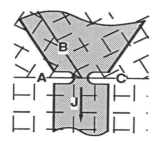

Fig. 1. Tricrystal microbridge with 30° misorientation angles.

We have also carried out experiments on cuprate superconducting tricrystal microbridges with a wide range of critical temperatures, including doped and ion-irradiated YBCO. The field-modulated critical current behavior approaches that expected for an ideal d-wave superconductor as the T_c and J_c are both suppressed in proton-irradiated tricrystal microbridges. Some of this behavior could be attributed to the fact that the tricrystal junction becomes further into the short junction limit as the critical current is reduced and the Josephson penetration length is increased. In addition, the magnitude of any real s-wave component might be reduced as the number of irradiation-induced defects, which act as oxygen traps, increases and the structure becomes more tetragonal.

In conclusion, phase-sensitive measurements with tricrystal can be used to probe the pairing state symmetries of a wide range of materials. Our ongoing efforts include an investigation of other cuprate superconductors, such as thallium-based cuprates.

Acknowledgements

This research was supported by the State of Texas through the Texas Center for Superconductivity and the Advanced Research Program, and by the Robert A. Welch Foundation.

References

1. C. C. Tsuei, J. R. Kirtley, M. Rupp, J. Z. Sun, A. Gupta, M. B. Ketchen, C. A. Wang, Z. F. Ren, J. H. Wang, and M. Bhushan, *Science* **271** (1996) 288 and references cited therein.
2. A. G. Sun, D. A. Gajewski, M. B. Maple, and R. C. Dynes, *Phys. Rev. Lett.* **72** (1994) 2267.
3. J. H. Miller, Jr., Q. Y. Ying, Z. G. Zou, N. Q. Fan, J. H. Xu, M. F. Davis, and J. C. Wolfe, *Phys. Rev. Lett.* **74** (1995) 2347.

II. EXPERIMENTS

Transport

NORMAL-STATE RESISTIVITY OF SUPERCONDUCTING $La_{2-x}Sr_xCuO_4$ IN THE ZERO-TEMPERATURE LIMIT

Yoichi Ando,[*] G. S. Boebinger, A. Passner, R. J. Cava
Bell Laboratories, Lucent Technologies, Murray Hill, NJ 07974, USA

T. Kimura, J. Shimoyama, K. Kishio
Department of Applied Chemistry, University of Tokyo, Hongo, Bunkyo-ku, Tokyo 113, Japan

> The low-temperature normal-state resistivities of underdoped $La_{2-x}Sr_xCuO_4$ were studied by suppressing the superconductivity with pulsed magnetic fields of 61 T. Over a wide temperature range below T_c, both in-plane and out-of-plane resistivities of single crystals were found to diverge logarithmically as $T/T_c \to 0$. The logarithmic divergence is accompanied by a nearly constant anisotropy ratio, ρ_c/ρ_{ab}. The low-temperature resistivities of similarly underdoped polycrystalline samples also diverge logarithmically, while more-underdoped insulating polycrystals show variable-range-hopping conduction.

1 Introduction

The unusual properties of the normal-state resistivity of high-T_c cuprates reflect the electronic structure that leads to high-T_c superconductivity. However, due to the extreme values of H_{c2}, the behavior of the normal-state resistivity at low temperatures has been largely hidden. To address this issue, we use a 61 T pulsed magnetic field to suppress the superconductivity in $La_{2-x}Sr_xCuO_4$ (LSCO) down to ^{3}He temperatures in both single crystals[1] and polycrystals.

2 Experiments

The single crystal LSCO samples are grown by the floating-zone method.[2] We studied two underdoped compositions, nominally $x = 0.08$ and 0.13, which have T_c of 20 K and 35 K, respectively. ρ_{ab} and ρ_c are measured in differently-oriented platelet-shaped crystals cut from the same rod, with dimensions chosen to facilitate each measurement. The crystals are typically $3 \times 2 \times 0.2$ mm^3. The polycrystals are prepared by solid-state reaction processing[3] and shaped typically to $3 \times 1 \times 0.3$ mm^3.

The resistivity is measured with an ac four-probe technique. For the pulsed experiments, a fast lock-in amplifier (driven at ~ 120 kHz) and a transient digitizer are used to record the data during the 100 msec magnet pulses. The ac-current density (as much as 5 A/cm^2 for the lowest-impedance sample) was chosen to avoid sample heating and non-ohmicity. Details of the experiment, including precautions to avoid eddy-current heating are in Ref.1.

Figure 1 shows the magnetic field dependence of ρ_{ab} for $x = 0.13$ crystal at a variety of temperatures from 1.4 K to 45 K. Figure 2 shows the temperature dependence of ρ_{ab} for $x = 0.13$ at fixed magnetic fields, determined from a series of 61-T pulses. These

---

[*]Permanent address: Central Research Institute of Electric Power Industry, 2-11-1 Iwato-kita, Komae, Tokyo 201, Japan

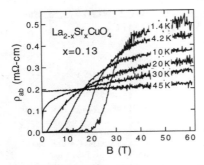
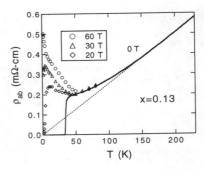

Figure 1: ρ_{ab} vs B of $x = 0.13$ single crystal at selected temperatures.

Figure 2: ρ_{ab} vs T in 0 T (solid line), 20 T, 30 T, and 60 T. Dotted line fits the linear-T part.

data for $x = 0.13$ are striking, because ρ_{ab} is metallic down to T_c and shows linear-T behavior above 150 K which extrapolates to a zero intercept (dotted line). However, when superconductivity is suppressed with high magnetic fields, a weakly insulating behavior is observed, which is well described by $\rho \sim \ln(1/T)$ for $T < 25$ K [Fig. 3(a)]. The ρ_c data for the same composition are shown in Fig. 3(b); it is surprising that ρ_c is best described by the same $\ln(1/T)$ divergence.

The logarithmic divergence is better demonstrated in more underdoped crystals, $x = 0.08$, where it extends over a temperature range of about one-and-a-half decades [Figs. 3 (c) and (d)]. Note that, once again, both ρ_{ab} and ρ_c show the logarithmic divergence. Since the 50-T and 60-T data superimpose, the normal-state magneto-resistance is negligible compared to the temperature dependence. We therefore find it reasonable to conclude that the $\ln(1/T)$ dependence mimics the behavior of the high-T_c electron system at zero magnetic field in the absence of superconductivity.

It should be noted that the anisotropy ratio ρ_c/ρ_{ab} (not shown), which is strongly

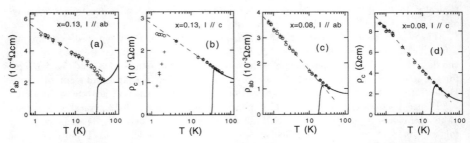

Figure 3: Resistivity vs $\ln T$ plots for the $x = 0.13$ and 0.08 single crystals. Circles, crosses, and solid lines are 60-T, 50-T, and 0-T data, respectively. Dashed lines are $\ln T$ fits to the low-temperature data.

261

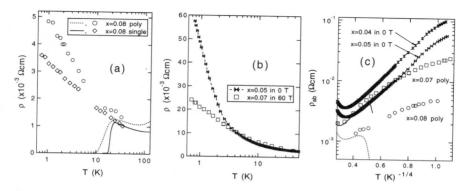

Figure 4: log T plot comparing (a) single and poly-crystals; (b) superconducting and nonsuperconducting samples. (c) Variable-range-hopping plot for the polycrystals. Open symbols are 60-T data.

temperature dependent at high temperatures, becomes essentially temperature independent over the entire temperature range of the $\ln(1/T)$ behavior in these high fields.[1] This implies that the fundamental mechanism that determines the low-temperature resistivity is the same for the ab and c directions when the resistivity diverges logarithmically.

To investigate the evolution of the temperature dependence upon further reducing the carrier density, we measured polycrystalline samples of $x = 0.08, 0.07, 0.05, 0.04$. The use of polycrystals is necessary, because $x = 0.08$ is currently the lower limit for single crystals.[2] The T_c (midpoint) of the $x = 0.08$ and 0.07 samples are 16 K and 11 K, respectively, while the $x = 0.05$ and 0.04 samples are nonsuperconducting. Figure 4 (a) shows the similarity of the resistivity of the $x = 0.08$ polycrystal and single crystal. Figure 4 (b) shows the striking difference between the temperature dependence of the normal-state resistivity for the $x = 0.07$ and 0.05 samples. This is better demonstrated in Fig. 4 (c), where the linearity of the $x = 0.04$ and 0.05 data on a $\ln \rho$ vs $T^{-1/4}$ plot is evidence for variable range hopping (VRH), as previously reported for nonsuperconducting LSCO.[4,5] Note that the data from the superconducting samples are nowhere linear in this plot. As might be expected, the VRH onset occurs near the 2D Mott limit, $\rho_{ab} \sim 10$ mΩcm.[1] These results suggest that the peculiar $\ln(1/T)$ dependence in the normal-state is closely related to the occurrence of superconductivity.

3 Discussion

The mechanism for the peculiar $\ln(1/T)$ dependence is not yet clear. While both Kondo effect and weak localization can give rise to logarithmic increases in resistivity, we tend to exclude both mechanisms, due to the extremely high magnetic field.[1] Other possibilities include:

Electron interactions in 2D systems[6] — This results in a suppression of the 2D density of states and a logarithmic correction to σ_{ab}. If ρ_c is determined by tunneling between 2D systems, which reflects the 2D density of states, ρ_c might show the same $\ln(1/T)$ behavior.[7]

Marginal Fermi Liquid Model[8] — This model predicts a leading $\ln(T_0/T)$ correction to the resistivity when quasiparticle-impurity scattering dominates conduction,[9] with T_0 the same in all directions. Also, the model conjectures a 3D insulating state at low temperatures; however, calculations to date give a $[\ln(1/T)]^2$ low-temperature divergence.[9]

2D Luttinger Liquid Model[10] — Incoherent inter-liquid conductivity has been predicted to follow a power law $\sigma_c(T) \sim T^{4\alpha}$, where α is the Luttinger liquid exponent. When localization in the Luttinger Liquid occurs, σ_{ab} could be proportional to σ_c.[11] The best fit of a power law to our data gives $4\alpha \simeq 0.33$ and 0.14 for $x = 0.08$ and 0.13, respectively, although the range of 4α should be limited to $0 < 4\alpha < 1/4$. As in the data, 4α is expected to be larger for more underdoped samples.

Phenomenological c Axis Resistivity Model[12] — This model proposes that the low-temperature ρ_c divergence results from suppression of the in-plane density of states, as measured by the Knight shift. Our data can be fitted with a reasonable set of parameters down to ~ 10 K,[13] below which the data diverge faster than the fit.

We thank E. Abrahams, B.L. Altshuler, P.W. Anderson, P.B. Littlewood, A.J. Millis, D. Pines, and C.M. Varma for helpful discussions.

References

1. Y. Ando, G. S. Boebinger, A. Passner, T. Kimura, and K. Kishio, *Phys. Rev. Lett.* **75**, 4662 (1995).
2. T. Kimura, K. Kishio, T. Kobayashi, Y. Nakayama, N. Motohira, K. Kitazawa, and K. Yamafuji, *Physica* C **192**, 247 (1992).
3. H. Takagi, R. J. Cava, M. Marezio, B. Batlogg, J. J. Krajewski, W. F. Peck,Jr., P. Bordet., and D. E. Cox, *Phys. Rev. Lett.* **68**, 3777 (1992).
4. B. Ellman, H. M. Jaeger, D. P. Katz, T. F. Rosenbaum, A. S. Cooper, and G. P. Espinosa, *Phys. Rev.* B **39**, 9012 (1989).
5. C. Y. Chen, E. C. Branlund, C.-S. Bae, K. Yang, M. A. Kastner, A. Cassanho, and R. J. Birgeneau, *Phys. Rev. Lett.* **51**, 3671 (1995).
6. B. L. Altshuler and A. G. Aronov, in *Electron-Electron Interactions in Disordered Systems*, ed. M. Pollak and A. L. Efros (North-Holland, 1985), pp.1-153.
7. B. L. Altshuler, private communication.
8. C. M. Varma, P. B. Littlewood, S. Schmitt-Rink, E. Abrahams, and A. E. Ruckenstein, *Phys. Rev. Lett.* **63**, 1996 (1989).
9. G. Kotliar, E. Abrahams, A. E. Ruckenstein, C. M. Varma, P. B. Littlewood, and S. Schmitt-Rink, *Europhys. Lett.* **15**, 655 (1991).
10. D. G. Clarke, S. P. Strong, and P. W. Anderson, *Phys. Rev. Lett.* **74**, 4499 (1995).
11. P. W. Anderson (private communication).
12. Y. Zha, S. L. Cooper, and D. Pines, preprint.
13. D. Pines and Y. Zha (private communication).

High Pressure Study of The T_c and Thermopower of Hg-based Cuprates

F. Chen, X.D. Qiu, Y. Cao, L. Gao, Y.Y. Xue and C.W. Chu

Department of Physics and Texas Center for Superconductivity at University of Houston
University of Houston, Houston, TX 77204-5932

Q. Xiong

Department of Physics and High Density Electronic Center
University of Arkansas, Fayetteville, AR 72701

The pressure effects on superconducting transition temperature T_c and thermopwer S were studied up to 16 GPa for $HgBa_2Ca_{m-1}Cu_mO_{2m+2+\delta}$ with m = 1–3 over a broad doping range. Direct measurement for a $HgBa_2Ca_2Cu_3O_{8+\delta}$ sample with $T_c = 115$ K shows that the charge transfer is negative and unusually small, *i.e.* ~ -0.0004 holes/$(CuO_2 \cdot GPa)$. The $T_c(P)$ over a broad doping levels, on the other hand, suggests that factors not being considered in the charge transfer model may play important roles. Site-selective cation-substitution shows that the local environment of Hg may be one of the factors.

1 Introduction

The charge-transfer induced by pressure P^1 and the van Hove singularity[2] have been used to interpret the large T_c-enhancement under pressure in $HgBa_2Ca_{m-1}Cu_mO_{2m+2+\delta}$, but contradictions exist[3]. To clarify the situation, $T_c(P)$ was measured up to 16 GPa over a broad n-range, and dn/dP was deduced from the measured dS/dP, where n is the carrier density and S is the thermopower. Our data demonstrate not only that the dn/dP of $HgBa_2Ca_{m-1}Cu_mO_{2m+2+\delta}$ is too small to contribute significantly to the T_c-enhancement, but also that other factors not being considered before may play important roles. We observed that the T_c-enhancement is very sensitive to substitutions at the Hg-site, suggesting the local environment of Hg is one of the factors.

2 Experiments

All samples were single phase polycrystalline with a sharp superconducting transition. Instruments for providing quasi-hydrostatic pressure (QHP) up to 18 GPa and hydrostatic pressure up to 1.6 GPa were reported before.[3,4] The thermopower was measured using a homemade apparatus[5] designed to reduce the pressure-medium effects. The pressure effects of the p-Chromel/Cu thermocouples, which were used to measure the temperature gradient across the sample, was calibrated.

3 Data and Discussion

The traditional charge transfer models[1] obtain $dT_c/dP = dT_c^{max}/dP - 2\alpha(n - n^{op}) \cdot (dn/dP)$ from the proposed $T_c = T_c^{max}[1-\alpha(n - n^{op})^2]$, and assume that both dT_c^{max}/dP and dn/dP are independent of P and n, where α and n^{op} are universal constants. Therefore, $dn/dP = -(\partial^2 T_c/\partial P \partial n)/(2\alpha)$ can be deduced from the n-dependence of dT_c/dP. dn/dP of $HgBa_2Ca_{m-1}Cu_mO_{2m+2+\delta}$ so obtained has been given previously as $\sim 0 \pm 0.0001$,[6] $+0.0008$,[4] and -0.0013 holes/$(CuO_2 \cdot GPa)$[7] for m = 1, 2 and 3 respectively (Inset of Fig. 1). These values are rather small, e.g., about one tenth of the $dn/dP \sim 0.011$ holes/$(CuO_2 \cdot GPa)$ in $YBa_2Cu_3O_{7-\delta}$.[8] However, these small dn/dP still contributes a significantly m-dependent term of $-\alpha T_c^{max}(P \cdot dn/dP)^2$, ~ 0, -7 and -15 K at 30 GPa for m = 1, 2 and 3 respectively, following the model in Ref. 1. Such predicted m-dependence seems to be too large based on the observed $\Delta T_c = T_c(P) - T_c(0)$, which is m-independent within 3 K (Fig. 1). Therefore, a direct measurement of dn/dP is desired.

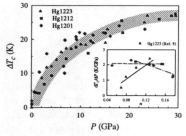
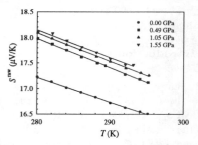

Figure 1: Universal ΔT_c vs pressure for optimal doped $HgBa_2Ca_{m-1}Cu_mO_{2m+2+\delta}$. Inset: dT_c/dP vs hole concentration for $HgBa_2Ca_{m-1}Cu_mO_{2m+2+\delta}$.

Figure 2: Thermopower raw data near room temperature under pressure for $HgBa_2Ca_2Cu_3O_{8+\delta}$ ($T_c = 115$ K).

We deduced dn/dP by measuring the thermopower S under pressure. It has been suggested that n vs S at 290 K obeys a universal trend in cuprates.[10] The proposed correlation has been tested in $HgBa_2Ca_{m-1}Cu_mO_{2m+2+\delta}$ with m = 1-3 at the ambient pressure.[5] dn/dP can be deduced, therefore, from $dS(290\,K)/dP$ if the proposed universal S vs n trend is hold under pressure. It should be note that a change of n usually causes a significant change of S in most metal and semiconductors where the proposed S vs n universality is not valid. An underdoped $HgBa_2Ca_2Cu_3O_{8+\delta}$ with $T_c \sim 115$ K was measured under hydrostatic pressure up to 1.6 GPa. Two Cu/p-Chromel thermocouples were used to measure the temperature difference ΔT with the Cu-wires also serving as the voltage leads. The thermopower raw data $S^{raw}(P)$ calculated assuming $S_{p-Chromel}(P) = S_{p-Chromel}(0)$ and $S_{Cu}(P) = S_{Cu}(0)$, will be related to the actual $S(P)$ as $[S^{raw}(P) - S_{Cu}(0)]/[S_{p-Chromel}(0) - S_{Cu}(0)] = [S(P) - S_{Cu}(P)]/[S_{p-Chromel}(P) - S_{Cu}(P)]$, where $S(P)$, $S_{p-Chromel}(P)$ and $S_{Cu}(P)$

are the thermopowers of the sample, p-Chromel and Cu under pressure respectively. The observed S^{raw} shifted parallel with the pressure (Fig. 2). The random data-scattering was better than 10 nV/K or 0.00002 holes/CuO_2, and $S_{Cu}(P) \approx S_{Cu}(0)$ is a good approximation. However, a large systematic uncertainty may be caused by $S_{p-Chromel}(P) - S_{Cu}(P)$. By carefully calibrating $S_{p-Chromel}(P) - S_{Cu}(P)$ against two Pt-thermometers, we reduce the total uncertainty to < 0.3 μV/(K·GPa) or 0.0007 holes/GPa.

The deduced dn/dP of the $HgBa_2Ca_2Cu_3O_{8+\delta}$ samples is -0.0004 ± 0.0007 holes/(CuO_2·GPa) using the universal $S(290)$ vs n trend. Although this value is not directly in contrast to that of ~ -0.0013 holes/(CuO_2·GPa) obtained from the charge transfer model, the smaller value of -0.0004 seems to be more reasonable based on the observed universal $\Delta T_c(P)$. This may suggest that the actual charge transfer is not even equal to $-(\partial^2 T_c/\partial P \partial n)/(2\alpha)$ if further investigations confirm our results.

dn/dP of $HgBa_2Ca_2Cu_3O_{8+\delta}$ has been calculated theoretically. Novikov[2] gave a rather large value of 0.027 holes/(CuO_2·GPa) over 0–9.4 GPa and Singh[11] reported a $dn/dP = 0.0022$ holes/(CuO_2·GPa). Both results are in disagreement with our data and further investigations are needed. Such negligible dS/dP can neither be accommodated with the proposed van Hove singularity[2]. S should change drastically when the Fermi surface approaches a singularity, which is in direct contrast with our data.

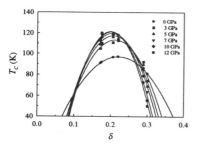
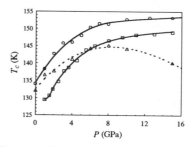

Figure 3: Isobaric plot of T_c vs carrier concentration for $HgBa_2CuO_{4+\delta}$. Inset: dT_c/dP for $HgBa_2CuO_{4+\delta}$ with different δ.

Figure 4: T_c vs QHP for $HgBa_2Ca_2Cu_3O_{8+\delta}$ (O), $(Hg_{0.8}Pb)Ba_2Ca_2Cu_3O_{8+\delta}$ ($\triangle$) and $Hg(Ba_{0.75}Sr_{0.25})_2Ca_2Cu_3O_{8+\delta}$ ($\square$).

To further verify the charge transfer models, T_c of $HgBa_2CuO_{4+\delta}$ was measured up to 16 GPa over a broad n range (Fig. 3). The isobaric T_c vs δ plot will only parallel shift with P, i.e. only T_c^{max} changes, under the charge transfer model[1] and accepting $dn/dP \sim 0$ as we have demonstrated. Although our data (Fig. 3) can be reasonably fitted with $T_c = T_c^{max} \cdot [1 - \beta(\delta - \delta^{op})^2]$, all three parameters T_c^{max}, δ^{op} and β vary with P. The value of β, for example, increases nearly 50% with P up to 16 GPa. The obtained $d(\delta^{op})/dP \sim -0.002$ oxygen/(CuO_2·GPa), which $\propto d(n^{op} - n)/dP$, is also too large to be accommodated with the observed dn/dP, suggesting a large negative dn^{op}/dP as proposed by us before.[3] These observations suggest that many factors not being considered in the charge transfer models play important roles in the pressure effects in $HgBa_2Ca_{m-1}Cu_mO_{2m+2+\delta}$.

To explore these factors, substitutions were carried out at the Ba and Hg sites for $HgBa_2Ca_2Cu_3O_{8+\delta}$. The optimal T_c at ambient pressure changed with these substitutions only moderately, *i.e.* ~ -2 K with 20% substitution of Hg by Pb and ~ -8 K with 25% substitution of Ba by Sr. However, the pressure effects of these substitutions were very different. The maximum T_c-enhancement under pressure in the Pb-substituted samples is only half of that of pure $HgBa_2Ca_2Cu_3O_{8+\delta}$ (Fig. 4). On the other hand, the enhancement remains almost unchanged, or even increases slightly with the Sr-substitution. Therefore, we relate the unusual pressure effects to the local environment of Hg, especially the linear coordination of O-Hg-O.

4 Conclusion

In conclusion, both T_c and S were measured for $HgBa_2Ca_{m-1}Cu_mO_{2m+2+\delta}$ at various doping levels. The results show that the charge transfer induced by pressure is very small and has negligible contribution to the T_c-enhancement under pressure. However, many factors not being considered in the charge-transfer models may play important roles. One of the factors may be the local environment of Hg.

Acknowledgments

This work is supported in part by NSF, EPRI, DoE, the State of Texas through Texas Center for Superconductivity at the University of Houston, and the T. L. L. Temple Foundation. This work is also supported by NFSF-DMR-9318946 at University of Arkansas.

Reference

1. J.J. Neumeier and H.A. Zimmermann, *Phys. Rev.* B **47**, 8385 (1993).
2. D.L. Novikov *et al.*, *Physica C* **222**, 38 (1994).
3. L. Gao *et al.*, *Phys. Rev.* B **50**, 4260 (1994).
4. F. Chen *et al.*, *Phys. Rev.* B **48**, 16047 (1993).
5. F. Chen *et al.*, submitted to *Phys. Rev.* B.
6. Y. Cao *et al.*, *Phys. Rev.* B **52**, 6854 (1995).
7. F. Chen *et al.*, *J. Appl. Phys.* **76**, 6941 (1994).
8. C. Murayama *et al.*, *Physica C* **185-189**, 40 (1991).
9. A.K. Klehe *et al.*, *Physica C* **223**, 313 (1994).
10. S.D. Obertelli *et al.*, *Phys. Rev.* B **46**, 14928 (1992).
11. D.J. Singh and W.E. Pickett, *Physica C* **233**, 237 (1994).

PUZZLING BEHAVIOR OF THE ELECTROTHERMAL CONDUCTIVITY OF HIGH-T_C SUPERCONDUCTORS

J. A. Clayhold, Y. Y. Xue, C. W. Chu
Texas Center for Superconductivity
University of Houston
Houston, Texas 77204-5932

J. N. Eckstein and I. Bozovic
Varian Research Center
3075 Hansen Way
Palo Alto, CA 94304

The electrothermal conductivity in the vortex state of a type-II superconductor probes only the normal excitations, both inside and outside the vortex core. Unlike the case of niobium, where the electrothermal conductivity is well-behaved, measurements in high-T_c superconductors have shown a surprising low-field anomaly. The anomaly is tentatively identified with vortex-core excitations, but, at present, neither the origin nor even the sign of the anomaly is understood.

The electrothermal conductivity, P, which appears in the linear response equation for the electrical current in the presence of both an electric field, $\vec{E}$, and a temperature gradient, $\vec{\nabla}T$,

$$\vec{j} = \sigma \vec{E} - P \vec{\nabla} T , \tag{1}$$

is a powerful probe of the transport of normal excitations in type-II superconductors [1]. P can be obtained experimentally from the ratio of the thermopower to the resistivity [1]. In contrast to the electrical conductivity, which, in general has contributions from three sources

$$\sigma_{total} = \sigma_s + \sigma_c + \sigma_{qp} , \tag{2}$$

with σ_s resulting from vortex motion, σ_c from core electrons, and σ_{qp} from the quasiparticles outside the core—the electrothermal conductivity only has contributions from the latter two sources [1]:

$$P = P_c + P_{qp} , \tag{3}$$

because only normal carriers can transport entropy parallel to the electric field. The selectivity of the electrothermal conductivity can be exploited to study the behavior of the normal-core electrons.

In conventional, low-temperature type-II superconductors, such as niobium, the dominant contribution to P comes from the core electrons. The core electrons in niobium can be regarded as normal, nonsuperconducting electrons and it was consequently observed experimentally [2] for niobium that $P_{core} = P_{normal}$. The electrothermal conductivity was the same above and below H_{c2}, as shown in Fig. 1.

The electrothermal conductivity of high-T_c materials is more interesting [1]. Results for a $YBa_2Cu_3O_{7-\delta}$ c-axis thin-film sample are shown in Fig. 2a. The thermal gradient was in the a-b plane and the magnetic field was parallel to the c-axis. The data at high magnetic field resemble the niobium data, being mostly constant. The low-field divergence of P, however, is an unexpected but apparently universal [1] feature of the high-T_c superconductors. This feature has also been seen [1] in samples of $Tl_2Ba_2CaCu_2O_{8+\delta}$ and $Bi_2Sr_2CaCu_2O_{8+\delta}$.

The mystery of P is the origin of the low-field divergence. As described below, it must be related to the properties of the normal core, because the last term in Eq. 3, the external-quasiparticle contribution, is known to be non-divergent and smoothly-varying with magnetic field in this temperature range [1]. The schematic breakdown of the two contributions to P is illustrated in Fig. 2b.

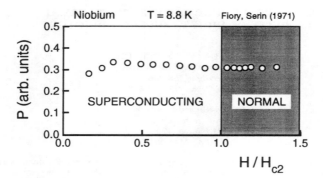

Figure 1: a) Magnetic-field dependence of the electrothermal conductivity, P, in niobium above and below H_{c2}, data from Ref. 2. The value is the same in both the superconducting and normal phases indicating that the vortex core excitations in niobium can be taken to be normal, non-superconducting electrons.

The puzzle is to explain why the core excitations of $YBa_2Cu_3O_{7-\delta}$ behave so differently from what was observed in niobium, where P was independent of the magnetic field.

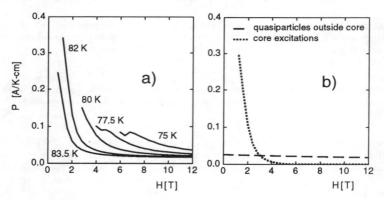

Figure 2: a) Magnetic-field dependence of the electrothermal conductivity, P, of the $YBa_2Cu_3O_{7-\delta}$ sample at various temperatures. b) The likely breakdown of P at $T = 82$ K into its constituent parts, the core and external-quasiparticle contributions.

The mystery was deepened by some surprising results [1] from a sample of $Bi_2Sr_2CaCu_2O_{8+\delta}$. The raw thermopower data displayed a *sign-change* as a function of magnetic field and temperature. The sign-change was interpreted [1] as evidence for two contributions to P, as indicated by Eq. 3. The sample, an epitaxial c-axis thin-film sample of $Bi_2Sr_2CaCu_2O_{8+\delta}$ with $T_c = 80$ K, was prepared by atomic layer-by-layer molecular beam epitaxy (ALL-MBE) [3]. The raw data for the in-plane mixed-state thermopower with $\vec{H} \parallel c$ are shown in Fig 3. At the highest temperature, $T = 76$ K, the thermopower reversed sign below $H = 2$ T. The sign reversal moved to higher magnetic fields as the temperature was reduced.

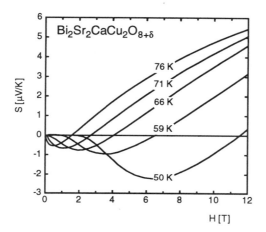

Figure 3: Thermopower of the $Bi_2Sr_2CaCu_2O_{8+\delta}$ sample as a function of the magnetic field at various temperatures below T_c, showing the sign-change with field and temperature.

The origin of the sign-change becomes clearer when we examine the behavior of the electrothermal conductivity. Data for P of the $Bi_2Sr_2CaCu_2O_{8+\delta}$ sample are shown in Fig. 4a. The similarity to the $YBa_2Cu_3O_7$ data of Fig. 2 are striking. The main difference is that the low-field divergence in $Bi_2Sr_2CaCu_2O_{8+\delta}$ is *negative* in sign. Just as in $YBa_2Cu_3O_7$, the electrothermal conductivity of $Bi_2Sr_2CaCu_2O_{8+\delta}$ appears to approach a constant, positive value in large magnetic fields. The high-field constant value of P is evident in the expanded scale of Fig. 4b.

To determine the origin of the low-field anomaly, it is useful to examine the electrothermal conductivity in more detail, which allows comparison with other recent mixed-state transport measurements. From elementary textbooks, the electrothermal conductivity of each species can be written in the Boltzmann form

$$P_i = \frac{e}{T} \int \frac{d\mathbf{k}}{4\pi^3} \left(-\frac{df}{d\epsilon}\right)(\epsilon - \mu)\mathbf{v}^2 \tau(\epsilon) , \qquad (4)$$

where $f(\epsilon)$ is the distribution function, μ is the chemical potential, and τ is the relaxation time. For the quasiparticles *outside* of the vortex core, e.g., at the nodes of the gap in $YBa_2Cu_3O_7$, it has been demonstrated recently from measurements of the Righi-Leduc effect [4] that the main effect of the external magnetic field is through the relaxation time, $\tau(\epsilon)$. The vortices are efficient scatterers of the external quasiparticles at high field and low-temperature. At the temperatures used in the present work, however, other, non-vortex sources of scattering were observed to dominate.

The low-field divergences of Figs. 2 and 4 increase faster than $1/H^2$ as the magnetic field is reduced and are thus inconsistent [1] with the kinematic reduction [4] of quasiparticle transport by scattering from vortices. It therefore seems more likely that the low-field anomaly is associated with the transport of core excitations. Nevertheless, the microscopic origin of the anomaly, and even its sign remain an unexplained mystery.

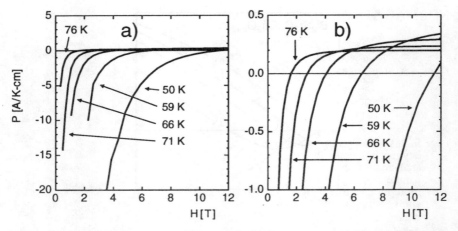

Figure 4: a) Magnetic-field dependence of the electrothermal conductivity, P, of the $Bi_2Sr_2CaCu_2O_{8+\delta}$ sample at various temperatures. b) Expanded view of the region of the sign-change, showing the asymptotically-constant, positive values of P at high-field.

Acknowledgments

Useful discussions with A. van Otterlo, G. Blatter, and C. S. Ting are gratefully acknowledged.

References

1. J. A. Clayhold, Y. Y. Xue, C. W. Chu, J. N. Eckstein, and I. Bozovic *Phys. Rev.* B **53**, 8681 (1996).
2. A. T. Fiory and B. Serin *Physica* **55** 73 (1971).
3. J. N. Eckstein and I. Bozovic, *Ann. Rev. Mat. Sci.* **25**, 679 (1995)
4. K Krishana, J. M. Harris, and N. P. Ong *Phys. Rev. Lett.* **73**, 610 (1995).

NORMAL STATE TRANSPORT PROPERTIES OF TL-2201 SINGLE CRYSTALS

A.M. HERMANN, W KIEHL and R. TELLO
Department of Physics
University of Colorado
Boulder, CO 80309-0390

ABSTRACT

In this paper, transport properties of Tl2201 single crystals as a function of oxygen doping (and corresponding variation of Tc) are presented and analyzed. Hall effect results are shown in detail. Measurements both parallel to and perpendicular to the ab-planes are presented. The ab-plane room-temperature thermoelectric power and Hall constants have differing signs and unusual temperature dependencies. Hall coefficients are temperature dependent and, for B $\perp$ ab-plane, the Hall data at optimal doping (high Tc) do not show the inverse Hall mobility - T^2 correlation "universality" found for most HTSC materials.

1. Introduction

Since its discovery[1], $Tl_2Ba_2CuO_x$ (Tl-2201) has been one of the most interesting high-temperature superconductor systems available for study. It is one of the simplest HTSC crystal structures containing only one Cu-O plane per formula unit[2]. The superconducting transition temperature, Tc, of this system can be tuned over a large temperature range (0 to 92 K) by simple oxygen or inert gas (N_2, He, Ar) annealing[3] (oxygen acts as a p-dopant). In this paper we discuss measurements of the Hall effect on Tl-2201 single crystals as a function of Tc (hole doping).

Hall coefficient measurements have previously performed on bulk Tl-2223[4,5,6], Tl-2212[5], and Tl-2201 [5,7], and on highly doped single crystals of Tl-2201[8,9] The Hall carrier densities observed in these materials are low in comparison to those seen in $YBa_2Cu_3O_7$. Kubo et al.[9] have made measurements of the temperature dependence of R_H on single phase Tl-2201 polycrystalline samples for various oxygen doping levels. Hall measurements on Tl-2201 single crystals for heavily over-doped samples (Tc<50K) have been reported by Manako et al.[8,10]. We report here measurements of two independent components of the Hall coefficient on Tl-2201 single crystals for a wide range of doping concentrations.

2. Experimental

The crystals were grown using a self-flux method.[11] Dimensions were approximately 1 mm x 0.5 mm x 0.1 mm. Selected crystals were analyzed by X-ray diffraction and microprobe analysis for structure and composition. Laue X-ray diffraction patterns verified the quality of the crystals and showed the unit cell to be 3.86 Å x 3.86 Å x 23.2 Å± 0.1 Å. The crystals used were flat plates and the X-ray data showed that the flat plane is the *ab*-plane. The superconducting properties were characterized by both ac susceptibility and four-probe resistivity versus temperature. The as-grown crystals had Tc's of ~ 85 K which could be lowered by simple oxygen annealing.[3]

A double ac Hall measurement method was used. The current and magnetic field were produced at different frequencies resulting in Hall signals at the sum and difference frequencies. By using a lock-in amplifier and adequate filters it was possible to measure very small Hall signals at these frequencies. An advantage of this type of Hall measurement is that thermoelectric and thermomagnetic effects are eliminated since the period of the Hall signal is short compared to the time it takes to set up the temperature gradients associated with these effects. Power requirements for the magnet determined our selection of 60 Hz for the magnetic field frequency f_m. A current frequency f_c of 85 Hz was chosen and the Hall voltage was measured at 25 Hz. No significant unwanted mixing in the circuit was detected. Contacts were made with DuPont Conductor Composition 4871N silver paint.

3. Hall Effect Measurement Results

Two independent components of the Hall tensor and the resistance were measured as a function of temperature for different oxygen doping levels. For measurements of R^H_{abc} the crystal's c-axis is aligned with the magnetic field while for R^H_{cab} it is perpendicular. The alignments were mechanical.

The Hall coefficient was computed using the expression

$$R_H = \frac{V_{Hrms} t}{\sqrt{2} I_{RMS} B_{RMS} f}$$

where V_{Hrms} is the RMS Hall voltage, t is the crystal thickness, I_{rms} is the RMS current and B_{rms} is the RMS magnetic field, and f is the factor[12] which compensates for the shorting of the Hall field that occurs when the length to width ratio is not greater than 3 or 4.

Measurements of R^H_{abc} are reported for the same crystal at four different levels of oxygen doping. The as-grown crystal was superconducting at T_{cmid}=88.5°K. The crystal was then annealed in oxygen to T_{cmid} = 70.0°K, 50.5°K and finally to a Tc less than the refrigerator's loaded cooling limit of 30°K.

Figure 1 shows the R^H_{abc} temperature dependence. The change of sign in the mixed state at low temperature is attributed to vortex motion. Hall coefficient temperature dependences for differing doping levels are compared in Figure 2. The in-plane data are in good agreement with single crystal data available in the literature (i.e. for intermediate-to-low Tc overdoped-crystals). The data for in-plane hole orbits give uniformly positive Hall coefficients.

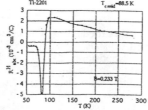

Fig. 1. R^H_{abc} vs. temperature data showing the sign change in the mixed state.

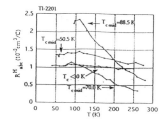

Fig. 2. Comparative measurements of R^H_{abc} vs. temperature for different oxygen doping levels.

Measurements of the temperature dependence of the R^H_{cab} component of the Hall coefficient for three different oxygen doping levels are shown in Figure 3. The midpoint transition temperatures are Tc$_{mid}$ of 87.5°K, 82.9°K, and 69.8°K. The sign is found to be n-type for the optimally doped samples and changes to p-type after a small amount of doping. A small linear decrease of the Hall coefficient with temperature is observed. This dependence contrasts sharply with the other component of the Hall coefficient. The magnitude of the Hall coefficient is surprisingly large (~0.8 cm^3/C at room temperature for Tc$_{mid}$=87.5°K).

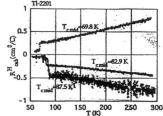

Fig. 3. R^H_{cab} vs. temperature for differing oxygen doping levels.

4. Discussion

With the universalities observed in the Seebeck coefficient, Hall coefficient, and critical temperature vs. doping, it is reasonable to consider that only one band is the major player for ab-plane transport properties.

As pointed out by Ong[13] we may follow Anderson et al.[14] and write Hubbard Hamiltonian consisting of Cu(3d$_{x^2-y^2}$) - O(2pσ) antibonding terms (leading to a wideband) and of on-site repulsive energy U terms as

$$H = -t\sum_{ij} c_{i\sigma}^+ c_{j\sigma^+} + U\sum_i n_{i\uparrow} n_{i\downarrow}$$

where $c_{i\sigma}^+$ creates an electron of spin σ at site i. We can discuss our ab-plane data in terms of the limits of the Hamiltonian in Eq (1) above. If U=0, Eq. 1 has the tight-binding solution in which holes occupy states (in a single band) up to mid-band (for only nearest-neighbor interactions included).

Kubo[15] has proposed a two-dimensional tight-binding model of the Cu-O plane assuming non-zero next-nearest neighbor interactions as well. The single band showed both electron-like and hole-like characteristics arising from the geometry of the band. His model showed that in areas around half filling, there can be a negative Seebeck coefficient (as we have observed previously[16]) and a positive Hall coefficient. The experimentally observed temperature-dependent R_H cannot be explained within this framework, however.

Anderson[17,18] has proposed that the cuprates are in the strong correlation limit (U>>t) in which the ground state (undoped) is an antiferromagnetic (Mott) insulator. Doping by oxygen or by site-selective substitutional atoms introduces holes in the upper Hubbard band.

A T^2 "universality" has been observed[19,20] in the Hall angle (proportional to reciprocal mobility) for a wide variety of cuprates. This T^2 dependence has been interpreted in the large U limit as being due to scattering processes with two separate relaxation times, one due to transport scattering, and one due to transverse Hall scattering. The T^2 "universality" is seen in our data for high doping (low Tc), but significant deviation from T^2 behavior is observed near optimal doping (high Tc). In fact, at optimal doping, $\cot\theta_H$ varies as T^3, as was noted in the infinite layer HTSC compounds[21]. It appears that no simple model is yet available to explain the normal state properties.

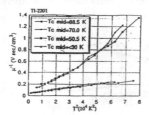

Fig. 4. Reciprocal mobility vs. T^2 for various oxygen doping levels (μ^{-1} is defined as ρ_{ab}/R^H_{cab})

Acknowledgment

We gratefully acknowledge the support of the Office of Naval Research under ONR grant number N00014-95-1-0910.

References

[1] Z.Z. Sheng and A.M. Hermann, *Nature* (London) **332** (1988) 55.
[2] C.C. Torardi, M.A. Subramanian, J.C. Calabrese, J. Gopalakrishnan, K.J. Morrissey, T.R. Askew, R.B. Flippen, U. Chowdhry and A.W. Sleight, *Science* **240** (1988) 631.
[3] A. Maignan, C. Martin, M. Huve, J. Provost, M. Hervieu, C. Michel and B. Raveau, *Physica C* **170** (1990) 350.
[4] J. Clayhold, N.P. Ong, P.H. Hor, and C.W. Chu, *Phys. Rev. B* **38** (1988) 7016.
[5] M.A. Tanatar, V.S. Yefanov, V.V. Dyakin, A.I. Akimov, and A.P. Chernyakova, *Physica C* **185**, (1991) 1247.
[6] C. Hohn, A. Dascoulidou, O. Maldonado, T. Zetterer, and A. Freimuth, *Physica C* **191** (1992) 354.
[7] Y. Kubo, Y. Shimakawa, T. Manako, and H. Igarashi, *Phys. Rev. B* **43** (1991) 7875.
[8] T. Manako, Y. Shimakawa, Y. Kubo, and H. Igarashi, *Physica C* **190** (1991) 62.

[9] Y. Kubo, Y. Shimakawa, T. Manako, T. Kondo and H. Igarashi, *Physica C* **185** (1991) 1253.
[10] T. Manako, Y. Shimakawa, and Y. Kubo, *Proceedings of the Materials Research Society Spring Meeting 1992*, San Francisco, CA.
[11] H.M. Duan, T.S. Kaplan, B. Dlugosch, and A.M. Hermann, *Physica C* **203** (1992) 257.
[12] I. Isenberg, B.R. Russell, and R.F. Greene, *J. Sci. Instrum.* **19** (1948(685.
[13] N.P. Ong, in *Physical Properties of Superconductors II*, ed. by D.M. Ginsberg, World Scientific, Singapore, (1990).
[14] P.W. Anderson, *Science* **235** (1987) 1196.
[15] Y. Kubo, *Phys. Rev. B* **50** (1994) 3181.
[16] W.Kiehl, H.M. Duan, and A.M. Hermann, *Physica C* **253** (1995) 271.
[17] P.W. Anderson, G. Baskaran, Z. Zou, and T. Hsu, *Phys. Rev. Lett* **58** (1987) 2790.
[18] P.W. Anderson, in *Frontiers and Borderlines in Many Particle Systems, Proceedings of theVarenna Summer School,* Varenna, Italy, (1987), ed. by R.A. Broglia and J.R. Schrieffer (North Holland, Amsterdam, 1988).
[19] Y. Kubo and T. Manako, *Physica C* **197** (1992) 2855.
[20] A. Carrington, A.P. Mackensie, C.T. Lin, and J.R. Cooper, *Phys. Rev. Lett.* 69 (1992) 2855.
[21] E.C. Jones, D.P.Norton, D.K. Christen and D.H. Lowndes, *Phys. Rev. Lett.* **73** (1994) 166.

PRESSURE EFFECT ON THE SUPERCONDUCTING AND NORMAL-STATE PROPERTIES FOR THE PBCO/PBCO SUPERLATTICE

J.G.LIN, M. L. LIN, AND H.C.YANG
Department of Physics, Taiwan National University, Taipei, R.O. C.

C. Y. HUANG
Center for Condensed Matter Sciences, Taiwan National University, Taipei, R.O.C.

Z. J. HUANG
*Texas Center for Superconductivity at University of Houston
Houston, Texas 77204-5932*

ABSTRACT

The pressure dependence of the resistivity and superconducting transition temperature for the $(YBCO)_n/(PBCO)_m$ superlattice with n/m = 48/36 and 120/120 has been measured, and the results have been compared with that of pure YBCO. A metal-semiconductor-like transition around 7 kbar has been observed for both superlattice, which phenomenun is absent in pure YBCO. We have attributed this observation to the pressure-enhanced localization effect at the interface of YBCO and PBCO layers.

1. Introduction

Since the discovery of high temperature superconductors (HTS) in 1986[1], an enormous amount of experimental and theoretical efforts have been devoted for understanding these materials. It is generally accepted that the charge carriers are holes in the CuO_2 layer, and the Fermi surface, if there is one, is expected to have little dispersion along the c axis. But, an important issue has often been raised; that is to what extent the charge transport is a property of the individual CuO_2 layer and how important the interlayer coupling is for the superconducting properties.[2]

Experimentally, studies of $YBa_2Cu_3O_7(YBCO)/PrBa_2Cu_3O_7(PBCO)$ superlatties show that the superconducting transition temperature T_c persists even in a unit cell of YBCO.[3-5] The superlattice $YBa_2Cu_3O_7/PrBa_2Cu_3O_7$ (YBCO/PBCO) has an advantage of being tailored into various numbers of YBCO and PBCO layers. With an increase in the thickness of the non-superconducting PBCO layer, the strength of the interlayer coupling decreases. Similar to the changing of the thickness of PBCO layer between YBCO layers, pressure could decrease the volume of the PBCO unit cell, and, consequently, increases the interlayer coupling between two YBCO layers. We, therefore, decided to carry a high pressure study on the superlattice films of $(YBCO)_n/(PBCO)_m$ with n and m varied, hoping to provide some useful information regarding the role of interlayer coupling in the mechanism of high-T_c superconductivity.

2. Experiments

The $(YBCO)_n/(PBCO)_m$ superlattices with n/m = 48/36 and 120/120 were prepared by an off axis RF magnetron sputtering system onto $SrTiO_3$ (001) substrates. The films consist of 10 repeating layers of n/m. The hydrostatic pressure up to 14 kbar was generated by the modified clamp technique, the equipment is composed mainly of two parts: (I) a high pressure cell constructed of BeCu and (II) a cylinder capsule made of Teflon. We use 3M Flourinant as our pressure medium.

The resistivity is determined by the standard four-probe technique. The typical dimension of the sample is 2000Å×2mm×5mm. Electrical contacts were established by attaching 0.05 mm Pt-wires to the sample with In-solder.

3. Results and discussion

The value of T_c is defined as the temperature at which the electrical resistivity drops to 50% of its extrapolated normal state value. Figure 1 shows the resistance R(T) for the superlattice with n/m = 48/36 under the pressures of 0, 3, 5, 7, and 9 kbar. The T_c-values are 87.4 K, 87.6 K, 87.7 K, 82.5 K and 81.5 K, respectively. The slopes of R vs. T in the normal state are positive at P = 0.1, 3 and 5 kbar but negative at 7 and 9 kbar.

Figure 2 shows the resistance R(T) for the superlattice with n/m = 120/120 under the pressures of 0, 3, 5, 7 and 9 kbar. The T_c-values are 88.4 K, 88.7 K, 88.5 K, 83.5 K and 83.0 K, respectively. The slopes of R vs T in the normal state are positive at P = 0, 3 and 5 kbar, and, are negative at 7 and 9 kbar.

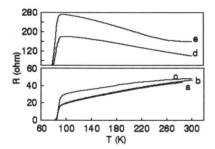

Fig. 1 Resistivity (R) vs Temperature (T) for (YBCO)n/(PBCO)m superlattice with n/m = 48/36 at the pressures ranging from 0 to 9 kbar.

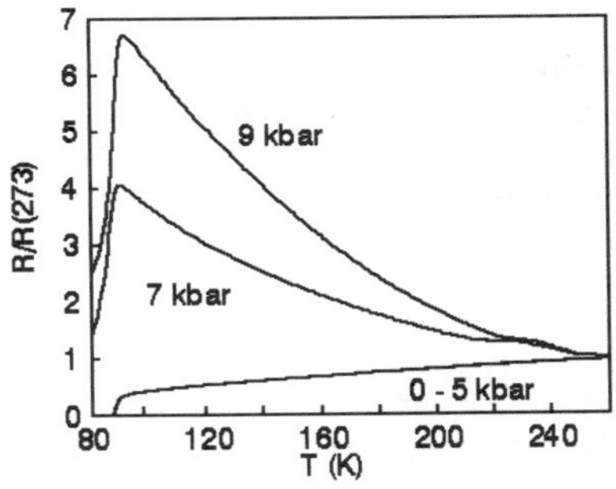

Fig. 2 Resistivity (R) vs Temperature (T) for (YBCO)n/(PBCO)m superlattice with n/m = 120/120 at the pressures ranging from 0 to 9 kbar.

Fig. 3(a), 3(b) and 3(c) show the variation of T_c under various pressure for the superlattice with n/m = 120/120, and 48/36, respectively. For pure YBCO, T_c linearly increases with pressure up to 14kbar; while for n/m = 120/120 and 48/36, it is observed that T_c initially increases with increasing pressure when the applied pressures are lower than 7Kbar. When P equals or larger than 7 kbar, T_c decreases with pressure for both superlattices.

Figures 4(a), 4(b) and 4(c) show dR/dT vs. P for the pure YBCO and the superlattices with n/m = 120/120 and 48/36, respectively. In Fig.4(a), dR/dP is positive up to 14 kbar. Fig.4(c) is similar to Fig.4(b). The common feature in Fig.4(b) and 4(c) is a metal to semiconducting-like transition at P = 7kbar. Many explanations have been put forth to explain the T_c dependence of the number of PBCO unit cell. The papers of Wu et al.[6] and Rajagopal et al.[7] used the proximity effect arguments, Wood [8,9] invoked spin-polaron model, while Liu et al.[10] proposed the model of the interlayer scattering effect.

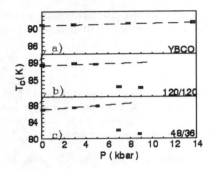 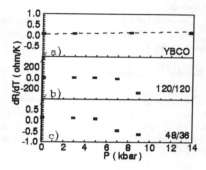

Fig. 3 T_c vs. pressure (P) for (a) pure YBCO, (b) n/m = 120/120 and (b) 46/36. The dash lines are the guidance for eyes.

Fig. 4 dR/dT vs. pressure (P) for (a) pure YBCO (b) n/m = 120/120 and (c) 46/36.

Accordingly, pressure shortens the c-lattice parameter of the unit cell between the CuO planes, thus, it is expected that the coupling should be enhanced with appling pressure. The proximity effect should also be enhanced with the pressure because that the leaking of the pair amplitude of the superconducting YBCO layer to the insulating PBCO layer depends on the distance between these layers. When pressure is lower than some critical point (below 7kbar), we have observed an increasing T_c with increasing P, which is consistent with the proximity effect. But for P equaling or higher than 7 kbar, T_c decreases with increasing P, which contradicts with the proximity effect or interlayer coupling model.

Based on our results, there exist a critical pressure which generates some dramatic effect on the transport property of the CuO_2 layer. The diffusion scattering model proposed by Berkowitz et al.[10] suggested an origin for the transport property of YBCO/PBCO superlattice. In this model, some structure defeats at the interface, act as scattering centers and provide an additional contribution to the total resistivity. We, thus, believe that this interface effect may be greatly enhanced when pressure exceeds some critical value. And the scattering process may evolve into the semiconducting-like localization of holes.

4. Conclusion

We have measured the pressure effect on T_c for $(YBCO)_n/(PBCO)_m$ superlattice with n/m = 48/60 and 48/36, and have observed a phase transition from the metallic to the semiconducting state for both samples. We have adopted the model of scattering at the interface of YBCO/PBCO to explain our observation. It is assumed that pressure enhances the interface-defect between YBCO and PBCO layers, thus inducing a hole-localization in CuO_2 plane. The critical pressure required for inducing this hole-localization may be around 7 kbar for our YBCO/PBCO superlattices.

Acknowledgments

The authors would like to thank Mr. L. M. Wang for providing the YBCO/PBCO superlattice. This work is supported in part by the National Science Council of the R. O. C. under the grant No. NSC-85-2112-M-002-022.

References

1. M. K. Wu, J. R. Ashburn, C. J. Torng, P. H. Hor, R. L. Meng, L. Gao, Z. J.Huang, Y. Q.Wang and C. W. Chu, Phys. Rev. Lett. **58**, 908 (1987).
2. S. L. Cooper and K. E. Gray, Physical Properties of High Temperature Superconductors IV, ed. Donald M. Ginsberg (Singapore, World Scientific, 1992).
3. D. H. Lowndes, D. P. Nortan, J. D. Budai, Phys. Rev. Lett. **65**, 1160 (1990)
4. Q. Li, X. X. Xi, X. D. Wu, A. Inam, S. Vadlamannati, W. L. Mclean, T. Venkatesan, R. Ramesh, D. M. Hwang, J. A. Martinez, L. Nazar, Phys. Rev. Lett. **64**, 3086 (1990).
5. Terashima, K. Shimura, Y. Banda, Y. Matsuda, A. Fujiyama, and S. Komiyama, Phys. Rev. Lett. **67**, 1362 (1991).
6. J. Z. Wu, C. S. Ting, W. K. Chu, and X. X. Yao, Phys. Rev. B **44**, 411 (1991).
7. A. K. Rajagopal and S. D. Mahanti, Phys. Rev. B **44**, 10210 (1991).
8. R. F. Wood, Phys. Rev. Lett. **66**, 829 (1991).
9. R. F. Wood, M. Monstoller, and J. F. Cooke, Physica C **165**, 97 (1990).
10. G. Liu, G. Xiong, G. Li, G. Lian, K. Wu, and S. Liu, Phys. Rev. B **49**, 15287 (1994).

ANOMALOUS RESISTIVITY IN THE MIXED STATE OF $CeRu_2$ [*]

J. Herrmann,[a] N. R. Dilley,[a,b] S. H. Han,[a] M. B. Maple[a,b,c]

[a] *Institute for Pure and Applied Physical Sciences, University of California, San Diego*
La Jolla, CA 92037-0360, USA
[b] *Department of Physics, University of California, San Diego*
La Jolla, CA 92093-0319, USA
[c] *Institute for Theoretical Physics, University of California, Santa Barbara*
Santa Barbara, CA 93106-4030, USA

Low frequency ac electrical resistivity ρ measurements, performed as a function of magnetic field H and temperature T on a polycrystalline sample of $CeRu_2$, revealed the development of a finite resistance anomaly in the superconducting state well below the upper critical field $H_{c_2}(T)$. The anomaly is very sensitive to the transport current density and appears to be due to a decrease and subsequent increase of fluxoid pinning as $H_{c_2}(T)$ is approached from within the mixed state.

1 Introduction

Research on the mixed state properties of the high T_c cuprate superconductors has revealed a rich variety of phases in the magnetic magnetic field H – temperature T plane.[1] This has rekindled interest in vortex phases and dynamics in low T_c superconductors such as the A15, magnetic rare earth and heavy fermion superconductors. A particularly interesting example is the binary compound $CeRu_2$ for which an irreversible feature in magnetization M vs. H isotherms was observed [2] between a low-field region (where the magnetization appears to be reversible) and the upper critical field H_{c_2}. The anomaly closely resembles the irreversible magnetization curves observed in superconducting compounds exhibiting a "peak effect" in the critical current density,[3] which involves an increase and subsequent decrease (peak) of the pinning of the flux line lattice (FLL) as the superconducting transition is approached from within the mixed state. Among the possible mechanisms that have been invoked to explain this magnetization anomaly are the formation of a spatially inhomogeneous superconducting Fulde-Ferrel-Larkin-Ovchinnikov (FFLO) state,[4] a softening of the shear modulus of the FLL,[6] matching effects of commensurate pinning and vortex structures, or collective pinning,[7] which can give rise to a peak in the pinning force when the correlation length approaches the lattice constant of the FLL.

We report electrical resistivity ρ measurements on $CeRu_2$ as a function of H and T that reveal the development of a finite resistance anomaly in the superconducting state well below the upper critical field $H_{c_2}(T)$ in the range $0 \leq \mu_0 H \leq 1.5$ T. The appearance of the resistive region seems to be due to a decrease and subsequent increase of fluxoid pinning as $H_{c_2}(T)$ is approached from within the mixed state and appears to be related to the irreversible behavior of the magnetization near $H_{c_2}(T)$ in $CeRu_2$ that is observed above 1 T.

---

[*]This work was supported by the U.S. Department of Energy under Grant No. DE-FG03-86ER-45230 at UCSD and in part by the National Science Foundation under Grant No. PHY94-07194 at UCSB.

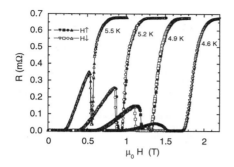

Figure 1: Magnetic field dependences of the ac electrical resistance R for increasing and decreasing field at different temperatures, illustrating the resistive anomaly well below the mean field superconducting transition. For $T < 5.4$ K, hysteresis develops at the transition into the resistive state. This behavior is qualitatively reproduced in the temperature dependences of R in different magnetic fields.

2 Results and Discussion

Figure 1 shows the electrical resistance $R(H)$ of a polycrystalline sample of CeRu$_2$, measured with a transport current density of 13 A cm^{-2} while ramping the magnetic field at a rate of 400 mT/min for different constant temperatures. The most striking feature of $R(H,T)$ is the appearance of a region with nonzero resistance well below the onset of the main superconducting transition for $T > 4.3$ K. A minimum in $R(H)$ reminiscent of the anomaly can be traced to temperatures as high as 5.9 K. The different characteristic fields associated with the anomaly are illustrated in Fig. 2 for a field sweep at $T = 4.5$ K. The $R(H)$ curves measured at higher temperatures ($T \geq 5.4$ K) are (within our experimental uncertainty) completely reversible. At lower temperatures, for which the resistance drops to zero below the onset field H_0 of the main transition, we observe that the subsequent transition into the state with finite resistance starts to exhibit hysteretic characteristics, which become more pronounced with decreasing temperature; for $T \leq 4.4$ K, the resistance remains zero for decreasing field in the entire field interval $H < H_0$.

The $R(H)$ results were confirmed by data obtained while monotonically sweeping the temperature at a rate of ~ 25 mK/min in a fixed magnetic field that clearly showed the presence of the anomaly and the hysteretic character of the transition into the resistive state, which becomes more pronounced with increasing field. During warming sweeps, we observed instabilities of the resistance and spontaneous warming of the sample for fields $\mu_0 H \geq 0.7$ T in the temperature region where the resistance first becomes nonzero. This behavior appears to be a manifestation of flux jumps in the sample.

The resistive anomaly is strongly dependent on the transport current I; it is more pronounced for higher I and gradually disappears for lower I. At the same time, the hysteresis in the isothermal $R(H)$ dependences becomes weaker with increasing I and stronger with decreasing I.

Figure 2 compares isothermal field sweeps of resistance and magnetization (measured in a Faraday magnetometer) at $T = 4.5$ K. The field interval $H_{2\uparrow} \leq H \leq H_0$ with $R = 0$ corresponds to the field interval in which the irreversible anomaly in $M(H)$ is observed. Although field hysteresis develops in the onset of the $M(H)$ anomaly at lower temperatures (see Fig. 3(a)), it is negligible in the temperature interval 4.6 K $< T <$ 5.4 K where the

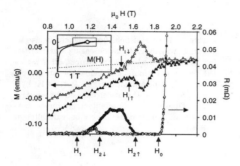

Figure 2: Magnetic Field H dependence of resistance R and bulk magnetization M at $T = 4.5$ K. The field interval where the anomaly in $M(H)$ is observed roughly corresponds to the field interval below the onset field H_0 in which R vanishes. The dashed line is a linear fit to the reversible normal state paramagnetic magnetization which extrapolates to zero. For decreasing H, the hysteresis in the resistive onset $H_{2\uparrow\downarrow}$ is much more pronounced than that in magnetization $H_{i\uparrow\downarrow}$. The inset shows the entire magnetization curve $M(H)$ for $H > 0$ with the anomaly indicated.

hysteresis in $R(H)$ is observed. Isothermal magnetization curves $M(H)$ obtained from Faraday magnetometer measurements reveal a small yet distinct irreversibility throughout the superconducting mixed state when the anomaly in $M(H)$ is present near H_{c_2}.

Figure 3 summarizes our results in the form of H-T phase diagrams. The size, location, and the hysteretic features of the resistive anomaly in the H-T plane are strongly dependent on the current density: with increasing current, the resistive region becomes more pronounced ($H_1(T)$ gets shifted to lower fields) and the maximum resistance attained increases, while the position of the "transition" line $H_{2\uparrow}(T)$ is less affected. The transition into the state with $\rho = 0$ at $H_{2\uparrow}$ corresponds to the onset of the irreversible magnetization at $H_{i\uparrow}$ over the entire temperature interval studied.

The strong dependence of our results on the transport current density suggests a correlation of the resistive anomaly with changes in the pinning properties. Small-angle neutron scattering measurements [8] performed at $T = 1.8$ K showed that the FLL in the mixed state below the magnetization anomaly ($H < H_i$) consists of aligned rigid bundles of parallel flux lines, whereas the diffracted intensity dropped and eventually vanished for $H > H_i$. Upon field cooling from the irreversible state, the bundle size decreased in both the direction parallel and perpendicular to the external magnetic field. In light of these findings, a possible scenario for the mixed state behavior of the resistivity is a transition from the regime of weak collective pinning to strong pinning of individual vortices at $H_{2\uparrow\downarrow}$ and $T_{2\uparrow\downarrow}$, respectively. In the former case, which is described by the collective pinning theory, [7] the translational order of the FLL is destroyed by weak pinning, and vortex bundles characterized by correlation lengths L_c and R_c parallel and perpendicular to the applied magnetic field are coherently pinned, where the bundle size depends on the defect concentration. In the strong pinning regime, the nature of the flux arrangement is determined by the size, distribution, and density of the pinning sites. Whether there exists an intermediate vortex glass phase with true zero resistivity as opposed to a strongly pinned vortex liquid with a finite flux-creep resistivity $\rho \sim \exp(-U/k_BT)$ due to the thermal activation of vortices across a random pinning potential of energy scale U,[9] remains to be investigated.

The distinct hysteretic character of the transition at $H_{2\uparrow\downarrow}$ and $T_{2\uparrow\downarrow}$ in $R(H,T)$ and of the onset of the magnetization anomaly at $H_{i\uparrow\downarrow}$ in $M(H)$ could be an indication that the transition the FLL undergoes is of first order. We note that hysteretic features similar

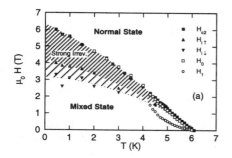
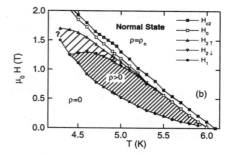

Figure 3: (a) Comprehensive phase diagram comparing $R(H)$ data for $J = 13.0$ A cm^{-2} (open symbols) with $M(H)$ data obtained in SQUID and Faraday magnetometers (solid symbols). The shaded area represents the irreversible anomaly in $M(H)$, with a lower bound H_i which becomes more hysteretic at lower temperatures. All boundaries are guides to the eye. (b) Detail of the H-T phase diagram obtained from $R(H)$ measurements. The various characteristic fields are defined in Fig. 2.

to those observed in the magnetization curves at $H_{i\uparrow\downarrow}$ have been reported.[10] Furthermore, pronounced magnetocaloric and magnetostriction anomalies have been observed which indicate that a first order transition occurs at H_i.[11] The pronounced hysteresis of a diamagnetic anomaly observed in the temperature dependence of the ac magnetic susceptibility χ',[12] as well as thermal hysteresis in isofield magnetostriction [13] were interpreted as indications of a first order transition associated with the formation of the FFLO state.

References

1. G. Blatter et al., Rev. Mod. Phys. **66**, 1125 (1994).
2. S. B. Roy, Phil. Mag. B **65**, 1435 (1993).
   A. D. Huxley et al., J. Phys.: Condens. Matter **5**, 7709 (1993).
   K. Yagasaki, M. Hedo, and T. Nakama, J. Phys. Soc. Jpn. **62**, 3825 (1993).
3. A. M. Campbell and J. E. Evetts, Adv. Phys. **21**, 327 (1972).
4. P. Fulde and R. A. Ferrell, Phys. Rev. **135**, A550 (1964).
   A. I. Larkin and Y. N. Ovchinnikov, Zh. Eksp. Teor. Fiz. JETP **47**, 1136 (1964).
5. K. Gloos et al., Phys. Rev. Lett. **70**, 501 (1993).
6. A. B. Pippard, Philos. Mag. **19**, 217 (1969).
7. A. I. Larkin and Y. N. Ovchinnikov, J. Low Temp. Phys. **34**, 409 (1979).
8. A. Huxley et al., Proceedings SCES'95, to be published.
9. D. S. Fisher, M. P. A. Fisher, and D. A. Huse, Phys. Rev. B **43**, 130 (1991).
10. H. Sugawara et al., Physica B **206–207**, 196 (1995).
11. H. Goshima et al., Physica B **206–207**, 193 (1995).
    M. Tachiki et al., Z. Phys. B, submitted.
12. T. Nakama et al., J. Phys. Soc. Jpn. **64**, 1471 (1995).
13. R. Modler et al., Phys. Rev. Lett. **76**, 1292 (1996).

TEMPERATURE DEPENDENCE OF THE UPPER CRITICAL MAGNETIC FIELD IN BiSrCuO and NdCeCuO

M. S. OSOFSKY, R. J. SOULEN, JR., S. A. WOLF
Naval Research Laboratory, Washington, DC 20375-5000

J. M. BROTO, H. RAKOTO, J. C. OUSSET, G. COFFE, S. ASKENAZY
Service National Des Champs Magnetique Pulses, 31077 Toulouse Cedex, France

P. OSWALD
Clarendon Laboratory, Oxford University, Parks Road, Oxford OX13PU, UK

P. PARI
Centre d'Etudes de Saclay, Service de Physique de l'Etat Condense, Laboratoire des Basses Temperatures, 91191 Gif-sur-Yvette, France

I. BOZOVIC, J. N. ECKSTEIN, G. F. VIRSHUP
Edward L. Ginzton Research Center, Varian Associates, Palo Alto, CA 94304-1025

J. COHN
Department of Physics, University of Miami, Coral Gables, FL 33124

X. JIANG, S. MAO, R. GREENE
University of Maryland, Center for Superconductivity Research, Department of Physics, College Park, Md 20742

ABSTRACT

The temperature dependence of the upper critical field, $H_{c2}(T)$, has been measured for thin $Bi_2Sr_2CuO_x$ and $Nd_{1.85}Ce_{0.15}CuO_x$ films in magnetic fields up to 35 Tesla and temperatures down to 70mK. $H_{c2}(T)$ for Biscco diverged anomalously as the temperature decreased while the NdCeCuO sample exhibited conventional behavior. Thermoelectric power measurements indicate that the Bisco sample is overdoped suggesting that the $H_{c2}(T)$ data can be explained by a pair-breaking theory of Ovchinnikov and Kresin.

Recent measurements of the temperature dependence of the upper critical field, $H_{c2}(T)$, in several high temperature superconductors have revealed an unusual divergence as the temperature approaches zero.[1,2,3] More recently, Ovchinnikov and Kresin[4] developed a theory that explains these results as being due to a pair-breaking effect which weakens when the pair-breaking moments order at low temperatures. In this paper we present $H_{c2}(T)$ and thermoelectric power (TEP) results for two systems with T_c near 20 K: $Bi_2Sr_2CuO_x$, which shows the divergent $H_{c2}(T)$ at temperatures down to 65 mK and fields up to 35 T, and $Nd_{1.85}Ce_{0.15}CuO_x$, which has conventional $H_{c2}(T)$ behavior.[5]

The preparation of the thin film samples[6,7] and details of the pulsed field measurements[3] are discussed elsewhere. In ambient magnetic field the BiSrCuO film the superconductive transition began at 19 K and was complete at 12 K. The resistive onset of the NdCeCuO film began at 22 K and the transition was complete at 19 K. When the magneto-resistance was measured, the films were oriented with the c-axis parallel to the applied field. The BiSrCuO film was measured with a 0.4 µA current and the NdCeCuO film with 1 and 10 µA currents.

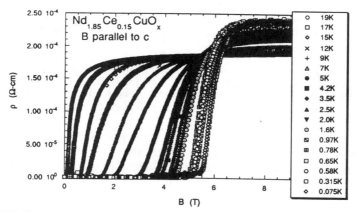

FIG 1. Resistivity, ρ as a function of magnetic field H at several temperatures T. The temperatures decrease monotonically from 19 K (far left-hand curve) to 75 mK (far right-hand curve).

Fig. 1 shows a series of ρ(H,T) curves for NdCeCuO for temperatures varying from 19 K down to 75 mK. These curves indicate that the transition widths are relatively insensitive to the applied field strength and that the quenched, normal state resistance increases below $T_C(H=0)$. We define H_{c2} for each curve as the magnetic field where the extrapolated normal state resistance and the tangent of the transition meet (see ref. 3).

Fig. 2 displays the $H_{c2}(T)$ data extracted from the R(H,T) data shown in Fig. 2 along with the BiSrCuO Hc2(T) curve from ref. 3. The line through the NdCeCuO data is the conventional WHHM curve.[3,8] The line through the BiSrCuO data is the two parameter fit to the pair-breaking theory of Ovchinnikov and Kresin. The Details of the theory are given in reference 4.

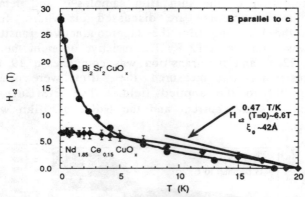

FIG. 2 Upper critical magnetic field $H_{c2}(T)$ data for NdCeCuO derived from Fig. 1 and BiSrCuO data from ref. 3. Solid curves: WHHM theory for NdCeCuO (estimated ξ_0=42Å in the ab plane) and the theory of Ovchinnikov and Kresin for BiSrCuO.

It has been suggested that, in at least some of the cuprates,[4] overdoping introduces impurities into the system which would be the source of this pair-breaking. Heat capacity measurements[9] confirm this in the TlBaCuO system which also had a diverging $H_{c2}(T)$.[2]

Obertelli *et al.* have shown that the room temperature value of the TEP is a reliable indicator of the doping of the cuprate superconductors.[10] We performed TEP measurements on samples with similar T_c to those in the $H_{c2}(T)$ study to get an estimate of the doping of the two systems. The TEP data for the BiSrCuO sample looks rather conventional with a large negative TEP (-12.5 μv/K) at room temperature indicating that the sample is strongly overdoped. From the results of ref. 10 we conclude that the T_c in optimally doped BiSrCuO is over 100 K. The NdCeCuO TEP data has an unusual kink at around 80 K. The room temperature value is -1.25 µV/K showing that the sample is almost optimally doped.

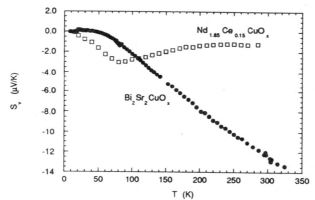

Figure 3. TEP of samples similar to those in figure 2.

These results are consistent with the TlBaCuO data suggesting that overdoping is crucial to the divergent $H_{c2}(T)$ phenomenon, perhaps through the introduction of magnetic impurities as suggested by Ovchinnikov and Kresin.

Research at the University of Maryland was supported by the National Science Foundation.

References.

1. Y. Dalichaouch, et al., *Phys. Rev. Lett.* **64**, 599 (1990).
2. A.P. Mackenzie et al, *Phys. Rev. Lett.* **71**, 1238(1993).
3. M. S. Osofsky, et al., *Phys. Rev. Lett.* **71**, 2315 (1993); M. S. Osofsky, et al., *J. Supercond.* **7**, 279 (1994).
4. Y. N. Ovchinnikov and V. L. Kresin, *Phys. Rev. B* **52**, 3075 (1995).
5. M. Hikata and M. Suzuki, *Nature* **338**, 635 (1989); M. Suzuki and M. Hikata, *Phys. Rev. B* **41**, 9566 (1990).
6. J. N. Eckstein, et al., *Thin Sold Films* **216**, 8 (1992); and references therein.
7. S. N. Mao et al., *Appl. Phys. Lett.* 61, 2357 (1992).
8. E. Helfand, N. R. Werthamer, and P. C. Hohenberg, *Phys. Rev.* **147**, 295 (1966); and references therein.
9. J. Wade et al., *J. Supercond.* **7**, 261 (1994).
10. S. D. Obertelli et al., *Phys. Rev. B* **46**, 14928 (1992).

NONLINEAR MICROWAVE SWITCHING RESPONSE OF BSCCO SINGLE CRYSTALS

T. JACOBS[1], BALAM A. WILLEMSEN[1,2] and S. SRIDHAR[1]

[1] *Physics Department, Northeastern University, Boston, MA 02115*
[2] *Rome Laboratory, Hanscom AFB, Bedford, MA 01731*

QIANG LI

Material Science Division, Brookhaven National Laboratory, Upton, NY 11973

G. D. GU and N. KOSHIZUKA

Superconductivity Research Laboratory, ISTEC 10-13, Shinonome I-chrome, Koto-ku, Tokyo, 135, Japan

Measurements of the surface impedance, Z_s, in $Bi_2Sr_2Ca_1Cu_2O_{8+\delta}$ single crystal with microwave currents flowing along the $\hat{c}$ axis show clear evidence of a step-like nonlinearity. The surface resistance switches between apparently quantized levels for microwave field strength changes of less than 1 mG. This non-linear response can arise from the presence of intrinsic Josephson junctions along the $\hat{c}$ axis of these samples driven by the microwave current.

The work described in this paper was motivated by a chance observation during the study of the microwave response of high quality $Bi_2Sr_2Ca_1Cu_2O_{8+\delta}$ (BSCCO) single crystals. While measuring the surface impedance, $Z_s(T)$, of these crystals, an unusual step-like switching feature was observed in the cavity resonances. (See Fig. 1) These effects are only observed in the presence of microwave currents induced along the $\hat{c}$ axis, and seem qualitatively consistent with the underdamped response of intrinsic Josephson junctions (JJs) in these samples.

The measurements are performed in a Nb cavity [1] operating in the TE_{011} mode at 10 GHz which is held at fixed temperature, $T \leq 4.2$ K. The sample sits on a sapphire rod at the center of the cavity so that its temperature can be varied independent of that of the cavity. The microwave field at the center of the cavity induces circulating microwave currents in the sample. By orienting the sample $\hat{c}$ axis $\parallel H_{rf}$ these currents flow entirely in the ab plane, while for $\hat{c} \perp H_{rf}$ current flows along the $\hat{c}$ axis as well as in the ab plane. All of the measurements reported here are for the latter geometry, and the $\hat{c}$ axis response dominates due to the large anisotropy.

The sample surface impedance, $Z_s = R_s + iX_s$, is typically determined from changes in the quality factor and resonance frequency of the perturbed cavity by $R_s = \Gamma(Q^{-1} - Q_b^{-1})$ and $\Delta X_s = \xi(f_b - f_0)$, where Γ and ξ are geometric factors, Q_b and f_b are the background Q and f_0. When the cavity resonance is Lorentzian, as is usually the case, f_0 is determined from the maximum of the peak, and the 3 dB (half power) width of the peak, Δf_{3dB}, gives $Q^{-1} = \Delta f_{3dB}/f_0$.

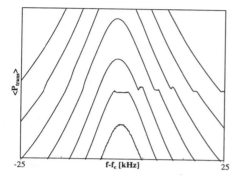

Figure 1: Typical non-linear response of BSCCO at 4 K presented for an 50 kHz frequency span centered at $f_c = 10.114738$ GHz. The traces presented were taken in 3 dB increments of the input power.

Careful examination of the cavity resonances revealed clear step-like deviations from Lorentzian shape (See Fig. 1), and thus from linear response. At low input powers the resonance appears Lorentzian, and can thus be used with the ab plane results to extract the temperature dependence for the $\hat{c}$ axis alone.[2] An important feature of the data which can clearly be seen in Fig. 1 is the existence of a characteristic microwave power level at which the switching occurs, which stays constant over a wide range of input power.

When the cavity resonances are clearly no longer Lorentzian the usual relationships between the 3 dB bandwidth, quality factor Q and R_s cease to hold, and we must resort to other methods to relate the measurements to sample properties. We have developed a method [3] which relates the complex transmission through the perturbed cavity, S_{21}, directly to changes in surface impedance ΔZ_s of the sample.

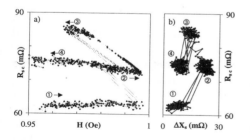

Figure 2: (a) $R_s(H_{rf})$ extracted from a resonance near $H_{rf\,switch}$(4 K) (b) $R_s(H_{rf})$ vs. $\Delta X_s(H_{rf})$

The results of our analysis applied to a resonance measured very near the initial onset of nonlinearities at 4 K is shown in Fig. 2a. The plot of R_s vs. H_{rf} shows that initially R_s is independent of H_{rf} as in the case of linear response, but an abrupt jump to increased dissipation occurs at a critical value of H_{rf}. This enhanced dissipation leads directly to the observed drop in H_{rf}. Interestingly the resistance almost appears quantized, being independent again of H_{rf} for $H_{rf} > H_{rf\,switch}$. The response is clearly hysteretic since the downward jump occurs at a different value of $H_{\rm rf}$, giving rise to the observed asymmetry of the resonance. It is obvious that these features can only be quantitatively analyzed after the extraction of the R_s and X_s vs. $H_{\rm rf}$. The changes in R_s are accompanied by simultaneous changes in X_s, as can be seen in Fig. 2b.

We were able to track $H_{\rm rf\,switch}$ as a function of temperature (See Fig. 3) in a number of samples up to ~ 70 K where the signal strengths become too weak to clearly determine it. For a simple one junction model, one expects that $H_{\rm rf\,switch} \propto 1/\lambda_c$, so for comparison $1/\lambda_c$ is presented on the right hand axis.

$H_{\rm rf\,switch} \sim 1$ Oe can be related to a critical current density by $J_c = H_{\rm rf}/\lambda_c(T)$, where $\lambda_c(T)$ is the $\hat{c}$ axis penetration depth. Using $\lambda_c(0) \sim 40\mu$m, as estimated from fitting $\lambda_c(T)$ data to a tunneling model [2], we get $J_{c\,\rm rf\,switch} = 200\,{\rm A/cm}^2$. This is well within the range obtained in mixed ac-dc experiments [4,5] with $70\,{\rm A/cm}^2 \leq J_{c\,\rm rf\,switch} \leq 1250\,{\rm A/cm}^2$ and corresponds to the value expected for optimally doped crystals with high T_c.

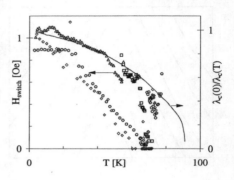

Figure 3: $H_{\rm rf\,switch}(T)$, corresponding to the onset of nonlinearities. A line representing $\lambda_c(0)/\lambda_c$ with $\lambda_c(0) = 40\mu$m from a tunneling model are plotted on the right axis for comparison with single junction expectations.

The ab plane critical current densities are much higher, $J_{c\,ab} \sim 7 \times 10^6$ A/cm^2.[6] The fact that similar switching behavior is not observed in our experiments when $H_{\rm rf} \parallel \hat{c}$ (i.e. when only ab plane currents are induced) is entirely consistent with this large anisotropy in J_c.

The picture which emerges from these measurements is consistent with the previous results on BSCCO, being a highly anisotropic material composed of CuO_2 superconducting layers in the ab plane, coupled weakly along the $\hat{c}$ axis. The presence of JJs along the $\hat{c}$ axis is a distinct possibility and behavior consistent with this has also been observed in microwave emission and Shapiro step measurements in the presence of dc-currents.[7]

It is important to point out that the experiments described here differ from typical microwave experiments on JJs, in that we measure *dynamic* losses as a function of pure ac drive in the absence of any dc current or field bias. This problem has only recently begun to be studied on experiments in manufactured JJs and associated calculations.[9,8,10]

All of these consider the case of an overdamped nonlinear resistively shunted junction (RSJ), which does not seem to be able to describe our present data on

BSCCO, even qualitatively. The switching response and the rather complicated hysteresis observed may indicate that one needs to consider an underdamped junction for BSCCO, by inclusion of a mass term in the calculations. The equation $(\ddot{\phi} + \beta\dot{\phi} + \sin\phi = i_{rf}\cos\omega t)$ which arises has many complex solutions,[11] including regions of chaos, where such switching behavior may be observed. This distinguishes the nonlinearities observed in BSCCO from those observed in the *ab* plane of twinned YBCO single crystals [12] which do seem to be very well described by the overdamped RSJ model.[10] Overall, such nonlinear microwave response with JJ like behavior appears to be a characteristic intrinsic feature of the cuprates.

Acknowledgments

This work was supported by NSF-DMR-9623720. We thank John S. Derov and Rome Laboratory, Hanscom AFB for encouragement and the loan of essential equipment and P. Kneisel for annealing the Nb cavity. Qiang Li is supported by the US Dept. of Energy under contract No. DE-AC02-76CH00016. T. Jacobs is presently at the University of Karlsruhe, Germany.

References

1. S. Sridhar and W. L. Kennedy, Rev. Sci. Instrum. **59**, 531 (1988).
2. T. Jacobs *et al.*, Phys. Rev. Lett. **75**, 4516 (1995).
3. T. Jacobs, B. A. Willemsen and S. Sridhar (to be published); T. Jacobs, Diploma Thesis, Universität Karlsruhe, Karlsruhe, February 1996.
4. R. Kleiner *et al.*, Phys. Rev. B **50**, 3942 (1994).
5. S. Luo, G. Yang, and C. E. Gough, Phys. Rev. B **51**, 6655 (1995).
6. B. D. Biggs *et al.*, Phys. Rev. B **39**, 7309 (1989).
7. R. Kleiner and P. Müller, Phys. Rev. Lett. **68**, 2394 (1992).
8. D. E. Oates *et al.*, Appl. Phys. Lett. **68**, 705 (1996).
9. L. M. Xie *et al.*, Bull. Am. Phys. Soc. **41**, 283 (1996).
10. S. Sridhar *et al.*, (in preparation)
11. S. H. Strogatz *Nonlinear dynamics and chaos: with applications to physics, biology, chemistry, and engineering* (Addison-Wesley, Reading 1994).
12. A. Erb, E. Walker, and R. Flükiger, Physica C **245**, 245 (1995).

THERMOELECTRIC POWER OF HIGH-T_c CUPRATES

J.L. TALLON

New Zealand Institute for Industrial Research, P.O. Box 31310, Lower Hutt, New Zealand.

Systematics in the thermoelectric power (TEP) provide a simple means to determine the doping state of any HTS material just from the room temperature TEP, S(290). It also provides important insights into metallicity of non-CuO_2 substructures, the 60K plateau, the pseudogap and vHS scenario.

1. Introduction

Nine years have passed since the first reported study[1] of the thermoelectric power (TEP) in high-T_c superconductors (HTS) and TEP studies continue to reveal important features of the physics of HTS[2]. Little progress, however, has been made in the intervening years towards understanding the unusual dependence of the TEP on temperature and hole concentration, p. It has, rather, provided insights into other properties of HTS materials and has most importantly provided a very convenient measure of the doping state, whether of single crystals, thin films or polycrystalline materials. It is these features which will be reviewed in the following.

2. Doping systematics in S(T)

In general, the temperature-dependent TEP, S(T), is linear in T with negative slope and has large positive values (400-500µV/K) at low p which fall rapidly with increasing p becoming negative in the heavily doped cuprates. This is illustrated by the three groups of curves[3] shown in Fig. 1 at (i) the insulator/metal transition (upper), (ii) optimal doping (middle) and (iii) the overdoped superconductor/metal transition (lower). The three upper curves in Fig 1 show S(T), for $YBa_2Cu_3O_{7-\delta}$ (Y-123), polycrystalline $Bi_2Sr_2CaCu_2O_{8+\delta}$ (Bi-2212) and single crystalline $Bi_2Sr_2CaCu_2O_{8+\delta}$. The fact that the latter two are almost identical shows that the TEP of sintered samples is just the a-b plane TEP and is much less sensitive to porosity and intergranular links than the electrical resistivity suggesting that thermal current is transported more efficiently through these boundaries than the electrical current. It is this fact which renders the TEP such a simple yet, as we shall see, useful measurement.

It has long been noted that the phase behaviour of the different cuprates is similar, all showing a Neel state at low doping and a narrow domain of superconductivity where T_c follows a roughly parabolic dependence on hole concentration. It is our view that all the cuprates have identical scaled phase curves, conveniently approximated by[4]

$$T_c(p)=T_{c,max}[1-82.6(p-0.16)^2] . \qquad (1)$$

This has been demonstrated[4] for $La_{2-x}Sr_xCuO_4$ (214) where p=x, for $La_{2-x}Sr_xCaCu_2O_6$ (2126) where p=x/2 and for $Y_{1-x}Ca_xBa_2Cu_3O_6$ where p=x/2. For $YBa_2Cu_3O_{7-\delta}$ and $Y_{1-x}Ca_xBa_2Cu_3O_{7-\delta}$ p was estimated from the bond valence sum parameter $V_-=2+V_{Cu}-V_{O2}-V_{O3}$. Each of the above compounds is found to follow the same phase curve and is well represented by equ. (1) (although we will see that the situation is more complex in the "60K plateau" region). The implication is that the three TEP curves shown in Fig. 1 at the insulator/superconductor transition, with S(T) rising to 80µV/K, all have the same hole concentration (p≈0.05) as do the

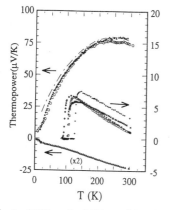
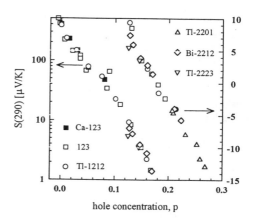

Fig. 1. S(T) at the M/I transition (upper curves), optimal doping (middle curves) and the S/M transition (lower curve).

Fig. 2. The p-dependence of the room-temperature TEP, S(290), for a variety of HTS cuprates.

set of four at optimum doping ($p \approx 0.16$) where S(T) typically falls from ~7 to ~$2\mu V/K$.

The fact that the TEP has the same temperature dependence for a wide variety of HTS materials at two key points on the phase curve suggests that S(T,p) has a common dependence upon p for all HTS. Working with this assumption the room temperature TEP, S(290), was taken as a convenient representation and is plotted against p in Fig. 2[3,4]. Where possible p is taken from the cation composition and defined by x or x/2, for Y-123 $p=V_-$, for $Tl_{0.5}Pb_{0.5}Sr_2Ca_{1-x}Y_xCu_2O_7$ p was taken as (0.5-x)/2 near optimum doping (x=0.2) and away from optimum doping, where either oxygen stoichiometry or Pb valency begins to alter, p was taken to vary as V_-. For the remaining compounds p was estimated from the ratio $T_c/T_{c,max}$ using equ. (1). The data is plotted as two separate data sets: on the underdoped side where the variation with p is exponential and on the overdoped side where the variation is approximately linear. The remarkable correspondence in S(290,p) amongst the rather diverse set of HTS materials has provided a simple and reliable means to determine the doping state of any HTS material in a measurement that takes just a couple of minutes. These materials include single, double and triple layer cuprates with Bi layers, Hg layers, single or double Tl layers or $CuO_{1-\delta}$ layers. The only exception to the correlation is 214 which closely follows the trend of the other cuprates at low doping but from p=x=0.12 deviates to higher values of S(290) and remains positive even for the most overdoped samples. This may arise from scattering from oxygen vacancies in the CuO_2 plane which are known to occur for higher Sr concentrations[5].

3. Metallic non-CuO_2 sublayers?

It has been suggested[6] that 214 is the ideal model system to study the TEP of the CuO_2 planes and that the other Bi, Tl and Hg based systems, as well as R-123, are complicated by the presence of BiO, TlO etc sublayers which are metallic and contribute to a negative thermopower in the overdoped region. It is difficult to imagine, however, that layers consisting of Bi_2O_2, $(Bi,Pb)_2O_2$, TlO, $Tl_{0.5}Pb_{0.5}O$, Tl_2O_2 and Hg all contribute identically to the TEP. Rather, it is more reasonable to deduce that these layers are not metallic, they do not contribute appreciably to S(T) and that it is 214 which is the exception to the trend of the other cuprates. 123 provides a good counter-example. Here the chains are known to be metallic as confirmed by transport, infrared, NMR and µSR studies[7] and they contribute a positive slope to $S(T)^2$. As oxygen is depleted from the chains their metallicity is destroyed and S(T) reverts to the behaviour typical of the other cuprates. By means of Ca substitution 123 can be substantially overdoped and then by removing oxygen it may be brought back to optimal doping. With 20% Ca substitution in 123 the optimal oxygen deficiency is as high as δ=0.38, the chains are inert and S(T) is found to be much the same as the other optimally doped cuprates shown in Fig. 1. Certainly, S(290) is not at all comparable to that for optimally doped 214 (≈20µV/K).

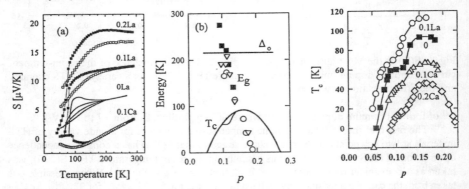

Fig. 3(a) Suppression of the pseudogap-induced TEP enhancement in $YBa_2Cu_4O_8$ due to Zn substitution. ■: 0% Zn. □: 3.7% Zn. (b) the p-dependence of T_g (■) and E_g from NMR (▼) and C_p (○).

Fig. 4. The p-dependence of T_c for 123 substituted with 0.1La, 0.0La, 0.1Ca and 0.2Ca. For clarity, the curves are displaced 20K apart.

4. Normal-state pseudogap

Fig. 3a shows the effect of Zn substitution on S(T) in La and Ca substituted $YBa_2Cu_4O_8$ (124). Above a temperature T_g S(T) is independent of Zn content (and we deduce that the Zn does not affect the hole concentration) but below T_g there occurs an enhancement in the TEP which is progressively diluted with increasing Zn content. T_g is seen to fall with increasing hole concentration rather like the decrease in the normal state pseudogap with doping. First referred to as a "spin gap" as observed in NMR and neutron scattering, the pseudogap was found from heat capacity (C_p) data[8] to be a gap in the total excitation spectrum occuring in the normal state above T_c. T_g and the pseudogap energy, E_g determined from both NMR and C_p are plotted in Fig. 3b as a function of p (which we determine from the ratio $T_c/T_{c,max}$ using equ. (1)). It is clear that the TEP, NMR and C_p all probe the same pseudogap which results in an enhancement in S(T) as the gap opens. The fact that Zn suppresses this enhancement without altering T_g is strong evidence that Zn locally suppresses the pseudogap leaving E_g unchanged in regions remote from a Zn atom. The fact that Zn does not affect S(T) above T_g while Zn is known to suppress both spin fluctuations and superconductivity suggests that the unusual TEP for the cuprates is unrelated to either scattering from spin fluctuations or to the pairing mechanism.

5. The "60K plateau"

Armed with S(290) as a convenient measure of the hole concentration we have determined the variation in p with δ and hence the phase curves $T_c(p)$ for a range of La and Ca substituted 123 samples and these are shown in Fig. 4. Notably the "60K plateau" is preserved in these plots indicating that this is primarily a p-dependent rather than δ-dependent feature and should probably be associated with the cusp which occurs at the same hole concentration, $p \approx 0.12$, in 214. For the 0.1La, pure 123, 0.1Ca and 0.2Ca substituted samples the plateaux occur at $\delta = 0.22$, 0.45, 0.62 and 0.73 respectively, confirming that they are not primarily associated with oxygen ordering at $\delta \approx 0.5$. We have identified the edge of the plateau with the point where the normal-state pseudogap energy, E_g equals the superconducting gap energy, Δ, as shown in Fig. 3b and the plateau probably arises due to phase separation there.

6. Thermoelectric power and van Hove singularity

As noted, S(T) for 123 differs from that of the other cuprates due to contributions from the chains which provide a positive slope in contrast to the prevailing negative slope. As a consequence the effect of unloading oxygen is to provide a progressive reversal of the slope as the chains are broken up. At the same time the doping state of the planes moves from overdoped to underdoped so that S(T) moves up from negative to positive values. This peculiar reversal of both sign and slope of S(T) in the neighbourhood of optimal doping has been taken as evidence for the progression of the Fermi energy across a van Hove singularity (vHS) resulting in T_c passing through a maximum and the reflection of S(T) about the S=0 axis. However, the above data makes it quite clear that this behaviour is peculiar to 123, it originates in the $CuO_{1-\delta}$ chains and is not generic. It was shown that by Ca substitution in 123 the reversal of slope can be displaced arbitrarily far into the overdoped region[9] and is therefore unrelated to optimal doping. This serves to fully confirm the picture we have presented of 214 being the exception to the trend of the other cuprates and the unusual S(T) for 123 deriving from additional chain contributions rather than from traversing a vHS. As a corollary, the search for metallised substructures in the cuprates, which could have major advantages in performance[7], should focus on departures of S(T) from the systematic trends which we have summarised here.

Thanks are due to my collaborators Dr G.V.M. Williams, Dr J.R. Cooper, C. Bernhard, P.S.I.P.N. de Silva, Dr J.W. Loram and Dr N.E. Flower.

1. A. Mawdsley et al., Nature **328**, 233 (1987).
2. J.L. Tallon *et al.*, Phys. Rev. Lett. **75**, 4114 (1995).
3. S. Obertelli *et. al.*, Phys. Rev. B **46**, 14928 (1992).
4. J.L. Tallon *et al.*, Phys. Rev. B **51**, 12911 (1995).
5. P.G. Radaelli et al., Phys. Rev. B **51**, 12911 (1995).
6. J.S. Zhou and J.B. Goodenough, Phys. Rev. B **51**, 3104 (1995).
7. J.L. Tallon *et al.*, Phys. Rev. Lett. **74**, 1008 (1995).
8. J.W. Loram *et al.*, Phys. Rev. Lett. **71**, 1740 (1993).
9. J.L. Tallon and C.Bernhard, Phys. Rev. Lett. **75**, 4552 (1995).

SCATTERING TIME: A UNIQUE PROPERTY OF HIGH-T_c CUPRATES

I. Terasaki, Y. Sato and S. Tajima
Superconductivity Research Laboratory, ISTEC,
1-10-13 Shinonome, Koto-ku, Tokyo 135, Japan

From the comparison with other layered superconductors, the T-linear in-plane resistivity and the semiconductor-like out-of-plane resistivity are found to be unique in high-T_c cuprates, which strongly suggests that scattering time is most anomalous in their physical properties. Our measurements have revealed anomalies in the scattering time such as the anomalously short scattering time along the c axis and the rapid enhancement of the in-plane scattering time in the mixed state.

1 Introduction

Since the discovery of high-T_c cuprates (HTSC), electron correlation in low-dimensional systems has become a central issue in recent solid-state physics. It has been long discussed whether the electronic states of HTSC are two-dimensional (2D) or three-dimensional,[1] and whether the Fermi-liquid (FL) picture is valid or not.[2] A difficulty in these questions is that the effective mass (m^*) of HTSC is highly anisotropic, from which apparent non-FL behavior arises. In fact, some theorists have proposed that the semiconducting out-of-plane resistivity (ρ_c) can be explained by highly anisotropic FL pictures.[1] Thus one has to find a quantity that is independent of the anisotropic m^*.

For the above purpose, layered superconductors where FL description is valid can be a good reference. Prime examples are organic BEDT salts[3] and Sr_2RuO_4.[4] They exhibit large anisotropies between ρ_{ab} and ρ_c, and show clear de-Haas-van-Alphen oscillation—direct evidence for the Fermi liquid. It should be emphasized that the T-linear ρ_{ab} is unique in HTSC, while the other two have $\rho_{ab} \propto T^2$. Another difference is that semiconducting ρ_c at low T is observed only in HTSC. These facts indicate that the T-dependence of resistivity and/or the scattering time (τ) are quite unique in HTSC.

One way to minimize the contribution of m^* is to study the least anisotropic HTSC, $YBa_2Cu_3O_y$ (YBCO). We have studied the charge dynamics of YBCO single crystals[5,6] and have proposed that the 2D nature of the electronic states of HTSC is characterized by τ, not by m^*. Here we present our recent results of transport and optical studies of YBCO, mainly focusing on (i) the anomalously short τ_c, and (ii) the rapid enhancement of τ_{ab} in the mixed state.

2 The out-of-plane conduction

In a previous paper,[5] we examined various expressions for ρ_c, and found that the best candidate is

$$\rho_c = C_1 \rho_{ab} + C_2/T, \qquad (1)$$

where C_1 and C_2 are constants. Equation (1) requires $\rho_c \cdot T$ to be a linear function of $\rho_{ab} \cdot T$, and actually, as shown in Fig. 1(a), the observed ρ_c and ρ_{ab} roughly obey Eq. (1). Since ρ_c

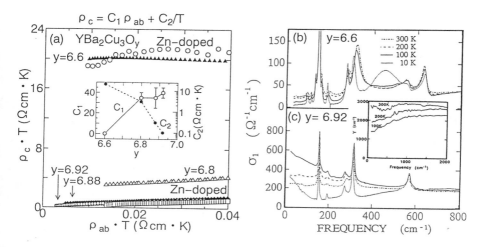

Figure 1: (a) $\rho_c \cdot T$ is plotted as a function of $\rho_{ab} \cdot T$. Zn-doped data for the 60-K phase is taken from Ref 7. The doping dependence of C_1 and C_2 is plotted in the inset. (b)(c) c-axis optical conductivity spectra for the 60-K and 90-K phases. The scattering rate γ ($\equiv 1/\tau_c$) of the 90-K phase is shown in the inset.

is expressed by the sum of two terms, we cannot regard σ_c ($=1/\rho_c$) of YBCO as a single component except for the two extreme cases where C_1 or C_2 is negligibly small.

The carrier-doping dependence of C_1 and C_2 is shown in the inset of Fig. 1(a). ρ_c of 60-K phase YBCO ($y=6.6$) has no coherent term ($C_1=0$), implying the breakdown of the Boltzmann transport. As doping proceeds, C_2 rapidly decreases with y, while C_1 is nearly constant once it becomes finite. Note that the Zn-doped data [7] are consistently plotted in Fig. 1(a), which means that the Zn doping modifies only the coherent part as $\rho_c = C_1(\rho_{ab} + \rho_0^{Zn}) + C_2/T$. In terms of C_1 and C_2, therefore, we can understand the Zn-doping in ρ_c, which increases ρ_c in the 90-K phases and remains ρ_c in the 60-K phase unchanged.

Let us look closer at the frequency dependence of the c-axis conduction. Figure 1(b) and 1(c) show $\sigma_c(\omega)$ for $y=6.6$ and $y=6.92$, respectively. Reflecting that ρ_c for $y=6.6$ is dominated only by C_2, its $\sigma_c(\omega)$ is not Drude-like, but rather is characterized by conductivity suppression below 800 cm^{-1}. On the other hand, the $y=6.92$ sample has a Drude-like conductivity, which is consistent with the fact that ρ_c is mainly determined by C_1. However, there still exists an anomaly; the evaluated $1/\tau_c(=\gamma)$ (1200 cm^{-1} at 100K) is, as shown in the inset of Fig. 1(c), 10 times larger than $1/\tau_{ab}(\sim k_B T)$. As a result, the out-of-plane conduction is in the dirty limit, while the in-plane conduction is in the clean limit. Such a large anisotropy in τ has never been seen in other layered materials, and would be possibly discussed in the framework of the RVB theory.

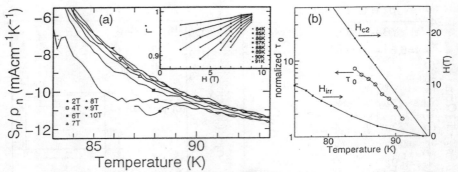

Figure 2: (a) S_n/ρ_n in various magnetic fields. The H dependence of Γ^* (normalized S_n/ρ_n at 10 T) is shown in the inset. (b) The enhancement of τ_{ab} below T_c and H_{irr} and H_{C2}

3 The enhancement of the in-plane scattering time in the mixed state

It is now established that τ_{ab} enhances rapidly below T_c,[8] which is a direct proof that the scattering mechanism is electronic. A further question is whether the enhancement of τ_{ab} occurs in the flux-flow regime. Although some implications are reported through the thermal conductivity and Hall conductivity measurements,[9] there still remain uncertainties. For example, Hall conductivity inevitably includes τ_{ab}^{tr} and τ_{ab}^{H} which gives different T-dependence of ρ_{xx} and ρ_{xy} above T_c.

Magneto-thermopower is a useful probe for τ_{ab}^{tr} in the mixed state. When the Hall angle is negligibly small, magneto-Seebeck coefficient (S_{xx}) along the CuO_2 plane is phenomenologically expressed as [10]

$$S_{xx} = S_n \rho_{xx}/\rho_n, \qquad (2)$$

where S_n and ρ_n are the Seebeck coefficient and resistivity of quasiparticles (QP), and ρ_{xx} is the resistivity below T_c. Accordingly, we can obtain S_n/ρ_n by measuring S_{xx} and ρ_{xx}.

Figure 2(a) shows thus obtained S_n/ρ_n in various magnetic fields.[11] Note that S_n/ρ_n depends not only on T, but also on H, which suggests that the scattering rate is modified in the flux-flow regime. We interpreted the H-dependence in terms of the scattering of QP by vortices, that is, $S_n/\rho_n \propto \tau_{ab}^{tr}(H,T)$ (Here we assume that the normal-state τ would naturally evolve to τ of QP below T_c). To see the H-dependence more clearly, we plotted $\Gamma^* \equiv S_n(10\text{ T})/S_n \cdot \rho_n/\rho_n(10\text{ T})$ in the inset of Fig. 2(a). The scattering by vortices is evidenced by the observed relation of $\Gamma^* \sim aH + b$.

Figure 2(b) shows the evaluated τ_{ab}^{tr} from Γ^*, which clearly shows its rapid enhancement in the flux-flow regime. Another advantage of the magneto-thermopower measurement is that the H_{c2} line can be obtained from the off-diagonal signal (the Nernst effect).[10] In addition, the irreversibility line can be evaluated from $\rho_{xx}=0$, and thus we can obtain a gross feature of the phase diagram simultaneously. The measurement results strongly suggest that the enhancement of τ occurs in the presence of the amplitude (not the phase) of the superconducting order parameter. This could be a clue for the elucidation of the scattering mechanism of HTSC.

4 Summary

Comparison between high-T_c cuprates and other layered superconductors has revealed that the scattering time characterizes their anomalous two-dimensional nature. The semiconducting ρ_c of the 60-K phase YBCO is dominated by conductivity suppression below 800 cm^{-1}, and is unlikely to obey the Boltzmann transport. Even the metallic ρ_c of the 90-K phase still shows the anomaly that the scattering time is ten times shorter along the c-direction than along the ab-direction. The in-plane scattering time rapidly enhances in the flux-flow regime, implying that the growth of the amplitude of superconducting order parameter is an origin of the enhancement.

Acknowledgments

Authors would like to thank Y. Yamada, and S. Shiohara for the technical assistance of the crystal growth of YBa$_2$Cu$_3$O$_7$. They appreciate S. Miyamoto, R. Hauff, and J. Schützmann for collaboration, and H. Mori for the discussion of organic superconductors. This work is partially supported by NEDO.

References

1. S. L. Cooper and K. E. Gray, in *Physical Properties of the High Temperature Superconductors IV*, edited by D. M. Ginsberg (World Scientific, Singapore, 1994), p. 61.
2. P. W. Anderson, Science **256** (1992) 1526.
3. M. Dressel, O. Klein, G. Grüner, K. D. Carlson, H. H. Wang, and J. M. Williams, Phys. Rev. B**50** (1994) 13603.
4. Y. Maeno, H. Hashimoto, K. Yoshida, S. Nishizaki, T. Fujita, J. G. Bednorz, and F. Lichtenberg, Nature **372** (1994) 532.
5. I. Terasaki, Y. Sato, S. Miyamoto, S. Tajima, and S. Tanaka, Phys. Rev. B**52** (1995) 16246; Physica C**235-240** (1994) 1413.
6. J. Schützmann, S. Tajima, S. Miyamoto, and S. Tanaka, Phys. Rev. Lett. **73** (1994) 174; S. Tajima, J. Schützmann, S. Miyamoto, I. Terasaki, Y. Sato, and R. Hauff, submitted to Phys. Rev. B.
7. K. Mizuhashi, K. Takenaka, Y. Fukuzumi, and S. Uchida, Phys. Rev. B**52** (1995) R3884.
8. D. A. Bonn, P. Dosanjh, R. Liang, and W. N. Hardy, Phys. Rev. Lett. **68** (1992) 2390; Phys. Rev. B**47** (1993) 11314.
9. J. M. Harris, N. P. Ong, P. Matl, R. Gagnon, L. Taillefer, T. Kimura, and K. Kitazawa, Phys. Rev. B**51** (1995) 12053.
10. For a review, R. P. Huebener, Supercond. Sci. Technol. **8** (1995) 189.
11. Y. Sato, I. Terasaki, and S. Tajima, Submitted to Phys. Rev. B.

SPIN GAP EFFECTS ON THE CHARGE DYNAMICS OF HIGH T_c SUPERCONDUCTORS

S. UCHIDA, K. TAKENAKA, K. TAMASAKU
Department of Superconductivity University of Tokyo Yayoi 2-11-16, Bunkyo-ku, Tokyo 113, Japan

The spin gap is a phenomenon characteristic of high-T_c cuprates in the underdoped regime. Remarkable changes are observed in resistivity, Hall effect, and optial conductivity, both in-plane and out-of-plane components, below a certain temperature which appears to correspond to the opening of the spin gap.

1 Introduction

An interplay between charges dynamics and spin fluctuations has been suggested by the recent transport and optical measurements on the Y based cuprates.[1-3] For the highly doped material (7-$y \sim 6.93$ in the YBa$_2$Cu$_3$O$_{7-y}$(Y123) system), the in-plane resistivity ρ_{ab} (or ρ_a) shows the T-linear behavior from above T_c to room temperature. On the other hand, ρ_{ab} of the underdoped crystals deviates from T linearity at low temperatures. The onset of the deviation, T^*, increases with decreasing oxygen content.[1]

There is a strong indication for the underdoped compounds that a gap opens at $q = Q$ (antiferromagnetic wavevector) in the spin excitation spectrum at temperature T_s well above T_c.[4,5] It is also suggested that the decreasing magnetic susceptibility, characteristic of the underdoped regime of all the HTSC, might be connected to a spin gap that opens near $q = 0$ (corresponding to the formation of spin siglets). The decrease of the magnetic susceptibility seems to start at temperature T^* higher than T_s. The anomalies in resistivity, start at $T \sim T^*$ and become substantial at $T \sim T_s$. It has been found that the deviation from the T-linear dependence in ρ_{ab} is correlated with the spin gap formation.

The characteristic temperature T^* is also identified for YBa$_2$Cu$_4$O$_8$ (Y124)[2] and La$_{2-x}$Sr$_x$CuO$_4$ (La214).[6] As in the case of Y123 T^* is a characteristic temperature separating different high- and low- temperature charge dynamics also in Y124 and La214.

2 Spin Gap Effects on the In-Plane Charge Transport

The deviation from the T-linear resistivity apparently correlates with the temperature dependence of the Hall coefficient (R_H).[1] While $R_H \propto 1/T$ for fully oxygenated samples, for reduced samples a deviation from the $1/T$ dependence becomes obvious, with a peak occurring at a temperature well above T_c. As in the case of $\rho_{ab}(T)$, the onset of the deviation from $1/T$ increases in temperature with decreasing oxygen concentration.

It can be seen in Fig. 1 that all the ρ_{ab} vs T curves for Y123 coincide by scaling T with T^* and $\rho_{ab}(T)$ with $\rho_{ab}(T^*)$; ρ_{ab} is linear in T above a certain temperature T^* and shows a stronger T dependence ($\rho_{ab} \sim T^\alpha$ with $\alpha \sim 2.5$ for any y) below T^*.

It is also shown in Fig. 1 that the T dependence of R_H shows a fairly good scaling as in the case of ρ_{ab}. For the lowest oxygen contents (6.45 and 6.58) a spurious component shows up in $R_H(T)$ at low temperatures which appears to be related with a fairly large residual resistivity in these compounds possibly arising from inhomogeneity in the crystals. Without this component the scaling becomes as good as that for ρ_{ab}.

We can demonstrate that the observed peak in $R_H(T)$ is connected with the change in the T dependence of ρ_{ab} below T^*. The T dependent R_H in HTSC[7] has long been a mystery and it is not fully understood even at present. Anderson[8] proposed a novel explanation by introducing a scattering time τ_H in addition to the scattering time τ which determines the in-plane resistivity. τ_H is assumed to be relevant to the Hall effect and directly connected to the Hall angle by $\cot\theta_H = 1/\omega_c\tau_H$ (ω_c : cyclotron frequency). The experimental result by Chien, Wang, and Ong[9] shows that $\tau_H^{-1} \sim T^2$ for the oxygenated Y123.

The data of $\cot\theta_H$ are best described by the T^2 law also for the underdoped compounds and no significant feature is seen at T^*. In terms of the two-scatteirng-time model this implies that $\tau_H^{-1} \sim T^2$ irrespective of presence of the spin gap. Then, it follows that the change in T dependence of τ below T^* is responsible for the peak in $R_H(T)$ appearing below T^*. Actually, as illustrated in Fig. 2, $\tau \sim T^{-1}$ and $\tau_H \sim T^{-2}$ lead to $R_H \sim \tau_H/\tau \sim T^{-1}$ at $T > T^*$, whereas $R_H \sim T^{0.5}$ at $T \ll T^*$ since $\tau \sim T^{-2.5}$ and $\tau_H \sim T^{-2}$, thus giving rise to a peak at $T \sim T^*$.

3 Out-of-Plane Charge Transport and Optical Conductivity Spectrum

The out-of-plane resistivity ρ_c is much more sensitive to the oxygen content than that of ρ_a, and the temperature dependence of ρ_c is characterized by a crossover from high-T metallic to low-T semiconducting regime. The crossover temperature, T_{c0}, decreases with increasing oxygen content.[10] From these results we notice that the crossover to non-metallic regime in ρ_c is obviously linked with the crossover from T-linear to nonlinear regime in ρ_a. The onset temperatures of the anomaly in ρ_a and ρ_c do not exactly coincide — T_{c0} is higher than T^*, perhaps because ρ_c is more sensitive to the spin gap.

The electronic component of the c-axis optical conductivity $\sigma_c(\omega)$ in the normal state of the underdoped compound is nearly ω-independent in the low energy region. $\sigma_c(\omega)$ is gradually depressed in the energy region, as the temperature is lowered below $\sim T^*$. This is in accordance with the semiconducting T dependence of ρ_c below T^* and seems to indicate a deepening of pseudogap with decreasing temperature in the normal state.[3]

The pseudogap is seen only in the c-axis spectrum, and the in-plane spectrum is dominated by a Drude peak at $\omega = 0$. It is found that the development of the pseudogap in $\sigma_c(\omega)$ is correlated with a rapid narrowing of the Drude width in $\sigma_{ab}(\omega)$,[11] consistent with the rapid decrease in ρ_{ab} below T^*. As a consequence, the Drude peak is isolated from the mid-IR band in the $\sigma_c(\omega)$ spectrum which is responsible for a low carrier density (possibly $n = x$) in the underdoped regime.

4 Summary

The normal-state charge transport of the underdoped cuprates show characteristic behaviors below T^* well above T_c. Below T^* the in-plane resistivity is reduced and concomitantly R_H shows a peak, whereas the out-of-plane resistivity is enhanced and becomes semiconducting.

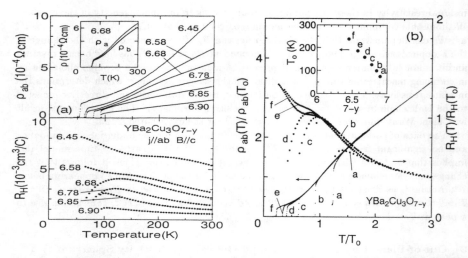

Figure 1: (a) The temperature dependence in-plane resistivity (ρ_{ab}) and the Hall coefficient (R_H) for YBa$_2$Cu$_3$O$_{7-y}$ with $0.10 \leq y \leq 0.55$. (b) The temperature is scaled with $T_0 \sim 0.7T^*$, and ρ_{ab} and R_H are rescaled with their values at T_0.

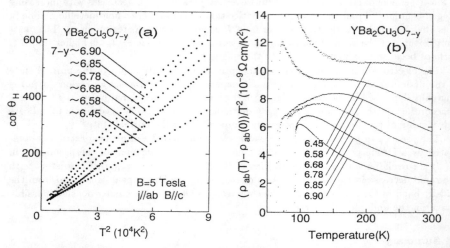

Figure 2: (a) The Hall angle cotθ_H as a function of T^2 for YBa$_2$Cu$_3$O$_{7-y}$. (b) The in-plane resistivity for YBa$_2$Cu$_3$O$_{7-y}$ plotted rescaled as $(\rho_{ab}(T) - \rho_{ab}(0))/T^2$ vs T, where $\rho_{ab}(0)$ is a residual resistivity as extrapolated from the high temperature data

This observation is suggestive of totally different mechanisms for ρ_a and ρ_c, but clearly indicates that the same physical effect, which shows up below T^* acts on the in-plane and c-axis charge transport in the opposite direction.

The characteristic temperature T^* coincides with that below which a gap opens in the spin fluctuations. The opening of the spin gap below T^* leads to the reduction in ρ_a due to the reduction in the scattering rate between charge carriers and spin fluctuations. In this context, ρ_c might be enhanced if the c-axis conduction were managed by interlayer hopping assisted by the spin fluctuations.

In the optical spectrum, a pseudogap is seen in the c-axis, while the in-plane spectrum is dominated by a Drude peak at $\omega = 0$ which shows a rapid narrowing below T^* and, as a consequence, is isolated from the mid-IR band. In this regard, the low carrier density in the underdoped regime is linked with the presence of the spin gap.

These results are generic to the underdoped cuprates which show quite universal behaviors in various properties. Contrary to the underdoped regime, highly doped cuprates exhibit non-universal behaviors, strongly dependent on material and doped hole density,[14] indicating that the highly doped cuprates are in the region of a rapid crossover from underdoped to (non-superconducting) overdoped regime.

Acknowledgements

This work was supported by a Mombusho Grant for Scientific Research on Priority Area, "Anomalous Metallic State Near the Mott Transition" and NEDO Grants for Advanced Industrial Technology Research and International Joint Research.

References

1. T. Ito, K. Takenaka, and S. Uchida, *Phys. Rev. Lett.* **70**, 3995 (1993).
2. B. Bucher *et al.*, *Phys. Rev. Lett.* **70** 2012 (1993).
3. C. C. Homes *et al.*, *Phys. Rev. Lett.* **71** 1645 (1993).
4. H. Yasuoka, T. Imai, and T. Shimizu, in *Strong Correlation and Superconductivity*, edited by H. Fukuyama, S. Maekawa, and A. P. Malozemoff (Springer, Berlin 1989) p.254.
5. J. Rossat-Mignod *et al.*, Physica C**185-189**, 86 (1991).
6. H.Y. Hwang *et al.*, *Phys. Rev. Lett.* **72** 2636 (1994) ; T. Nakano *et al.*, Phys. Rev. B**49** 16000 (1994).
7. S. Uchida *et al.*, Jpn. J. Appl. Phys. **26**, L440 (1987); N. P. Ong *et al.*, *Phys. Rev.* B**35**, 8807 (1987).
8. P. W. Anderson, *Phys. Rev. Lett.* **67**, 2992 (1991).
9. T. R. Chien, Z. Z. Wang, and N. P. Ong, *Phys. Rev. Lett.* **67**, 2088 (1991).
10. K. Takenaka *et al.*, *Phys. Rev.* B**50** 6534 (1994).
11. K. Takenaka *et al.*, unpublished ; L. D. Rotter *et al.*, *Phys. Rev. Lett.* **67**, 2741 (1991).
12. Y. Fukuzumi *et al.*, *Phys. Rev. Lett.* **76**, 684 (1996).

THERMAL CONDUCTIVITY OF HIGH-TEMPERATURE SUPERCONDUCTORS IN A MAGNETIC FIELD

CTIRAD UHER
Department of Physics, University of Michigan
Ann Arbor, Michigan 48109

ABSTRACT

This paper gives a brief review of the experiments aimed at studying the effect of external magnetic field on the thermal conductivity of high-temperature superconductors.

1. Introduction

The entry of a material into the superconducting domain below its transition temperature T_c and the formation of a gap in the energy spectrum have a profound effect on the distribution function, the entropy content, and the scattering power of the charge carriers as well as on the mean-free path (mfp) of phonons, the other essential heat-carrying entity. In high temperature superconductors (HTS) this leads to a rise in the thermal conductivity below T_c which culminates in a pronounced peak near $T_c/2$. This striking behavior is observed in all hole-type CuO_2-plane superconductors and in all their structural forms[1]. Whether the effect is due to phonons or quasiparticles, is the source of much controversy[2-4].

It was hoped originally that the use of a magnetic field might shed new light on the role of carriers and phonons in the thermal transport of HTS. As it turns out, while the interpretation of the data remains equivocal, the studies have opened up new and interesting insight into the role of vortices as scattering centers and uncovered fascinating features which relate to fundamental issues concerning the nature of superconductivity in HTS. After first noting the trend in the conventional superconductors I provide a brief review of the existing measurements on HTS.

2. Conventional Superconductors

External magnetic field $H > H_{c1}$ drives a superconductor into the mixed state characterized by the presence of Abrikosov vortices which can be viewed as tubes of radius ξ containing bound excitations not too different from normal electrons. The density of vortices increases with field and they interact, setting up a triangular vortex lattice. Although the group velocity of the bound excitations is small and they do not contribute to the heat flux, the influence of vortices on the heat transport is nevertheless dramatic. Close to T_c the heat is carried by unbound excitations (uncondensed electrons) and their interaction with vortices determines the overall behavior. Two processes are possible: quasiparticles with mfp l_e scatter on vortices having the effective cross section a, resulting[5] in the thermal resistance

$$W_e(B) = W_e(0) [1 + (l_e a/\Phi_o)B] \qquad (1)$$

The second contribution might arise from the Andreev process[6] where an incoming electron-like quasiparticle scatters on the modulation of the order parameter $\Delta(\mathbf{r})$ and transforms into an outgoing hole-like quasiparticle. This process obviously drastically alters the heat flow because of the near reversal of the group velocity. There is no general theory available which covers the entire range of magnetic fields but it is clear that beyond a certain field strength the scattering ought to weaken because the maximum value of $\Delta(\mathbf{r})$ is smaller as the vortices start to overlap, and tunneling between vortices should reverse the trend and drive the thermal conductivity toward its normal state value κ_n as $B \rightarrow H_{c2}$. How the thermal conductivity actually attains its normal state value depends critically[7,8] on how large is l_e in relation to the coherence length ξ.

Phonon-vortex scattering can be studied only at sufficiently low temperatures where the lattice carries most of the heat. It has two components: one due to a purely geometrical scattering and the other one due to phonons scattering on bound quasiparticles within the vortex core. For phonons to scatter, their wavelength must be comparable to the crossection of the vortex. Phonon scattering intensifies with increasing vortex density as[9]

$$W_p(B) = W_p(0) [1 + \alpha \, l_p(B/H_{c2})] \qquad (2)$$

Here α is the average scattering diameter of the vortex, and l_p is the phonon mfp. Eventually, as the vortices come close together, the electrons start to tunnel between them and with a further increase in the field strength the conductivity turns up and approaches the normal state as $B \rightarrow H_{c2}$. Typical behaviors of the conductivity are sketched in Fig.1. It is important to note that Eqs. (1) and (2) predict the same field dependence and thus the magnetic field, on its own, cannot distinguish between the effects due to quasiparticles and those due to phonons.

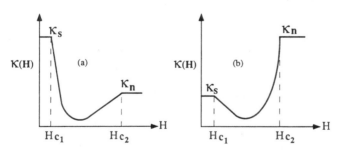

Figure 1 : Effect of magnetic field on the thermal conductivity of conventional superconductors.
(a) Behavior of a typical superconducting alloy.
(b) Behavior of a "clean" superconductor ($l_e > \xi$).

3. High Temperature Superconductors

The first indication that field does affect the thermal transport of HTS came from the work of Zhu et al.[10] on sintered YBCO. Precision measurements had to await single crystals of reasonable size. Such measurements confirmed[11-14] a strong influence (30% reduction near the conductivity peak) the field has on the heat transport of HTS below T_c. The data also displayed interesting hysteresis effects and it was shown[14] that the time evolution of the remnant field can be used profitably in the study of vortex dynamics. By rotating the magnetic field from the B $\perp$ ab-plane to B $\parallel$ ab-plane orientation it became clear that vortices scatter far more effectively when they lie perpendicular to the CuO_2 plane than when they fit between the planes. Further detailed studies[15] on a variety of perovskite structures with vortices parallel to the CuO_2 planes have revealed a close relation between thermal resistance and the dimensionality of superconductivity. In marked contrast to the conventional superconductors, all experimental evidence indicates that field dependence of the thermal conductivity in HTS is substantially sublinear as if with the increasing density of vortices their ability to limit heat flow was weakening, Fig.2. Richardson et al.[16] proposed and used a stretched exponential given by Eq. (3) which yielded an excellent fit to their data

$$W_v(B) = cB \exp[-p\, B^{1/4}] \qquad (3)$$

at all temperatures and fields; this functional form was adopted in other studies. Mechanisms put forward to explain the sublinear field dependence vary. Richardson et al. assumed improved perfection of the vortex lattice in increasing field and phonons traveling in the Bloch wave fashion as leading to less intensive scattering. Bougrine et al.[17] explained the sublinear behavior in their Fe-doped YBCO as arising from a combined scattering of phonons on free vortices and on vortices pinned by dislocation cores of twin boundaries. This latter scatterer yields a hierarchy of stretched exponential field dependences with the exponent reflecting the type of defect. Sublinear field dependence of the thermal conductivity in HTS is clearly in discord with strictly linear variations predicted for phonon-vortex or electron-vortex scattering in conventional superconductors by Eqs. (1) and (2).

All the above investigations explicitly assume that heat transport is due to phonons. Soon after it became known[18,19] that quasiparticles in HTS possess unusually long relaxation times below T_c, attempts were made to reevaluate the nature of the thermal conductivity by

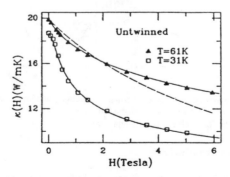

Figure 2: Field dependence of the thermal conductivity of an untwinned crystal of YBCO. Solid lines are fits by Eq. (3). The dashed line is an unsuccessful fit assuming functional dependence of Eq. (2). (Adapted from Ref. 16.)

highlighting the influence of quasiparticles. The likelihood that below T_c there exists a significant quasiparticle contribution to the thermal transport has led to inquiry as to how this transport channel is affected by the presence of vortices. Making exactly the same assumptions regarding the vortices as in their earlier work but considering quasiparticles rather than phonons as heat carriers, the authors of Ref.17 reconfirmed[20] the stretched exponentional form of the field dependence. Weighing their two studies equally further substantiates the fact that it is difficult to identify experimental conditions which unequivocally distinguish between the two competing heat conducting entities, quasiparticles and phonons. Nevertheless, attempts to separate the thermal conductivity into its components have been made. Yu et al.[21] combined conventional formulas for the phonon-vortex and electron-vortex scattering, and assumed that one of the two channels (phonons) has a negligible field dependence. The remaining, field dependent contribution was taken as due to quasiparticles. With this, somewhat prejudged assignment, they separated the conductivity into its components. The result merely demonstrates that it is possible to model the transport with quasiparticles as the "active" channel. Apart from the fact that the proponents of the phonon-dominated transport can make just as convincing a claim in favor of phonons, it also seems unrealistic to assume that phonons are left completely intact by the presence of vortices. A more likely scenario is that both quasiparticles and phonons participate and contribute to the unusual features observed in the heat transport of HTS and, because of their similar functional dependence on temperature and field, one must explore nontraditional transport techniques in order to ascertain the relative weight of each. Attempts along these lines were made recently and turned out to be very successful.

Krishana et al.[22] made use of the Righi-Leduc effect (thermal equivalent of the Hall effect) in twinned crystals of YBCO whereby an in-plane thermal current and a magnetic field applied perpendicular to the CuO_2 planes give rise to an in-plane but transverse thermal flow. The physical origin of the effect is asymmetric scattering to which unbound quasiparticles are subjected as they encounter circulating currents around the vortex. Since phonons are scattered symmetrically by the vortices, they are filtered out by this process and the transverse thermal conductivity κ_{xy} is thus given entirely by quasiparticles without any phonon background. It is easy to confirm the quasiparticle nature of this effect simply by reversing the magnetic field, which in turn reverses the direction of the transverse thermal current. It is remarkable that such a robust effect is detected down to 15K, implying that a significant fraction of the unbound quasiparticles exists down to very low temperatures. This substantiates the early findings of uncondensed carriers at subkelvin temperatures.[23,24] From the initial slope of κ_{xy} vs. H the authors calculate the quasiparticle mfp and this input allows them to estimate the in-plane component of the thermal conductivity due to quasiparticles. The temperature dependence of the in-plane total thermal conductivity together with the quasiparticle contribution are shown in Fig.3. Comparing the overall rise in $\kappa^{tot}(T)$ below T_c with the quasiparticle conductivity $\kappa^e(T)$, it is seen that quasiparticles are an important contributor to the heat transport in the superconducting domain, accounting for about one half of the anomalous rise in the thermal conductivity below T_c. This elegant and effective approach to determine the respective contributions to the overall thermal transport will surely find much appeal in the community. One can look forward to interesting studies on a

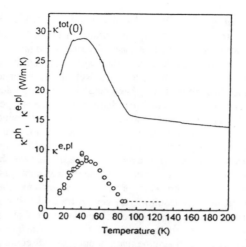

Figure 3: Temperature dependence of the total in-plane thermal conductivity (solid line) and the quasiparticle thermal conductivity contribution (open circles). (Adapted from Ref. 22.)

variety of HTS structures the aim of which is to ascertain quasiparticle and phonon properties in this otherwise difficult temperature domain.

In an experiment that may seem similar to the one in Ref. 22 but conceptually is entirely different, Yu et al.[25] attempted to detect a signature of the gap function symmetry from the behavior of the off-diagonal component of the thermal conductivity tensor in an untwinned crystal of YBCO. The off-diagonal component is believed to arise from quasiparticle scattering on vortices via the Andreev process. A fundamental difference with regard to the experiment of Ref. 22 is the orientation of magnetic field. While Krishana et al. place the field perpendicular to the CuO_2 planes and thus make use of the transverse thermomagnetic effect, Yu et al. orient the field strictly in-plane. From the angular dependence of the tensorial components of the thermal conductivity and their degree of modulation the authors conclude that their data are consistent with either a d-wave or an extended s-wave gap function. While the experiment is clever, its interpretation may not be unique due to the particular shape of the crystal. As pointed out by Klemm et al.[26], a square-shaped crystal necessarily leads to anomalous flux pinning and demagnetization effects which in turn result in an anomalous flux entry pattern well into the sample interior. A combination of periodicity of these flux patterns with a particular layout of thermocouple junctions on the sample gives results which, according to Klemm et al., are probably unrelated to the order parameter symmetry and are most likely due to a "corner effect". Clearly the measurements should be repeated with different sample geometry and location of thermocouples to clarify the origin of the effect. Furthermore, as with all unexpected experimental features of HTS, one should check for their uniqueness by performing the same experiment on Nb or other suitable conventional superconductor.

Studies of the thermal conductivity of HTS have proved to be very rewarding in terms of exciting new results and on a number of occasions provided an early indication of the exotic nature of superconductivity, unusual quasiparticle dynamics and unexpected phonon properties. Application of magnetic field in the measurements of the thermal conductivity have brought into focus the role of vortices as scattering centers for both phonons and quasiparticles. The results are very intriguing and undoubtedly will stimulate further experimental and theoretical inquiry.

4. Acknowledgments

It is my pleasure to acknowledge usefull discussions with Prof. Joshua Cohn. The work was supported by the ONR Grant No. N00014-92-J-1335.

5. References

1. C. Uher, in *Physical Properties of High-Temperature Superconductors*, D. M. Ginsberg, ed. (World Scientific, Singapore, 1992), Vol. 3, p. 159, also J. Supercond. **3**, *337* (1990).
2. C. Uher, Y. Liu, and J. F. Whitaker, *J. Supercond.* **7**, 323 (1994).
3. R. C. Yu, M. B. Salamon, J. P. Lu, and W. C. Lee, *Phys. Rev. Lett.* **69**, 1431 (1992).
4. J. L. Cohn, V. Z. Kresin, M. E. Reeves, and S. A. Wolf, *Phys. Rev. Lett.* **71**, 1657 (1993).
5. R. M. Cleary, *Phys. Rev.* **175**, 587 (1968).
6. A. F. Andreev, *Sov. Phys. JETP* **19**, 1228 (1964).
7. C. Caroli and M. Cyrot, *Phys. Kondens. Mater.* **4**, 285 (1965).
8. K. Maki, *Phys. Rev.* **158**, 397 (1967).
9. W. F. Vinen, E. M. Forgan, C. E. Gough, and M. J. Hood, *Physica* **55**, 94 (1971).
10. D. Zhu, A. C. Anderson, T. Friedmann, and D. Ginsberg, *Phys. Rev.* **B41**, 6605 (1990).
11. T. T. M. Palstra, B. Batlogg, L. F. Schneemeyer, and J. V. Waszczak, *Phys. Rev. Lett.* **64**, 3090 (1990).
12. V. V. Florentiev, A. V. Inyushkin, A. N. Taldenkov, O. Melnikov, and A. Bykov, in *Progress in High Temperature Superconductivity*, R. Nicolsky, ed. (World Scientific, Singapore, 1990), Vol. 25, p. 462.
13. N. V. Zavaritsky, A. V. Samoilov, and A. A. Yurgens, *Physica* **C180**, 417 (1991).
14. S. D. Peacor, J. L. Cohn, and C. Uher, Phys. Rev. **B43**, 8721 (1991).
15. A. V. Inyushkin, A. N. Taldenkov, S. Yu. Shabanov, L. N. Demyanets, and T. G. Uvarova, *J. Supercond.* **7**, 331 (1994).
16. R. A. Richardson, S. D. Peacor, F. Nori, and C. Uher, *Phys. Rev. Lett.* **67**, 3856 (1991).
17. H. Bougrine, S. Sergeenkov, and M. Ausloos, *Solid State Commun.* **86**, 513 (1993).
18. D. A. Bonn, P. Dosanjh, R. Liang, and W. N. Hardy, *Phys. Rev. Lett.* **68**, 2390 (1992).
19. M. C. Nuss, P. M. Mackiewich, M. L. O'Malley, E. H. Westerwick, and P. B. Littlewood, *Phys. Rev. Lett.* **66**, 3305 (1991).
20. S. Sergeenkov and M. Ausloos, *Phys. Rev.* **B52**, 3614 (1995).
21. F. Yu, M. B. Salamon, V. Kopylov, N. Kolesnikov, and A. Herman, *Physica* **C235-240**, 1489 (1994).
22. K. Krishana, J. M. Harris, and N. P. Ong, *Phys. Rev. Lett.* **75**, 3529 (1995).
23. U. Gottwick, R. Held, G. Sparn, F. Steglich, H. Rietschel, D. Ewert, B. Renker, W. Bauhoffer, S. von Molnar, M. Wilhelm, and H. E. Hoenig, *Europhys. Lett.* **4**, 1183 (1987).
24. J. L. Cohn, S. D. Peacor, and C. Uher, *Phys. Rev.* **B38**, 2892 (1988).
25. F. Yu, M. B. Salamon, A. J. Leggett, W. C. Lee, and D. M. Ginsberg, *Phys. Rev. Lett.* **74**, 5136 (1995).
26. R. A. Klemm, A. M. Goldman, A. Bhattacharya, J. Buan, N. E. Israeloff, C. C. Huang, O. T. Valls, J. Z. Liu, R. N. Shelton, and U. Welp, submitted for publication.

THERMOPOWER OF THE CUPRATES UNDER HIGH PRESSURE

J.-S. ZHOU and J. B. GOODENOUGH

Materials Science & Engineering, University of Texas at Austin
Austin, TX 78712

The temperature dependence of the thermopower of the intergrowth copper oxides is $\alpha(T) = \alpha_0 + \delta\alpha(T)$. The enhancement factor $\delta\alpha(T)$ disappears in polaronic conductors; a $\delta\alpha(T)$ having a $T_{max} \approx 140$ K is found in the superconductors, and a $\delta\alpha(T) > 0$ extends into the overdoped regime. Pressure experiments show a strong correlation between T_C and $\delta\alpha(T)$ indicative of a common underlying mechanism. In underdoped $YBa_2Cu_3O_{6.7}$, the $Cu(1)O_x$ chains do not contribute to $\alpha(T)$, and pressure transfers electrons from the CuO_2 sheets to $Cu(1)O_x$ chains, which increases T_C and $\delta\alpha(T)$. In fully oxidized $YBa_2Cu_3O_{6.96}$ and $YBa_2Cu_4O_8$, the more conductive chains give a negative $\delta\alpha(T)$ that dominates the positive $\delta\alpha(T)$ of the CuO_2 sheets, and pressure increases the superconductive-pair concentration n_s in the chains, but not in the sheets; it also increases the negative component in $\delta\alpha(T)$ relative to any change in the positive component. A c- axis vibration observed with Raman spectroscopy supports a vibronic coupling as the underlying mechanism responsible for $\delta\alpha(T)$ and the inducing of superconductivity in the chains.

The cuprate superconductors exhibit not only a higher transition temperature of superconductivity, but also unique normal-state properties. A resistivity with linear temperature dependence extends to highest temperatures without saturation on approaching the mobility limit imposed by the Boltzmann transport theorem [1]. The Hall coefficient R_H is temperature dependent to high temperatures and its scaling temperature is outside the Debye temperature range [2]. The thermopower α and its temperature dependence provide more information about the normal state. α can be separated into two terms, $\alpha = \alpha_0 + \delta\alpha$, where α_0 is nearly temperature-independent but doping-dependent, $\delta\alpha$ is an enhancement that gives a negative slope in α vs T below room temperature until a maximum at $T_{max} \approx 140$ K. We have demonstrated a correlation between the enhancement $\delta\alpha$ and the superconductive transition by introducing oxygen vacancies while keeping the same oxidation state in the CuO_2 sheets [3]. The oxygen vacancies suppress the superconductive transition and the enhancement $\delta\alpha$ as well. Tallon et al [4] also found the same relation by substituting Zn for Cu. However, these studies alone do not prove a relation between superconductivity and the $\delta\alpha$ enhancement as vacancies or impurities could suppress the enhancement and T_C independently. In this paper, we present thermopower measurements under hydrostatic pressure; the results firmly establish the relationship between the superconductive transition and the enhancement $\delta\alpha$.

The thermopower measurements under high pressure were carried out with a self-clamped device. A differential T-type thermocouple 20 μm in diameter was used to determine the temperature gradient generated by an internal heater. The Cu-lead contribution was carefully measured and subtracted from the results. The set-up was thoroughly tested; for example, it gives the correct thermopower of Cu and Pt metals. The pressure inside the chamber was monitored by a manganin coil. The modification of the thermopower due to pressure loss during cooling could therefore be corrected; it is plotted together with the raw data. The thermopower measurements under high magnetic field were also performed on a home-built apparatus. Pt resistors were used to monitor the temperature in the probe. The Cu-lead contribution was not subtracted out from this measurement.

Fig.1 shows $\alpha(T)$ under different pressures for three $La_{2-x}Sr_xCuO_4$ samples: an underdoped $x = 0.13$, an optimally doped $x = 0.15$, and an overdoped $x = 0.22$. In this system there is no charge reservoir, and a pressure-independent chemistry guarantees a pressure-independent hole concentration in the CuO_2 sheets. T_c behaves differently under pressure for the orthorhombic and tetragonal phases. A $dT_c/dP > 0$, i.e. 0.2 K/kbar for $x = 0.13$ and 0.1 K/kbar for $x = 0.15$ in the orthorhombic phase, and a $dT_c/dP = 0$ for $x = 0.22$ in the tetragonal phase are well documented [5]. In the orthorhombic samples with $dT_c/dP > 0$, pressure clearly increases the enhancement term $\delta\alpha$. On the other hand, the $x = 0.22$ tetragonal sample with $dT_c/dP = 0$ shows no pressure dependence of $\alpha(T)$. Therefore, the superconductive transition and the enhancement $\delta\alpha(T)$ are correlated with each other.

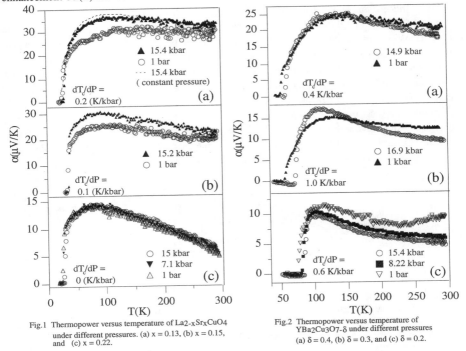

Fig.1 Thermopower versus temperature of $La_{2-x}Sr_xCuO_4$ under different pressures. (a) $x = 0.13$, (b) $x = 0.15$, and (c) $x = 0.22$.

Fig.2 Thermopower versus temperature of $YBa_2Cu_3O_{7-\delta}$ under different pressures (a) $\delta = 0.4$, (b) $\delta = 0.3$, and (c) $\delta = 0.2$.

In the 123 system, it has been argued that pressure induces hole transfer from the CuO_x charge reservoir layer to the CuO_2 sheets. Given this assumption, the α_0 term should decrease under pressure. In Fig.2, we plotted the results for three underdoped samples in the 123 system. In these underdoped samples, the conduction from segmented chains is negligible. The thermopower, in this case, should reflect that from the CuO_2 sheets. Pressure indeed reduces α_0 in these samples, but it increases the enhancement $\delta\alpha$ above 100 K and T_c as well. The sample with $\delta = 0.3$ shows the largest $dT_c/dP = 1$ K/kbar and also $d\delta\alpha/dP$; this composition is located at the transition from the 60 K phase to the 90 K phase. In the sample $\delta = 0.2$, $\delta\alpha$ appears to be smaller, which is caused by a chain contribution. For $\delta \leq 0.1$, the chain contribution becomes dominant [7]. The relation between dT_c/dP and $d\delta\alpha/dP$ in these underdoped 123 samples is similar to that of the $La_{2-x}Sr_xCuO_4$ system.

In $YBa_2Cu_4O_8$ (124), a naturally stoichiometric system, double Cu-O chains are fully established. The conduction from double chains have been reported [6], and Cooper pairs are condensed out of the Cu-O double chains [8]. Fig.3 shows the thermopower as a function of temperature under pressure. The temperature dependence of the thermopower looks different from other cuprates. The minimum at 150 K may be attributed to the influence from double chains. Indeed, an enhancement $\delta\alpha$ in the negative direction from Cu-O chains has been deduced for the $Y_{1-x}Ba_{2-x}La_xCu_3O_{7-\delta}$ system [7]. Pressure reduces α_0, which means that hole transfer from double chains to CuO_2 sheets takes place. Moreover pressure deepens further the minimum at 150 K, which would indicate an increase of the enhancement $\delta\alpha$ from double chains under high pressure. In the Y124 system, the charge transfer mechanism alone may not be sufficient to interpret the large coefficient $dT_c/dP = 0.66$ K/kbar as a relatively high value $dT_c/dP = 0.25$ K/kbar was found even in optimally doped $Y_{0.8}Ca_{0.2}Ba_2Cu_4O_8$ [9]. High pressure may increase the pair concentration n_s in the double chains as well as inducing charge transfer. Given this situation, T_c increases with increasing total pair concentration n_s, which is consistent with μSR data [10]. Therefore the enhancement $\delta\alpha$ from the double chains can be treated as an indicator of pair concentration in the chains.

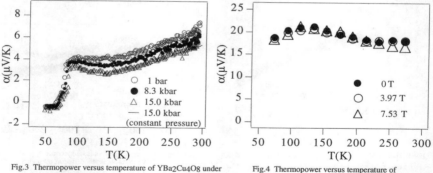

Fig.3 Thermopower versus temperature of $YBa_2Cu_4O_8$ under high pressure

Fig.4 Thermopower versus temperature of $La_{1.85}Sr_{.15}CuO_4$ under high magnetic field

With this clarification of the relation between T_c, $\delta\alpha$ and n_s, we turn to the possible origin of the enhancement $\delta\alpha$. The calculation based on the van Hove singularity model gives an enhancement [11], but the relation between $\delta\alpha$ and T_c is just opposite to what we have observed. Within the marginal Fermi liquid model, Kumar [12] obtained a negative slope in α vs T, but he failed to give an enhancement $\delta\alpha$. Tallon *et al* [4] have suggested that the spin gap found in the NMR data for underdoped samples may account for $\delta\alpha$. In $La_{1.85}Sr_{.15}CuO_4$, however, the spin gap from NMR opens right above T_c whereas the onset of the enhancement $\delta\alpha$ is near room temperature [3]. Moreover, the thermopower in this sample is magnetic-field independent up to 30 T at three fixed temperatures. We have performed the thermopower measurement as a function of temperature under a magnetic field. The results, Fig.4, show that $\delta\alpha$ is unchanged within our device resolution up to 7.5 T.

We have already pointed out [3] that it is difficult to account for an enhancement $\delta\alpha$ with a $T_{max} > 100$ K by a phonon-drag effect due to electron interactions with acoustic phonons. However, Raman spectra [14] for Y124 show that an oxygen in-phase vibration mode of energy 627 K along the c axis saturates at 5 Gpa whereas other modes vary linearly with pressure. At the same pressure (5 GPa), T_c becomes saturated against pressure also [15]. Therefore we could assume that this optical vibrational mode involves superconductive pair formation. In the case of phonon drag, a peak appears at 0.2-0.3 θ_D in the curve of α *vs* T, where θ_D is Debye temperature. Taking the same format into the optical-phonon domain, with a characteristic energy of 627 K, we obtain an enhancement due to this phonon having a $T_{max} \approx 134$ K, which is compatible with our results. We conclude that strong electron coupling to optical mode vibrations is the underlying mechanism relating T_c, $\delta\alpha$, and n_s.

This work is supported by the National Science Foundation.

References

1. H. Takagi *et al*, *Phys. Rev. Lett.* **69** (1990) 2975
2. H.Y. Hwang *et al*, *Phys. Rev. Lett.* **72** (1994) 2636
3. J.-S. Zhou and J.B. Goodenough, *Phys. Rev.* **B51** (1995) 3104
4. J.L. Tallon *et al*, *Phys. Rev. Lett.* **75** (1995) 4114
5. J.-S. Zhou, H. Chen and J..B. Goodenough, *Phys. Rev.* **B49** (1994) 9084
   N. Yamada and M, Ido, *Physica* **C203** (1992) 240
6. B. Bucher *et al* , *Phys. Rev. Lett.* **70** (1993) 2012
7. J.-S. Zhou, J.P. Zhou, J.B. Goodenough and J.T. McDevitt, *Phys. Rev.* **B51** (1995) 3250
8. D.N. Basov *et al*, *Phys. Rev. Lett.* , **74** (1995) 598
9. T. Miyatake *et al*, *Physica* **C167** (1990) 297
10. J.L. Tallon *et al*, *Phys. Rev. Lett.* , **74** (1995) 1008
11. D.M. Newns *et al*, *Phys. Rev. Lett.* **73** (1994) 1695
12. D. Kumar, *J. Phys. Condens. Matter*, **5** (1993) 8277
13. R.C. Yu *et al*, *Phys. Rev.* **B37** (1988) 7963
14. N. Watanabe *et al*, *Phys. Rev.* **B49** (1994) 9226
15. J.J. Scholtz *et al*, *Phys. Rev.* **B45** (1992) 3077

BEHAVIOR OF ScN AND ScS CONTACTS UNDER MICROWAVE IRRADIATION

A.B. AGAFONOV, D.A. DIKIN, V.M. DMITRIEV, A.L. SOLOVJOV

B. Verkin Institute for Low Temperature Physics and Engineering
National Academy of Sciences of Ukraine,
47 Lenin Ave., Kharkov, 310164, UKRAINE

Measurements of the I–V characteristics and their derivatives of ScN and ScS metallic weak links prepared by point contact and break–junction technique and made of Y(123), Bi(2212), Tl(2212) have been performed in the temperature range from 4.2 to 100 K. Contacts were irradiated by microwave field of wide frequency range $10^5 \div 2 \cdot 10^{10}$ Hz. Temperature dependencies and behavior under irradiation of the critical and excess currents are reported. Observed effects point to the presence of Cooper–pair as well as quasiparticles photon assisted charge transport. We revealed the clear evidence of existence in these structures of nonlinear and nonequilibrium phenomena typical for conventional weakly coupled structures, such as planar microbridges or bridges of variable thickness. But high–T_c samples indicate much more complicated picture of the quasiparticles dynamics. That is the subject of our discussion.

1 Objects of investigation

We report the results of our study of two types of weak links. The first type is point contact formed by niobium needle with point diameter $\simeq 1\mu$m and high-T_c superconductor anvil (single–phase ceramic crystallites): Nb–Tl$_2$Ba$_2$CaCu$_2$O$_x$ with critical temperature $T_c \cong 100$K, Nb–YBa$_2$Cu$_3$O$_x$ ($T_c \cong 84$K), Nb–Bi$_2$Sr$_2$CaCu$_2$O$_x$ ($T_c \cong 90$K). The second type is break junction made of YBa$_2$Cu$_3$O$_x$–single crystal ($T_c \cong 88 \div 92$K) narrowed in the middle to the bridge of some tens of μm and then broken in the cryostat. The contacts had IVC's of plain metallic type, big value of excess current (I_{exc}) at high bias voltages. Measurements were performed by standard 4–probes current–biased technique.

2 Experimental results and discussion

For ScN point contacts Nb–Tl(2212) we revealed that under the microwave irradiation ($f \cong 12$GHz) at temperatures $T/T_c \simeq 0.5$ excess current oscillates with increasing of power level P. According to the theory [1] at high bias voltages (eV $\gg \Delta$, T) at ScN contacts IVC under irradiation could be written as:

$$I(P) = V/R_n + I_{exc} = V/R_n + J_0^2(Z) \cdot I_0 \cdot \mathrm{sign} V,$$

where I_0 is excess current in absence of irradiation, $J_0^2(Z)$ – Bessel function of parameter $Z = (eV_\omega/\hbar\omega) \sim \sqrt{P}$, V_ω – ac voltage induced by microwave power. So, I_{exc} in this case oscillates with P increasing as square of Bessel function. Our experimental data fit well predicted [1] dependence. Such related to Andreev reflection nonlinear IVC behavior could be used for detecting and mixing of electromagnetic signals.

At higher temperatures we did not observed I_{exc} oscillations. But at temperatures near $T_c(T/T_c \simeq 0.9)$ and voltages at the contact $V > \Delta$ we revealed increase of I_{exc} at some microwave power so that $I_{exc}(P)$ exceeded I_{exc} at P = 0. That means an enhancement of contact superconducting properties. Similar behavior of critical current was observed earlier on Nb–Nb [2] and Nb–Sn point contacts near T_c. At ScS Nb–Y(123) contacts and temperature $T \ll T_c$ (T = 4.2 K) we also observed I_{exc} and I_c stimulation that is impossible for conventional superconductors. The theory of superconductivity stimulation effects was developed by Eliashberg [3] and Aslamazov, Larkin [4] for conventional superconductors. But these theories have some strong restrictions on superconductor and weak link characteristics that very difficult to realize and control in high-T_c materials.

For ScS break junctions we observed IVC typical for conventional superconducting bridges of variable thickness with clear evidence of quasiparticle nonequilibrium behavior: two values of normal resistance and critical temperature, big excess current and linear dependence of $I_{exc}R_n$ at T close to T_c, intrinsic voltage steps.

The observation of such effects, associated with nonequilibrium processes, in temperature range from T_c to $T \ll T_c$ suggests more complicated dynamics of quasiparticles in HTS than in conventional superconductors. The nonequilibrium dynamics of quasiparticles in HTS may be due to some peculiarities of this material. We suppose that there are mechanisms providing an existence of nonfreezing quasiparticles at low temperatures and delocalization of quasiparticles under microwave irradiation.

Some our experimental results (contacts' R_d constancy for ScN in wide temperature range from $T/T_c \cong 0.8$ to 0; deviation to the lower values of Δ and $I_{exc}R_n$–product from BCS temperature dependence for ScS point contacts and break junctions) point in favor of these assumptions.

3 Conclusion

The observed phenomena are consequence of complicated nature of the quasiparticle spectrum and transport properties of HTS. An interpretation of stimulating properties of the external irradiation in the spirit of known for conventional superconductors mechanisms requires comprehensive studies and particular care.

Acknowledgments

Presentation of this paper became possible thanks to the financial support of Workshop Organizing Committee, NSF, research group of Dr.Paul C.W.Chu and International Foundation "Renaissance" (Ukraine) sponsored by J.Soros Foundation.

References

1. A. F. Volkov and A. V. Sergeev, Zh.Teor.Fiz. **51**, 1716 (1981).
2. V. M. Dmitriev *et al*, Fiz. Kondens. Sost. **issue 28**, 3 (FTINT AN USSR, 1976).
3. G.M. Eliashberg, Pis'ma v ZhETF **11**, 186 (1970).
4. A.G. Aslamazov and A.I. Larkin, Zh.Eksp. and Teor.Fiz. **36**, 2109 (1987).

TEMPERATURE DEPENDENCE OF RADIOFREQUENCY ABSORPTION AND CRITICAL CURRENT DENSITY IN HIGH-T_c YBaCuO THIN SAMPLES

N.T. CHERPAK, G.V. GOLUBNICHAYA, E.V. IZHYK,
A.YA. KIRICHENKO, I.G. MAXIMCHUK, A.V. VELICHKO
Institute of Radiophysics and Electronics, National Academy of Sciences
12 Acad. Proskura street, 310085 Kharkov, Ukraine

The temperature and field dependences of radiofrequency (rf) absorption in both YBaCuO ceramic plate and thin epitaxial film in presence of size effect are analyzed. The rf absorption in ceramic is shown not to be described satisfactory in terms of the model of critical state or that of thermal assisted magnetic flux creep (TAFC). However the peculiarities of the absorption in thin film can be explained within the scope of TAFC model. The temperature dependence of the absorption was used to determine the temperature one of critical current density $J_c(T)$. It has been found that S-N-S type of weak links leads the behavior $J_c(T)$.

While studying the energy absorption temperature dependence of high-T_c superconducting (HTS) YBaCuO samples in radiofrequency (rf) range (1-15 MHz) the well pronounced peak at $T_m < T_c$ is revealed under condition that the penetration depth of electromagnetic field in the normal state exceeds the sample thickness d (size effect).[1,2] On the other hand impact of temperature T and magnetic field (dc and ac) on rf absorption was studied in a number of works, but quantitative data are not sufficient for identification of the physical models for the absorption in HTS. Study of size effect as a function of extrinsic factors occurs a good approach in this aspect. Another aspect of the size effect is a possibility to determine the T-dependence of critical current J_c near T_c.

In our work the T-dependence peculiarities of rf absorption in HTS ceramic and epitaxial film under conditions of the size effect are analyzed. On this basis the conclusion is drawn about applicability of various physical models of rf absorption, about $J_c(T)$ and nature of intergranular weak links in the studied samples.

The measurement technique used is based on dependence of the resonant circuit Q-factor versus HTS sample impedance properties.[1] In the work some other author's data are also involved for the analysis.

For the ceramic the study of the sample thickness dependence of the absorption maximum has carried out. We emphasize that frequency dependence $T_m(f)$ is weak (logarithmic) function and is the same for both the critical state (CS) and thermal assisted magnetic flux creep (TAFC) models. At the same time these models result in distinguished dependence of T_m against d. Within the framework of CS model the meaning of T_m is given by expression $H_{rf} = 0.971\mu_0 J_c(T_m)d$ and within the scope of TAFC model we have $\delta_{sk}(T_m) = d/2$, where $\delta_{sk} = 2\rho_f/(\omega\mu_0)^{1/2}$ and ρ_f is the sample resistance in magnetic flux regime[3]. But results of the experimental data analysis for ceramic sample (d=1.2, 3.3, 3.8 nm) with using the CS and TAFC models do not witness in favor of either of models. We believe the TAFC model is preferable for description of high temperature part of the absorption curve ($T > T_m$) and

CS model gives better the position of the absorption maximum. In contrast to the ceramic the empirical expression for the film [4] $(1 - T_m/T_c)^n \times (T_c/T_m) = aH_{rf}$ may be explained in the frame of TAFC model at $n = 1.5$ and $a = [H_e \ln(f_0/f)]^{-1}$ where f_0 and H_e are constants.

In works [5,6] a relation between J_c and hysterical losses in HTS has been established. For disc geometry the losses are

$$\frac{P}{P_{max}} = \frac{\chi''(T)}{\chi''_m} = \left\{ \frac{2\chi_0}{\pi} \int \left\{ -S(x) + (1 - \cos\theta)S\left[\frac{x}{2}(1 - \cos\theta)\right] \right\} \sin\theta\, d\theta \right\} / (0.241\chi_0),$$

where $S(x) = (1/2x)[\cos^{-1}(1/\cosh x) + \sinh x/\cosh^2 x]$, $x = H_{rf}/H_d$, $\chi_0 = 8R/(3\pi d)$, R is the sample radius, where $H_d = J_c d/2$ whence we can restore dependence $J_c(T)/J_d$, where $J_c = 2H_d/d$. The temperature dependence of the absorption for ceramic with $d = 3,3$ mm was used. The fitting function used was $J_c(T) = J_{c0}[(T_c - T)/(T_c - T_0)]^n$ where $J_{c0} = J_c(T=0)$. The best approximation is obtained at $J_{c0}/J_d = 1.89$, $n = 1.52$, $T_0 = 80.4$ K ($T_c = 91.3$ K). Dependence $J_c(T) \simeq (T_0 - T)^{3/2}$, characteristic for both "dirty" HTS and percolative ones, testifies in favor of realization of S-N-S type weak link contacts in the samples studied.

Thus, the rf absorption in ceramic is shown not to be described satisfactory in terms of critical state model or thermal assisted magnetic flux creep. However peculiarities of the absorption in thin films can be explained within the scope of the TAFC model. The temperature dependence of rf absorption has allowed to determine the temperature dependence of the critical current density $J_c(T)$. The authors has found that dependence $J_c(T)$ in ceramic is driven by S-N-S type weak links.

References

1. E.V. Izhyk et al, Sov. Tech. Phys. Lett. (USA) **15**, 2478 (1989).
2. N.T. Cherpak et al, Physica C **180**, 280 (1991).
3. V.B. Geshkenbein et al, Phys. Rev. B **43**, 3748 (1991).
4. W. Xing et al. Ac susceptibility investigation rf YBaCuO thin films, *(private commun.)*
5. Yu.F. Revenko et al, Sov. Tech. Phys. Lett. (USA) **14**, 909 (1988).
6. J.R. Clem, A. Sanches, Phys. Rev. B. **50**, 9355 (1994).

NORMAL STATE MAGNETORESISTANCE AND HALL EFFECT IN LaSrCuO THIN FILMS

F.F. BALAKIREV, I.E. TROFIMOV, S. GUHA, P. LINDENFELD
Department of Physics and Astronomy, P.O.Box 849, Rutgers University, Piscataway, NJ 08855, USA

We report measurements of the magnetoresistance, $\Delta\rho(H)/\rho_0$, and Hall effect of thin films of $La_{2-x-y}Sr_xNd_yCuO_4$ between T_c and 300 K. The specimens were grown on $LaSrAlO_4$ substrates by laser ablation. At high temperatures the orbital part of the magnetoresistance is found to be proportional to the square of the tangent of the Hall angle, but there are large deviations at temperatures that are lower although not yet in the fluctuation regime. Measurements of compounds with Nd substituted for some of the Sr show the same characteristic T-dependence of the magnetoresistance, while the behavior of the Hall angle is quite different.

High-temperature superconductors (HTS) demonstrate unusual transport properties, including a linear temperature dependence of the resistivity[1] and a cotangent of the Hall angle, $cot\theta_H$, proportional to T^2 (Ref. 1,2). One proposal that has been made to explain these phenomena is for quasiparticles with two different lifetimes, τ_{tr} and τ_H, for the longitudinal and transverse currents[3]. Recent experiments on magnetoresistance (MR) give support for this model[4], and show that the orbital MR in HTS cuprates violates Kohler's rule, and that $\Delta\rho(H)/\rho_0$ is proportional to τ_H^2. While data on YBCO show good agreement with this proportionality, the data on LSCO deviate at low temperatures. We have measured the MR and Hall effect on several LSCO samples to check the validity of the model.

C-axis oriented single-crystal thin films of $La_{2-x-y}Sr_xNd_yCuO_4$ were prepared by laser ablation on $LaSrAlO_4$ substrates. The sample resistance and Hall voltage were measured using a DC four-probe method with resolution better than 2×10^{-6}, and temperature stability better than 1 mK.

We investigated two specimens with y=0 and with x=0.15 and 0.135, and one specimen with x=0.1 and y=0.15. For the MR the measurements were made both in the transverse (**H** ∥ c-axis) and in the longitudinal (**H** ∥ **j**) geometry. For T more than 10 K above the superconducting transition temperature (T_c) the magnetoresistance is proportional to B^2 up to the highest field (10 T) in both directions for all measured samples, while at lower T it has the form characteristic for the fluctuation region. The transverse MR is larger than the longitudinal by more then one order of magnitude, indicating that the spin-related MR, if present, is much smaller then that due to the orbital effects. While the two samples near optimal doping exhibit a linear resistance (Fig.1a) and $cot\theta_H$ proportional to T^2 (Fig. 1b), the Nd-doped sample shows a lower T_c, an upturn in resistance, and also in $cot\theta_H$, at low temperatures. Despite quite

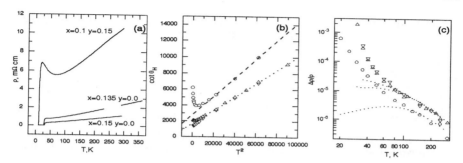

Figure 1: **(a)** Resistivity vs. temperature. **(b)** $cot\theta_H$ vs. T^2 for $La_{2-x-y}Sr_xNd_yCuO_4$: ▽ x=0.15, y=0; △ x=0.135, y=0; ○ x=0.1, y=0.15. **(c)** $\Delta\rho/\rho$ at 1 Tesla vs. T: ▽ x=0.15, y=0; △ x=0.135, y=0; ○ x=0.1, y=0.15. The dashed lines show the fit to the equation in the text.

different behavior of ρ and θ_H, the MR of all three samples shows an almost identical temperature dependence (Fig. 1c). We have analyzed our MR data with the formula $\Delta\rho/\rho = (\omega_c\tau_H)^2(\tau_{tr}/\tau_H - 1)$. This temperature dependence differs from that used in Ref. 4 by a factor of $(\tau_{tr}/\tau_H - 1)$. It may be seen from Fig. 1c, that deviations from the equation start at temperatures too high to be attributed to superconducting fluctuations. For the Nd doped sample the deviation begins at $T \sim 6T_c$. Although the model of the two lifetimes describes the data well above 100 K, it does not seems to do so at lower temperatures.

Acknowledgments

This work was supported by NSF grants 93-05860 and 95-01504, and by the Rutgers research Council.

References

1. G. Xiao, P. Xiong, M.Z. Cieplak, *Phys. Rev. B* **46**, 8687 (1992).
2. T. R. Chien, Z. Z. Wang, and N. P. Ong, *Phys. Rev. Lett.* **67**, 2088 (1991).
3. P. W. Anderson, *Phys. Rev. Lett.* **67**, 2092 (1991).
4. J.M. Harris, Y.F. Yan, P. Matl, N.P. Ong, P.W. Anderson, T. Kimura, and K. Kitazawa, *Phys. Rev. Lett.* **75**, 1391 (1995).

GROUND STATE OF SUPERCONDUCTING La$_{2-x}$Sr$_x$CuO$_4$ IN 61-TESLA MAGNETIC FIELDS

G. S. Boebinger, Yoichi Ando,* A. Passner
Bell Laboratories, Lucent Technologies, Murray Hill, NJ 07974, USA

K. Tamasaku, N. Ichikawa, S. Uchida
Superconductivity Research Course, University of Tokyo, Yayoi, Bunkyo-ku, Tokyo 113, Japan

M. Okuya, T. Kimura, J. Shimoyama, K. Kishio
Department of Applied Chemistry, University of Tokyo, Hongo, Bunkyo-ku, Tokyo 113, Japan

Using a 61 T pulsed magnet to suppress superconductivity, we measure the low-temperature normal-state in-plane (ρ_{ab}) and c-axis (ρ_c) resistivities of $La_{2-x}Sr_xCuO_4$ (LSCO) single crystals with Sr concentrations, x, ranging from 0.08 to 0.22. Underdoped samples exhibit a previously reported logarithmic divergence of both ρ_{ab} and ρ_c. As the carrier concentration is increased in the underdoped regime, the divergence becomes weaker until an insulator-to-metal (I-M) transition occurs simultaneously in ρ_{ab} and ρ_c at optimal doping ($x \sim 0.16$). At the I-M transition, the magnitude of the sheet resistance per CuO_2 layer is well below both the 2D Mott limit and $h/4e^2$.

Because upper critical fields in the high-T_c cuprates are extremely large, a systematic study of the normal-state at low temperatures requires intense pulsed magnetic fields. Normal state resistivity measurements of LSCO single crystals over a wide range of carrier concentrations reveal an insulator to metal transition occurring at $x \sim 0.16$, the Sr concentration corresponding to optimum T_c.

The single crystal LSCO samples are grown by the floating-zone method for a variety of Sr concentrations.[1,2] Samples with $x = 0.08, 0.13, 0.15$ (not shown), 0.17, 0.18 (not shown), and 0.22 were grown by the Kishio group, while samples with $x = 0.12$ and 0.15 were grown by the Uchida group. ρ_{ab} and ρ_c are measured in differently-oriented platelet-shaped crystals cut from the same rod. During the 100 msec magnet pulses, the four-probe resistivity is measured at a fixed temperature using a $\sim$120 kHz lock-in and a transient digitizer. Experimental details are given elsewhere.[3,4]

Figure 1a shows the temperature dependence of ρ_{ab} for five representative crystals. Differences between the data above T_c at 0 T (lines) and 60 T (symbols) are due to superconducting fluctuations and the magnetoresistance of the normal-state. Although not evident on this semi-log plot, ρ_{ab} shows linear-T behavior above 50K for all samples with $x \geq 0.15$ and above 150 K for the $x = 0.13$ sample. The underdoped samples exhibit a previously reported logarithmic divergence of both ρ_{ab} and ρ_c.[3,4] For the $x = 0.13$ and 0.15 samples, even though ρ_{ab} at 0 T is metallic down to T_c, clearly insulating behavior is observed in the $T/T_c \to 0$ limit, once superconductivity is suppressed by the intense magnetic field. In contrast, the $x = 0.17$ and 0.22 samples show metallic ρ_{ab} down to our lowest experimental temperatures, where the onset of the superconducting transition is evident. Figure 1b shows the temperature dependence of ρ_c for seven crystals. For ρ_c, there is an

---

*Permanent address: Central Research Institute of Electric Power Industry, 2-11-1 Iwato-kita, Komae, Tokyo 201, Japan

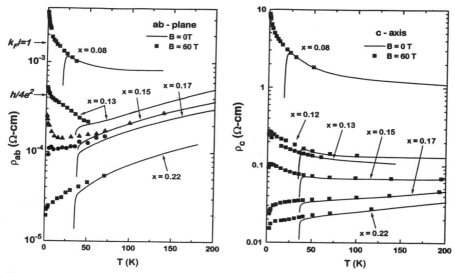

Figure 1: (a) ρ_{ab} and (b) ρ_c vs T at 0 T (lines) and 60 T (symbols) for a variety of Sr concentrations, x.

I-M transition occuring between $x = 0.15$ and 0.17, just as for ρ_{ab}. We therefore conclude that there is an I-M transition in the normal state of the high T_c cuprates which occurs at $x \sim 0.16$, the Sr concentration corresponding to optimal superconductivity.

Note the arrows in Fig. 1a which indicate that the magnitude of ρ_{ab} at the I-M transition is smaller than both (1) the 2D Mott limit, $hc_o/\rho_{ab}e^2 = k_F l = 1$, where $c_o = 6.5\text{Å}$ is the CuO layer spacing; and (2) the universal value for a 2D superconductor-insulator transition, $h/4e^2$ per CuO layer, recently reported for underdoped high T_c cuprates.[5]

We thank E. Abrahams, B.L. Altshuler, P.W. Anderson, P.B. Littlewood, A.J. Millis, D. Pines, and C.M. Varma for helpful discussions.

References

1. T. Kimura, K. Kishio, T. Kobayashi, Y. Nakayama, N. Motohira, K. Kitazawa, and K Yamafuji, *Physica C* **192**, 247 (1992).
2. Y. Nakamura and S. Uchida, *Phys. Rev.* B **47**, 8369 (1993).
3. Y. Ando, G. S. Boebinger, A. Passner, T. Kimura, and K. Kishio, *Phys. Rev. Lett.* **75**, 4662 (1995)
4. Y. Ando, G. S. Boebinger, A. Passner, R. J. Cava, T. Kimura, J. Shimoyama, and K. Kishio, in this volume.
5. Y. Fukuzumi, K. Mizuhashi, K. Takenaka, and S. Uchida, *Phys. Rev. Lett.* **76**, 684 (1996).

THE EFFECT OF Sr IMPURITY DISORDER ON THE MAGNETIC AND TRANSPORT PROPERTIES OF $La_{2-x}Sr_xCuO_4$, $0.02 \leq x \leq 0.05$

R. J. Gooding, N. M. Salem
Dept. of Physics, Queen's University, Kingston, ON Canada K7L 3N6

R. J. Birgeneau, F. C. Chou
Dept. of Physics and the CMSE, MIT, Cambridge, MA U.S.A. 02139

We discuss the properties of the intermediately doped $LaSrCuO$ compounds, a system which is known to behave like a canonical spin–glass phase at low temperatures. Arguments are made in support of the existence of disorder–induced meandering rivers of charge connecting impurity centres, thus producing small clusters of essentially undoped antiferromagnetic spins. We detail how these clusters result in the anomalously small Curie constant measured in the temperature interval just above the spin–glass transition.

Experimental studies in the doping region of $La_{2-x}Sr_xCuO_4$ separating the antiferromagnetic (AFM) insulating phase ($0 \leq x \leq 0.02$) from the superconducting phase ($0.05 \leq x \leq 0.2$), *viz.* in the intermediately doped spin–glass regime ($0.02 \leq x \leq 0.05$), have begun to produce a consistent picture of this unusual system: It is known that at low temperatures this phase behaves like a canonical spin–glass phase.[1] Also, an unusual scaling of the spin correlation length is found,[2]

$$\xi^{-1}(x,T) = \xi^{-1}(x,0) + \xi^{-1}(0,T) \quad , \tag{1}$$

an anomalously small Curie constant is found in the temperature range above the spin–glass transition,[1] and the magnetic correlations seemingly remain commensurate.[2] Earlier neutron–scattering[3] and La NQR experiments[4] interpreted their complicated results in terms small domains of antiferromagnetically correlated spins. Lastly, transport measurements of this compound[2,5] displayed (i) a logarithmic conductance with temperature and (ii) an isotropic, negative magnetoresistance, both of which are difficult to interpret in tandem with the above–mentioned magnetic data.

One theoretical approach that yielded a $\xi(x,T)$ consistent with Eq. (1) was based on explicitly including the effects of the disordered Sr impurities, and then examining the spin texture that results when carriers in the system are localized in the regions around each impurity.[6] We have reexamined the spin texture that led to this spin correlation length, and have found a configuration similar to that of a cluster spin glass, *viz.*, similar to that suggested in Refs. 3 and 4. Then, in this report we focus on the small Curie constant found in susceptibility experiments,[1] as well as reporting new measurements on better samples, and show that a cluster spin glass must also have a small Curie constant, further supporting the existence of such a novel phase of matter.

The simplest configuration of impurities that serves to illustrate how the domains of a cluster spin glass are formed is three Sr impurities that are non–colinear. Figure 1 depicts a representative configuration of spins — note that in the region bounded by the lines connecting the impurities, the AFM order parameter is rotated relative to that of the background spins. When many impurities are included, this physics repeats itself for differing trios and quartets and ... of Sr impurities, and the resulting spin texture is that of small clusters of spins with boundaries defined by the meandering paths that connect the Sr sites. The boundaries correspond to disordered spin states that would be desirable, relative to the antiferromagnetically clusters, for the carriers to move in. This serves to explain the absence of incommensurate correlations: the carriers do become mobile as the temperature is increased, but they move along the disordered domain walls thus effectively being expelled from the commensurately ordered cluster. (A similar effect was proposed to result from the electronic phase separation phenomenology of Emery and Kivelson[7] — here we find that it is the disorder which leads to the existence of these rivers of charge and the associated expulsion of the carriers from the clusters.)

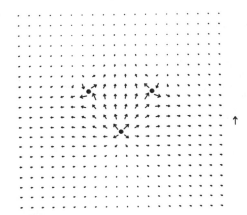

Figure 1: Three Sr impurities (filled circles) and the ground state spin texture that they induce (open boundary conditions are used). The effect of the impurities on the spins is introduced via the interaction discussed in Ref. 6. The projection of the spins onto the xy plane is shown — a legend at the right shows a spin rotated completely into the plane and pointing along the y direction. For clarity, all spins on B sublattice sites are flipped, an operation that would make an AFM look like a FM. The spins in the region bounded by the impurities are antiferromagnetically correlated, but the order parameter is pointing in the y direction. Outside of this region the order parameter asymptotically approaches its bulk direction, namely, it points along the z axis.

Measurements of the Curie constant just above the spin–glass transition in Ref. 1 for a $x = 0.04$ crystal found that only 0.5 % of the Cu spins contributed to the susceptibility. Preliminary measurements on crystals (grown from 5N–99.999 % pure original compounds in air using a crucible–free, contamination–free, travelling solvent float–zone method) of $x = 0.025$ and 0.035 found Curie constants corresponding to 0.13 % and 0.35 % of the Cu spins contributing, respectively. These small values, and the increasing Curie constant with Sr concentration, may be derived for a cluster spin glass in a straightforward fashion: As a simple picture of such spin configurations, let each domain be of a $L \times L$ geometry with $L = a_0/\sqrt{x}$, a_0 being the lattice constant. Then, noting that the quantum–mechanical ground state for such a cluster is a $S = 0$ singlet for L even, and a $S = \frac{1}{2}$ doublet for L odd, and that the gaps to the first excited states are large (e.g., 0.3 J for a 32–square lattice, J ($\approx$ 1500 K for $LaSrCuO$) being the AFM exchange constant), as the first approximation we take the singlet clusters to be magnetically inert (in the temperature interval over which the susceptibility was measured). Thus, for a lattice with N Cu spins, the susceptibility arises from $N/(2L^2)$ spin-$\frac{1}{2}$ magnetic units — this immediately leads an upper bound for the Curie constant of $\frac{1}{2}x$ of the total number of Cu spins, consistent in both magnitude and trend with experiment. (A somewhat more rigorous treatment, including the excitations into the first excited states, yields $\frac{3}{8}x$.)

Details of the effect of quantum fluctuations on these spin textures, and a full theoretical treatment of the magnetoresistance of mobile carriers in a cluster spin glass, both explicitly accounting for the critical role played by Sr impurity disorder, will be presented elsewhere.

1. F. C. Chou, et al., Phys. Rev. Lett. **75**, 2204 (1995), and references therein.
2. B. Keimer, et al., Phys. Rev. B **46**, 14034 (1992).
3. S. M. Hayden, et al., Phys. Rev. Lett. **66**, 821 (1991).
4. J. H. Cho, et al., Phys. Rev. Lett. **46**, 3179 (1992).
5. N. W. Preyer, et al., Phys. Rev. B **44**, 407 (1991).
6. R. J. Gooding and A. Mailhot, Phys. Rev. B **48**, 6132 (1993).
7. S. Kivelson and V. J. Emery, 1994 LANL Proceedings.

THE NORMAL STATE TRANSPORT PROPERTIES IN PURE, Pr- AND Ca-DOPED $RBa_2Cu_3O_{7-\delta}$ SYSTEMS

WEIYAN GUAN
Department of Physics, Tamkang University, Tamsui, Taiwan, R. O. C.

We report detailed studies of the normal state resistivity and Hall-effect in pure, Pr- and Ca-doped $RBa_2Cu_3O_{7-\delta}$ systems. We find a linear temperature dependence of the normal state resistivity ρ_n and the Hall number n_H above T_C. This fact seems to be the most puzzling: *The higher the carrier density, the smaller the conductivity!* At a constant temperature ρ_n and n_H are dependent on ionic radius of rare earth r_{R3+}, viz. the larger R^{3+} ionic radius, the larger ρ_n ($\rho_n \propto r_{R3+}$), but the lower n_H ($n_H \propto 1/r_{R3+}$). For a fixed temperature the relation of $\rho_n \propto 1/n_H$ is confirmed for pure, Pr- and Ca-doped $RBa_2Cu_3O_{7-\delta}$ with different R ion. *The higher the carrier density, the higher the conductivity.*

In the simple Drude model, one expects $\rho = m^*/ne^2\tau$, where m^* is the effective electron mass, and $1/\tau$ is the electron scattering rate. However, for most of the high-T_C superconductors both ρ and $n_H \propto T$, then, $\rho \propto n_H$ instead of $\rho \propto 1/n$. This fact seems to be the most puzzling in the normal state transport properties of the high-T_C cuprates. In a single-band Fermi liquid system, the Hall number, $n_H = V/eR_H$ would be the number of carriers per unit cell V. For the high-T_C superconductors, due to its T-dependence and the complex band structure, n_H may not represent the actual carrier concentration. However, the value of n_H at a particular T correlates rather consistently with the carrier concentration determined by other methods. n_H scales with actual carrier concentration.

In our recent papers we have reported detailed studies of the normal state resistivity and the Hall-effect of bulk $R_{0.8}Pr_{0.2}Ba_2Cu_3O_{7-\delta}$ (R=Nd, Eu, Gd, Dy, Er, Yb)[1] and $R_{0.9}Ca_{0.1}Ba_2Cu_3O_{7-\delta}$ (R=Nd, Gd, Ho, Tm)[2] systems. The results indicate that ρ and n_H are linearly dependent on temperature. At a constant temperature the normal state resistivity increases linearly with increasing R^{3+} ion size, r_{R3+} (Fig. 1), but the Hall numbers n_H monotonically decrease with increasing r_{R3+}: $1/n_H \propto r_{R3+}$ (Fig. 2).

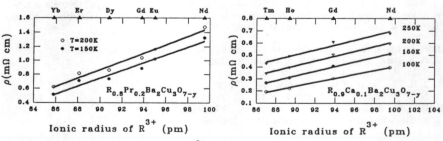

Fig.1 Normal state resistivity ρ vs. R^{3+} ionic radius in Pr- and Ca-doped $RBa_2Cu_3O_{7-y}$

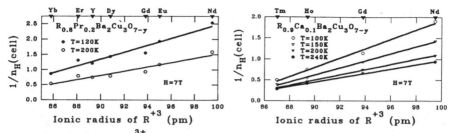

Fig.2 $1/n_H$ vs. R^{3+} ionic radius in Pr- and Ca-doped $RBa_2Cu_3O_{7-y}$

In Fig.3 we present the correlation between ρ and $1/n_H$. For a fixed temperaturethe the relation of $\rho \propto 1/n_H$, is confirmed. The experimental data fall on a common straight line though the origin of coordinates for all pure, Pr- and Ca-doped $RBa_2Cu_3O_{7-y}$ systems.

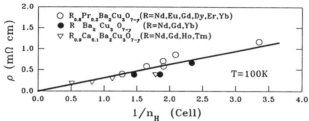

Fig.3 Normal state resistivity ρ vs. $1/n_H$ at T=100K in pure, Pr- and Ca-doped $RBa_2Cu_3O_{7-y}$ systems

The cotangent of the Hall angle follows a universal T^2-dependence: $\cot\theta_H = \alpha T^2 + C$. In Pr-doped systems α and C remain constant.[1] However, in Ca-doped systems,[2] α decreases with increasing $r_{R^{3+}}$, while leaving C unchanged. In an unified picture of Hall effect parameter C is a measure of the in-plane impurity scattering rate. It is not surprising that C is independent on R ion substituted in Y-site. The results demonstrate that the increase of ρ with increasing $r_{R^{3+}}$ is due to the decrease of the number of mobile holes n_H, but not due to the increase of the impurity scattering.

The observed R^{3+} ion size dependence of the Hall number n_H in bulk $R_{0.9}Ca_{0.1}Ba_2Cu_3O_{7-y}$ systems at a fixed Ca concentration reveals that the change of number of the mobile holes in these systems is not only connected to the hole generation by the substitution of Ca^{2+}, but also connected to the hole localization caused by disorder and lattice compression.

This study is supported by the National Science Council, R.O.C.

References
1. J. C. Chen, Yunhui Xu, M. K. Wu, and Weiyan Guan, *Phys. Rev.* **B53** (1996).
2. Weiyan Guan, J. C. Chen and S. H. Cheng, (Unpublished).

ANISOTROPY OF THERMAL CONDUCTIVITY OF YBCO AND SELECTIVELY DOPED YBCO SINGLE CRYSTALS

P.F. HENNING, G. CAO, and J.E. CROW
*Department of Physics and National High Magnetic Field Laboratory, Florida State University,
1800 E. Dirac Dr., Tallahassee, FL 32306 USA*

The anisotropy of the thermal conductivity of single crystals of YBCO has been studied. Along the a-b plane, the thermal conductivity has the expected $1/T$ behavior at high temperatures followed by a sharp upturn at T_c and a peak near $T_c/2$. Along the c-axis the thermal conductivity shows $1/T$ behavior at high temperatures. However, K_c does not display a break in slope at T_c and peaks near 20 K. This behavior is compared to the behavior of heavily doped $YBa_2(Cu_{1-x}Zn_x)_3O_7$ and $Y_{1-x}Pr_xBa_2Cu_3O_7$ single crystals, which have a-b plane thermal conductivities similar to that observed along the c-axis in pure YBCO.

The dramatic rise at T_c and a peak a factor of two over the normal state in the a-b plane thermal conductivity of YBCO has spurred much examination, both theoretical and experimental[1]. In this study, we examine two regimes of the thermal conductivity in the YBCO family. First, we examine the response of the thermal conductivity to the introduction of a small amount of impurities into YBCO single crystals via oxygen and Pr doping. In this case, the charge carrier contribution to K_{ab} strongly dictates the magnitudes and positions of the observed peaks below T_c. The peaks are shifted to higher temperatures and are reduced in magnitude with increasing oxygen vacancies and Pr doping. This is consistent with a theory of charge carrier thermal conductivity in which the quasiparticle scattering rate, gamma, has its relaxation below T_c softened by the introduction of impurities[2]. Second, we examine the thermal conductivity along the c-axis of a pure YBCO single crystal and K_{ab} of more heavily doped Zn and Pr single crystals. K_c in pure YBCO has no break in slope at T_c in contrast to the same crystal's K_{ab}. The magnitude of K_c in the normal state is reduced relative to K_{ab}, $K_c/K_{ab} = .2$, and its temperature dependence exhibits uninterrupted $1/T$ behavior down to its peak at 20 K. Similarly, in the superconducting heavily Pr and Zn doped single crystals along the a-b plane, no change in slope at T_c and peak at 20 K are observed. The behavior in this second regime can be qualitatively understood in terms of phonon thermal conductivity limited by phonon and impurity scattering, along with a decreasing contribution from charge carriers as T_c is decreased.

The crystals used in this study were grown by a self-flux technique in our laboratory. Further details are provided elsewhere[3]. The thermal conductivity along both the a-b plane and c-axis was measured by a standard steady-state technique[4]. Figure 1 shows K_{ab} of two pure YBCO single crystals at various stages of oxygenization. A fully oxygenated $Y_{.92}Pr_{.08}Ba_2Cu_3O_7$ single crystal with $T_c=88.5$ K is shown on the same plot. K_{max}/K_{T_c} is driven down quickly as the position T_{max}/T_c of the peak is driven to higher values, and abruptly reaches saturation at $K_{max}/K_{T_c}=1.15$, and $T_{max}/T_c = .45$. Figure 2 shows K_c of a pure YBCO single crystal, along with K_{ab} of three fully oxygenated doped YBCO single crystals, $Y_{.85}Pr_{.15}Ba_2Cu_3O_7$, $Y_{.65}Pr_{.35}Ba_2Cu_3O_7$ and $YBa_2(Cu_{.97}Zn_{.03})_3O_7$. In all cases, there is no break in K at T_c and peak at $T_{max} = 20$ K. The general features displayed by

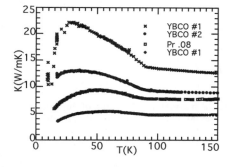

Figure 1: K_{ab} of two YBCO single crystals and a $Y_{.92}Pr_{.08}Ba_2Cu_3O_7$ single crystal.

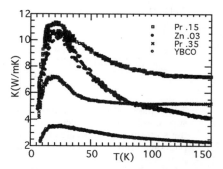

Figure 2: K_c of a pure YBCO single crystal, and K_{ab} of $Y_{1-x}Pr_xBa_2Cu_3O_7$ (x=.15,.35), and $YBa_2(Cu_{.97}Zn_{.03})_3O_7$ single crystals.

K_{ab} in the heavily doped Pr and Zn YBCO and K_c of YBCO are similar to that reported for $Ba_{1-x}K_xBiO_3$[5] and $Nd_{2-x}Ce_xCuO_4$[6]. The thermal conductivity of a superconductor is the sum of the charge carrier thermal conductivity K_e and the phonon thermal conductivity K_p. If increased impurity scattering causes a decrease in T_c, gamma will be a function of T_c and not just scale with the reduced temperature. Here, T_{max}/T_c will rise with the introduction of impurities with the falloff of T_c[7]. For the phonons, a maximum in the thermal conductivity can arise from two sources: directly from the electron-phonon scattering rate when the condensation of charge carriers at T_c causes an abrupt upturn from the reduction of this scattering or indirectly from the competition of scattering rates. The second source of a peak in K_p will not be sensitive to T_c. Peaks in K_e can only reasonably arise from the onset of superconductivity. It is therefore reasonable to ascribe the peaks in K_c of YBCO, and in K_{ab} of the heavily Pr and Zn doped YBCO to phonons and not to an artifact of pair condensation. In contrast, the trend of K_{ab} of the crystals in Figure 1 cannot be reasonably described by the BRT theory, as reasonable scattrering rates lead to monotonic decreases in T_{max} as T_c falls. K_e on the other hand, under reasonable choices of gamma, mirrors the trend seen in these lightly doped samples. A more quantitative analysis will be published elsewhere.

References

1. C. Uher in *Physical Properties of High Temperature Superconductors III*, ed. D.M. Ginsberg (World Scientific, Singapore, 1992).
2. P. J. Hirschfeld and W.O. Putikka, (preprint, 1996).
3. G. Cao *et al*, Physica **B206**, 749 (1995).
4. J. L. Cohn *et al*, Physica **C192**, 435 (1992).
5. Baoxing Chen *et al*, Phys. Rev. **B51**, 6171 (1995).
6. J. L. Cohn *et al*, Phys. Rev. **B46**, 12053 (1992).
7. P. F. Henning *et al*, J. Supercond. **8**, 453 (1995).

THERMAL CONDUCTIVITY OF HIGH-T_c SUPERCONDUCTORS

M. HOUSSA AND M. AUSLOOS
S.U.P.R.A.S, Institut de Physique B5, Université de Liège,
B-4000 Liège, Belgium

The in-plane electronic contribution $\kappa_{e,ab}$ to the thermal conductivity of high-T_c superconductors was calculated using a variational method. We took into account the scattering of heat carrying electrons by point defects as well as by acoustic phonons. We considered both isotropic s-wave and anisotropic $d_{x^2-y^2}$-wave gap parameter symmetries. The electronic density of states was chosen to be relevant for 2D and 3D systems and a van Hove singularity was included if necessary. The peak structure observed in single crystals of $YBa_2Cu_3O_{7-\delta}$ and $Bi_2Sr_2CaCu_2O_8$ could be well reproduced with very reasonable values of the physical parameters. However, the very low temperature behavior of κ in these materials was found to be incompatible with an s-wave gap parameter but could be explained by considering a gap parameter of $d_{x^2-y^2}$-wave type.

The behavior of the in-plane electronic contribution to the thermal conductivity $\kappa_{e,ab}$ of $YBa_2Cu_3O_{7-\delta}$ and $Bi_2Sr_2CaCu_2O_8$ compounds can be described using a variational method along the lines of a two fluid model, considering the scattering of electrons by point defects and acoustic phonons[1,2].

The quasiparticle energy spectrum is assumed to be[3]

$$E(\vec{k}) = \sqrt{(\varepsilon(\vec{k})-\varepsilon_F)^2 + \Delta(\vec{k})^2} \qquad (1)$$

where $\varepsilon(\vec{k})$ is the electronic spectrum in the normal state, ε_F the Fermi energy and $\Delta(\vec{k})$ the superconducting gap parameter. To account for the anisotropic electronic structure of high-T_c compounds, we consider either a Lawrence-Doniach like spectrum[4] or a strictly two dimensional spectrum[5] with a saddle point located at the Fermi energy, corresponding to a logarithmic van Hove singularity in the density of states. On the other hand, the gap parameter is chosen to be of either isotropic s-wave symmetry[1] or anisotropic $d_{x^2-y^2}$-wave symmetry[2].

The experimental results on an $YBa_2Cu_3O_{7-\delta}$ single crystal[6] and a $Bi_2Sr_2CaCu_2O_8$ single crystal[7] are shown in Fig.1 (a) and (b) respectively. The solid lines are theoretical fits to the data using the Lawrence-Doniach model for $YBa_2Cu_3O_{7-\delta}$ and two dimensional saddle point model for $Bi_2Sr_2CaCu_2O_8$ and using a $d_{x^2-y^2}$-wave gap parameter. They are in excellent agreement with the data. It can be shown that this characteristic peak structure observed on the thermal conbductivity of high-T_c materials can also be reproduced by an s-wave gap parameter[1] with slightly different physical parameters somewhat less realistic in the s-wave case than in the d-wave case symmetry.

However, the very low temperature behavior of $\kappa_{e,ab}$ in these materials removes the possible controversy. The data are shown in the insets of Fig.1 (a) and (b) on log-log plots. One can see that $\kappa_{e,ab}$ behaves like power laws in both materials at very low temperatures. Since $\kappa_{e,ab}$ exponentially vanishes at low temperature in an s-wave

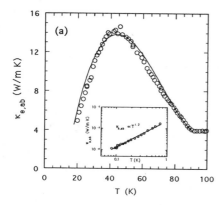 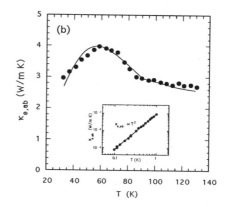

Fig.1 Temperature dependence of the in-plane electronic thermal conductivity $\kappa_{e,ab}$ of (a) an $YBa_2Cu_3O_{7-\delta}$ single crystal from ref.6 and (b) a $Bi_2Sr_2CaCu_2O_8$ single crystal from ref. 7. Insets : very low temperature behavior of $\kappa_{e,ab}$ on log-log plots.

superconductor[2], this power law behavior seems only consistent with a gap parameter of $d_{x^2-y^2}$-wave type. The physical argument is found in the mean free path temperature dependence and in the different behavior of the quasiparticle density of states at low temperature wich is either an exponential decrease (s-wave case) or a power law decrease (d-wave case).

We conclude that the analysis of the peak structure of the thermal conductivity of high-T_c superconductors can be well described by considering an electronic model with realistic band structures. Besides, the very low temperature behavior of $\kappa_{e,ab}$ allows us to conclude that the gap parameter symmetry is most probably of d_wave type.

Acknowledgements

Part of this work has been financially supported through the Inpulse Program on High Temperature Superconductors of Belgium Federal Services for Scientific, Technological and Cultural (SSTC) Affairs under contract SU/02/013 and the ARC 94-99/174 contract of the Ministery of Higher Education and Scientific Research through the University of Liège Research Council.

References

1. M. Houssa and M. Ausloos, *J. Phys. C : Condensed Matter* **8**, (1996) 2043.
2. M. Houssa and M. Ausloos, *Physica C* **257**, (1996) 321.
3. P.G. De Gennes, *Superconductivity of metals and alloys*, ed. W.A. Benjamin (New York, 1966), p.111.
4. W.E. Lawrence and S. Doniach, *Proc. of the 12th Int. Conf. on Low Temperature Physics* (Kyoto, 1970), ed. E. Kanda (Tokyo) p. 361.
5. D.M. Newns, C.C. Tsuei, R.P. Huebener, P.J.M. van Bentum, P.C. Pattnaik, and C.C. Chi, *Phys. Rev. Lett.* **73**, (1994) 1695.
6. R.C. Yu, M.B. Salamon, J.P. Lu, and W.C. Lee, *Phys. Rev. Lett.* **69**, (1992) 1431.
7. P.B. Allen, X. Du, L. Mihaly, and L. Forro, *Phys. Rev. B* **49**, (1994) 9073.

ANGULAR DEPENDENCE OF THE c-AXIS NORMAL STATE MAGNETORESISTANCE IN $Tl_2Ba_2CuO_6$

N.E.HUSSEY, J.R.COOPER, I.R.FISHER, A.P.MACKENZIE and J.M.WHEATLEY
IRC in Superconductivity, University of Cambridge, Madingley Road, Cambridge, CB3 0HE, U.K.

On rotating the field B within the ab-plane, the out-of-plane transverse magnetoresistance of overdoped $Tl_2Ba_2CuO_6$ ($T_c \leq 25K$) exhibits a striking anisotropy with four-fold symmetry, reflecting the anisotropy of the in-plane mean free path. The T-dependence of this anisotropy, however, cannot account for the observed T-dependence of the Hall coefficient.

1. Introduction

It is widely thought that an understanding of the unusual variation of the Hall coefficient R_H with temperature T and hole concentration p is an important step towards the correct microscopic theory of high-T_c superconductivity. Although the systematic behaviour of $R_H(T, p)$ is well established experimentally, the theoretical situation is still unresolved and several contrasting theoretical approaches have emerged. These include the spinon-holon model with two distinct relaxation times[1] and more conventional pictures where a single scattering rate varies strongly around the Fermi surface FS[2]. Measurements of the angular dependence of the c-axis transverse MR (i.e. with $B//ab$) should be able to probe the anisotropy of the in-plane mean free path and thus help to distinguish between these different points of view. Here we report the first angular dependent MR studies of overdoped $Tl_2Ba_2CuO_6$ ($T_c \leq 25K$) for which both $d\rho_{ab}/dT$ and $d\rho_c/dT$ are positive.

2. Results

The transverse MR is positive for $I//c$ and $I//ab$, and in both cases it varies as $\rho(B) = \rho_0 + \Delta\rho_2 - \Delta\rho_4$, where $\Delta\rho_2$ and $\Delta\rho_4$ are positive and proportional to B^2 and B^4 respectively[3]. The $\Delta\rho_2/\rho_0$ terms have essentially the same T dependence for both current directions but the c-axis MR is ≈ 6 times the corresponding in-plane MR[3]. On rotating B within the ab-plane, a striking anisotropy is observed in the c-axis MR, which is dominated at low T by a term having four-fold symmetry (fig. 1). The maximum MR occurs for B along [110] i.e.

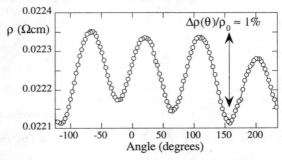

Fig. 1. Angular dependence of the c-axis transverse magnetoresistance of overdoped $Tl_2Ba_2CuO_6$ ($T_c \approx 25K$) at $T = 32.1K$ and $B = 13T$. ρ_0 is the zero-field resistivity at 32.1K.

at 45° to the Cu-O-Cu bonds. The magnitude of the anisotropy $A = \rho(\theta=\pi/8) - \rho(\theta=0)$ scales as B^4 and decreases rapidly with increasing T. As a first step we have calculated all components of the conductivity tensor σ_{ij} for an open FS with a small cosine dispersion along the c-axis using the relaxation time approximation, the Jones-Zener expansion to order B^4 and smoothly varying four-fold anisotropies in τ, k_F and v_F, represented by; $\tau^{-1}(\theta) = \tau_0^{-1}(1 + \varepsilon\cos^2 2\theta)$, $k_F(\theta) = k_F(1 + \alpha\sin^2 2\theta)$ and $v_F(\theta) = v_F(1 + \beta\sin^2 2\theta)$, where θ is the in-plane angle between k and the a (or b) axes[3]. Using typical values of $\alpha = 0.2$ and $\beta = 0.1$ from band structure calculations, we calculated the anisotropy of the transverse MR $(A/\rho_0)/(\Delta\rho_4/\rho_0)$ as a function of ε, the anisotropy of τ in the ab-plane (fig. 2). Several of the experimental features of the in- and out-of-plane MR are consistent with this simple model[3] and the measured T dependence of the MR anisotropy corresponds to ε increasing from 0.29 to 0.44 between 30 and 120K. However, the measured $R_H(T)$ requires larger changes than those that account for the T-dependence of the MR anisotropy as shown in fig. 3. One possible new ingredient which may account for this discrepancy is the two-lifetime model in which the Boltzmann transport equation can be written as[4] $g_k = \tau_{tr}eE.v_k(-\partial f_o/\partial E) - \tau_H ev_k \times B.(\partial g_k/\partial k)$; $\tau_H^{-1}(T)$ is proportional to the inverse Hall angle (i.e. $\cot\Theta_H \approx AT^2 + B$) while $\tau_{tr}^{-1}(T) \propto \rho_{ab}(T)$. In this model, $\Delta\rho_{ab}^{(2)}/\rho_{ab}^{(0)} \propto \tau_H^2$ and $R_H \propto \tau_H/\tau_{tr}$. It is also plausible within this model that ρ_c should obey Kohler's rule as observed[3]. Namely, $v_{//}$ is large so the Lorentz force with $B//ab$ and $I//c$ drives the carriers parallel to I and thus the same τ governs ρ_c and the c-axis MR. In this picture, however, there is still difficulty in accounting for the magnitude[3] of $\Delta\rho_{ab}/\rho_{ab}$ unless it is assumed that the FS is in fact small, or there is considerable anisotropy in τ_H ($\varepsilon \sim 1$-2). It should be stressed, finally, that the details of the Bloch-Boltzmann analysis outlined here could be altered if the c-axis dispersion varied with $k_{//}$ as suggested by band structure calculations[5].

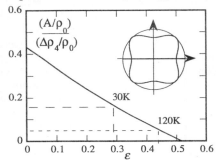
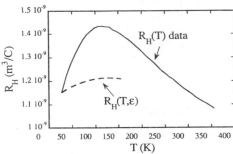

Fig. 2 Dependence of the MR anisotropy on ε. Dashed lines represent the measured MR anisotropy at 30K and 120K. Inset is the in-plane variation of τ for Tl$_2$Ba$_2$CuO$_6$ within this picture for $\varepsilon = 0.4$.

Fig. 3 $R_H(T)$ of Tl$_2$Ba$_2$CuO$_6$ ($T_c \approx 20$K) Solid line is measured R_H. Dashed line is R_H estimated within the present model from the T-dependence of the c-axis MR anisotropy.

References

1) P.W.Anderson, *Physical Review Letters* **67** 2092 (1991).
2) see, for example, A.Carrington et al., *Physical Review Letters* **69** 2855 (1992).
3) N.E.Hussey et al., *Physical Review Letters* **76** 122 (1996).
4) J.M.Harris et al., *Physical Review Letters* **75** 1391 (1995).
5) J.M.Wheatley, *these proceedings*

THERMOELECTRIC POWER OF SUPERCONDUCTING ALLOYS
YNi$_2$B$_2$C and LuNi$_2$B$_2$C

J. H. Lee, Y. S. Ha, Y. S. Song, Y. W. Park
Department of Physics, Seoul National University,
Seoul 151-742, Korea

Y.S. Choi
Samsung Display Devices Co., Ltd., Energy Resources Center,
Suwon, Kyungkido, 445-970, Korea

Thermoelectric Power of LnNi$_2$B$_2$C(Ln=Y, Lu) is presented as a function of temperature. The superconducting transition temperature is 14.5K for Ln=Y, and 15.5K for Ln=Lu. TEP has a nonlinear temperature dependence, which was assumed to have the mixed valent character. Some parameters appearing in mixed valence figure seem to have reasonable values.

The superconducting transition was identified in our samples by the resistivity measurements. The resistivity at room temperature for these samples is of the order of 100 Ωcm which is the same as those reported.[1] The normal state resistivity shows nearly linear temperature dependence above 50K for all these samples and slope is approximately 0.4 $\mu\Omega$cm/K. The resistivity starts to drop to zero at 16K and 15K for LuNi$_2$B$_2$C and YNi$_2$B$_2$C, respectively.

The TEP data(Fig.1) show the superconducting transition at temperature specified by the resistivity data. TEP at room temperature is -5.1 μV/K and -4.7 μV/K for LuNi$_2$B$_2$C and YNi$_2$B$_2$C respectively. Nearly the same temperature dependence of TEP in YNi$_2$B$_2$C and LuNi$_2$B$_2$C is not surprising in view of the obvious YNi$_2$B$_2$C-LuNi$_2$B$_2$C band similarities[2]. The negative sign of TEP indicates that the charge carriers are electron-like. The diffusion thermopower is given by the Mott formula and since electronic states at the Fermi level are mostly populated by d-states derived from Ni and narrow d-band electrons are much less mobile than p electrons, we can expect the Mott $(s-d)$-like scattering process[3] could be involved in diffusion thermopower, in which case the sign of S_d will be negative considering that the energy derivative at the Fermi level of d-density of state is negative from the calculated band structure.[1,2]

Within the measured range, TEP shows a highly nonlinear temperature dependence, and furthermore it looks as if it is temperature independent above approximately 130 K. Band structure ensures that these materials are good metals,[4] but the so-called 'phonon drag peak' which can be seen in conventional metals, is not observed at low temperares. The thermoelectric power will be highly nonlinear at low temperatures if the electron-phonon interaction is large, even in the absence of phonon drag.[5] It was suggested that the electron-phonon interaction enhances the diffusion thermopower of metals at low temperatures, the size of the enhancement factor being the usual $(1 + \lambda)$ mass enhancement as seen, for example, in low temperature specific heat. But estimated value of $\lambda \geq 8$, which is unreasonably high compared with 2.7 suggested by resistivity data[4] and 1.2 obtained from specific heat measurement.[6] Therefore it is hard to say that this nonlinear behavior of TEP

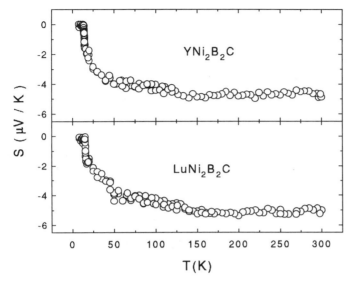

Figure 1: TEP of LnNi$_2$B$_2$C (Ln=Y, Lu).

is due to electron-phonon mass enhancement.

As another approach, it is interesting when we compare the present system with the mixed valence compounds. TEP of the mixed valence system (e.g. CeNi$_x$) was calculated in reference 7:

$$S = \frac{AT}{B^2 + T^2}$$

The initial idea is that superimposed on a broad band there is a peak in the density of states near the Fermi level. This resonance gives the characteristic temperature dependence of S. The parameter $A \sim E_0 - E_F$, where E_0 is the center of resonance and $B^2 \sim (E_0 - E_F)^2 + \Gamma^2$ where Γ is the width of the resonance. we have fitted the data with the above expression adding a linear term αT where the αT term is the normal band contribution. It is noticible that T/(S-αT) is a linear function in T^2 in the whole temperature range except near Tc(Fig.2). The obtained values of α, $E_0 - E_F$, and Γ are as follows: $\alpha \sim -0.00874 \mu V/K^2$, $E_0 - E_F \sim -5.1$K, $\Gamma \sim 207$K for LuNi$_2$B$_2$C and $\alpha \sim -0.0091 \mu V/K^2$, $E_0 - E_F \sim -3.8$K, $\Gamma \sim 156$K for YNi$_2$B$_2$C. These are reasonable values comparable to those obtained for intermetallic compounds CeNi and CeNi$_2$. The remaining question is whether the present intermetallic compounds can be approached in view of the mixed valency. Ni 3d-derived conduction electron near the Fermi level can contribute to electrical conduction and the mixed valency can be viewd as the partial filling of the 3d shell of the ionized Ni ion. The

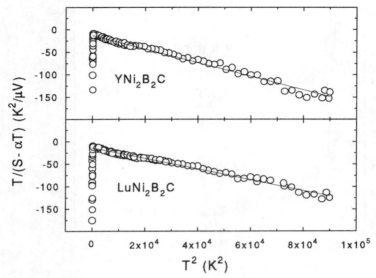

Figure 2: Plot of T^2 vs. $T/(S-\alpha T)$.

NMR study[8] showed that Ni atoms indeed carry localized moments at the NMR time scale and YNi_2B_2C is identified as antiferromagnetic spin fluctuating material. It is expected that the localized moments of Ni sites will fluctuate with the antiferromagnetic correlations, from which the valence fluctuation could be derived, supporting the mixed valence figure. But since the ionic states of Ni atom sites are not clear, it seems too early to interpret these data with the mixed valency.

References

1. A. Fujimori, K. Kobayashi, T. Mizokawa, K. Mamiya, A. Sekiyama, H. Eisaki, H. Takagi, S. Uchida, R.J. Cava, J.J. Krayewski, and W.F. Peck Jr., *Phys. Rev.* **B50**, (1994) 9660.
2. L.F. Mattheiss, *Phys. Rev.* **B49**, 13279 (1994).
3. N.F. Mott, *Phil. Mag.* **26**, 1249 (1972).
4. W.E. Pickett and D.J. Singh, *Phys. Rev. Lett.* **72**, 7302 (1994).
5. A.B. Kaiser, *Phys. Rev.* **B37**, 5924 (1988).
6. J.S. Kim, W.W. Kim, and G.R. Stewart, *Phys. Rev.* **B50**, 3485 (1994).
7. U. Gottwick, K. Gloos, S. Horn, F. Steglich and N. Grewe, *J. Magn. Magn. Mat.* **47 & 48**, 536 (1985).
8. T. Kohara, T. Oda, K. Ueda, Y. Yamada, A. Mahayan, K. Elankumaran, Zakir Hossain, L.C. Gupta, R. Nagarayan, R. Vijayaraghavan, and Chandan Mazumdar, *Phys. Rev.* **B51**, 3985 (1995).

ANOMALOUS DIELECTRIC EFFECT OF $La_{2-x}Sr_xCuO_4$ FILM AT $x = 1/4^n$

M. SUGAHARA, X.-Y. HAN, H.-F. LU, S.-B. WU,
N. HANEJI, H. KANEDA and N.YOSHIKAWA
Faculty of Engineering, Yokohama National University, Yokohama, 240 Japan

The resistivity of LSCO materials drops sharply at $x \approx 4^{-n}$ from low up to room temperature. Anomalous dielectric response of LSCO film at $x \approx 4^{-n}$ and $x \approx 2 \times 4^{-n}$ is found along c axis. The capacitance C_t of multi-layer structure Pd/LSCO/STO/Pd increases at the x values exceeding the capacitance C_{STO} of Pd/STO/Pd structure from low to room temperature, where effective dielectric constant of the conductive normal state is considered to be negative. The C-V measurement reveals the existence of "critical voltage" in the anomalous dielectric property.

1. Introduction

We found that the resistivity ρ of $La_{2-x}Sr_xCuO_4$ (LSCO) materials drops sharply at $x \approx 4^{-n}$ from low up to room temperature [1,2] with appurtenant anomalies at $x \approx 2 \times 4^{-n}$. There is the following similarities between the ρ drop in LSCO and the quantum Hall effect [3] in a 2D system at density $\sigma = (eB/2\pi\hbar)j$ (j, integer). (a) Both phenomena occur at regular intervals of carrier densities. (b) The systems have 2D character. (c) In the ideal limit the resistance R_{xx} decreases sharply at the special carrier densities.

We propose the following model [1,4,5] for the carrier state at $x \approx 4^{-n}$ and $x \approx 2 \times 4^{-n}$ based on the similarities (a), (b) and (c). (i) 2D carriers in LSCO are originally in pairing state. (ii) Each hole pair occupies one of the quantized *localized steady states* named "n'th order hole pair" (HPn) [1] with quantized area $2 \times 4^n S_{CuO_2}$ (S_{CuO_2}, the area of a CuO_2 unit) when $x \approx 4^{-n}$ and $4^n S_{CuO_2}$ when $x \approx 2 \times 4^{-n}$. (iii) Each CuO_2 layer is spatially replete with a mode of HPn's. (iv) This *spatially ordered state* is a macroscopic quantum state with properties dual to superconductivity where pairs are *ordered in momentum space*. (v) External charge introduced into the material is quantized in the form of charge quanta. (vi) The material is expected to reveal "complete dielectricity".

If the above model is true, we should observe some dielectric anomaly in the charge motion along c axis of LSCO at $x \approx 4^{-n}$ and $x \approx 2 \times 4^{-n}$. We conducted the following experimental study to find the dielectric anomaly.

2. Anomalous Capacitive Characteristic of LSCO Film along c Axis

The inset of Fig. 1 shows two capacitor types. One, C_t, has a double layer structure consisting of an insulator and a LSCO layer of thickness d_{LSCO} between two metal (Pd) electrodes. (i) The insulator layer consists of (100) $SrTiO_3$ (STO) single crystal of thickness $d_{STO} = 1$ mm, which acts as the substrate for growth of the c-axis oriented LSCO layer in the case of C_t structure. (ii) The area of the Pd electrodes and the LSCO layer is $S = 5 \times 5$ mm^2.

We get $C_{STO} = \varepsilon_{STO} S / d_{STO}$ and $C_t = C_{STO} / [1 + (d_{LSCO} / d_{STO}) (\varepsilon_{STO} / \varepsilon_{LSCO})]$. Whether LSCO is a simple conductor, or simple dielectric, we find $C_t / C_{STO} \leq 1$.

Figure 1 and 2 shows the observed x dependence of C_t / C_{STO} at low temperature and room temperature. $C_t / C_{STO} \geq 1$ is found in the measured range of x, and C_t / C_{STO} increases at around the doping levels $x \approx 4^{-n}$ and $x \approx 2 \times 4^{-n}$. The anomalous C_t increase reveals that the effective dielectric constant ε_{LSCO} is negative especially at the special doping levels.

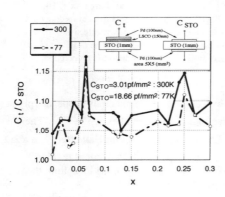

Fig. 1: x dependence of C_t at $0 \leq x \leq 0.25$. Fig. 2: x dependence of C_t at $0.25 \leq x \leq 0.6$.

Besides the above x dependence we observed at $x \approx 4^{-n}$ and $x \approx 2 \times 4^{-n}$ (1) C_t increase with d_{LSCO} increase and (2) the existence of critical voltage in C-V measurement at 4.2 K.

3. Conclusion

We found anomalous dielectric response of LSCO film at at $x \approx 4^{-n}$ and $x \approx 2 \times 4^{-n}$ along c axis. The properties are explainable as the macroscopic quantum feature dual to superconductivity in the hole pair repletion state.

References
1. M. Sugahara, *Jpn.J.Appl.Phys.* **31**, 324 (1992).
2. M. Sugahara and J.-F. Jinag, *Appl.Phys.Lett.* **63** (1993) 255; Physica B, **194-196**, 2177 (1994).
3. K. von Klitzing, G. Dorda and M. Pepper, *Phys. Rev. Lett.* **45**, 494 (1980).
4. M. Sugahara, X.-Y. Han, H.-F. Lu, S.-B. Wu, N. Haneji, H. Kaneda and N. Yoshikawa, *Jpn. J. Appl. Phys.* **35**, 793 (1966).
5. M. Sugahara, to be published in *Bulletin of the Faculty of Engineering, Yokohama Natl. Univ.*, Vol. 45, March 1996.

INVESTIGATION OF THE MICROWAVE POWER HANDLING CAPABILITIES OF HIGH-T_c SUPERCONDUCTING THIN FILMS

J. WOSIK, L. M. XIE, D. LI, P. GIERLOWSKI, J. H. MILLER, Jr., S. A. LONG.

Texas Center for Superconductivity at the University of Houston, Houston, TX 77204 USA

We have measured the microwave power dependence of the surface resistance R_s of $YB_2Cu_3O_x$ (YBCO) thin films up to very high microwave power levels, using both frequency and time domain techniques. Our measurements show, for the first time, that heating effects contribute significantly to the nonlinearity of the surface resistance at higher microwave power levels. We are currently using a non-linear RSJ model, combined with a vortex motion model, to interpret the remaining contributions to the nonlinear behavior of the surface resistance.

1. Introduction

Recent years have seen outstanding technological progress in the deposition and characterization of high quality high temperature superconducting (HTS) films. These developments have led to significant progress in the design and fabrication of HTS microwave devices. In particular, prototype HTS components for cellular and satellite communications have been demonstrated. Existing performance at low power levels is sufficient in some such applications, but high power handling capability is required in many other applications. It has been reported that HTS thin films exhibit highly nonlinear behavior in the power-dependent surface resistance,[1,2] which can lead to the generation of harmonics and to intermodulation effects. In spite of the great body of experimental data on the power dependence of HTS surface resistance reported in literature, no fundamental understanding of the observed behavior has been achieved. Therefore, it is of considerable importance to determine whether or not the experimentally observed nonlinearity in the power dependence of R_s is intrinsic, and, if not, if the quality of HTS films should be improved. In this paper, we present a study of the electromagnetic response of a dielectric cavity with two YBCO films to incident microwave power as high as 50 W.

2. Experimental Techniques

Thin YBCO films were deposited onto $LaAlO_3$ substrates using a substrate scanning laser ablation technique. Measurements were done in the TE_{011} mode of a dielectric cavity at 14 GHz. Two 20 mm by 20 mm YBCO films were placed inside the cavity on both sides of a dielectric disc. The cavity was kept cold by placing it in a specially designed gas flow cryostat for the high rf power measurements. An HP vector or scalar analyzer was used to measure the Q in the frequency domain using a single 100 ms time sweep (the steady state method). A microwave pulsed system with a 150 W Varian TWT amplifier was used for the time domain measurements. Both the unloaded Q and frequency were calculated by analyzing the decay signal from the cavity after the square pulse of incident power was switched off (the transient method).[3] The pulse duration was sufficiently long for the steady state (in which the cavity is fully charged) to be reached.

3. Results

The primary manifestations of R_s nonlinearity are a reduction in Q and a shift in resonant frequency. Such behaviors can be seen in Fig. 1, in which several resonant curves of the cavity are shown for input power levels ranging from 14 dBm to 47 dBm. At higher rf power than 47 dBm, the films in the cavity were driven into the normal state. The output power was measured as a function of time without changing the cavity arrangment. This enabled us not only to determine the loaded Q of the cavity from the decay of the output signal, but also to study the dynamics of the electromagnetic response of the cavity. The results of these measurements are shown in Fig. 2. From this figure we can conclude that, at input power levels higher than 35 dBm, significant heating of the film takes place.

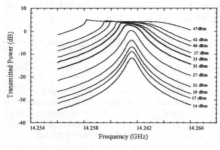

Fig. 1. Transmitted power versus frequency for input power levels ranging from 14 dBm to 47 dBm is measured. Cavity operated at 70 K.

The output power, in the absence of any heating effects, would normally stay constant after the cavity was charged until the pulse was switched off (see the low power results). At higher power levels, the peaks in the response, followed by plateaus, indicate that the dissipated rf power causes the temperature of the YBCO films to rise to a steady value before the cavity reaches its steady state.[4] The films are still superconducting but have larger surface resistance, which results in a shortening of the decay time and thus a decrease in Q.

In Fig. 3, a plot of 1/Q (calculated from the cavity decay), which is proportional to R_s, versus the square root of the output power (which is proportional to the microwave current in the films) is shown for temperatures of 20 K, 50 K, and 70 K. The maximum rf magnetic field H_{rf} reached in these measurements is estimated to be about 100 Gauss. Our results show that a complete interpretation of the $R_s(H_{rf})$ behavior requires one to include the effects of heating on the observed R_s nonlinearity.

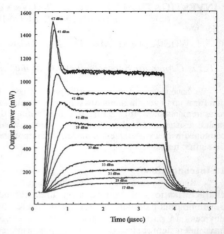

Fig. 2. The resonator pulse response. Transmitted power versus time for input powers ranging from 17 dBm to 47 dBm. The pulse duration was 3.5 μsec.

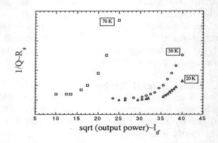

Fig. 3. Nonlinear surface resistance of a YBCO film versus microwave current amplitude is shown for three different temperatures.

5. Acknowledgements
This work was supported by the Texas Higher Education Coordinating Board (ARP grants 003652-238 and 003652-227), by the State of Texas via Texas Center for Superconductivity at University of Houston, and by the Robert A. Welch Foundation (E-1221).

6. References
1. D. Oates, P. P. Nguyen, G. Dresselhaus, M. S. Dresselhaus, C. W. Lam, and S. M. Ali, J. of Superconductors 5, 361(1992).
2. for the review see T. B. Samoilova, Supercon. Sci. Technol. 8, 259 (1995).
3. J. P. Turneaure and I. Weissman, *J. Appl. Phys.* **39**, 4417 (1968).
4. A. N. Rheznik, A. I. Smirnov, and M. D. Chernobrovtseva, *Superconductivity: Physics, Chemistry, Technology*, **6**, 186 (1993).

II. EXPERIMENTS

Spectroscopy

THERMODYNAMIC EVIDENCE ON THE SUPERCONDUCTING AND NORMAL STATE ENERGY GAPS IN $La_{2-x}Sr_xCuO_4$

J.W.LORAM, K.A.MIRZA, J.R.COOPER, N.ATHANASSOPOULOU and W.Y.LIANG
IRC in Superconductivity, University of Cambridge, Madingley Road, Cambridge CB3 0HE, U.K.

We present electronic specific heat $\gamma(T)$ and susceptibility $\chi(T)$ data for $La_{2-x}Sr_xCuO_4$ ($0 \leq x \leq 0.45$). Below T_c γ exhibits a d-wave dependence $\gamma = \alpha T$, where α is independent of x in the underdoped region but increases rapidly with overdoping. The normal state Wilson ratio $S/\chi T$ (where S is the electronic entropy) is close to the value for a non-interacting Fermi liquid. The number of states within $E_F \pm 35$meV increases by ~0.8 per hole up to x~0.25, leading to peak in the DOS. For x<0.24 a gap Δ_n develops at E_F in the quasiparticle spectrum, the observed initial dependences $\chi, \gamma \sim T/\Delta_n$ possibly indicating d-wave symmetry for the normal state gap.

$La_{2-x}Sr_xCuO_4$ (LSCO) has single CuO_2 planes and can be hole doped across the entire generic phase diagram from AF insulator to superconductor to normal metal. We have made a detailed comparison of the electronic specific heat coefficient $\gamma \equiv C^{el}/T$ and bulk susceptibility χ for 20 values of x in the range $0 \leq x \leq 0.45$. Full details of sample preparation and the determination of γ using differential calorimetry will be presented elsewhere. The susceptibility results are in good agreement with previous studies[1,2].

The behaviour of γ in the superconducting region is shown in Fig 1. On emerging from the spin glass phase a weak superconducting anomaly is first observed at x=0.06. With increasing x (hole doping) the magnitude of the specific heat jump continues to increase into the overdoped region (x>0.16) and shows clear evidence of bulk superconductivity well into the tetragonal phase (x>0.2), in agreement with the conclusions of Nagano et al[3]. T_c is zero for $x \geq 0.30$. Below T_c γ is well represented by an initial linear

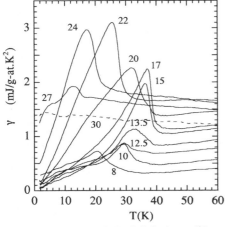

Fig 1. γ for $La_{2-x}Sr_xCuO_4$. Labels show x%Sr

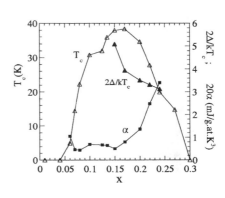

Fig 2. T_c, α and $2\Delta(0)/k_BT_c$ for LSCO

dependence $\gamma=\gamma(0)+\alpha T$, with coefficient α shown in Fig 2. This linear dependence was first demonstrated for the LSCO system by Momono et al[4] and attributed to a d-wave order parameter, and Chen et al[5] have observed d-wave gap anisotropy in the Raman spectrum of $La_{1.83}Sr_{.17}CuO_4$. The initial slope α is insensitive to x for $x \leq .15$ but increases rapidly at higher x, though $\gamma(0)$ remains low up to x=0.24. Taking $\alpha=3.14\gamma_n k_B/\Delta$ for a d-wave superconductor with maximum gap Δ we find that the gap parameter $2\Delta/k_B T_c$ decreases from 5.2 to 3.1 between x=0.15 and 0.24 (Fig 2). This may reflect increased pair breaking[6], or may possibly be related to the filling of the normal state gap (see below).

The normal state behaviour of γ, S/T and S are shown in Figs 3, 5 and 7, where

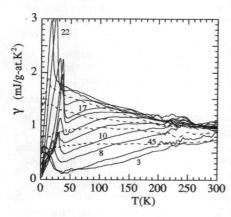

Fig 3. γ vs T for
x=3,5,8,10,12.5,13.5,15,17,20,22,24 %Sr
Broken curves (top to bottom) 27,30,35,45 %Sr

Fig 4 $d(\chi T)/dT$ vs T for
x=3,6,8,10,12.5,13.5,15,17,20,22,24 %Sr
Solid curves (top to bottom) 27,30,35,45 %Sr

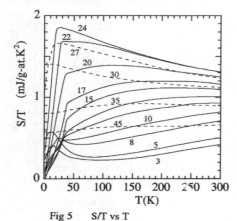

Fig 5 S/T vs T

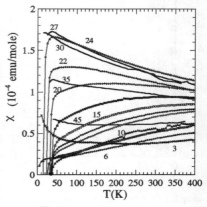

Fig 6 χ vs T

$S(T)=\int^T \gamma(T')dT'$ is the electronic entropy. $d(\chi T)/dT$, χ and χT are shown in Figs 4, 6 and 8. Strikingly similar T-dependences are observed for each pair (γ,$d(\chi T)/dT$), (S/T,χ) and (S,χT), showing that the total spin+charge spectrum over all q (from S) and the q=0 spin spectrum (from χ) have a similar energy dependence. In addition comparison of Figs 3,5 and 7 with Figs 4,6 and 8 shows that the magnitudes are related by a factor close to the free electron Wilson ratio $a_0=\pi^2/3(k_B/\mu_B)^2$ (10^{-4} emu/mole≡1.05mJ/g.at.K^2 for LSCO). Indeed a plot of S/T vs χ for equal x at 40K gives $S/T=0.96a_0(\chi-\chi_0)$, where $\chi_0= -0.3.10^{-4}$ emu/mole may be attributed to non-spin contributions to χ. These results suggest that the low energy excitations are predominantly those of conventional fermions, as found previously for the $YBa_2Cu_3O_{6+x}$ system[7,8]. For weakly interacting fermions (WIF) the weight functions in Fermi surface integrals over the single particle density of states (DOS) correspond closely for each pair, and in this case S and χT simply count the number of orbitals within the Fermi window $\sim(E_F\pm 2k_BT)$ in units of $\ln(4)k_B$ and $\mu_B^2/2$ respectively, whilst S/T and χ reflect the DOS averaged over the same energy window. We conclude that

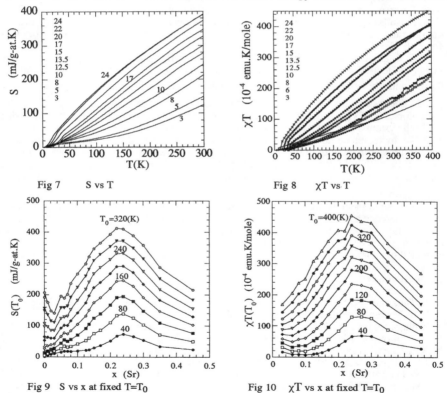

Fig 7   S vs T

Fig 8 χT vs T

Fig 9   S vs x at fixed T=T_0

Fig 10 χT vs x at fixed T=T_0

the substantial T-dependences of S/T and χ etc are determined primarily by the energy dependence of the single particle density of states g(E) in the vicinity of E_F, and that any mass renormalisation affects both properties similarly. From the closeness of the Wilson ratio to its free electron value it is evident that charge and spin excitations make comparable contributions to the low temperature entropy[8].

From Figs 3-6 it appears that a gap or minimum in the DOS at E_F for low x develops with hole doping into a peak at x~0.25, and that the DOS away from E_F is insensitive to x. As noted above the number of states within the Fermi window ~$(E_F \pm 2k_B T)$ can be deduced from plots of S vs T or (assuming WIF) from χT vs T. For both properties (Figs 7,8) the curves for each x are parallel above ~200K. Such positive parallel shifts with x imply additional states located in a narrow energy range relative to the width of the window [8].

Plots of S and χT vs x at fixed temperature T_o (Figs 9,10) show that the number of extra states within a *fixed* Fermi window ~$(\pm 2k_B T_o)$ increases linearly with x to x~0.25, then decreases at a comparable rate. The slopes dS/dx and d(χT)/dx, when expressed in the units ln(4)k_B and $\mu_B^2/2$ respectively appropriate for WIF, both show that the number of orbitals within an energy window $E_F \pm 34$ meV (T_o=200K) increases at a rate of ~0.8/Sr atom. We have found a similar result for $YBCO_{6+y}$ where the number increases as ~0.3/oxygen[8]. These results suggest that ~1 additional state per added hole is located within this narrow energy range around E_F for both systems. For LSCO the extra states de-localise from E_F for x>0.25.

We finally consider the shape of the DOS close to E_F. Figs 3 and 4 suggest that in the normal state g(E) is close to zero at E_F for low x (≤ 0.15). The linear T-dependences above T_c of γ and d(χT)/dT suggest an energy dependence g(E)~ $|E-E_F|/\Delta_n$, where the normal state gap Δ_n decreases with hole density[9]. We have observed[8] a similar linear T-dependence in underdoped $YBCO_{6+y}$ raising the possibility of a d-wave normal state gap with line nodes for both systems. For LSCO the gap fills above x=0.15 to give a single peak at E_F for x~0.24 which then collapses to a flat DOS at x=0.45. An alternative interpretation is that g(E) is not symmetric about E_F but has, for example, a van Hove singularity which moves through E_F with hole doping. A difficulty here is that where g(E_F) is small (eg for x$\leq$0.15) the chemical potential will be strongly T-dependent, preventing[8] the large observed increase in γ,χ with T.

Our results strongly suggest the possibility of precurser pairing without long range phase coherence as an explanation for the normal state gap. It may be significant that the normal state gap starts to fill at around optimum T_c (x~0.16) and vanishes as T_c falls to zero at higher x. If the superconducting and normal state gaps were not related T_c might be expected to increase up to x~0.25 due to the large increase in low energy spectral weight.

References

1. H.Takagi,T.Ido,S.Ishibashi,M.Uota,S.Uchida,Y.Tokura, *Phys. Rev* **B40** (1989) 2254
2. J. B. Torrance et al. *Phys. Rev.* **B40** (1989) 8872
3. T.Nagano,Y.Tomioka,Y.Nakayama,K.Kishio,K.Kitazawa,*Phys.Rev.***B48**(1993) 9689
4. N.Momono,M.Ido,T.Nakano,M.Oda,Y.Okajima,K.Yamaya,*Physica* **C233** (1994) 395
5. X.K.Chen,J.Irwin,H.Trodahl,T.Kimura,K.Kishio, *Phys. Rev. Letts.* **73** (1994) 3290
6. J.L.Tallon,G.V.M.Williams,C.Bernhard,N.E.Flower,J.W.Loram, to be published.
7. J.W.Loram,K.A.Mirza,J.Wade,J.R.Cooper,W.Y.Liang,*Physica* **C235**-240 (1994)134
8. J.W.Loram,J.R.Cooper,K.A.Mirza,N.Athanassopoulou,W.Y.Liang, to be published.
9. B.Batlogg,H.Hwang,H.Takagi,R.Cava,H.Kao,J.Kwo,*Physica* **C235**-240 (1994)130

NEUTRON SCATTERING MEASUREMENTS ON $YBa_2Cu_3O_{7-\delta}$

H.A. Mook,[1] P. Dai,[1] F. Doğan,[2] K. Salama,[3] G. Aeppli,[4] and M.E. Mostoller[1]

[1] *Oak Ridge National Laboratory, Oak Ridge, Tennessee 37831-6393*
[2] *University of Washington, Seattle, Washington 98195*
[3] *Texas Center for Superconductivity, University of Houston, Houston, Texas 77204*
[4] *NEC, 4 Independence Way, Princeton, NJ 08540*

Polarized and unpolarized neutrons have been used to study the excitations in $YBa_2Cu_3O_{7-\delta}$. The magnetic excitations are found to be centered at (π,π) in the reciprocal lattice and are sharply peaked in energy at 41 meV in the superconducting state and broad in energy in the normal state. The intensity is found to result from coupled bilayers in both the normal and superconducting state. A new set of incommensurate one-dimensional phonon excitations has been discovered in scans that integrate over energy along planes in reciprocal space. These are found to be very sharply defined in momentum space and are thought to stem from $2k_F$ effects for the Cu-O chains.

1 Magnetic Excitations

Magnetic inelastic neutron scattering is a powerful technique which gives direct information about the behavior of the electrons in materials. This information is of particular value for high-temperature superconductors where electron correlation effects may be responsible for some of the unusual physical properties of the material. One of the best studied superconductors is $YBa_2Cu_3O_{7-\delta}$(YBCO) [1-7]. Many years ago, Rossat-Mignod et al. [1] reported enhanced scattering in the superconducting state in the region of 41 meV although other features in the scattering were reported as well. Phonon scattering is found to be stronger than the magnetic scattering in many cases for YBCO and thus polarized beams are needed to distinguish the magnetic scattering. [2] The polarized measurements showed that the only scattering found in the superconducting state was centered at 41 meV and that lower energy scattering disappeared as the superconducting gap opened. The normal state scattering first appeared to have a small peak at 41 meV as well[2] but improved measurements have shown the normal state scattering is generally featureless in energy, but still centered at (π,π) in the reciprocal lattice. Fong et al. [4] suggested from unpolarized measurements that the normal state scattering is unobservable, but we[5] have shown it is readily found if polarized neutrons are used. Fig. 1(a) & (b) show polarized measurements of the magnetic excitations in YBCO at 100 K and 10 K. The open points are the spin flip scattering and thus show the magnetic intensity directly. The filled points were taken with the HF-VF technique[3] which determines the background directly. The HF-VF measurements are very time consuming and were done at selected points to determine the background. We have found from high resolution unpolarized measurements that the 41 meV peak found in the superconducting state is very narrow in energy and we estimate that its width is as small as 2.5 meV.

The polarized measurements shown in Fig. 1 were taken with a wide resolution of 12.5 meV to obtain sufficient signal to observe the scattering in the normal state. A comparison of the height of scattering in the normal and superconducting state thus must take the energy resolution into consideration. The scattering in the normal state is found to be broad in energy and is not affected substantially by the resolution. Assuming 2.5 meV for

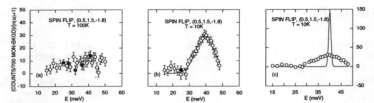

Figure 1: Polarized beam measurements at (0.5,1.5,-1.8). The open circles are the spin flip scattering. The closed circles are obtained by the HF-VF method. (a) shows the normal state scattering at 100 K. (b) is the scattering at 10 K. (c) shows the 10 K scattering where the solid line is the 41 meV peak corrected for resolution assuming a 2.5 meV intrinsic width.

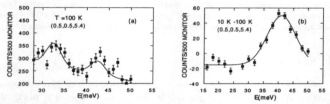

Figure 2: (a) shows measurements at (0.5,0.5,5.4) at 100 K which is essentially the phonon scattering. The solid line is the fit from the shell model calculation. (b) is the temperature subtracted scattering which shows the 41 meV scattering. The solid line is a Gaussian fit.

the width of the 41 meV peak would result in a height shown in Fig. 1(c) where the solid line shows a peak of the same area as that given in the polarized measurements.

The 41 meV peak has attracted considerable attention and a number of theoretical attempts [7] have been made to explain its origin. The absolute value of the integrated intensity of the 41 meV peak can be obtained if the magnetic scattering is compared carefully to the phonon scattering. The cross section for phonon scattering is given by:

$$\frac{d^2\sigma}{d\Omega dE} = 2.09 \frac{k_f}{k_i} (\sum_l e^{-i\mathbf{q}\cdot\mathbf{l}} \frac{[\mathbf{Q}\cdot\sigma_l(\mathbf{q})]e^{-W_l(\mathbf{Q})}b_l}{\sqrt{M_l}})^2 \frac{1}{\omega_q}(1+n(\omega)) \quad \text{barns/unit cell} \quad (1)$$

where $\mathbf{Q}$ is the momentum transfer, b_l the nuclear scattering amplitudes of the atoms at site l, σ_l the polarization vectors for each branch, M_l the mass of the atom at the site l, $n(\omega)$ the population factor, and W_l the Debye-Waller factors for the atoms. If the frequency ω is given in meV and the masses in amu the units are in barns/unit cell. In order to study both magnetic and non-magnetic scattering, we have made our measurements for $\mathbf{Q} = (0.5, 0.5, 5.4)$ where the phonon modes are not pure modes and have a complex displacement pattern. The polarization vectors then must be calculated and we have done this using a shell model for the lattice dynamics. The shell model uses parameters that give a good account of the optical and neutron measurements on the phonons in YBCO.

Fig. 2(a) shows the unpolarized data at 100 K which is largely the phonon scattering as the normal state magnetic scattering is small on the scale of the phonon scattering.

The line shows a fit to the data using the result of the theoretical phonon scattering cross-section convolved with the experimental resolution. There are a number of phonon modes that involve all the atoms in the unit cell in the energy range shown. The contributions from most of the modes are small, with the largest contributions coming from modes that involve c-axis displacements of the oxygen plane atoms and the oxygen chain atoms. The calculation gives a good account of the measurements so that the theoretical phonon cross section can be relied upon to obtain an absolute scale for the magnetic scattering from the same sample.

For unpolarized neutron scattering from a system of unaligned Heisenberg spins, the cross-section is:

$$\frac{d^2\sigma}{d\Omega dE} = \frac{1}{\pi}\frac{k_f}{k_i}r_0^2\frac{1}{2}f^2(\mathbf{Q})(1+n(\omega))\chi''_{\alpha\alpha}(\mathbf{q},\omega)/\mu_B^2 \quad (2)$$

where r_0 is 5.4×10^{-13} cm, and $f(\mathbf{Q})$ is the magnetic form factor. Fig. 2(b) shows the 41 meV peak obtained by taking the difference between 10 K and 100 K. Polarized measurements have shown that the phonon scattering does not change much between 100 K and 10 K. The 100 K magnetic scattering is quite small and energy-independent [Fig. 1(a)] so it doesn't affect the peak height determined for the 41 meV excitation. This measurement then gives the magnetic scattering which can be compared to the phonon scattering.

Since as far as we can determine, the $\hbar\omega_0 = 41$ meV peak is arbitrarily narrow we can rewrite the susceptibility at two-dimensional zone-boundary points $(0.5, 0.5, Q_c)$ as:

$$\chi''_p/\pi = \frac{1}{2}\chi' \sin^2(\pi Q_c/d)(\delta(\hbar\omega - \hbar\omega_0) - \delta(\hbar\omega + \hbar\omega_0))\hbar\omega_0 \quad (3)$$

where the bilayer spacing is $d = 3.43$ Å and Q_c is in Å^{-1}. The absolute intensity calibration provided by the phonons yields a value of 4.3 ± 0.13 μ_B^2/eV/unit cell for the real part χ' of the magnetic susceptibility associated with the 41 meV resonance.

Now that the size of the 41 meV peak is available in absolute units we expect it will be easier to choose between various suggestions for the origin of the 41 meV peak. In choosing between various scenarios for the origin of the peak it must be remembered that the peak results from a coupled bilayer excitation. We find this to be the case as well for the normal state scattering.

2 Incommensurate Fluctuations

We have also performed a series of investigations to examine scattering from the Cu-O chains. These are one-dimensional in nature so that the scattering from them can be distinguished from other sources of scattering as it should appear as sheets in reciprocal space. We have made scans that integrate in energy along planes of scattering that are perpendicular to the **b***(**a***) direction. Since the sample is heavily twinned we cannot distinguish **a*** from **b***. Such scans yield very sharp incommensurate peaks that occur between the standard Bragg reflections.

Fig. 3 shows some of these scans, obtained with an incident energy of 18.93 meV. The satellite peaks narrow as the temperature is reduced and also reduce in intensity. The solid

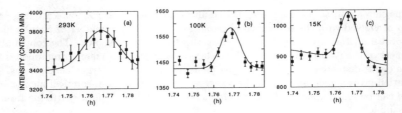

Figure 3: Excitations observed by integrating over planes of scattering perpendicular to $b^*(a^*)$. These are thought to stem from a dynamic charge density wave associated with the Cu-O chains.

line is a least squares fit to the data using a Gaussian and a sloping background. The peaks are found to be centered at 1.767 along b^* and have a width of 0.01 in b^* which is consistent with the size of the resolution. Additional smaller peaks are found at 0.767 and at 1.233 which represent the same points in reciprocal space. We have checked a number of other directions in reciprocal space and have confirmed the one-dimensional origin of the peaks in Fig. 3. Calculations have also been made that simulate the experiment using the shell model for the phonon positions and structure factors. These show that no structure from the normal phonons that can account for the satellites. Likewise, identical measurements on $YBa_2Cu_3O_{6.15}$ show no satellites either.

The position of the satellites is consistent with calculated values[8] for $2k_F$ for the Cu-O chains. Since the satellite scattering is inelastic in nature and has the temperature and Q dependence of phonon scattering we propose that it stems from an uncondensed charge density wave associated with the Fermi surface of the Cu-O chains. It is perhaps surprising that the scattering is so narrow in momentum space as the chain bands are expected to hybridize with other bands to some degree. We expect magnetic excitations at the chain $2k_F$ position as well and are in the process of studying these excitations.

The research at Oak Ridge was supported by US DOE under Contract No. DE-AC05-84OR21400 with Lockheed Martin Energy System, Inc.

References

1. J. Rossat-Mignod et al., Physica **169B**, 58 (1991).
2. H.A. Mook et al., Phys. Rev. Lett. **70**, 3490 (1993).
3. H.A. Mook et al., Physica **213B**, 43 (1995).
4. H.F. Fong et al., Phys. Rev. Lett. **75**, 321 (1995).
5. P. Dai et al., preprint (1995).
6. P. Bourges et al., Phys. Rev. B **53**, 876 (1996).
7. There have been a number of suggestions for the origin of the 41 meV peak. See for instance, Ref. 4, E. Demler and S.C. Zhang, Phys. Rev. Lett. **75**, 4126 (1995); D.Z. Liu et al., ibid. **75**, 4130 (1995); I.I. Mazin and V.M. Yakovenko, ibid. **75**, 4134 (1995); V. Barzykin and D. Pines, Phys. Rev. B **52**, 13585 (1995).
8. W.E. Pickett et al., Science **255**, 46 (1992).

HIGH-ACCURACY SPECIFIC-HEAT STUDY ON $YBa_2Cu_3O_7$ AND $Bi_2Sr_2CaCu_2O_{8.2}$ AROUND T_c IN EXTERNAL MAGNETIC FIELDS

A. SCHILLING*
*Lawrence Berkeley Laboratory and Department of Chemistry,
University of California, Berkeley, CA 94720, USA*

O. JEANDUPEUX*, C. WAELTI*, H. R. OTT*, A. VAN OTTERLOO[†]
*Laboratorium für Festkörperphysik, ETH Hönggerberg, 8093 Zürich, Switzerland
†Institut für Theoretische Physik, ETH Hönggerberg, 8093 Zürich, Switzerland

We deduce an upper limit for a latent heat related to a possible first-order phase transition at the irreversibility line of our investigated $YBa_2Cu_3O_7$ single crystals, $L < 0.05$ k_BT per vortex per layer in $\mu_0H = 7$ T parallel to the c axis. The high resolution in our specific-heat $C_p(T)$ data makes it also possible to analyse the second-order specific-heat discontinuities at T_c of $YBa_2Cu_3O_7$ and $Bi_2Sr_2CaCu_2O_{8.2}$ in more detail. Combining these results with corresponding magnetization $M(H,T)$ data, we conclude that the 3D–XY model for describing the phase transition to superconductivity at T_c does not reproduce in a satisfactory way the experimental data for magnetic fields exceeding $\mu_0H \approx 1$ T.

1. Introduction

We modified the standard experimental configuration of the differential thermo-analysis (DTA) technique to obtain accurate specific-heat $C_p(H,T)$ data on cuprate superconductors [1]. We thereby achieve a relative accuracy $\delta C_p/C_p < 4\times10^{-3}$/mg, with a negligibly small instrumental broadening in temperature. The measurements have been done on single crystals of $YBa_2Cu_3O_7$ ($T_c = 90.5 - 91.4$ K) and on a textured film of $Bi_2Sr_2CaCu_2O_{8.2}$ ($T_c = 91$ K), in magnetic fields up to $\mu_0H = 7$ T applied parallel to the c axes of the crystallites.

2. Investigations near the irreversibility line of $YBa_2Cu_3O_7$

2.1. Upper limit for a latent heat at the irreversibility line

It is assumed that in clean samples of cuprate superconductors a liquefaction of the vortex lattice occurs at the irreversibility line $H_{irr}(T)$, well below the fluctuation dominated upper critical-field $H_{c2}(T)$ crossover. Theoretical and experimental work supports that this melting corresponds to a first-order phase transition associated with a latent heat of the order of $L \approx k_BT$ per vortex per superconducting layer [2-5]. It has been seriously questioned, however, whether the experimentally detected anomalies seen in the *reversible* region of the H-T phase diagram (i.e., at $T \neq T_{irr}$) are due to a real thermodynamic phase transition or not [6].

The theory of operation of the here used DTA technique [1] implies that a first-order transition should manifest itself in an increase of the temperature difference between the sample (with total heat capacity C_p) and a reference specimen by $\Delta T = L/C_p$ (see Fig. 1, inset). Inserting material parameters for calculating ΔT we obtain $\Delta T = k_BTB/C_p\Phi_0 s$ for $L = k_BT$ /vortex/layer (where $k_B=1.38\times10^{-23}$ J/K, $\Phi_0=2.07\times10^{-15}$ Vs, B is the magnetic induction, and

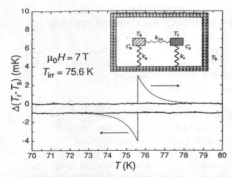
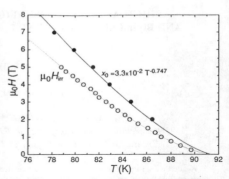

Fig. 1: DTA signal between a YBa$_2$Cu$_3$O$_7$ crystal (T_s) and a Cu reference (T_r). The line corresponds to a latent heat $L = k_B T$/vortex/layer at T_{irr} (see text).

Fig. 2: H-T diagram of YBa$_2$Cu$_3$O$_7$, showing H_{irr} and the coordinates of magnetization (x_0) and specific-heat features (filled circles) described in the text.

$s = c \approx 12$ Å is the distance between the layers). In Fig. 1, we present corresponding $\Delta T(T)$ data that we obtained on a 16.8 mg YBa$_2$Cu$_3$O$_7$ crystal in $\mu_0 H = 7$ T, for both heating and cooling the sample in the field. Distinct shifts in ΔT are clearly absent. Similar data in smaller fields down to 0.5 T around the magnetically measured $H_{irr}(T)$ [1] (see Fig. 2) did not reveal any clear features at T_{irr} either. From these data we can calculate an upper limit $L < 0.05$ k$_B T$ per vortex per layer in $\mu_0 H = 7$ T, which increases as $1/H$ when we reduce the external magnetic field H.

2.2. Considerations about the mixed state specific heat near the irreversibility line

Upon freezing, the vortex lattice can no longer adapt to the still significant T variations of the penetration depth $\lambda_{ab}(T)$ and $H_{c2}(T)$, respectively. In thermodynamic equilibrium, these variations give rise to a C_p/T term proportional to $-B \partial^2 [\lambda_{ab}^{-2} \ln(\alpha \mu_0 H_{c2}/B)]/\partial T^2$, where α is a weakly T and H dependent constant of the order of unity [7]. If this term changes below $H_{irr}(T)$, one may expect a certain change in the measured specific heat when crossing the boundary, irrespective of the order of the phase transition. In a similar way, the magnetization $M(H,T)$ should change its slope around $H_{irr}(T)$.

In Fig. 3 we show $M(T)$ data from another YBa$_2$Cu$_3$O$_7$ crystal for selected magnetic fields H, plotted as $y = MH^{-0.5}$ vs. $x = tB^{-0.747}$ (with $t = T/T_c - 1$), thus accounting for a 3D-XY scaling (see below). Indeed, $y(x)$ exhibits a slope change by $\approx 10\%$ at $x_0 \approx 3.3 \times 10^{-2}$ T$^{-0.747}$. Similar slope changes in $M(T)$ have been recognised in Refs. 4 and 5 at approximately the same coordinate x_0. We do not see any jump in $y(x)$ at x_0, however. The coordinate x_0 is *in the apparently reversible region* of the H-T diagram (see Fig. 2). Fig. 4 shows the measured specific-heat difference $(C_p(7T) - C_p(6T))/T$, together with an estimate of what we expect if the measured slope change in $y(x)$ at x_0 were due to a corresponding change in the *equilibrium magnetization*. Large anomalies in C_p/T at $T_0 = T_c(x_0 B^{0.747} + 1)$, as expected from thermodynamic relations, are not visible in our heat-capacity data. This leads us to conclude that the seemingly sharp, large slope change in $y(x)$ at x_0 does not represent an intrinsic equilibrium-magnetization $M(T)$ feature of YBa$_2$Cu$_3$O$_7$.

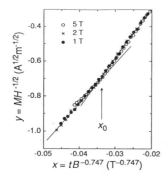
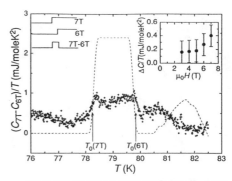

Fig. 3: Magnetization M of $YBa_2Cu_3O_7$ in 3D-XY scaling variables. The lines are to show the kink at x_0. Corresponding C/T curves were generated from such data using thermodynamic relations (Fig. 4).

Fig. 4: C/T difference for $YBa_2Cu_3O_7$ in $\mu_0H = 6$ and 7 T. The dashed line shows our expectations according to $C/T = \partial^2/\partial T^2 \int MdH$. The sizes of the C/T anomalies at $T_0(H)$ are plotted in an inset.

Nevertheless, smaller features in the specific heat, just near the detection limit of our apparatus, are reproducibly discernible at $T_0(H)$ ($\Delta C_p/T \approx 0.3$ mJ/mole K^2, see Fig. 4 as an example). The location of these $C_p(T)$ anomalies in the H-T diagram is displayed in Fig. 2. At present we are not able to draw final conclusions about the physical origin of these features. They may indeed suggest a certain change in the specific heat of $YBa_2Cu_3O_7$ when crossing a transformation line of the vortex matter at $tB^{-0.747} \approx 3.3 \times 10^{-2}$ T$^{-0.747}$.

3. Measurements near the $H_{c2}(T)$ crossover

For an interpretation of our thermodynamic data near $H_{c2}(T)$ we do not only rely on an apparent scaling collapse of the data onto single curves as predicted by certain theories. It turns out that most descriptions for the $H_{c2}(T)$ crossover that require such a scaling property, produce satisfactory scaling plots of measured physical quantities in suitable scaling variables [8,9].

In Fig. 5 we show $C_p(H,T)/T$ of $Bi_2Sr_2CaCu_2O_{8.2}$ and $YBa_2Cu_3O_7$, respectively, for external magnetic fields up to $\mu_0H = 7$ T parallel to c. The high resolution of these data allows us to calculate accurately the derivatives $\partial[C(H) - C(0)]/\partial T$ and $\partial C/\partial H$, respectively, which are insensitive to the presence of a phonon contribution to C_p/T, thereby representing quantities that are related to the respective electronic specific heats only. Inspired by the apparently good scaling of $M(H)$ of $YBa_2Cu_3O_7$ according to the 3D-XY model description of the critical region [4,9,10] (see Fig. 3), we accordingly plotted the specific-heat data as $[C(0) - C(H)]/B^{0.005}$ vs. $tB^{-0.747}$ in Fig. 6 [9,10]. At first sight, the scaling seems to be satisfactory. Nevertheless the data below T_c cannot exactly collapse to a universal curve because the mixed state specific heat of $YBa_2Cu_3O_7$ increases with increasing H in the temperature range of interest (see Fig. 5). Moreover, the low-field data for $\mu_0H < 2$ T strongly deviate from the scaling curve suggested by the high-field data [9]. This is particularly important because the validity of the 3D-XY model can be justified only if the correlation length $\xi_{XY} \approx \xi_0 t^{-\upsilon}$ (where $\upsilon_{3DXY} \approx 2/3$ is the critical exponent) does not exceed the magnetic length scale $(\Phi_0/\pi H)^{1/2}$, which, together with the Ginzburg criterion for the critical region, $t \leq Gi \approx 10^{-2}$, requires $H \leq 2H_{c2}(0)Gi^{2\upsilon} \approx 1$ T [9].

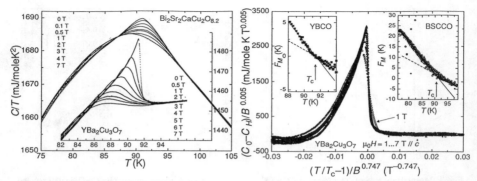

Fig. 5: Specific heats $C(H,T)/T$ of YBa$_2$Cu$_3$O$_7$ and Bi$_2$Sr$_2$CaCu$_2$O$_{8.2}$ around T_c.

Fig. 6: 3D-XY scaling plot for YBa$_2$Cu$_3$O$_7$. The insets show $F_M(T)$ data with lines that demonstrate that the slopes $dF_M/dT = v^{-1}$ are not constant.

Further support for the failure of the 3D-XY description to explain thermodynamic data in large magnetic fields comes from studying $F_C(T) = 2H(\partial C/\partial H)/\partial[C(H) - C(0)]/\partial T$ and $F_M(T) = 2H(\partial M/\partial H - 1/2M/H)/(\partial M/\partial T)$, respectively. In the critical region both F_C and F_M must vary as $F_C = F_M = (T_c - T)/v$ if $|\alpha/2v| \ll 1$ (where $\alpha_{3DXY} \approx -0.007$) [9]. In Fig. 6 we plotted F_M for Bi$_2$Sr$_2$CaCu$_2$O$_{8.2}$ and YBa$_2$Cu$_3$O$_7$, respectively, to demonstrate that there is no universal, T independent critical exponent $v = (dF_M/dT)^{-1}$ for $\mu_0 H > 0.75$ T. Below T_c, $v \approx 2/3$, but above T_c, the data rather imply $v > 1$. We have reported [9] the same discrepancy for $F_C(T)$ of YBa$_2$Cu$_3$O$_7$. These observations seem to clearly invalidate the claim that the 3D-XY scaling hypothesis is applicable for describing the transition to superconductivity in these compounds in external magnetic fields larger than $\mu_0 H \approx 1$ T, which is not at all obvious from focusing only on the apparent collapse of scaling plots of thermodynamic data onto single curves.

Acknowledgements

We thank to B. Heeb and Th. Wolf for providing the samples. This work was supported by the Schweizerische Nationalfonds zur Förderung der Wissenschaftlichen Forschung.

References

1. A. Schilling and O. Jeandupeux, *Phys. Rev. B* **52**, 9714 (1995).
2. R. E. Hetzel *et al.*, *Phys. Rev. Lett.* **69**, 518 (1992).
3. E. Zeldov *et al.*, *Nature* **375**, 373 (1995).
4. R. Liang *et al.*, *Phys. Rev. Lett.* **76**, 835 (1996).
5. U. Welp *et al.*, preprint.
6. D. E. Farrell *et al.*, submitted to *Phys. Rev. B*.
7. L. N. Bulaevskii and M. P. Maley, *Phys. Rev. Lett.* **71**, 3541 (1993).
8. M. Roulin *et al.*, submitted to *Physica* C.
9. O. Jeandupeux *et al.*, submitted to *Phys. Rev. B*.
10. M. B. Salamon *et al.*, *Phys. Rev. B* **47**, 5520 (1993).

PHONON ANOMALY IN HIGH TEMPERATURE SUPERCONDUCTING $YBa_2Cu_3O_{7-\delta}$ CRYSTALS.

R. P. SHARMA[2], Z. ZHANG[1], J. R. LIU[1], R. CHU[1], T. VENKATESAN[2] AND W. K. CHU[1]

[1]Texas Center for Superconductivity, University of Houston, Houston, TX-77204. [2]Center for Superconductivity Research, University of Maryland, College Park, MD-20742.

Investigations of lattice vibrations and small displacement of atoms (u) from their regular lattice sites in $YBa_2Cu_3O_{7-\delta}$ are made by Ion Channeling, a technique which provides a direct real space probe of uncorrelated atomic displacements as small as 0.01 Å. By comparing the results obtained in two high quality single crystals of the above material, one superconducting (T_c=92.5K) and the other non superconducting, a large deviation from normal thermal behavior is seen in the former case over the temperature range 300K to 30K. The deviation is mainly associated with Cu-O rows, Y-Ba rows have shown normal Debye like behavior. The observed decrease in u is much faster in this case as the temperature is lowered below 300K, there is almost no variation between 180K and 220K indicating a possible phase transition in this region followed by a sudden drop across T_c (between 100K and 80K) and then saturation in the u value below T_c. The value of this constant amplitude of lattice vibration is almost the same as the zero point vibration.

Despite significant work the role of phonons in high temperature superconductors remains unclear. The isotope effect [1], specific heat [2], elastic constant [3] and several other measurements [4] have yet to provide definite answers to the behavior of phonons near T_c in these materials. Raman scattering studies in $YBa_2Cu_3O_{7-\delta}$ (YBCO) samples, point towards possible phonon softening [5] in the vicinity of T_c. Neutron and x-ray diffraction [6] investigations have provided very accurate information to a precision of 10 fm regarding the structure of these materials. However, they are relatively insensitive to local uncorrelated atomic displacements such as small structural distortions or thermal vibrations. This is due to the fact that the diffraction data are analyzed within the constraints of a particular model based on the symmetry of the crystal space group. Random displacements from the ideal atom positions, whether static or thermal, contribute only to diffuse scattering (which is not included in the model). Furthermore, several neutron diffraction studies and also neutron inelastic scattering investigations have indicated the possibility of structural and vibrational anomalies [7] in the vicinity of T_c.

In the present work the ion channeling technique is used to investigate the phonon behavior and any other atomic displacement related to static or dynamic distortion, such as the Jahn Teller type, in $YBa_2Cu_3O_{7-\delta}$. The critical angle for channeling to occur is dependent on the incident ion energy, the atomic numbers of the projectile and target, the interatomic spacings, electron screening potential and most important for the present study, any displacements (static or vibrational) of the atoms from their regular lattice sites. It is important to note that as mentioned above neutron and x-ray diffraction are relatively insensitive to local uncorrelated atomic displacements in the sample. By comparing the channeling results in superconducting and non superconducting samples we find large deviation from normal thermal behavior in the former case associated only with CuO rows.

The samples used in the experiment were two [001] oriented $YBa_2Cu_3O_{7-\delta}$ single crystals, 4 to 5 mm^2 in area and 100μm thick. One of the crystals had $\delta < 0.1$, $T_c = 92.7$ K, while the other had $\delta=0.77$ and was non-superconducting.

Ion channeling measurements were carried out using a well-collimated beam (0.5

mm diameter and < 0.01° divergence) of 1.5 MeV He+ ions. The sample was mounted with a thermally conducting epoxy on a 3 axis goniometer having an angular resolution of 0.01°. The target holder was thermally insulated from the goniometer and could be cooled down to 33 K via a flexible Cu braid attached to a closed-cycle refrigeration unit. The target system was surrounded by two co-axial copper cryoshields, the outer one at 83 K ,while the inner one was maintained at 25 K. In this way the effective pressure of condensible gases was reduced to a negligible level; the background pressure in the target chamber was maintained at ~2x10^{-8} torr. The specimen temperature could be varied by a small 25W heater mounted on the back of the target holder, and could be maintained within ±2 K at any desired temperature between 33 and 300 K. The backscattered particles were analyzed using an annular surface barrier detector of 300 mm^2 active area with a 4 mm diameter central hole which was mounted along the beam axis at a distance of 5 cm from the target. Thus good statistics for backscattered (170°) counts at a dose of only 3 nC of the incident He-ions is obtained. Since YBa$_2$Cu$_3$O$_{7-\delta}$ single crystals are very susceptible to radiation damage, it was important to keep the incident beam dose to a minimum.

The crystal was first aligned and the angular channeling scans were made by measuring the Rutherford backscattering (RBS) yield as a function of the tilt angle about the [001] axis of the sample at several different temperatures. In order to avoid radiation damage, the 0.5 mm diameter beam spot was shifted to a new position on the specimen after every incident ion dose of 300 nC. Figure 1 shows examples of [001] axial channeling angular scans made with the RBS yield as a function of specimen tilt angle at

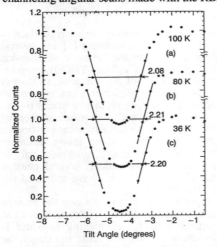 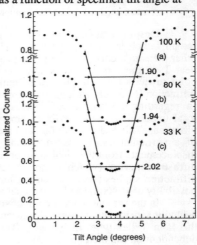

Fig.1 The [001] axial channeling angular scans made in YBa$_2$Cu$_3$O$_{7-\delta}$ (T$_c$=92.7K). A 5% increase in FWHM is seen across T$_c$, while practically no change in the b and c scans.

Fig. 2 Same as Fig.1 except the crystal is non-superconducting. There is no abrupt change between a and b, and a small increase from 80K to 33K, as expected from a normal thermal behavior.

temperatures of 100, 80 and 33 K. The RBS yield contains the signals emanating from the three constituents, Cu, Y and Ba of the superconducting sample and cover a depth of ~ 400 nm from the sample surface for the Cu signal . Thus the information from both the Cu-O and Y-Ba rows is mixed in the data collected. Similar scans from a non- superconducting sample are shown in Fig. 2. The large reduction in counts, to <3% of the random yield when the [001] axis of the specimen is aligned with the incident beam direction, demonstrates a very high-quality, single crystal specimen. A sudden increase of about 6.5% in the FWHM of the channeling scans in the superconducting case (Fig.1) is clearly seen when cooling through T_c between 100 and 80 K, confirming our earlier results [8]. Such an increase is absent in the case of the nonsuperconducting sample.

The plots of the FWHM of the channeling angular scans made with Cu-Y-Ba signals from the respective superconducting and non superconducting samples as a function of temperature, 300 down to 33K, are shown in Fig.3. Also shown in this figure is a similar plot obtained with the signals only from Y-Ba rows of the superconducting sample (Fig. 3c). It is evident that there is a gradual increase in the value of the FWHM as the temperature is lowered in the non superconducting sample (Fig.3a) as well as for the Y-Ba rows in the superconducting case, indicating that the thermal vibration amplitude is decreasing with temperature as expected. However for the Cu-Y-Ba signals in the superconducting sample there is an anomaly in the variation of the FWHM as a function of temperature (Fig.3b).

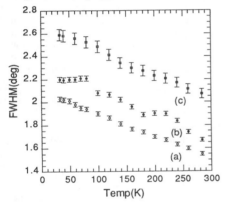

Fig. 3 A plot of the FWHM versus T
(a) non superconducting, (b) Y-Ba-Cu rows and
(c) Y-Ba rows in Superconducting sample

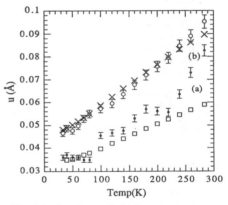

Fig. 4 A plot of u vs. T. (a) Superconducting (•) Experimental, (▫) Calculated. (b) Non Superconducting (o) Experimental, (×) Calculated.

The lattice vibration amplitude or the atomic displacements (u) as extracted from the measured FWHM of the channeling angular scans, using the continuum model [9] for channeling, with corrections based upon the Monte Carlo computer simulation of Barrett [10] are shown as points with error bars in Fig.4a and b both for superconducting and non superconducting YBCO crystals respectively. The average values of respective atomic numbers and lattice spacing are used in this calculation. The calculated values of the thermal vibration amplitude (u) based on Debye model for respective cases are also shown as squares and crosses for both samples. The normal trend of the decrease in the thermal vibrational amplitude as a function of decreasing temperature is clearly seen in the case of non superconducting sample (Fig.4b). Similar trend for Y-Ba rows in the superconducting specimen has also been obtained. However the u values obtained from the combined Y-Ba-Cu signals (Fig.4a) in the superconducting case, differ considerably from the calculated ones throughout the temperature range, an effect clearly associated with the Cu-O rows.

The absence of any variation in the u values (Fig.4a) within the experimental error in the superconducting state of the sample, indicates the existence of a temperature independent small amplitude <3.5 pm, in the available wide temperature range (33K to 80K). Using the relation $<u^2>=3\hbar\omega_D{}^2/8\pi\rho v^3$ (where $\hbar\omega_D = k_B\theta_D$, k_B is Boltzman's constant, θ_D is the Debye temperature, v is the velocity of sound and ρ is the mass density), the zero-point vibration in the superconducting YBCO is obtained as 3.2 pm which is approximately the same as the measured 3.5 pm between 30 and 80K. The longitudinal velocity of sound in YBCO along (110) in the plane (a,b) has been estimated to be 7×10^5 cm/sec by M. Saint-Paul et al. [11] and is used in the above calculation. This behavior is quite complex indicating that the material is in a more ordered state and a small constant amplitude of lattice vibration persists.

It appears that the structural fluctuation starts right from room temperature. In dilatation experiments [12] a clear change in the thermal expansion coefficient in the a and b direction across Tc and some instabilitiy along the c-axis around 200K is seen. In the channeling experiments also a similar anomaly is seen, but the effect is an order of magnitude larger than mentioned above, giving a clear indication of dynamic distortion. Thus there exists a combination of static and dynamic distortion followed by a stable state in the superconducting region. In the neutron inelastic scattering experiments [7], a shift of the apical O4 atom in the <110> direction in the a-b plane in a random fashion has been suggested. If this is true, this shift of the apical O(4) of the O(4)-Cu(2)-O(2,3) pyramid can create a <110> buckling motion and an instability in the structure which may stabilize as the superconducting state is reached. The main cause of such a fluctuation may be related to charge transfer between Cu-O planes and the chains as explained by Khomskii et al [13].

We thank Lynn Rehn, Boyed Veal and A. Paulikas from Argonne National Lab. for providing high quality YBCO crystals and useful suggestions. We also acknowledge F. Wellstood, R. Greene, K. B. Ma and P. H. Hor for helpful discussions. This work is supported by NSF Grant # DMR 9404579 at the University of Maryland and by the state of Texas through the Texas Center for Superconductivity at the University of Houston.

References

1. B. Battlog, R. J. Cava, A. Jayaraman, R. B. van Dover, G. A. Kourouklis, S. Sunshine, D. W. Murphy, L. W. Rupp, H. S. Chen, A. White, K. T. Short, A. M. Mujsce and E. A. Rietman, Phys. Rev. Lett. 58, 2333 (1987)
2. W. C. Lee, K. Sun, L. L. Miller, D. C. Johnston, R. A. Kumm, S. Kim, R. A. Fisher and N. E. Phillips, Phys. Rev. B43, 463 (1991).
3. X. D. Shi, R. C. Yu, Z. Z. Wang, N. P. Ong and P. M. Chaikin, Phys. Rev. B39, 827 (1989).
4. J. Mustre de leon, S. D. Conrodson, I. Batistic and A. R. Bishop, Phys. Rev. Lett. 65,1675 (1990); Brinkmann, Physica 153-155C, 75 (1988).
5. C. Thomsen, M. Cardona, B. Gegenheimer, R. Liu and A. Simon, Phys. Rev. B37, 9860(1988); B. Friedl, C. Thomsen, M. Cardona, Phys. Rev. Lett. 65, 915 (1990).
6. R. P. Sharma, F. J. Rotella, J. D.Jorgensen and L. E. Rehn, Physica C174, 409(1991).
7. M. Arai, K. Yamada, Y. Hideka, S. Itoh, Z. A. Bowden, A. D. Taylor and Y. Endoh, Phys. Rev. Lett. 69, 359(1992); H. A. Mook, M. Mostoller, J. A. Harvey, N. W. Hill, B. C. Chakoumakos and B. C. Sales, Phys. Rev. Lett. 65, 2712 (1990).
8. R. P. Sharma, L. E. Rehn, P. M. Baldo and J. Z. Liu, Phys. Rev. Lett. 62, 2869(1989) and Phys. Rev. B40, 11396 (1989).
9. J. Lindhard, K. Dan. Vidensk. Selsk. Mat. Fys. Medd. 34, No. 14 (1965);
10. J. H. Barrett, Phys. Rev. B3, 1527 (1971).
11. M. Saint-Paul, J. L.Tholence, H. Noel, J. C. Levet, M. Potel and P. Gougeon, Solid State Comm. 69,1161(1989).
12. C. Meingast, O. Kraut, T. Wolf, H. Wuhl, A. Erb and G. Muller-Vogt, Phys. Rev. Lett. 67, 1634 (1991)
13. Daniil I. Khomskii and Feodor V. Kusmartsev, Phys. Rev. B46, 14245(1992).

ELECTRODYNAMICAL PROPERTIES OF HIGH TC SUPERCONDUCTORS STUDIED WITH POLARIZED ANGLE RESOLVED INFRARED SPECTROSCOPY

D. van der Marel, J. Schützmann, H. S. Somal, and J. W. van der Eb
*Materials Science Centre, Laboratory of Solid State Physics,
University of Groningen, Nijenborgh 4, 9747 AG Groningen, The Netherlands*

Using infrared spectroscopy at grazing angle of incidence we study the electrodynamical properties of high temperature superconductors. We review some of our experiments where transverse polarized light is absorbed by longitudinal optical modes with their mode of oscillation perpendicular to the plane. This is particularly useful for the study of the plasmons and phonons perpendicular to the plane, and allows us to study in detail the c-axis dynamical properties of flux grown single crystals for which usually no samples with large dimensions in the c-direction exist.

1 Introduction and motivation

Recently P. W. Anderson pointed out[1], that for single layer superconductors the following correlation should exist between the bare Josephson plasmon energy (in in units of meV) and T_c (in K) if superconductivity is caused by the Anderson-Chakraverty interlayer-tunneling mechanism:

$$\hbar\omega_J = 2.9\ T_c N(0)^{1/2} d^{1/2} a^{-1}$$

where a (in $\mathring{A}$) is the in-plane lattice parameter, d (in $\mathring{A}$) is the spacing between CuO_2 planes, and $N(0)$ (in eV^{-1}) is the density of states at the Fermi energy per unit of CuO_2. For all cuprate superconductors $N(0)$ is approximately 1 eV, as follows e.g. from specific heat data. Experimentally one observes the plasma resonance at a reduced value $\omega_J/\sqrt{\epsilon_S}$ due to screening. In the cuprates this reduction is a factor 3 to 5 depending on the compound considered. This relation between measurable quantities is a unique feature of this mechanism, and thus provides an experimental test of this theory. Using the above expression we constructed the following table:

Compound	T_c (K)	a ($\mathring{A}$)	d ($\mathring{A}$)	$\hbar\omega_J$ (meV)	λ_c (μm)
$Bi_2Ba_2CuO_6$	12	3.86	11.57	30	6.6
$Nd_{2-x}Ce_xCuO_4$	24	3.95	6.035	43	4.5
$La_{2-x}Sr_xCuO_4$	32	3.79	6.64	63	3.1
$Tl_2Ba_2CuO_6$	85	3.86	11.57	216	0.91
$Hg_1Ba_2CuO_5$	98	3.86	9.51	225	0.88

For optimally doped $La_{2-x}Sr_xCuO_4$ the value of the *unscreened* Josephson plasmon energy is 25-30 meV [4], which is not too far below the prediction based on the Anderson-Chakravarty model. For the other systems no direct observations of the Josephson-plasmon energy have been reported yet, possibly due to the fact that samples of sufficient thickness for conventional normal incidence reflectivity experiments are not available.

2 The PARIS method

In this paper we discuss the reflection properties at grazing angles of incidence of anisotropic materials. In particular we consider the situation where the light is p-polarized, *i.e.* with the electric field vector parallel to the reflection-plane, and where the dielectric tensor component of the material along the crystal surface is metallic-like (Reϵ is large and negative). In this case the absorptivity A_p displays a series of resonance peaks at frequencies corresponding to the longitudinal optical modes with polarization perpendicular to the sample surface[2]. Using the Fresnel equations

$$\frac{A_p |n_x| \cos\theta}{2(2-A_p)} = \frac{\text{Im}(\alpha_p)}{1+|\frac{\alpha_p}{n_x \cos\theta}|^2} \quad \text{with} \quad \alpha_p(\omega) = e^{i\eta}\sqrt{1-\frac{\sin^2\theta}{\epsilon_z}}$$

we obtain the pseudo-loss function $\text{Im}(\alpha_p)/(1+|\alpha_p/n_x \cos\theta|^2)$ directly from the experimental data, without the need of a Kramers-Kronig analysis. In this expression θ is the angle of incidende with the surface normal (the z-direction), $\epsilon_i = n_i^2$ is the dielectric tensor component along x_i, and $\eta \equiv \pi/2 - \text{Arg}(n_x)$. If $|\epsilon_z| \gg \sin^2\theta$, this is the loss function $\text{Im}(-e^{i\eta}/\epsilon_z)$ with a Fano-type phase factor. The limiting behaviour $\eta = 0$ is reached for a superconductor or for a metal with $\omega\tau \gg 1$. In the low frequency limit of a metal $\eta = \pi/4$. For a metal with θ sufficiently far below the critical angle we have $|\alpha_p| \ll 1$, so that the pseudo-loss function becomes $\text{Im}(\alpha_p)$. Note that for grazing angles of incidence A_p is enhanced with a factor $1/\cos\theta$. We recently took advantage of this fact to study the in-plane conductivity of $La_{2-x}Sr_xCuO_4$ below T_c in detail.[3]

3 The pseudo-loss function of $Tl_2Ba_2CuO_6$

Let us now consider $Tl_2Ba_2CuO_6$ ($T_c = 85$ K). To simulate the pseudo-loss function at a grazing angle of incidence we use the following parameters for the electronic c-axis dielectric function: $\epsilon_\infty = 4, \hbar\omega_{pc} = 200$ (meV), $\rho_{DC} = 1$ (Ωcm), and transverse (longitudinal) optical phonons at 16.0 (17.7), 51.2

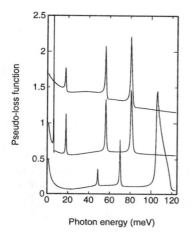 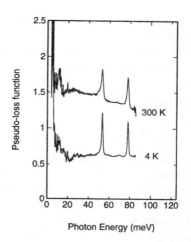

Figure 1: Simulation of the pseudo loss-function for $Tl_2Ba_2CuO_6$. Top curve: 300 K. Middle curve: 4 K using BCS theory. Lower curve: 4K, with $\omega_J = 200$ meV.

Figure 2: Experimental plot of the pseudo-loss function from ab-plane surface of $Tl_2Ba_2CuO_6$ (T_c=85 K) using p-polarized light with $\theta = 80°$.

(55.9) and 74.6 (80.3) meV with $\hbar/\tau = 0.6 meV$. The result of this simulation is displayed in Fig. 1. In the normal state the electronic contribution is overdamped and does not give rise to an additional zero-crossing of ϵ'. If we model the superconducting state with BCS theory, as is shown in the middle curve of Fig. 1, a Josephson plasmon appears at 6 meV due to transfer of spectral weight of $\sigma(\omega)$ in the gap-region to the δ-function at $\omega = 0$. The effect of the appearance of an unscreened Josephson plasma energy of 200 meV in the superconducting state is shown in the lower curve: All longitudinal modes are now of mixed phonon/plasmon character, and the longitudinal phonons are pushed to a higher frequency compared to the normal state.

In Fig. 2 the experimentally measured pseudo-loss function is displayed above and below the phase transition. The half-width of the loss-peaks is $1/\tau_{ph} + 4\pi\sigma_e\epsilon_\infty^{-1}S_{ph}/(\epsilon_\infty + S_{ph})$, where τ_{ph} is the intrinsic phonon life-time, S_{ph} the oscillator strength, and σ_e the electronic optical conductivity. Hence the linewidth of the longitudinal phonons can be used to obtain an upper-limit of the electronic contribution to the optical conductivity in the c-direction. From the experimental data we obtain an upper-limit for σ_c of 1 S/cm near the two dominant longitudinal phonon-peaks. From comparison with Fig. 1 we conclude that the shift of longitudinal frequencies below T_c, as well as the occurance of an extra peak at 48 meV, are both absent in the experiment. This implies that the unscreened Josephson plasmon energy is below 20 meV, or $\lambda_c > 10\mu m$ in this material.

4 Conclusions

The method of polarized angle dependent infrared spectroscopy was used to measure the c-axis infrared properties of thin single-crystalline platelets. The Josephson plasmon energies were calculated for a number of single layer compounds adopting an expression suggested by Anderson, and compared to experimental values for $La_{2-x}Sr_xCuO_4$ and $Tl_2Ba_2CuO_6$. The agreement with $La_{2-x}Sr_xCuO_4$ is within 50 percent. For $Tl_2Ba_2CuO_6$ no shift in longitudinal phonon frequencies was observed below T_c. This may be taken as an indication, that, as is also the case in conventional BCS theory, the electronic part of the dielectric function remains almost unaffected on an energy scale larger than $3.5k_BT_c$.

Acknowledgments

We gratefully acknowledge numerous discussions with P. W. Anderson and stimulating comments by A. J. Leggett, and N.N. Kolesnikov for supplying $Tl_2Ba_2CuO_6$ crystals. This investigation was supported by the Netherlands Foundation for Fundamental Research on Matter (FOM) with financial aid from the Nederlandse Organisatie voor Wetenschappelijk Onderzoek (NWO).

1. P. W. Anderson, Science **268**, 1154 (1995), and private communication.
2. D. van der Marel, B. J. Feenstra, and A. Wittlin, Phys. Rev. Lett. **71** 2676 (1993); Jae Kim, B.J. Feenstra, H.S. Somal, Wen Y. Lee, A.M. Gerrits, A. Wittlin, D. van der Marel, Phys. Rev. B, 49 (1994) 13065-13069
3. H.S. Somal, J.H. Kim, D. van der Marel *et al*, Phys. Rev. Lett. **76** 29 february, (1996).
4. K. Tamasaku, Y. Nakamura, and S. Uchida, Phys. Rev. Lett. **69** (1992) 1455; J. H. Kim, A. Wittlin, D. van der Marel *et al.*, Physica C **247**, 297 (1995).

HIGH-PRESSURE RAMAN STUDY OF THE MERCURY-BASED SUPERCONDUCTORS AND THE RELATED COMPOUNDS

IN-SANG YANG, HYE-GYONG LEE, NAM H. HUR*, SUNG-IK LEE**

Department of Physics, Ewha Womans University, Seoul 120-750, Korea;
*Korea Research Institute of Standards and Science, Taejon 305-606, Korea;
**Department of Physics, Pohang Institute of Science and Technology, Pohang 790-390, Korea

We report pressure-induced effects on the Raman frequencies of the apical oxygen (O_A) A_{1g} modes in the Hg-O_A-Cu bonds of the mercury-based superconductors and modes in some related compounds. The rates of increase in the force constants for O_As in the mercury-based superconductors are found to be strongly correlated with their T_c vs P behaviors, respectively. Study on the Hg-1212 samples annealed in various conditions shows that the O_δ content does not affect the pressure dependence of the O_A A_{1g} modes. The major cause of the increase of the Raman frequencies of the O_A A_{1g} modes is attributed to the Hg-O_A bondstength and its change.

1. Introduction

The transition temperature (T_c) of the mercury-based superconductors increases by ~ 30 K under high pressure.[1] Investigating the pressure-induced changes in the local environment and the electronic structures of the mercury-based superconductors would be helpful in understanding the high temperature superconductivity of the material.

In this article, we present results of micro-Raman study of the A_{1g} modes of the apical oxygens in HgBa$_2$Ca$_{n-1}$Cu$_n$O$_{2+2n+\delta}$, Hg-12(n-1)n, superconductors and the impurity modes under pressure upto 10 GPa. Our results show that the Raman frequencies of the O_A A_{1g} modes increase for all the superconductors as pressure is applied. The increase rate is found to be nearly the same for Hg-1201, Hg-1212 and Hg-1223 below $P \approx 5$ GPa, above which the rate for Hg-1223 becomes much larger than those for Hg-1201 and Hg-1212.

2. Experiments

The mercury compounds were synthesized by the precursor method.[2,3] Magnetic susceptibility and resistance versus temperature measurements of these samples showed that there exists only one superconducting phase in each of the samples. The onset temperatures for superconductivity are 92 K and 134 K for Hg-1201 and Hg-1223, respectively. Complete transition to zero resistance state occurs within 3 K. For Hg-1212, the onset temperatures were 112 K, 120 K, and 124 K for Ar-annealed, as-sintered, and O$_2$-annealed samples, respectively.

Polycrystalline samples were loaded at room temperature into a gasketed diamond anvil cell without particular pressure medium. It is important to use a pressure medium in maintaining hydrostatic conditions at high pressure. However, using the sample itself as the pressure medium was the best choice available in this experiment, since any other pressure media would result in unbearable deterioration of the already weak Raman signal-to-noise ratio. The dimensions of the sample chamber were approximately 300 μm in diameter and 150 μm in thickness. Pressures were determined by the well known ruby fluorescence technique.[4] The band widths and the splitting of the R_1 and R_2 lines of the fluorescent spectra of the ruby chips were utilized to ensure the quasihydrostatic condition.

Raman spectra were excited by the 514.5 nm line of the Ar-ion laser through a metallurgical microscope with ultra-long-working-distance objective lenses. Specific crystal(s) in the cell was(were) focused onto, and the back-scattered light from it(them) was dispersed by the Jobin Yvon U1000 double monochromator. The exact orientation of the crystals under pressure is not known.

3. Results and Discussion

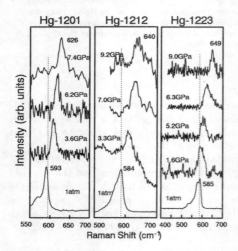

Fig. 1: Typical Raman spectra of Hg-1201, Hg-1212, and Hg-1223 at high pressures.

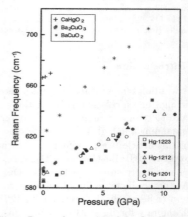

Fig. 2: Raman frequencies of the O_A A_{1g} modes of Hg-1201 (circles), Hg-1212 (triangles), and Hg-1223 (squares) along with the strong peaks of several impurity phases at high pressures.

Figure 1 shows typical Raman spectra of the Hg-1201, Hg-1212, and Hg-1223 samples under different pressures, respectively. For the superconducting samples in the pressure cell, only the strong peaks of the O_A A_{1g} modes are distinguishable due to poor signal-to-noise ratios and strong fluorescence of diamond anvils. The samples do have impurity phases like $Ba_2CuO_{3+\delta}$, $CaHgO_2$, and $BaCuO_2$. The spectral shape and the relative intensities of several modes, and their behavior under high pressure were taken into account in distinguishing superconducting phases from the impurity phases.[5]

Figure 2 shows the peak positions of the O_A A_{1g} modes of Hg-1201 (circles), Hg-1212 (triangles) and Hg-1223 (squares) along with those of strong peaks in the range of 580 - 700 cm^{-1} from $Ba_2CuO_{3+\delta}$ (#), $CaHgO_2$ (+), and $BaCuO_2$ (*). Different types of circles, triangles and squares represent different batches of samples, indicating reproducibility of the data.

The O_A A_{1g} peak positions of the Hg-1212 samples lie almost same as those from the Hg-1201 and Hg-1223 samples. Moreover, it is notable that the pressure dependence of the Raman mode is independent of the oxygen content, therefore independent of the values of T_c at ambient pressure. We investigated the effect of O_δ content on the behavior of the O_A A_{1g} Raman frequency of Hg-1212 under pressure. The O_A A_{1g} peak frequencies of the Hg-1212 shown in Fig. 2 are for the samples as sintered (open triangles), annealed in oxygen (up triangles), and annealed in argon gas (down triangles). The pressure dependence of the O_A A_{1g} mode of Hg-1212 samples as-sintered is the same as that of Hg-1212 samples that were annealed in oxygen or in argon gases. The same seems to be true for the Hg-1201 and Hg-1223. In fact, the data for Hg-1201 and Hg-1223 are from samples that were made by slightly different heat treatments. The dependence of T_c on pressure depends on the oxygen contents for most oxide superconductors. However, the T_c vs P behavior of the mercury-based superconductors above 2 GPa is essentially independent of the initial values of T_c, as long as the values of T_c are not too far off the optimal values.[6]

In measurements on the Tl-doped Hg-1223, it has been shown that the occupancy of O_δ-site is not affecting the Raman frequencies of the O_A A_{1g} mode.[7] Therefore, the increase of the frequency of the O_A A_{1g} Raman mode under increasing pressure is not related with the possible change in the O_δ content of the Hg-1223 samples under pressure. That is, the change of the frequency of the O_A A_{1g} mode is independent from the occupancy of O_δ-site, but might be directly related with the bondstrength of the Hg-O_A-Cu bond.

To probe the pressure dependence of the O_A A_{1g} mode, we plotted our Raman data in (ω^2/ω_0^2) vs P for Hg-1201, Hg-1212, and Hg-1223 (Fig. 3), where ω_0 is the Raman frequency of corresponding samples at ambient pressure. ω^2 represents effectively the force constants of the bonds between O_A and its neighboring atoms. The rates of the increase of (ω^2/ω_0^2) for both Hg-1223 and Hg-1201 are the same upto $P \approx 5$ GPa. For Hg-1212, it is also linear throughout the measured pressure range, but the slope of it is slightly higher than that of Hg-1201 and Hg-1223 at pressures ≤ 5 GPa. Whereas, the Hg-1223 shows a sharp increase of the slope above $P \approx 5$ GPa and the slope becomes larger than that of Hg-1212. At this pressure (≈ 5 GPa), we note slope changes in T_c vs P for the Hg-1223 system. Shown in the inset of Fig. 3 are T_c data for Hg-1223 from electrical measurements by Gao and Chu et al.[1] and from magnetic measurements by Takahashi et al.[8]. In fact, $P \approx 5$ GPa corresponds to the pressure where the level crossing of the Hg-O_A orbital energy (E_{Hg-O_A}) and Fermi energy E_F of Hg-1223 occurs.[9] This may mean a strong correlation of $\Delta(\omega/\omega_0)^2$ with the change in the electronic structure of the mercury-based superconductors. Our Raman results indicate that the change in the Hg-O_A-Cu bondstrength is not due to the possible change in the O_δ content, but might be due to charge transfer from the CuO_2 planes. A scheme to explain this high pressure behavior of Hg-1223 is drawn in Fig 4. At above 5 GPa, the Hg-O_A-Cu bond acquires charges from the CuO_2 planes resulting in higher doping-state of the planes, which may lead to *decreasing dT_c/dP*.

In conclusion, we observed effects of pressure on the Raman frequencies of the apical oxygen O_A A_{1g} modes of the $HgBa_2Ca_{n-1}Cu_nO_{2n+2+\delta}$ (n=1,2,3) superconductors, respectively. Our results show that the Raman frequencies of the O_A A_{1g} modes increase significantly as the pressure is increased, and their increase rate are strongly correlated with the changes of T_c and the electronic structure of the system. The major cause of the increase of the Raman frequency of the mode is attributed to the strengthened Hg-O_A-Cu bonds, which might be due to the possible charge transfer from the CuO_2 planes into the Hg-O_A bonds.

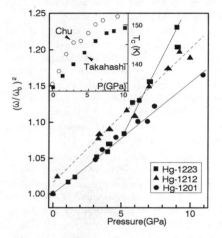

Fig. 3: Raman peak frequencies in (ω^2/ω_0^2) vs P for Hg-1201 (circles), Hg-1212 (triangles), and Hg-1223 (squares). The lines are guides to the eye. Inset shows the T_c vs P data of Hg-1223.[1,8]

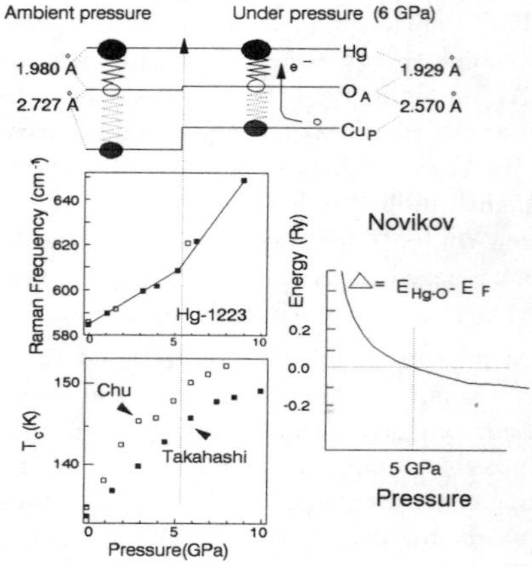

Fig. 4: A scheme to explain the high pressure behavior of Hg-1223. Lattice parameters indicated are from Ref.[10].

Acknowledgements

We thank professor Seung-Joon Jeon for the essential help in the high-pressure work. The financial support from the fund through Basic Science Research Institute Program, Ministry of Education, Project No. BSRI-95-2428 is gratefully acknowledged.

References

1. L. Gao et al., Phys. Rev. B. **50**, 4260 (1994).
2. N. H. Hur et al., Physica C **234** 19 (1994).
3. Sergey Lee et al., J. Mat. Chem. **4**, 991 (1994).
4. H. K. Mao et al., J. Geophys. Res. **91**, 4673 (1986).
5. In-Sang Yang et al., Physica C **222**, 386 (1994); In-Sang Yang et al., Phys. Rev. B **51**, 644 (1995).
6. L. Gao, private communication.
7. In-Sang Yang et al., Phys. Rev. B **52**, 15 078 (1995).
8. H. Takahashi et al., Physica C **218**, 1 (1993).
9. D. L. Novikov et al., Physica C **222**, 38 (1994).
10. A. R. Armstrong et al., Phys. Rev. B. **52**, 15 551 (1995).

ION CHANNELING STUDIES IN YBCO THIN FILM AT LOW TEMPERATURE

XINGTIAN CUI, Z. ZHANG, QUARK CHEN, J.R. LIU and W.K. CHU

Texas Center for Superconductivity at the University of Houston
Houston, TX 77204-5932

High quality epitaxially grown c-oriented YBCO thin films have been used for ion channeling investigations. An abrupt increase in the width of the channeling angular scan was observed when temperature changes from above Tc to below Tc, which does not follow the normal behavior described by Debye or Einstein model.

Introduction

Phonon behavior of HTS across T_c is highly interesting since it could help shed some light on the mechanism of high T_c superconductivity. Recently, several reports[1-3] showed the existence of anomalies in the motions of Cu and O atoms in bulk single crystalline YBCO, ErBaCO and $Bi_2Sr_2CaCu_2O_8$ across Tc. As compared with bulk single crystals, thin film sample has the advantages of large area and the flexibility in choosing the crystal orientations. In this paper, we report a RBS channeling work on high quality YBCO thin films measured in a wild temperature range. From an ion channeling angular scan, we can measure the full width at half maximum (FWHM) as a function of temperature for each constituent element, which would reflect the phonon behaviors of different modes.

Experimental

The high quality epitaxially grown c-oriented YBCO thin films were deposited on $SrTiO_3$ single crystal substrates by pulsed laser ablation technique. The typical thickness of the films was 7000-9000 Å and Tc was about 91K. χ_{min} near surface region at 100K was below 5%.

Thin film sample was mounted on a cold stage, which was cooled by a closed cycle cryocooler. The temperature of the sample can be varied from above room temperature down to 35K. The transition temperature of YBCO thin film was monitored in situ by resistance measurement. The 1.5MeV He ion RBS-channeling experiments were carried out using 1.7MV tandem accelerator at TCSUH. The backscattered He ions were registered in a ring type Si detector, which was placed 5-6 cm away from the sample and was aligned with the beam line. The beam divergence was less than 0.01 degree. The goniometer that manipulates the sample has an angular resolution better than 0.01 degree. Great attention had been paid to the radiation dose to avoid radiation damage which may degrade the sample quality.

Two energy intervals on the RBS spectra were used to obtain the integrated intensity, each covering 15-20 channels of ion energies which correspond to a film thickness of 1500-2000Å. One interval has only Y, Ba backscattering counts, while the other has Cu, Y and Ba backscattering counts. Each angular scan was conducted at different temperatures.

Results and Discussion

Fig.1 shows the FWHM of the channeling angular scans, which was the Y and Ba signals of the RBS spectra. No anomalous change in FWHM across the Tc was observed. The variation of FWHM with temperature follows Debye-like behavior. However, in Fig.2, we can clearly see an abrupt 6% change in FWHM of the channeling angular scans across Tc for signals from Cu,Y and Ba, while there is only a 2% change for the Y and Ba signals. This indicates that the anomalous behavior is due to the contribution of the Cu signals. This observation is in agreement with the results reported by the Argonne group on bulk YBCO and O. Mayer et al. on thin film[1,2].

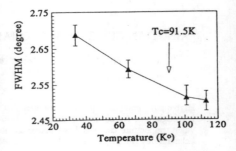

Fig.1 FWHM of Ba and Y channeling angular scans as function of temperature.

The magnitude of lattice vibration amplitude is directly related to the FWHM of the channeling angular scan[4]. The abrupt changes of FWHM reflect the abnormal phonon behavior of Cu atoms along a-b plane, although, we still could not distinguish based on this technique whether the abnormal phonon behavior of Cu arises from the Cu-O planes or from the Cu-O chains.

This is a progress report of an ongoing project on superconducting and non-superconducting oxides, which aims at studying the vibration amplitude of various atoms as a function of temperature.

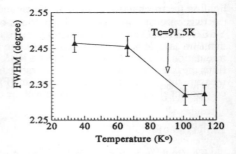

Fig.2 FWHM of Cu, Ba and Y channeling angular scans as function of temperature.

Acknowledgments

The authors would like to thank Dr. R.P. Sharma and Dr. K.B. Ma for useful discussions.

References

1. R.P. Sharma, L.E. Rehn, P.M. Baldo and J.Z. liu, *Phys. Rev.* **B 38** (1988) 9287.
2. J.Remmel, O.Meyer, J.Geerk, J.Reiner and G. Linker, *Phys. Rev.* **B 48**(1993) 16168
3. H.A. Mook, M. Mostoller, J.A. Harvy, N.W. Hill, B.C. Chakoumakos and B.C.Sales, *Phys. Rev. Lett* . **6 5** (1990) 2712
4. L.C. Feldman, J.W. Mayer and S.T. Picraux, *Material Analysis by Ion Channeling*, New York, Academic Press (1982).

ELECTRONIC RAMAN SCATTERING OF YBa$_2$Cu$_4$O$_8$ AT HIGH PRESSURE

T. Zhou, K. Syassen, and M. Cardona

Max-Planck-Institut für Festkörperforschung, Heisenbergstr. 1, D-70569 Stuttgart, Germany

J. Karpinski and E. Kaldis

Laboratorium für Festkörperphysik, ETH Zürich-Hönggerberg, 8093 Zürich, Switzerland

>We have investigated the electronic Raman scattering in the normal state of YBa$_2$Cu$_4$O$_8$ at hydrostatic pressure up to 9 GPa. In all three polarization configurations measured, the continuum is changed from a strongly temperature dependent Bose-Einstein form to very weakly temperature dependent by increasing pressure. We can explain our results within a model recently proposed by Varma, who suggests there is a crossover from the Fermi-Liquid to the so-called Marginal-Fermi-Liquid behavior when the superconducting cuprates are tuned toward optimal doping.

The electronic Raman scattering continuum observed in YBa$_2$Cu$_3$O$_7$ and other superconducting cuprates [1] exhibits several unusual features, particularly its weak dependence on frequency and temperature in the normal state. This is quite different from the behavior observed in doped semiconductors [1], where the dependence of the low frequency Raman scattering efficiency on T is mostly determined by the Bose-Einstein factor $1 + n_\omega = [1 - exp(-w/T)]^{-1}$. Varma [2] explained this anomaly by assuming that the quasipartical scattering rate is linearly dependent on ω and T. This phenomenological approach is called Marginal Fermi Liquid (MFL) model. Recently, Donovan et al. [3] reported that in YBa$_2$Cu$_4$O$_8$ (Y124) samples at ambient pressure, all of the five components of the electronic Raman continuum exhibit the "usual" Bose-Einstein-like behavior in the normal state as the temperature is decreased. On the other hand, Y124 is of particular interest at high pressure, for its T_c increases with pressure at a rate of 5.5 K/GPa below 3 GPa and saturates above 6 GPa [4]. It is just these two particular properties of Y124 that motivated us to investigate the electronic Raman scattering in Y124 at high pressure.

Figure 1 shows two sets of spectra measured at 1.5 and 6 GPa in the xx polarization configuration. Each set consists of two spectra measured at 300 and 110 K, respectively. The background of the 1.5 GPa spectrum clearly shows a strong temperature dependence. When both spectra are corrected for the Bose-Einstein factor, the agreement between the two spectra is remarkable, except for some redistribution of the intensity of the phonons. This behavior is very similar to that observed by Donovan et al. [3] in Y124 at ambient pressure. On the contrary, the 6 GPa spectra show a very weak temperature dependence. In this case, the temperature behavior is very similar to that observed in YBa$_2$Cu$_3$O$_7$ at ambient pressure [1]. Similar behaviors are also observed in yy and xy configurations.

In order to quantify the observation described above, we multiply n(ω) by α where α is a fit parameter, and renormalize the 110 and 300 K spectra at the same pressure with αn(ω) in order to make them agree with each other. Figure 2 exhibits the pressure dependence of the fitted value of α for the three different polarization configurations. At pressures below 1.5 GPa, α is about 1, the value expected for the conventional Fermi-Liquid metals and heavily doped semiconductors. With increasing pressure, α decreases and gradually tends to remain constant above 6 GPa for all the three configurations measured.

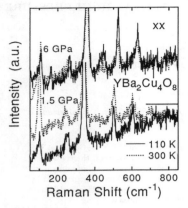

Figure 1.

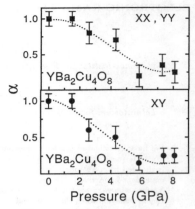

Figure 2.

Bucher et al. [5] measured the temperature dependent resistivity $\rho(T)$ along the a axis in Y124 at ambient pressure. They found a kink around 160 K, while the curve is linear both above and below 160 K. This kink is suggested to be related with some low-energy excitation gap [5]. Furthermore, various experiments suggest that Y124 can be tuned from underdoped towards optimally doped upon increasing pressure [4,6]. In a recent developement of the Maginal-Fermi-Liquid (MFL) model [7], Varma reached at the conclusion that MFL behavior is to be expected only for $T \geq G(x)$ with a crossover to Fermi-Liquid (FL) at lower temperature. $G(x)$ is a characteristic temperature of the order of that at which the resistivity kink mentioned above occurs, x is the doping concentration. When x tends to the optimal doping value, $G(x) \to 0$. According to this scenario, at low pressure the Y124 sample should be underdoped with $G(X) \sim 160$ K, therefore MFL behavior is expected approximately above 160 K and FL behavior below 160 K. Figure 1 shows that at 1.5 GPa the background of the 300 K spectrum is rather flat, thus suggesting MFL behavior, while a drastic background decrease in the low frequency regime is observed for the 110 K spectra, contrary to the MFL behavior. Above 6 GPa the sample should be nearly optimally doped since T_c is almost at its highest value. The MFL model therefore predicts that $G(x) \to 0$ and the corresponding behavior appears in the whole temperature range. The 6 GPa spectra of Fig. 1 taken at both 300 and 110 K show that the elctronic Raman continuum is weakly frequency and temperature dependent, a result that is consistent with the MFL model. Thus the updated MFL model [7] is in qualitative agreement with our electronic Raman results for the normal state.

References

1. M. Krantz et al., J. Low Temp. Phys. **99**, 205 (1995).
2. C.M. Varma, et al, Phys. Rev. Lett. **63**, 1996 (1989).
3. S. Donovan et al., J. Superconductivity **8**, 417 (1995).
4. B. Bucher et al, Physica. C **157**, 478 (1989).
5. B. Bucher et al, Phys. Rev. Lett. **70**, 2012 (1993).
6. T. Miyatake et al, Physica. C **167**, 297 (1990).
7. C.M. Varma, Phys. Rev. Lett. **75**, 898 (1995).

RAMAN STUDY OF $HgBa_2Ca_{n-1}Cu_nO_{2n+2+\delta}$ (n=1, 2, 3, 4 AND 5) SUPERCONDUCTORS

Xingjiang Zhou and M. Cardona

Max-Planck-Institut für Festkörperforschung, Heisenbergstr. 1, D-70569 Stuttgart, Germany

C. W. Chu and Q. M. Lin

Department of Physics and the Texas Center for Superconductivity at the University of Houston, Houston, Texas 77204-5932, U.S.A.

S. M. Loureiro and M. Marezio

Laboratoire de Cristallographie, CNRS-UJF, BP 166, 38042 Grenoble Cedex 09, France

Polarized micro-Raman scattering measurements have been performed on the five members of the $HgBa_2Ca_{n-1}Cu_nO_{2n+2+\delta}$ (n=1, 2, 3, 4 and 5) high T_c superconductor family. A systematic evolution of the spectrum, which mainly involves oxygen-related phonons around 590, 570, 540 and 470 cm^{-1}, with the increasing number of CuO_2 layers, has been observed. Local laser annealing measurements clearly demonstrate that all these phonons are closely related to interstitial oxygen in the HgO_δ planes. The origin of the spectrum evolution with the number of CuO_2 layers lies in the variation of interstitial oxygen content.

Raman scattering has already been employed to probe phonon structures for the first three members of $HgBa_2Ca_{n-1}Cu_nO_{2n+2+\delta}$ superconductors.[1,2,3,4] We have performed polarized micro-Raman measurements on the five members of the series and found a systematic spectral evolution with increasing number of CuO_2 layers (see Fig. 2 in ref.[5]). We also performed local laser annealing measurements to investigate the change of Raman spectrum with the variation of excess oxygen content, δ.

Figure 1 shows the Raman spectra of Hg-1201 before and after laser annealing in vacuum with a laser power of 30 mW for 2 hours. The 570 cm^{-1} peak is found to vanish after annealing. The 590 cm^{-1} peak also displays a significant linewidth narrowing with FWHM decreasing from 7 to 5 cm^{-1}. The disappearance of the 570 cm^{-1} peak after annealing strongly suggests that this mode is related to the interstitial oxygen. It also confirms that the remaining single 592 cm^{-1} peak definitely corresponds to the only A_{1g} mode of the apical oxygen in Hg-1201.

Figure 2 shows the spectra of Hg-1234 before and after different stages of annealing in vacuum. With decreasing oxygen content the 470 and 540 cm^{-1} peaks initially decrease strongly in intensity. Then the peak at 570 cm^{-1} gradually shifts from 567 cm^{-1} towards a higher frequency (580 cm^{-1}). New peaks appear at the higher frequency wing of the 570 cm^{-1} peak, first at 588 cm^{-1} and then the 592 cm^{-1} peak. The latter increases in intensity upon further annealing and eventually becomes dominant. This evolution is quite similar to that with the number of CuO_2 layers.[5] This clearly indicates that the origin of the spectral evolution with the number of CuO_2 layers lies in the variation of interstitial oxygen content.

The evolution of the 470, 540, 570 and 590 cm^{-1} peaks with decreasing oxygen content suggests that all these modes are closely related to the interstitial oxygen. In Hg-based compounds, the interstitial oxygen may create new vibrational modes in two ways. The

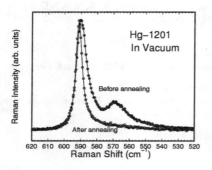 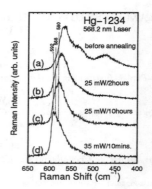

Figure 1: zz-polarized Raman spectra of Hg-1201 before and after local laser annealing with 30 mW laser power for 2 hours in vacuum.

Figure 2: zz-polarized Raman spectra of Hg-1234 measured before and after annealing in vacuum at different stages.

first way corresponds directly to their own vibrations. The second way is for the interstitial oxygen in the HgO$_\delta$ plane to affect the vibration of apical oxygens and thus create frequency-shifted, more or less localized apical oxygen modes.

The laser annealing measurements have demonstrated that the 592 cm^{-1} mode represents the vibration of the apical oxygen without neighboring interstitial oxygens. A careful analysis of the 570 cm^{-1} peak indicates that it seems to be composed of a series of narrower structures centered at 592, 588, 580, 571 and 557 cm^{-1}. The possible decomposition suggests that it is due to frequency-shifted vibrations of the apical oxygens affected by neighboring interstitial oxygens: in this case the interstitial oxygen would tend to lower the vibration frequency of the neighboring apical oxygen. The identified fine structures would then represent vibrations of apical oxygens in different enviroments of neighboring interstitial oxygens.

The 540 and 470 cm^{-1} peaks are found to be strongly correlated and they show different behaviors from the higher energy peaks. They can be assigned to local vibrations of the interstitial oxygens. The 540 cm^{-1} mode is likely to represent the vibration along the c direction while the 470 cm^{-1} mode may correspond to vibrations within the a-b plane.

References

1. Y. T. Ren *et al*, Physica C **217** 273 1993.
2. N. H. Hur *et al*, Physica C **218** 365 1993.
3. M. C. Krantz *et al*, Phys. Rev. B **50** 1165 1994.
4. I. S. Yang *et al*, Physica C **222** 386 1994.
5. M. Cardona, X. J. Zhou and T. Strach, these proceedings.

II. EXPERIMENTS

Magnetic

MAGNETIC AND STRUCTURAL PROPERTIES AND PHASE DIAGRAMS OF $Sr_2CuO_2Cl_2$ AND LIGHTLY-DOPED $La_{2-x}Sr_xCuO_{4+\delta}$

D.C. JOHNSTON, F. BORSA[a], J.H. CHO[b], L.L. MILLER, B.J. SUH, D.R. TORGESON[c]
Ames Laboratory and Department of Physics and Astronomy,
Iowa State University, Ames, Iowa 50011, USA

P. CARRETTA, M. CORTI, A. LASCIALFARI
Dipartimento di Fisica "A. Volta", Universita di Pavia, I-27100 Pavia, Italy

R.J. GOODING, N.M. SALEM, K.J.E. VOS
Department of Physics, Queen's University, Kingston, Ontario, Canada K7L 3N6

F.C. CHOU
Center for Materials Science and Engineering,
Massachusetts Institute of Technology, Cambridge, MA 02139, USA

The evolution of the structural and magnetic properties with doping are reviewed for the systems $Sr_2CuO_2Cl_2$, $La_2CuO_{4+\delta}$ and $La_{2-x}Sr_xCuO_4$. In the first system, a Heisenberg to XY-like crossover is found upon cooling towards the Néel temperature $T_N = 256$ K. In the second system, macroscopic phase separation occurs for $0.01 < \delta < 0.06$, and superconducting compounds occur with $\delta > 0.06$. In the third system, in the antiferromagnetic region $0 < x < 0.02$ the effective spins of the localized (below ~ 30 K) doped holes freeze into a spin-glass state below $T_f \approx (815\ K)x$, whereas at higher temperatures the doped holes appear to become delocalized and segregate into stripes separating undoped domains.

1 Introduction

The evolution of the magnetic and structural properties of the insulating parent cuprate compounds as they are doped into the metallic and superconducting state continues to be an important area of research. Understanding this evolution and its relationship to the microscopic distribution of the doped holes may offer insights into interactions important in the superconducting materials. In addition, the nature of this evolution in strongly correlated electron systems is interesting in its own right. In this article, we will briefly review some of the recent progress made in attempting to address this issue in the La_2CuO_4-type systems. We will start with a brief review of the magnetic properties of undoped La_2CuO_4, augmented by recent observations on the related system $Sr_2CuO_2Cl_2$. The structural and magnetic properties of the oxygen-doped $La_2CuO_{4+\delta}$ and Sr-doped $La_{2-x}Sr_xCuO_4$ systems will then be discussed in turn.

---

[a] Also at Dipartimento di Fisica "A. Volta", Universita di Pavia, I-27100 Pavia, Italy
[b] Present address: Research Center for Dielectric and Advanced Matter Physics, Pusan National University, 30 Jangjun-Dong, Kumjeong-Ku, Pusan 609-735, South Korea
[c] Deceased

2 La$_2$CuO$_4$ and Sr$_2$CuO$_2$Cl$_2$

As is well-known, La$_2$CuO$_4$ is an antiferromagnetic (AF) insulator which exhibits nearly ideal two-dimensional (2D) Heisenberg magnetic behavior above the Néel temperature $T_N \approx 325$ K,[1] as measured by neutron scattering techniques. However, interpreting the magnetic data in the vicinity of T_N is problematic because of the large Dzyaloshinskii-Moriya anisotropy term in the spin Hamiltonian resulting from the tetragonal to orthorhombic transition at $T_o \approx 530$ K.[2] This in turn causes a sharp peak in the magnetic susceptibility χ at T_N, which is not expected for the Heisenberg model.

The compound Sr$_2$CuO$_2$Cl$_2$, which contains the same CuO$_2$ layers as in La$_2$CuO$_4$, remains tetragonal down to 10 K, exhibits AF ordering at $T_N \approx 250$ K,[3,4] and offers the opportunity of examining the magnetic properties of undistorted CuO$_2$ layers. The primary anisotropy is an XY-type anisotropy, so to leading order the spin Hamiltonian is given by

$$\mathcal{H} = J\sum_{i,j}(\mathbf{S}_i \cdot \mathbf{S}_j - \alpha_{XY} S_i^z S_j^z) \qquad (1)$$

where $\alpha_{XY} = 1.4 \times 10^{-4}$ and $J/k_B = 1450$ K.[5] We carried out high precision ^{35}Cl NMR and $\chi(T)$ measurements on single crystals to explore anisotropy effects near T_N.[6] The ^{35}Cl nuclear spin-lattice relaxation rate ($2W$) measurements for the field parallel to the c-axis, $H \parallel c$, showed a divergence upon approaching $T_N = 256$ K from above. Since dynamic scaling should hold near T_N, one expects $2W_c \propto \xi_\perp^{z-\eta}$, where $\xi_\perp$ is the 2D AF correlation length in the CuO$_2$ planes and z and η are critical exponents. Above ~ 300 K, we find an exponential behavior $2W_c \propto \xi_\perp^{z-\eta} \propto \exp[(2310\text{ K})/T]$ as expected for the 2D square-lattice Heisenberg antiferromagnet, whereas upon closer approach to T_N, $2W_c \propto \xi_\perp^{z-\eta} \propto \exp[3080\text{ K})/(T-231\text{ K})]^{1/2}$, consistent with expectation for XY-like behavior. Thus, the system shows a crossover from 2D Heisenberg to 2D XY-like behavior upon approaching T_N from above.

3 La$_2$CuO$_{4+\delta}$

The parent compound La$_2$CuO$_4$ can be doped into the metallic state either by anion or cation doping. Very soon after the beginning of the high-T_c field, it was found that excess oxygen can be inserted at moderate temperatures under oxygen atmosphere, with the amount inserted increasing with increasing oxygen pressure.[7] Later, a novel room-temperature electrochemical oxidation technique was discovered which allowed precise control over the oxygen content.[7] Samples prepared using this latter technique were examined by neutron diffraction, and the phase diagram for $\delta < 0.06$ was determined.[8] A miscibility gap is found for the approximate range $0.01 < \delta < 0.06$ at low T, in which the T_N remains constant at ≈ 250 K. This T_N is due to the oxygen-poor phase in the phase-separated samples. In addition, superconductivity is observed in the adjoining oxygen-rich phase with $T_c \approx 32 - 34$ K. The details of the phase diagram at higher doping levels $\delta > 0.06$ are at present unclear. A series of "staged" compounds is found, in which the excess oxygen atoms are located in layers which are separated by n (stage $n = 2$ to 7) CuO$_2$ layers,[9] but the compositions and T_c's of these phases are as yet unknown.

4 La$_{2-x}$Sr$_x$CuO$_4$

In the La$_{2-x}$Sr$_x$CuO$_4$ system, T_o decreases nearly linearly with x from 530 K for $x = 0$ and goes to zero at $x \approx 0.21$. The T_N decreases much more rapidly and disappears by $x = 0.02$. Bulk superconductivity occurs for $0.06 < x < 0.27$. Our interest here is in the AF regime $0 \leq x < 0.02$. Previous ^{139}La NQR measurements indicated that below T_N, the effective doped-hole spins freeze below a temperature $T_f = (815 \text{ K})x$, and that this doped-hole spin-glass state coexisted with the long-range AF order of the Cu spin system.[10] Theoretical calculations indicated that the effective spins of the doped-holes corresponded to the transverse components of the nearby Cu spins, and that these are actually the entities which freeze below T_f.[11] In the calculations, the doped holes were assumed to be localized at temperatures near T_f, in agreement with inferences from the NQR data.

In our more recent work, the static local internal magnetic fields below $T_N(x)$ were determined both by ^{139}La NQR measurements and by μSR measurements.[12] Both measurements of the internal field were found to track each other versus both T and x, indicating that both measurements are probing the sublattice magnetization $M^\dagger(x,T)$. Above ~ 30 K, $M^\dagger(T)$ decreased uniformly with x. However, below this T, $M^\dagger(x, T = 0)$ becomes nearly independent of x. We attribute this anomalous behavior below ~ 30 K to localization of the doped holes below this T.[12]

The $M^\dagger(x,T)$ data above 30 K were fitted by power law behaviors, and extrapolated to $T = 0$ to obtain the zero temperature values that $M^\dagger(x)$ would have exhibited in the absence of hole localization. We then modeled the system assuming that the mobile doped holes formed 1D stripes in the CuO$_2$ planes which effectively decoupled adjacent undoped domains from each other. After including a weak interplanar coupling to induce 3D AF order, and solving for the resulting $M^\dagger(x, T = 0)/M^\dagger(0,0)$ ratio using conventional spin-wave theory, excellent agreement with the values extrapolated from above 30 K was obtained. We conclude that the hole-stripe model is a viable candidate for explaining the observed data above 30 K. Further support for this physical model was recently obtained from nonlinear sigma model and renormalization group calculations,[13] where the predicted $M^\dagger(x, T = 0)/M^\dagger(0,0)$ and $T_N(x)/T_N(0)$ ratios were both found to be in agreement with the data.

5 Concluding Remarks

Macroscopic phase separation occurs in the La$_2$CuO$_{4+\delta}$ system, which is allowed by the high mobility of the excess oxygen atoms. We infer from comparison of experiment with theory that a related inhomogeneous doped-hole distribution occurs in the lightly-doped La$_{2-x}$Sr$_x$CuO$_4$ system above ~ 30 K; here macroscopic phase separation does not occur presumably because the Sr ions are immobile and their random positions are quenched in during synthesis. Although such phase separation and doped-hole inhomogeneities have been widely predicted to occur by an electronic mechanism,[14] further quantitative microscopic predictions are needed for comparison with experiment. Other important questions are whether and to what extent the interactions responsible for the phase separation or doped-hole inhomogeneities at low doping levels are important to the T_c and other physical properties in the metallic compositions.

Acknowledgments

Ames Laboratory is operated for the U.S. Department of Energy by Iowa State University under Contract No. W-7405-Eng-82. The work at Ames was supported by the Director for Energy Research, Office of Basic Energy Sciences. The work at Pavia was supported by Instituto Nazionale Fisica della Materia. One of us (R.J.G.) acknowledges support by NSERC of Canada.

References

1. For a review, see R.J. Birgeneau and G. Shirane, in *Physical Properties of High Temperature Superconductors I*, ed. D.M. Ginsberg (World Scientific, Singapore, 1989), pp. 151-211.
2. For a review, see D.C. Johnston, *J. Magn. Magn. Mater.* **100**, 218 (1991).
3. L.L. Miller *et al*, *Phys. Rev.* B **41**, 1921 (1990).
4. D. Vaknin *et al*, *Phys. Rev.* B **41**, 1926 (1990).
5. M. Greven *et al*, *Phys. Rev. Lett.* **72**, 1096 (1994); *Z. Phys.* B **96**, 465 (1995).
6. B.J. Suh *et al*, *Phys. Rev. Lett.* **75**, 2212 (1995).
7. For a review, see D.C. Johnston *et al*, in *Phase Separation in Cuprate Superconductors*, ed. E. Sigmund and K.A. Müller (Springer-Verlag, Heidelberg and Berlin, 1994), pp. 82-100.
8. P.G. Radaelli *et al*, *Phys. Rev.* B **49**, 6239 (1994).
9. B.O. Wells *et al*, *Z. Phys.* B (to be published).
10. F.C. Chou *et al*, *Phys. Rev. Lett.* **71**, 2323 (1993).
11. R.J. Gooding, N.M. Salem and A. Mailhot, *Phys. Rev.* B **49**, 6067 (1994).
12. F. Borsa *et al*, *Phys. Rev.* B **52**, 7334 (1995).
13. A.H. Castro Neto and D. Hone, preprint (1996).
14. See, e.g., V.J. Emery and S.A. Kivelson, *Physica* C **209**, 597 (1993).

ANISOTROPY PROPERTIES OF HIGH-T_c MICROCRYSTALS BY A MINIATURIZED TORQUEMETER

C. ROSSEL, P. BAUER

IBM Research Laboratory, Zurich Research Laboratory, 8803 Rüschlikon, Switzerland

D. ZECH, J. HOFER, M. WILLEMIN, H. KELLER

Physik-Institut der Universität Zürich, 8057 Zürich, Switzerland

J. KARPINSKI

Laboratorium für Festkörperphysik, ETH Zürich, 8093 Zürich, Switzerland

The torquemeter is an ideal tool to investigate the pinning and anisotropy properties of high-T_c superconductors. For this purpose we have developed a miniaturized torquemeter based on Si piezoresistive cantilevers with a high sensitivity of $\Delta \tau \simeq 10^{-14}$ Nm. Magnetic moments as small as $m \simeq 10^{-14}$ Am^2 can be measured in a field of 1 T on microcrystals of mass ≤ 1 μg. Angular torque data on $Hg_1Ba_2Ca_3Cu_4O_{10}$ are presented.

1 Introduction

The power of torquemagnetometry in the study of the anisotropic magnetic properties of the high-T_c cuprates has been widely demonstrated.[1-5] Investigation of the properties of the magnetic phase diagram — pinning, melting of the vortex lattice, role of anisotropy and thermal fluctuations — is of fundamental interest for itself and in view of technical applications. High-quality single crystals are necessary to determine such fundamental parameters as coherence length ξ, penetration depth λ, and effective mass anisotropy ratio γ. The fact that the best single crystals are usually the smallest places stringent conditions on the sensitivity of the measuring devices. We have thus developed a new miniaturized and highly sensitive torque magnetometer based on Si piezoresistive cantilevers and used it to investigate the magnetic properties of microcrystals of $Hg_1Ba_2Ca_3Cu_4O_{10}$ and $Bi_2Sr_2Ca_1Cu_2O_8$.[6,7]

2 Principle of the piezoresistive torquemeter

The device is based on the accurate measurement of the deflection Δz of a silicon piezoresistive cantilever produced by the torque $\vec{\tau} = \vec{m} \times \vec{B}$ generated by the magnetic moment m of a microcrystal glued to its free extremity. The measurements are taken by increasing the magnetic field B or by rotating

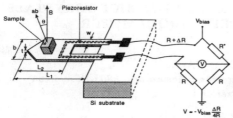

Figure 1: Schematic of piezoresistive torquemeter. R* can be replaced by a bare lever for drift compensation.

it with respect to the a-b plane. We used levers of thickness 4 μm, length $L_1 \simeq 165$ μm and width $b \simeq 90$ μm fabricated by Park Scientific Instruments.[8] As shown in Fig. 1, the change in piezoresistance ΔR of the lever due to its bending is measured via a Wheatstone bridge. The sensitivity, given by the percentage change in resistance per Å of deflection Δz, can be measured or calculated based on its geometry and its elastic properties.[6] It is found to be $R^{-1}\Delta R/\Delta z = \beta \pi_L E t(L_1 + L_2)/2(L_1^3 - L_2^3) \simeq 9 \cdot 10^{-7}$ Å^{-1}, where π_L is the longitudinal piezoresistive coefficient for $\langle 110 \rangle$ Si, E the Young's modulus, $\beta \leq 1$ a correction coefficient and L_1, L_2, and t the dimensions of the lever displayed in Fig. 1. As described in more detail in Ref. 6, the lowest measurable torque is given by the accuracy of measuring the deflection. In principle the limiting factors are the various noise sources (thermal Johnson noise, $1/f$ noise) of the piezoresistor and measuring circuit. With appropriate care (temperature stability, band width etc.) and a differential measurement setup, a resolution in Δz of 0.1 Å is achievable,[9] which corresponds to a torque sensitivity of $\Delta \tau \simeq 2 - 3 \cdot 10^{-14}$ N/m at room temperature. This allows the measurements of magnetic moments of $m \simeq 10^{-14}$ Am^2 in a field of 1 Tesla, which is about three orders of magnitude smaller than in commercial SQUID magnetometers.

3 Effective mass anisotropy of HgBa$_2$Ca$_3$Cu$_4$O$_{10}$ (Hg-1234)

Angular-dependent torque measurements were performed on a high-quality single crystal of Hg-1234 ($T_c^{\text{onset}} = 113$ K) at several temperatures below T_c.[7] This crystallite had dimensions of $60 \times 60 \times 10$ μm^2 and a mass of only $m \simeq 380$ ng. One example of the torque signal $\tau(\theta)$, measured at $T = 109.77$ K and $B = 0.5$ T, is shown in Fig. 2. The torque monitored for increasing and decreasing angles was found to be fully reversible over the entire measured angular range ($\pm$ 90°), except close to the a-b plane ($\theta \leq 0.4°$) where a small hysteresis curve with a pronounced peak at $\theta = 0.1°$ is observed. The shape of the reversible magnetic torque $\tau(\theta)$ can be analyzed by means of the 3D anisotropic continuum London model[10] through the equation

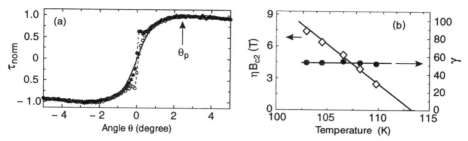

Figure 2: (a) Angle-dependent torque of a Hg-1234 microcrystal at $T = 109.77$ K and $B = 0.5$ T. θ: angle between direction of field B and a-b plane. Solid line: fit with Eq. (1). (b) Temperature dependence of anisotropy γ and critical field ηB_{c2}^c.

$$\tau(\theta) = \frac{\Phi_0 V B}{16\pi\mu_0 \lambda_{ab}^2} \frac{\gamma^2 - 1}{\gamma} \frac{\sin 2\theta}{\epsilon(\theta)} \ln \frac{\gamma \eta B_{c2}^c}{B\epsilon(\theta)} \qquad (1)$$

where $\epsilon(\theta) = \sqrt{\cos^2(\theta) + \gamma^2 \sin^2(\theta)}$ is the angular scaling function, B_{c2}^c the upper critical field along the c-axis, Φ_0 the flux quantum and η a parameter close to unity (here $\eta \simeq 0.7$). By fitting the data with Eq. (1), the anisotropy parameter $\gamma = 51.4(5)$ and the upper critical field $\eta B_{c2}^c = 2.43(6)$ T can be derived. The anisotropy is independent of temperature in the 10 K range below T_c with a mean value of $\gamma = 52$, and $B_{c2}(T)$ is linear with a slope $\eta(dB_{c2}^c/dT) = -0.80(6)$ T/K. From the extrapolated field $B_{c2}(0) = 0.7 \, (dB_{c2}^c/dT)_{T_c} \cdot T_c$, the in-plane coherence length $\xi_{ab}(0) = \sqrt{\Phi_0/2\pi B_{c2}^c(0)} = 1.8$ nm can be derived, in good agreement with previous reports on Hg-based cuprates.

The anisotropy value of Hg-1234 fits well with the general trend of increasing 2D character as one moves from YBCO-123 ($\gamma = 5$-8), and YBCO-124 ($\gamma = 12$) to BSCCO-2212 ($\gamma = 150$-400). This trend is also visible in the position of the respective irreversibility/melting line of these four compounds, as plotted in Fig. 3. Larger anisotropy induces larger thermal fluctuations as described by the position fluctuations of vortices $\langle u^2(T) \rangle \propto ((T\lambda_{ab}^2)/\sqrt{B}) \cdot \gamma$. In fact the position and shape of the vortex lattice melting line is usually determined by the Lindemann criterion $\langle u^2(T) \rangle \simeq c_L^2 a_0^2$, and the transition line takes the form $B_m(T) \propto (c_L^4 \Phi_0^2/\gamma^2 T_c^2)(T_c/T - 1)^2$ in the intermediate-field regime.[11] Here $c_L \simeq 0.1 - 0.4$, and $a_0 = \sqrt{\Phi_0/B}$ is the intervortex distance. The melting field $B_m(T)$ is thus lowered through the increasing anisotropy γ.

The deviations of the experimental reversible torque signal from the 3D London model close to the $a - b$ plane (Fig. 2) can be explained[7] appropriately by the 2D Lawrence–Doniach model,[12] which takes into account the discrete na-

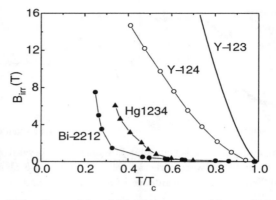

Figure 3: Irreversibility lines of four high-T_c cuprates of increasing anisotropy γ.

ture of the vortex lines (pancakes) due to the weak interlayer coupling present in this material.

4 Conclusion

With one example, it is shown that torque magnetometry can be successfully performed on very small, high-quality crystals by using piezoresistive cantilevers. This technique opens the door to many detailed studies, for example on the nature and order of the melting transition line[13] or on the fourfold symmetry of d-wave superconductivity in the a-b plane.[14]

1. D.E. Farrell et al., Phys. Rev. Lett. **61**, 5158 (1989).
2. L. Fruchter and I.A. Campbell, Phys. Rev. B **40**, 5158 (1989).
3. B. Janossy et al., Physica C **170**, 22 (1990).
4. J.C. Martinez et al., Phys. Rev. Lett. **69**, 2276 (1992).
5. D. Zech et al., submitted to Phys. Rev. B (1995).
6. C. Rossel et al., submitted to J. Appl. Phys. (1995).
7. D. Zech et al., Phys. Rev. B – Rapid Commun. (in press, 1996).
8. Park Scientific Instruments, Sunnyvale, CA 94089, USA.
9. M. Tortonese et al., Appl. Phys. Lett. **62**, 834 (1993).
10. V.G. Kogan et al., Phys. Rev. B **38**, 7049 (1988).
11. G. Blatter et al., Rev. Mod. Phys. **66**, 1125 (1995).
12. L.N. Bulaerskii, Phys. Rev. B **44**, 910 (1991).
13. E. Zeldov et al., Nature **375**, 373 (1995).
14. T. Ishida et al., preprint (1996).

ELECTRONIC STUCTURE AND MAGNETIC PROPERTIES OF $RENi_2B_2C$(RE=Pr, Nd, Sm, Gd, Tb, Dy, Ho, Tm, Er)

Z. ZENG, DIANA GUENZBURGER, E. M. BAGGIO-SAITOVITCH
Centro Brasileiro de Pesquisas Físicas
Rua Dr. Xavier Sigaud 150, 22290-180, Rio de Janeiro, RJ, Brasil

D.E. ELLIS
Department of Physics and Astronomy and Materials Research Center,
Northwestern University, Evanston, IL 60208, U.S.A.

ABSTRACT

The electronic stucture and magnetic properties of $RENi_2B_2C$ (RE=rare earth) are studied using first principles density-functional theory within the embedded cluster model. Spin-polarization of the conduction electrons by the RE moments is examined in detail, and related to the interplay between superconductivity and antiferromagnetic order.}

1. Introduction

The recent discovery of superconductivity in rare-earthquaternary compounds $RENi_2B_2C$ (RE = rare-earth)[1] has inspired much work regarding the interplay of superconductivity and magnetism. Although the compounds $RENi_2B_2C$ can be made across the whole RE series, only for the later rare-earths they are superconducting[2].

The compounds in which magnetism coexists with superconductivity are for RE = Dy, Ho, Er and Tm with antiferromagnetic (AFM) order among the RE-C layers.

We have performed self-consistent spin-polarized first-principles electronic structure calculations for large embedded clusters (73 atoms) representing the compounds $RENi_2B_2C$ (RE = Pr, Nd, Sm, Gd, Tb, Ho and Tm). The Discrete Variational method (DVM) was employed, as described extensively in the literature[3,4]. The local density approximation was used, with the exchange-correlation potential derived by von Barth-Hedin[5].

2. Results

The calculated spin magnetic moments μ on the RE atom are 2.40 for Pr, 3.62 for Nd, 5.83 for Sm, 6.65 for Gd, 3.16 for Ho and 0.83 for Tm (in μ_B). Most of this moment is due to the 4f orbital, although small contributions from the 5d (predominantly), 6s and 6p are present. The small 5d moments align ferromagnetically with 4f, and the 5p antiferromagnetically. The 6s moments are negligibly small or zero in all cases. A small moment develops on the C atoms by polarization; this is negative for the lighter rare-earths up to Gd, then turns positive for Ho and Tm. Small positive moments are present on B; on Ni the moments are zero due to the magnetic symmetry.

A pronounced difference in the conduction electrons spin polarization among the different rare earth compounds could be seen from the spin density, which is spatially more extended for lighter rare earths. The AFM polarization of the C atoms in lighter rare earth compounds is characterizing a superexchange-type interaction between rare earth and C. For heavier rare earth compounds like $HoNi_2B_2C$ and $TmNi_2B_2C$, the spin moments on C change sign. The polarization on B is very small.

The polarization field created by rare-earth spins along lines joining neighbor atoms in the sequence between the two RE–C layers (antiferromagnetic layers) RE↑–C–B–Ni–B–C–RE↓, for all compounds is investigated. Even Pr, which has a considerably smaller 4f spin moment ($2.37\mu_B$) than Ho ($3.14\mu_B$), polarizes the conduction electrons more effectively than the latter, due to a larger radius of the 4f orbital. Furthermore Sm ($\mu(4f)=5.79\mu_B$) polarizes the C atoms considerably more than Gd which has a larger 4f moment ($6.57\mu_B$). In fact, the C polarization due to Gd scales with that due to Nd, with only $3.59\mu_B$ for the 4f spin moment, but with a larger extension of 4f orbital. Hybridization between the 4f and 5d orbitals of the RE, which is more pronounced for the earlier rare-earths, was found to be an inportant factor to explain these differences.

In view of these results, we propose that the differences found in the spin polarization of the conduction electrons along the Lanthanide series in the $RENi_2B_2C$ compounds are relevant in explaining the coexistence or not of magnetism and superconductivity.

Acknowledgments

Calculations were performed at the Cray YMP of the Supercomputing Center of the Universidade Federal do Rio Grande do Sul. The work of DEE was supported in part by the MRL program of the Materials Research Center of Northwestern University, award #DMR-9120521.

References

1. R.J. Cava, et al., Nature, **367**, 252 (1994).
2. R.J. Cava, et al., Phys. Rev. B **49**, 12 384 (1994).
3. D.E. Ellis and G.S. Painter, Phys. Rev. B**2**, 2887(1970); D.E. Ellis, Int. J. Quant. Chem. Suppl. **2**, 35 (1968).
4. D. Guenzburger and D.E. Ellis, Phys. Rev. B **51**,12519 (1995), and references therein.
5. U. von Barth and L. Hedin, J. Phys. C **5**, 1629 (1972).

THE MAGNETIC PROPERTIES OF THE QUATERNARY INTERMETALLICS RNiBC (R=Er, Ho, Tb, Gd, Y)

Julio C. T. Mondragón, E. M. Baggio-Saitovich
CBPF, R. Xavier Sigaud 150, 22290-180, Rio de Janeiro, RJ. Brazil

M. El Massalami
Intituto de Física, UFRJ. caixa postal 68.528 Rio de Janeiro, RJ, Brazil

ABSTRACT

The structural and magnetic features of the RNiBC series of intermetallic compounds (R=Y, Er, Ho, Tb and Gd) were studied by XRD, resistivity and magnetic measurements, their properties are compared with those of the corresponding RNi_2B_2C compounds. The a-parameter reflects the lanthanide contraction (similar to RNi_2B_2C) while c-parameter remains almost constant (different of RNi_2B_2C). This structural features reflect into different magnetic behavior.

1. Introduction

The structure of the intermetallic series RNiBC (R=Y, rare earth) is isomorphous to LuNiBC[1] wherein the nonmagnetic Ni_2B_2 sheets are alternately stacked on double NaCl-type RC-layers, one more RC-layer that on the RNi_2B_2C series[1]. Thus, the physical features of the derived RNiBC are expected to be drastically modified. In this communication, we study the magnetic and transport properties of the RNiBC (R=Er, Ho, Tb, Dy, Gd) and compare their behavior with the corresponding RNi_2B_2C compounds. For to study the system RNiBC we prepared the samples by standard arc-melting method[2] and characterized by Co-Kα X-ray powder diffraction, four point dc resistivity (1.2 K<T<300K) and ac-susceptibility (500 Hz, 1 Oe, 1.5K<T<100K).

2. Results and discussion

Despite the samples were not single phase, as also seen in ac-susceptibility data (Fig.2), their structural features show that on the insertion of larger R, the a-parameter and the unit cell volume are increased, however the c-parameter remains quasi-constant.

The resistivity data for R= Er, Ho, Dy, Tb, Gd (see Fig. 1 for typical examples) reveals a metallic behavior at high temperatures. No superconductivity is observed, as expected, due to the low density of state at the Fermi level[4] and pair breaking effects. On approaching T_N, the resistivity shows a T-independent saturation features (except for R=Gd which was communicated before[6]). The transition temperatures T_C obtained from the magnetic susceptibility, except for R=Gd, Tb, seem to scale with de Gennes factor ErNiBC is collinear ferromagnetic along c-axis with T_C=4.5K[5]) in contrast to the RNi_2B_2C series (where the de Gennes scaling of T_N is strictly obeyed[3]).

Although T_N values of the other magnetic members, e.g. R=Tm, are needed for a valid generalization, we may speculate that these deviation from de Gennes-type behavior

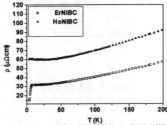

Fig.1 R.-vs-T for ErNiBC and HoNiBC reveals typical metallic behavior up to 100K in both cases, and an anomaly at 4.7 K for Er.

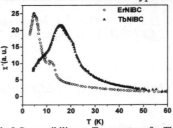

Fig.2 Susceptibility-vs-Temperature for TbNiBC ErNiBC with T_N about 18K for Tb and 4.7K for Er, with a clear Curie Weiss behavior.

(beyond R larger than Dy) is attributed to a weakening in the interlayer interactions (which determine the strength of T_N) due to the separation between the RC layers. The T_c values for Ho and Er are in agreement with recent neutron diffraction measurements[5].

In summary, it is found that the insertion of an extra RC layers in the structure of the quaternary RNi_2B_2C series modifies drastically the structural, transport and magnetic features of the derived RNiBC compounds. As an example, the monotonic decrease of the a-parameter and the monotonic increase, of the c-parameter and the monotonic increase of T_N are not observed in RNiBC (strictly followed in $RNi_2B_2C^3$).

The difference in the metric behavior of c-parameter is related to the opposite influence of the lanthanide contraction on the basic structural building blocks: on increasing the R-size, the width of RC is increased while that of the Ni_2B_2 is decreased. On the other hand, it is argued that the variation in the spatial separation of the RC layers leads to a weakening of the RKKY-type interlayer interactions and thus destroy the monotonic increase of T_N with the de Gennes factor.

3. References

1. R.J. Cava, H. Takagi, H.W. Zandbergen, J.J. Krajewski, W.P.Peck,.T. Siegrist, B. Batlogg, R.B. van Dover, R.J. Felder, K. Mizuhashi, J.O. Lee, H. Eisaki, S. Uchida, *Nature* **367** (1994) 252.
2. T.E. Grigereit, J.W Lynn, Q. Huang, A. Santoro, R.J. Cava, J.J. Krajewski, and W.F. Peck, *Phys. Rev. Lett.* **73**(1994) 2756..
3. M. El Massalami, S.L. Bud'ko, B. Giordanengo, E.M. Baggio-Saitovitch, *Physica C* **244** (1995) 41.
4. L. F. Mattheis, *Phys. Rev. B* **49**(1994)
5. L.J.Chang, submitted to *J. Phys. Condensed Matter*.
6. M. El Massalami, B. Giordanengo, J. Mondragón, E.M. Baggio-Saitovitch, A. Takeuchi, J. Voiron and A. Sulpice, *J. Phys. Cond. Matter*, **7**(1995)10015.

MÖSSBAUER STUDIES ON OXYANIONS SUBSTITUTED RELATED Y-Ba-Cu-O SYSTEM

ANGEL BUSTAMANTE DOMINGUEZ
Universidad Nacional Mayor de San Marcos, Facultad de CienciasFísicas, Apdo. Postal 14-0149, Lima - Perú.

R.B. SCORZELLI and E. BAGGIO-SAITOVITCH
Centro Brasileiro de Pesquisas Físicas, Rua Dr. Xavier Sigaud 150, Urca, CEP 22290-180, Rio de Janeiro, Brazil.

ABSTRACT

The partial replacement of copper by CO_3^{2-}, BO_3^{3-}, SO_4^{2-} and PO_4^{3-} oxyanions in the superconducting system $YBa_{2-x}Sr_xCu_3O_{7-\delta}$ occurs mainly on the Cu(1) sites and can stabilize the Ba free phase $YSr_2Cu_3O_{7-\delta}$ at normal conditions. Here we report on the results obtained by X-ray diffraction and Mössbauer Spectroscopy (MS) studies of samples containing BO_3^{3-}, SO_4^{2-} and PO_4^{3-} group and doped with 0.3% ^{57}Fe.

1. Introduction

Superconducting cuprates containing oxyanions of carbon, phosphour, sulphur, boron and other elements were synthesized recently[1]. According to the structural studies, ligth elements substitute copper atoms in chains of Y-Ba-Cu-O crystal structure forming oxyanion groups. The introduction of the oxyanions, which occur on the "chain sites", stabilize the Ba free phase $YSr_2Cu_3O_{7-\delta}$, that can only be prepared at high pressures[2].

X-ray powder diffraction analysis shows that these compounds are isostructural to the superconductor $YBa_2Cu_3O_{7-\delta}$ showing a reduction of the lattice parameters due to the incorporation of the oxyanion. In the present paper we report, on the results obtained by X-ray diffraction (XRD) and Mössbauer studies on three representative samples containing (SO_4^{2-}), (PO_4^{3-}) and (BO_3^{3-}) group and doped with 0.3% ^{57}Fe:

S2FEN= $[Y_{0.84}Sr_{0.16}]Sr_2Cu_{2.77}Fe_{0.01}(SO_4)_{0.22}O_{6.12}$

P2FEN= $[Y_{0.70}Ca_{0.30}]Sr_2Cu_{2.78}Fe_{0.01}(PO_4)_{0.21}O_{6.16}$

B2FEN= $[Y_{0.85}Ca_{0.15}](SrBa)\{Cu_{2.49}Fe_{0.01}(BO_3)_{0.50}\}O_{5.50}$

2. Discussion

The X-ray structural analysis showed that the samples S2FEN and P2FEN have orthorhombic (Pmmm) structure with lattice parameters: a=3.8266Å, b=3.8429Å and c=11.2638Å, showing a reduction of the c/a and c/b values as compared to the

Y-Ba-Cu-O system, while for the sample B2FEN (which contains Ba) showed a tetragonal (P4/mmm) structure with lattice parameters a=3.8469Å and c=11.2449Å.

The figure 1 shows the MS at room temperature (RT) for the samples S2FEN, P2FEN and B2FEN. All spectra were fitted with four doublets, corresponding to Fe species in different local environments. Species A (QS=1.98mm/s, IS=0.04mm/s), B (QS=1.20mm/s, IS=0.03mm/s), C (QS=0.30mm/s, IS=0.30mm/s), D (QS=1.56mm/s IS=-0.18mm/s) and F (QS=0.55mm/s, IS=0.30mm/s) are similar to species observed in the Y-Ba-Cu-O system[3,4]. The hyperfine parameters of the species that can be correlated with the incorporation of the oxyanions (D', H and G) are listed in table 1 and can be considered as a fingerprint of the stabilized Ba free phase $YSr_2Cu_3O_{7-\delta}$ containing oxy-anion.

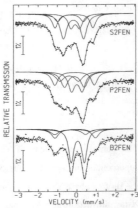

Fig. 1 Mössbauer spectra at RT for samples S2FEN, P2FEN and B2FEN.

Table 1.

[57]Fe Mössbauer parameters for samples S2FEN, P2FEN and B2FEN at RT. IS = Isomer Shift relative to iron metal (mm/s), QS = Quadrupole Splitting (mm/s), Γ = Linewidths (mm/s), A = Relative Areas (%).

Samples	S2FEN-(D')				P2FEN-(H)				B2FEN-(G)			
	IS	QS	Γ	A	IS	QS	Γ	A	IS	QS	Γ	A
	-0.097	1.33	0.40	13	-0.166	0.97	0.54	20	0.181	1.54	0.50	12

References
1. P. Slater and C. Greaves, *Physica* C **215**, 191 (1993), and references therein.
2. B. Okai, *Japanese Journal of Applied Physics*, **29**, L2180 (1990).
3. E. Baggio-Saitovitch, I. Souza Azevedo, R.B. Scorzelli, H. Saitovitch, S.F. da Cunha, A.P. Guimarães,P.R.Silva and A. Takeuchi, *Phys. Rev.* B **37**, 7967 (1988).
4. P. Boolchand, and D. McDaniel, *Hyperfine Interactions*, **72**,125 (1992).

MAGNETIC PROPERTIES OF SOME HIGH-TEMPERATURE SUPERCONDUCTORS

C. Y. HUANG AND J. G. LIN
Center for Condensed Matter Sciences, National Taiwan University, Taipei, Taiwan, R.O.C.

P. H. HOR, R. L. MENG, AND C. W. CHU
*Texas Center for Superconductivity at University of Houston
Houston, Texas 77204-5932*

In this paper, we have reviewed the early results of our magnetic measurements for the early-made high-temperature superconducting oxides, including Y-Ba-Cu-O, La-Sr-Cu-O, and La-Ba-Cu-O. It is interesting to note that many magnetic properties (such as paramagnetic Meissner effect) observed in the poor-quality polycrystaline samples are consistent with those in the high-quality single crystals measured recently, indicating the close relevence between some magnetic properties and magnetic impurities or weak -links.

In the early stage of the discovery of high temperature superconductors (HTS) in 1986,[1] most materials were impure polycrystaline samples. Later, the quality of samples were much improved, made into single crystals, films, or even superlattices. However, many salient magnetic properties were shown to have already appeared in poor-quality samples in comparison with those demonstrated later in high-quality samples. In this paper, we will give some examples to address this point.

Table 1 lists our data of coherence lengths (ξ) and effective penetration depths (λ_{eff}), extracted from our early measurements,[2-4] In comparison with that obtained by other groups later,[5] it is apparent that our data for early-made HTS are consitent with those for better samples.

Table 1. The list of our data obtained in 1987

Sample	ξ(nm)	λ_{eff}(nm)	Reference
$(La_{0.9}Sr_{0.1})_2CuO_4$	4.1	64	2
$(La_{0.9}Ba_{0.1})_2CuO_4$	6.7	---	3
$Y_{1.2}Ba_{0.8}CuO_4$	14	---	4

Figure 1 shows[6] the magetization (M) with respect to temperature (T) at various cooling fields (H_{cool}) for $Y_{1.2}Ba_{0.8}CuO_4$, which contains only $\sim$ 30% of $YBa_2C_3O_{6+\delta}$ and mainly the insulating green phase, Y_2BaCuO_4. Clearly, M is diamagnetic right below T_{cm} ($\approx$ 90 K) at all H_{cool}. The 30-G data (denoted by solid circles) show a jump from a negative value at 87 K to a positive value at 85 K and M increases with decreasing T. For H_{cool} = 40 G, the switching takes place at 83 K. As H_{cool} increases, the switching temperature (T_s) decreases (Fig. 2). In the inset of Fig. 2, the dependence of T_s on H_{cool} is shown. For H_{cool} > 87 G, the switching disappears. Apart from the switching, the data for warming are identical to those for cooling within experimental errors. As shown in Fig. 2, M at low temperature is much greater than the paramagnetic magnetization above T_{cm}. In our ealier paper,[6] this switching phenomenon and the related paramagnetic Meissner effect (PME) were interpreted to be attributed to spin-glass like frustrated, weakly linked superconducting clusters present in the impure Y-Ba-Cu-O sample.

Recently, similar but less pronounced results of the diamagnetic to paramagnetic magnization-transitions were also observed in Bi-oxide,[7] Tl-oxides,[8] and $YBa_2Cu_3O_7$.[9] The proposed interpretations include: (1) trapped flux,[8] (2) orbital paramagnetic moment due to the

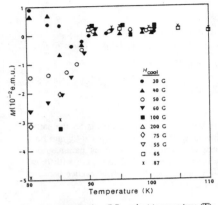

Fig. 1 Magnetization (M) against temperature (T) for $Y_{1.2}Ba_{0.8}CuO_4$ cooled in the magnetic field, H_{cool}.

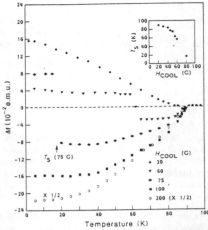

Fig. 2 M vs. T for $Y_{1.2}Ba_{0.8}CuO_4$ cooled in the magnetic field, H_{cool}. The inset shows the dependence of the switching Temperature (T_s) on H_{cool}.

supercurrent,[7] and (3) spontaneous current loop caused by π-junctions.[9] In terms of the π-junction model, when the field or temperature is large enough to destroy π-junctions, the magnization becomes diamagnetic (no longer paramagnetic). At low temperature, the π-junctions recover and the sample is paramagnetic again. This argument is consistent with our M-T data. Some authors assumed the occurrence of the π-junctions, and hence PME, to be related to d-wave superconductivity. In our case, in addition, because of abundant impurities, there could be many magnetic impurities at some grain boundaries, thus leading to the formation of π-junctions.

In summary, we have demonstrated that the PME obtained from our earlier impure samples are even more pronounced than some recent observations employing much better samples. The connection of the positive magnetization below T_c to π-junctions calls for a need to investigate this effect on the samples with strong-magnetic impurites.

References

1. J. G. Bednorz and K. A. Muller, Revs. Mod. Phys. **60**, 585 (1988); M. K. Wu, J. R. Ashburn, C. J. Torng, P. H. Hor, R. L. Meng, L. Gao, Z. J. Huang, Y. Q. Wang and C. W. Chu., Phys. Rev. Lett. **58**, 908 (1987).
2. E. Zirngieble, J. D. Thompson, C. Y. Huang, P. H. Hor, R. L. Meng, C. W. Chu, and M. K. Wu, Appl. Phys. Commun. **7**, 1 (1987).
3. E.Zirngiebl, J.O.Willis, J. D. Thompson, C. Y. Huang, J. L. Smith , Z. Fisk, P. H. Hor, R. L. Meng, C. W. Chu and M. K. Wu, Solid State Commun. **63**, 721 (1987).
4. P.H.Hor, R. L. Meng, J. Z. Huang, C. W. Chu and C. Y. Huang, Appl. Phys, Commun. **7**, 129 (1987).
5. K. G. Vandervoort, U. Welp, J. E. Fessler, H. Claus, G. W. Crabtree, W. K. Kwok A. Umezawa, B. W. Veal, J. W. Downey, and A. P. Paulikas, Phys. Rev. B **43**, 13042 (1991).
6. P. H. Hor, R. L. Meng, C. W. Chu, M. K. Wu, E. Zirngiebl, J. D. Thompson and C. Y. Huang, Nature 326, 669 (1987) .
7. W. Braunisch, N. Knauf, V. Fataev, S. Neuhausen, A. Grutz, A.Kock, B.Roden, D. Fh0mskii, and D. Wohlleben, Phys. Rev. Lett. **68**, 1908 (1992).
8. F. J. Blunt , A. R. Perry, A. M. Campbell, and R. S. Liu , Physica C **175**, 539 (1991).
9. S. Riedling, G. Brauchle, R. Lucht, F. Röherg, and H. V. Lohneysen , Phys. Rev. B **49**, 13283 (1994).

MÖSSBAUER STUDIES OF $RE_{1.85}Sr_{0.15}CuO_4$ T'-PHASE

ADA LÓPEZ, M.A.C. DE MELO, D. SÁNCHEZ, I. SOUZA AZEVEDO, E. BAGGIO-SAITOVITCH
Centro Brasileiro de Pesquisas Físicas, Rua Xavier Sigaud, 150
22290-180 Rio de Janeiro - Brazil

F.J. LITTERST
TU - Braunschweig-Mendelssohnstr 3
38106 -Braunschweig -Germany

^{57}Fe Mössbauer studies have been performed in $^{57}Fe:RE_{1.85}Sr_{0.15}CuO_4$ samples (RE= Gd, Eu, Nd, and Pr) at room temperature and 4.2K. Several iron species are observed corresponding to different oxygen coordination. Moreover Cu/Fe spin dynamic was detect, which can be explained by frustration processes.

1. Introduction

Many studies of high-T_C superconductors have focused on the modification of the magnetic properties as the number of carriers is varied. The $(La_{1-x}Gd_x)_{1.85}Sr_{0.15}CuO_4$ system changes from a superconducting to a paramagnetic and finally to an antiferromagnetic material with increasing Gd concentration, the local $Cu-O_2$ units being an octahedron (T-phase), a pyramid (T*-phase), and a square (T'-phase), respectively[1]. We have studied the Mössbauer spectra of the T'-phase using ^{57}Fe, ^{119}Sn and ^{151}Eu isotopes as probe in the samples $RE_{1.85}Sr_{0.15}CuO_4$, where RE are the rare-earth Gd, Eu, Pr and Nd. Here we will focus mainly on the magnetism and the spin dynamic in the T'-phase observed by ^{57}Fe Mössbauer spectroscopy.

Polycrystalline samples have been prepared according to the standard solid state reaction method and the quality control was done by powder X-ray diffraction. The Mössbauer spectra have been recorded at 295 and 4.2K temperatures on all samples.

2. Results and Discussion

The room temperature (RT) ^{57}Fe Mössbauer spectra on Gd and Eu samples indicate that there are three sites present, presumably associated with the *ideal* T* fivefold pyramidal site (C), and the *ideal* T'fourfold site (E) and an interstitial oxygen site (D)[2]. The very small relative intensities of C sites (ca. 10% of all intensity) show that the samples are in the T'-phase, as confirmed from our X-ray diffraction experiments, which show only a single phase. The relative intensities of D and E Mössbauer subspectra are rare-earth and iron concentration dependent (see Figure 1).

At 4.2K no clear magnetic hyperfine pattern is found for Gd and Eu samples doped with 1% ^{57}Fe, only rather a broad line. The spectral lineshapes can be described by relaxation processes due to spin flip-flop. The relaxation rates γ and the hyperfine magnetic field B_{hf} are shown in Table 1. The Gd samples with 2 and 3 % ^{57}Fe show the same spectral lineshapes. For samples above 5% ^{57}Fe fully developed magnetic hyperfine patterns are formed which yield an angle θ between B_{hf} and the electric field gradient (Table 1), i.e. the moments are canted. This different behavior observed for the samples with high iron concentration indicates that the relaxation effects in the samples with low iron concentration are not due to Fe-Fe, but rather to Fe-Cu magnetic interactions.

The relaxation effects observed in $RE_{1.85}Sr_{0.15}CuO_4$ samples are seen in terms of frustration of an antiferromagnetic order in the Cu planes caused by the Cu-Cu ferromagnetic interaction by magnetic O^-. The presence of a magnetic oxygen ion between two adjacent Cu ions provides an effective ferromagnetic interaction between the Cu ions, that frustrates the antiferromagnetic order. The competing exchange interactions lead to the observed spin flip-flop of the iron moments as observed by µSR in the $La_{2-x}Sr_xCuO_4$ system[3]. These frustration effects are not caused by the onset of the three dimensional interactions corresponding to the magnetic Gd ions between the Cu-O planes, since these same frustration effects are also observed in this system with nonmagnetic Eu ions.

For the samples with RE=Nd and Pr different Mössbauer spectra are observed. Our X-ray diffraction results show a small abundance of a different phase, apparently, formed from RE and Fe. Attempts are being done to improve the sample preparation in order to clear out this point.

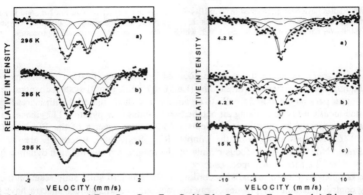

Fig. 1- Mössbauer spectra: a) $Eu_{1.85}Sr_{0.15}Cu_{0.99}Fe_{0.01}O_4$, b) $Gd_{1.85}Sr_{0.15}Cu_{0.99}Fe_{0.01}O_4$ and c) $Gd_{1.85}Sr_{0.15}Cu_{0.95}Fe_{0.05}O_4$

Table 1.- Mössbauer hyperfine parameters (* θ = 45°).

Sample	T (K)	D site			E site		
		ΔEq (mm/s)	B_{hf} (Tesla)	γ (1/s)	ΔEq (mm/s)	B_h (Tesla)	γ (1/s)
$Eu_{1.85}Sr_{0.15}Cu_{0.99}Fe_{0.01}O_4$	295	0.70	-	-	1.70	-	-
	4.2	0.90	45.0	8.1×10^8	1.70	34.0	4.3×10^7
$Gd_{1.85}Sr_{0.15}Cu_{0.99}Fe_{0.01}O_4$	295	0.70	-	-	1.70	-	-
	4.2	0.90	45.0	8.1×10^8	1.70	34.0	4.3×10^7
$Gd_{1.85}Sr_{0.15}Cu_{0.95}Fe_{0.05}O_4$	295	0.90	-	-	1.70	-	-
	15	0.90*	49.0	$<10^6$	1.70	34.0	9.7×10^7

References
1. Gang Xiao, Marta Z. Cieplak, and C.L. Chien *Phys. Rev.* **B 40** (1989) 4538.
2. E.Baggio-Saitovitch, M.A.Márquez, D.R.Sanchez and S.G.Garcia *Physica C* **235-240** (1994) 865.
3. G.M.Luke L.P.Le, B.J.Sternlieb, Y.J.Uemura, J.H.Brewer, R.Kadono, R.F.Kiefl, S.R.Kreitzman, T.M.Riseman, C.E.Stronach, M.R.Davis, S.Uchida, H.Takagi, Y.Tokura, Y.Hidaka, T.Murakami, J.Gopalakrishnan, A.W.Sleight, M.A.Subramanian, E.A.Early, J.T.Markert, M.B. Maple, and C.L.Seaman ., *Phys. Rev* **B42**, (1990) 7981.

OBSERVATION OF A PAIR-BREAKING FIELD AT THE Ni SITE IN NON-SUPERCONDUCTING RENi$_2$B$_2$C

D.R. Sánchez, M.B. Fontes, S.L. Bud'ko, E. Baggio-Saitovitch
Centro Brasileiro de Pesquisas Físicas
Rua Xavier Sigaud 150 Urca, CEP 22290-180. Rio de Janeiro, Brazil.

and

H. Micklitz
Physikalisches Institut, Universität zu Köln, Germany.

ABSTRACT

A transferred magnetic hf field is observed in TbNi$_2$B$_2$C (T<15K) and HoNi$_2$B$_2$C (5.5K>T>4.2K) resulting from the non collinear antiferromagnetic spin structure of the RE moments. However, no magnetic hyperfine field (hf) is observed at the ^{57}Fe nucleus in the superconducting compounds DyNi$_2$B$_2$C and ErNi$_2$B$_2$C down to 4.2K. We conclude that this transferred magnetic hf field (B_{hf}) acts as a pair-breaking field at the Ni planes.

1. Introduction

The rare earth (RE) nickel borocarbides are intermetallic layered compounds which offer the possibility to study the interplay between superconductivity and magnetism. Mössbauer spectroscopy on ^{57}Fe doped RENi$_2$B$_2$C, in principle, offers the possibility to use ^{57}Fe as a local probe. Since it is well established that Ni has no magnetic moment in RENi$_2$B$_2$C, it is very likely that Fe sitting at a Ni-site has also no magnetic moment.

Polycrystalline samples of RE(Ni$_{0.99}$^{57}Fe$_{0.01}$)$_2$B$_2$C, RE= Tb, Dy, Ho, Er, were prepared by argon arc melting and characterized by X-ray diffraction and AC susceptibility. ^{57}Fe Mössbauer effect (ME) spectra were taken with the RENi$_2$B$_2$C samples in a variable temperature helium cryostat and the ^{57}Co:Rh source at room temperature.

2. Results and Discussion

The Mössbauer spectra for all the samples in the paramagnetic state consist of a main doublet with a RE dependent quadrupole splitting, reflecting the structural changes. Temperature dependent Mössbauer spectra show a B_{hf} below T_N only for HoNi$_2$B$_2$C (5.5K>T>4.2K) and TbNi$_2$B$_2$C (T<15K) samples.

In Fig. 1 we show T-dependent spectra corresponding to HoNi$_2$B$_2$C sample: starting with a quite symmetric spectrum at T~5.5K, the spectra change their shape for 5.3K>T>4.2K and recover the symmetric shape for T=3K. The temperature dependence of B_{hf} for TbNi$_2$B$_2$C and HoNi$_2$B$_2$C is given in Fig. 2.

The observation of a B_{hf} at Fe (Ni) site in the RENi$_2$B$_2$C is due to the non cancellation of the transferred hiperfine field induced by the four nearest RE neighbors.

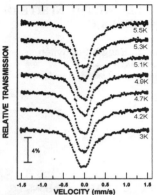

FIG. 1. ^{57}Fe Mössbauer spectra of HoNi$_2$B$_2$C in the temperature region 6K>T>3K

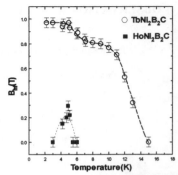

FIG. 2. Temperature dependence of magnetic hf field B_{hf} at the ^{57}Fe nucleus in TbNi$_2$B$_2$C and HoNi$_2$B$_2$C as obtained from least-square fittings to the spectra as shown in Fig. 1.

In the case of DyNi$_2$B$_2$C the simple collinear antiferromagnetic spin structure[1] will result in a zero transferred B_{hf}. An incommensurate modulated spin arrangement is observed: along the a (b)-axis for ErNi$_2$B$_2$C [2], and along the c-axis as well as along the a (b)-axis (6K≥ T ≥4.7K) for HoNi$_2$B$_2$C [3]. However, a transferred B_{hf} is observed only in the HoNi$_2$B$_2$C ^{57}Fe:ME measurements.

For TbNi$_2$B$_2$C the magnetic spin structure, to our knowledge, has not yet been determined. However, a weak ferromagnetic behavior was already observed for T<8K [4]. This phase transition is also detected by our Mössbauer measurement as a slope change in the plot of B_{hf}-T near 8K (see fig. 2).

We observe a B_{hf} in the case of TbNi$_2$B$_2$C below T_N~15K (see fig. 2) and for HoNi$_2$B$_2$C between T_{N1}~6K and T_{N2}~4.7K. At the same time it is known that superconductivity does not appear at all in TbNi$_2$B$_2$C and disappears for HoNi$_2$B$_2$C in the temperature region T_{N1}>T>T_{N2} (reentrance behavior).

Taking these two facts together it is very reasonable to assume that the observed B_{hf} at the ^{57}Fe nucleus acts as a pair-breaking field at the Ni site which leads to the suppression of superconductivity in some of the RENi$_2$B$_2$C compounds.

3. References

1. P. Devernagas, J. Zarestky, C. Stassis, A.I. Goldman, P.C. Canfield, and B.K. Cho, *Physica B* **212**, 1 (1995).
2. J. Zarestky, C. Stassis, A. I. Goldman, P.C. Canfield, P. Devernagas, B.K. Cho, and D.C. Johnston, *Phys. Rev.* **B51**, 678 (1995).
3. A. I. Goldman, C. Stassis, P.C. Canfield, J. Zarestky, P. Devernagas, B.K. Cho, and D.C. Johnston, *Phys. Rev.* **B50**, 9668 (1994).
4. B. K. Cho, P.C. Canfield, and D.C. Johnston (to be published in *Phys. Rev. B*).

II. EXPERIMENTS

Flux Dynamics

COMPARATIVE STUDY OF VORTEX CORRELATION IN TWINNED AND UNTWINNED $YBa_2Cu_3O_{7-\delta}$ SINGLE CRYSTALS

F. de la CRUZ, D. LÓPEZ, E. F. RIGHI, G. NIEVA*.
Centro Atómico Bariloche, 8400 S. C. de Bariloche, R. N., Argentina

Vortex correlation in the c crystallographic direction in YBCO single crystals has been studied by means of transport measurements. The solid-liquid phase transition in untwinned single crystals is followed by a discontinuous change from a vortex structure that is correlated in the c direction to a non correlated liquid in all directions. Correlated vortex liquids in the c direction are only possible when extended defects such as twin boundaries stabilize a finite temperature-dependent vortex correlation in the field direction. In this case the liquid can be characterized by vortex lines with a diverging temperature dependent length.

The originally proposed existence of new phase transitions [1] in the $H-T$ phase diagram of the high temperature superconductors has received strong experimental and theoretical support [2]. On the other hand, the nature of the ground state of the vortex solid phase and the characteristics of the liquid state in the oxide superconductors are expected to be determined by the anisotropy of the superconductor, the disorder introduced by thermal fluctuations [3] and by the particular characteristics of the pinning potentials [4].

In this paper we review transport measurements using the transformer contact configuration in twinned and untwinned $YBa_2Cu_3O_{7-\delta}$ (YBCO) single crystals [5]. The results show qualitatively different characteristics of the liquid and solid vortex structures in clean and twinned samples. The presence of twin boundaries not only changes the nature of the transition from the solid to the liquid state but modifies the correlation function of the vortex structure in the c direction.

Plotted in Fig. 1 are the top and bottom resistances defined as V_{top}/I and V_{bottom}/I, where I is the injected current, for both a twinned and an untwinned sample. The applied field in the c direction is 40 kOe. Similar results have been obtained for other crystals and fields [6]. The sharp drop of the resistance of the untwinned sample ($\Delta T(10\% - 90\% \simeq 200mK$) at $T_m(H)$ has been associated [7] with a melting transition. In agreement with previous results [7] $T_m(H)$ is the temperature below which the $I-V$ characteristics are non linear. Thus, above $T_m(H)$ the response is Ohmic and $R_{top} > R_{bottom}$, whereas below $T_m(H)$ $R_{top} = R_{bottom}$ [6,8] for currents higher than the critical current in the ab planes. This is shown in Fig. 2 where we have plotted $\Delta V = V_{top} - V_{bottom}$ as a function of the applied current in the close vicinity of $T_m(H)$.

The data in Fig. 1 and Fig. 2 demonstrate that for $T > T_m(H)$ there is no vortex velocity correlation in the c direction. The non linear behavior for $T < T_m(H)$ and the results in both figures make evident that the vortex pinning is effective once the solid is formed and the phase correlation in the c direction is established. The differences in the behavior of the twinned and untwinned samples can be seen from the comparison made in Fig. 1. In the case of twinned samples the $I-V$ characteristics in the ab plane are linear for $T > T_i(H)$. For practical purposes $T_i(H)$ is the temperature below which $\rho_{ab} = 0$. For $T > T_i(H)$ there is a vortex liquid [5] with velocity correlation in the c direction [5] up to

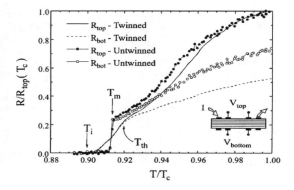

Figure 1: R_{top} and R_{bottom} normalized by $R_{top}(T_c)$ as a function of reduced temperature for a twinned and an untwinned crystal $30\mu m$ and $35\mu m$ thick respectively, in an applied magnetic field of 40 kOe parallel to the c-axis. Inset: electrode configuration.

a temperature $T_{th}(H)$. That is, below T_{th} the vortex system is superconducting in the c direction while dissipative in the ab planes, while above T_{th} the vortex system is dissipative in all directions.

It is interesting to emphasize the qualitatively different vortex response in the two types of samples. Figure 1 shows that the disorder introduced by twin boundaries lowers the temperature [6,8] where the vortices become pinned, $T_i(H)$, with respect to the solidification temperature in the clean sample ($T_i < T_m$). It is also evident that the c-axis phase correlated liquid in the twinned sample persists above T_m, where the liquid in the clean sample is already phase uncorrelated in the out-of-plane direction. This shows that the correlated pinning potential [4] of the twin boundaries localizes vortex lines on length scales longer than the thickness of the sample for $T < T_{th}$. On the other hand, the disorder induced by the same potential transforms the character of the phase transition and depresses the temperature at which the vortex system becomes pinned. [a] In this sense the data strongly suggest that the twin boundaries stabilize a disentangled vortex state [6,8,5], where the long-range topological order of the vortex position is suppressed.

The liquid in the twinned samples at $T < T_{th}$ is stiffer than that of the clean ones at the same reduced temperatures. Thus, vortices become localized in the quasi-two-dimensional twin potential. As we will show later, the liquid or solid in the twinned samples is less stiff than the crystalline vortex solid at the same reduced temperatures, despite the presence of the correlated twin potential localizing the vortices in the c direction. Within this picture the collective vortex response at a given reduced temperature against the non homogeneous Lorentz force is stronger in the crystalline solid than either in the c-axis correlated liquid or

---

[a]Simulations in three-dimensional Josephson networks have shown that the transition at T_m changes from first order to a continuous one at $T_i < T_m$ with the introduction of "twin boundaries" (E. Jagla and C. Balseiro, unpublished).

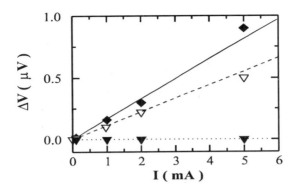

Figure 2: $\Delta V = V_{top} - V_{bottom}$ as a function of current at different temperatures near T_m. The external field is 30 kOe. Solid diamonds: 85.87 K; open triangles: 85.74 K; solid triangles: 85.15 K. For this sample $T_m(30\ kOe) = 85.50 K$.

in the disordered solid, for the same reduced temperature. This has been demonstrated by measuring the current necessary to induce flux cutting[5] J_{cut} in twinned samples ($T < T_{th}$) and in the untwinned ones for $T < T_m$ (it should be realized that $J_{cut} = 0$ for $T > T_m$). We found[6] that J_{cut} changes discontinuously at T_m, as can be inferred from Fig. 2, while it decreases continuously to zero[5] at T_{th}. In fact, J_{cut} is so large in the crystalline vortex solid that it exceeded the maximum current ($\sim 20\ mA$) we were able to feed without sample heating. The different behavior of J_{cut} has important consequences in the behavior of the electrical resistance in the c direction[6] in both type of samples.

It is surprising to see that $V_{top} = V_{bottom}$ at $T_m(H)$. We did not expect this behavior in a first order transition. On the other hand if the difference in voltage dropped to zero continuously down to T_m it would mean that the first order phase transition could be triggered by the establishment of phase correlation in the c direction. The other possibility (the most likely one as will be shown) is that there is a finite voltage difference at $T_m(H)$ too small to be detected within our experimental sensitivity. To solve this problem we measured samples of different thicknesses. In Fig. 3 we compare the sample thickness dependence of the two temperatures, T_i and T_{th} , characterizing the twinned samples and T_m for the untwinned ones. We see that while T_{th} is thickness dependent[8] (see Fig. 3a) T_m is independent of the sample thickness, as observed in Fig. 3b. Measurements[6] of V_{top} and V_{bottom} in a thick untwinned sample made evident that ΔV drops to zero at T_m from a finite value.

In summary, the results of Fig. 3 show that the phase coherence in the c direction at the first order phase transition is established discontinuously. As a consequence, coherence in the ab and c directions is established at the same thickness independent temperature T_m. The thickness dependence of T_{th} can be interpreted as a result of a vortex correlation length in the liquid state diverging at $T_i(H)$. When the correlation length coincides with the

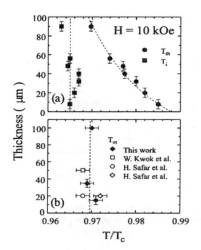

Figure 3: Thickness dependence of the characteristic temperatures discussed in the text: (a) T_{th} and T_i; (b) T_m. The solid lines are guides to the eye.

thickness of the sample V_{top} becomes equal to V_{bottom}. The results in this paper and those in Ref. 3 strongly support that in an infinite sample of twinned material the vortex solid has a Bose glass ground state[4]. In the clean case the first order transition is associated with the simultaneous establishment of phase coherence in the ab and c directions, in agreement with recent theoretical work[9].

References

[*] Member of CONICET Argentina.
1. P. L. Gammel *et al.*, Phys. Rev. Lett. **61**, 1666 (1988).
2. G. Blatter *et al.*, Rev. Mod. Phys. **66**, 1125 (1994).
3. D. H. Huse, M. P. A. Fisher, and D. S. Fisher, Nature **358**, 553 (1992).
4. D. R. Nelson and V. M. Vinokur, Phys. Rev. B **48**, 13060 (1993).
5. F. de la Cruz, D. López, and G. Nieva, Philos. Mag. B **70**, 773 (1994); H. Safar *et al.*, Phys. Rev. Lett. **72**, 1272 (1994).
6. D. López *et al.* (unpublished); D. López *et al.*, this conference
7. W. K. Kwok *et al.*, Phys. Rev. Lett. **69**, 3370 (1992); H. Safar *et al.*, Phys. Rev. Lett. **69**, 824 (1992); W. K. Kwok *et al.*, Physica B **197**, 579 (1994).
8. D. López *et al*, Phys. Rev. B (in press).
9. D. Domínguez, N. Grønbech-Jensen, and A. R. Bishop, Phys. Rev. Lett. 75, 4670 (1995); R. Šášik and D. Stroud, Phys. Rev. Lett. **72**, 2462 (1994); G. Carneiro, Phys. Rev. Lett. **75**, 521 (1995).

First- and Second-Order Vortex-Lattice Phase Transitions in $Bi_2Sr_2CaCu_2O_8$

B. Khaykovich[1], T. W. Li[2], M. Konczykowski[3], D. Majer[1], E. Zeldov[1], and P. H. Kes[2]

[1] *Department of Condensed Matter Physics, Weizmann Institute of Science, 76100 Rehovot, Israel*

[2] *Kamerlingh Onnes Laboratorium, Rijksuniversiteit Leiden, 2300 RA Leiden, The Netherlands*

[3] *CNRS, URA 1380, Laboratoire des Solides Irradiés, École Polytéchnique, 91128 Palaiseau, France*

> Local magnetization measurements are used to study the vortex-lattice phase transitions in $Bi_2Sr_2CaCu_2O_8$ crystals with various oxygen stoichiometry. The first-order phase transition line at elevated temperatures shifts upward for more isotropic over-doped samples. The local measurements at lower temperatures reveale another sharp, probably second-order transition at the onset of second magnetization peak. The two phase transition lines merge at a multicritical point forming apparently a continuous phase transition line that is anisotropy dependent.

In high-temperature superconductors (HTSC) the nature of the different vortex phases and the thermodynamic transitions between them are of fundamental interest and subject of substantial recent theoretical and experimental efforts [1,2,3,4,5]. It is generally accepted that the high anisotropy plays a crucial role in the richness of the phase diagram of HTSC. We have investigated the effect of anisotropy on the first-order vortex-lattice phase transition in $Bi_2Sr_2CaCu_2O_8$ (BSCCO) that was observed recently as a sharp step in the local magnetization [4]. Another intriguing feature of the phase diagram of many HTSC crystals and BSCCO in particular is the anomalous second magnetization peak at lower temperatures. The associated increase of magnetization with magnetic field has been attributed to surface barrier effects [6], crossover from surface barrier to bulk pinning [7], sample inhomogeneities [8], dynamic effects [9], and 3D to 2D transitions [10]. Our investigation of the local vortex dynamics indicates that a very abrupt upturn in the bulk critical current occurs at the onset of the second magnetization peak in BSCCO. We postulate that this behavior is triggered by an underlying thermodynamic phase transition of the flux-line lattice. Furthermore, for the different anisotropy crystals the two phase transition lines are found to form apparently one continuous transition line that changes from first to possibly second order at a critical point. The existence of such a phase transition at low temperature at flux densities on the order of H_{c1} is unprecedented and calls for new theoretical and experimental studies.

The experiments were carried out on several as-grown BSCCO crystals [11,12] with typical dimensions of $700 \times 300 \times 10 \ \mu m^3$ and on two crystals [12] which were heat treated in order to change the oxygen stoichiometry [13]. The as-grown crystals [11,12] show practically identical phase diagrams. The over-doped crystal (annealed at 500°C in air) has $T_c \simeq 83.5$ K and the optimally-doped crystal (annealed at 800°C) has $T_c \simeq 89$ K. As shown in [13] annealing in air between 500°C and 800°C reversibly changes the oxygen stoichiometry. With increasing oxygen content both the c-axis and T_c decrease linearly with the oxygen concentration. It was further shown [14] that the ratio between the c-axis and ab-plane resistivities at 100 K also depends strongly on the oxygen contents. A significant reduction of ρ_c/ρ_{ab} was reported when going from optimally to strongly over-doped. These results convincingly show that also the anisotropy in BSCCO depends on oxygen contents and the crystal becomes more

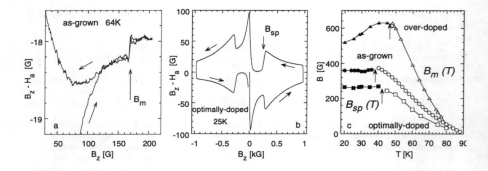

Figure 1: Local magnetization loops $B_z - H_a$ measured in the central part of BSCCO crystals as a function of *local* magnetic induction ((a) and (b)). At elevated temperatures a first-order transition is manifested by equilibrium magnetization step at B_m (a). (b) The second magnetization peak B_{sp} at lower temperature T=25K is shown. (c) First-order transition lines B_m (empty symbols) together with the second-order transition $B_{sp}(T)$ (filled symbols) for over-doped ($\triangle$), as-grown ($\bigcirc$), and optimally-doped BSCCO crystal ($\square$). $B_{sp}(T)$ was defined as the point of the steepest drop of the local magnetization peak on decreasing field as indicated by arrow in Fig. 1b. The arrows indicate the position of the critical point.

isotropic in going from optimally to over-doped. Although oxygen content influences the value of the critical current, in the higher temperature and low field region that is of interest here no observable change in pinning was obtained[13]. The as-grown crystals[11,12] have doping level close to optimal ($T_c \simeq 90$ K), but the exact oxygen stoichiometry is not known due to the large vertical temperature gradient during the growth and *in situ* annealing process[12]. The local magnetization measurements were performed in applied field $H_a||z||c$-axis using arrays of $10 \times 10 \mu m^2$ GaAs/AlGaAs Hall sensors[15].

Figures 1a and 1b illustrate typical local magnetization loops as measured by the sensors. The first-order transition[4] is clearly seen as a small positive sharp step in $B_z - H_a$ at elevated temperatures (Fig. 1a). Figure 1c shows the mapping of the first-order transition as a function of temperature for three crystals of different oxygen stoichiometry, i.e. different anisotropy. We find that the transition line shifts significantly with anisotropy and moves to higher fields for more isotropic over-doped crystals. In addition, in all the crystals the first-order transition terminates abruptly at a sample dependent critical point (indicated by arrows) in the range of 40 to 50 K[4].

We discuss now the behavior at temperatures below the critical point. As demonstrated in Fig. 1b, a large second magnetization peak is observed at fields above $B_{sp}(T)$. Analysis of the field profiles across the sample shows a transition from smooth dome-shaped profiles due to surface and geometrical barriers at low fields to Bean profiles due to onset of significant critical currents at fields above B_{sp}[16]. Several models have been proposed to explain the origin of this enhancement of critical current[8,9,10]. All these models describe crossover processes in which the width of the transition region is on the order of the characteristic field at the transition. In global measurements the observed onset of the second peak is broad since the magnetization builds up gradually with H_a and reaches a maximum only when a fully penetrated new critical state is obtained. In addition, the global measurements average over regions of different values of B, and therefore different values of $J_c(B)$, across the sample. The use of Hall-sensors, on the other hand, lifts this limitation. In the Bean model [17] the field gradient dB_z/dx is proportional to J_c and hence can be used for a direct evaluation of the behavior of J_c. A very sharp drop of local J_c in the second peak region on decreasing H_a was recently demonstrated [5], which occurs whenever the local field reaches $B_{sp}(T)$ regardless of the position. The width of the transition is *locally* very sharp. It is about 5 G $\simeq$ 1.5% of B_{sp} when measured by smaller sensors ($3 \times 3 \mu m^2$) and about 20 G if measured by sensors of $10 \times 10 \mu m^2$. We therefore infer that the sharp change in pinning at the onset of the second peak is driven by an underlying thermodynamic second-order phase transition with a width which is at least two orders of magnitude narrower than the characteristic field B_{sp}. We cannot exclude, though, that this is a weakly first-order transition with a magnetization step that is below our experimental resolution or obscured by the onset of bulk pinning. To the best of our knowledge there is currently no theoretical explanation for the existence of a sharp vortex-lattice phase transition at low temperatures and fields on the order of H_{c1}[1].

Figure 1c shows the second peak line $B_{sp}(T)$ for three crystals along with $B_m(T)$. The position of B_{sp} is strongly anisotropy dependent [10]. We find that it shifts to higher fields to the same extent as the shift of $B_m(T)$. Moreover, the first-order transition terminates at a temperature below which the second magnetization peak develops for each of the samples[5,18]. $B_{sp}(T)$ and $B_m(T)$ merge at two sides of the same, sample dependent, critical point on the $B - T$ phase diagram for the various anisotropies. This finding is a strong indication that the two lines represent in fact one continuous vortex-lattice transition that changes from a first-order to probably a second-order (or a weakly first-order) at a tricritical point as the temperature is decreased. This interpretation is consistent with the fact that a first-order vortex-lattice phase transition is not expected to terminate at a simple critical point and should be followed by a second-order phase transition due to involved symmetry breaking. The amplitude of the second magnetization peak decreases as the temperature is increased, and the peak becomes not very well-defined close to the critical point probably due to rapid relaxation of the bulk current at higher temperatures. This is the reason for the apparent gap in the data in a narrow temperature interval around the critical point (Fig. 1c).

The exact nature of the various vortex phases and the phase transitions in BSCCO is still unresolved. The neutron diffraction data [3] suggest an ordered Abrikosov lattice in the entire low field phase, therefore one possible scenario would be that $B_m(T)$ is a simultaneous melting and decoupling transition into a pancake liquid, whereas $B_{sp}(T)$ is a transition into a decoupled pancake solid. One may thus anticipate existence of another vertical phase transition or crossover line that separates the two high-field phases. Presence of such a vertical depinning line was recently reported [15]. This line shows no sharp features but it seems to terminate at the same critical point as discussed above, which implies existence of a multicritical point. A similar phase diagram was recently proposed on the basis of Monte-Carlo simulations [19].

In conclusion, we find that a weekly pinned vortex lattice undergoes a sharp, probably second-order, phase transition into a strongly pinned state as the field is increased at low temperatures. At elevated temperatures a first-order transition into unpinned liquid state is observed. The two transitions seem to merge into one continuous phase transition line with a multicritical point at intermediate temperatures. This entire phase transition is significantly shifted to higher fields as the anisotropy of BSCCO crystals is reduced.

Acknowledgments

This work was supported by the Ministry of Science and the Arts, Israel, and the French Ministry of Research (AFIRST), by the Glikson Foundation, by Minerva Foundation, Munich/Germany, by contract CT1*CT93-0063 from the Commission of the European Union, and by the Dutch Foundation for Fundamental Research on Matter (FOM).

References

1. for a recent review see G. Blatter et al., Rev. Mod. Phys. **66**, 1125 (1994).
2. H. Safar et al., Phys. Rev. Lett. **69**, 824 (1992); W.K. Kwok et al., Phys. Rev. Lett. **69**, 3370 (1992); W.Jiang et al., Phys. Rev. Lett. **74**, 1438 (1994); S. L. Lee et al., Phys. Rev. Let. **71**, 3862 (1993); H. Pastoriza et al, Phys. Rev. Lett. **72**, 2951 (1994); H. Safar et al., Phys. Rev. Lett. **70**, 3800 (1993).
3. R. Cubbit et al., Nature **365**, 407 (1993).
4. E. Zeldov et al., Nature **375**, 373 (1995).
5. B. Khaykovich et al., Phys. Rev. Lett. (in press).
6. V.N. Kopylov et al., Physica C **170**, 291 (1990).
7. N. Chikumoto et al., Phys. Rev. Lett. **69**, 1260 (1992).
8. M. Daeumling et al., Nature **346**, 332 (1990).
9. L. Krusin-Elbaum et al., Phys. Rev. Lett. **69**, 2280 (1992); Y. Yeshurun et al., Phys. Rev. B **49**, 1548 (1994).
10. G. Yang et al., Phys. Rev. B **48**, 4054 (1993); T. Tamegai et al., Physica C **223**, 33 (1993); K. Kishio et al., in "Proc. 7th Intnl. Workshop on Critical Currents in Superconductors" ed. H.W. Weber, World Sci. Pub., Singapore, p. 339 (1994).
11. N. Motohira et al., J. Ceram. Soc. Jpn. Int. Ed. **97**, 994 (1989).
12. T.W. Li et al., J. Crys. Grow. **135**, 481 (1994).
13. T.W. Li et al., Physica C **224**, 110 (1994); ibid **257** (in press).
14. L. Forro, Phys. Letters A **179**, 140 (1993); Y. Kotaka et al., Physica C **235-240**, 1529 (1994).
15. E. Zeldov et al., Europhys. Lett. **30**, 367 (1995).
16. D. Majer et al., Physica C **235-240**, 2765 (1994).
17. C.P. Bean, Phys. Rev. Lett. **8**, 250 (1962); E. Zeldov et al., Phys. Rev. B**49**, 9802 (1994); E. H. Brandt and M. V. Indenbom, Phys. Rev. B**48**, 12893 (1993).
18. T. Hanaguri et al., Physica C **256**, 111 (1996).
19. R. Sasik and D. Stroud, Phys. Rev. B **52**, 3696 (1995).

ON THE THICKNESS DEPENDENCE OF IRREVERSIBILITY LINE IN $YBa_2Cu_3O_{7-\delta}$ THIN FILMS

PABLO MENEZES and J. ALBINO AGUIAR
Departamento de Física, Universidade Federal de Pernambuco, 50670-901 Recife-PE, Brasil

In this paper we analyse the thickness dependence of the irreversibility line of $YBa_2Cu_3O_{7-\delta}$ thin films. We show that is possible to fit the irreversibility line data with the scalling function $H_{irr}=A(1-t)^\alpha$, where $t=T_{IRR}/T_C$ and A and α are adjustables parameters for all thickness studied. We obtain α values varying from 1.0 to 1.6 when the film thickness changes from 10000 to 200 Å. We argue that surface pinning effects can be responsible for this behavior.

1. INTRODUCTION

One of the most widely studied features of vortex behavior in high-temperature superconductors is the experimentally observed irreversibility line (IL) in the H-T phase diagram[1]. The irreversibility line H_{irr} divides the H-T plane into two different regions: one at high fields and temperatures where characteristic magnetic properties are reversible, and the other at low fields and temperatures where a flux pinning[2] or a surface or edge barrier is active[3], yielding a nonzero critical current and hysteretic magnetic response. A related phenomenon is the large (logarithmic) magnetic relaxation observed below the IL[1]. The large relaxation implies a rapid approach to the reversible equilibrium state which appears above the IL. The irreversibility line is found in single crystals and may be determined by several techniques[3].

In this paper we analyse the thickness dependence of the IL using the data of Civale et al[4] obtained in $YBa_2Cu_3O_{7-\delta}$ thin films. These authors found that thinner films show a depressed irreversibility line. By using a de Almeida-Thouless[5] type of scalling function to fit all the data we argue that this behavior can be attributed to the increase of disorder introduced by surface deffects and/or by strains, which can lead to surface pinning effect.

2. EXPERIMENTAL

In this article we use the data of Civale et al[4] for the IL of thin films. They produced $YBa_2Cu_3O_{7-\delta}$ thin films by laser ablation. The films had thickness between 100 and 10 000Å and were epitaxially grown on $(100)SrTiO_3$ substrates with the Cu-O planes parallel to the substrate surface. X-ray analysis confirm the orientation and purity of the films. The irreversibility line was measured using an ac susceptibility technique. A dc magnetic field was applied parallel to the c axis of the film. The maximum of imaginary part of the susceptibility (χ'') was taken as the definition of the irreversibility line.

3. DISCUSSION

In figure 1 we present the irreversibility line H_{irr} defined by the maximum in χ'' for $YBa_2Cu_3O_{7-\delta}$ laser-ablated films of different thickness as function of 1-t. Here t is the reduced temperature $t=T_{irr}/Tc$. The data were colleted from from the work of L. Civale et al[4]. It is clear that thinner films show a depressed irreversibility line. Note that deppression of $H_{irr}(t)$ starts at films with thickness

~1000 Å. In their work Civale et al[4] discuss the various theoretical scenarios that can used to explain the results, such flux-creep[2], vortex glass[6] order to analyse the data we took the following procedure. We fit the thickness dependence of the IR with a scaling of de Almeida-Thouless[5] type $H_{irr}=A(1-t)^{\alpha}$ when A and α are adjustable parameters, and $t=T_{irr}(H)/Tc(0)$. The fitting curves are show in figure 1.

For the parameter α we obtain value varying from 1.0 to 1.6 when the film thickness changes from 200 Å to 10000 Å. The change in the crossover expoent α associated with the irreversibility line is a indication of the increase of the disorder in the system. As shown by Soares et al[8] for an Ising spin glass in the presence of a gaussian randon field the crossover expoent can assume value between 1/2 and 3/2 depending on the magnitude of the randon field. We believe that in our case we can attribute the origin of this disorder to deffects or strain induced at the film surface, which can lead to surface pinning reducing the irreversibility. Indeed, recently Konczy- kowski et al[3] have shown that surface damage leads to a reduction in the irreversibility in $YBa_2Cu_3O_7$ crystal.

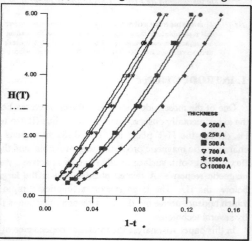

figure 1. Irreversibility line $H_{irr}(1-t)$ for YBa2Cu3O7-δ laser-ablated films of differents thickness. The lines are fitting to the experimental data using the scaling $H_{irr}(t)=A(1-t)^{\alpha}$ as explained in the text.

ACKNOWLEDGMENTS

We thank the Brasilian Agencies CNPq and FINEP for financial support.

REFERENCES

1. K.A. Müller, M. Takashige and J.G. Bednorz, Phys. Rev. Lett. **58**, 1183 (1987).
2. Y. Yeshurum, A.P. Malozenoff, Phys. Rev. Lett. **60**, 2202 (1988).
3. M. Konczykowski et al, Phys. Rev. B **43**, 13707 (1991).
4. L. Civale, T.K. Worthington, A. Gupta, Phys. Rev. B **43**, 5425 (1991).
5. J.R.L. de Almeida and D.J. Thouless, J. Phys. A **11**, 983 (1978).
6. M.P.A. Fisher and D. Huse, Phys. Rev. Lett. **62**, 1415 (1989).
7. D.R. Nelson and H.S. Seung, Phys. Rev. B **39**, 9153 (1989).
8. R.F. Soares et al, Phys. Rev. B **50**, 6151 (1994).

MAGNETO-OPTIC IMAGING OF MELT-TEXTURED YBa$_2$Cu$_3$O$_{6+x}$ BICRYSTALS

MICHAEL B. FIELD, ANATOLY POLYANSKII, ALEX PASHITSKI
AND DAVID C. LARBALESTIER
Applied Superconductivity Center, University of Wisconsin-Madison, 1500 Engineering Drive, Madison, WI, 53706 USA

APURVA PARIKH AND KAMEL SALAMA
Texas Center for Superconductivity and Mechanical Engineering, University of Houston, Houston, TX, 77204, USA

ABSTRACT

Magneto-optic imaging was used to observe the flux penetration in bulk scale 5°[100] and 11°[13 3 -13]YBa$_2$Cu$_3$O$_{6+x}$ grain boundaries. No preferential penetration was seen in the strongly coupled boundary while the mixed-behavior boundary exhibited inhomogeneous penetration.

1. Introduction

Magneto-optic (MO) imaging can identify weak and strong coupling regions in high temperature superconductors[1], as shown here for two melt-textured bicrystals. In earlier work we have shown that the coupling across grain boundaries (GB) may be inhomogeneous[2] and we accordingly studied one strongly coupled, low angle bicrystal (5°[100] misorientation) and one with mixed coupling (11°[13 3 -13]), seeking to see if MO imaging could reveal differences in their behavior.

2. Experimental Details

Samples were made by the melt-out, melt process technique[3] and then thinned by polishing to reveal their grain structure. Individual bicrystals were isolated by laser cutting, and their crystal misorientation determined from x-ray pole figures. Transport measurements of J_c were made and the bicrystals were then MO imaged[2] in a microscope/cryostat system[4] with Bi-doped YIG films placed directly on the polished flat bicrystal surface. The field was applied perpendicular to the sample surface.

3. Results and Summary

Figures 1 and 2 compare the real space surface and MO images of the two bicrystals. No preferential flux penetration was seen at the GB of the 5°[100] bicrystal but there was substantial penetration at cracks (some of which have been shown to contain insulating phases[5]), and larger 211 particles. The J_c data of Figure 3 are consistent with the GB being strongly coupled. The images of the 11°[13 3 -13] bicrystals are much more inhomogeneous and do show that flux penetrates preferentially

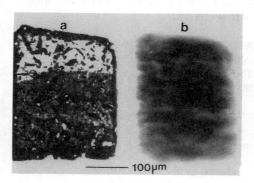
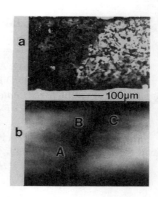

Fig. 1 a) Polarized light micrograph of the 5°[100] bicrystal b) reflected light MO image at 15K in 40mT. C axis is 9° from surface plane. Flux penetrates preferentially along cracks between platelets and at large 211 particles but not at the grain boundary.

Fig. 2 a) Polarized light micrograph of the 11°[13 3 -13] bicrystal b) reflected light MO image at 15K in zero field after field cooling in 40mT. Regions A, B, and C pin flux better than adjacent segments of the boundary

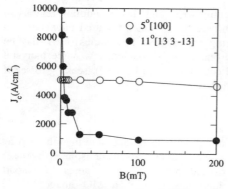

Fig.3 Cross-boundary J_c vs B, 77K

at the boundary, consistent with the stronger field dependence of J_c seen in Fig. 3. There appear to be strong pinning bridges at A, B and C. Correlation of the two images shows that the 211 particles crossing the GB seem to act as weak points. These two images are consistent with an increase in electromagnetic coupling inhomogeneity as the misorientation angle increases.

This work was funded by the NSF-MRG Program DMR-9214707

4. References

1. D.C. Larbalestier, et al. *J. of Metal*, 20, December 1994.
2. M.B. Field, et al., *unpublished*, (1996).
3. A.S. Parikh, et al., *Supercond. Sci. Technol.*, **7**, (1994) 455.
4. V.K. Vlasko-Vlasov, et al., in *Springer Series in Materials Science*, **23** (1993) 111.
5. K.B. Alexander, et al., *Phys Rev* **B**, **45(10)**, (1992) 5622.

VORTEX PHASE TRANSITION IN $Bi_2Sr_2CaCu_2O_y$ SINGLE CRYSTAL

H. IKUTA, S. WATAUCHI, J. SHIMOYAMA, K. KITAZAWA, and K. KISHIO
Department of Superconductivity, University of Tokyo
Hongo 7-3-1, Bunkyo-ku, Tokyo 113, Japan

The ab-plane resistivity and the magnetization under magnetic fields applied along the c-axis direction were measured on a $Bi_2Sr_2CaCu_2O_y$ single crystal. The oxygen contents were carefully controlled and an overdoped single crystal with T_c=78.3 K was prepared for this study. We found distinct step structures in the temperature dependence both of the ab-plane resistivity and the magnetization. The positions of the step structures observed resistively and magnetically coincided well on the H-T diagram.

Over the past few years, the magnetic phase diagram of the mixed state of high-temperature superconductors (HTSC's) has been of much interest. It has been proposed that a solid-to-liquid transition of the flux-line-lattice (FLL) exists near the superconducting transition temperature (T_c). Recently, Zeldov *et al.* have observed a distinct step in the local flux density of a $Bi_2Sr_2CaCu_2O_y$ (BSCCO) single crystal.[1] Because magnetization is a thermodynamic quantity, this result strongly supports the first-order FLL melting picture. On the other hand, there has been no report regarding the corresponding resistive anomaly for BSCCO so far. Therefore, we have measured the resistivity on an overdoped BSCCO single crystal and observed for the first time for BSCCO a step structure in the low-resistivity region, corresponding to the FLL melting transition.

The crystal used in this study was grown by the floating-zone technique as described elsewhere.[2] The as-grown crystal was annealed at 800°C for three days to remove structural inhomogeneity. Electrodes were prepared on the crystal using gold paste and the crystal was fired then at 800°C for 1 hour. Thereafter, the crystal was sealed in a quartz tube with appropriate amount of partial oxygen gas and annealed at 400°C for three days. The partial pressure of oxygen during the heat treatment was about 2.1 atm. After the heat treatment, the crystal was quenched to room temperature. As reported previously, this process results into high-quality samples with oxygen contents corresponding to the carrier-overdoped state.[3] The resistivity was measured by the four-probe AC method along the crystalline a-axis direction using an AC resistance bridge (Linear Research, LR-700). The transport AC current was 3 mA which corresponded to a current density of 52 A/cm^2. The zero-field-cooled (ZFC) and field-cooled (FC) magnetization curves were measured by a SQUID magnetometer (HOXAN HSM-2000X). T_c was determined magnetically to be 78.3 K (onset) and ΔT_c=1.1 K (10-90 %).

Figure 1 shows the temperature dependence of the resistivity and the derivative of the magnetization with respect to the temperature (dM/dT) measured on the same crystal. Due to the correction factor of the magnet of our SQUID system, the applied field was slightly different for resistivity and magnetization measurements as indicated in the figure. It can be seen that there is a rather fast drop of resistivity with decreasing the temperature at the low end of the transition curve as indicated by an arrow-head (T_1). By further decreasing the temperature, the resistivity fell for almost one order of magnitude to the

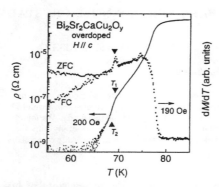

Figure 1: The temperature dependence of the resistivity and the derivative of the magnetization with respect to the temperature measured on the same overdoped BSCCO single crystal. The applied fields were 200 Oe and 190 Oe, respectively.

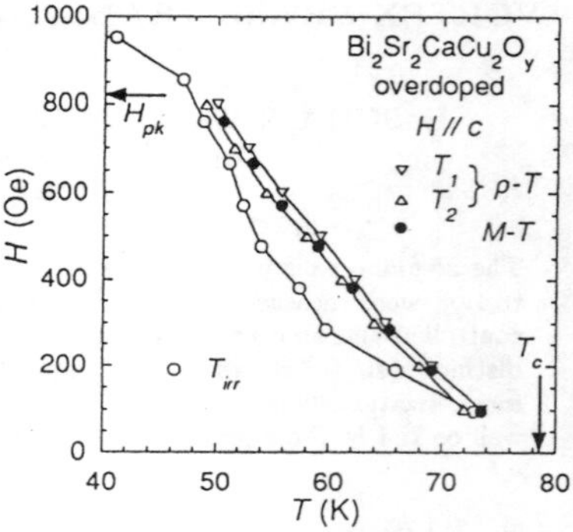

Figure 2: The phase diagram determined from the resistivity and the magnetization data measured on the same overdoped BSCCO single crystal.

point denoted by T_2, and decreased rather slowly again for further temperature decrease. Such a step-like structure at the end of the resistive transition has not been reported for BSCCO in the literature. Furthermore, we have also observed a distinct peak for both ZFC and FC dM/dT curves as shown in the figure. This peak of dM/dT corresponds to a step-like increase of magnetization with temperature and indicates the FLL melting transition.

Figure 2 is the H-T diagram determined from the resistivity and the magnetization data. Bisides the locations of the step-like structures, the irreversibility line is also plotted in the figure. T_c and the location of the so-called "secondary-peak" are shown by the arrows. The figure shows that the melting lines determined from the resistivity and the magnetization measurements coincides noticeably well. We can also see that the structures observed in the magnetization and the resistivity data could be resolved for applied fields up to about 820 Oe, the so-called "secondary-peak" field of this crystal, and that the melting line is well separated from the irreversibility line.

In conclusion, we have measured the resistivity and the magnetization on an overdoped BSCCO single crystal. We have observed a well defined FLL melting transition in both of these measurements. The positions of the melting lines on the H-T diagram determined by both methods coincided quite well.

References
1. E. Zeldov et al., Nature **375**, 373 (1995).
2. N. Motohira et al., J. Ceram. Soc. Jpn. **97**, 994 (1989).
3. J. Shimoyama et al., Proc. 6th US-Japan Workshop on High-T_c Superconductors, (World Scientific, Singapore, 1994) p. 245.

Three dimensional vortex fluctuation in $HgBa_2Ca_{0.86}Sr_{0.14}Cu_2O_{6-\delta}$

Mun-Seog Kim[a] and Sung-Ik Lee
Department of Physics, Pohang University of Science and Technology, Pohang 790-784, South Korea

Seong-Cho Yu
Department of Physics, Chungbuk National University, Cheongju 360-763, South Korea

Nam H. Hur
Korea Research Institute of Standards and Science, Taedok Science Town, Taejon 305-600, South Korea

Abstract
This study measures the temperature dependence of reversible magnetization of grain-aligned $HgBa_2Ca_{0.86}Sr_{0.14}Cu_2O_{6-\delta}$ high-T_c superconductor with external magnetic fields parallel to the c-axis. The fluctuation of vortices was quantitatively interpreted for the first time from the 3-dimensional vortex fluctuation model.

It is well-known that vortex fluctuation effect plays an important role in the magnetization of highly anisotropic superconductors, i.e., the magnetization curves $M(T, H)$ significantly deviate from the prediction of the mean-field theory[2] and are field-independent at the characteristic temperature T^*. These behaviors could be explained by the BLK model, i.e., the entropy contribution to the free energy due to thermal distortions of vortices, suggested by Bulaevskii, Ledvij, and Kogan[1].

This paper reports the experimental results on reversible magnetization for grain aligned $HgBa_2Ca_{0.86}Sr_{0.14}Cu_2O_{6-\delta}$ (Hg-1212). Field-induced magnetic fluctuations are clearly observed near $T_c(H)$, and a crossover of magnetization curves $M(T)$ for various magnetic fields was found at $T^* = 114.5$ K, which indicates the vortex fluctuation effect.

Figure 1 shows temperature dependence of reversible magnetization for various external magnetic fields. The inset of Fig. 1 shows details of $M(T)$ near the crossover temperature. According to Bulaevskii et al., the reversible magnetization is as follows:

$$-M(T) = \frac{\phi_0}{32\pi^2 \lambda_{ab}^2(T)} \ln \frac{\eta H_{c2}}{eH} - \frac{k_B T}{\phi_0 R} \ln \frac{16\pi k_B T \kappa^2}{\alpha \phi_0 R H \sqrt{e}}, \qquad (1)$$

---

[a] Also Department of Physics, Chungbuk National University, Cheongju 360-763, South Korea

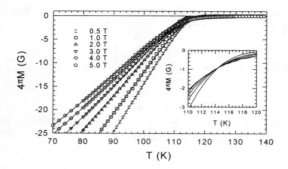

Figure 1: Temperature dependence of reversible magnetization $4\pi M(T)$ with the theoretical curves(solid line) derived from the model of Hao et al.. Inset: $4\pi M$ vs. T near T^*

where $e = 2.718...$, η and α are constants of order unity, R is effective interlayer spacing for quasi-2D system. Even though a superconductor is an anisotropic 3D system, all expressions are retained, if R is replaced by $C\xi_c(T)$, where C is a numerical factor of the order unity[1]. From the 3D-BLK model, the lower limit of coherence length along the c-axis $\xi_c(0) \simeq 2$ Å and the anisotropy ratio $\gamma \leq 7.7$ were obtained, which implies that this sample is anisotropic 3-dimensional superconductor as $Y_1Ba_2Cu_3O_{7-\delta}$. The confident evidence supporting to be 3D system can be found from good 3D scaling behavior of high-field magnetization around $T_c(H)$ as a function of $[T - T_c(H)]/(TH)^{2}$.[3]

Acknowledgments

We wish to express appreciation for the financial support of the Korean Ministry of Education, Basic Science Research Centers of Pohang University of Science and Technology, Agency of Defense Development of Korea, and Korea Science and Engineering Foundation.

References

1. L. N. Bulaevskii, M. Ledvij, and V. G. Kogan, Phys. Rev. Lett. **68**, 3733 (1992).
2. V. G. Kogan *et al.*, Phys. Rev. Lett. **70**, 1870 (1993).

DISCONTINUOUS ONSET OF THE C-AXIS VORTEX CORRELATION AT THE MELTING TRANSITION IN $YBa_2Cu_3O_{7-\delta}$

D. LÓPEZ, E. F. RIGHI, G. NIEVA*, F. DE LA CRUZ
Centro Atómico Bariloche, 8400 S. C. de Bariloche, R. N., Argentina

W. K. KWOK, J. A. FENDRICH, G. W. CRABTREE
Materials Science Division, Argonne Nat. Lab., Argonne, IL 60439

L. PAULIUS
Western Michigan University, Kalamazoo, MI 49008

Transport measurements in untwinned $YBa_2Cu_3O_{7-\delta}$ single crystals using a multiterminal contact configuration have demonstrated that the long-range vortex correlation in the c direction is discontinuously established at the liquid-solid transition. The stiffness of vortices against current-induced cutting becomes nonzero abruptly at the same temperature. These observations mean that in twin-free samples there is no evidence of a disentangled vortex liquid.

Recent transport measurements using the flux transformer configuration [1] in twinned $YBa_2Cu_3O_{7-\delta}$ (YBCO) crystals have shown that longitudinal superconductivity in the vortex liquid phase is lost at a characteristic temperature $T_{th}(H)$ well above the irreversibility line. For temperatures below T_{th} it has been shown [1] that for applied currents less than $I_{cut}(T, H)$ the vortex liquid moves with out-of-plane velocity correlation.

In this paper we show that in clean (untwinned) YBCO crystals the vortex liquid is always uncorrelated. Reducing temperature, we observe that at the melting temperature $T_m(H)$ the elastic properties of the vortex structure change abruptly. Our samples are high-quality YBCO crystals prepared and characterized as described in Ref. 1.

Shown in Fig. 1(a) are the linear-response voltages V_{top} and V_{bottom} as a function of temperature for un untwinned sample $100\mu m$ thick with an external magnetic field of 20 kOe parallel to the c-axis. The fact that the voltages are different on both faces of the crystal is evidence that the vortex motion above T_m (see arrow) has no out-of-plane velocity correlation. Fig. 1(b) is a plot of the temperature dependence of the linear-regime voltage in the c direction V_c for the same sample and magnetic field. The out-of-plane resistance shows a discontinuous drop to zero at the same temperature $T_m(H)$ where the vortex liquid freezes into a solid. The results show an abrupt onset of longitudinal superconductivity at $T_m(H)$, indicating that in clean samples the vortex solid melts into a highly entangled vortex liquid [2]. This result is in agreement with recent theoretical predictions [3]. The sharp drop observed in V_c is consistent with the finite observed difference $V_{top}(T_m) - V_{bottom}(T_m)$.

Measurements of $V_{top}(I)$ and $V_{bottom}(I)$ in the vicinity of T_m show that the current necessary for flux cutting I_{cut} rises discontinuously from zero ($T > T_m$) to a finite value larger than the maximum current we can apply without heating the sample [4]. Since it has been shown [5] that $I_{cut} \propto c_{44}$ (tilt modulus), the jump in I_{cut} means that the solid melts into a liquid with $c_{44} = 0$.

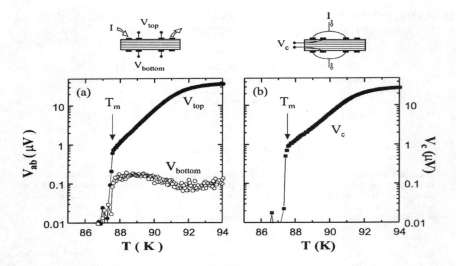

Figure 1: (a) V_{top} and V_{bottom} as a function of temperature for a untwinned $YBa_2Cu_3O_{7-\delta}$ single crystal. (b) Temperature dependence of the voltage in the c direction for the same sample. The applied magnetic field is 20 kOe parallel to the c-axis. The diagram above each figure shows the respective contact configuration.

In summary, the melting transition in clean YBCO crystals separates an ordered vortex solid with finite tilt modulus from an entangled vortex liquid with linear out-of-plane resistance.

References

[*] Member of CONICET Argentina.
1. D. López *et al*, Phys. Rev. B (in press) ; F. de la Cruz, D. López, and G. Nieva, Philos. Mag. B **70**, 773 (1994); H. Safar *et al.*, Phys. Rev. Lett. **72**, 1272 (1994).
2. D. R. Nelson, Phys. Rev. Lett. **60**, 1973 (1988).
3. R. Šašik and D. Stroud, Phys. Rev. Lett. **72**, 2642 (1994); A. Schönenberger, V. Geshkenbein, and G. Blatter, Phys. Rev. Lett. **75**, 1380 (1995); D. Domínguez, N. Grønbech-Jensen, and A. R. Bishop, Phys. Rev. Lett. **75**, 4670 (1995).
4. D. López *et al* (unpublished); F. de la Cruz *et al*, this conference.
5. D. López *et al.*, Phys. Rev. B **50**, 9684 (1994).

RESISTIVE TRANSITIONS OF HTS UNDER MAGNETIC FIELDS: INFLUENCE OF FLUCTUATIONS AND VISCOUS VORTEX MOTION

E.SILVA, R.MARCON
*Dipartimento di Fisica "E.Amaldi" and INFM, Università di Roma Tre,
Via della Vasca Navale 84, 00146 Roma, Italy*

R.FASTAMPA, M.GIURA
Dipartimento di Fisica and INFM, Università "La Sapienza", 00185 Roma, Italy

V.BOFFA, S.SARTI
ENEA CRE-Frascati, 00044 Frascati, Roma, Italy

We present resistive transitions in high magnetic fields (6-14T) in YBCO films. Through a scaling approach we show that superconducting fluctuations are responsible for the resistance drop around T_c. At low temperature, viscous vortex motion, with an activation term diverging at the glass-liquid line, is the main source for the dissipation. A phenomenological model incorporating all those contributions accurately describes the experimental data.

In HTS the broadening of the transitions and the vanishing of the measured resistivity below the irreversibility line are the main dissipative phenomena in magnetic fields. The first feature is usually ascribed to the strong superconducting fluctuations, while the second is often connected to the existance of a frozen configuration of vortices, eventually leading to a vortex-glass phase with zero linear resistivity. In this paper we present measurements of the resistive transitions at high magnetic fields on YBCO films. The data are interpreted within a phenomenological model [1] which takes into account both superconducting fluctuations and viscous vortex motion. The two regimes can be separately identified, thus avoiding possible ambiguities in determining the fitting parameters. The data are described with great accuracy by the model.

We measured resistive transitions in two $YBa_2Cu_3O_{7-\delta}$ films, with the magnetic field applied along the c axis (perpendicular to the film plane). Field strength ranges from 4 to 14 T. 1500 Å thick films are etched on 3 mm-long, 100 μm wide strips. Four-probes, lock-in detection (20 Hz) is used, with 5 nV sensitivity. Current density is 10 A/cm². T_{c0} is chosen as the inflection point of the zero-field resistive transition.

In the employed model, the fluctuation conductivity, $\Delta\sigma(H,T)$, enters into the total resistivity as:

$$\rho_{normal+fluctuation} = \frac{1}{\sigma_n(T) + \Delta\sigma(H,T)} = \frac{\rho_n(T)}{1 + \Delta\sigma(H,T)/\sigma_n(T)} \quad (1)$$

where $\rho_n(T) = 1/\sigma_n(T)$ is the normal-state resistivity. The fluctuation conductivity has a particularly useful expression at high fields, where a lowest-Landau-level (LLL) scaling can apply. In a three dimensional superconductor, with zero-field transition temperature T_{c0} and zero-temperature upper critical field H_{c20}, one can write: [2]

$$\frac{\Delta\sigma(H,T)}{\sigma_n} = \left[\frac{t^2}{h}\right] F_{3D}(x_{3D}) \quad (2)$$

where $x_{3D} = (ht)^{-2/3}(t-1+h)$, $h=H/H_{c20}$, $t = T/T_{c0}$.

The limiting form of the scaling function at $x_{3D} \ll 0$, $F_{3D} \sim -x_{3D}$, is the flux-flow. If the scaling can be performed, one can extract numerically the scaling function, and obtain $\Delta\sigma(H,T)$. In the insert of Fig.1 we show the successful scaling ($T_{c0} = 89.6$ K, $H_{c20} = 140$ T), and we plot the scaling function. In both films we find that a reduced flux-flow, $F_{3D} = -4x_{3D}$, matches better the experimental scaling function.

The vortex motion is introduced through a phase-slip model.[3] One can write down the complete expression for the resistivity:

$$\rho(H,T) = \rho_n(T) \frac{1}{1 + \Delta\sigma(T,H)/\sigma_n(T)} \left[I_0(\gamma(H,T))\right]^{-2} \quad (3)$$

where the activation term is taken to diverge at the vortex-glass transition:[4]

$$\gamma(H,T) = \frac{A(H)}{T - T_g(H)} \left[1 - (T/T_d(H))^4\right]^{1/2} \quad (4)$$

The total fits (Fig.1, main panel) are obtained by the full expression, Eq.(3), where the fitting parameters are (see Eq.(4)) A, T_g, T_d. We obtain that, from 4 to 14 T, $A \approx 200$ K (roughly constant); $T_g(H)$ reproduces the usual shape of the irreversibility line; $T_d(H)$ is linear with H, approaching T_{c0} at zero field.

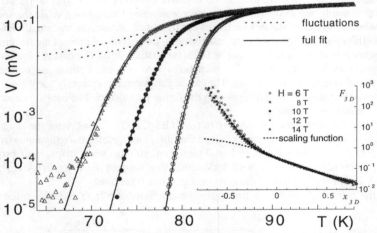

Fig.1 Fluctuational scaling according to Eq.(2) (insert) and fit according to Eqs. (3) and (4). Data on a second sample exhibit the same features.

References

1. S.Sarti, R.Fastampa, M.Giura, E.Silva, and R.Marcon, *Phys. Rev.* B **52**, 3734 (1995)
2. S.Ullah and A.T.Dorsey, *Phys. Rev.* B **44**, 262 (1991)
3. V.Ambegaokar and B.I.Halperin, *Phys. Rev. Lett.* **25**, 1364 (1969)
4. M.Giura, R.Marcon, E.Silva, and R.Fastampa, *Phys. Rev.* B **46**, 5753 (1992)

MEISSNER HOLES IN REMAGNETIZED SUPERCONDUCTORS

V.K.VLASKO-VLASOV[1,2], U.WELP[2], G.W.CRABTREE[2], D.GUNTER[2], V.I.NIKITENKO[1],
V.KABANOV[1], L.PAULIUS[3]

[1] *Institute for Solid State Physics, 142432 Chernogolovka, Russia*
[2] *Argonne National Lab, 9700 South Cass Av., Argonne, IL 60439*
[3] *Western Michigan University, Kalamazoo, MI 49008*

Flux distributions on different faces of YBCO single crystals remagnetized by unidirectional and rotating fields are studied using advanced magneto-optical techniques. Unusual structures corresponding to the appearance of strong current concentration along certain fronts are found. At such fronts the vortex lines bend into closed loops which then collapse and form flux free cylinders. Strong magnetization currents along these Meissner holes result in essential flux redistribution and development of dynamical instabilities in the samples.

Remagnetization processes in superconductors are usually considered within the critical state model where the sample is infinitely long in the field direction, vortices are straight, and the remagnetization front separating oppositely magnetized regions is planar or cylindrical. In this paper we show that in any finite superconductor an inevitable curvature of vortices results in 3D remagnetization structures and formation of flux free cylinders (Meissner holes) which strongly affect the magnetic induction and current distributions and induce instabilities in the vortex motion.

Flux distributions on different faces of YBCO crystals were imaged using iron-garnet magneto-optical indicator films with in-plane anisotropy[1]. The crystal surface to be imaged was oriented normal to the optical axis of a polarizing microscope and the value of the component of the local induction normal to the surface was inferred from the image intensity. The presence of flux free cylinders was revealed by strong local perturbation of flux patterns at fields both normal and parallel to the plane of the crystal plates.

Fig.1 shows the magnetic image of the **ab**-plane of a 25 μm thick crystal remagnetized in a normal field. The flux in the center trapped after application of the initial up field is bright, the flux entering from the edges after partial remagnetization by a down field is dark. The remarkable feature of the field pattern is the locally higher flux density on either sides of the remagnetization front. This concentration of vortices on the front strongly suggests a local increase of the current along it which would produce similar increase in the adjacent flux density. Such a local current would arise from bending, closure, and collapse of vortices at the front. In a finite size sample, lines of field form closed loops centered at the remagnetization front. Closed vortices feel a collapsing force ϵ/R due to their line tension ϵ, where R is the radius of the loop. The collapsing force is opposed by the pinning force $F_p = \Phi_o J_c$ which permits collapse only for loops of radius $R_c < \epsilon/F_p$.[2] Inside this radius, the superconductor is in the Meissner state and a Meissner current flows along the cylindrical boundary of the Meissner hole with the vortex state. The total Meissner current is $I_M \approx (H_{c1}/\lambda)2\pi R_c \lambda$ which can be large and cause a dramatic redistribution of flux at the remagnetization front, especially if pinning is weak.[3].

Meissner holes first form on the faces of the crystal parallel to the field direction (Fig.2-

3), and move to the interior as the remagnetizing field is increased. The most unexpected structures are nucleated by the rotation of the field in the plane of the sample. Fig.4 shows the distribution of the normal component of the field for the crystal initially magnetized in its **ab**-plane, after the field has been rotated within this plane by $90°$. Two bright curved lines appear, representing the current flowing along the Meissner holes. These curved lines change shape as the field is further rotated, appearing successively on different sides of the sample. Fits of the observed field profiles to those expected for currents flowing along the remagnetization front confirm the Meissner hole model. [3]

The lines of Meissner holes are unstable to sharp bends in arbitrary directions, similar to the current pinch instability in plasmas. These instabilities produce intricate and unusual field structures which should strongly affect the ac losses of superconductors in numerous applications.

The work was supported by the US DOE, BES-Materials Science under contract #W-31-109-ENG-38 (UW,GWC,LP), the NSF STCS under contract #DMR91-20000 (DG), and ISF (grants #RF100 and RF1300)(VKV-V,VIN,VK).

1. L.Dorosinskii et al., *Physica C* **203**, 183 (1992).
2. A.M.Campbell and J.E.Evetts *Critical Currents in Superconductors*, (1972).
3. present authors, *to be published*

MEASUREMENT OF THE TOTAL TRANSVERSE FORCE ON MOVING VORTICES IN YBCO FILMS

X.-M. Zhu, E. Brändström, B.Sundqvist, G. Bäckström
Department of Experimental Physics, Umeå University, S-901 87, Umeå, Sweden

We have designed a mechanical experiment to measure the total transverse on moving vortices in type II superconductors. The vortices are driven into motion by the vibration of a permanent magnet above a relatively large superconductor. The transverse force is measured by monitoring the motion of the supercoducting film mounted on a quartz plate.

The transverse force on a moving vortex in a type II superconductor remains an unsettled issue. Generally speaking, there are two kinds of opinions. It is either shown to be essentially the same as its classical one, with the fluid density replaced by the superconducting electron density [1], or almost zero with possible different directions in a realistic superconductor with disorders [2]. From the phenomenological point of view, the time dependent Ginzburg-Landau equation gives zero transverse force while the non-linear Schrödinger equation gives one similar to the classical case [3].

We have designed a mechanical experiment to directly measure the transverse force on the superconductor when the vortices are driven into motion. We have a 1cm×1cm superconducting film glued on top of a quartz plate, with the dimension $l/w/d = 20/10/0.4mm$ and coated with Au strips, which is clamped at its other end. Above the superconductor, we vibrate a thin mu-metal plate which is attached to a permanent magnet. The mu-metal plate concentrates the magnetic flux such that the moving magnetic flux is confined to the central area of the superconducting film. The direction of its vibration is oriented to be perpendicular to the norm of the quartz plate within half a degree. The quartz plate is driven into motion by the magnetic interaction between the vortices of the superconductor and the moving magnet. We have monitored the motion of the quartz plate capacitively in the same way as in a vibrating reed experiment [4], for two opposite magnetic field directions. A typical set of data is shown in Fig.1. There are contributions from both the magnetic field dependent transverse force and the magnetic field independent pinning force due to the slight misalignment. We have adjusted the temperature and the magnetic field distribution to minimize the effect of pinning. In the optimized situation, the effect from pinning is negligible.

The amplitude of the transverse force is obtained by calibrating the motion of the quartz plate glued with the superconductor to a known driving force, provided by applying an ac voltage between an electrode and the quartz plate. This calibration also helps to determine the direction of the transverse force. Our results show that the transverse force has the same order of magnitude and the same sign as given by Ao and Thouless [1].

The transverse force inferred from a Hall measurement is usually two to three orders of magnitude smaller than the one obtained by our measurement and may carry a different sign [5]. We believe, however, our results are not incompatible with the Hall measurement, because the Hall anomaly can be caused by the motion of vacancies and vacancy-like dislo-

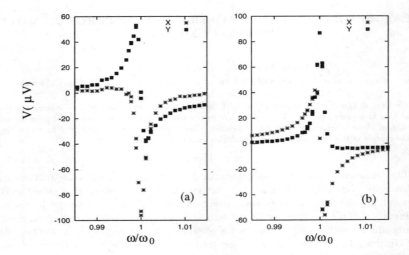

Figure 1: The in-phase (X) and out-of-phase (Y) component of the voltage on the capacitor monitoring the vibration of superconductor for the magnetic field pointing down (a) and up (b) towards the superconductor.

cations in the vortex lattice, or by any other kinds of effect due to the vortex correlation. Our experiment, on the other hand, is independent of the vortex correlation.

Acknowledgments

This work is financially supported by the Swedish Natural Science Research Council (NFR).

References

1. P. Ao, D.J. Thouless, Phys. Rev. Letts. **70**, 2158(1993).
2. For a review, see E. H. Brandt, Rep. Prog. Phys. **58**, 1465(1995).
3. P. Ao, D.J. Thouless, and X.-M. Zhu, Mod. Phys. Letts **B9**, 755(1995).
4. For a review, see P. Esquinazi, J. Low Temp. Phys. **85** 233(1991).
5. Eg. S.J. Hagen et. al, Phys. Rev. **B47**, 1064(1993).

II. EXPERIMENTS

Occurrence

PHASE SEPARATION AND STAGING BEHAVIOR IN $La_2CuO_{4+\delta}$

R.J. BIRGENEAU[1], F.C. CHOU[1], Y. ENDOH[2], M.A. KASTNER[1], Y.S. LEE[1], G. SHIRANE[3],
J.M. TRANQUADA[3], B.O. WELLS[1], K. YAMADA[2]

[1] *Department of Physics and Center for Material Science and Engineering, Massachusetts Institute of Technology, Cambridge, MA 02139, USA*
[2] *Department of Physics, Tohoku University, Aramaki Aoba, Sendai 980-77, Japan*
[3] *Department of Physics, Brookhaven National Laboratory, Upton, NY 11973, USA*

Neutron scattering and magnetization measurements have been performed on three single crystals of $La_2CuO_{4+\delta}$. For two crystals with $\delta \sim 0.02$ and 0.03 phase separation of the intercalated oxygen occurs below 280 K whereas stage ordering occurs below 255 K. A third crystal exhibits no phase separation and a higher stage ordering temperature. All three show evidence for an in-plane ordering around 210 K. We propose a phase diagram which summarizes our observations.

Among the family of superconducting copper-oxides, $La_2CuO_{4+\delta}$ and $La_{2-x}Sr_xCuO_4$ have particularly simple structures. $La_2CuO_{4+\delta}$ is of great interest because its doping concentration may be well controlled and the oxygen dopants are mobile to temperatures as low as 200 K, in contrast to the quenched disorder resulting from the fixed Sr^{2+} ions in $La_{2-x}Sr_xCuO_4$. Recent experiments by Wells et al.[1] have shown that in the oxygen-rich phase of $La_2CuO_{4+\delta}$ the oxygen intercalants order into layers parallel to the copper oxide planes. This paper extends the previous work with further experiments on three large single crystals of $La_2CuO_{4+\delta}$. In particular, we examine two crystals in the doping regime where phase separation occurs.

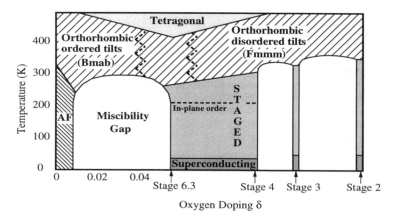

Figure 1: Proposed phase diagram for $La_2CuO_{4+\delta}$.

The data in this paper together with that from Wells et al.[1], measured from seven crystals in total, suggest the phase diagram in Figure 1. At low temperatures and low doping, $La_2CuO_{4+\delta}$ exhibits three dimensional antiferromagnetic order. For a range of dopings ($\delta \simeq 0.012 - 0.055$),[2,3] below room temperature the sample phase separates into oxygen-rich and oxygen-poor phases. In oxygen-rich samples, which are superconducting below ~ 35 K, the interstitial oxygen dopants order one-dimensionally along the **c** direction, tending to segregate into planes parallel to the CuO_2

sheets and regularly spaced every n CuO_2 host layers. We call this configuration stage n, adopting the nomenclature used for intercalated graphite. As we show below, for crystals which phase separate, the phase separation temperature ($T_{PS} \simeq 280$ K - 290 K) is higher than the stage ordering temperature ($T_{SO} \simeq 255$ K).

All three single crystals used in this experiment were grown at MIT using the travelling-solvent-floating-zone technique. This method does not employ a crucible, hence contamination is reduced, and the crystals appear to be more homogenous than those grown by other means. The crystals are electrochemically oxygenated as described in Ref.1; different oxygen dopings are produced by varying the duration of the electrolysis.

The magnetization, measured with a SQUID magnetometer, at high fields reveals the weak ferromagnetism associated with the antiferromagnetic transition of the undoped phase, as shown in Figure 2(a). From the size of the moment we conclude that the doped crystals are 20% and 45% oxygen-rich by volume. From the phase diagram previously presented by Radaelli et al[3] for powder $La_2CuO_{4+\delta}$ we infer corresponding macroscopic oxygen concentrations of $\delta \sim 0.02$ and $\delta \sim 0.03$ respectively. Shielding measurements shown in Figure 2(b) demonstrate that all three doped crystals exhibit Meissner diamagnetism and hence have an oxygen-rich fraction that is superconducting. T_c for the $\delta \sim 0.02$ and 0.03 crystals is about 32 K, while for the crystal labeled stage 4.4, T_c is about 33 K. It is evident therefore that the two crystals with $\delta \sim 0.02$ and 0.03 have at low temperatures a superconducting fraction coexisting with an antiferromagnetic fraction.

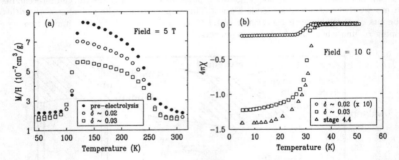

Figure 2: Magnetization of samples as a function of temperature. (a) High field reveals the weak ferromagnetic transition of the oxygen-poor phase. (b) Low field measurements of the shielding signal after slow cooling at zero field. The applied field is along the c direction.

To determine the temperature below which the single crystals phase separate, we performed scans about nuclear Bragg reflections to monitor changes in the lattice constants as a function of temperature. Figure 3(a) shows that at high temperatures there is only one (0,0,6) Bragg peak signifying a single homogenous phase. At low temperature there are two Bragg peaks, corresponding to two different phases with different c-axis lattice constants; the phase with the longer c-axis lattice constant is the oxygen-rich phase and the other is oxygen-poor. The data in Figure 3 are for the $\delta \sim 0.03$ sample. The relative intensities of the two peaks at low temperatures provide a second measure of the fractions of sample in the two phases. We find that about 45% of the sample is oxygen-rich, consistent with the magnetization data. From fits to the data, we find that phase separation occurs below $T_{PS} \simeq 290$ K. This is a lower phase separation temperature than that for the powder samples as studied by Radaelli et al[3]. The $\delta \sim 0.02$ sample also phase separates with $T_{PS} \simeq 280$ K. However, the stage 4.4 sample does not phase separate signifying that it is doped beyond the miscibility gap.

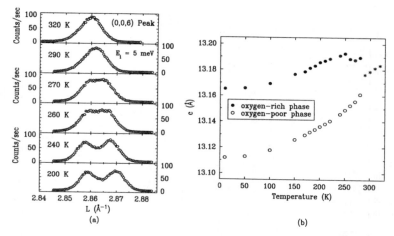

Figure 3: Crystal with $\delta \sim 0.03$, about 45% oxygen-rich. (a) Scans over a fundamental nuclear Bragg peak. (b) Temperature dependence of the lattice constants obtained from scans like those in (a).

A range of staging numbers from $n = 2$ to 7 have been observed in $La_2CuO_{4+\delta}$. Staging is evidenced by incommensurate peaks displaced along L on either side of the Bmab superlattice peaks which originate from the tilting of the CuO_6 octahedra in the orthorhombic phase. In a homogenous staged sample, the commensurate Bmab superlattice peaks disappear altogether being replaced by the two incommensurate peaks. These staging superstructure peaks result from antiphase domain boundaries at which the CuO_6 octahedra tilt direction reverses. This reversal is presumed to be caused by the intercalated oxygen layers, similar to the model proposed by Tranquada et al. for $La_2NiO_{4+\delta}$.[4] For the two samples doped in the miscibility gap, we find that the oxygen-rich phase has $n = 6.3$ at low temperatures. For the sample that was doped beyond the miscibility gap, we find that $n = 4.4$, independent of temperature. This suggests that n is simply a function of the amount of oxygen in the oxygen-rich phase.

Figure 4 presents data taken on the $\delta \sim 0.02$ sample. The temperature dependence of the lattice constants in the top panel of part (a) shows phase separation occuring below $T_{PS} \simeq 280$ K. The lower panel shows the temperature dependence of the 2D integrated intensity in the (0,K,L) plane of the (0,1,4.15) staging peak from fits to a 2D lorentzian squared. We see that the onset of stage ordering in this sample occurs at $T_{SO} \simeq 255$ K. The staging number evolves from ~ 7.1 to 6.3 with decreasing temperature. The sample with $\delta \sim 0.03$ follows the same behavior. Examining the staging intensity upon slow cooling and slow warming (220 K to 260 K in 25 hours), we find 5 K hysteresis. In the stage 4.4 sample, stage ordering occurs at a higher $T_{SO} \simeq 290$ K. However the sample remains orthorhombic, and there is no commensurate Bmab superlattice peak up to at least 330 K, suggesting that above 290 K the tilts are disordered.

An apparent drop in the intensity of the staging peak occurs for temperatures below 210 K, as first reported by X. Xiong et al.[5] This results from the establishment of an in-plane modulation, in which the staging peak itself splits into two superlattice peaks along H. Scans along H through the (0,1,4.15) staging peak are shown in Figure 4(b) taken on warming. From the width of the single staging peak at 220 K, we estimate a staging domain size of about 83 lattice constants along a. As the crystal cools, the staging peak intensity diminishes and two peaks rise on either side. The two

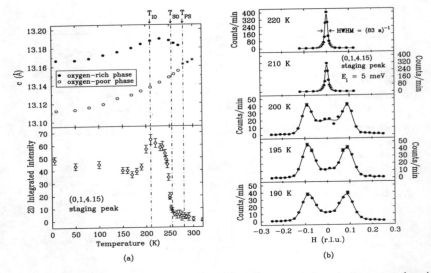

Figure 4: Crystal with $\delta \sim 0.02$, about 20% oxygen-rich. (a) The upper panel shows the temperature dependence of the lattice constants; the lower panel shows the temperature dependence of the 2D integrated intensity of the (0,1,4.15) staging peak. (b) Scans along H through the (0,1,4.15) staging peak. The solid line for the 220 K and 210 K scans is a fit to a lorentzian convolved with the resolution function. Note the change of scale for the vertical axis. The solid line for the lower temperature scans is a fit to three 3D gaussians.

new peaks appear in coexistence with the initial staging peak, implying a first-order transition. The displacement of the superlattice peaks from the original staging peak position suggests that they are caused by a modulation along H with a periodicity of about 11.5 **a**. Thus, this in-plane ordering occurs within a staging domain. No new satellite peaks are found along the in-plane K direction. A similar drop in staging peak intensity is seen in the $\delta \sim 0.03$ sample and, to a much lesser extent, in the stage 4.4 sample, suggesting that this H modulation also occurs in these crystals.

Acknowledgements

The work at MIT was supported by the MRSEC Program of the National Science Foundation under award number DMR 94-00334 and by the NSF under award number DMR 93-15715. Research at Brookhaven National Laboratory was carried out under contract No. DE-AC-2-76CH00016, Division of Material Science, U.S. Department of Energy.

References

1. B.O. Wells et al., Z. Phys. B(in press).
2. J.D. Jorgenson et al., Phys. Rev. B **38**, 11337 (1988).
3. The values for δ are taken from P.G. Radaelli et al., Phys. Rev. B **49**, 6239 (1994).
4. J.M. Tranquada et al., Phys. Rev. B **50**, 6340 (1994).
5. X. Xiong et al., Phys. Rev. Lett.(in press) and in this volume.

X-RAY SEARCH FOR CDW IN SINGLE CRYSTAL $YBa_2Cu_3O_{7-\delta}$

P. WOCHNER,[1] E. ISAACS,[2] S. C. MOSS,[3]
P. ZSCHACK,[4] J. GIAPINTZAKIS,[5] D. M. GINSBERG[5]

[1]*Physics Department, Brookhaven National Laboratory, Upton, NY 11973*

[2]*AT&T Bell Laboratories, Murray Hill, NJ 07974*

[3]*Physics Department and Texas Center for Superconductivity, University of Houston, Houston, TX 77204-5506*

[4]*ORAU, Brookhaven National Laboratory, Upton, NY 11973 (currently at ANL/APS, Argonne, IL 60439)*

[5]*Physics Department, University of Illinois, Urbana, IL 61801*

ABSTRACT

Recently, H. L. Edwards et al.[1] observed, in STM experiments at 20K, modulations in the CuO chain layer of cold-cleaved single crystals of $YBa_2Cu_3O_{7-\delta}$ which they interpreted as a possible charge density wave (CDW). Since X-ray scattering is an ideal tool for the study of static or dynamic lattice displacements, we performed a synchrotron X-ray study at beamline X14 at the NSLS of BNL on a high quality single crystal of $YBa_2Cu_3O_{7-\delta}$, which was mainly single domain with a spacially well localized volume fraction of other twin orientations of roughly 10%. Appropriate scattering configurations were chosen to enable observations of longitudinal or transverse CDWs with polarization either in the chain direction, $\parallel$ <010> or $\perp$ to it in <001>. The X-ray energy of 16keV allowed us to reach large momentum transfers to increase the sensitivity to lattice displacements. In none of our scans, which definitely covered the case of a 1-dimensional longitudinal CDW with propagation in the b direction as proposed by Edwards et al., did we find intensity other than the main Bragg peak(s) and the twin reflections. We therefore suspect that the STM finding may be a surface-induced phenomenon.

1. Introduction

Following the lead of Edwards et al.[1] who found strong evidence with the STM for a static (i.e. pinned) charge density wave (CDW) at the 001 surface of a single crystal of $YBa_2Cu_3O_{7-\delta}$, we have performed an extensive synchrotron X-ray search for this effect which, with X-rays, is invariably seen as a ("static") longitudinal modulation of the structure in which the resultant atomic displacements are parallel to the wave vector of the proposed CDW. Essentially we have been guided in our study by the findings of Edwards et al.[1] where they show real-space STM images of a cold-cleaved (001) surface of the $YBa_2Cu_3O_{7-\delta}$ crystal which reveal the Cu-O chain plane to be corrugated along the chains (the b-axis of the orthorhombic crystal) with a wavelength of ~1.3nm. These corrugations persist over several wavelengths but are only weakly correlated between chains. The

Fourier transform of a well-defined image therefore shows somewhat cigar-shaped and diffuse "satellites", sharper in the b*-direction but elongated normal to it (along a* and, by implication, c*). Through a number of tests they identified the observed modulation as a Fermi-surface-induced 1D CDW whose period was quite plausibly given by a known Fermi surface spanning length ($2k_F$) in the direction in question. The observed CDW's were perforce pinned both because they would otherwise not be seen in the STM and because observed oxygen vacancies could act as pinning centers.

The present study was thus made rather straightforward. For normal X-ray scattering the assignment of static vs. dynamic is not possible. However, the scattering is not off the CDW itself but invariably off the attendant displacement wave or mass density wave (MDW) resulting from the correlated core readjustments to the CDW. We therefore expect to measure satellite(s) about the average Bragg peaks which appear at $(2\pi)/\Lambda$ where Λ is the CDW wavelength. Actually, if the MDW is well-defined, even if it is sinusoidal, higher order satellites can be expected.[2] However, their absence is usually attributed to defects and variations in Λ. Without entering seriously into the scattering formalism we may note that for such (static or dynamic) displacive waves, the scattered intensity shows the following proportionality:[2]

$$I_{MDW} \propto |F|^2 \cdot e^{-2M} (\mathbf{Q} \cdot \boldsymbol{\varepsilon}_{MDW})^2$$

where $|F|$ is the structure factor for the Bragg peak in question, $2M$ is a combined static and dynamic Debye-Waller factor, $\mathbf{Q}$ is the diffraction vector ($\hbar\mathbf{Q}$ is the momentum transfer in the experiment) and $\boldsymbol{\varepsilon}_{MDW}$ is the unit polarization of the MDW, usually longitudinal or transverse. When the propagation vector of the MDW, $\mathbf{q}_{MDW} \parallel \boldsymbol{\varepsilon}_{MDW}$, the wave is longitudinal; when $\mathbf{q}_{MDW}$ is $\perp \boldsymbol{\varepsilon}_{MDW}$, it is transverse.

In the present case we shall be looking at **q** vectors in a radial direction about the (080) position (i.e. longitudinal) where $\boldsymbol{\varepsilon}_{MDW} \parallel \mathbf{q}_{MDW}$. For a transverse wave our principal experiments give $\boldsymbol{\varepsilon}_{MDW} \perp \mathbf{q}_{MDW}$, and thus $\boldsymbol{\varepsilon}_{MDW} \perp \mathbf{Q}$, and $\mathbf{Q} \cdot \boldsymbol{\varepsilon}_{MDW} = 0$! We thus measure only the longitudinal CDW contribution along the b-axis (b* reciprocal axis or k-direction using the conventional hkℓ indices for our reciprocal lattice notation). We chose to measure at (080) because at this large value of Q the observed effect, $(\mathbf{Q} \cdot \boldsymbol{\varepsilon})^2$, will be much enhanced, if present, while the value of $2M$ is not so large as to cancel this advantage.

2. Experiment

Our study was done on a nominally single domain crystal of $YBa_2Cu_3O_{7-\delta}$ (0.5mm x 2.0mm x 0.06mm) from a batch with T_c's of ~91K, a transition width of <0.1K and an oxygen content $0 \leq \delta \leq 0.05$. The measured mosaic spread was less than 0.1°, although the wings of an azimuthal (transverse) scan through a Bragg peak will always show some sharp structure with synchrotron resolution. The crystal was mounted on a glass fiber in a He-filled, Be-walled, can in an evacuated displex cryostat Be chamber which was offset-mounted on a Huber 4-circle diffractometer on line X14 at the National Synchrotron Light Source (NSLS) at Brookhaven. The temperature was maintained at 23-24K. No analyzer crystal was used but a solid state detector was employed to remove fluorescence scattering ($CuK\alpha$). The incident energy was 16keV ($\lambda=0.7744$Å) which is below the K-edge of Y ($\lambda=0.7277$Å) to leave a background relatively clean of parasitic scattering.

We show in Fig.1 a schematic of the twin structure to be expected in the reciprocal space of an orthorhombic crystal. The notations (1,2) and (3,4) refer to domains in which the a and b axes are reversed. From (1) to (4) or (2) to (3) we have twin related spots. As we explore the (080) reflection at lower intensity contours we can expect to see, not only weaker mosaic blocks, but the twinning and domain structure of Fig.1 even though our crystal consists mainly of one untwinned domain (~90%). As we shall see, however, the extensive structure is nowhere in evidence at the predicted positions of the CDW satellites ($\Lambda=13$Å, b=3.89Å): namely at a value of $\Delta k=\pm 0.3$ from any 0k0 Bragg peak.

In Fig.2 we show a radial line scan through the major twin reflection at 080 where the central Bragg reflection is several orders of magnitude off scale and the twin reflection appears as a cut through its mosaic tail. Figures 3(a) and (b) show contour maps of the a*-b* plane covering two ranges: 3(a) shows contours running from 4.6×10^4 down to 10^3 counts. Figure 3(b) shows a much more diffuse (weaker) range covering 2×10^3 to 900 counts. At $\Delta k=\pm 0.3$, i.e. at k=8.3 and 7.7, there is no evidence in either plot for a diffuse structure parallel to the h-axis. The structure in Fig.3(a) results from a major twin domain centered on 080 with its twin arranged as in (2-3) of Fig.1. The second (weaker) pair (1-4) is shifted, as in Fig.1 to higher h by $\Delta h=0.1$. The split in k is only ~0.15. In other words, over an intensity range covering the weakest background scattering there is no evidence for the CDW (MDW) structure suggested by the results of Ref.1.

Figure 3(c) completes this picture with contour maps in the b*-c* plane about (080). Again there is the expected twin structure displaced in k by $\Delta k \sim 0.15$ and showing extensive mosaic structure associated, again, with weak wings in this case along the c* direction (ℓ) which is also transverse to the radial <0k0> direction, but with no diffuse peaks ("cigars") along ℓ at k=8.3 or 7.7!

3. Conclusions

Based on our data to date, we conclude that the observations of Edwards et al. are related to a surface phenomenon and do not represent a bulk CDW. This result is, however, at variance with the recent report by Mook and co-workers at this conference in which their integrated neutron intensities along the suggested sheet of diffuse scattering at $\Delta k \sim 0.3$ about (020) shows a peak. We have, at this time, no explanation for this discrepancy.

Acknowledgments

This work was supported at Houston by the Texas Center for Superconductivity (TCSUH) and the NSF on DMR-92-08420; at BNL by the Division of Materials Research, US DOE on DE-AC-02-76CH00016; and at Illinois by the NSF on DMR-91-20000 through the Science and Technology Center for Superconductivity.

References

1. H. L. Edwards, A. L. Barr, J. T. Markert and A. L. de Lozanne, Phys. Rev. Lett. 73, 1154 (1994).
2. R. W. James, "Optical Principles of the Diffraction of X-rays," Oxbow Press (Woodbridge, CT, 1962); B.E. Warren, "X-ray Diffraction," Dover Publ. (NY, 1990).
3. H. A. Mook, P. Dai, G. Aeppli, F. Dogan and K. Salama (these proceedings).

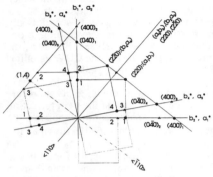

Fig. 1 Schematic reciprocal lattice for a twinned crystal with 2 domains.

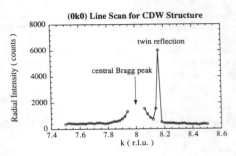

Fig. 2 Radial scan through major 080 reflection with no indication for peaks at 7.7 or 8.3.

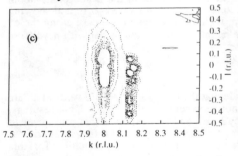

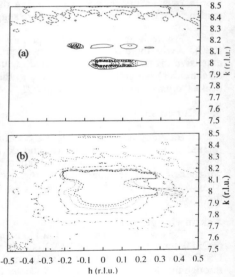

Fig. 3 (a) and (b): Contour maps in the a*-b* (hk0) plane: (a) contours run from 4×10^4 to 10^3 counts; (b) runs from 2×10^3 to 900; (c) contour map in b*-c* (0kℓ) plane. There is no evidence for CDW structure at $\Delta k = \pm 0.3$ (see text).

SUPERCONDUCTING PROPERTIES OF Nb THIN FILMS

ANA LUÍZA V.S. ROLIM, J.C.C. DE ALBUQUERQUE, E.F. DA SILVA JR,
J.M. FERREIRA AND J. ALBINO AGUIAR
DEPARTAMENTO DE FÍSICA, UNIVERSIDADE FEDERAL DE PERNAMBUCO
50670-901 RECIFE-PE, BRASIL

Niobium thin films with thickness between 100-1000nm have been deposited by dc magnetron sputtering on dieletric substrates (glasses). The magnetic field-temperature (H-T) diagram, obtained from zero field cooling (ZFC) and field cooling (FC) cycles, reveals a strong dependence of the irreversibility line as a function of film thickness. Thinner films suppress the irreversibility line. We attribute this behavior to surface pinning effect.

1. INTRODUCTION

For many years thin films of transition-metal elements have been of great interest due to the possibility of having high-T_c superconductor and also because of their good materials properties for use in superconducting electronics. The use of these materials has not been successfull due to the degradation of the superconductivity in the surface layer. The critical temperature (T_c) is quite sensitive to small amounts of dissolved oxygen[1], to strain[2], evaporation rate and substrate temperature[1]. As one reduces the film thickness the depression in T_c can be attributed to proximity effect, lifetime broadening of density of state and to localization[3,4]. Recently, due to the observation of an irreversibility line (IL) in high temperature superconductors[5] and to the debate on their origin[7], the study of Nb films has received more attention[6]. In this paper we present results of T_c measurement and IL determination of Nb films. We show that T_c has a decrease approximately linear with the reciprocal of the film thickness and that the IL is strongly dependent on the film thickness.

2. EXPERIMENTAL

Glass substrates used in this study were cleaned with a solution of HF priori to the deposition. The substrates were rinsed successively in acetone and flowing deionized water. They were dried in a N_2 flow and then mounted in the substrate holders. The deposition chamber was also cleaned with HF solution before the deposition. The Nb targets were cutted from high purity Nb bar obtained after six fussion with electronic beam. The Nb films were deposited by magnetron dc sputtering using a commercial system. The Ar plasma pressure was of 2.3µbar and the power 300W. A plasma etching was done priori making the films used in this work. Using a commercial SQUID magnetometer, the temperature dependence, of both, ac susceptibility and dc magnetization as a function of dc magnetic field were measured. The dc magnetization was recorded using zero field cooling (ZFC) and fiel cooling (FC) cycles. The irreversibility temperature (T_{irr}) was identified as been that where the difference between ZFC and FC curves were less then the mean square deviation of the measurement.

3. RESULTS AND DISCUSSIONS

The diamagnetic transition of the films studied were very sharp. The critical temperatures (T_c) show an approximate linear decrease with the reciprocal of the film thickness. The thicker film 1000nm has $T_c=9.2K$ and the thinner one (102.4nm) has T_c of 8.2K. This is in agreement with both, proximiy effect and Ebisawa-Fukuyama-Maekawa[8] theory of localization. No attempt was done to fit the data with these theories.

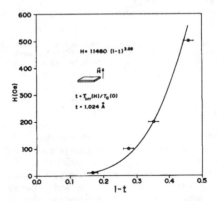

Fig.1 Irreversibility line $H_{irr}(t)$ for a 1.024 A Nb film as a function of the reduced temperature $t=T_{irr}/T_c$. The continuous line is a fitting to the data using the scalling $H_{irr}(t)=A(1-t)^\alpha$ as explained in the text.

The magnetic-field-temperature (H-T) phase diagram for the films studied were constructed from the data obtained from ZFC-FC cycles. In Fig.1 we show the H-T diagram for a 102.4nm film. Thicker films present similar diagram. The data was fitted using the scalling function $H=A(1-t)^\alpha$ were $t = T_{irr}^{(H)} / T_c^{(0)}$ and α and A are adjustable parameters. For the 102.4nm thick film we found $\alpha=3.9$. For a 1000nm film $\alpha=1.3$. These results show that the irreversibility line is depressed in thinner films. Many theoretical models are consistent with these results[7], but, more experimental data are need before attempting to make any quantitative comparison. We believe that surface damage due to deffects and strain induced in thinner films, can be the responsible for the observed behavior. This is in agreement with the results of Konczykoski et al.[9] in $YBa_2Cu_3O_7$ crystals. They found that damaging of the surface by electron irradiation reduces the irreversibility.

ACKNOWLEDGMENTS

We thanks Brazilian Agencies FINEP and CNPq for financial support and, CEMAR/ FAENQUIL at Lorena-SP for providing the Nb targets.

REFERENCES

1. J.R. Rairden and C.A. Neugebauer, Proc. IEEE, 52, 1234 (1964).
2. W. De Sorbo, Phys. Rev. 135, A1190 (1964).
3. S.I. Park and T.H. Geballe, Physica 135B, 108 (1985).
4. M.S.M. Minhaj et al, Phys. Rev. B49, 15235 (1994)
5. K.A. Müller, M. Takashige and J.G. Bednorz, Phys. Rev. Lett. 58, 1183 (1987).
6. M.F. Schmidt, N.E. Israeloff and A.M. Goldman, Phys. Rev. Lett. 70, 2162 (1993).
7. L. Civale et al., Phys. Rev. B 43, 5425 (1991) and references therein.
8. H. Ebisawa, H. Hukuyama, and S. Maekawa, J. Phys. Jpn. 54, 2257 (1985).
9. Konczykowski et al, Phys. Rev. B 43, 13707 (1991).

Pressure Effects on T_c of $HgBa_2Ca_{n-1}Cu_nO_{2n+2+\delta}$ with $n \geq 4$

Y. Cao, X. D. Qiu, Q. M. Lin, Y. Y. Xue and C. W. Chu
Texas Center for Superconductivity and Department of Physics at University of Houston, Houston, Texas 77004-5932

The initial pressure effect dT_c/dP and the maximum T_c-enhancement under pressure ΔT_c^{max} of $HgBa_2Ca_{n-1}Cu_nO_{2n+2+\delta}$ was measured for $1 \leq n \leq 6$. Both dT_c/dP and ΔT_c^{max}, which are almost n-independent for $n < 4$, show a drop with the increase of n at $n \sim 4$. This observation might be attributed to the pressure-induced inhomogeneous-distribution of carriers between the inner and outer CuO_2 layers.

The pressure effect on the $HgBa_2Ca_{n-1}Cu_nO_{2n+2+\delta}$ [Hg-12(n-1)n] series drew much research interest, especially after the record-high T_c up to 164 K was achieved under 30 GPa for Hg-1223 samples [1]. One striking feature is the universal T_c enhancement for nearly optimally-doped Hg-1201, -1212 and -1223 samples [1-3]. The present study is to determine whether this universality will hold as $n \geq 4$.

High quality Hg-12(n-1)n samples with $4 \leq n \leq 6$ were successfully synthesized and characterized. The detailed sample preparation will be published elsewhere [4]. T_c was determined resistively by the four-lead method using a Linear Research LR-400 bridge. The hydrostatic pressure environment up to 1.6 GPa was generated inside a Teflon cup in a Be-Cu high pressure clamp, using 3M fluorinert as the pressure medium, and the quasi-hydrostatic pressure environment up to 18 GPa was obtained in a Bridgman anvil arrangement with solid steatite as the medium.

Resistance was measured as a function of temperature under hydrostatic pressures up to 1.5 GPa on several Hg-1234 samples. By changing the synthesis and annealing conditions, we can adjust the T_c of the samples between 80 and 124K. In this doping region, $dT_c/dP \sim 1.2$ K/GPa is nearly independent of oxidation, and is considerably smaller than those of the optimally-doped Hg-1201, -1212 and -1223 samples.

Similar measurements were also done on the $n = 5$ and 6 compounds. The obtained dT_c/dP's are quite close: $dT_c/dP \sim 1.0$ K/GPa for Hg-1245 with $T_c \sim 110$ K, and ~ 1.1 K/GPa for Hg-1256 with $T_c \sim 103$ K. dT_c/dP also is insensitive to oxidation/reduction annealing for these compounds in this low pressure range. The n-dependence of dT_c/dP is shown in Figure 1, where a step-like drop appears at $n = 4$ even after considering the sample-to-sample data scattering. The quasi-hydrostatic pressure effect on Hg-1234 and -1245 was also measured up to 15 GPa. Unlike the results of the first three members of the Hg-12(n-1)n series, there was a much smaller T_c enhancement and a lower saturation pressure.

These simultaneous decrease of both dT_c/dP and ΔT_c^{max} with $n \geq 4$ was accompanied by a drop in the T_c of the as-synthesized samples at ambient pressure. The nonmonotonic T_c-n correlation has been interpreted to be the result of a non-homogeneous distribution of carriers among the CuO_2 layers in compounds with $n \geq 4$ [5]. The electrostatic forces at the carriers, which may also cause carrier-concentration difference between different

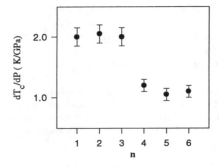
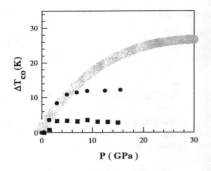

Fig. 1 dT_c/dP vs number of CuO_2 layers for $HgBa_2Ca_{n-1}Cu_nO_{2n+2+\delta}$ under hydrostatic pressure

Fig. 2 ΔT_c vs P for $HgBa_2Ca_{n-1}Cu_nO_{2n+2+\delta}$
Shaded: n= 1, 2 and 3; Circle: n= 4
Square: n = 5

CuO_2 layers, will increase with pressure. Therefore, a pressure-induced carrier-redistribution may occur to enhance the carrier inhomogeneous distribution, and a negative term, which is significant only at n > 3, might be added into the effect of pressure on T_c.

In summary, we observed a step-like decrease of the initial dT_c/dP and a drastic drop in ΔT_c^{max} in Hg-12(n-1)n with n ≥ 4. We tentatively attribute that to a pressure-induced carrier-inhomogeneity between the inner and outer CuO_2 layers.

Acknowledgment

*This work is supported in part by NSF Grant No DMR 95-00625, EPRI RP-8066-04, DoE Grant No DE-FC-48-95-R810542, the State of Texas through the Texas Center for Superconductivity at University of Houston and the T. L. L. Temple Foundation.

Reference

1. L. Gao et al., Phys. Rev. B **50** 4260 (1994)
2. C.W. Chu et al., Nature 365 (1993) 323, and L. Gao et al., Phil. Mag. Lett. **68** 345 (1993)
3. A. K. Klehe et al., Physica C**223** 345 (1994)
4. Q. M. Lin, Y. Y. Sun, Y. Y. Xue, C. W. Chu, these proceedings
5. M. Stasio et al., Phys. Rev. Lett. **64** 2827 (1990)
6. F. Chen, X. D. Qiu, Y. Cao, L. Gao, Y. Y. Xue, C. W. Chu, these proceedings

LOW TEMPERATURE SCANNING TUNNELING MICROSCOPY AND SPECTROSCOPY OF THE CuO CHAINS IN $YBa_2Cu_3O_{7-x}$

D. J. DERRO, T. KOYANO, HAL EDWARDS, A. BARR,
J. T. MARKERT, AND A. L. DE LOZANNE
Department of Physics, University of Texas,
Austin, TX 78712-1081, USA

We present data on the CuO chains of cold-cleaved single crystals of the high-temperature superconductor $YBa_2Cu_3O_{7-x}$. An energy gap (about 25meV) is clearly observed on this surface. The energy gap disappears near oxygen vacancies. By utilizing Current Imaging Tunneling Spectroscopy (CITS) and logarithmic derivative analysis, we have obtained a spatial map of the gap over the surface.

In the past we transferred $YBa_2Cu_3O_{7-x}$ (YBCO) thin films into our Scanning Tunneling Microscope (STM) without exposure to atmosphere. But because we could not achieve atomic resolution nor reproducible spectroscopic results,[1] we developed a technique for cleaving single crystals of YBCO *in situ* at low temperatures (~20K). Using this procedure we were able to achieve atomic resolution, reproducible spectroscopy, and a reasonable work function all simultaneously under the same measurement conditions. The most interesting layer of YBCO shows the CuO chains with some oxygen vacancies. Superconductivity is generally thought to reside in the CuO planes in YBCO. Yet the electronic structure of the CuO chains may obscure or modify the superconducting nature of the CuO planes. Therefore, in order to understand the mechanism of superconductivity in YBCO it may be important to study the CuO chains.

Initially, we performed STM on YBCO to obtain images of the CuO chains and single-point Scanning Tunneling Spectroscopy (STS) to obtain I(V) curves which displayed a reproducible energy gap of 20-30 meV ($2\Delta/kT_c$=6-8).[2] The STM images showed a 1.3nm electronic modulation along the CuO chains, about three times the unit cell length in that direction.[3] We asserted previously that these modulations could be due to a charge density wave because we performed reversed-bias STM imaging of the chains and the 1.3nm corrugations changed sign under bias polarity reversal. Another plausible interpretation is that these modulations can be caused by Friedel oscillations.

Since we first observed these modulations, an important question has been whether this is exclusively a surface phenomenon. Very recently, similar modulations have been observed in the bulk by neutron diffraction, albeit with a slightly longer periodicity of 1.6 nm.[4] The change in periodicity can be due to the different electronic environment of the chains on the surface compared to those in the bulk. On the other hand, a careful search by x-ray diffraction has not detected any such modulations along the chains.[5] While it is too early to know with certainty the solution to this apparent contradiction, a possible explanation is that these modulations are due to a spin density wave which is therefore not visible by x-ray scattering. A second explanation would be that Friedel oscillations depend strongly on the number of scattering centers (oxygen vacancies in our model), so that the ability to detect these modulations would be strongly sample dependent. Clearly, more studies of this phenomenon are desirable.

Most recently, we have used CITS to study the detailed nature of the spectroscopy of the CuO chains. CITS works by simultaneously performing STM imaging and STS. We have shown that the superconducting energy gap is destroyed near oxygen vacancies by comparing I(V) curves far from and near an oxygen vacancy.[6] We have also used the logarithmic derivative to analyze the CITS data where $R(V) \equiv dlnI/dlnV = (dI/dV)/(I/V)$.[7] Now to interpret the logarithmic derivative images, we note that near the edge of the energy gap, dI is large and I is small so that R(V) is large and a bright region will correspond to a region with an energy gap. The logarithmic derivative over a normal region has a value equal to one because the I(V) curve is linear. Shown below in figure 1 is one such logarithmic derivative image that highlights this feature: notice how the lower left region is bright meaning there is a definite energy gap here and the upper right region is dark where there is not a definite energy gap.

An important conclusion from the cross section in figure 1 is that the superconducting properties change quite rapidly, within about 2 nm. This is consistent with the coherence length in the a-b plane, which is about 1 nm, because all spatial variations of the order parameter should be longer than or equal to the coherence length.

This work is supported by the Texas Advanced Technology Program (3658-172), the R. A. Welch Foundation (Grant Numbers F-1047 and F-1191) and the National Science Foundation (Grant Number DMR-9158089).

References
1. A.L. de Lozanne, E. Ogawa, R.M. Silver, A.B. Berezin, and S. Pan, in *Science and Technology of Thin Film Superconductors*, McConnel & Wolf, eds. (Plenum Press, New York, 1988) pp. 111-119.
2. H. L. Edwards, J. T. Markert, and A. L. de Lozanne, *Phys. Rev. Lett.* **69**, 2967 (1992).
3. H. L. Edwards, A. L. Barr, J. T. Markert, and A. L. de Lozanne, *Phys. Rev. Lett.* **73**, 1154 (1994).
4. Mook et al., these proceedings.
5. Moss et al., these proceedings.
6. H. L. Edwards, D.J. Derro, A.L. Barr, J. Markert, and A. de Lozanne, *Phys. Rev. Lett.* **75**, 1387 (1995).
7. H. L. Edwards, D. J. Derro, A. L. Barr, J. T. Markert, and A. L. de Lozanne, *J. Vac. Sci. Technol.*, to appear May/June (1996).

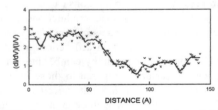

Figure 1: Logarithmic derivative image of CITS data. The cross section is taken through the diagonal shown in the image. The vertical scale in the cross-section is arbitrary, so we have normalized it to 1 in the normal region.

Zn and Ni Impurities in 123 Materials

D. Goldschmidt and Y. Eckstein

Crown Center for Superconductivity and Department of Physics

Technion, Haifa, Israel

It is generally accepted that in the most investigated 123 material $YBa_2Cu_3O_7$ (YBCO), impurities of Ni and Zn replace mainly copper atoms in the CuO_2 plane[1]. The effect is, however, quite different for those two impurities. Non magnetic Zn is by far more effective in reducing T_c than magnetic Ni. This was taken as a strong indication for the AF spin fluctuation mechanism for superconductivity in cuprates[2].

In this Workshop we present data of another 123 material which was obtained by replacing the Yttrium atoms by Ca and La and at the same time some of the Ba by La in a way that yields charge compensation[3]. In this way $(Ca_xLa_{1-x})(Ba_{1.75-x}La_{0.25+x})Cu_3O_7$, (CaLaBaCuO) was formed. Notice that the average charge of the cations excluding Cu is 7.25 (in YBCO it is 7) and is independent of x. This makes this family isoelectronic[4], that is, the average charge concentration remains constant. Nevertheless, it was found that changing x changes T_c. By thermopower (TEP) measurements it was shown that this is due to doping (change of carrier density) suggesting that the interaction does not change[4,5]. Being tetragonal this 123 family of materials also offers the possibility of investigation free of the effects of ordered chains.

We investigated the $x = 0.4$ material of this family, i.e. $(Ca_{0.4}La_{0.6})(Ba_{1.35}La_{0.65})$ $(Cu_{1-z}M_z)_3O_y$ with regards to Zn and Ni impurities. M stands for Ni or Zn. The simplest assumption to make is that both impurities replace the copper atoms in the same manner as in YBCO. These materials form high purity single phase. In these materials getting the optimum doping, where T_c is maximal, was easily achieved. As a matter of fact for 8% Ni and higher these materials can be made completely overdoped, i.e., $T_c = 0K$ (Fig. 1). An interesting point is that in the underdoped region the TEP as function of z (impurity concentration) is constant for a given oxygen concentration y. This suggests that in this case the doping does not change, but that Zn and Ni impurities cause pair breaking, confirming previous suggestions on the effect of these impurities in YBCO[1,6].

The main result is shown in Fig. 2 where the critical temperature as function of the impurity concentration at optimal doping is presented. It is clearly seen that YBCO and CaLaBaCuO do not behave in the same manner. The difference in T_c depression between both impurities is much smaller in this material than in YBCO. In fact, in CaLaBaCuO, Zn and Ni induce almost the same T_c depression. Numerically, in YBCO dT_c/dz is $-3K/\%$ and $-13K/\%$ for Ni and Zn, respectively. It is $-7K/\%$ and $-9/\%$ for Ni and Zn in CaLaBaCuO.

The question is what is the reason for the difference between these two 123 materials (YBCO and CaLaBaCuO)? One would expect the behavior to be the same in both 123 materials, provided that both impurities substitute for Cu in the planes in the same manner in both cuprates, and that the planes are the only source for superconductivity[2]. We think it is unlikely that site distribution of the impurities (e.g., between planes and chains) is different for both materials. However, verification of the actual site of Zn and Ni in CaLaBaCuO (e.g. by neutron diffraction or NMR) is essential to clear this point. It is also possible that the ordered chains in YBCO, which are missing in CaLaBaCuO, make different contributions to the impurity effect in both materials. After all, these chains may also contribute to superconductivity[7].

References

1. A.V. Mahajan, H. Alloul, G. Collin, and J.F. Marucco, Phys. Rev. Lett. **72**, 3100 (1994).
2. P. Monthoux and D. Pines, Phys. Rev. **B49**, 4261 (1994) and this Workshop.
3. D. Goldschmidt, A. Knizhnik, Y. Direktovitch, G.M. Reisner, and Y. Eckstein, Phys. Rev. **B49**, 15928 (1994).
4. D. Goldschmidt, Y. Direktovitch, A. Knizhnik, and Y. Eckstein, Phys. Rev. **B51**, 6739 (1995).
5. D. Goldschmidt, A. Knizhnik, Y. Direktovitch, G.M. Reisner, and Y. Eckstein, Phys. Rev. **B52**, 12982 (1995).
6. H. Alloul, P. Mendels, H. Casalta, J.F. Marucco, and J. Arabski, Phys. Rev. Lett. **67**, 3140 (1991).
7. J.L. Tallon, APS March '96 Meeting, St. Louis.

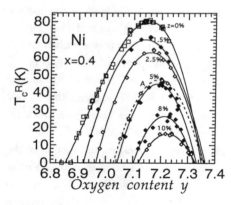

Fig. 1: T_c vs y for various Ni concentrations

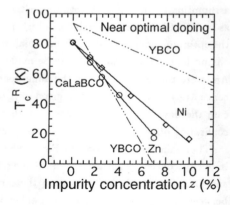

Fig. 2: The dependence of T_c^{max} on impurity concentration

PHYSICAL PROPERTIES OF INFINITE-LAYER AND T'-PHASE COPPER OXIDES

J. T. MARKERT, B. C. DUNN, A. V. ELLIOTT, K. MOCHIZUKI, AND R. TIAN
Department of Physics, The University of Texas at Austin
Austin, TX 78712, USA

> We report the effects of isoelectronic and n-type doping in the infinite-layer copper-oxide materials; we also mention briefly attempts at p-type doping of both the infinite-layer and T'-phase materials. Steric effects due to the introduction of isoelectronic Ca in $Sr_{0.90-x}Ca_xLa_{0.10}CuO_2$ cause a smooth and rapid reduction of T_c. Comparison with the 2-1-4 T'-phase materials suggests a universal behavior. The concentration dependence of T_c in $Sr_{1-x}La_xCuO_2$ is more like that of the hole-doped materials, suggesting that the unusual high-concentration onset of superconductivity in the 2-1-4 T' phase (at $x \approx 0.14$) is related to defects in that structure.

There are only two known electron-doped copper oxide superconductors: the 2-1-4 T' phase (e.g., $Nd_{2-x}Ce_xCuO_{4\pm y}$) and the infinite-layer materials (e.g., $Sr_{1-x}La_xCuO_2$). Both compounds have square-planar coordinated copper ions, i.e., ideally they have no apical oxygens. Recently, the T' phase has been shown[1] to have appreciable interstitial oxygen occupancy of the apical site. It is thus interesting to compare electron doping in that system, with its unusual onset of superconductivity at the relatively high concentration $x \approx 0.14$, with the same n-type doping in the infinite-layer material, where data indicate[2] that oxygen occupancy is ideal and stoichiometric (no interstitials).

Polycrystalline samples of $Sr_{1-x}La_xCuO_2$ and $Sr_{0.90-x}Ca_xLa_{0.10}CuO_2$ were prepared by high pressure synthesis in sealed platinum capsules (25 kbar at ~1040°C for 2 hr) after several preliminary grindings and firings at 900°C under ambient conditions. These conditions were previously found[3] to provide optimal properties. Data are reported for single-phased samples only; for $x > 0.12$, impurity phases were observed, confirming previous estimates of the solubility limit.[3]

In Fig. 1 are shown magnetization data for samples of $Sr_{0.90-x}Ca_xLa_{0.10}CuO_2$ with

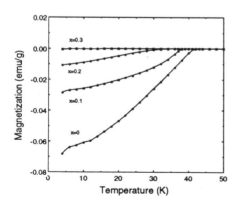
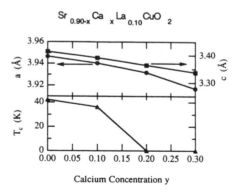

Fig. 1. Magnetization data for various x for $Sr_{0.90-x}Ca_xLa_{0.10}CuO_2$.

Fig. 2. Lattice constants and T_c as functions of x for $Sr_{0.90-x}Ca_xLa_{0.10}CuO_2$.

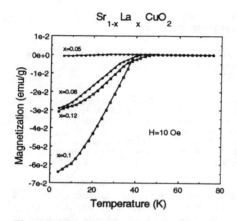
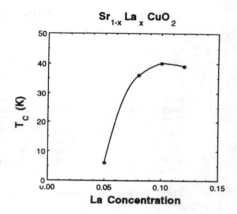

Fig. 3. Magnetization data for various x for $Sr_{1-x}La_xCuO_2$.

Fig. 4. T_c versus x for $Sr_{1-x}La_xCuO_2$.

various values of x. The superconducting volume fraction decreases with decreasing lattice constant; in Fig. 2, where T_c corresponds to 5% of ideal diamagnetism, superconductivity has essentially disappeared at $x \approx 0.2$, where the in-plane lattice constant of $a = 3.92$ Å. This is identical to the value observed in the electron-doped 2-1-4 T' phase,[4] suggesting a universal Cu-O bond tension required for electron doping.

In Fig. 3 are shown magnetization data for samples of $Sr_{1-x}La_xCuO_2$. Evidently, superconductivity occurs for even low concentrations ($x \approx 0.05$), in sharp contrast to the 2-1-4 T' phase, where, presumably due to the oxygen interstitials discussed above, superconductivity is suppressed, typically until $x \approx 0.14$. In Fig. 4 is shown explicitly the behavior of T_c (here, when χ is 1% of $-1/4\pi$). It is interesting that in the infinite-layer materials, where the bond distances are too small to permit interstitial occupancy, the concentration dependence is very similar to that of the hole-doped materials, although the encountered solubility limit prevents access to the overdoped region.

We have also attempted to hole dope both the 2-1-4 T' phase and the infinite-layer structure. For the smallest ambient-pressure T' phase series, $Gd_{2-x}M_xCuO_{4-y}$, we attempted the alkaline earth substitutions M = Sr, Ca, and Mg. The solubility limit for M = Sr is $x \approx 0.10$, but the M = Ca substituent is apparently insoluble; however, surprisingly, the even smaller M = Mg doping appears to have a larger maximum solubility, in the range $x \approx 0.2$–0.3. However, iodometric titration indicated an oxygen deficiency $y \approx 0.5x$, i.e., essentially zero net hole doping. That this structure does not seem amenable to hole doping may indicate insufficient bond compression or that apical oxygens are required.

For the infinite-layer phase, a somewhat larger range of bond compression is accessible. Therefore, we have attempted hole doping with various high-pressure syntheses in the series $Sr_{1-x}M_xCuO_2$ and $Sr_{1-x-y}Ca_xM_yCuO_2$ for M = Na and K. To date, no superconductivity has been observed in these materials.

This work was supported by the National Science Foundation under Grant No. DMR-9158089 and the Robert A. Welch Foundation under Grant No. F-1191.

References
1. P. G. Radaelli et al., Phys. Rev. B **49**, 15,322 (1994).
2. J. D. Jorgensen et al., Phys. Rev. B **47**, 14,654 (1993).
3. J.L.Cobb, A.Morosoff, L.Stuk, and J.T.Markert, Physica C **194-196**, 2247 (1994).
4. C. E. Kuklewicz and J. T. Markert, Physica C **253**, 308 (1995).

X-RAY SINGLE CRYSTAL STRUCTURE ANALYSIS OF $HgBa_2Ca_{n-1}Cu_nO_{2n+2+\delta}$ COMPOUNDS

H. SCHWER, J. KARPINSKI, K. CONDER, R. MOLINSKI, G.I. MEIJER
Laboratory of Solid State Physics, ETH Hönggerberg,
8093 Zürich, Switzerland

C. ROSSEL
IBM Research Division, Zurich Research Laboratory,
8803 Rüschlikon, Switzerland

ABSTRACT

The crystal structures of the homologous series $Hg_xPb_{1-x}Ba_2Ca_{n-1}Cu_nO_{2n+2+\delta}$ (n = 2 - 5, x = 0.5 - 1) have been determined and refined by X-ray single crystal diffraction. Pb-doped crystals have transition temperatures up to 129 K, but contain stacking faults. Lead and the excess oxygen atom are shifted off their ideal sites. Yttrium of the crucible is incorporated to 50 % at the Ca site in HgPb-1212, but not in other members of the HgPb-12(n-1)n family. Crystals without Pb have a Hg-deficiency but contain much less stacking faults. Structural parameters and bondlengths vary in a systemtic way as a function of the number n of CuO_2 layers in the structure.

1. Experimental

Single crystals of $Hg_xPb_{1-x}Ba_2Ca_{n-1}Cu_nO_{2n+2+\delta}$ (n = 2 - 5, x = 0.5 - 1) were grown with a gas phase high pressure technique from a PbO flux[1] at an Ar pressure p = 10 kbar. The samples were cooled down from 1050 - 1070 °C to 860 °C and annealed for two hours at that temperature. Superconducting transition temperatures range from T_c = 70 K (HgPbY-1212) to 129 K (HgPb-1234). The crystals are black plates up to 1 x 1 x 0.02 mm^3 with (001) as main faces. Single crystals without traces of intergrowth, twinning or peak broadening have been measured on a Siemens P4 X-ray single crystal diffractometer at room temperature, using MoKα radiation (λ = 0.71073 Å). The structures of HgPbY-1212, HgPb-1223, -1234, -1245 and of Hg-1223 crystals have been refined and typical results for each type are presented in table 1.

2. Results

2.1 HgPb-1223, -1234, -1245

The crystal structures were refined to R < 0.04. All HgPb-1223, 1234, and 1245 crystals showed characteristical peaks in the Fourier maps, which are interpreted as effects of single stacking faults of material with one more or one less $CaCuO_2$ unit than the main phase[2,3]. The electron densities of those peaks should be proportional to the number of stacking faults. The peaks were attributed to "ghost atoms" and by constraining their occupancies the same value, it was possible to refine the amount of stacking faults in HgPb-12(n-1)n crystals, which is usually 2 - 5 %.

The basal z = 0 plane of the crystal structure contains the excess oxygen atom at (.5, .5, 0) and Hg in the origin; Hg is substituted by 20 - 50 % Pb. All crystals showed shifts of the Pb and the excess oxygen atoms off their ideal positions by about 0.3 - 0.6 Å of the crystal structure. This is due to average Pb-O bondlengths (≈ 2.3 Å) which are too short to match the Hg-12(n-1)n unit cell (≈ 2.72 Å).

2.2 HgPbY-1212

All 1212 crystals grown in Y_2O_3 crucibles incorporated 50 % Y at the Ca site in the crystal structure. The distribution of Y was randomly, because no superstructure could be found. The lattice parameter a is larger and c is smaller than in undoped crystals and ceramic material. Almost no stacking faults were found, but the transition temperature is lower than in undoped material. Y incorporation was only observed in 1212, not in other members of the HgPb-12(n-1)n family.

2.3 Hg-1223

Usually, the Hg site is not fully occupied by mercury, and impurities like C and Cu are incorporated even to large extents[2-5]. In our undoped crystals, the Hg concentration was refined to about 80 %. Again, these crystals contained much less stacking faults than the Pb doped ones. The as-grown crystals had very small excess oxygen concentrations of about 5 %.

2.3 Trends in bondlengths

Bondlengths in 12(n-1)n materials indicate systematic trends with increasing number n of CuO_2 layers: Ba moves towards the basal plane, and the apex oxygen atom O(1) approaches to the CuO_2 plane. Buckling of the CuO_2 layers decreases from the outer planes to the central one, indicating the approach to the ideal $CaCuO_2$ structure[1] in the central part of the crystals structures.

Table 1. Refinement results of $Hg_xPb_{1-x}Ba_2Ca_{n-1}Cu_nO_{2n+2+\delta}$ crystals (space group P4/mmm).

	HgPbY-1212	HgPb-1223	Hg-1223	HgPb-1234	HgPb-1245
a [Å]	3.8630(4)	3.8580(2)	3.8577(3)	3.8530(5)	3.8529(3)
c [Å]	12.4683(9)	15.801(2)	15.715(1)	18.968(3)	22.172(2)
stacking faults	0	1.9(2) %	0	3.9 %	4.4 %
R, R_w [%]	3.40 3.10	2.70 2.50	2.73 2.35	2.18 1.92	3.3 2.57

References

1. J. Karpinski, H. Schwer, et al., *Physica* C **234** (1994) 10.
2. H. Schwer, J. Karpinski, et al., *Physica* C **243** (1995) 19.
3. H. Schwer, J. Karpinski, et al., *Physica* C **254** (1995) 7.
4. M. Kopnin, E.V. Antipov, et al, *Physica* C **243** (1995) 222.
5. J. Akimoto, Y. Oosawa, et al., *Physica* C **242** (1995) 360.

IN-PLANE ORDERING IN PHASE-SEPARATED AND STAGED SINGLE CRYSTAL $LA_2CUO_{4+\delta}$

X. XIONG,[1] P. WOCHNER,[2] and S. C. MOSS[1]

[1]*Physics Department and Texas Center for Superconductivity, University of Houston, Houston, TX 77204-5506*

[2]*Physics Department, Brookhaven National Laboratory, Upton, NY 11973*

ABSTRACT

A phase-separated and well-staged single crystal, $La_2CuO_{4.015}$, was studied by neutron scattering. Evidence for a periodic one-dimensional modulation of the in-plane order was observed with a splitting of the staging satellites $(0,k,\ell\pm0.16$; with k odd and ℓ even) along the a* direction through a long annealing from 213K to 185K. Within the above temperature range the a*-modulated phase and the unmodulated phase, both with staging along the c* direction, coexist. The modulation period decreases on cooling and is 69Å at 185K. We suggest that an ordering into laterally separated (vertically offset), and thus phase shifted, domains is responsible for the splitting of the staging satellites. These split peaks are broad (FWHM is 0.09 in r.l.u. along h), indicating a distributed mixture of the in-plane domains. Calculations of the structure factor reproduce the experimental results and lead to a complete picture of the equilibrium domain structure in this superconductor.

Wells et al.[1] have shown in single crystal neutron scattering studies of $La_2CuO_{4+\delta}$ that the oxygen-rich phases are staged on cooling as also occurs in the well-documented $La_2NiO_{4+\delta}$ studies of Tranquada et al.[2] We summarize here our evidence for a reversible temperature-induced in-plane superstructure formation along the a-axis after staging in $La_2CuO_{4+\delta}$.[3]

The onset temperature of phase separation (T_{PS}) into oxygen-poor ($\delta\sim0$) and oxygen-rich ($\delta\sim0.05$) phases and the Néel temperature (T_N) of the oxygen-poor phase (Bmab) were determined to be 270K and 260K, respectively, in our single crystal. Typical ℓ-scans of the (0 1 4) reflection at 240K and 224K after slow coolings are shown in Fig.1. The central peak (reduced in intensity by 1/60 in this figure) is the (0 1 4) reflection of the Bmab phase while both left and right peaks are first-order staging satellites of the oxygen-rich phase. At 250K, when the satellites first appear on cooling, their position corresponds to stage 7, but is temperature-dependent, as in the insert of Fig.1, and saturates below 213K at 6.5.[3] On heating, the satellites show ~10K hysteresis. The temperature dependence of the integrated intensity of an ℓ-scan through $(0,1,4+\Delta)$ ($\Delta\approx0.150$-0.153) reveals, below 213K, an intensity drop coinciding with a lock-in of the staging.[3]

This intensity decrease on cooling is due to a further (in-plane) splitting of the staging satellites along the a* direction. High resolution h-scans on the satellite (0,1,4.153) at several temperatures are presented in Ref.4. We name the new phase the "a*-modulated phase" after the splitting along h and refer to "unmodulated phase" before the h-splitting, but both of retain their staging along the c-axis. The development of the a*-modulated phase and the coexistence of these two phases may be clearly seen in Ref.4 at 200.5K and 194.5K. The modulation period decreases on cooling which must be attributed to in-plane diffusion of the interstitial oxygens. The average modulation period along the a-axis is ~13 unit cells, or ~69Å from the fitting results at 185K.

We suggest a simple model as shown in the insets of Fig.2 to interpret the observed h-splitting on the staging satellites. In model (A), the interstitial oxygens are distributed randomly along the a-axis while still staged along the c-axis but with a smearing of the boundaries to smooth the tilt reversal. In model (B), the interstitial oxygens are segregated laterally to form vertically offset domains that remain individually staged. The modeling of these results based on (A) and (B), were obtained by a structure factor calculation convoluted with the instrumental resolution[4] and are compared to experimental data in Fig.2. At 213K model (A) shows only a central staging peak as expected for this temperature. In Fig.2b a random mixture of domains offset up to (up or down) three layers is considered because of the broad h-split width (0.09 in r.l.u. at 185K) and the calculated result is in excellent agreement, including the missing higher-order peaks. The detailed calculations for these long striped domains are presented in Ref.4.

Acknowledgments

This work was supported at Houston by the Texas Center for Superconductivity (TCSUH) and the NSF on DMR-9208420 and at BNL by the Division of Materials Research, US DOE contract number DE-AC-02-76CH00016.

References

1. B. Wells, et al. (submitted).
2. J. M. Tranquada et al., Phys. Rev. B 50, 6340 (1994).
3. X. Xiong et al., Phys. Rev. Lett (in press).
4. P. Wochner et al. (manuscript in preparation).

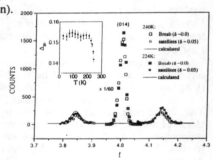

Fig.1 The first order satellite pair in a slow cooling to 240K and to 224K in the oxygen-rich phase (δ~0.05); the central peak (at 1/60 intensity) is the reflection (0 1 4) of the Bmab phase (δ~0.0). The line is calculated using a simple staging model for each temperature. The insert shows the temperature dependence of the satellite peak position.

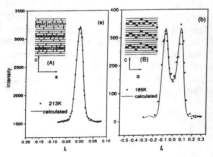

Fig. 2 a) The experimental data and calculated results based on model (A) at 213K. In the schematic inset the solid lines are CuO layers and the circles are interstitial oxygens. b) A calculated result based on model (B) using a Hendricks-Teller model along the a-axis. The schematic inset corresponds to model (B).

III. THEORY

III. THEORY

Mechanisms of Superconductivity

ANTIFERROMAGNETIC REAL-SPACE SCENARIO FOR THE CUPRATES

ELBIO DAGOTTO

Department of Physics, and National High Magnetic Field Lab, Florida State University, Tallahassee, FL 32306, USA

A model of weakly interacting hole quasiparticles for the cuprates is briefly reviewed. Strong antiferromagnetic correlations dress the holes inducing unusual features in the quasiparticle dispersion. Of special importance is the presence of flat bands that resemble those found in photoemission experiments. These flat bands produce a large density of states (DOS) that enhances the critical temperature once hole attraction is included in the problem. We study the influence of a finite hole density on these results. It is shown that even for short antiferromagnetic correlation lengths, d-wave remains the dominant channel for superconductivity. The chemical potential crosses a large peak in the DOS as we move away from half-filling. The possibility of s-wave superconductivity (SC) in the overdoped regime is also discussed.

Some of the theories proposed for the cuprates use antiferromagnetic correlations to induce superconductivity at finite hole density and low temperatures. In this context, it is believed that short-range antiferromagnetism is enough to produce pairing. The hole attraction is better visualized in real-space as the sharing of spin polarons. In this paper we review a "real-space" scenario based on antiferromagnetism, which was proposed recently by A. Nazarenko, A. Moreo and the author,[1] mainly concentrating on recent results that addressed the stability of the scenario against hole doping.[2]

The presence of "flat" regions in the normal-state q.p. dispersion is a remarkable feature of the phenomenology of hole-doped high temperature superconductors.[3] These flat bands are located around momenta $\mathbf{p} = (\pi, 0)$ and $(0, \pi)$, and at optimal doping they are $\sim$ 10meV below the Fermi energy, according to angle-resolved photoemission (ARPES) studies.[3] Antiferromagnetic (AF) correlations likely play an important role in the generation of these features, as has been suggested by studies of holes in the 2D t-J and Hubbard models at and away from half-filling.[4,5] The large peak in the DOS induced by the flat regions can be used to boost T_c. This leads to a natural explanation for the existence of an "optimal doping" which in this framework occurs when the flat regions are reached by μ. We will refer to these ideas as the "Antiferromagnetic van Hove" (AFVH) scenario.[1] Superconductivity in the $d_{x^2-y^2}$ channel is natural in the AFVH scenario due to the strong AF correlations.[6] The interaction of holes is better visualized in *real space* i.e. with pairing produced when dressed holes share a spin polaronic cloud, as in the spin-bag mechanism.[7] This real-space picture holds even for a small AF correlation length, ξ_{AF}. Previous scenarios have also used van Hove (vH) singularities in the band structure to increase T_c,[8] but d-wave SC is not natural in this context unless AF correlations are included.

To obtain quantitative information from these intuitive ideas, holes moving with a dispersion calculated using one hole in an AF background, $\epsilon_{AF}(\mathbf{p})$, and interacting through a nearest-neighbors (NN) attractive potential, that mimics the sharing of spin polarons, have been previously analyzed.[1] Note that the hole attraction could also be induced by a phononic mechanism,[9] but here we will only concentrate on AF-induced pairing. Within a

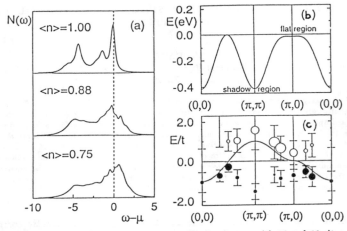

Figure 1: (a) $N(\omega)$ for the 2D t-J model averaging results for clusters with 16 and 18 sites, at $J/t = 0.4$ and for the densities indicated. The δ-functions were given a width $\eta = 0.25t$. Similar results were found at other values of J/t; (b) q.p. energy vs momentum obtained at half-filling.[4] The result shown, $\epsilon_{AF}(p)$, is a good fit of Monte Carlo data on a 12×12 cluster at $J/t = 0.4$; (c) q.p. dispersion vs momentum at $\langle n \rangle = 0.87$ and $J/t = 0.4$ using exact diagonalization of 16 and 18 sites clusters.[11] The open (full) circles are IPES (PES) results. Their size is proportional to the peak intensity. The solid line is $\epsilon_{NN}(p)$.

rigid band filling of $\epsilon_{AF}(p)$ and using a BCS formalism, $d_{x^2-y^2}$ superconductivity dominates with $T_c \sim 100K$ caused by the large DOS implicit in the hole dispersion. $2\Delta/kT_c \sim 5$ was reported in this context.[1] The existence of d-wave SC in this model can be understood simply from the two body problem.[10] The size of the pairs is just of a few lattice spacings. These tight pairs lead us to believe that a region of *preformed pairs* in the normal state should exist in the cuprates.

To study the presence of a large DOS peak in the dispersion of the t-J model at finite hole density, $N(\omega)(= \sum_p A(p,\omega))$ has been calculated numerically in systems with strong AF correlations.[2] In Fig.1a, $N(\omega)$ for the 2D t-J model obtained with exact diagonalization (ED) techniques is shown at several densities. At half-filling, a large DOS peak caused by flat regions in the q.p. dispersion appears at the top of the valence band, as discussed before.[4] Weight exists at energies far from μ i.e. the large peak carries only part of the total weight. The maximum in the DOS is *not* reached at the top of the valence band but at slightly smaller energies.[4] As $\langle n \rangle$ decreases, the peak is now much broader but it remains well-defined. At $\langle n \rangle = 0.87$, μ is located close to the energy where $N(\omega)$ is maximized. If a source of hole attraction exists in this system, a SC gap would open at μ, and the resulting T_c could be large. For $\langle n \rangle \sim 0.75$, μ moves to the left of the peak, where T_c should become smaller. Then, as $\langle n \rangle$ is reduced away from half-filling in the 2D t-J model Fig.1a suggests that μ travels across a broad DOS peak providing a natural definition for the "underdoped, optimal and overdoped" densities.

Now let us discuss the p-dependence of the q.p. band obtained from $A(p,\omega)$. Results

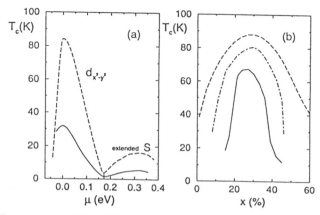

Figure 2: (a) T_c vs μ obtained with the BCS gap equation, with $\mu = 0$ as the chemical potential corresponding to the saddle-point. The dashed line corresponds to $\epsilon_{AF}(p)$, and the solid line to $\epsilon_{NN}(p)$. Note the presence of both $d_{x^2-y^2}$-wave and extended s-wave SC; (b) T_c for d-wave SC vs the percentage x of filling of the q.p. band (*not* of the full hole spectrum). The dashed line is the same as in (a). The solid line corresponds to the BCS gap equation result making zero the weight Z_p of states in the q.p. dispersion that are at energies from the saddle point larger than 2.5% of the total bandwidth (i.e. including states only in a window of energy $\sim 125K$ around the flat regions). The dot-dashed line is the same but using a window of $\sim 250K$.

are already available in the literature.[5,11] In Fig.1b,c, the q.p. dispersion is shown at $\langle n \rangle = 1$, and at $\langle n \rangle \sim 0.87$ for the t-J model.[11] Note that upon doping the small q.p. bandwidth and the flat regions remain well-defined, inducing a large DOS peak (Fig.1a). However, the region around $p = (\pi, \pi)$ has changed substantially, i.e. the AF induced region clearly observed at $\langle n \rangle = 1$ has reduced its intensity and considerable weight has been transferred to the inverse PES region (Fig.1c). Actually, the q.p. dispersion at $\langle n \rangle \sim 0.87$ can be fitted by a tight-binding nearest-neighbors (NN) dispersion with a small effective hopping.[12,5,2] The changes in the quasiparticle dispersion could be interpreted using dressed quasiparticle operators. However, the difference between Figs.1b and 1c could correspond to an intrinsic change in the q.p. dispersion as ξ_{AF} decreases. This is the less favorable case for the AFVH approach, and thus we should analyze this possibility in detail. For this purpose, we applied the BCS formalism to a model with a low density of q.p.'s having a dispersion $\epsilon_{NN}(p)/eV = -0.2(cosp_x + cosp_y)$, to reproduce the dominant features of Fig.2b. As before, we include a NN attraction $V = -0.6J$ between q.p.'s,[1] with $J = 0.125eV$, for both dispersions $\epsilon_{AF}(p)$ and $\epsilon_{NN}(p)$. Solving numerically the gap equation, T_c is shown in Fig.2a. SC at $T_c \sim 80 - 100K$ in the $d_{x^2-y^2}$ channel appears naturally if the AF-dispersion is used.[1] If, instead, $\epsilon_{NN}(p)$ is used, the flat bands present in this narrow dispersion produce $T_c \sim 30K$ which is still large. Even more remarkable is the fact that the $d_{x^2-y^2}$ character of the SC state is maintained. This result can be understood noticing that a combination of $\epsilon_{NN}(p)$ with an attractive NN potential effectively locates us in the family of "t-U-V" models with U repulsive and V attractive, where it is known that for a "half-filled" band the dominant SC state is $d_{x^2-y^2}$-wave. In other words, when μ is at the flat region in Fig.1c it approximately corresponds to a "half-filled" q.p. band, leading to a $d_{x^2-y^2}$-wave SC state (Fig.2a). Then, even if the q.p. dispersion changes with doping near the Q point, such an effect does not alter the main qualitative features found in previous studies.[1]

The robust T_c mostly arises from the influence of the flat bands in the spectrum. Actually, we analyzed T_c using $\epsilon_{AF}(p)$ but modulating the contribution of each momentum with a p-dependent weight Z_p in the one particle Green's function. We considered the special case where Z_p is zero away from a window of total width W centered at the saddle point, which is located in the flat bands region. Inside the window W, $Z_p = 1$. Results are shown in Fig.2b, for d-wave SC. Note that even in the case where W is as small as just 5% of the total bandwidth (itself already small of order $2J$), T_c remains robust $\sim 70K$.

We finish this paper with an unexpected prediction[2] obtained from $\epsilon_{AF}(p)$ and $\epsilon_{NN}(p)$, when μ reaches the bottom of these bands (which should *not* be confused with the bottom of the full hole spectrum since a large amount of weight lies at energies lower than those of the q.p. band). In this regime, the BCS analysis applied to any of the two dispersions shows that extended s-wave SC dominates over $d_{x^2-y^2}$-wave SC in the "overdoped" regime (Fig.2a). This occurs when the q.p. band is nearly empty which, from Fig.1a, corresponds to an overall density of $\langle n \rangle \sim 0.7$. This symmetry change of the SC state can be understood based again on the analogy with the "t-U-V" model. Other recent theories and experiments have also discussed the possibility of s-wave SC in the overdoped regime.[13]

Acknowledgments

E. D. is supported by grant NSF-DMR-9520776.

References

1. E. Dagotto, A. Nazarenko and A. Moreo, Phys. Rev. Lett. **74**, 310 (1995).
2. A. Nazarenko, S. Haas, J. Riera, A. Moreo and E. Dagotto, preprint.
3. D.S. Dessau et al., Phys. Rev. Lett. **71**, 2781 (1993).
4. E. Dagotto, A. Nazarenko and M. Boninsegni, Phys. Rev. Lett. **73**, 728 (1994).
5. N. Bulut, D. J. Scalapino and S. R. White, Phys. Rev. B **50**, 7215 (1994).
6. D. J. Scalapino, Phys. Rep. **250**, 331 (1995).
7. A.Kampf and J.R.Schrieffer,Phys.Rev.B **41**,6399(1990).
8. C. C. Tsuei et al., Phys. Rev. Lett. **65**, 2724 (1990); R. S. Markiewicz, J. Phys. Condens. Matt. **2**, 6223 (1990); A. A. Abrikosov et al., Physica C **214**, 73 (1993).
9. A. Nazarenko and E. Dagotto, Phys. Rev. B **53**, R2987 (1996).
10. A. Nazarenko, et al., preprint.
11. A. Moreo et al., Phys. Rev. B **51**, 12045 (1995).
12. E. Dagotto, F. Ortolani and D. Scalapino, Phys. Rev. B **46**, 3183 (1992).
13. J. Ma, et al., Science **267**, 862 (1995); B. E. C. Koltenbah and R. Joynt, preprint; E. Dagotto et al., Phys. Rev. B**49**, 3548 (1994).

CHARGE INHOMOGENEITY AND HIGH TEMPERATURE SUPERCONDUCTIVITY

V.J. EMERY
Dept. of Physics, Brookhaven National Laboratory, Upton, NY 11973-5000, USA

S.A. KIVELSON
Dept. of Physics, UCLA, Los Angeles, CA 90095, USA

When the condensation of a gas of fermions into a (self-bound) liquid state is frustrated by the long-range Coulomb interaction, the consequence is a large local fluctuation of the charge density, together with pairing on a high energy scale. The competition between these two effects at long length scales determines the nature of the ordered state at low temperatures. Evidence for the central role of this competition in determining the physical properties of the high temperature superconductors is provided by the delicate interplay of superconductivity, charge and spin ordering, and structural phase transformations in the La_2CuO_4 family of materials. There the gas-liquid transition corresponds to the phase separation of holes doped into an antiferromagnetic insulator. Because of the low superfluid density and poor conductivity, the critical temperature for the superconducting transition in underdoped and optimally doped materials is governed by the onset of phase coherence and not by the pairing energy scale.

1 Introduction

The fundamental driving force for high temperature superconductivity and for many of the unusual properties of the cuprates is the frustration of the motion of holes in an antiferromagnet. *Neutral holes* are unstable to phase separation into an antiferromagnetically ordered insulating region and a hole-rich metallic region [1,2]. This is the best compromise between the hole kinetic energy and the exchange interaction between the spins. Since superexchange is itself a kinetic process, phase separation minimizes the total zero-point energy of the strongly-correlated system. *For charged holes* the long-range Coulomb interaction frustrates phase separation, unless the dopants are mobile enough to counterbalance the effect of the long-range Coulomb interaction, as in the case of photodoping or oxygen doping [3,4]. Of course this does not mean that phase separation is without consequence. Between the spinodals, the instability of the neutral system (which is signalled by a negative compressibility) is converted by the Coulomb interaction into charge inhomogeneity on intermediate time scales or length scales, and especially into ordered or fluctuating charge stripes [5,6]. As we shall see, the other consequence is local pairing on a high energy scale. Typically, charge ordering and superconducting phase coherence compete at long length scales, although they may coexist in specific regions of the phase diagram [7].

There is extensive experimental evidence to show that: 1) Phase separation is the fate of neutral holes in an antiferromagnet [3,4]. 2) The La_2CuO_4 family displays ordered [8] or fluctuating [9,10,11] charge stripes together with antiphase spin domains. 3) Charge order and superconductivity compete with each other [12,13,14].

The existence of charge-density wave fluctuations adds a new dimension to discussions of the physics of the high temperature superconductors which, almost from the outset, have been dominated by the interplay of antiferromagnetism and superconductivity. The clearest evidence for this behavior is to be found in the La_2CuO_4 family although, as we shall see, angle resolved photoemission data on other materials may also be understood from this point of view. Nevertheless the three phenomena are simply different manifestations of a common underlying physical theme, and the best way to investigate any one of them is to find the material in which it shows up the most clearly.

The charge stripes are topological defects, which form antiphase domain walls in the antiferromagnetic order of the background spins [8]. We have argued [15] that this kind of *"topological doping"* is a common feature of correlated insulators such as polyacetylene (for which the defects are solitons) and the high temperature superconductors, and that it is responsible for the rapid restoration of the symmetry of the ground state upon doping; non-topological defects would produce much more modest reduction in the magnitude of the order parameter, proportional to the dopant concentration.

2 Evidence for Stripes.

Recent neutron scattering studies [8] of $La_{1.6-x}Nd_{0.4}Sr_xCuO_4$ with $x = 0.12$ found a succession of transitions; first to the low temperature tetragonal (LTT) structure, then to a charged-ordered state (charge-stripes) and finally, at a slightly lower temperature, to a period-doubling magnetically-ordered state. We conclude that the transitions are driven by frustrated kinetic phase separation because the charge order appears first and the relatively low charge density along the stripes (one doped hole per two Cu ions) allows the holes to be mobile. The alternative route to stripe order, (a Hartree-Fock, Fermi surface instability [16]) predicts a single transition to a charge and spin ordered insulating state, with one doped hole per Cu, twice as large as observed.

$La_{1.6-x}Nd_{0.4}Sr_xCuO_4$ with $x = 0.12$ does not exhibit bulk superconductivity. But there are substantial low-energy stripe fluctuations in optimally-doped, superconducting samples of $La_{2-x}Sr_xCuO_4$; inelastic neutron scattering experiments [9,10,11] have found peaks at energies as low as a few meV and in essentially the same positions in k-space as in (ordered) $La_{1.6-x}Nd_{0.4}Sr_xCuO_4$. Clearly the stripe correlations are an essential part of the physics of these materials.

The single-particle properties of a disordered stripe phase [17] also account for the peculiar features of the electronic structure of high temperature superconductors observed by angle-resolved photoemission spectroscopy (ARPES) on hole-doped high temperature superconductors. In particular, the spectral function of holes moving in a disordered stripe background reproduces the experimentally-observed shape of the Fermi surface and the existence of nearly dispersionless states in its neighborhood. The essential ingredient in this interpretation of the data is a background of slowly-fluctuating stripes whose dynamics is determined by collective effects (the competition between phase separation and the long-range Coulomb force), rather than the single-particle behavior.

3 The Mechanism

As might be expected, an appreciation of the electronic structure and the underlying physics also leads to a natural mechanism of high temperature superconductivity. When the holes on the stripes tunnel into the intervening region, which has a low density of doped holes and strong antiferromagnetic correlations, they experience the very same effective attraction that would have caused them to coalesce (phase separate) in the absence of the longer range Coulomb interaction. In fact, the holes moving on a stripe form a quasi one-dimensional electron gas in an "active" environment which has low-energy excitations in the spin (and possibly charge) degrees of freedom. The environment renormalizes the kinetic energy of the holes and mediates effective interactions between them [18]. We have found several such processes that lead to pairing, *even though the basic Hamiltonian contains only repusive interactions* [18]. In particular, when two holes hop into the antiferromagnetic environment, they may form a bound state, which typically has a large binding energy and a low mobility, and cannot, by itself, lead to high temperature superconductivity. However, the holes in the stripe are able to combine this large binding energy with their own mobility and achieve a high superconducting transition temperature [18].

4 Phase Fluctuations

It is possible to define a characteristic temperature for superconducting phase ordering T_θ^{max} which is proportional to the phase stiffness $V_0 = (\hbar c)^2 d/16\pi e^2 \lambda^2(0)$ where $\lambda(0)$ is the zero-temperature penetration depth and d is a the spacing between the CuO_2 planes. We have shown [19] that, if $T_c \ll T_\theta^{max}$, then phase fluctuations are relatively unimportant and T_c is likely to be close to the mean-field transition temperature T^{MF} predicted by BCS theory. On the other hand, if $T_\theta^{max} \sim T_c$, then the value of T_c is determined primarily by phase ordering, and T^{MF} is simply the characteristic temperature below which local pairing becomes significant.

The value of T_θ^{max}/T_c for the cuprate superconductors is of order unity [19], which places them in a quite different category from conventional superconductors, for which the ratio can be as large as 10^5. Indeed, T_c is predominantly determined by phase fluctuations in underdoped high temperature superconductors and by the mean-field transition temperature T^{MF} in overdoped materials such as Tl 2201 ($T_\theta^{max}/T_c \geq 2$). Optimally doped materials, such as $YBa_2Cu_3O_{7-\delta}$ and $La_{2-x}Sr_xCuO_4$, with δ and x in the neighborhood of 0.05 and 0.15 respectively, are in the crossover region between the two.

When T_c is much smaller than T^{MF}, the existence of pairing without phase coherence should be manifested as a pseudogap in the the temperature range $T_c < T < T^{MF}$. Such a pseudogap has been observed in underdoped samples of high temperature superconductors, such as $YBa_2Cu_3O_{7-\delta}$ and $Bi_2Sr_2CaCu_2O_{8+x}$ and in the stoichiometric material $YBa_2Cu_4O_8$. The pseudogap has been seen in spin [20], charge [21,22,23], and single-electron [24] properties. Moreover, the central peak in the a-axis optical conductivity (which becomes a δ-function in the superconducting state) already narrows above T_c [25] in $YBa_2Cu_4O_8$ and underdoped $YBa_2Cu_3O_{7-\delta}$. The prediction [19] that T_c is supressed by phase fluctuations in organic conductors has been confirmed by the observation [26] of pseudogap behavior in the nuclear spin relaxation rate. Taken as a whole, these experiments strongly support the existence of a phase fluctuation regime in the cuprate and organic superconductors.

Acknowledgements: We have benefitted from discussions with John Tranquada on all aspects of this problem and especially on charge ordering in the nickelates and cuprates. This work was supported in part by NSF grant #DMR-93-12606 and by the Division of Materials Sciences, U.S. Department of Energy, under contract DE-AC02-76CH00016.

References

1. V.J. Emery and S.A. Kivelson, and H-Q Lin, *Phys. Rev. Lett.* **64**, 475-478 (1990).
2. S.A. Kivelson and V.J. Emery, in *Strongly Correlated Electronic Materials: The Los Alamos Symposium 1993*, ed. by K.S. Bedell, *et al.* (Addison Wesley, Redwood City, 1994) p. 619.
3. V.J. Emery and S.A. Kivelson, Physica C **209** 594 (1993).
4. *Phase Separation in Cuprate Superconductors*, eds. K. A. Müller and G. Benedek (World Scientfic Singapore, 1993); *Proc. of the Second and Third International Confs. on Phase Separation in Cuprate Superconductors*, Cottbus, Germany, 1993 and Erice, Sicily, July 1995.
5. U. Löw *et al. Phys. Rev. Lett.* **72**, 1918 (1994).
6. L. Chayes *et al.*, Physica **A**, in press.
7. V.J. Emery and S.A. Kivelson, to be published.
8. J. Tranquada *et al.*, *Nature* **375**, 561 (1995) and to be published.
9. S-W. Cheong, *et al.*, *Phys. Rev. Lett.* **67**, 1791-1794 (1991).
10. T.E. Mason, *et al.*, *Phys. Rev. Lett.* **68**, 1414-1417 (1992).
11. T. R. Thurston, *et al.*, *Phys. Rev.* B **46**, 9128-9131 (1992).
12. B. Büchner it et al., *Phys. Rev. Lett.* 73, 1841 (1994).
13. M. Nohara it et al., *Phys. Rev. Lett.* 70, 3447 (1993).
14. C-H. Lee it et al., Sendai preprint.
15. V.J. Emery and S.A. Kivelson, Sissa preprint cond-mat/9603009, to be published in Synthetic Metals.
16. J. Zaanen and O. Gunnarsson, *Phys. Rev.* B **40**, 7391 (1989); papers cited in ref. 2.
17. M.I. Salkola, V.J. Emery and S.A. Kivelson, *Proc. of the Third International Conf. on Phase Separation in High T_c Superconductors*, Erice, July, 1995 (in press) and submitted to *Physical Review Letters*.
18. V.J. Emery and S.A. Kivelson, O. Zachar, to be published.
19. V.J. Emery and S.A. Kivelson, *Nature* **374**, 434 (1995).
20. M. Mehring, Appl. Magn. Reson. **3**, 383-421 (1992).
21. D.N. Basov, *et al.*, *Phys. Rev.* B **50**, 3511-3514 (1994).
22. P. Wachter, B. Bucher, and R. Pittini, *Phys. Rev.* B **49**, 13164-13171 (1994).
23. D.N. Basov, *et al.*, *Phys. Rev. Lett.* **74**, 598-601 (1995).
24. D. S. Marshall *et al.*, (Stanford University Preprint) submitted to *Phys. Rev. Lett.*.
25. D.N. Basov *et al.*, McMaster University preprint.
26. H. Mayaffre *et al.*, *Europhys. Lett.*, **28**, 205 (1994).

SUPERCONDUCTIVITY MEDIATED BY SCREENED COULOMB FIELD

J. D. FAN and Y. M. MALOZOVSKY

Department of Physics, Southern University
Baton Rouge, Louisiana 70813, USA
e-mail: phjola@lsuvax.sncc.lsu.edu

ABSTRACT

A graphic functional derivative method based on Ward's identity is introduced to obtain the contributions to the interparticle interaction. The pairing mechanism mediated by the screened Coulomb field and discussed in this article is an extension of the conventional Cooper pairing. The scattering diagrams from two methods are the same and are used to calculate the two particle interaction kernel K and hence the vertex function Γ. The transition temperature T_c at which instability occurs is obtained by the pole condition of $\Gamma^{-1} = 0$. It has been found that T_c may be quite high in a layered two-dimensional Fermi gas and gives rise to the bell shape. Of more importance is that T_c is related to the dielectric constant, electron band mass, doping density and interlayer distance.

1. A Perturbative Treatment of the Vertex Part

The single-particle self-energy with the spin indices incorporating the vertex correction can be represented by [1]

$$\Sigma_\sigma(p) = i \sum_{\sigma'} \int \frac{d^4k}{(2\pi)^4} G_{\sigma'}(p-k) D(k) \Gamma_{\sigma'\sigma}(p,k), \qquad (1)$$

where p and k are four-dimensional momenta, $G_\sigma(k)$ is the dressed electron Green's function, $\Gamma_{\sigma\sigma'}(p',k)$ the three-point vertex function, and $D(k)$ is the boson Green's function satisfying Dyson's equation $D(k) = D_0(k) + D_0(k)\widetilde{\chi}(k)D(k)$ with $D_0(k)$ being the zeroth-order boson Green's function and $\widetilde{\chi}_{\sigma\sigma'}(k) = -i \int \frac{d^4p'}{(2\pi)^4} G_\sigma(p') \Gamma_{\sigma\sigma'}(p',k) G_{\sigma'}(p'-k)$ being the irreducible response function. For example, $D_0(k) = V(k)$ in the case of the Coulomb interaction, while $D_0(k) = D_{0ph}(k,\omega) = 2g_0^2(k)\omega_0(k)/(\omega^2 - \omega_0^2(k) + i\delta)$ in the case of the electron-phonon interaction.

The interaction kernel satisfies Ward's identity [1], $K^{\sigma\sigma'}(p,p') = -i \, \delta\Sigma_\sigma(p)/\delta G_{\sigma'}(p')$. To evaluate this kernel the iteration procedure is used. First of all, take the self energy in the random phase approximation (RPA), i.e. let $\Gamma_{\sigma\sigma'}(p,k) = \delta_{\sigma\sigma'}$ and $\widetilde{\chi}(k) = \chi_0(k)$ in Eq. (1), where $\chi_0(k) = -i \sum_\sigma \int \frac{d^4q}{(2\pi)^4} G_\sigma(q) G_\sigma(q-k)$. Using the relation $dG_\sigma(p)/dG_{\sigma'}(p') = \delta(p-p')\delta_{\sigma\sigma'}$, the kernel in the RPA can thus be derived as

$$K^{\sigma\sigma'}(p,p') = K_1^{\sigma\sigma'}(p,p') + K_2^{\sigma\sigma'}(p,p'), \qquad (2)$$

where

$$K_1^{\sigma\sigma'}(p,p') = D(p-p')\delta_{\sigma\sigma'} \qquad (2a)$$

and

$$K_2^{\sigma\sigma'}(p,p') = -i \int \frac{d^4q}{(2\pi)^4} D^2(p-q) G_\sigma(q) [G_{\sigma'}(p'-p+q) + G_{\sigma'}(p'+p-q)] \qquad (2b)$$

with the boson propagator D(k) in the RPA. The diagrams for these interactions are shown in Fig. 1. If the bisection (cutting) of any electron line is, however, introduced to represent the procedure of the graphic functional derivative of $\delta G_\sigma(p)/\delta G_{\sigma'}(p') = \delta(p-p') \delta_{\sigma\sigma'}$, it is seen that the cutting at a on the top graph of Fig. 1 reproduces Fig. 1(a), the cuttings at b and c of the particle-hole bubble with consideration of particle-hole excitations yield the other two sub-graphs. Using the cutting technique one more time, one can obtain the diagrams issued on the basis of Fig. 1(a, b & c). Therefore, the introduced graphic derivative method provides a way to obtain step by step the sequential diagrams of the interaction kennel in any order based on Ward's identity. This method presents an approach to the desired level beyond the RPA and avoid the arbitrariness in the selection of diagrams.

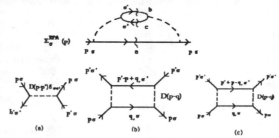

Fig. 1 Diagrams of the interaction kernel K(p,p') beyond the RPA

2. A New Pairing Mechanism

Two electrons above the Fermi sea can only come from the Fermi sea and simultaneously leave two holes in it. From the concept of the Landau Fermi liquid, quasielectron and quasihole appear and disappear in pairs [1]. This implies that scattering between electron and electron (e-e), electron and hole (e-h) as well as hole and hole (h-h) should be taken into account. Nevertheless, only the e-e scattering was considered in the original Cooper pairing due to the assumption of a quiescent Fermi sea. [2]

A positronium molecule-like (bi-exciton) model is suggested to understand the pairing between e-e (h-h) in k-space. Two positronium-like atoms, each of which consists of one electron and one hole, may form a positronium molecule-like structure, in which e-e (h-h) are tightly bound together due to the resonating valence bond (RVB) analogous to the treatment by Heitler and London [3] in a hydrogen molecule.

With this picture in mind one may construct a complete pattern of the scattering diagrams as shown in Fig. 2. Note that there are two possibilities for the final momentum of an electron (hole) after scattering: it is either equal to or different from the initial. For the single scattering there is only one possibility because of the momentum conservation: different initial and final momenta. Three diagrams in Fig. 2(a) correspond to the e-e, e-h and h-h scattering and are identical. One can take either of them in calculations. For the double scattering, however, one has the e-e, e-h and h-h scattering with both the same Fig. 2(b) and different Fig. 2(c) initial and final momenta. It is also easy to see that the ladder approximation of Fig. 2(a) will reproduce diagrams in Fig. 2(c) which can thus be excluded from the basic diagrams. Since the h-h scattering is exactly identical with the e-e scattering, the third diagrams in Fig. 2 (b) should be omitted in the consideration of the basic diagrams. Therefore, the basic diagrams are composed of those chosen from one of the diagrams in Fig. 2(a) and the first two in Fig. 2(b) and represented in Fig. 2(d). In regard to the triple and multiple scattering, it is apparent that they can be reproduced by the ladder approximation of Fig. 2(d). It is seen that Fig. 2(d) is the same as Fig. 1, which was obtained in the first step beyond the RPA by using Ward's identity one time.

3. Superconductivity

The diagrams in Fig. 2 (d) or Fig. 1 can be used to calculate [4] the two-particle interaction kernel in the charge channel, $K_c(p,p')$, with four-dimensional momenta $p = (\mathbf{p},\omega)$ and $p' = (\mathbf{p}',\omega')$. It is found that

$$K_c(\mathbf{p},i\omega;\mathbf{p}',i\omega') = D(\mathbf{p}-\mathbf{p}') + 2\sum_q D^2(\mathbf{p}-\mathbf{q}) \frac{n_F(\xi_q) - n_F(\xi_{q-p+p'})}{\xi_q - \xi_{q-p+p'} - i(\omega-\omega')}, \qquad (3)$$

$$- 2\sum_q D^2(\mathbf{p}-\mathbf{q}) \frac{1 - n_F(\xi_q) - n_F(\xi_{p+p'-q})}{i(\omega+\omega') - \xi_q - \xi_{p+p'-q}}$$

where $D(\mathbf{p}-\mathbf{p}')$ is taken to be frequency-independent in the particle-hole channel and $n_F(\omega) = [\exp(\omega/T) + 1]^{-1}$ is the Fermi distribution function. The three point vertex part in the charge density channel Γ_c is given by

$$\Gamma_c(p,k) = 1 + i \int \frac{d^4 p'}{(2\pi)^4} K_c(p,p') G(p') G(p'-k) \Gamma_c(p',k). \qquad (4)$$

Taking $\Gamma_c(p',k) = 1$ in Eq. (4) and substituting Eq. (3) into Eq. (4), we obtain that

$$\Gamma_c^{(1)}(\mathbf{p},i\omega;\mathbf{k},i\omega') = 1 + \Delta\Gamma_c^{(1)}(\mathbf{p},i\omega;\mathbf{k},i\omega'), \qquad (5)$$

where

$$\Delta\Gamma_c^{(1)}(\mathbf{p},i\omega;\mathbf{k},i\omega') \approx -\sum_{\mathbf{p}'} \chi^0_{\mathbf{p}'\mathbf{k}}(\omega') K_c(\mathbf{p},i\omega;\mathbf{p}',\xi_{\mathbf{p}'-\mathbf{k}}+i\omega') \qquad (5a)$$

with

$$\chi^0_{\mathbf{p}\mathbf{k}}(\omega) = [n_F(\xi_p) - n_F(\xi_{p-k})] / (\xi_p - \xi_{p-k} - i\omega). \qquad (5b)$$

Now, substituting Eq. (5) into Eq. (4), we can calculate $\Gamma_c(p,k)$ up to the second order, then similarly to third order and so on to the n-th order expressed in terms of $\Delta\Gamma_c^{(1)}(p,k)$. The expression of $\Gamma_c^{(n)}(p,k)$ is a series of $\Delta\Gamma_c^{(1)}(p,k)$ and $\Gamma_c(p,k) = \lim_{n\to\infty} \Gamma_c^{(n)}(p,k)$. Making the summation of the series, the vertex function can be evaluated as

$$\Gamma_c(p_F,\omega;\mathbf{k},\omega') \approx \frac{1 - [B(0) - B(\omega)]\chi_0(\mathbf{k},\omega')}{1 + [A - B(0)]\chi_0(\mathbf{k},\omega')}, \qquad (6)$$

where

$$B(\omega) = \langle D^2(\mathbf{p}-\mathbf{q})\rangle \sum_q \frac{1 - 2n_F(\xi_q)}{\omega - 2\xi_q + i\delta} \quad \text{and} \quad A = \frac{1}{2}[\langle D(\mathbf{p}-\mathbf{p}')\rangle - \langle D^2(\mathbf{p}-\mathbf{q})\rangle N_F]. \qquad (6a)$$

Here $\langle D(\mathbf{p}-\mathbf{p}')\rangle N_F = 2\alpha F(\alpha,\zeta)/\pi$ and $\langle D^2(\mathbf{p}-\mathbf{p}')\rangle N_F = 2\alpha \widetilde{F}(\alpha,\zeta) e^2/\kappa p_F$, averaged over the Fermi surface, with $\alpha = e^2/\kappa v_F$, $\zeta = 2p_F c$, and two integrals $F(\alpha,\zeta)$ and $\widetilde{F}(\alpha,\zeta)$, evaluated for a layered 2D system elsewhere. [4]

The critical temperature T_c at which a pole first appears in Γ_c can be calculated with $\Gamma_c^{-1}(p_F,0;0,0) = 0$. This is analogous to the Cooper instability [5]. It turns out that

$$T_c \approx 1.13 \, \varepsilon_F \exp\{-1/\lambda_{eff}(\alpha,\zeta)\} = 1.13 \, E^*_B \, \alpha^{-2} \exp\{-1/\lambda_{eff}(\alpha,\zeta)\}, \qquad (7)$$

where

$$\lambda_{eff}^{-1}(\alpha,\zeta) = \frac{1}{\alpha \widetilde{F}(\alpha,\zeta)} \left[\frac{\pi}{\alpha} - F(\alpha,\zeta) + \alpha \widetilde{F}(\alpha,\zeta)\right], \qquad (7a)$$

is the effective interelectron coupling constant, and $E^*_B = e^4 m_b / 2\kappa^2$ is the effective Bohr energy. Fig. 3 shows a numerically calculated result of Eq. (7) and the bell-shape curve of T_c vs. α is apparent.

It deserves a mention that the first diagram in Fig. 2(d) plus its ladders construct the conventional Cooper instability. [5] Other two diagrams with equal initial and final momenta supply two more ladder summations which do not exist in the original Cooper instability. In the high density limit and isotropic 3D, however, their contributions to the kernel, the second and third terms in Eq. (3), are found to be negligible. This confirms the validity of the original Cooper assumption of a quiescent Fermi sea for conventional superconductors. It is not difficult to note the transition temperature calculated from Eq. (7) is related to the dielectric constant κ and the electron band mass m_b, the carrier density n_s, associated with α, and the interlayer distance c. Eq. (7) gives rise to both high and low T_c for a layered 2D lattice depending on the physical, chemical and structural parameters, and is consistent with the earlier results by the authors. [6]

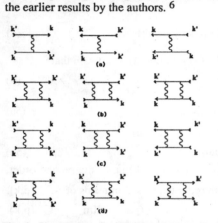

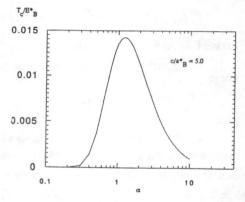

Fig. 2. Single scattering (a); double scattering with the same, (b), and different, (c), initial and final states; and three basic diagrams (d).

Fig. 3. The dimensionless transition temperature T_C vs. α. interelectron coupling constant.

3. Acknowledgment

This work is supported by DoE through Science Engineering Alliance under grant No. DE-FG05-94ER25229. The numerical calculations were performed on the CRAY supercomputer provided by NCSA at UIUC.

4. References

1. A. A. Abrikosov, L. P. Gor'kov and L. E. Dzyaloshinskii, "Methods of Quantum Field Theory in Statistical Physics," Ed. Richard A. Silver man, Prentice-Hall, Inc. 1963.
2. Leon N. Cooper, Phys. Rev. 104, 1189 (1956).
3. W. Heitler and F. London, Z. Physik, 44, 455 (1927).
4. Y. M. Malozovsky and J. D. Fan, Superconductor Science and Technology, IOP, submitted, 1996.
5. G. D. Mahan, "Many-Particle Physics," Second Edition, Plenum Press, NY, 1990.
6. Y. M. Malozovsky and J. D. Fan, Physica C 231, 63 (1994).

INSIDE HT$_c$ SUPERCONDUCTORS: AN ELECTRONIC STRUCTURE VIEW

A.J. Freeman, D.L. Novikov

Science and Technology Center for Superconductivity, Department of Physics and Astronomy, Northwestern University, Evanston,
IL, 60208-3112, USA

From the earliest days of high-T_c research, electronic structure theory has successfully revealed the physical and chemical properties of the normal state of the cuprates. In this paper, a review is given of some striking predictions of electronic structure theory about the role of van Hove singularities (vHs) on the electronic structure and properties of the newest and highest temperature superconducting cuprates at ambient and high pressures. The results provide possible strong evidence for the role of vHs in the superconductivity of quasi-2D high T_c systems. They thus serve to call attention to their role not only in enhancing T_c through large increases in N(E_F) and Fermi surface areas, but also in possibly providing support for vHs based excitonic pairing mechanisms for superconductivity. Further, this information, derived from the understanding - expressed as an "empirical rule" - gained from the related role of doping and pressure, is used to investigate other likely systems for being made into high T_c superconductors.

1 Introduction

Electronic structure theory has been an active and important part of the high T_c story since late 1986. Its predictions still remain valid and increasingly demonstrate that the normal state properties of the cuprates can be understood as conventional metals (i.e., Fermi liquids). A number of experiments, notably photoemission, de Haas-van Alphen and positron annihilation, have confirmed in detail and with excellent agreement the predictions of band theory on the Fermi surfaces and energy bands near E_F. This has served to firmly establish a Fermi liquid picture as the basis for developing the theory of high T_c superconductivity. Recently, there has developed mounting evidence for the possible important role of van Hove singularities (vHs) on the superconducting properties of the cuprates. From the first calculations of the electronic structure of the high T_c cuprates [1,2], vHs were expected and found to play an important role. While most obvious in La 214, where the single Cu-O$_2$ layer yields the now well-known 2D $dp\sigma$ band crossing E_F and a clearly separated vHs, the appearance and role of vHs in the other high T_c cuprates has not been clearly apparent until recently.

Here, we review briefly our recent work [3] about the role of vHs on the electronic structure and properties of the newest and highest T_c cuprates. The calculations were performed within the local density approximation using the full-potential linear muffin-tin orbital method [3].

2 vHs in different structures

2.1 The infinite layered superconductor $(Sr_{1-x}Ca_x)_{1-y}CuO_2$

Superconductivity at 110 K in the infinite-layer (IL) compound with x=0.3 and y=0.1, was reported [4]. We have investigated [5] in detail the band structure, chemical bonding and density of states (DOS) and resulting properties of $(Sr_{1-x}Ca_x)CuO_2$ for x=0.0, 0.3 and 1.0, which we correlated with varying composition y of the divalent Ca/Sr vacancies. We showed that a highly important role is played by a strong 2D vHs for $y > 0$ that strongly affects the Fermi surface (FS) topology as discussed by Lifshitz [6]. As is typical of all cuprates, the single free-electron-like valence band originating from the Cu 3d and O 2p states crosses E_F and gives rise to a simple Fermi surface. The vHs at the X and R points in the Brillouin zone contributes to the enhancement of the DOS close to E_F. Assuming a rigid-band approximation, a doping of x=0.3 causes E_F to move exactly onto the vHs. At this point, the change of the topology of the FS may be accompanied by various anomalies in thermal and electronic properties [6] and possible changes in T_c as well [7].

We also investigated [8] the effect of pressure on the electronic structure and properties of the IL. Our predicted [8] compressibility parameters are in excellent agreement with the later measurements [9]. The most important band structure change with pressure (obtained using different volumes of the unit cell) is the movement of the vHs (at the R-point) with respect to E_F. The decrease of volume, which is mainly due to the c-axis shortening, leads to a raising of the vHs towards the Fermi level as opposed to hole doping which lowers E_F onto vHs. This may provide a clue to understanding the nontrivial pressure dependence of T_c in high-T_c superconductors.

2.2 $HgBa_2Ca_{n-1}Cu_nO_{2n+2+\delta}$

Following close on to IL observations, the discovery of HTSC (T_c=94 K) in $HgBa_2CuO_{4+\delta}$ (Hg-1201), turned attention away from the more complex cuprates based on Y, Bi and Tl to these simpler single layer high T_c materials.

The electronic structure and FS [10,11] of Hg-1201 shows again the dominant role of a major vHs. As in the IL case, the single free-electron-like band crossing E_F gives rise to the simple FS, which again has the shape of a rounded square, centered at the M point in the (ΓXM) plane and around the A point in the top (ZRA) plane. Assuming that a rigid-band approximation is valid for the doping region considered, we calculated the FS[10,11] corresponding to the doping level at which the maximum critical temperature (T_c=95 K) is reached. In this case, E_F hits the vHs at R in the top (ZRA) plane; the diameter of the FS in the (ΓXM) plane is increased but the nesting character stays almost unchanged, while in the (ZRA) plane the former rounded square Fermi surface becomes pinched, touches the BZ boundaries and serves to maximize the FS area. This change in FS may be accompanied by various kinds of electronically driven anomalies[6]. It also leads to a kind of "empirical" rule, that the maximum in T_c for hole doped high-T_c cuprates is expected when the area of the FS reaches its maximum value.

While we have focused on Hg-1201, the results for Hg-1212 and 1223 are very similar[11], but they are "self-doped" structures that move the Cu-O $dp\sigma$ antibonding band away from half-filling. This band pattern is reminiscent of the case for the Tl- based superconductors[12] — the crystal analog of the Hg-materials — where the Tl 6s-O 2p band is always found to be located below E_F. Our results lend strong support to our earlier finding that the vHs may play a dominant role, and call attention to vHs based pairing mechanisms for superconductivity - as emphasized in the recent work of Markiewicz[13].

2.3 $Sr_2CuO_2F_{2+\delta}$ and $(Ca_{1-x}Na_x)_2CuO_2Cl_2$

This discovery[14,15] opened a new class of HTSC compounds with the apical oxygen substituted by other elements (F, Cl). Thus, we have investigated[16] the role of the lack of apical oxygens on the band structure and DOS of the parent compounds, ($Sr_2CuO_2F_2$, $Sr_2CuO_2Cl_2$ and, $Ca_2CuO_2Cl_2$) and their electronic properties which we correlate with varying degree of "hole doping". These display the same features as found in the high-T_c cuprates (strong two-dimensionality, a low DOS and a simple Fermi surface in the form of a rounded square). Since a major 2D vHs exists near E_F, we again assume[5,10,11] that its position determines the optimum hole doping level needed to achieve the highest possible T_c. From our results, we argue that $Sr_2CuO_2F_{2+\delta}$ with δ=0.6, which is the estimated composition synthesized[14], is heavily overdoped and that proper doping may help achieve better superconducting properties.

Since the optimum doping level for $(Ca_{1-x}Na_x)_2CuO_2Cl_2$[15] is unknown, we estimated this important parameter. From our results we propose the possibility of achieving HTSC by monovalent metal substitutions for Ca or Sr with an optimum composition of $M_{0.35-0.38}M'_{1.65-1.62}CuO_2X_2$, where M is a monovalent metal, M' = Ca or Sr and X = F or Cl. In such a case, from the total DOS at the vHs we can expect a higher T_c for $M_{0.35-0.38}Sr_{1.65-1.62}CuO_2X_2$ rather than for the compounds considered.

3 Possible experiments on vHs affected properties

As indicated above, for a wide range of HTSC materials there is an apparent correlation between T_c as a function of doping (and/or pressure) and a maximum in the DOS located close to E_F, with T_c depending strongly on the vHs passing close to or through E_F; the latter case is known as an electronic topological transition (ETT) [6]. The direct proof of this hypothesis would be experimental observations of properties of HTSC materials that are sensitive to ETT. These include thermopower, bulk modulus, phonon frequencies and thermal expansion coefficients [17].

In fact, we have recently investigated [18] effects of pressure on the A_{1g} phonon frequency in Hg-1201. Theoretically determined zero pressure lattice parameters and phonon frequencies were found in good agreement with experiment. An ETT was found to occur when the vHs was shifted close to the vicinity of E_F by pressure which caused considerable phonon softening.

4 Acknowledgments

This work was supported by the NSF (Grant No. DMR 91-20000)

References

1. J. Yu, A.J. Freeman et al., *Phys. Rev. Lett.*, 58:1035, 1987.
2. J.H. Xu et al., *Phys. Rev. Lett.*, 120:489, 1987.
3. M. Methfessel, *Phys. Rev. B*, 38:1537, 1988.
4. M. Azuma et al., *Nature*, 356:775, 1992.
5. D. L. Novikov et al., *Physica C*, 210:301, 1993.
6. I. M. Lifshitz. *Sov. Phys. - JETP*, 11:1130, 1960.
7. L. Dagens. *J. Phys. F*, 8:2093, 19987.
8. D. L. Novikov and A. J. Freeman. *Physica C*, 219:246, 1993.
9. H. Shaked et al., *Phys. Rev. B*, 50:12752, 1994.
10. D. L. Novikov and A. J. Freeman. *Physica C*, 212:233, 1993.
11. D. L. Novikov and A. J. Freeman. *Physica C*, 216:273, 1993.
12. J. Yu, et al., *Physica C*, 152:273, 1988.
13. R. S. Markiewicz, *Physica C*, 217:381, 1993.
14. M. Al-Mamouri et al., *Nature*, 369:382, 1994.
15. Z. Hiriri et al., *Nature*, 371:139, 1994.
16. D. L. Novikov et al., *Phys. Rev. B*, 51:6675, 1994.
17. V. I. Nizhankovski et al., *Sov. Phys. - JETP Lett.*, 59:733, 1994.
18. D. L. Novikov et al., *Phys. Rev. B.* in print.

PAIR-BREAKING AND THE UPPER LIMIT OF T_C IN THE CUPRATES

VLADIMIR Z. KRESIN
*Lawrence Berkeley National Laboratory, University of California,
Berkeley, CA 94720*

STUART A. WOLF
*Naval Research Laboratory,
Washington, D.C. 20375-5000*

and

YU N. OVCHINIKOV
*Landau Institute for Theoretical Physics,
Moscow, Russia, 11733V*

ABSTRACT

There is an "intrinsic" value of T_C which is similar for all high Tc cuprates and corresponds to the absence of all pair-breaking. This value represent an upper limit of the critical temperature for "classical" cuprates. Pressure induced doping allows T_C to be raised above the the usual maximum.

High Tc oxides are doped superconductors and their properties depend strongly on carrier concentration n; this parameter did not play an essential role for conventional metallic superconductors. The dependence $T_C(n)$ has a maximum, that is T_C decreases in the overdoped region. The question arises: what is the reason for such decrease? An analysis of various properties of the overdoped state leads to the conclusion that this decrease is caused by the pair-breaking effect.

The influence of magnetic impurities (pair-breaking, gaplessness at some concentration of magnetic impurities, depression of T , etc.) was studied previously [1], see also the reviews [2]. The interaction between localized magnetic moments and Cooper pairs which are in the singlet state, destroys the pair correlation. It is important to understand that this process is accompanied by spin-flip scattering (such scattering provides the conservation of total spin).

The existence of such pair-breaking in the cuprates is supported by a number of experiments. Indeed, heat capacity measurements (see e.g., [3]) which show low temperature Shottky anomalies are direct experimental evidence for the presence of magnetic impurities.

Magnetic impurities provide the pair-breaking effect. Pair breaking, has been used to explain μSR spectroscopy data [4], i.e. the depression in T_C in the overdoped state was accompanied by an increase in the "normal" component. Pair-breaking leads to a depression in T_C [1] and, eventually, at some value of the concentration of magnetic impurities $n_M=n_{M;cr.}$, to the total supression of superconductivity. This suppression is proceeded by a gapless state. A gapless spectrum has also been observed in [3,4]. Therefore, the scenario, based on pair-breaking effect by magnetic impurities, has strong experimental support.

According to our approach, the observed value of T_C is affected by two competing factors: 1. Increase in the carrier concentration n; this is a favorable factor (T_C increases with increasing n) , and 2. Pair-breaking effect of magnetic impurities which depresses T_C. At the " optimum" doping level which corresponds to the maximum observed value of T_C (e.g. $T_{C;.opt.}=90K$ for the 2201 Tl-based cuprate), we are dealing with a crossover of these two trends. The "intrinsic" value of T_C corresponds to the "maximum" level of doping, and is equal to the value of the critical temperature for this doping in the absence of pair-breaking. We think that this value represent the upper limit of T_C for the system.

The decrease in T_C and the concommitant decrease in H_{C2} allowed a direct measurement of the dependence of the critical field on temperature. Such measurements were performed on the 2201 Tl-and Bi-based cuprates[5,6]. Contrary to the conventional picture [7], the critical field displays positive curvature over the entire temperature range with a sharp increase in the region near T=0 K. According to the physical picture, developed in [8], spin-flip scattering leads to a depression of the superconducting state, and this is reflected in relatively small values of T_C and H_{C2} (near T_C). In this temperature region, the magnetic impurities can be treated as independent [1], and the spin-flip scattering by the impurities provides conservation of total spin. However, at low temperatures (in the region T=1K), because of the correlation of the magnetic moments, the trend to ordering of the moments becomes important, and this trend frustrates the spin-flip scattering. Pairing becomes less depressed, and this leads to a large increase in the value of H_{C2} and, correspondingly, to a positive curvature in H_{C2} vs T (for a more detailed discussion see [8]). Note that a similar effect has been observed in the Sm-Ce-Cu-O, La-Sr-Cu-O and, recently, in Y-Ba-Zn-Cu-O systems [9].

Based on the study [8], which analyzes in detail $H_{C2}(t)$ one can obtain the value of critical temperature in the absence of magnetic impuritiesFor the Tl 2201 compound one can obtains $T_{C;m}{}^0=160K$. It is remarkable that this value greatly exceeds the experimental value $T_{C;m}=90K$, observed at the optimum ambient pressure doping. Therefore, we conclude that the material contains magnetic impurities even at optimum doping, and the value of the critical temperature $T_{C;m}{}^{dop.}$ is depressed relative to the 'intrinsic" value $T_{C;m}{}^0$ which is equal to $T_{C;intr.}=160K$. One can carry out a similar analysis for the 2201 Bi-based cuprate with use of the data [6]. A detailed calculation leads to a similar value $T_{c;intr.} = 160K$.

Recently measurements of $H_{c2}(T)$ for the underdoped Y-Ba-Zn-Cu-O system were described in [9]. The behavior of the critical field appears to be similar to that of the overdoped cuprates studied in [5],[6]. This case corresponds to the "dirty" case where the mean free path was about 12 angstoms. One can evaluate the value of the "intrinsic" Tc for this material. It is remarkable that for the YBCO compound the value of $T_{c;intr.}$ appears to be close to that for other cuprates. A similar value can be obtained also for the Sm-Ce-Cu-O.

The calculations shows that four different cuprates (Tl- and Bi-based cuprates, YBCO, SmCeCuO) have a similar value of the "intrinsic" T_C which is in the region of 160-170K. We think that this coincidence is not occasional. Indeed, all these cuprates, have an important common feature: they contain the same structural unit, the Cu-O plane. This unit in each of the compounds is doped by a similar doping channel (reservoir-apical oxygen-CuO plane), and the system of dopants-holes is superconducting. We think that 160-170 K represents the upper limit of Tc for such systems.

Qualitatively, we are dealing with one high T_C superconductor, namely, the Cu-O plane, and this superconductor, depending on many factors, such as the doping level, structure of the sample, strength of the coupling, phonon spectrum, oxygen content, etc. can have a different value of critical temperature in the range from $T_C = 0K$ up to $T_{c;intr.} = 160\text{-}170K$.

The maximum value of T_C for the Hg-based cuprate is not far from our estimated limit. One can ask: what makes this material almost ideal? We speculate that this is related to the fact that the Hg-based cuprate, even at maximum T_C, is characterized by a small amount of oxygen in the Hg-O layer, and, according to our scenario, by a small concentration of magnetic pair-breakers; the doping occurs because of the non-stoichiometry of the Hg, Pb, and Ba.

Of course, the conclusion about the upper limit of T_C in cuprates does not mean the end of the road to higher T_C. This conclusion is concerned with this class of materials only, that is, the "classical" copper oxides. If we are interested in the further increase in T_C, then ,probably, it is time to look at different class of materials.

In discussing the location of the magnetic moments, one should note that localized magnetic moments can be formed on the apical oxygen site; there is a possibility of the formation of a paramagnetic radical O2-, (this involves the apical oxygen and the additional oxygen in the Tl-O (Bi-O) layers.

External pressure decreases the c-lattice parameter which increases the charge transfer and, therefore, T_C, without adding magnetic moments.

We can make a specific prediction. If we take an underdoped 2201 Tl-based cuprate with Tc=80K, then the applied pressure can raise T_C up to approximately 105K, that is noticeably higher then the usually observed value $T_{C;max}$=90K.

In conclusion, the analysis of magnetic and other properties of non-stoichiometric cuprates, including the behavior of T_C upon overdoping, gaplessness, and the temperature dependence of the critical field, allows us to introduce a new parameter, " intrinsic" T_C; its value appears to be similar for different cuprates and lies in the region of T_C =160-170K. This value represent the upper limit of T_C for this class of materials.Since the " intrinsic" value is depressed by pair-breaking, one can raise T_C above the maximum ambient pressure measured value by pressure induced doping.

The authors are grateful to D. Gubser for fruitful discussions. The research of VZK is supported by the US Office of Naval Research contract No.N00014-95-F0006.

References.

1. A. Abrikosov and L. Gor'kov, *Sov. Phys.- JETP* 12 ,1243 (1961); P. De Gennes, *Phys. Cond. Mat.* 3, 79 (1964) ;
2. K.Maki, *Superconductivity*, R. Parks, Ed., p. 1035, Marcel Dekker,NY (1969) ; P. de Gennes, *Superconductivity in Metals and Alloys*, Benjamin, NY (1966); D. Saint-James, G. Sarma,and E.Thomas, *Type II Superconductors*, Pergamon, Oxford (1969); A. Abrikosov, *Fundamentals of the Theory of Metals*, North-Holland, Amsterdam (1988)
3. N. Phillips, R. Fisher, and J. Gordon, in *Progress in Low-Temperature Physics* 13, p. 267, D. Brewer, Ed., North-Holland, The Netherlands (1992); A. Junod, in *Studies of High Temperature Superconductors*, A. V. Narlikar, Ed., Nova Science,Publ., NY (to be published); J.Wade, J. Loram,K. Mirza,J. Cooper,J. Tallon, *J.of Super.*, 7, 261 (1994).
4. C. Niedermayer,C. Bernhard, U. Bunninger, H. Gluckler, J. Tallon, E. Ansaldo, J. Budnick *Phys. Rev. Lett.* 71, 1764 (1993) ; *J. of Super.* 7, 165 (1994); J. Tallon ,J. Loram, *J. of Super.*, 7,15 (1994); J. Tallon et al., *Phys. Rev. Lett.* 74, 1008 (1995).
5. A. P. Mackenzie et al., *Phys. Rev. Lett.* 71, 1938 (1993).
6. M. Osofsky et al.,, *Phys. Rev. Lett.* 71,, 2315 (1993).
7. L.Gor'kov, *Sov. Phys.-JETP*, 10 ,593 (1960); E. Helfand and N.R..Werthamer, *Phys. Rev. Lett.* 13, 686 (1964)
8. Yu. Ovchinnikov, V. Kresin, *Phys. Rev. B*52, 3075 (1995); preprint.
9. Y.Dalichaouch et al., *Phys. Rev. Lett.*, 64, 599 (1990); M. Suzuki and M. Hikita, *Phys. Rev. B*44, 249 (1991); D. Walker et al. ,ibid. B51, 9375 (1995).

WHAT DOES D-WAVE SYMMETRY TELL US ABOUT THE PAIRING MECHANISM?

K. LEVIN, D. Z. LIU and JIRI MALY

James Franck Institute, University of Chicago, Chicago, IL 60637, USA

In this paper we argue that d-wave symmetry is a general consequence of superconductivity driven by repulsive interactions. Van Hove (or flat band) effects, deriving from the two dimensionality of the CuO_2 plane are important in stabilizing this state. By extending the original Kohn-Luttinger picture to a 2 D lattice, we find that the screened Coulomb term has important wave vector structure which leads to $d_{x^2-y^2}$ superconductivity

There is a growing body of evidence to support the claim that the pairing symmetry in the cuprates is most probably d-wave wave[1], although some inconsistencies remain. These observations lead naturally to the question: what constraints does this information provide on the detailed nature of the pairing mechanism? In this paper we investigate this important question based on the Kohn Luttinger picture[2] in which it was argued that the gap equation

$$\Delta_{\mathbf{k}} = -\sum_{\mathbf{k'}} V(\mathbf{k}-\mathbf{k'}) \left(\Delta_{\mathbf{k'}} \frac{\tanh \beta E(\mathbf{k'})}{2E(\mathbf{k'})} \right) \qquad (1)$$

could be satisfied for repulsive interactions $V > 0$, provided V contained some significant variation with wave-vector. A reasonable choice for $V(\mathbf{q})$ is the screened Coulomb interaction $V_{eff}(\mathbf{q},\omega) = V_o(\mathbf{q})/\epsilon(\mathbf{q},\omega)$ which in coordinate space exhibits the well known Friedel oscillations; thus there are regions of negative sign and superconductivity can take advantage of this attraction by forming anisotropic Cooper pair states.

The magnetic pairing scenario[1], which has been widely discussed for the cuprates, may be viewed as a simple extension of the Kohn Luttinger picture where the repulsive interaction (in wave vector space) is chosen to be the magnetic susceptibility with peaks around the $(\pi/a, \pi/a)$ point. As was shown by several different groups, this particular form for the wave-vector dependence in $V(\mathbf{q})$ leads to a $d_{x^2-y^2}$ pairing symmetry.

While this scenario evidently explains the pairing symmetry, it is not at all clear whether it is the appropriate mechanism for the superconductivity in the cuprates. Our group[3] has extensively investigated this question based on the constraints imposed on the magnetism by neutron data[4] in the copper oxides, and have inferred that the magnetism appears too weak to be responsible for the high superconducting transition temperatures found in the cuprates.

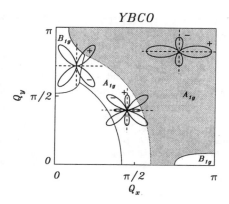

Figure 1: Phase diagram indicating gap symmetry for YBCO-like bandstructure.

Where, then, does the $d_{x^2-y^2}$ symmetry originate? To address this question we first build intuition by analyzing a simple "generic" model which contains variable wave vector structure. This wave vector structure is crucial for anisotropic superconductivity. Consider $V(k_x, k_y)$ given by

$$V(k_x, k_y) = \frac{\lambda}{[1 - J_o(\cos(k_x \pm Q_x) + \cos(k_y \pm Q_y))]^2} \quad (2)$$

which is taken to be repulsive ($\lambda > 0$) and to have maxima at variable Q_x, Q_y with peak widths characterized by J_o. The magnetic pairing scenario is a special case of this more general interaction with $Q_x = Q_y = \pi/a$. As expected from the Kohn Luttinger picture, if this interaction is substituted into Eq(1), a variety of anisotropic pairing states will arise. The various allowed symmetries[5] are plotted in Figure 1 as a function of Q_x, Q_y. Here the electronic structure is taken to be that of YBCO, near optimal stoichometry where the Van Hove singularities (or flat band regions) are in close proximity to the Fermi surface.

An important conclusion of this study is that the largest fraction of phase space is associated with $d_{x^2-y^2}$ pairing. Thus, one may conclude that this symmetry should *not* be related to any particular wave vector structure. It appears to be a robust consequence of superconductivity driven by repulsive interactions. What stabilizes this $d_{x^2-y^2}$ pairing state over other candidate states? There are two important effects: (1) it is a state with a minimal number of sign changes in the gap function, (as compared to the eight lobe s-state (A_{1g}) also shown) and consequently yields higher T_c's for most slowly varying interactions and (2) it is a state which takes maximal advantage of

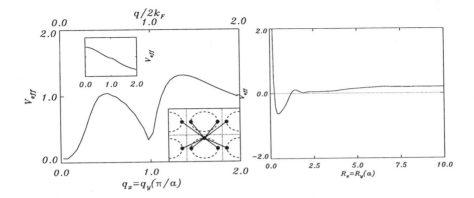

Figure 2: (a) Screened Coulomb interaction for LSCO model bandstructure. Upper left inset: V_{eff} for 3D jellium. Lower right inset: demonstrates how Van Hove effects are enhanced by Umklapp processes. (b) Screened Coulomb interaction in real space in the static limit.

the high density of states associated with the flat band regions or "Van Hove" effects. Moreover, these flat bands are observed experimentally in a variety of cuprates[6]. It should be stressed that our early work[7] has noted that correlation effects contribute in an important way to pin the flat bands near the Fermi energy, so that these effects should not be interpreted as simple (i.e., one electron) Van Hove singularities.

Despite these suggestive results, a more microscopic picture is clearly desirable. The most natural source for a repulsive interaction which may drive the superconductivity in the cuprates, is the direct Coulomb interaction. Moreover, it is essential to our work that we begin with the full long range Coulomb interaction, rather than the Hubbard model approximation. We have, therefore, generalized the Kohn Luttinger calculation based on the screened Coulomb interaction to the case of a two dimensional, tight binding lattice. Here we find two important effects play a role in our analysis: local field and Van Hove contributions. Local field effects lead to a matrix form for the dielectric constant[8], so that higher Brillouin zone contributions or Umklapp processes are included.

In Figure 2a we plot our results for the screened Coulomb interaction for LaSrCuO (at optimal stoichiometry) and, as a point of comparison, for three dimensional jellium. The results for the YBaCuO family are found to be similar. The effect of the flat band regions enters not only in the density of states effects discussed in the context of the simple generic model in Figure 1, but also directly into the interaction itself. These large density of states regions cause a dip in the interaction associated with enhanced screening. This dip is then followed by a maximum, slightly above the (π/a, π/a) point. Moreover these Van Hove effects are enhanced further through higher zone or Umklapp processes as shown in the inset. Finally, and most importantly, the superconducting instability associated with this wave vector structure is indeed, the $d_{x^2-y^2}$ state.

In Figure 2b we plot the screened Coulomb interaction in co-ordinate space in the static limit. This figure shows the expected Friedel oscillations, which (in the Kohn Luttinger language) may be viewed as driving the superconductivity. The characteristic energy scale beyond which the Friedel oscillations become significantly altered, so that the d-wave pairing is no longer stable, is of the order of 4t (where t is the nearest neighbor hopping). This cut-off thus represents a sizeable fraction of the plasma frequency.

What about the characteristic size of T_c? We are in the process of introducing corrections to the mean field equation which include phase fluctuation effects and which will enable us to compute T_c more reliably. These phase fluctuation effects are manifested in "pseudo-gap" behavior away from optimal stoichiometry. It is only when such a scheme is in hand that we can with some confidence make further progress on the pairing mechanism.

In summary, this paper has demonstrated the generality of the $d_{x^2-y^2}$ pairing symmetry and has suggested an alternate route to d-wave pairing via the long range Coulomb interaction. Whether this is all or only a component of the pairing is too soon to say, but it is clear that direct Coulombic effects will act in concert with any other underlying d-wave pairing mechanism.

This work is supported by the National Science Foundation (DMR 91-20000) through the Science and Technology Center for Superconductivity.

1. See, for example, D. Scalapino (preprint) for a review.
2. W. Kohn and J.M. Luttinger, Phys. Rev. Lett **15**, 524 (1965).
3. R.J. Radtke et al, Phys. Rev. B **46**, 11975 (1992).
4. H.A. Mook et al, Phys. Rev. Lett. **70**, 3490 (1993).
5. J. Maly, D. Z. Liu and K. Levin, Phys. Rev. B **53**, 6786 (1996).
6. See for example, K. Gofron et al Phys. Rev. lett. **73**, 3302 (1994); D. King et al Phys. Rev. Lett. **73**, 3298 (1994).
7. K. Levin et al, Physica C **175**, 449 (1991).
8. L.J. Sham, Phys. Rev. B **6**, 3584 (1972); R.P. Gupta and S.K. Sinha, Phys. Rev. B **3**, 2401 (1971).

SPIN FLUCTUATIONS, MAGNETOTRANSPORT AND $d_{x^2-y^2}$ PAIRING IN THE CUPRATE SUPERCONDUCTORS

DAVID PINES

Department of Physics, University of Illinois,
1110 West Green Street, Urbana, IL 61801-3080 USA

I use a nearly antiferromagnetic Fermi liquid description of planar excitations to review recent experimental and theoretical work on spin fluctuation excitations in the cuprate superconductors which establishes magnetic scaling behavior in the normal state and reconciles the results of neutron and NMR experiments on $La_{2-x}Sr_xCuO_4$ and $YBa_2Cu_3O_{6-x}$ systems. I describe the results of model NAFL calculations which make evident the causal relationship between the magnetic quasiparticle interaction, normal state magnetotransport, and the transition at high temperatures to a superconducting state with $d_{x^2-y^2}$ pairing, and discuss further recent experiments which support that pairing state.

The proposal that in the cuprate superconductors it is the magnetic interaction between planar quasiparticles which is responsible for their anomalous normal state behavior and the transition at high temperatures to a superconducting state with $d_{x^2-y^2}$ pairing was made over six years ago;[1] it was based on results obtained in the first generation of nuclear magnetic resonance experiments and their explanation using a phenomenological one-component description which demonstrates the close approach of $YBa_2Cu_3O_7$ and $La_{1.85}Sr_{0.15}CuO_4$ to antiferromagnetism.[2] It led to a nearly antiferromagnetic Fermi liquid (NAFL) description of normal state behavior [a review of work prior to 1995 may be found in Ref. (3)] based on the ansatz[4] that the effective magnetic interaction between planar quasiparticles mirrors the highly anisotropic momentum dependence of the spin-spin response function, $\chi(\mathbf{q},\omega)$, measured in NMR experiments. For systems in which the spin excitation spectrum peaks at the commensurate wave vector, $\mathbf{Q} = (\pi,\pi)$, $\chi(\mathbf{q},\omega)$ can be written in the form[2]

$$\chi(\mathbf{q},\omega) = \chi_{\text{MMP}}(\mathbf{q},\omega) + \chi_{\text{FL}}(\mathbf{q},\omega) \tag{1}$$

where the anomalous contribution, $\chi_{\text{MMP}}(\mathbf{q},\omega)$, introduced by Millis, Monien and Pines (1990, hereafter MMP) is,

$$\chi_{\text{MMP}}(\mathbf{q},\omega) = \frac{\chi_Q}{1+(\mathbf{Q}-\mathbf{q})^2\xi^2 - i(\omega/\omega_{\text{SF}})} \equiv \frac{\alpha\xi^2}{1+(\mathbf{Q}-\mathbf{q})^2\xi^2 - i(\omega/\omega_{\text{SF}})} \tag{2}$$

and χ_{FL} is a parametrized form of the normal Fermi liquid contribution, which is wave vector independent over most of the Brillouin zone,

$$\chi_{\text{FL}}(q,\omega) \cong \frac{\chi_0(T)}{1-i\pi\omega/\Gamma} \tag{3}$$

modified to take into account the temperature dependence of the bulk susceptibility, $\chi_0(T)$. Subsequent experiments on the ^{17}O and ^{63}Cu spin-lattice relaxation rates, when combined with measurements of the ^{63}Cu spin-echo decay rate, made possible the determination of the temperature-dependent AF correlation length, ξ, and the frequency, ω_{SF}, of the relaxational mode.[5] The AF correlations are comparatively long-range ($\xi(T_c) \geq 2a$, for YBa$_2$Cu$_3$O$_7$, $\xi(T_c) \sim 7a$ for La$_{1.85}$Sr$_{0.15}$CuO$_4$), while ω_{SF} is typically in the 10-20 mev range.[6] Quite remarkably, the parameters ω_{sf} and ξ are not independent for a wide range of temperatures and hole concentrations, but rather display scaling behavior,

$$\omega_{SF} = \text{const}/\xi^z \tag{4}$$

where one has a crossover from $z = 2$ to $z = 1$ behavior at the same temperature, T_{cr}, at which $\chi_0(T)$ begins to decrease as T is decreased.[6]

As both sample quality and INS (inelastic neutron scattering) techniques improved, there emerged a number of apparent contradictions between the spin fluctuation spectrum deduced from NMR experiments and that measured directly in an INS experiment.[7] Fortunately, these have now been resolved. Zha et al.[7] have shown that if one modifies (a) the phenomenological nuclear hyperfine Hamiltonian, and (b) χ_{MMP} to reflect the presence of four incommensurate peaks at $\mathbf{Q}_i$ produced by the quasiparticle Kohn anomalies, then the one-component, $\chi_{MMP}(\mathbf{q},\omega)$ obtained from fits to NMR experiments is in excellent quantitative agreement with the recent INS measurements of the absolute intensity of $\chi(\mathbf{q},\omega)$ by Aeppli et al. on La$_{1.86}$Sr$_{0.14}$CuO$_4$[8] which show, inter alia, $\xi = 2a$ at $T_{cr} \sim 300K$, $\chi_{Q_i} = 350$ states/eV and $\omega_{SF} \sim 9$ meV at 35K, and by Fong et al. on YBa$_2$Cu$_3$O$_7$.[9]

Perhaps the strongest single constraint on candidate descriptions of normal state behavior in the cuprate superconductors is the appearance of pseudogap and pseudoscaling behavior. Analysis of NMR experiments on $^{63}T_1$ and $^{63}T_{2G}$ shows that there are, in general, three distinct regimes of normal state behavior.[6]

- a *mean field regime*, found for $T > T_{cr}$, in which the influence of antiferromagnetic correlations may be described in the RPA, ω_{SF} displays mean-field, non-universal $z = 2$ behavior, ξ^{-2} is linear in temperature, $(^{63}T_1 T/T_{2G}^2)$ is independent of temperature, and $\chi_0(T)$ increases as T decreases.

- a *pseudoscaling regime*, found for $T_* \leq T \leq T_{cr}$, where $(^{63}T_1 T/T_{2G})$ is independent of temperature, ω_{SF} displays non-universal $z = 1$ scaling behavior, both ω_{SF} and ξ^{-1} vary linearly with temperature, and $\chi_0(T)$ decreases linearly as T decreases.

- a *pseudogap regime*, found for $T_c \leq T \leq T_*$, in which $\chi_0(T)$, ξ, and ω_{SF} display behavior which is qualitatively similar to what happens when one gets a quasiparticle gap through the long range order which accompanies superconductivity or itinerant antiferromagnetism. Thus $\chi_0(T)$ changes character at T_*, and as T decreases, ω_{SF}, after passing through a minimum, increases as T decreases, while ξ and T_{2G} become independent of temperature. Since there is no long range order, "pseudogap" rather than "gap" is the proper description of quasiparticle behavior in this regime.

Barzykin and I found that T_{cr} decreases rapidly with increased doping and proposed that at T_{cr}, $\xi \sim 2$. With this ansatz, one then is able to set the scale for the temperature variation of ξ measured in T_{2G}. Thus for YBa$_2$Cu$_3$O$_{6.92}$, one has $T_{cr} \sim 150K$, while for

YBa$_2$Cu$_4$O$_8$, confirmation of our predicted crossover has come in very recent work by Curro et al.[10], who find $z = 1$ behavior below ~ 485K, and $z = 2$ behavior from 485K to ~ 700K.

Pseudoscaling and pseudogap behavior is not confined to the magnetic properties of the normal state, but is found as well in the charge response. Thus, for YBa$_2$Cu$_3$O$_{6.63}$ below T_*, the planar resistivity ceases to exhibit a linear in T behavior, while the Hall effect likewise changes character at this temperature;[11] a similar conclusion applies for YBa$_2$Cu$_4$O$_8$.[12] In the La$_{2-x}$Sr$_x$CuO$_4$ system, for $x \geq 0.15$, the Hall coefficient, $R_H(T)$, follows scaling behavior, with a characteristic temperature which exhibits the same strong dependence on doping level that is found for $T_{\rm cr}$.[13] For this same system, with $x \leq 0.14$, the resistivity ceases to be linear in T below T_*.[14] Thus the crossover temperatures, $T_{\rm cr}$ and T_*, which Barzykin and I identify in low frequency magnetic measurements possess direct counterparts in transport measurements.

A major challenge for the NAFL approach has been explaining anomalous transport in an applied magnetic field, where experiments show that for optimally doped materials the Hall resistivity is a strong function of temperature, yet the cotangent of the Hall angle has a very simple behavior: $\cot\theta_{\rm H} = {\rm A} + {\rm B}T^2$. Quite recently, Stojkovic and I have investigated the temperature dependence of the normal state Hall effect and magnetoresistance in YBa$_2$Cu$_3$O$_7$ using the NAFL model experiment-based planar quasiparticle interaction,

$$V_{\rm mag}^{\rm eff}(\mathbf{q},\omega) = g^2 \chi(\mathbf{q},\omega) \sigma_1 \cdot \sigma_2. \tag{5}$$

and $\chi(\mathbf{q},\omega)$ determined from fits to the NMR experiments.[15] We obtain a direct (nonvariational) numerical solution of the Boltzmann equation and find that the highly anisotropic scattering at different regions of the Fermi surface brought about by the momentum dependence of the interaction, Eq. (5), gives rise to the measured anomalous temperature dependence of the resistivity and Hall coefficient while yielding the quadratic temperature dependence of the Hall angle observed for both clean and dirty samples. We have extended our calculations to both the underdoped and overdoped systems, and find that changes in the effective interaction and the Fermi surface with doping and temperature are capable of bringing about the systematic changes in $\sigma_{xx}(T)$ and $\sigma_{xy}(T)$ which are observed as one passes from optimally doped systems such as La$_{1.85}$Sr$_{0.15}$CuO$_4$ to either strongly overdoped or underdoped systems.

Let me now turn to the superconducting transition and pairing state. The NAFL strong coupling (Eliashberg) calculations, carried out in collaboration with Philippe Monthoux[16,17] of the normal state properties and T_c for YBa$_2$Cu$_3$O$_7$ using the model experiment-based planar quasiparticle interaction, Eq. (5), showed that when the momentum dependence of the quasiparticle interaction is taken into account, although lifetime effects do lead to a dramatic reduction in T_c, for parameters appropriate to YBa$_2$Cu$_3$O$_7$ a superconducting transition into a $d_{x^2-y^2}$ pairing state occurs at $T_c \sim 90$K for comparatively modest values of the coupling constant g. For this same range of coupling constants the calculated resistivity and optical properties in the normal state were in quantitative agreement with experiment, so that a bridge was built between the measured anomalous normal state charge response, the anomalous normal state spin response, and $d_{x^2-y^2}$ superconductivity at high temperatures. Our calculations provided for YBa$_2$Cu$_3$O$_7$ a proof of concept of the magnetic mechanism and the NAFL approach. However, since our nearly antiferromagnetic

Fermi liquid approach *predicted unambiguously that the pairing state of* $YBa_2Cu_3O_7$ *must possess* $d_{x^2-y^2}$ *symmetry*, experimental detection of that pairing state was a *necessary* condition for the magnetic mechanism to explain high T_c. Initially only NMR experiments on $YBa_2Cu_3O_7$, showed the characteristic $d_{x^2-y^2}$ signature of a linear dependence on T of the Knight shift and a T^3 dependence of the ^{63}Cu spin-lattice relaxation rate we had predicted.[5] However, as Leggett has told you at this meeting, with one exception, subsequent experiments using a variety of techniques and materials, support this pairing assignment for the hole-doped cuprate superconductors.

As the $d_{x^2-y^2}$ nature of the order parameter has become well established, advocates of other than the magnetic mechanism have begun to find ways of obtaining the pairing state from their mechanism of choice. Thus it is natural to inquire whether there is a "smoking gun" for the magnetic mechanism. Monthoux and I have shown that there is.[17] Because $\xi(T_c)$ is typically a few lattice spacings, any imperfection which destroys local magnetic order acts as a unitary (strong) scatterer in its influence on the low temperature properties, while for a pairing potential of magnetic origin, our model calculations showed that because it both scatters at the unitary limit *and* changes the pairing potential, such an impurity will exceed the unitary limit in its influence on T_c. Thus the magnetic mechanism is confirmed by measurements which show that impurities, such as Zn, which change the local planar magnetic order exert a considerable influence on both T_c and the low temperature superconducting properties of the cuprate superconductors, while those such as Ni, which do not appreciably change the local magnetic order, have much less influence on these properties.[18].

A simple physical picture of why the magnetic interaction yields $d_{x^2-y^2}$ pairing may be obtained in the limit of long correlation lengths, where it is straightforward to pursue the consequences of a pairing potential, $V_{\text{eff}}(\mathbf{q}, 0)$ which has four incommensurate peaks at $\mathbf{Q}_i = [\pi(1 \pm \delta), \pi(1 \pm \delta)]$.[19] The pairing potential in configuration space is,

$$V_{\text{eff}}(x,0) = \frac{g^2 \chi_{\mathbf{Q}_i}}{2} (-1)^{n_x + n_y} [\cos(\pi \delta n_x) + \cos(\pi \delta n_y)] \quad (6)$$

where $\mathbf{x} \equiv (n_x, n_y)$ describes a point on the lattice, and n_x and n_y are integers. The interaction is thus maximum and repulsive at the origin, while at the nearest-neighbor sites $(\pm 1, 0)$ and $(0, \pm 1)$, it is attractive, with

$$V_{\text{eff}} = -\frac{g^2 \chi_{\mathbf{Q}_i}}{2}[1 + \cos(\pi \delta)]. \quad (7)$$

Evidently, the interaction is most attractive when $\cos(\pi\delta) = 1$, *i.e.* when $\delta = 0$ and one has a commensurate spin-fluctuation spectrum. At longer distances, V_{eff} is cut off by the finite correlation length that this simple model ignores, so the main contribution really comes from the nearest-neighbor term (especially for YBCO$_7$ with $\xi \simeq 2$).

Monthoux and I emphasize that it is this on-site repulsion, plus the effective attraction between the nearest neighbors, which is responsible for d-wave pairing; in this pairing state, quasiparticles, by virtue of their $l = 2$ relative angular momentum, avoid sampling the on-site repulsion while taking advantage of the attraction between nearest neighbors to achieve superconductivity. That the pairing state must be $d_{x^2-y^2}$ follows from the fact that along the diagonals in configuration space (where $n_x = \pm n_y$), the effective pairing potential, Eq. (6), is always repulsive for $n_x \leq \xi$, and not too large a discommensuration, δ; hence it is energetically favorable to place the nodes of the gap parameter,

$$\Delta(T) = \Delta_0(T)(\cos k_x a - \cos k_y a) \tag{8}$$

along these same diagonals, since for that pairing state the effective repulsion, which would otherwise be deleterious for superconductivity, is rendered harmless.

The above simple model calculation explains why as long as the $\mathbf{Q}_i$ span the Fermi surface, which appears to be the case for all cuprate superconductors, an effective magnetic interaction which possesses incommensurate peaks will be less effective in bringing about superconductivity than a magnetic interaction which is peaked at $\mathbf{Q}$. It also explains why Hubbard calculations of T_c show a T_c which is maximum near half-filling, falling off as the doping increases, since an elementary calculation for the Hubbard model shows that $\delta = 0$ at half filling, and increases with increased doping.

Where do matters stand? First, thanks to the superb experimental work by many people on increasingly better samples, there now exist some forty-odd "strong constraints" on candidate descriptions of the normal state, which space prohibits me from presenting here. Second, given the near-consensus on $d_{x^2-y^2}$ pairing (and the ad-hoc character of the arguments for a non-magnetic origin of this pairing), I would anticipate that both theorists and experimentalists alike will increasingly focus their attention on describing and deriving the three distinct normal state phases I discuss above. The finding that $\xi(T_{cr}) = 2$ argues strongly in favor of a magnetic origin of pseudoscaling and pseudogap behavior. Experimental evidence in support of this point of view comes from the NMR experiments which show that Zn impurities suppress the pseudogap, while Ni impurities do not. From a Fermi liquid perspective, the only way pseudoscaling and pseudogap behavior can come about is through a non-linear feedback of the magnetic interaction on the quasiparticles which are in turn the source of that interaction. Monthoux and I have developed a Fermi-liquid description of pseudoscaling, and have reviewed the non-linear feedback effects at work above T_{cr} and when the system goes superconducting, and then considered how similar effects in the normal state could lead to pseudoscaling and pseudogap behavior below T_{cr}.[20] Eliashberg calculations of strong coupling effects on the quasiparticle lifetime explain the $z = 2$, mean-field behavior;[17] however below T_{cr}, vertex corrections, which can be safely neglected above T_{cr}, must be responsible for pseudoscaling and pseudogap behavior. Finding a first-principles, microscopic derivation of pseudoscaling and pseudogap behavior remains a major unsolved problem. A significant first step has been taken by Chubukov and his collaborators, who find that for the underdoped systems where the pseudogap effects play a significant role at low temperatures, one begins to lose those parts of the Fermi surface which are most strongly coupled by the magnetic interaction.[21] This provides a natural explanation for the very recent ARPES results of Marshall et al. on the underdoped Bi system.[22] However, much, much more work remains to be done.

Acknowledgments

I should like to thank my collaborators and colleagues for stimulating discussions on these and related topics. The research described here has been supported by NSF grants to the Frederick Seitz Materials Reseach Laboratory (NSF DMR89-20538) and the Science and Technology Center for Supercomputing (NSF DMR91-20000).

1. D. Pines, *Physica* B **163**, 78 1990.
2. A. Millis, H. Monien, and D. Pines, *Phys. Rev. B* **42**, 167 (1990).
3. D. Pines, in *High Temperature Superconductors and the C^{60} System*, (Gordon & Breach), ed. H.-C. Ren, 1 (1995).
4. P. Monthoux, A. Balatsky, and D. Pines, *Phys. Rev. B* **46**, 14803 (1992).
5. C. P. Slichter, in *Strongly Correlated Electron Systems*, ed. K. S. Bedell *et al.* (Addison-Wesley, NY), 427 (1995).
6. V. Barzykin and D. Pines, *Phys. Rev. B* **52**, 13585 (1995).
7. Y. Zha, V. Barzykin, and D. Pines, *Phys. Rev. B* in the press (1996).
8. G. Aeppli *et al.*, preprint.
9. H. Fong *et al.*, preprint.
10. N. Curro *et al.*, private communication (1996).
11. T. Ito *et al.*, Phys. Rev. Lett. **70**, 3995 (1993).
12. B. Bücher *et al.*, Phys. Rev. Lett. **70**, 2012 (1993).
13. H. Y. Hwang *et al.*, *Phys. Rev. Lett.* **72**, 2636 (1994).
14. T. Nakano *et al.*, *Phys. Rev. B* **49**, 16000 (1994).
15. B. Stojkovic and D. Pines, *Phys. Rev. Lett.* **76**, 811 (1996).
16. P. Monthoux and D. Pines, *Phys. Rev. B* **47**, 6069 (1993).
17. P. Monthoux and D. Pines, *Phys. Rev. B* **49**, 4261 (1994).
18. Y. Kitaoka, *J. Phys. Chem. Solids* **54**, 1385 (1993).
19. D. Pines and P. Monthoux, *J. Phys. Chem. Solids* **56**, 1651 (1995).
20. P. Monthoux and D. Pines, *Phys. Rev. B* **50**, 16015 (1994).
21. A. Chubukov *et al.*, *Phil. Mag.* in the press (1996).
22. D. S. Marshall *et al.*, preprint (1996).

PAIRING MECHANISM IN THE TWO–DIMENSIONAL HUBBARD MODEL

D.J. SCALAPINO
*Department of Physics, University of California, Santa Barbara,
CA 93106-9530, USA*

Here we discuss Quantum Monte Carlo results for the magnetic susceptibility, single–particle spectral weight and the irreducible particle–particle interaction vertex of the two–dimensional Hubbard model. In the doped system, as the temperature is lowered below $J = 4t^2/U$, short–range antiferromagnetic correlations develop. These lead to a narrow low–energy quasiparticle band with a large Fermi surface and a particle–particle vertex which increases at large momentum transfer, which favor $d_{x^2-y^2}$–pairing.

A variety of experiments on the high T_c cuprates can be interpreted in terms of a gap with dominant $d_{x^2-y^2}$ symmetry [1]. However, the implications of this with respect to the nature of the underlying pairing mechanism remains an open question. Here we review some results which have been obtained for the two–dimensional Hubbard model, which has been found to exhibit $d_{x^2-y^2}$–like pairing fluctuations. Our aim is to examine in this particular case the structure of the quasiparticle spectrum and the interaction in order to gain insight into the mechanism which leads to $d_{x^2-y^2}$ pairing correlations in this model.

The two-dimensional Hubbard Hamiltonian provides an approximate model of a CuO_2 layer:

$$H = -t \sum_{\langle ij \rangle, s} \left(c_{is}^\dagger c_{js} + c_{js}^\dagger c_{is} \right) + U \sum_i n_{i\uparrow} n_{i\downarrow}. \qquad (1)$$

Here $c_{is}^\dagger$ creates an electron of spin s on site i, and $n_{is} = c_{is}^\dagger c_{is}$ is the occupation number for spin s on site i. The one–electron transfer between near–neighbor sites is t, and U is an onsite Coulomb energy. The bare energy scale is set by the bandwidth $8t$ and the effective Coulomb interaction U, which are both of order electron volts. Near half–filling, electrons on neighboring sites tend to align antiferromagnetically so as to lower their energy by the exchange interaction $J = 4t^2/U$. This interaction is of order a tenth of an electron volt and, as we will see, sets the energy, or temperature scale, below which antiferromagnetic (AF) correlations, the low–energy structure in the single–particle spectral weight, and the pairing interaction develop.

While Monte Carlo [2] and Lanczos [3] calculations for a 4 × 4 lattice find that two holes added to the half–filled Hubbard ground state form a $d_{x^2-y^2}$ bound state, and density matrix renormalization group calculations [4] find that $d_{x^2-y^2}$–like pairs are formed on two–leg Hubbard ladders, it is not known what happens for the two–dimensional Hubbard model. It is possible that on an energy scale of order $J/10$, a $d_{x^2-y^2}$ superconducting state forms. However, this may well require modifications of the model, such as an additional near–neighbor $\Delta J \mathbf{S}_i \cdot \mathbf{S}_j$ term or possibly a next–near–neighbor hopping t'. Nevertheless, it is known that as the temperature is reduced below J, $d_{x^2-y^2}$ pairing correlations develop in the doped two–dimensional Hubbard model, and here we will examine why this happens.

At half–filling, $\langle n_{i\uparrow} + n_{i\downarrow} \rangle = 1$, the 2D Hubbard model develops long–range antiferromagnetic order as the temperature goes to zero. In the doped case, strong short–range AF correlations develop as the temperature decreases below J. This is clearly seen in the temperature dependence of the wave vector dependent magnetic susceptibility

$$\chi(\mathbf{q}) = \frac{1}{N} \sum_{\boldsymbol{\ell}} \int_0^\beta d\tau \, \langle m_{i+\ell}^-(\tau) m_i^+(0) \rangle e^{-i\mathbf{q}\cdot\boldsymbol{\ell}}. \qquad (2)$$

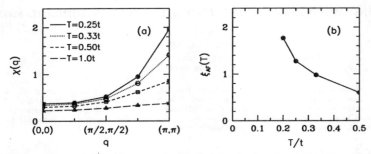

Figure 1: (a) Magnetic susceptibility $\chi(\mathbf{q})$ versus $\mathbf{q}$ along the $(1,1)$ direction for various temperatures. (b) Temperature dependence of the AF correlation length ξ_{AF} in units of the lattice spacing a. These results are for an 8×8 lattice with $U/t = 4$ and a filling $\langle n \rangle = 0.875$.

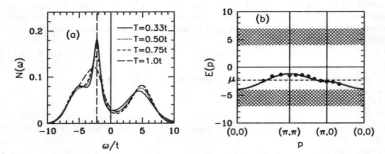

Figure 2: (a) Evolution of the single–particle density of states with temperature for $U/t = 8$ and $\langle n \rangle = 0.875$. (b) Dispersion of the quasiparticle peak in the spectral weight versus $\mathbf{p}$. The solid points mark the low–energy peaks of $A(\mathbf{p},\omega)$ shown in Fig. 3, and the solid curve represents an estimate of the quasiparticle dispersion using these data and Lanczos results for $\mathbf{p}$ near $(0,0)$. The broad darkened areas represent the incoherent spectral weight in the upper and lower Hubbard bands. The horizontal dashed line denotes the chemical potential μ.

Here $m_i^+ = c_{i\uparrow}^\dagger c_{i\downarrow}$ and $m_{i+\ell}^-(\tau) = e^{H\tau} m_{i+\ell}^- e^{-H\tau}$, where $m_{i+\ell}^-$ is the hermitian conjugate of $m_{i+\ell}^+$. Monte Carlo results for $\chi(\mathbf{q})$ versus $\mathbf{q}$ along the $(1,1)$ axis for an 8×8 lattice with $U/t = 4$ and a filling $\langle n \rangle = 0.875$ are shown in Fig. 1(a). As the temperature decreases below $J = 4t^2/U$, significant short–range dynamic AF correlations evolve. Fig. 1(b) shows the AF correlation length ξ_{AF} versus T. Here, ξ_{AF}^{-1} is defined as the half-width at half-maximum of $\chi(\mathbf{q})$.

As these AF correlations develop, the single-particle spectral weight [5]

$$A(\mathbf{p},\omega) = -\frac{1}{\pi} \operatorname{Im} G(\mathbf{p}, i\omega_n \to \omega + i\delta) \qquad (3)$$

and the density of states $N(\omega) = \frac{1}{N} \sum_\mathbf{p} A(\mathbf{p},\omega)$ also change. Figure 2(a) shows $N(\omega)$ for $U/t = 8$ and $\langle n \rangle = 0.875$. As the temperature is lowered, a peak appears on the upper edge of the lower Hubbard band. This peak arises from a narrow quasiparticle band shown in the single–particle spectral weight $A(\mathbf{p},\omega)$ of Fig. 3 and plotted as the solid curve in Fig. 2(b). As the momentum

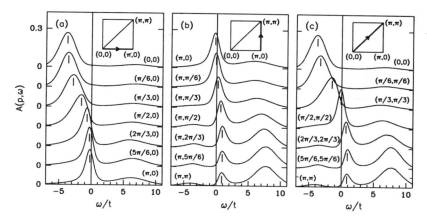

Figure 3: Single–particle spectral weight along various cuts in the Brillouin zone is shown for $U/t = 8$ and $\langle n \rangle = 0.875$ on a 12×12 lattice at $T = 0.5\,t$.

p goes towards the Γ point $(0,0)$, we believe that the quasiparticle peak is obscured by the lower Hubbard band because of the resolution of the maximum entropy technique which we have used. Indeed, at the Γ point a separate quasiparticle peak is found from Lanczos exact diagonalization on a 4×4 lattice [6]. Note that this implies a large hole–like Fermi surface. As clearly evident in the spectral weight shown in Figs. 3(a) and (b), the quasiparticle dispersion is anomalously flat near the $(\pi, 0)$ corner. As discussed by various authors, this reflects the influence of the AF correlations on the quasiparticle excitation energy. It is clear that the peak structure in $N(\omega)$ also arises from the short–range AF correlations and is a many–body effect rather than simply a non–interacting band Van Hove singularity.

Monte Carlo calculations [7] have also been used to determine the singlet irreducible particle–particle vertex $\Gamma_{\rm IS}(p', -p', p, -p)$ in the zero center–of–mass momentum and energy channel which gives the effective pairing interaction. Here $p = (\mathbf{p}, i\omega_n)$. In Fig. 4(a), $\Gamma_{\rm IS}(q = p - p')$ is plotted for $\mathbf{q}$ along the $(1,1)$ direction and $i\omega_n = i\omega_{n'} = i\pi T$, corresponding to zero Matsubara energy transfer. Comparing Figs. 1(a) and 4(a), one clearly sees that the structure of the interaction and $\chi(\mathbf{q})$ are similar, both reflecting the development of the AF correlations as T is reduced below J.

Given the Monte Carlo results for the irreducible particle–particle vertex $\Gamma_{\rm IS}(p', -p', p, -p)$ in the zero energy and center–of–mass momentum channel, and the single–particle Green's function $G(\mathbf{p}, i\omega_n)$, the Bethe–Salpeter equation for the particle–particle channel is

$$\lambda_\alpha \phi_\alpha(p) = -\frac{T}{N} \sum_{p'} \Gamma_{\rm IS}(p, -p, p', -p') |G(p')|^2 \phi_\alpha(p'). \qquad (4)$$

Here, the sum on p' is over both $\mathbf{p}'$ and $\omega_{n'}$. In the parameter regime that the Monte Carlo simulations are carried out, the leading singlet eigenvalue is in the $d_{x^2-y^2}$ channel. Fig. 4(b) shows the temperature dependence of the $d_{x^2-y^2}$ eigenvalue.

The development of both the low–energy quasiparticle dispersion and the peak in the singlet particle–particle vertex at large momentum transfers arises from the growth of short–range AF correlations as T decreases below J. As is known [8], for the large Fermi surface associated with the observed quasiparticle dispersion, a particle–particle vertex which *increases* at large momentum

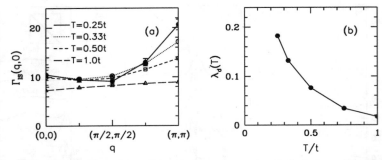

Figure 4: (a) Singlet irreducible particle–particle vertex for zero energy transfer $\Gamma_{IS}(\mathbf{q}, i\omega_m = 0)$ versus $\mathbf{q}$ along the (1,1) direction. As the temperature decreases below $J = 4t^2/U$, the strength of the interaction is enhanced at large momentum transfer. Note the similarity to $\chi(\mathbf{q})$ in Fig. 1(a). (b) Temperature dependence of the $d_{x^2-y^2}$ eigenvalue of the Bethe–Salpeter equation. These results are for $U/t = 4$ and $\langle n \rangle = 0.875$.

transfer favors $d_{x^2-y^2}$ pairing. Note that the tendency for $d_{x^2-y^2}$ pairing does not require a particularly sharp or narrow peak in $\Gamma_{IS}(\mathbf{q})$ for $\mathbf{q} = (\pi, \pi)$, but rather simply sufficient weight at large momentum transfers. Thus it is the strong short–range AF correlations which lead to the formation of $d_{x^2-y^2}$ pairing correlations in the Hubbard model.

Acknowledgments

The work reviewed here was carried out with N. Bulut and S.R. White. I would also like to acknowledge many useful discussions with J.R. Schrieffer. This work was supported in part by the National Science Foundation under grant No. DMR92–25027. The numerical computations reported in this paper were carried out at the San Diego Supercomputer Center.

References

1. J.R. Schrieffer, *Solid State Commun.* **92**, 129 (1995); D.J. Scalapino, *Phys. Reports* **250**, 329 (1995); D. Van Harlingen, *Rev. Mod. Phys.* **67**, 515 (1995).
2. E. Dagotto, A. Moreo, R.L. Sugar, and D. Toussaint, *Phys. Rev.* B **41**, 811 (1990).
3. A. Parola, S. Sorella, M. Parrinello, and E. Tosatti, *Phys. Rev.* B **43**, 6190 (1991).
4. R.M. Noack, S.R. White, and D.J. Scalapino, *Phys. Rev. Lett.* **73**, 882 (1994); R.M. Noack, S.R. White, and D.J. Scalapino, *Europhys. Lett.* **30**, 163 (1995).
5. N. Bulut, D.J. Scalapino, and S.R. White, *Phys. Rev.* B **50**, 7215 (1994).
6. E. Dagotto, A. Moreo, F. Ortolani, J. Riera, and D.J. Scalapino, *Phys. Rev. Lett.* **67**, 1918 (1991); E. Dagotto, F. Ortolani, and D.J. Scalapino, *Phys. Rev.* B **46**, 3183 (1992).
7. N. Bulut, D.J. Scalapino, and S.R. White, *Phys. Rev.* B **50**, 9623 (1994).
8. D.J. Scalapino, *Physica* C **235–240**, 107 (1994).

A POSSIBLE PRIMARY ROLE OF THE OXYGEN POLARIZABILITY IN HIGH TEMPERATURE SUPERCONDUCTIVITY.

M. Weger[1], M. Peter[2] and L.P. Pitaevskii[3]

1) Racah Institute of Physics, Hebrew University, Jerusalem
2) DPMC, Geneva University, 24 quai Ernest Ansermet, Geneve 4
3) Department of Physics, Technion, Haifa

Abstract.
  The ionic dielectric constant in superconducting cuprates is found to be very large at low frequencies, $\epsilon_0 \approx 40$ with a dispersion around $\omega_t \approx 15$ meV. This dielectric constant has a profound effect, the material being a "doped semiconductor" rather than a "conventional" metal. Since $\epsilon_{ion} > \epsilon_{el}$ (except for very small q values), the electron-electron interaction is shielded by the ionic polarizability, and the electronic Thomas-Fermi parameter κ_{TF} is very small. This causes the electron-phonon matrix element to be abnormally large for small values of $\omega(\omega < \omega_t)$ and for small (but not extremely small) values of $q(q \approx \kappa_{TF})$. As a result, the McMillan coupling constant λ is abnormally large at small energies. We calculate the superconducting transition temperature T_c and other properties by solving the Eliashberg equations in a "very strong coupling" scenario. The maximum value of T_c in a 2-D system is given approximately by: $(T_c)_{max} = 0.1(e^2/2a_c)\sqrt{m/M} \cdot 1/\sqrt{\Delta\theta} \cdot (\omega_t/\Omega)$, where $2a_c$ is the width of the metallic layer (a_c is the Bohr radius of the oxygen $2p\sigma$ orbitals in the c-direction), m the band mass, M the oxygen mass, $\Delta\theta$ the scattering angle ($\approx \kappa_{TF}/k_F$), and Ω the frequency of the Cu-O stretching vibrations (≈ 40 meV). This gives a maximum value of about 200 K for parameters characteristic of the cuprates. The small value of $\Delta\theta \approx 0.3$ radians indicates forward-scattering of the electrons by the phonons. This causes a near-degeneracy between superconducting states with d-wave and s-wave symmetry, the d-wave state being favored in "clean" samples. Under certain (but not all) conditions, the isotope effect is greatly compensated.
  Therefore, we believe that it is possible to account for high-temperature superconductivity by the BCS phonon-mediated interaction.

1. INTRODUCTION.

When high-T_c superconductivity was discovered in YBCO [1], it was believed that the "conventional" BCS mechanism, with a phonon-mediated interaction, can not account for the superconductivity in this material, and several alternative mechanisms were proposed; some of which being very original and sophisticated. One reason why the "conventional" mechanism was not taken seriously, is that since the work of McMillan [2], it was believed that the maximum value of T_c that can be obtained by it is limited to about 20-30 K. This "saturation" of T_c follows from McMillan's expression for T_c: $T_c = \frac{\Omega}{1.3} \cdot \exp\left(-\frac{1.04(1+\lambda)}{\lambda - \mu^*(1+0.64\lambda)}\right)$. Ω is the phonon frequency, λ the McMillan electron-phonon coupling constant, given by $\lambda = I^2 n(E_F)/M\Omega^2$ where I is the electron-phonon matrix element, and $\mu^* = \mu/[1 + \mu ln(E_F/\Omega)]$ where μ is the bare Coulomb interaction. This expression applies for $\lambda < 2$; for larger values of λ, no "universal" formula applies, but some approximations, like the KGL formula [3] $T_c = 0.18 \cdot \Omega\sqrt{\lambda/(1 + 2.6 \cdot \mu^*)}$ give a reasonable approximation.

Values of T_c for some "reasonable" parameters are shown in Table I. If we arbitrarily set $\mu^* = 0.1$, a high value of T_c is possible, but since the bare μ, given by $n(E_F)U$ is large (≈ 3), and E_F is small (0.5-1 eV), this selection is not reasonable. In particular, the value of $T_c = 164K$ obtained in HgBa$_2$Ca$_2$Cu$_3$O$_8$ [4] requires a value of λ in excess of 10.

TABLE I

	Ω	λ	μ*	T$_c$
realistic:	400K	2	0.35	21 K
too high Ω:	700K	2	0.41	27 K
excitonic:	1200K	2	0.57	11 K
wrong μ*:	600K	3	0.1	89 K

Such a large value of λ was not considered seriously, for several reasons: (i) Band structure calculations yield $\lambda \approx 1.5$ [5]. (ii) Such a large value of λ would give rise to a very large electronic specific heat, a very large resistivity in the normal state, superconducting tunneling curves with very broad maxima [6], large broadening and pulling of phonon frequencies [7], etc. and such phenomena are not observed. In particular, on the basis of the normal state resistivity, it was stated that $\lambda < 0.1$ [8]. (iii) The isotope effect is very small in optimally-doped samples [9].

The saturation of T_c pointed out by McMillan, and by Cohen and Anderson [10], and many others for 30 years, is thus a very fundamental property. If it does not hold for the cuprates, a very "basic" physical reason must exist.

We suggest that such a basic physical cause indeed exists, and it is the large ionic dielectric constant ϵ_{ion} of these perovskites, which replaces the electronic dielectric constant ϵ_{el} ($\approx 1 + \kappa_{TF}^2/q^2$, where κ_{TF} is the Thomas-Fermi screening parameter) as the main screening agent. The cuprates are "doped semiconductors" rather than "conventional" metals.

A "back-of-the envelope" estimate of the increase in T_c by this cause is as follows: For large λ, $T_c \approx 0.18\Omega\sqrt{\lambda} = 0.18I\sqrt{n(E_F)/M}$. I is given approximately by: $I(q) = \langle \psi_k | \nabla V | \psi_{k+q} \rangle /(1 + \kappa_{TF}^2/q^2)$ [11,12]. κ_{TF} is given by: $\kappa_{TF}^2 = 4\pi e^2 n(E_F)/\epsilon_{ion}$. In "conventional" metals, $\epsilon_{ion} \approx 1$ and $\kappa_{TF}^2 \approx (2-3)k_F^2$. In cuprates, $\epsilon_{ion} \approx 40$ [13] and therefore $\kappa_{TF}^2 \ll k_F^2$. As a result, the matrix element I is about 3-4 times bigger, and so is T_c. The enhancement of I is particularly large near the Mott transition, since there $\kappa_{TF}^2/k_F^2 \approx r_s$ attains its largest possible value. Therefore, while in a "conventional" metal the highest possible T_c is given by: $(T_c)_{max} = 0.1 E_F \sqrt{m/M}$, in a "doped semiconductor" ($\epsilon_{ion} > 20$), it is given by: $(T_c)_{max} = 0.1 E_F \sqrt{m/M} \cdot r_s$ [11,12].

2. EXTREMELY INHOMOGENEOUS DIELECTRIC.

There is an obvious weakness in the above estimate. Namely, ϵ_{ion} is expected to screen the electron-ion potential as well as the electron-electron interaction, leading to an electron-ion potential given by: $V_{screened}(q) = \frac{4\pi Ze^2/\epsilon_{ion}}{q^2 + (\kappa_{TF}^0)^2/\epsilon_{ion}}$ $[(\kappa_{TF}^0)^2 = 4\pi e^2 n(E_F)]$, which is **not** enhanced over the unscreened potential ($\epsilon_{ion} = 1$) and is even **reduced** for $q \neq 0$ (Fig.1a).

The flaw in this argument is subtle.

Cuprates are extremely **inhomogeneous** dielectrics. For a homogeneous dielectric, $\epsilon_{ion}(r_1, r_2) = \epsilon_{ion}(r_1 - r_2)$ and the Fourier transform $\epsilon_{ion}(q)$ enters into the expression for $V_{screened}(q)$. In an inhomogeneous dielectric, ϵ_{ion} is a function of **two** variables, and cannot be reduced to a function of a single variable $\epsilon_{ion}(q)$. The two variables can be chosen as two position variables, r_1, r_2 or two momentum variables, q and G, (the reciprocal lattice vector): $V(q) = \Sigma \frac{\rho(q+G)}{q^2} \cdot \frac{1}{\epsilon(q,G)}$. (The terms with $G \neq 0$ can be regarded as "Umklapp" terms). We prefer to use one momentum variable q, and one position variable r_2, where: $\frac{1}{\epsilon(q,r_2)} = \int \frac{1}{\epsilon(r_1-r_2,r_2)} \cdot e^{iq(r_1-r_2)} d^3(r_1-r_2)$, and therefore: $V_{ihd}(q) = \int \frac{\rho(r_2)}{q^2} \cdot \frac{1}{\epsilon(q,r_2)} d^3 r_2$ (ihd stands for inhomogeneous dielectric). $V_{ihd}(q)$ is thus a functional of the charge distribution $\rho(r_2)$.

For an ion situated in the center of a hollow sphere of radius R, in a dielectric with a dielectric constant ϵ_0, $V_{ihd}(q)$ is given by (n is the density of conduction electrons): $V_{ihd} = n\frac{4\pi e^2}{q^2}[1 - (1 - 1/\epsilon_{ion}(\omega))]\frac{\sin qR}{qR}$. We write: $V_{ihd}(q) = \frac{4\pi Ze^2}{q^2}\{\frac{1}{\epsilon(q,r_2)}\}_{r_2}$ and regard this as a definition of the response function, for this specific charge distribution.

For the potential induced by the conduction electrons, spread-out more-or-less uniformly over the atomic cell, we write likewise: $V_{ee}(q) = \frac{4\pi e^2}{q^2} <\frac{1}{\epsilon(q,r_2)}>_{r_2}$, where now the response function is different, because of the different distribution of charge that induces the potential. The response functions $\{\}$ and $<>$ are plotted in Fig. 1b for parameters appropriate for the cuprates. We see that for $q \to 0$, both reduce to $1/\epsilon_0$, and for large values of q/k_F, both approach 1. But in the region $0.15 < q/k_F < 1.5$, there is a "window", where $\{\}$ approaches 1, while $<>$ is still close to $1/\epsilon_0$.

The electron-ion potential, screened by the conduction electrons (as well as by the dielectric medium), is given by:

$$V_{ihs}(q) = \frac{4\pi Ze^2 \{1/\epsilon_{ion}(q,r_2)\}_{r_2}}{q^2 + 4\pi e^2 n(E_F) <1/\epsilon_{ion}(q,r_2)>_{r_2}}$$

(ihs stands for inhomogeneous dielectric screened). $V_{ihs}(q)$ is plotted in Fig.1a. For $q \to 0$, it attains the value $Z/n(E_F)$ which is required by rather general arguments [14]. But for small values of q, it rises sharply and attains a sharp maximum at $q \approx \kappa_{TF} (\approx 0.3 k_F$ for parameters appropriate for the cuprates).

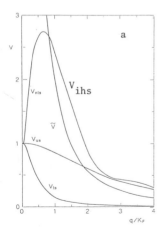
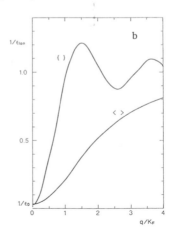

Figure 1

(a) The electron-ion potential.
V_{us}: Without ionic screening.
V_{ls}: Screening by homogeneous dielectric.
V_{ihs}: Inhomogeneous dielectric screening.

(b) The inverse ionic dielectric constant.
$\{\}$ Electron-ion interaction.
$<>$ Electron-electron interaction.

3. DIMENSIONAL EFFECT.

The electron-phonon matrix element $I(q,\omega)$ given by $qV_{ihs}(q,\omega)$ is seen to possess a very large, sharp peak at a small value of q ($q \approx 0.3 k_F$) (Fig.1a). The McMillan λ is given by integration of $I^2(q,\omega)$ over the FS. Thus, there is an important dimensional effect. In 3D, the area of the FS corresponding to the small q values is small (πq^2); as a result, the increase in λ is not very large. In 2D, the FS is a cylinder of length $2\pi/c$, and the area of the FS (for the small q-values) is given by $q \cdot 2\pi/c$ (Fig.2). Therefore, the small q-values contribute much more and λ actually diverges as $\Delta\theta = \kappa_{TF}/k_F \approx q/k_F$ goes to zero. Calculating the averages over the FS is straightforward, and the expressions for the maximum T_c are: 3D: $(T_c)_{max} = 0.04 r_s E_F \sqrt{m/M} \cdot \omega_t/\Omega$; 2D: $(T_c)_{max} = 0.1 \frac{e^2}{2a_c} \sqrt{m/M} \cdot 1/\sqrt{\Delta\theta} \cdot \omega_t/\Omega$. a_c is the Bohr radius of the oxygen $2p\sigma$ orbitals in the c-direction, and m is the band mass.

In the calculation of T_c, we also take into account the ω-dependence of $V_{ihs}(q,\omega)$, which is cut-off at ω_t, the cutoff frequency of the ionic dielectric constant, which in the Lyddane-Sachs-Teller theory is given by: $\epsilon_{ion}(\omega) = \epsilon_\infty + \frac{\epsilon_\infty - \epsilon_0}{\omega^2/\omega_t^2 - 1}$, ω_t in the cuprates is approximately 15 meV [13].

Since $\omega_t/\Omega \approx 0.4$, $1/\sqrt{\Delta\theta} \approx 2$, we may approximate: $(T_c)_{max} \approx 0.1 \frac{e^2}{2a_c} \sqrt{m/M} \approx$ 200K. This expression is extremely simple; even simpler than McMillan's expression for T_c, since the phonon frequency drops out, and we don't have to "guess" λ. The maximum T_c is seen to be the inverse thickness of the metallic layer, which in the cuprates is very small, due to the small width of the oxygen $2p\sigma$ orbitals.

The ω_t/Ω term in the expression for T_c cancels the isotope effect of the phonon mode Ω (≈ 40 meV); the factor $1/\sqrt{\Delta\theta} \cdot \omega_t/\Omega$ partially cancels the isotope effect of ω_t (since $\epsilon_0 \propto \omega_t^{-2}$, and therefore $\Delta\theta$ increases with ω_t). We consider here the "clean" limit; when the elastic scattering rate exceeds ω_t, these expressions no longer apply.

We considered solutions of the Eliashberg equations for this scenario, of a very large λ with a small cutoff [11]. The small value of $\Delta\theta$ favors a gap $\Delta(k)$ with d-wave symmetry, even for the phonon-mediated interaction [15].

REFERENCES.
1) M.K. Wu, J.R. Ashburn, C.T. Torng, P.H. Hor, R.L. Meng, L. Gao, Z.J. Huang, Y.Q. Wang, C.W. Chu, *Phys. Rev. Lett.* **58** (1987) 908.
2) W.L. McMillan, *Phys. Rev.* **167** (1968) 331.
3) V.Z. Kresin, H. Gutfreund, W.A. Little, *Solid. State Commun.* **51** (1984) 335.
4) L.Gao, Y.Y. Xue, F. Chen, Q. Ziong, R.L. Meng, D. Ramirez, C.W. Chu, J.H. Eggert, H.K. Mao, *Phys. Rev. B* **50** (1994) 4260.
5) C.O. Rodriguez, A.I. Liechtenstein, I.I. Mazin, O. Jespen, O.K. Anderson, M. Methfessel, *Phys. Rev. B* **42** (1990) 2692. S. Massidda, J. Yu, K.T. Park, A.J. Freeman, *Physica C* **176** (1991) 159.
6) P.B. Allen and D. Rainer, *Nature* **349** (1991) 396.
7) R. Zeyher and G. Zwicknagl, *Z. Phys. B* **78** (1990) 175.
8) S. Martin, A.T. Fiory, R.M. Fleming, L.F. Schneemeyer, J.R. Waszszak, *Phys. Rev. B* **41** (1990) 846.
9) J.P. Franck in *"Physical Properties of High Temperature Superconductors IV"*, D.M. Ginsberg, Ed., World Scientific Singapore (1994) p.189.
10) M.L. Cohen and P.W. Anderson in *"Superconductivity of d- and f-band Metals"*, D.H. Douglas, Ed., AIP New York 1972.
11) M. Weger, B. Barbiellini, T. Jarlborg, M. Peter, G. Santi, *Ann. Phys.* **4** (1995) 431.
12) M. Weger, *Acta Physica Polonica A* **87** (1995) 723.
13) J. Humlicek, A.P. Litvinchuk, W. Kress, B. Lederele, C. Thomsen, M. Cardona, H.U. Habermeier, I.E. Trofimov, W. Konig, *Physica C* **206** (1993) 345. D. Reagor, A. Ahrens, S.W. Cheng, A. Migliori, Z. Fisk, *Phys. Rev. Lett.* **62** (1989) 2048.
14) V. Heine, P. Nozieres, J.W. Wilkins, *Phil. Mag.* **13** (1966) 741.
15) G. Santi, T. Jarlborg, M. Peter, M. Weger, *J. Superconductivity* **8** (1995) 405. *Physica C* (1996) (in press).

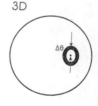
3D

2D

ABSENCE OF EXCHANGE SCATTERING BY CUPRATE-PLANE IMPURITIES IN HIGH–TEMPERATURE SUPERCONDUCTORS

HOWARD A. BLACKSTEAD

Physics Department, University of Notre Dame, Notre Dame, Indiana 46556 U.S.A.

and

JOHN D. DOW

Department of Physics and Astronomy, Arizona State University, Tempe, Arizona 85287-1504, U.S.A.

ABSTRACT

The exchange-scattering difference between Ni and Zn does not break Cooper pairs in most high-temperature superconductors, because the impurities are in the cuprate planes and the superconducting condensate is primarily in the charge-reservoirs, beyond the range of exchange-scattering.

1. Ni and Zn pair-breaking rates are the same

The four main experimental facts concerning Ni and Zn doping of high-temperature superconductors are:[1] (i) In each of the major high-temperature superconductors, except in $Nd_{2-z}Ce_zCuO_4$ (Nd214) and *possibly* in $YBa_2Cu_3O_7$ (Y123), Ni and Zn break Cooper pairs at the same rate and *suppress* T_c *the same*. The same amounts of Ni and Zn impurities drive T_c to zero. Hence the exchange scattering, which corresponds to the difference between Ni and Zn, has *no perceptible effect* on the critical temperature T_c. The obvious interpretation of this fact is that the superconducting condensate lies outside the range of the impurities' exchange scattering. The impurities cannot be assigned to sites within the superconducting condensate because it is inconceivable that magnetic exchange scattering could have *no effect* on the magnetically paired charge carriers of the Cooper pairs in these Meissner-effect superconductors. (ii) The main exception to same-rate-scattering by Ni and Zn occurs in Nd214, the one high-temperature superconductor whose cuprate-plane Cu-sites (the Ni and Zn impurity sites) are nearest-neighbors to interstitial dopant oxygen in the charge-reservoirs. Hence the cuprate-plane impurities break Cooper pairs in the charge-reservoirs. In Nd214, Ni and Zn dopants act as in an ordinary superconductor: Ni is a very strong pair-breaker, about six times as strong as Zn. (iii) The other purported exception to same-rate-scattering by Ni and Zn occurs in Y123, which does require more Ni than Zn to destroy superconductivity, but there is scant evidence that the Zn and Ni occupy the same sites: either one of the two inequivalent cuprate-planar Cu sites, Cu–O chain sites, or sites in microphases such as NiO. Observation of Zn being a stronger pair-breaker than Ni on the same site would be extremely strong evidence for a spin-fluctuation mechanism of Cooper-pairing. However, Y123 is infamous for its stoichiometry variations in the Cu–O chains, and compounds nearly identical to Y123, with reduced stoichiometry variations (including some Y123 homologues and $YBa_2Cu_4O_8$) have cuprate-plane Ni and Zn scattering the same — suggesting that the observed Y123 pair-breaking reflects different site-occupancies by the impurities, not different scattering strengths. (iv) The overall trend from one superconductor to another in the pair-breaking matrix element M for Ni and Zn, is that M dies out exponentially with the distance d between the cuprate-plane Cu-site occupied by the Zn or Ni impurity and the nearest oxygen ion in a charge-reservoir, $M \propto \exp(-d/d_0)$, where $d_0 \approx 3$ Å. (In Nd214, there is an additional short-ranged exchange-scattering contribution to M for Ni.) Hence the superconductivity is primarily in the charge-reservoir regions, not in the cuprate-planes (in which case M would be independent of d).

These four facts show clearly that *the primary superconductivity is in the charge-reservoirs* beyond the range of the short-ranged exchange-scattering by cuprate-plane impurities (except in Nd214), and only weakly perturbed by the long-ranged potentials of remote Ni or Zn impurities.[1] We doubt that *any* cuprate-plane theory can explain these facts.

2. Consequences

Other consequences of charge-reservoir superconductivity include: (i) Because the trivalent, magnetic ions of the lighter, larger-radius rare-earths are highly soluble on Ba-sites adjacent to the charge-reservoirs, while the heavier rare-earths are not, replacement of Y in Y123 by large-radius Ce, Pr, or Cm should break pairs and destroy the superconductivity, while replacement of Y with the heavier, smaller rare-earths should leave T_c at ≈ 90 K — as observed.[1] (ii) Pr123, if prepared under conditions that minimize Pr occupancy of Ba-sites (low temperatures, low oxygen concentrations), should contain grains that superconduct — as observed.[2] (iii) Magnetic rare-earth ions on Y-sites are so remote from the superconducting charge-reservoirs that they do not break pairs or adversely affect T_c — as observed.[1]

High-temperature superconductivity has remained unsolved for a decade because the primary superconductivity is in the charge-reservoirs, not in the cuprate-planes, as so many people were led to believe.

3. Acknowledgments

We thank the Office of Naval Research, the Air Force Office of Scientific Research, and the Department of Energy for their generous support (N00014-94-10147, AFOSR-F49620-94-10163, and DE-FG02-90ER45427).

4. References

1. H. A. Blackstead and J. D. Dow, *Philos. Mag.* **73**, 223 (1996), and references therein; *Phys. Rev.* **B 51**, 11830 (1995); *Proc. SPIE* **2397**, 617 (1995).
2. H. A. Blackstead, D. B. Chrisey, J. D. Dow, J. S. Horwitz, A. E. Klunzinger, and D. B. Pulling, *Phys. Lett.* **A 207**, 109 (1995).

HALL-EFFECT SCALING AND CHEMICAL EQUILIBRIUM IN NORMAL STATES OF HIGH-T_c SUPERCONDUCTORS

A.Kallio, and K.Honkala

Department of Physical Sciences, Theoretical Physics, University of Oulu, Linnanmaa, FIN-90570 Oulu, Finland

Abstract

We propose a simple treatment of boson fermion chemical equilibrium reaction $B^{2+} \rightleftharpoons 2h^+$ suitable for understanding the normal states of superconductors such as 123 where the bosons may continue to exist as preformed pairs for $T > T_c$ and gradually decay into pairing holes. The densities of bosons and fermions, given by a function $f(T)$, are temperature dependent. A proper treatment in analogy with ionization using the grand potential $\Omega(T,V,\mu)$ leads to reduction of free boson contribution to entropy and specific heat near T_c. Large free boson contribution immediately above T_c, not seen in experiments, has been the main argument against boson models. The treatment explains in a simple fashion also the Hall coefficient scaling law observed experimentally by Hwang et al. in $LSCO$. The logarithmic behaviour of resistivities ρ_{ab} and ρ_c in strong magnetic field by Ando et al is shown to come from bosons which then are 3D rather than 2D. We propose a "rational" principle how to manufacture superconductors with maximal T_c.

One of the most remarkable recent discoveries is the Hall coefficient scaling[1] in $La_{2-x}Sr_xCuO_4$ and other compounds[2]. It has been shown that for all x one can define a scaling temperature $T^*(x)$ such that all the temperature dependent curves for Hall coefficient $R_{ab}(x)$ can be collapsed into one curve in the scaling variable $t = T/T^*$ in the form

$$A(t) = \frac{R_{ab}(T) - R_\infty(x)}{R_{ab}(T^*)}, \qquad (1)$$

which is universal in variable t. The scaling temperature $T^*(x)$ drops roughly linearly from 600 K for $x \sim 0.07$ to 100 K for $x \sim 0.25$, where $T_c(x)$ vanishes. The purpose here is to show that the spectator fermion model (SFS) proposed recently[3,4] to explain the properties of the impurity system $YBa_2Cu_{3-x}M_xO_7$ can explain in a simple way the universality in Eq. (1). The SFS model proposes existence of bosons and their decay products, the pairing fermions, in chemical equilibrium governed by a boson breaking function $f(T)$ conserving the charge such that

$$N_B(T) = N_0 f(T),$$
$$N_h(T) = 2N_0(1 - f(T)), \qquad (2)$$

where N_B and N_h are the particle numbers for bosons and fermions following from the equilibrium reaction $B^{2+} \rightleftharpoons 2h^+$. The normalization is such that $N_B(0) = N_0, f(0) = 1$ and $f(\infty) = 0$. The reason for introducing the auxiliary function f is the fact that at least in the normal state it gives directly the carrier densities and contains in a simple way the charge conservation. Assuming that pairing holes constitute major charge carriers in the ab-direction the Hall coefficient for $T > T_{BL}$ is simply given by

$$R_{ab}(T) = \frac{R_\infty}{1 - f(T)} \qquad (3)$$

and by Eqs. (2) we obtain for the scaling function

$$A(t) = [1 - f(1)] \frac{f(t)}{1 - f(t)} \qquad (4)$$

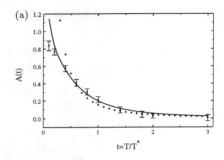 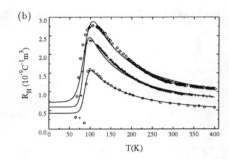

Figure 1: (a) The experimental scaling curve for $A(t)$ for $La_{2-x}Sr_xCuO_4$ (error bars): theory from Eqs.(4-7) with diamonds ($\alpha = 0.4$ $\beta = 0$) and line ($\alpha = 0.4$ $\beta = 0.15$). (b) Theoretical values of R_H compared with the experimental Hall coefficient for $YBa_2Cu_3O_{7-\delta}$ from ref.2.

with $A(1) = f(1)$. With $f(t) \to 0$ for large t this already explains qualitatively the scaling curve, which ends at $t_{BL} = T_{BL}/T^*(x)$ where T_{BL} ($> T_C$) is the boson delocalization temperature. Below T_{BL} also bosons will contribute to the Hall coefficient and the curves bend down below the "spin gap" where the fraction of localized bosons is given by $(1 - \xi(T))f(T)$ where ξ is step function like: $\xi(T > T_{BL}) = 0$, $\xi(T \ll T_{BL}) = 1$. Requiring the equilibrium condition $\mu_B = 2\mu_h$ gives the concentration product rule

$$\frac{f(2-f)}{(1-f)^2} = \varphi(T), \tag{5}$$

where $\varphi(T)$ is the chemical equilibrium constant. The solution for f is

$$f(T) = 1 - \frac{1}{\sqrt{1+\varphi(T)}} = 1 - (1 + \varphi(T))^{-1/2} \tag{6}$$

and $f(T) \propto \varphi(T)$ for large T. working with grand potential for bosons and fermions $\Omega = \Omega_h + \Omega_B - 2N_0\mu_h$ one obtains equilibrium constant [5]

$$\varphi(t) = \frac{\alpha}{t^2 + 2\beta t + 3\beta t \ln \frac{d}{t}}, \tag{7}$$

with α and β parameters and $d = eT_{BL}/T^*$. Most other experimental quantities can now be calculated with the simple rule: The temperature dependence is given by density dependence. This includes resistivity (n^{-1}), NMR –rate (n^2), specific heat (n). The free boson contribution to specific heat $3k_B(N_0/V)$ is cut by factor 20 in YBCO, but is not in the heavy fermion cases, where also the temperature dependence (A/T) of the giant linear coefficient agrees with experiments. Also the Hall coefficient in heavy fermion –cases exhibit the SFS behaviour of Fig. 1b.

1. H. Y. Hwang, et al., Phys. Rev. Lett. **72**,(1994) 2636.
2. A. Carrington et al., Phys. Rev. **B 48**,(1993) 13051.
3. A. Kallio et al., Physica **C 219**, (1994) 340-362.
4. A. Kallio, V. Apaja, and S. Pöykkö Physica **B 210**,(1995) 472.
5. A. Kallio, V. Sverdlov, and K. Honkala Preprint **No: 118** (1995)

PAIRING INSTABILITY AND ANOMALOUS RESPONSES IN AN INTERACTING FERMI GAS

Y. M. MALOZOVSKY AND J. D. FAN

Department of Physics, Southern University
Baton Rouge, Louisiana 70813, USA

ABSTRACT

The t-matrix (the two-particle vertex part) approach is used to investigate the pairing instability induced either by a given attractive interaction or by the particle-hole excitations in the case of the repulsive forces between particles. The effect of the vertex corrections on the charge and spin responses of a Fermi gas is also considered. It is shown that interaction in the particle-hole channel induces multipair excitations. It leads to the appearance of the anomalous term $Im\chi \sim -\omega/T$ in both the charge and spin responses of the Fermi gas in addition to the normal response $Im\chi \sim -\omega/v_F k$.

1. The t- matrix (the two-particle vertex part)

Carrying out the ladder summation of the exchange graphs and non-exchange ones in which all particles in the Fermi sea are taken into account,[1] it can be shown that the t-matrix for the case of either a given attractive interaction or the repulsive interaction incorporating the particle-hole excitations satisfies the following equation [2-3]

$$t(p,p') = \tilde{t}(p,p') \pm i \int \frac{d^4q}{(2\pi)^4} \tilde{t}(p,q) \, G(q) \, G(p'+p-q) \, t(q,p'), \qquad (1)$$

where (+) stands for the case with a given interaction and (−) corresponds to the case when the particle-hole excitations are taken into account. In Eq. (1) $t(p,p') = t(p,p',k)_{|k\to 0}$ is the effective t-matrix which is equivalent to the two-particle vertex part $I(p_1,p_2;p_3,p_4)$, i.e., $t(p_1,p_2,k) = I(p_1,p_2;p_3,p_4)|_{p_3=p_1+k, p_4=p_2-k}$. In Eq. (1) $\tilde{t}(p,p')$ is the irreducible kernel of interaction that in the zeroth-approximation coincides with the simple first-order vertex, namely, $\tilde{t}(p,p') = D(p-p')$, where $D(p-p')$ is the instantaneous interaction or interaction with a cutoff energy $\omega_c \ll \varepsilon_F$. Carrying out the frequency summation in Eq. (1), the solution to it can be derived as

$$t(\mathbf{Q},\omega_Q) = \pm \lambda \left(1 + \lambda \sum_{q,|\xi_q| \le \omega_c} \frac{1 - n_F(\xi_q) - n_F(\xi_{p'+p-q})}{\omega_Q - \xi_q - \xi_{p'+p-q} + i\delta} \right), \qquad (2)$$

where (−) corresponds to the case of a given attractive interaction with $\lambda = -\langle D(\mathbf{p-p'},0)\rangle > 0$ being the interaction constant, (+) is related to the case of a repulsive interaction when the particle-hole excitations are included with $\lambda = \langle D(\mathbf{p-p'},0)\rangle > 0$, where $\langle .. \rangle$ means the average over the Fermi surface. In Eq. (2) $\omega_Q = \omega + \omega'$ and $\mathbf{Q} = \mathbf{p'+p}$ are the total energy and momentum of the system of the two particles, respectively. Eq. (2) leads to the well-known result concerning the instability of the normal state in the case of a given attractive interaction and new result relating to the pairing instability caused by the particle-hole excitations if the interaction is repulsive, i.e., $\lambda > 0$.

489

2. The charge and spin responses in an interacting Fermi gas

The charge $\tilde{\chi}_c(k)$ and spin $\tilde{\chi}_s(k)$ responses follow from the solution of equations for $\tilde{\chi}_{c,s}(k) = -2i\int G(p')G(p'-k)\Gamma_{c,s}(p',k)d^4p'/(2\pi)^4$, the irreducible response functions with $\Gamma_{c,s}(p,k)$ being the three-point vertex part. The three point vertex parts in the charge and spin channels are given by $\Gamma_{c,s}(p,k) = 1 + i\int t_{c,s}(p,p')G(p')G(p'-k)\Gamma_{c,s}(p',k)d^4p'/(2\pi)^4$, where $t_{c,s}(p,p')$ are the t-matrices Eq. (1) in the charge and spin channels, respectively. In the case of the instantaneous interaction (i.e., $D(\mathbf{p}-\mathbf{p}',0)$ is a frequency-independent interaction) or when the cutoff energy of the interaction $\omega_c \sim \varepsilon_F$, the well known Migdal theorem is violated, and the vertex corrections essentially affect the irreducible response. In the case of the frequency-independent interaction, the real and imaginary parts of the charge $\tilde{\chi}_c(k)$ and the spin $\tilde{\chi}_s(k)$ irreducible responses, respectively, are evaluated as

$$\mathrm{Re}\tilde{\chi}_{c,s}(\mathbf{k},\omega)_{\omega<v_Fk} = -\frac{2N_F}{1 + F_0^{s,a}(\omega)}, \tag{3}$$

and

$$\mathrm{Im}\tilde{\chi}_{c,s}(\mathbf{k},\omega)_{\omega<v_Fk} = -\frac{\pi N_F}{[1 + F_0^{s,a}(\omega)]^2}\left(\frac{\omega}{v_Fk} \pm \tilde{\lambda}_{c,s}^2(\omega)N_F^2\tanh\frac{\omega}{2T}\right), \tag{4}$$

where N_F is the density of states per spin, and $(-)$ stands for the model with a given interaction and $(+)$ corresponds to the case when the particle-hole excitations are taken into account. In Eqs.(3) and (4) $F_0^s(\omega)$ and $F_0^a(\omega)$ are the frequency-dependent Landau's Fermi liquid parameters in the charge and spin channels ($F_0^s(\omega)$ is the spin-symmetric parameter, and $F_0^a(\omega)$ is the spin-asymmetric parameter), respectively. In Eq. (4) $\tilde{\lambda}_c(\omega) = F_0^s(\omega)/N_F$ and $\tilde{\lambda}_s(\omega) = F_0^a(\omega)/N_F$ are the effective frequency-dependent coupling parameters in the charge and spin channels, respectively, which appear as a result of the pair correlations induced either by a given interaction or by the particle-hole excitations. In the case of a given attractive interaction $F_0^{s,a}(\omega) = \lambda N_F/(1 - \lambda N_F\ln(\varepsilon_F/\max\{\omega,T\})$, the Landau Fermi liquid parameters in the charge and spin channels are the same, and positive ($F_0^{s,a}(\omega) > 0$), with $\lambda = -\langle D(\mathbf{p}-\mathbf{p}',0)\rangle$ being the interaction constant as before. In the case of the repulsive interaction when the particle-hole excitations are included into consideration, $F_0^s(\omega) \neq F_0^a(\omega)$, and they are determined by the equations $F_0^s(\omega) = -\lambda N_F/(1 -\lambda N_F\ln(\varepsilon_F/\max\{\omega,T\})$ for the Landau parameter in the charge channel, and $F_0^a(\omega) = -\lambda N_F/(1+\lambda N_F\ln(\varepsilon_F/\max\{\omega,T\})$ for the Landau parameter in the spin channel. In the case of the repulsive interaction $F_0^s(\omega)$ and $F_0^a(\omega)$ are negative ($F_0^{s,a}(\omega) < 0$), and $|F_0^s(\omega)| > |F_0^a(\omega)|$. It deserves a mention that the anomalous term in the irreducible charge response ($\mathrm{Im}\chi \sim -\omega/T$) in the case when the particle-hole excitations are included exhibits a negative sign in contrast to the positive sign in the case of a given interaction.

This work is supported by DoE under grant No. DE-FG05-94ER25229.

1. V.M. Galitskii, Sov. Phys. JETP 34,104(1958).
2. L.P.Gorkov and T.K. Melik-Barkhudarov, Sov. Phys. JETP 13,1018(1961).
3. J.R. Schrieffer, Physica 26, S124,(1960).

INTER-BAND THEORY OF SUPERCONDUCTORS: RESOLUTION OF OBSERVED S AND D-WAVE TUNNELING WITH ISOTROPIC S-WAVE PAIRING

JAMIL TAHIR-KHELI
Beckman Institute 139-74, California Institute of Technology, Pasadena, CA 91125, USA

A fundamentally different, yet conceptually and computationally simple theory is presented for high temperature superconductivity. We assume the existence of two bands, one with predominantly planar $d_{x^2-y^2}$ character on the copper sites and the other with predominantly p_z character with lobes pointing normal to the CuO planes on the oxygen sites. An attractive phonon coupling across the bands is assumed with a purely isotropic (s-wave) attraction. Thus, a Cooper pair consists of a $k \uparrow$ electron from one band and a $-k \downarrow$ electron from the other band. Due to the different masses of the two bands, Cooper pairs can carry current, or equivalently, are not invariant under time-reversal. This leads to a dramatic change in the standard picture of Josephson tunneling. Inter-band pairing makes the orbital nature of the bands contribute to the tunneling current. It is shown that the four key experiments designed to measure the phase of the gap: 1.) YBCO tri-crystals (d-wave) by Tsuei et al., 2.) YBCO-Pb corner junctions (d-wave) by Wollman et al., 3.) hexagonal YBCO junctions (s-wave) by Chaudhari et al., and 4.) YBCO-Pb c-axis tunneling (s-wave) by Dynes et al., are all a natural consequence of inter-band pairing. Finally, the temperature dependence of the Knight shifts in YBCO$_7$ and YBCO$_{6.63}$ and their completely different c-axis optics are qualitatively explained by the matrix element for electric-dipole transitions between the two bands.

1 Inter-Band Pairing Model

Assume an attractive coupling *across* two bands one with predominantly $d_{x^2-y^2}$ character on the Cu sites and the other with predominantly p_z character with lobes pointing normal to the CuO planes on the O sites. The bands are assumed to occupy the same phase space and have coincident Fermi surfaces at optimal doping. This assumption can be relaxed, but in this paper we will see how far we can go with the simplest possible assumptions about the bands.

Then, the possibility exists of optical absorption down to $\omega = 0$ from one band to the other so long as the relevant electric-dipole matrix element is non-zero. Experimentally, such optical absorption exists for light polarized along the c-axis for the "lower" T$_c$ materials, YBa$_2$Cu$_3$O$_{6.63}$ and La$_{2-x}$Sr$_x$CuO$_4$ but not for fully doped YBa$_2$Cu$_3$O$_7$. No strong absorption exists for light polarized in the CuO planes for any of these materials.

The d and p bands above can explain these facts with the added assumption that for YBCO$_7$ there is very little to no p_σ character of the $d_{x^2-y^2}$ on the O sites and no d_{xz}, d_{yz} character of the p_z band on the Cu site. The creation of oxygen vacancies in the chains causes some hybridization making the matrix element non-zero.

The key difference between inter-band pairing and BCS-like intra-band pairing is that inter-band Cooper pairs are *not* invariant under time-reversal although the complete Hamiltonian is time-reversal invariant. This dramatically alters the interpretation of Josephson tunneling. With intra-band pairing, the single electron tunneling matrix element gets mod-squared because both the electron and its time-reversed partner tunnel together. Thus, all phase information determining the supercurrent is contained in the phases of the gap functions on both sides of the junction.

With inter-band pairing, the single electron tunneling matrix elements are different for the two electrons since they come from different bands leading to a phase contribution from these matrix elements on top of the contribution from the gap functions. We define the simplest possible phases for the band wavefunctions and assume an isotropic s-wave attraction. This leads to a pure s-wave gap. Thus, the Josephson supercurrent is controlled solely by the phases of the band orbital matrix elements.

The four key tunneling experiments,[1-4] two leading to an "s-wave" gap and the other two to a "d-wave" gap can all *simultaneously* be explained by inter-band pairing. For the hexagonal YBCO (s-wave) of Chaudhari et al.,[3] the key feature leading to an s-wave result is the mismatch of the CuO planes across the grain boundary due to the use of a MgO seeding layer for the YBCO hexagon. This naturally leads to the dominant tunneling of d electrons to p electrons and p to d electrons. The single electron matrix element thus get mod-squared and we are left with "s-wave" behavior. On the other hand, for the YBCO tri-crystal experiment of Tsuei et al.,[2] the CuO planes align across the grain boundary forcing tunneling of d to d and p to p. In this case the d-d orbital overlap simulates the effect of a "d-wave" gap.

The temperature dependence of the Knight shift can be understood by essentially the same matrix elements that we used to understand the low-energy optical absorption. This is due to the orbital (Van Vleck) paramagnetism. Usually the orbital paramagnetism is temperature independent because the energy differences with non-zero matrix element in the expression for the orbital paramagnetism χ are either very small compared to kT or else very large compared to kT. With inter-band pairing as we have assumed this is no longer the case and the contribution to the Knight shift from the nuclear orbital coupling can be temperature dependent. It is not hard to see from the expression for χ that the shift must increase with increasing temperature as experimentally observed.

Finally, for $YBCO_7$ all of the relevant matrix elements will be zero for the same reasons as the optics. For $YBCO_{6.63}$ and LASCO, the matrix elements are zero for magnetic fields along the c-axis and non-zero for fields in the CuO planes as observed.

Acknowledgments

The author would like to thank Jean-Marc Langlois, Jason Perry, Siddharth Dasgupta, Tony Leggett and especially Bill Goddard for numerous discussions and encouragement. This research was not supported by any funding agency whatsoever.

References

1. D.A. Wollman *et al.*, *Phys. Rev. Lett.* **71**, 2134 (1993)
2. C.C. Tsuei *et al.*, *Phys. Rev. Lett.* **73**, 593 (1994).
3. P. Chaudhari *et al.*, *Phys. Rev. Lett.* **72**, 1084 (1994).
4. A.G. Sun *et al*, *Phys. Rev. Lett.* **72**, 2267 (1994).

THE CONNECTIONS OF THE EXPERIMENTAL RESULTS OF UNIVERSAL STRESS EXPERIMENTS AND OF THERMAL EXPANSION MEASUREMENTS AND THE MECHANISMS OF MICROSCOPIC DYNAMICS PROCESS ON CUO2 PLANES

DAWEI ZHOU

General Superconductor, Inc. 1663 Technology Avenue, Alachua FL 32615, USA

The experimental results of universal stress on the superconducting transition in YBCO by U. Welp et. al (Phys.Rev. Lett. 69, 2130, 1992) indicates that T_C is linear in p_i with slopes dT_C/dp_a =-2.0 K/Gpa, dT_C/dp_b =+1.9 K/GPa, and dT_C/dp_c =-0.3 K/GPa. These results confirmed that the small hydrostatic pressure dependence of T_C in YBCO is due to a cancellation of large and opposite effective effect in the a-b plane. This indicated that the pressure dependence of T_C and the relative sign change for the a- and b- axis compression may reflect the symmetry of 4-particle (electrons or electrons with hole) or higher order quantum tunneling cyclic loops (QTCLs) on the CuO_2 plane in high T_c superconductor. Because of the existence of mixed valences on the CuO_2 plane, the electron/holes in high T_c superconductor are weekly localized and strongly correlated thus able to form QTCL. The QTCLs have k=0 in the center of the mass frame, and a non-zero angular momentum. Each QTCL has total spin S=0. Each QTCL can contain two or more Cooper pairs in real space and in k-space. This dynamics is consistent with observations of very short coherence length on the CuO_2 plane. This is consistent with the thermal expansion measurement results (C. Meingast et. al Phys. Rev. lett. 67, 1634, 1991) that indicates that superconductivity favors asymmetric (b=a) CuO_2 plane.

1. Experimental Result

C. M. Meingast et. al. [1] have observed a large a-b anisotropy of the expansivity anomaly at T_c in untwinned $Y_1Ba_2Cu_3O_7$ sample. The onset of superconductivity is accompanied by highly anisotropic jumps of the expansivities in the a-b plane. This leads to a reduction of the orthohombic splitting below T_c, which suggests that superconductivity favors a symmetric (b=a) CuO_2 plane. Little effect is seen along the c axis.

U. Welp et. al. [2] have observed the effect of uniaxial stress on the superconducting transition in YBCO sample. U. Welp et. al. present the first direct measurement of T_c as a function of uniaxial stress P_i (i=a,b,c) along the crystal axes for an untwinned YBCO single crystal. T_c is linear in P_i with slopes dT_c/dP_a=-2.0+-0.2 K/GPa, dT_c/dP_b=+1.9=-0.2 K/GPa, and dT_c/dP_c=-0.3=-0.1 K/GPa. This result is consistent with C. Meingast's result.

2. Quantum Tunneling Cyclic Loop (QTCL) Model [3]

The strong correlation between the coherent particle motions will establish a macroscopic phase relation. On the CuO_2 planes the electrons and holes is viewed as a cyclic permutation of electrons (and holes) exchanging positions around a square determined by the Cu-atoms. At the microscopic level, those coherent four or more electrons and hole quantum tunneling cyclic loops (QTCLs) probably play a key role int he high-Tc superconductivity.

On the CuO_2 planes of high-Tc superconductors, the electrons and holes become weakly localized, therefore they can form coherent motion on the CuO_2 planes. There are four oxygen atoms around one Cu atom (in 3-D it would be six) and each Cu atom can supply at most three electrons to their nearest neighbors. This means that there can be no stable valence bond between the Cu atoms and the oxygen atoms. The Cu electrons are weakly localized (not free) and can pass across the oxygen bridges to complete the quantum tunneling. The microstructure of the CuO_2 planes of high-Tc superconductors established a weakly localized state of electrons and holes, this is the foundation of high-Tc superconductivity. Because of the planar structure of the high-Tc superconductors, coherent quantum tunneling cyclic loop (QTCL). Many 4-particle exchange permutations, forming a quantum tunneling cyclic loop (QTCL). Many 4-particle QTCLs can form to large size QTCL. The QTCLs have total K=0 in the center of mass frame, and non-zero angular momentum. Each QTCL has total spin S=0 to minimize the free energy. The Cyclic permutations and translations of QTCLs will transport electrons or holes from one site to another site coherently, and therefore without collision. The effective mass of QTCL will determine that a high-Tc superconductor whether belong to n-type or p-type high-Tc superconductor. For n-type high-Tc superconductor, QTCL has smaller effective mass, and the long distance QTCL has smaller effective mass, and the long distance QTCL translation is available. Cooper pairs will not only exist in K space, also exist in real space, exist inside QTCLs. This can be easily understood that why the coherent length is so short (about 20 angstroms), and is matched with the size of QTCL. Above Tc, QTCLs will exist but only as disconnected small clusters, and there will be no superconductivity. The connectivity of QTCLs will be destroyed by increasing the temperature and the magnetic field. The establishment of a connectivity between neighboring QTCLs at Tc and Hc can lead to a collective QTCL state--high-Tc superconducting state. The current carriers, electrons or holes will be transported by the coherent motion of QTCLs, collective QTCL rotation or collective QTCL translation. Increasing the temperature (or the applied magnetic field) will decrease the density of the QTCLs and destroy the connectivity, leading to a percolation transition to the normal state at the critical temperature. At Tc, the tendency to establish a connectivity between neighboring QTCLs will lead to tendency to reduce the orthorhombic splitting.

3. Conclusion

Thermal expansion and uniaxial stress experiment results are consistent with the QTCL model that proposed four or more particle quantum tunneling cyclic loops (QTCLs) probably is microscopic origin of the high temperature superconductivity.

4. Reference

1. C. Meingast et. al. PRL, 67, 1634, 1991
2. U. Welp et. al. PRL, 69, 2130, 1992
3. Dawei Zhou and N.S. Sullivan, Physica C, 162-164, 813, 1989

III. THEORY

Strongly Correlated Aspects

Quasiparticles in a Strongly Interacting Regime

A. Ferraz[1]

International Centre of Condensed Matter Physics
Universidade de Brasilia - C P 04667 - 70919-970 Brasilia-DF - Brazil

Fermi liquid theory is not appropriate to describe the normal phase of the high-T_c superconductors. Using the renormalization group approach we show that if the electronic effective mass m^* is sufficiently large the Fermi liquid is unstable for $d = 2$. In higher dimensions if the Fermi surface is not sharply defined and if its position is not fixed in momentum space there may exist quasiparticles of different nature in the physical system. We construct a model which takes explicit account of this feature. We show that if the 'low-energy' fields are integrated out the resulting effective model has a coupling function which can be large and negative. This leads to a breakdown of quasiparticle representation and to pairing of single particles at high energies.

The Fermi liquid is the reference model for conduction electrons in a normal metal. Unfortunately there is a large amount of experimental evidence indicating that the normal phase of high-T_c superconductors does not fit in this framework [1]. For these materials there are several physical quantities which have either the wrong energy or wrong temperature dependence. One example of this is the linear variation of the inverse of the relaxation time $\tau^{-1} \sim E$, at low energy as shown by the inelastic scattering measurements of infrared, Raman and ARPES. This behavior implies that the charge renormalization parameter $Z = (1 - \partial \Sigma (p_F, E) / \partial E|_{E=0})^{-1}$, where Σ is one-particle self-energy at the Fermi surface S_F, is either infinitesimally small or identically zero. This parameter measures the amount of single-particle character in the electronic excitations of the interacting system. It estimates how much single particle behavior exists among the collective state background. If $Z \cong 0$ the one-particle spectral function $A(\mathbf{p}, \epsilon)$ has no central peak. As a result if quasiparticles exist they are either of more extended nature or the Fermi surface has no fixed position and is not sharply defined in momentum space. Both these conditions violate Fermi liquid theory. We analyze these two possibilities in this present work.

Consider initially spinless quasiparticles in d ($d > 1$) spatial dimensions. If we discard interaction terms which are irrelevant in a renormalization group sense as we approach the Fermi surface their effective action S is given by [2]

$$S = \int dt d^d p \Psi^\dagger (p,t) \left[i\partial_t - \frac{1}{2m^*} (p^2 - p_F^2) \right] \Psi (p,t) - \frac{1}{4} U \int dt \Pi_i d^d p_i \delta^{(d)} (\Sigma_i p_i) \Psi^\dagger (\mathbf{p_1},t) \Psi (\mathbf{p_2},t) \Psi^\dagger (\mathbf{p_3},t) \Psi (\mathbf{p_4},t). \tag{1}$$

Here $|\mathbf{p_i}| \geq p_F$, the Fermi momentum, U is the quasiparticle coupling constant and m^* their effective mass. . If we now make the change of variables $\mathbf{p} \to \mathbf{p}/\sqrt{m^*}$ and rescale the fermion fields the effective lagrangian L for the quasiparticles is in convenient units ($\hbar = 1$) given by

$$L = \int d^d p \Psi^\dagger (i\partial_t - \frac{1}{2} (p^2 - \bar{p}_F^2)) \Psi - \frac{1}{4} m^{*\frac{d}{2}} \bar{u} \Lambda^\epsilon \int \Pi_i d^d p_i \delta^{(d)} (\Sigma_i p_i) \Psi^\dagger (\mathbf{p_1},t) \Psi (\mathbf{p_2},t) \Psi^\dagger (\mathbf{p_3},t) \Psi (\mathbf{p_4},t), \tag{2}$$

where Λ denotes a scaled momentum unit, $\bar{p}_F = p_F/\sqrt{m^*}$, $\bar{u}$ is a dimensionless coupling constant and $\epsilon = 2 - d$. It is therefore useful to establish the physical properties of L around the critical dimension $d_c = 2$.

Consider the scattering of two particles with opposite momenta as described by the one-particle irreducible four-fermion function $\Gamma^{(4)}(p_0)$ in one-loop order. Using the Feynman rules for L we get

$$\Gamma^{(4)}(p_0) = -i \frac{m^{*\frac{d}{2}} \bar{u} \Lambda^\epsilon}{4} - \frac{\left(-im^{*\frac{d}{2}} \bar{u} \Lambda^\epsilon\right)^2}{8} \int_\mathbf{q} \int_{q_0} \frac{1}{(q_0+p_0-\bar{\epsilon}+i\delta)(-q_0+p_0-\bar{\epsilon}+i\delta)}, \tag{3}$$

where $\int_{q_0} = \int_{-\infty}^{\infty} dq_0/2\pi$, $\int_{\mathbf{q}} = \int_{p_F \leq |\mathbf{q}|} \frac{d^d q}{(2\pi)^d}$, $\bar{\epsilon} = \frac{1}{2}\left(\mathbf{q}^2 - \bar{p}_F^2\right)$ and p_0 is measured with respect to $\mu(\bar{p}_F) = \bar{p}_F^2/2$.

Evaluating the integral over q_0 and the integrals over angles in d dimensions this reduces to $\Gamma^{(4)}(p_0)$ becomes

$$\Gamma^{(4)}(p_0) = -i\frac{m^{*\frac{d}{2}}\bar{u}\Lambda^{\epsilon}}{4} + i\frac{\left(m^{*\frac{d}{2}}\bar{u}\Lambda^{\epsilon}\right)^2}{4(4\pi)^{\frac{d}{2}}\Gamma\left(\frac{d}{2}\right)}\int_{\bar{p}_F}^{\Lambda}\frac{dq\, q^{d-1}}{q^2 - 2E} \quad (4)$$

where $E = p_0 + \frac{1}{2}\bar{p}_F^2$ and Λ is now a varying high energy cut-off.

Suppose that the quasiparticles are nearly localized and as a result the ground state of the electron fluid is close to suffer a transition to an insulating state. If this is the case we assume that the quasiparticle effective mass m^* is large enough to produce $\bar{p}_F \cong 0$. Since $m^* \sim Z^{-1}$ this also implies that the electron fluid is such that $Z \cong 0$. In this regime the single particle states don't differ considerably from the collective state background and they may exist far from $p = p_F$ in momentum space. Thus we may have $\Lambda > 0$ despite the fact that $\bar{p}_F \cong 0$.

Let us now define the renormalized dimensionless coupling u by the condition $\Gamma_R^{(4)}(p_0 = 0) = -i\frac{m^* u \Lambda^{\epsilon}}{4}$. However at this energy value, for $\bar{p}_F \cong 0$ and in the vicinity of $d = 2$, $\Gamma^{(4)}(0)$ becomes

$$\Gamma^{(4)}(0) = -i\frac{m^* u \Lambda^{\epsilon}}{4}\left[1 + \frac{m^* u}{4\pi\epsilon} + ...\right] \quad (5)$$

Clearly as $d \to 2$, $\Gamma^{(4)}(0)$ diverges and we need to regularize the theory against this. The way to proceed is to add a counter term to match the original lagrangian and to render the theory finite. We do this defining the bare coupling constant $m_0^* U_0$ to establish the renormalization prescription above. Thus when $\epsilon \to 0$ we write

$$m_0^* U_0 = m^* \bar{u} \Lambda^{\epsilon}\left[1 - \frac{m^* u}{4\pi\epsilon} + ...\right] \quad (6)$$

If we differentiate this equation with respect to Λ for a fixed $m_0^* U_0$ we obtain the renormalization group equation

$$\beta(m^* u, \epsilon) = -\epsilon m^* u - \frac{(m^* u)^2}{4\pi} \quad (7)$$

where $\beta(m^* u, \epsilon) = \Lambda \partial(m^* u)/\partial\Lambda$.

If we integrate this differential equation at $\epsilon = 0$ it follows that[3]

$$m^* u(\Lambda) = \frac{m^* u(\Lambda_0)}{1 - \frac{m^* u(\Lambda_0)}{4\pi}\ln\left(\frac{\Lambda_0}{\Lambda}\right)} \quad (8)$$

where Λ_0 is some upper bound value for the scaled momentum. Clearly if Λ decreases $m^* u(\Lambda)$ increases until we leave the domain of validity of perturbation theory for $\frac{m^* u(\Lambda_0)}{4\pi}\ln\left(\frac{\Lambda_0}{\Lambda}\right) \to 1$. Note that if $m^* u(\Lambda_0)$ is large enough this occurs at a scale $\Lambda \cong \Lambda_0 \exp\left(-4\pi/m^* u(\Lambda_0)\right) \equiv \Lambda_0^*$. The singularity in $m^* u(\Lambda)$ signals the transition to a non-metallic state. However the validity limits of perturbation theory in this regime indicates that we either should include contributions from higher order terms or consider a new effective model for the quasiparticles at this higher energy scale. Let us explore this last possibility.

If there are single particle states at high energies compared with $\mu(p_F)$ and if the electron fluid is such that $Z \cong 0$ it is most likely that the Fermi surface is not sharply defined nor its position is fixed in momentum space For simplicity let us suppose that there are two positions $|\mathbf{k}| = p_F$ and

$|\mathbf{k}| = \widetilde{p}_F$, with $p_F > \widetilde{p}_F$, around which single particles are long-lived and well defined in momentum space. Since these particles are located in thin shells around the Fermi surfaces at both $|\mathbf{k}| = p_F$ and $|\mathbf{k}| = \widetilde{p}_F$ it is convenient to use the momentum decomposition $\mathbf{p} = \mathbf{k} + \widehat{n}(\mathbf{k})\, l$, where $|\mathbf{k}| = p_F$, or $|\mathbf{k}| = \widetilde{p}_F$, $n(\mathbf{k})$ is the unit vector normal to S_F at $\mathbf{k}$ and l is the width of the very thin shell for the corresponding group of quasiparticles. In this way if we restrict ourselves to $d = 3$ and again neglect spin degrees of freedom the lagrangian L for these spinless single particles becomes[4]

$$L = \int_{|\mathbf{k}|=p_F} \frac{d^2k}{(2\pi)^3} \int_0^\Lambda dl \Psi^+(\mathbf{p},t) \left[i\partial_t - \tfrac{p_F}{m^*}l\right] \Psi(\mathbf{p},t)$$
$$+ \int_{|\mathbf{k}|=\widetilde{p}_F} \frac{d^2k}{(2\pi)^3} \int_0^{\widetilde{\Lambda}} dl \Psi^+(\mathbf{p},t) \left[i\partial_t - \tfrac{\widetilde{p}_F l}{m^*}\right] \Psi(p,t) \qquad (9)$$
$$-\tfrac{1}{4} U \int \prod_{i=1}^4 \frac{d^3\mathbf{p_i}}{(2\pi)^3} \delta^{(3)}\left(\sum_i \mathbf{p}_i\right) \Psi^+(\mathbf{p}_1,t) \Psi(\mathbf{p}_2,t) \Psi^+(\mathbf{p}_3,t) \Psi(\mathbf{p}_4,t)$$

where

$$\Psi(\mathbf{p},t) = \Psi_1(\mathbf{p},t) + \Psi_2(\mathbf{p},t) \qquad (10)$$

with

$$\Psi_1(\mathbf{p},t) = \begin{cases} \Psi(\mathbf{p},t), & for\ |\mathbf{k}| = \widetilde{p}_F, 0 \leq l \leq \widetilde{\Lambda} \\ 0, otherwise \end{cases} \qquad (11)$$

and

$$\Psi_2(\mathbf{p},t) = \begin{cases} \Psi(\mathbf{p},t), & for\ |\mathbf{k}| = p_F; 0 \leq l \leq \Lambda \\ 0; otherwise \end{cases} \qquad (12)$$

The physical model is specified by the functional integral W where

$$W = \int D\Psi D\Psi^+ \exp i \int dt (L_0 + L_{int})$$
$$= \int D\Psi_1 D\Psi_1^+ D\Psi_2 D\Psi_2^+ \exp i \int dt \left\{ \begin{array}{c} L_0[\Psi_1, \Psi_1^+] + L_0[\Psi_2, \Psi_2^+] \\ + L_{int}[\Psi_1 + \Psi_2, \Psi_1^+ + \Psi_2^+] \end{array} \right\} \qquad (13)$$

To consider what happens at a higher energy scale we integrate out the 'low-energy' modes Ψ_1 and Ψ_1^+ and generate a new effective model for the 'high-energy' fields Ψ_2 and Ψ_2^+. By doing this we define the effective action S_{eff}

$$\exp i S_{eff}\left[\overline{\Psi},\overline{\Psi}^+\right] = \exp i S\left[\Psi_2,\Psi_2^+\right] \int D\Psi_1 D\Psi_1^+ e^{iS_0}$$
$$\left[1 + \sum_{l=1}^\infty \frac{i^l}{l!} \frac{\int D\Psi_1 D\Psi_1^+ e^{iS_0} (\Delta S_{int})^l}{\int D\Psi_1 D\Psi_1^+ e^{iS_0}}\right] \qquad (14)$$

where $\Delta S_{int} = S_{int}[\Psi_1 + \Psi_2, \Psi_1^+ + \Psi_2^+] - S_{int}[\Psi_2, \Psi_2^+]$. The precise relation between the effective action and the original model is established perturbatively. The general form of S_{eff} is nevertheless given by

$$S_{eff}\left[\overline{\Psi},\overline{\Psi}^+\right] = \int_{p_0} \int_\mathbf{p} \Psi_2^+(\mathbf{p},p_0)$$
$$[p_0 - \mu(p_F) + \Sigma_1(p_0) - (\epsilon(\mathbf{p}) - \mu(p_F) + \Sigma_1(p_0))] \Psi_2(\mathbf{p},p_0) \qquad (15)$$
$$-\tfrac{1}{4} \int_{p_0} \prod_{i=1}^4 \int_{\mathbf{p_i}} U(p_0) \delta^{(3)}\left(\sum_i \mathbf{p}_i\right) \Psi_2^+(\mathbf{p}_1,p_0) \Psi_2(\mathbf{p}_2,p_0) \Psi_2^+(\mathbf{p}_3,p_0) \Psi_2(\mathbf{p}_4,p_0)$$

with $p_0 \geq \mu(p_F) - \Sigma_1(p_0)$, with Σ_1 being the correction to the chemical potential produced by the Ψ_1 fields and where $U(p_0)$ is the interaction function directly obtained from the one-particle irreducible $\Gamma_1^{(4)}(p_0)$ .Using the expansion(15) we can calculate $\Sigma_1(p_0)$ and $U(p_0)$ at any order in perturbation theory.

In this momentum thin shell configuration the most important diagrams are those whose structure constrains each pair of particles to have opposite momenta. Thus, in one-loop order ,Σ_1 is constant shift that can be neglected and if we take $\mu(p_F) = \mu(\widetilde{p}_F) + \frac{\widetilde{p}_F \widetilde{\Lambda}}{m^*}$ it then follows that [4]

$$U(p_0) = U - \frac{U^2}{8\pi^2}\widetilde{m}^*\widetilde{p}_F \ln\left(\frac{p_0 - \mu(p_F)}{p_0 - \mu(p_F) + \frac{\widetilde{p}_F\widetilde{\Lambda}}{m^*}}\right) + ... \qquad (16)$$

Once again we need to regularize the model against the logarithmic divergence at $p_0 = \mu(p_F)$ .If we use the same strategy as before we define the bare coupling U_0 such that

$$U_0 = U + \frac{U^2}{8\pi^2}\widetilde{m}^*\widetilde{p}_F \ln\left(\frac{\mathcal{E}}{\frac{\widetilde{p}_F\widetilde{\Lambda}}{m^*}}\right), \qquad (17)$$

with $\mathcal{E} = p_0 - \mu(p_F)$ and from this we obtain that

$$U(\mathcal{E}) = \frac{U(\mathcal{E}_0)}{1 - \frac{U(\mathcal{E}_0)}{8\pi^2}\widetilde{m}^*\widetilde{p}_F \ln\left(\frac{\mathcal{E}_0}{\mathcal{E}}\right)} \qquad (18)$$

where $\mathcal{E}_l \leq \frac{p_F\Lambda}{m^*}$ is again some upper bound energy value .Here the same non-perturbative trend is observed for the interaction function . Moreover for $\mathcal{E} \leq \mathcal{E}^* = \mathcal{E}_0 \exp\left(-8\pi^2/\widetilde{m}^*\widetilde{p}_F U(\mathcal{E}_0)\right)$, $U(\mathcal{E})$ can be large and negative . Thus it is possible for the $\overline{\Psi}\Psi$ to attract each other and to form a bound state . If this occurs the single particle may condense into pairs and become superconducting at this higher energy range.

In conclusion we consider the low lying excited states of a strongly interacting Fermi fluid with a charge parameter $Z \cong 0$. For $d \geq 2$ we show that if the quasiparticle cut-off $\Lambda \gg p_F$, the non-interacting Fermi momentum , there is an unlimited growth of m^*u , as we scale down the energy from some upper bound value. This signals the breakdown of Fermi liquid theory in this regime. Since the Fermi surface is not sharply defined we assume that its position is not fixed in momentum space. Considering that quasiparticles are long-lived in the vicinity of the Fermi surface there may exist single particle states of different character in the physical system . We take this into account constructing a model for quasiparticles located in different sectors of momentum space. We integrate out the 'low-energy' modes and derive an effective action for the 'high-energy' fermion fields. Using the renormalization group theory we then show that the coupling function for the resulting effective model can be large and negative in an appropriate energy range. If this is the case the quasiparticles attract each other and form bound states. This behavior may lead to the condensation of single particles into pairs and may produce superconductivity at higher energies .

References

[1] P.W.Anderson *The Central Dogmas* in *Career in Theoretical Physics*, World Scientific (1994)

[2] R.Shankar, *Physica A* **177** (1991) 530; *Rev. Mod. Phys* **66** (1994) 129 ; J. Polchinski , *TASI Lectures* (1992); G. Benfatto and G. Gallavotti, *J. Stat. Phys.* **59** (1990) 54; J. Feldmann and E. Trubowitz, *Helv. Phys. Acta* **63** (1990) 157; **64** (1991) 213

[3] A.Ferraz and X.Xue, *submitted for publication* (1996)

[4] A. Ferraz and Y.Ohmura-*submitted for publication* (1996)

HOLE SPECTRUM IN THE THREE BAND MODEL

L. P. GOR'KOV

National High Magnetic Field Laboratory, Florida State University,
1800 E. Paul Dirac Dr., Tallahassee, FL 32306-4005
and
L.D. Landau Institute for Theoretical Physics, Russian Academy of Sciences,
117334 Moscow, Russia

P. KUMAR

Physics Department, University of Florida, Gainesville, FL 32611

We consider the energy spectrum of a single hole in the framework of the three band model for CuO_2 plane. The hole Hamiltonian is derived perturbatevly up to the terms quadratic in the hopping matrix element between the Cu- and O- site, as usual. We get, however, some more insight into structure of the hole bands and their renormalization due to interactions of the antiferromagnetic background, by making use of the explicit symmetry of the resulting Hamiltonian. The dispersion law of two bands and polaronic effects are discussed. The spin cloud surrounding the hole has internal degrees of freedom. Two bands, one with a considerable dispersion and the second one, which is almost dispersionless, touch each other at the Γ point.

In this presentation we address again the issue of hole doping in cuprates. There is now a consensus that, at least for small concentrations, holes go into the oxygens' manifold of the CuO_2-planes, while the copper sites, $Cu^{2+}(d^9)$, preserve their spins. Therefore, the approach to the problem from the side of ionic states for both components, Cu^{2+} and O^{2-}, seems to be a resonable starting point. In what follows, we accept the simplest version of the three band model [1,2]: the (electronic) vacuum for copper is the configuration d^{10}; it is separated by the energy $\Delta = E_d - E_0 > 0$ from the p_x, p_y oxygen orbitals. Each Cu contains one hole (d^9), Cu^{2+}. This model produces the charge transfer gap (CT) insulator ground state in the antiferromagnetic state, well-documented in numerous optical experiments (see e.g. [3,4] and references therein). The CT-gap for most cuprates is $\Delta=1.5$–1.8 eV. The electronic states of oxygens and coppers are coupled via tunneling matrix elements $(-1)^\delta t_0$, where the sign is imposed by the relative orientation of the given p_x or p_y-oxygen orbital with respect to the representation, $d_{x^2-y^2}$, chosen for the Cu-site [5]. To confine doped holes on the oxygen network it is assumed that $t_0 < \Delta$. The emerging effective band is of order of $W \sim t_0^2/\Delta$, and the hole moves perturbatively along the oxygen network across copper sites.

The recent ARPES experiments in $Sr_2CuO_2Cl_2$ [6] have stimulated numerous theoretical attempts to reproduce the observed hole spectra. Most of them are based on the t-J model (e.g., [7]), but some come back to a more explicit use of the three band model [8]. The common simplification used in all these approaches is, however, that a combination of oxygen orbitals on which the hole resides, is chosen without rigorous justification [5,9]. This is discussed below, and, in an alternative view, it shows that the resulting picture reveals some new qualitative features.

The holes on the Cu-sites are described by the Hubbard term in the total Hamiltonian:

$$\hat{H}_{Cu} = -\Delta \sum_{i,\sigma} \hat{d}^+_{i\sigma}\hat{d}_{i\sigma} + U \sum \hat{n}_{i,d\uparrow}\hat{n}_{i,d\downarrow} \tag{1}$$

with a very large $U > 0$ to exclude the Cu^{3+}-orbitals. The oxygen orbitals are represented by operators $(\hat{X}^+_i, \hat{X}_i)$ and $(\hat{Y}^+_i, \hat{Y}_i)$. The unit cell carries the index $i = (i_x, i_y)$ assigned to the Cu-site. Oxygen orbitals X_i and Y_i are located on the x- and y-sides of the ith cell. The hybridizing part, H_{hyb}, has the form [5]:

$$H_{\text{hyb}} = +t_0 \sum_i \{\hat{d}^+_i(\hat{X}_i + \hat{Y}_i - \hat{X}_{i_x-1,i_y} - \hat{Y}_{i_x,i_y-1}) + \text{h.c.}\} \tag{2}$$

The elementary second order (in t_0) processes which would result in hole bands are now given by the following term:

$$h' = \frac{t_0^2}{\Delta}\{\hat{X}^+_i \hat{P}_i(\hat{Y}_i - \hat{X}_{i_x-1,i_y} - \hat{Y}_{i_x,i_y-1})\} \tag{3}$$

linking the four oxygen sites (the direct overlap between oxygen sites is omitted). Here,

$$\hat{P}_i = \frac{1}{2}(1 + \hat{\sigma} \cdot \hat{S}_i) \tag{4}$$

is the permutation operator between holes on an oxygen and the Cu-site ($\hat{\sigma}$ and $\hat{S}_i$ are both the Pauli matrices). Thus, tunneling produces <u>two</u> terms in (3): a "free motion" and an "exchange coupling" to the local spins. The Heisenberg exchange term

$$H_{\text{ex}} = J \sum_{i,i+\delta} \vec{S}_i \cdot \vec{S}_{i+\delta} \tag{5}$$

accounts for the nearest neighbors interactions of local spins. In the above model with $t_0 < \Delta$, the exchange between local spins would be of order of t_0^4/Δ^3. In practice, J$\simeq$0.15eV while $\Delta \simeq$1.8eV. Therefore, the assumption $t_0 << \Delta$ may have only qualitative meaning.

Before assuming any specific state for the quantum local spins, introduce the momentum representation for all operators:

$$\hat{X}_p = \frac{1}{\sqrt{N}} \sum_i \hat{X}_i e^{-ipr_i} \tag{6}$$

(and similarly for $\hat{Y}_p$ and $\hat{S}_p$.) After substitution of (6) into (3), the spin-independent part of the hole Hamiltonian can be diagonalized. Correspondingly, the total Hamiltonian for holes and spins may be re-written in the form:

$$H_{\text{tot}} = \sum J(q)\hat{S}_q \cdot \hat{S}_{-q} + W\{\sum_{p,\sigma} \varepsilon_1(p)\hat{a}^+_{p\sigma}\hat{a}_{p\sigma} + \sum_{p,\sigma} \varepsilon_2(p)\hat{b}^+_{p\sigma}\hat{b}_{p\sigma} + \sum_{p,q}(\hat{\psi}^+_p \hat{I} \hat{\psi}_{p-q})\} \tag{7}$$

The shorthanded notations in (7) are:

$$W = \frac{t_0^2}{2\Delta}; \quad \varepsilon_1(p) = -1; \quad \varepsilon_2(p) = -1 + 2(\sin^2\frac{p_x}{2} + \sin^2\frac{p_y}{2}) \tag{8}$$

The operator $\hat{I}$ is the 2×2 matrix acting on the column of operators, $\hat{a}$ and $\hat{b}$:

$$\hat{\psi}_p = \{\begin{matrix}\hat{a}_p \\ \hat{b}_p\end{matrix}\} \tag{8a}$$

The last term in (7) describes the hole interactions with local spins in the new basis:

$$\hat{I} \equiv \hat{I}(p,q,p-q)(\hat{\sigma}\cdot\hat{S}_q) \tag{8b}$$

The expressions for $\hat{I}(p,q,p-q)$ are rather cumbersome due to the tetragonal symmetry. We write down the most important, as we shall see, component:

$$I_{aa}(p,q,p-q) = -z_p^{-1}z_{p-q}^{-1}[\sin\frac{p_y}{2}\sin\frac{p_y-q_y}{2}\cos\frac{q_x}{2} + (x\to y)] \tag{8c}$$

Finally, one has

$$z_p^2 = \sin^2\frac{p_x}{2} + \sin^2\frac{p_y}{2} \tag{8d}$$

The Hamiltonian (7) has only two energy scales, J and W (J<<W at $t_0 \ll \Delta$.) For convenience of the analysis it will be useful to introduce an arbitrary parameter, λ, so that in (7) and (8b,c):

$$\hat{I} \Rightarrow \hat{I}\lambda \equiv \lambda\hat{I} \tag{9}$$

The $\hat{a}$-band is infinitely degenerate in absence of interactions. It is helpful, however, to relate $\hat{a}_p$ to the $\hat{X}$ and $\hat{Y}$ operators in the real space:

$$\hat{a}_p = \frac{i}{2z_p}e^{-i(p_x+p_y)/2}\sum_j \frac{e^{-ipr_j}}{\sqrt{N}}\{\hat{X}_j - \hat{X}_{j_x j_y+1} - \hat{Y}_j + \hat{Y}_{j_x+1,j_y}\}$$

$$\equiv \frac{i}{2z_p}e^{-i(p_x+p_y)/2}\sum_j \frac{e^{-ipr_j}}{\sqrt{N}}\hat{L}_j \tag{10}$$

Note that each configuration $\hat{L}_j$ is contacting four copper sites. The coupling I_{aa} in (8c) describes non-local interactions between $\hat{L}_j$'s and local spins on different sites.

Assuming $\lambda \ll 1$ in (9) makes it possible to perform a perturbative analysis of the (7). For that we may assume that the local spins are in the AF phase:

$$\langle S_{i,z}\rangle = \mu(-1)^i \tag{11}$$

where μ is the experimentally measured staggered moment ($\mu < 1$). Excitations of the spin sub-system are then characterized by the magnon spectrum ($\omega_0 \propto J$). It is easy to verify that the perturbative series for the $\hat{b}$-operators, interacting with magnons, proceeds as expansion in powers of the parameter λ itself.

The dispersive corrections to the energy of the $\hat{a}$-operators expand in powers of

$$\lambda \frac{W}{J} \gg \lambda; \text{(at } t_0 \ll \Delta\text{)} \tag{12}$$

This shows that the problem of holes interacting with spins, is first to be resolved for the dispersionless branch, ε_1, in (8). (Note that $\langle S_{(\pi,\pi)} \rangle$ from (11) merely folds the $\hat{a}$-branch twice onto the new magnetic zone).

In absence of a kinetic energy term (no dispersion) the motion of the $\hat{a}$-band hole results in a polaron formation at any λ. At $\lambda W/J \gg 1$ the physics is well described in terms of "ferropolarons"; at $\lambda \overline{W/J}$ small or of order of unity, the polaron is of an atomic size, such as a single cofiguration L_i in (10) (or a linear superposition of few L_i's). Note that because L_i itself interacts with four spins, such a polaron may have localized spin excitation on the energy scale of order of W itself.

The two branches, $\hat{a}$ and $\hat{b}$, in (8) touch each other at the Γ-point. This degeneracy is due to the symmetry and is not lifted in the presence of interactions with spin excitations. Therefore at low enough doping holes would go into the polaronic band where they are localized even by a weak disorder. For the $\hat{b}$-band polaronic effects may be absent at a proper choice of $W/J \sim 1$. Conductivity in such a system takes place via termal excitation of localized holes into the $\hat{b}$-band. This results in low temperature resistivity behavior

$$\rho(T) \propto T^{-\alpha}; \quad (\alpha = 1 \div 2).$$

Acknowledgements

This work was supported by the NHMF Laboratory through NSF cooperative agreement # DMR-9016241 and the State of Florida.

References

[1] C. M. Varma et al., *Solid State Comm.* **62**, 681 (1987).
[2] V. J. Emery, *Phys. Rev. Lett.* **58**, 2794 (1987).
[3] J. P. Falck et al., *Phys. Rev.* B **49**, 6246 (1994).
[4] S. Uchida et al., *Phys. Rev.* B **43**, 7942 (1991).
[5] F. C. Zhang and T. M. Rice, *Phys. Rev.* B **37**, 3759 (1988).
[6] B. O. Wells et al., *Phys. Rev. Lett.* **74**, 964 (1995).
[7] A. Nazarenko et al., *Phys. Rev.* B **51**, 8676 (1995).
[8] O. A. Starykh et al., *Phys. Rev.* B **52**, 12534 (1995).
[9] V. J. Emery and G. Reiter, *Phys. Rev.* B **38**, 4547, 11938 (1988).

Gauge Theory of the Normal State Properties of High-Tc Cuprates

Naoto Nagaosa
*Department of Applied Physics, University of Tokyo,
Tokyo 113, Japan*

Some recent developements on gauge theory for the normal state properties of high-Tc cuprates are reported. These include (1) effects of nonmagnetic impurities and (2) interlayer charge dynamics.

1 Introduction

Anomalous normal state properties are the most striking features in high-Tc cuprates, which can be understoood only when the strong correlations between electrons are taken into account. One of the theoretical framework is the gauge theoretical methods, where the strong on-site repulsion is taken into account as the local constraint that no double occupancy is allowed. This is closely related to the idea of introducing the local coordinates according to the spin fluctuation field. Then the connection between the neighboring local coordinates play an important role which is nothing but the gauge field. Then this method is particularly powerful when the spin system is not ordered but in the quantum liquid state, i.e., RVB state [1]. More explicitely this is formulted in the slave boson method where the elctron operator $C_{i\sigma}$ is expressed in terms of the product of the fermion (spinon) operator $f_{i\sigma}$ and the boson (holon) operator $b_i^\dagger$ as $C_{i\sigma} = f_{i\sigma} b_i^\dagger$ with local constraint $\sum_\sigma f_{i\sigma}^\dagger f_{i\sigma} + b_i^\dagger b_i = 1$.

The canonical model for the strongly correlated electron system is the t-J model whose Hamiltonian is given as

$$H = -\sum_{ij,\sigma} t_{ij} C_{i\sigma}^\dagger C_{j\sigma} + J \sum_{<ij>} \vec{S}_i \cdot \vec{S}_j \tag{1}$$

where $<ij>$ is the nearest neighbor pair. The mean field theory for this model has been developed by several authors [2] by introducing the order parameters $\chi_{ij} = <\sum_\sigma f_{i\sigma}^\dagger f_{j\sigma}>$, $\chi_{ij}^B = <b_i^\dagger b_j>$, $B_i = <b_i>$, and $\Delta_{ij} = <f_{i\uparrow} f_{j\downarrow}>$. χ and χ^B correspond to the hopping of the spinons and holons, respectively. B is the bose condensation order parameter and Δ is the spinon pairing. In terms of these four order parameters, the phase diagram in the plane of hole concentration x and the temperature T is devided into five regions. The first is the high temperature phase where all of the four are zero and the motion of the particles is incoherent. The most relevant to the normal state in the optimally hole concentration is the uniform RVB state where only χ and χ^B is non-zero while the others are zero. This uniform state is unstable against the spinon pairing Δ and the bose condensation B. In the underdoped region, the spinon pairing state without the bose condensation is possible which is identified as the spin gap state. The pseudo-gap exists for the triplet excitations while the system is not yet superconducting in this state. On the other hand in the overdoped region, a state is possible where only the bose condensation occurs. This state is nothing but the Fermi liquid state. Then the anomalous normal state properties coming from the composite nature of the electron (spin-charge separation) are realized when the holon is not bose condensed. If both Δ and B are non-zero, the superconductivity occurs. We have developed the gauge field theory for the uniform RVB state by taking into account the fluctuation around the mean field state [3]. We will describe below some recent advances along this line, i.e., the nonmagnetic impurity effects in the underdoped spin gap state and the interlayer charge dynamics of the uniform RVB state.

2 Nonmagnetic Impurities in the Spin Gap State

The effects of the nonmagnetic impurities, i.e., Zn, replacing Cu in the conducting planes show the following anomalous features in the underdoped region.

1. The formation of the magnetic moments due to Zn has been revealed by magnetic susceptibility [4], NMR [5], μSR [6] and EPR [7], whose number is roughly proportional to the Zn concentration and almost full moment appears in the underdoped region. The magnitude of the induced local moments is strongly dependent on the hole concentration x and becomes smaller or even vanishes as x increases [5].

2. The residual resistivity can be described in terms of the classical expression in terms of the Boltzmann transport theory [8], i.e.,

$$\rho_{\text{res}} = 4(\hbar/e^2)(n_{\text{imp}}/n)\sin^2\delta_0 \tag{2}$$

which is independent of the effective mass of the carriers and is determined only by the phase shift δ_0, the impurity concentration n_{imp} and the carrier density n It has been found that Zn acts as a strong sctterer with unitary limit, i.e., $\delta_0 = \pi/2$. The interesting thing is that the carrier number n is the hole concentration x in the underdoped region, and rather rapidly crosses over to the electron number $1 - x$ in the overdoped region.

We show in this section that these two aspects can be explained in a unified way from the viewpoint of the RVB theories. Although the total system is not superconducting, the spinon system is superconducting and the transport properties are determined by the holon according to Ioffe-Larkin law [9]. Therefore the residual resistivity ρ_{res} is that of the holon, and determined by the holon concentration x. Actually the criterion for the validity of Boltzmann expression eq.(2) is that the sheet conductance is much larger than the quantum conductance e^2/h, which is satisfied for the Zn concentration we are now interested in. Therefore it is reasonable that the holons, which are not superconducting or localized, give the observed resistivity. However it may appear difficult to explain the localized moment because the kinetic energy of the spinon is larger than that of holons and more robust against localization. Also the Ioffe-Larkin law tells that the spinon resistivity will dominate if the states near the Fermi energy are localized and spinon system is insulating. Actually these two dilemma can be simultaneously resolved by the spinon pairing with d-wave symmetry. In the case of d-wave pair, the density of states for the quasi-particle is reduced near the Fermi energy as $D(E) \propto |E - E_F|$, which favors the localization. This is compared with the boson density of states which is constant even near the bottom of the band in two-dimensions. This problem has been already analyzed by Lee [10] in the context of localized states in d-wave superconductos. Applying his argument, it can be shown that the number of strongly localized states roughly coincides with the number of nonmagnetic impurities. When the repulsive interaction between the fermions are again taken into account, it is expected that these localized states near the fermi level are singly occupied and give rise to the Curie moments proportional to the impurity density [11]. When the states of the quasi particle not the spinon itself are localized, the spinon system still remains superconducting and its resistivity is zero. Hence the total resistivity is still determined by the holons, and the conclusion on the resistivity above remains unchanged.

3 Interlayer Charge Dynamics

Recently there appeared several experiments on the physical properties of the high-T_c cuprates along the c-axis ,i.e., perpendicular to CuO_2 planes [12]. The d.c.resistivity ρ_c, infrared optical conductivity $\sigma_c(\omega)$, thermopower S_c, Hall constant R_H^c with the magnetic field H parallel to the plane, and the magnetoresistance have been studied revealing anomalous features. According to RVB scenarios with spin-charge separation, a spinon and a holon have to recombine to form a physical electron in order to hop between adjacent layers, and dissociate again into a spinon and a holon. Hence the conduction mechanism is expected to be quite different between inter- and intralayer.

In this section we study several physical properties based on the uniform RVB state [13]. We will treat the hopping matrix element t_c between adjacent layers as a small perturbation. Because the physical electron hops between layers, the c-axis conductivity σ_c, thermopower S_c, and thermal conductivity κ_c are expressed in terms of the physical electron Green's function $G(k,\varepsilon)$. In the uniform RVB state, the spectral function $A_{\vec{k}}(\varepsilon) = -\pi^{-1} Im G(k,\varepsilon)$ consists of the quasi-particle-like peak and the incoherent background [3]. We obtain the following results.

$$\sigma_c \sim e^2 (\frac{t_c}{J})^2 x [ax(\frac{J}{T})^{1/2} + b(\frac{T}{J})^{1/2}] \tag{3}$$

where a and b are dimensionless constants of order unity. Then the conductivity along the c-axis shows insulating behavior as $\sigma_c \sim T^{1/2}$ for $T \gg T_{B.E.} \cong xJ$. Similarly the thermopower S_c and the thermal conductivity κ_c are obtained as

$$S_c = \frac{1}{T} \frac{\varphi_c}{\sigma_c} \simeq \frac{1}{T} \frac{|\mu_B|}{e} = \frac{1}{e} \ln \frac{m_B T}{2\pi x} \tag{4}$$

$$\kappa_c = \frac{1}{T} \left(\sigma_c^E - \frac{\varphi_c^2}{\sigma_c} \right) \sim \frac{T^{3/2}}{e^2} \left[\ln \frac{m_B T}{2\pi x} \right]^2 \tag{5}$$

where we have assumed $T \gg T_{B.E.}$ and hence $|\mu_B| \gg T$. The thermopower is positive and its magnitude is of the order of $k_B/e \sim 90\mu V/K$ when we make the Boltzmann constant k_B explicit in eq.(4), which is in qualitative agreement with the experiment [12]. This positive sign comes from the incoherent background which extends only for $\varepsilon < 0$, which means that the "hole" carries the energy between the layers.

We now turn to the Hall constant R_H^c with the electric field $E(\| a)$ and the magnetic field $H(\| b)$ parallel to the layer. The composition rule for R_H^c is different from those of the in-plane quantities, and is given by $R_H^c = R_H^{cB} + R_H^{cF}$ where R_H^{cB} (R_H^{cF}) is the Hall constant assuming that the current in the plane is carried by the boson (fermion). These are obtained in the high frequency limit as $R_H^{cB} = \frac{2\eta_B}{ex}$, and $R_H^{cF} = -\frac{2\eta_F}{e(1-x)}$ where the dimensionless factors η_B, η_F are given by $\eta_B \sim \frac{\langle k^2 \rangle_{boson}}{k_F^2} \sim \frac{T}{J} \ll 1$ and $\eta_F = \langle e^{ik_x} \rangle_{F.S.}$. η_F is almost temperature independent but depends on the hole concentration x. For example $\eta_F \to 0$ as $x \to 0$ assuming the tight-binding band with only nearest neighbor hopping. Although one can not discuss about the d.c. Hall constant R_H^c, it is noted that the positive contribution R_H^{cB} is suppressed by the small factor $\eta_B \sim T/J \ll 1$ in contrast with the in-plane R_H which is almost determined by the boson Hall constant $R_H^B \cong \frac{1}{ex}$. Therefore $R_H^c = R_H^{cB} + R_H^{cF}$ is determined by the subtle balance between the reduced positive contribution R_H^{cB} and the negative one R_H^{cF}, and its magnitude is expected to be small and less than $\frac{1}{e}$ which is consistent with experiments [12].

References

1. P.W.Anderson, Science **235**, 1196 (1987).
2. G.Baskaran, Z.Zou, and P.W.Anderson, Solid State Commun. **63**, 973 (1987); Y.Isawa, S.Maekawa, and H.Ebisawa, Physica B**148**, 391 (1987); P.W.Anderson, Z.Zou, and T.Hsu, Phys. Rev. Lett. **58**, 2790 (1987); Y.Suzumura, Y.Hasegawa, and H.Fukuyama, J. Phys. Soc. Jpn. **57**, 2768 (1988).
3. N.Nagaosa and P.A.Lee, Phys. Rev. Lett. **64**, 2450 (1992); Phys. Rev. B**45**, 966 (1992); P.A.Lee and N.Nagaosa, Phys. Rev. B**46**, 5621 (1992).
4. G.Xiao, M.Z.Cieplak, J.Q.Xiao, and C.L.Chien, Phys. Rev. B**35**, 8782 (1987).
5. H.Alloul et al., Phys. Rev. Lett. **67**, 3140 (1991); A.V.Nahajan et al., Phys. Rev. Lett. **72**,3100 (1994)..
6. P.Mendels et al., Phys. Rev. B**49**, 10035 (1994).
7. A.M.Finkelstein, V.E.Kataev, E.F.Kukovitskii, and G.B.Teitel'baum, Physica C**168**, 370 (1990).
8. S.Uchida, unpublished.
9. L.B.Ioffe and A.I.Larkin, Phys. Rev. B**39**, 8988 (1989).
10. P.A.Lee, Phys. Rev. Lett. **71**, 1887 (1993).
11. N.Nagaosa and T.K.Ng, Phys. Rev. B**51**, 15588 (1995).
12. T.Ito et al., Nature(London) **350**, 596 (1991); K.Tamasaku, Y.Nakamura and S.Uchida, Phys. Rev. Lett. **69**, 1455 (1992), **70**,1533(1993); S.L.Cooper et al., Phys. Rev. Lett. **70**, 1533 (1993); C.C.Homes et al., Phys. Rev. Lett. **71**, 1645 (1993); J.M.Harris, Y.F.Yan, and N.P.Ong, Phys. Rev. B**46**, 14239 (1992); Y.F.Yan, P.Matl, J.M.Harris, and N.P.Ong, Phys. Rev. B**52**, R751 (1995).
13. N.Nagaosa, Phys. Rev. B**52**, 10561 (1995).

NOVEL CUPRATE MATERIALS

T. M. RICE and B. NORMAND
Theoretische Physik, ETH-Hönggerberg, CH-8093 Zürich, Switzerland

We discuss some examples of new copper-oxide compounds which raise challenging problems in the theory of strongly-correlated electrons and low-dimensional quantum magnetism. We concentrate on the "ladder" compounds, which are intermediate between one and two dimensions, particularly on the hole-doped ladder system $LaCuO_{2.5}$.

1 Introduction

All high-T_c cuprate superconductors at present contain as the key structural element infinite CuO_2 planes lightly doped away from the stoichiometric insulating phase. Lately, there has been much interest in exploring modifications to these CuO_2 planes, e.g. through a breaking up into weakly-coupled ladders, or arrays of strongly-coupled chains weakly coupled to eachother. So far this initiative has not yielded new superconducting compounds, but it has led to new forms of magnetic insulator. These new systems are especially interesting because they are intermediate between isolated chains, which can be well analyzed theoretically, and the two-dimensional planes which have proven so difficult to handle. This crossover between one and two dimensions is not at all smooth, and has many subtle aspects which are not yet fully understood. Here we will reserve further discussion of this point, and refer the reader to the recent review by Dagotto and Rice.[1] Rather we will concentrate on the analysis of some recently-synthesized cuprate systems.

The most extensively studied examples of "ladder" materials are the layered compounds $Sr_nCu_{n+1}O_{2n+1}$. The $n = 1$ member consists of weakly-coupled 2-leg ladders, and studies of its magnetic properties confirm the existence of a sizeable spin gap, as predicted theoretically (for details see the contribution to this conference of M. Takano, and Ref. 1). This behavior contrasts with that of the $n = 2$ member, which consists of 3-leg ladders, and therefore is predicted to behave as a single chain, i.e. to show quasi-long-range antiferromagnetic order with gapless excitations. Indeed, recent experiments confirm this behavior, and even show a transition to true long-range antiferromagnetic order.[2] Unfortunately, to date it has not been possible to dope these systems.

Recently however, a new 2-leg ladder system has been synthesized by Hiroi and Takano.[3] This is the high-pressure phase of $LaCuO_{2.5}$, which they could dope with holes by substituting Sr for La. This doping was found not to induce superconductivity, but instead a transition from insulating to metallic behavior only at a relatively large Sr concentration (≥ 0.15). Since hole-doped 2-leg ladders are expected to be examples of d-wave RVB states, it is important to examine the electronic properties of this system in some detail. A new aspect has very recently been uncovered,[4] namely a transition to apparently 3-dimensional antiferromagnetic order in the undoped parent insulator, with a value of $T_N \simeq 120K$ - an observation which points to strong interladder coupling. In this report we will examine this issue in more detail.

The interladder magnetic coupling is determined by the interladder hopping matrix element for the uppermost valence bands. In the next section we discuss the interpretation of the electronic bandstructure, using a tight-binding model with a single, Cu-centred orbital. The size of the hopping matrix element between orbitals on neighboring ladders is the quantity which determines the interladder superexchange coupling. We then use a mean-field analysis to examine the stability of the spin liquid phase against long-range antiferromagnetic order.

2 Electronic and Magnetic Structure of LaCuO$_{2.5}$

2.1 Tight-Binding Fit to the LDA Bandstructure

The bandstructure of LaCuO$_{2.5}$ has been computed by Mattheiss,[5] using the Local Density Approximation (LDA) method. Of primary interest is the dispersion of only the highest occupied valence bands, which we fit to a tight-binding model based on the single, copper-centered orbital in each planar CuO$_4$ unit. Restricting the set of hopping matrix elements to those between nearby copper sites, we obtain an effective one-band model, albeit with four mixed levels (one from each Cu atom in the unit cell). From the fit one may deduce the ratio t'/t, of inter- and intraladder overlap, and thus estimate the ratio J'/J of the magnetic interactions, by using the superexchange result $J \simeq 4t^2/U$. The details of this procedure are given elsewhere.[6]

The tight-binding Hamiltonian may be written with only short-ranged hopping matrix elements t_{ij}, and the maximal set which we need to achieve a reasonable fit includes n.n. intraladder parameters, t_r for hopping along a rung and t_z for hopping along a leg, n.n.n. parameters t'_r and t'_z for hopping across a square plaquette in the ladder, and along two leg bonds, respectively. and finally, a parameter t_s for hopping from a copper atom on one ladder to its neighbor on the adjacent ladder, and also the n.n. analog t'_s for transfer between atoms on adjacent ladders with a relative displacement of one leg bond. We take the intraladder rung and leg terms to have the same value $t_r = t_z = t \simeq 0.4$eV, as in planar CuO$_2$, which establishes the energy scale of the bands.

The Hamiltonian may be reexpressed in reciprocal space and diagonalized to yield the dispersion relations of the four energy bands

$$\epsilon_{\mathbf{k}} = \pm \left[\tilde{t}_r^2 + 4\tilde{t}_s^2 \cos^2 \tfrac{1}{2}k_x \pm 4\tilde{t}_r \tilde{t}_s \cos \tfrac{1}{2}k_x \cos \tfrac{1}{2}k_y \right]^{1/2} - 2t_z \cos k_z - 2t'_z \cos 2k_z, \qquad (1)$$

where $\tilde{t} = t + 2t' \cos k_z$. The bands are shown in Fig. 1(a) for a series of high-symmetry lines in the Brillouin zone. Each of the chosen transfer integral parameters has a specific role in fixing the band separation at the high-symmetry points, and so these may be adjusted to give a good fit to the LDA results. The final parameter choice shown in Fig. 1(a) consists of intraladder overlaps which are very close to those in the CuO$_2$ plane, and the relatively large interladder coupling $t_s = \tfrac{1}{2} t_r$.

Estimating the superexchange interaction as above, we conclude that the interaction between spins on neighboring ladders has a magnitude $J' \simeq 0.25 J$, where J is the intraladder magnetic coupling of both rung and leg spins. Because J' is an appreciable fraction of J, it is clear that the spin interactions in LaCuO$_{2.5}$ have significant three-dimensional character.

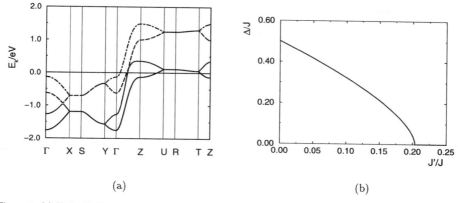

Figure 1: (a) Tight-binding bandstructure for the parameter set which appears closest to the LDA results of Mattheiss.[5] The parameters chosen are $t_z = t_r = t = 0.4\text{eV}$, $t'_z = \frac{1}{6}t_z$, $t'_r = -\frac{1}{5}t_r$, $t_s = \frac{1}{2}t_r$ and $t'_s = \frac{1}{5}t_s$. (b) Spin gap Δ as a function of the ratio between interladder and intraladder magnetic couplings J'/J, calculated for a three-dimensional system of isotropic ladders at $T = 0$, in the mean-field approximation from the starting point of dimerized rungs.

2.2 Mean-Field Analysis of the Spin Ground State

A mean-field analysis of the spin state for ladder systems was introduced by Gopalan et al.[7] Here we use a similar method to study the evolution of the spin liquid state with increasing interladder coupling $\lambda' J$.[6] The key quantity characterizing the spin-liquid ground state is the spin gap, the minimum energy of the triplet magnon excitations, which is found to occur at the wavevector $\mathbf{k}_M = (0, 0, \pi)$. The value of λ' where Δ is driven to zero will give the transition from the spin liquid state, where the spin orientation fluctuates with a time-averaged value of zero and with short-range correlations primarily at $\mathbf{k}_M$, to a magnetically ordered state with period $\mathbf{k}_M$. This wavevector corresponds to a simple antiferromagnetically aligned spin pattern in the 4-atom unit cell.

In the limit of no interladder coupling a value of $\Delta_0 = 0.501J$ is obtained for an isotropic 2-leg ladder. This mean-field result is in very good agreement with the result $\Delta_0 = 0.504J$ of numerical studies [8] by the Density Matrix Renormalization Group technique. In fact this agreement is largely serendipitous, but suggests that the mean-field approximation is reasonable. The spin gap obtained for the three-dimensionally coupled ladder system is shown in Fig. 1(b) as a function of the interladder coupling λ'. We see immediately that the spin gap decreases monotonically, with the transition point at $\lambda' = 0.203$. Comparison with the result of the previous section leads us to conclude that the LaCuO$_{2.5}$ system lies very close to the quantum critical point marking the phase transition from spin liquid to antiferromagnetic order.

3 Conclusion

In this brief report we have concentrated on the electronic properties of one particular material, $LaCuO_{2.5}$. This forms part of an overall strategy to look at novel cuprate compounds containing Cu^{2+} ions. Even as undoped insulators, such cuprates can form unusual magnetic structures described essentially by $S = \frac{1}{2}$ Heisenberg models. Quantum effects are very strong in such low-dimensional magnets, and as discussed above can lead to novel ground states and magnetic properties. Even more exciting is the possibility of finding ways to dope such materials. After ten years, the phenomenon of high-T_c superconductivity remains confined to cuprates containing infinite CuO_2 planes as the essential structural element. It is still of great interest to find out if this is an intrinsic limitation, or if there exist other structural forms which also can support high-T_c superconductivity. The CuO_4 squares which make up the CuO_2 planes can be assembled in many different ways, which allows for great flexibility in the synthesis of cuprate materials, as reviewed in detail by Müller-Buschbaum.[9] Novel cuprates can be expected to continue to pose new problems in the theory of strongly-correlated electrons and quantum magnetism in the coming decade.

Acknowledgments

We are grateful to Y. Kitaoka, M. Sigrist, M. Takano and especially Z. Hiroi for helpful discussions, and to the Swiss National Fund for financial support.

References

1. E. Dagotto and T. M. Rice, Science, **271**, 618 (1996).
2. M. Azuma, Z. Hiroi, M. Takano, K. Ishida and Y. Kitaoka, Phys. Rev. Lett. **73**, 3463 (1994).
3. Z. Hiroi and M. Takano, Nature **377**, 41 (1995).
4. S. Matsumoto, Y. Kitaoka, K. Ishida, K. Asayama, Z. Hiroi, N. Kobayashi and M. Takano, Phys. Rev. Lett., submitted.
5. L. F. Mattheiss, Solid State Commun. **97**, 751 (1996).
6. B. Normand and T. M. Rice, cond-mat/9603054 (1996).
7. S. Gopalan, T. M. Rice and M. Sigrist, Phys. Rev. B **49**, 8901 (1994).
8. S. White, R. Noack and D. J. Scalapino, Phys. Rev. Lett. **73**, 886 (1994).
9. H. Müller-Buschbaum, Angew. Chem. Int. Ed. Eng. **30**, 723 (1991).

PHASE STRING, SUPERCONDUCTIVITY, AND SPIN DYNAMICS IN THE T-J MODEL

Z.Y. WENG, D.N. SHENG, AND C.S. TING
TCSUH, University of Houston, Houston, TX 77204-5932, USA

The motion of an injected hole always leaves a trace of spin mismatches on the antiferromagnetic spin background. In contrast to previous belief, it has been rigorously shown recently that such a string effect *cannot* be repaired through low-energy spin flips. A series of new properties previously unknown to the $t-J$ model emerge after an appropriate treatment of this nonlocal string effect. We report a new phase diagram of the $t-J$ model at small doping concentration in which a continuous evolution from an AF insulating phase to a metallic phase is obtained. A gap is shown to be opened up at finite doping in the spinon spectrum, which determines an upper limit of a d-wave superconducting transition temperature and, at the same time, plays a central role in shaping the low-lying spin dynamics. The present theory well explains some of most important experimental measurements in the high-T_c cuprates.

1 Introduction

In this article we reexamine the following simple and old question: If holes are injected into a two-dimensional (2D) antiferromagnetic (AF) insulator, what kind of state(s) can be obtained at finite doping concentration and what are the associated physical properties? This problem is closely related to the high-T_c cuprates, and is to be studied here based on one of the simplest nontrivial model—the $t-J$ model.

One may start with a 2D Ising spin array with one doped hole as an example. Such a spin configuration will be disordered by the mobile hole on its hopping path, which is known as the spin "string" defect. It is well-known that this spin string defect can be easily repaired by spin flip process such that the hole still behaves like a quasiparticle in the long-wavelength, low-energy limit. This simple picture has become a general belief about the weakly-doped antiferromagnet for many years. However, there is an important point has been overlooked here. Namely, the real spins under consideration are quantum spins which have three non-commutable components. In other words, the above-mentioned "string" defect actually involves three spin directions, and the question whether such a spin string defect can be repaired by spin flips simultaneously *in all the three directions* has not been carefully addressed before. To reconsider this issue, we start with the ground state of the quantum Heisenberg model, where spin $x-y$ components are taken care of by the phases of the quantum states (with the z axis as the spin quantization direction) which satisfy the so-called Marshall sign rule[1] (e.g., every down spin at the A-sublattice is associated with an extra (-1) sign). Then it is easy to find that an injected hole's motion on such a spin background will mess up the Marshall signs, creating a string of extra signs in addition to the "correct" Marshall-sign assignment. It can be rigorously proven[2] that such a phase string is not repairable through the spin-flip process since the latter always respects the Marshall sign rule at low energy. Therefore, the previous understanding on the lightly-doped antiferromagnet has a serious flaw, and one has to re-think the whole problem over again.

The existence of non-repairable phase strings poses a fundamental challenge to any perturbative-minded methodology for the $t-J$ model. It implies a nonlocal topological-like characteristics of the solution, as the phase string always leads to a nontrivial Berry phase when a hole goes through a closed path back to the original location. To avoid losing this long-distance topological feature hidden in the $t-J$ model, one may explicitly incorporate the corresponding topological effect into the wavefuntion (basis) so that in the consequent new representation of the $t-J$ Hamiltonian the singular phase string effect is replaced by some topological phases more manageable perturbatively. According to Ref.2, such a new exact representation of the $t-J$ model has the following form:

$$H_{t-J} = -t \sum_{\langle ij \rangle} \left[e^{iA_{ij}^f} h_i^\dagger h_j \right] \left[\sum_\sigma e^{i\sigma A_{ji}^h} b_{j\sigma}^\dagger b_{i\sigma} \right] + H.c. - \frac{J}{2} \sum_{\langle ij \rangle \sigma,\sigma'} \left[e^{i\sigma A_{ij}^h} b_{i\sigma}^\dagger b_{j-\sigma}^\dagger \right] \left[e^{i\sigma' A_{ji}^h} b_{i\sigma'} b_{j-\sigma'} \right], \quad (1)$$

in which the phase string is replaced by the so-called topological phases A_{ij}^f and A_{ij}^h satisfying

$$\sum_C A_{ij}^h = \pi \sum_{l \in C} n_l^h; \qquad \sum_C A_{ij}^f = \pi \sum_{l \in C} \left(\sum_\sigma \sigma n_{l\sigma}^b - 1 \right), \quad (2)$$

for an arbitrary path C (here n_l^h and $n_{l\sigma}^b$ are holon and spinon number operators). And we have a new decomposition form $c_{i\sigma} = \tilde{h}_i^\dagger \tilde{b}_{i\sigma}(-\sigma)^i$ where $\tilde{h}_i^\dagger = h_i e^{i/2 \sum_l Im \ln(z_i - z_l)(\sum_\alpha \sigma n_{l\alpha}^b + 1)}$ and $\tilde{b}_{i\sigma} = b_{i\sigma} e^{-i/2 \sum_l Im \ln(z_i - z_l)\sigma n_l^h}$. Here both $h_i^\dagger$ and $b_{i\sigma}$ are bosonic operators and $\tilde{h}_i^\dagger$ and $\tilde{b}_{i\sigma}$ satisfy the so-called mutual statistics to guarantee the electron operator to be fermionic. Similar decomposition has been previously derived[4] based on the slave-boson formalism.

2 Mean-field solution

At half-filling Eq.(1) reduces to the Heisenberg model in the Schwinger-boson representation. In this case, Schwinger-boson mean-field (MF) theory with order parameter $\Delta_0 = \sum_\sigma \langle b_{i\sigma}^\dagger b_{j-\sigma}^\dagger \rangle$ is a fairly good solution.[3] If we make an analogy of it with the BCS theory, then at finite doping the phase A_{ij}^h in the superexchange term of (1) plays a role as if each holon carries a quantized π flux as seen by spinons, resembling the "vortex phase" in the BCS theory. Thus one may generalize the Schwinger-boson MF theory to finite doping by employing the Bogoliubov-de Gennes scheme for the vortex phase after introducing a generalized order parameter $\Delta_{ij} = \left\langle \sum_\sigma e^{i\sigma A_{ij}^h} b_{i\sigma}^\dagger b_{j-\sigma}^\dagger \right\rangle$. It turns out that without the presence of original string effect both superexchange and hopping processes can be simultaneously optimized in this new MF scheme and the corresponding MF Hamiltonian is given by

$$H_{t-J}^{MF} = \sum_{m\sigma} E_m \gamma_{m\sigma}^\dagger \gamma_{m\sigma} - t_h \sum_{<ij>} e^{iA_{ij}^f} h_i^\dagger h_j + H.c. + const, \quad (3)$$

where the new spinon operator γ is introduced through the Bogoliubov transformation $b_{i\sigma} = \sum_m \left(u_{m\sigma}(i)\gamma_{m\sigma} - v_{m\sigma}(i)\gamma_{m-\sigma}^\dagger \right)$ with $u_{m\sigma}(i) = u_m w_{m\sigma}(i)$ and $v_{m\sigma}(i) = v_m w_{m\sigma}(i)$.

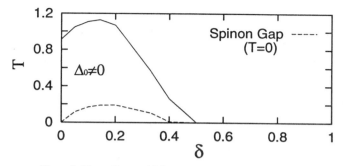

Figure 1: Phase diagram of the present MF solution of the $t - J$ model.

Here the single-particle wavefunction $w_{m\sigma}(i)$ satisfies

$$\xi_m w_{m\sigma}(i) = -\frac{J}{2} \sum_{j=n.n.(i)} \Delta_{ij} e^{-i\sigma A_{ji}^h} w_{m\sigma}(j) \qquad (4)$$

with $u_m = 1/\sqrt{2}\left(\frac{\lambda_m}{E_m}+1\right)^{1/2}$ and $v_m = 1/\sqrt{2}\left(\frac{\lambda_m}{E_m}-1\right)^{1/2} sgn(\xi_m)$. The spinon spectrum E_m and the effective hopping integral t_h will be determined after solving the self-consistent MF equations. At finite doping if we treat the flux lines described by A_{ij}^h as a uniform flux, then a uniform $\Delta_{ij} = \Delta_0$ solution may be found which gives a phase diagram shown in Fig. 1. Here we have a new phase which extends from the hall-filling to a finite doping concentration of holes, characterized by a nonzero Δ_0 which determines short-range AF correlations. A prominent feature in this phase is that a gap is opened up in the spinon spectrum E_m due to doping effect, whose zero-temperature scale is marked in Fig. 1 by a dashed curve. Such a spinon gap will play a central role in determining the superconducting transition temperature as well as anomalous spin dynamics as will be discussed below.

3 Superconducting condensation

Superconducting condensation is characterized by $\Delta_{ij}^p = \langle c_{i\uparrow}^\dagger c_{j\downarrow}^\dagger \rangle \neq 0$. For simplicity, we consider i and j to be nearest-neighboring sites. We find at $T = 0$, $\Delta_{ij}^p \propto \langle h_i^\dagger \rangle \langle h_j^\dagger \rangle \Delta_0$ due to the Bose-condensation of holons. And the pairing-symmetry: $\frac{\Delta_{i,i+\hat{x}}^p}{\Delta_{i,i+\hat{z}}^p} = -1$ is d-wave like. Since Δ_0 is always finite in this phase, the Kosterlitz-Thouless transition T_c^{KT} of holons will determine the superconducting transition at finite temperature. Here holons are subjected to the vector potential A_{ij}^f in (2) which represents π ($-\pi$) flux quanta bound to spinons. Spinon pairs may be broken up at finite temperature to carry free $\pm\pi$ flux quanta as seen by the holons. If the number of excited free spinon becomes comparable to the holon number, there will be simply no enough condensed holons to produce screening current around all the free $\pm\pi$ flux quanta carried by the unpaired spinons, and then it leads to the destruction of the Bose-condensation of holons. Thus, an upper limit for the superconducting transition

T_c may be estimated based on this criteria as $E_g^s/T_c > \ln 3 = 1.09$, where E_g^s is the ($T = 0$) spinon gap in the present MF case. Above T_c, the topological phase A_{ij}^f also provide a scattering-source which leads to anomalous transport properties of holons. Those anomalies have been discussed elsewhere,[4] which are consistent with experimental measurements in the high-T_c cuprates.

4 Spin dynamics

In the Schwinger-boson formalism, the Bose-condensation of spinons will give rise to an antiferromagnetic long-range ordering (AFLRO) at half-filling.[3] With the opening of a spinon gap at finite doping, such a type of AFLRO is prevented in favor of a d-wave superconducting condensation (previous section). Thus, our MF state provides a continuous evolution from an AF insulator (with AFLRO at $T = 0$) to a metallic phase with short-range AF correlations and a low-temperature superconducting condensation.

Like the spinon gap itself, spin dynamics is strongly modified by the doping due to the frustration effect caused by phase strings. For example, the NMR spin-lattice relaxation rate $1/T_1$ calculated by the present theory for the Cu site shows a strong doping-dependence at low-T, while they coincide together at high-T. The uniform spin susceptibility, behaving like $A + BT$ at $T > E_g^s$, also exhibits a strong doping-dependence in the coefficient A. Such behaviors qualitatively describe what have been observed in the one-layer $LSCO$ compounds, and the doping-dependences are very difficult to understand in the conventional theory where holes just locally distort the surrounding spin environment. $1/T_1T$ also shows drastically distinctive behaviors at planar Cu and O sites: a strong non-Korringa T-dependence at the former site in contrast to a more-or-less conventional Korringa law at the latter site. Similar difference shown in the cuprate superconductors has been usually interpreted as due to strong commensurate AF fluctuations. However, there are some subtle points in the present case. When we directly calculate the spin-spin correlation function at long distance, there is always a phase-string contribution leading to a shift of the AF peak to four incommensurate peaks at $(\pi \pm 2\pi\delta, \pi)$ and $(\pi, \pi \pm 2\pi\delta)$, etc., in the momentum space. But $1/T_1$ at O site only involves the on-site and nearest-neighboring sites spin-spin correlations where the phase string effect is absent. Therefore, both inelastic neutron-scattering and the NMR data found in the $LSCO$ compounds[5] can be easily reconciled within the present theoretical framework.

Acknowledgments This work is supported partially by a TARP grant under No.3652182 and a grant from the Robert A. Welch foundation, and by TSCUH.

References

1. W. Marshall, Proc. Roy. Soc. (London) **A232**, 48 (1955).
2. Z.Y. Weng, *et al.*, TCSUH preprint No. 95133.
3. D.P. Arovas and A. Auerbach, Phys. Rev. **B38**, 316 (1988).
4. Z.Y. Weng, *et al.*, Phys. Rev. **B52**, 637 (1995).
5. R.E. Walstedt, *et al.* Phys. Rev. Lett. **72**, 3610 (1994).

SPIN-SUSCEPTIBILITY OF STRONG CORRELATED BANDS IN FAST FLUCTUATING REGIME

M.V. EREMIN, S.G. SOLOVJANOV, S.V. VARLAMOV
Department of Radiospectroscopy and Quantum Electronics, Kazan State University, Lenin str.,18, Kazan 420008, Russia

I.M. EREMIN
Department of Theoretical Physics, Kazan State University, Lenin str.,18, Kazan 420008, Russia

We have derived an expression for the spin-susceptibility for the singlet correlated band in HTSC using a Hubbard-like theory. The fast spin fluctuating regime ($< s_z > \leq 1/2$) has been investigated. A new interesting feature of the obtained susceptibility function is that the Stoner-like factor is very sensitive to the temperature and to the hole doping level.

In our calculations we start from the Hubbard-like Hamiltonian for the CuO_2 plane in an external magnetic field:

$$H = \varepsilon_d \sum X_i^{\sigma,\sigma} + U_{pd} \sum X_i^{pd,pd} + \sum t_{ij}^{(1)} X_i^{pd,\sigma} X_j^{\sigma,pd} + \sum t_{ij}^{(2)} X_i^{\sigma,0} X_j^{0,\sigma} + \quad (1)$$

$$\sum t_{ij}^{(12)}(-1)^{\frac{1}{2}-\sigma}(X_i^{pd,\sigma} X_j^{0,\sigma} + X_i^{\sigma,0} X_j^{\sigma,pd}) - g\beta H_z \frac{1}{2}\sum(X_i^{\uparrow\uparrow} - X_i^{\downarrow\downarrow}) \quad . \quad (2)$$

Using the usual Hubbard theory it is easy to get the following relations for thermodynamical averages of anticommutators:

$$\begin{aligned} <\{X_i^{pd,\uparrow}, X_i^{\uparrow,pd}\}> &= \tfrac{1}{2} + \tfrac{\delta}{2} + <s_z>, & <\{X_i^{\uparrow,0}, X_i^{0,\uparrow}\}> &= \tfrac{1}{2} - \tfrac{\delta}{2} + <s_z>, \\ <\{X_i^{pd,\downarrow}, X_i^{\downarrow,pd}\}> &= \tfrac{1}{2} + \tfrac{\delta}{2} - <s_z>, & <\{X_i^{\downarrow,0}, X_i^{0,\downarrow}\}> &= \tfrac{1}{2} - \tfrac{\delta}{2} - <s_z>, \end{aligned} \quad (3)$$

where $\delta = <X^{pd,pd}>$ is the number of Zhang-Rice singlets per one copper site and $<s_z>$ is the thermodynamical expectation value of copper spins. These relations allowed us to calculate $<s_z>$ self-consistently under the assumption $<s_z> \leq 1/2$ (fast fluctuating regime) and, after that, the susceptibility.

In particular, when the chemical potential is located in the upper Hubbard-like band on gets from

$$<X_k^{pd,\uparrow} X_k^{\uparrow,pd}> = <X_k^{pd,\downarrow} X_k^{\downarrow,pd}>, \quad (4)$$

the following expression:

$$\chi_1(\theta,\delta) = (1+\delta)^2 \chi_p(\theta)/[4\delta + \Lambda_1(\theta,\delta) - Z_1(\theta,\delta)] \quad . \quad (5)$$

Here $\chi_p(\theta)$ is a typical expression for the Pauli-like susceptibility

$$\chi_p(\theta,\delta) = \frac{1}{2}(g\mu_b)^2 \sum \left[\left(\frac{E_{1k}-\varepsilon_k^d}{E_{1k}-E_{2k}}\right)\frac{\partial f(E_{1k})}{\partial E_{1k}} + \left(\frac{\varepsilon_k^d - E_{2k}}{E_{1k}-E_{2k}}\right)\frac{\partial f(E_{2k})}{\partial E_{2k}} \right] \quad . \quad (6)$$

E_{1k} and E_{2k} are the energies of the singlet correlated and lower copper Hubbard bands, respectively, as obtained in the Hubbard I approximation. We refer to [1] for a derivation and description of the energy dispersion in the present model. $\Lambda_1(\theta,\delta)$ and $Z_1(\theta,\delta)$ are temperature θ and doping level δ functions:

$$\Lambda_1(\theta,\delta) = \frac{(1+\delta)^2}{2N}\Sigma\Phi_k[f(E_{1k}) - f(E_{2k})] \quad , \quad (7)$$

$$\Phi_k = \left\{ \frac{t_k^{(1)} + t_k^{(2)}}{E_{1k} - E_{2k}} + \frac{\left(\varepsilon_k^{pd} - \varepsilon_k^d\right)\left[2\delta\left(t_k^{(12)}\right)^2 - \left(\varepsilon_k^{pd} - \varepsilon_k^d\right)\left(t_k^1 + t_k^2\right)\right]}{(E_{1k} - E_{2k})^3} \right\}, \quad (8)$$

$$Z_1(\theta, \delta) = \frac{(1+\delta)^2}{2} \sum \left[\left(\frac{E_{1k} - \varepsilon_k^d}{E_{1k} - E_{2k}}\right) F_{1k} \frac{\partial f(E_{1k})}{\partial E_{1k}} + \left(\frac{\varepsilon_k^d - E_{2k}}{E_{1k} - E_{2k}}\right) F_{2k} \frac{\partial f(E_{2k})}{\partial E_{2k}} \right], \quad (9)$$

$$F_{jk} = t_k^{(1)} - t_k^{(2)} \mp \left\{ \left[2\left(t_k^{12}\right)^2 \delta - \left(\varepsilon_k^{pd} - \varepsilon_k^d\right)\left(t_k^{(1)} + t_k^{(2)}\right)\right] / (E_{1k} - E_{2k}) \right\}, \quad (10)$$

where the upper sign is for $j = 1$ and the lower for $j = 2$.

The numerical results are shown on Fg. 1. The solid and dash-dotted lines correspond to

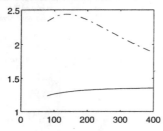 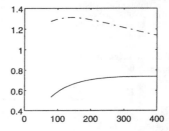

Figure 1: Temperature dependence of the spin susceptibility for χ_1 (left) and χ_p (right) for two different fillings.

values of the chemical potential near the bottom of the band and the singularity peak in DOS, respectively.[1] An interesting feature of the function $Z_1(\theta, \delta)$ is that the sign can change upon doping. This function is negative (~ -0.5) when the chemical potential is near the bottom of the band, and becomes positive when approaching to the top of it. The temperature dependence of $\chi_1(\theta, \delta)$ (suscept.) are given on Fig. 1a. For comparison, on Fig. 1b we show the temperature dependence of $\chi_p(\theta, \delta)$ (Pauli suscept.). We have found that $\Lambda_1(\theta, \delta)$ is small, varying from -0.005 to -0.02 .

When the chemical potential is in the lower Hubbard-like band the susceptibility calculated self-consistently from

$$<s_z> = \frac{1}{2N} \sum \{<X_k^{\uparrow,0}, X_k^{0,\uparrow}> - <X_k^{\downarrow,0}, X_k^{0,\downarrow}>\}, \quad (11)$$

is given by

$$\chi_2(\theta, \delta) = (1-\delta)^2 \chi_p(\theta) / \left[4 <X^{0,0}> -\lambda_2(\theta, \delta) - Z_2(\theta, \delta)\right], \quad (12)$$

where $<X^{0,0}>$ is the so-called vacuum average or number of Cu^+ ions per one copper site. For $t^{(12)} = 0$ our formula completely agrees with that of Hubbard and Jain.[2] We stress the fact that the expressions for $\chi_1(\theta, \delta)$ and $\chi_2(\theta, \delta)$ are not sensitive to Green's functions decouplimg scheme. In particular for the L. Roth's decoupling scheme [3] some corrections appeared forthe energy dispersion and for the functions Φ_k, F_{1k} and F_{2k} only.

We are grateful to R. Markendorf for useful discussions.

References

1. M.V. Eremin, S.G. Solovjanov, S.V. Varlamov, *Phys. Chem. Solids* **56**, 1713 (1995).
2. J. Hubbard and K.P. Jain, *J. Phys.* C (Proc. Phys. Soc.), Ser. 2, V. 1 (1968).
3. L.M. Roth, *Phys. Rev.* **184**, 451(1969).

SHORT–RANGE ANTIFERROMAGNETIC CORRELATIONS AND THE PHOTOEMISSION SPECTRUM

N. BULUT
Department of Physics, University of California, Santa Barbara,
CA 93106-9530, USA

We present Quantum Monte Carlo results on the antiferromagnetic correlations and the one–electron excitations of the doped two–dimensional Hubbard model. These results are helpful in interpreting the NMR, neutron scattering and photoemission experiments on the layered cuprates.

NMR and neutron scattering experiments have provided clear evidence that the superconducting layered cuprates have short–range and low–frequency antiferromagnetic (AF) correlations. In addition, the photoemission experiments have shown that the one–electron excitations are heavily damped and strongly renormalized. Perhaps, the simplest model to describe these properties is the two–dimensional Hubbard model on a square lattice,

$$H = -t \sum_{\langle i,j \rangle, \sigma} (c^\dagger_{i\sigma} c_{j\sigma} + c^\dagger_{j\sigma} c_{i\sigma}) + U \sum_i c^\dagger_{i\downarrow} c_{i\downarrow} c^\dagger_{i\uparrow} c_{i\uparrow}, \tag{1}$$

where t is the hopping matrix element, U is the onsite Coulomb repulsion, and $c_{i\sigma}$ annihilates an electron of spin σ at site i. Here, we present Quantum Monte Carlo (QMC) results on the magnetic fluctuations and the one–electron properties of the doped 2D Hubbard model.

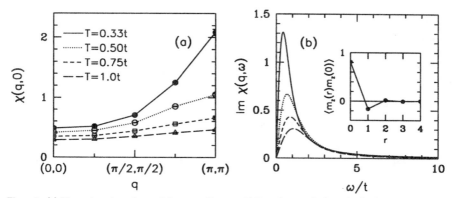

Figure 1: (a) Momentum dependence of the magnetic susceptibility $\chi(\mathbf{q}, \omega=0)$ along the $(1,1)$ direction at various temperatures. (b) Frequency dependence of the spin–fluctuation spectral weight $\mathrm{Im}\,\chi(\mathbf{q},\omega)$ for $\mathbf{q} = (\pi, \pi)$ at the same temperatures as in Fig. 1(a). Inset: Real–space structure of the equal–time magnetization–magnetization correlation function at $T = 0.33t$.

Using QMC simulations[1] and numerical analytic continuation methods[2] we have calculated the staggered magnetic susceptibility $\chi(\mathbf{q},\omega)$. The results that we will present here are for $U = 8t$ and $1/8$ doping on an 8×8 lattice. In addition, we use units such that $t = 1$ and $\mu_B = 1$. In Figure 1(a), the momentum dependence of $\chi(\mathbf{q}, \omega = 0)$ is plotted along the $(1,1)$ direction. Figure 1(b) shows the spectral weight $\mathrm{Im}\,\chi(\mathbf{q},\omega)$ versus ω for $\mathbf{q} = (\pi,\pi)$. In the inset of Fig. 1(b), the equal–time

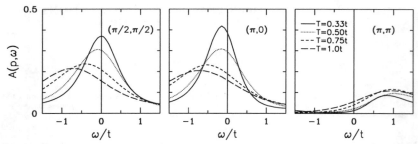

Figure 2: One–electron spectral weight $A(\mathbf{p},\omega)$ versus ω at various temperatures for $\mathbf{p} = (\pi/2, \pi/2)$, $(\pi, 0)$ and (π, π).

magnetization–magnetization correlation function is plotted as a function of distance along the $(1, 0)$ direction. These figures show that there are short range AF correlations in the 2D Hubbard model near half–filling. NMR T_1^{-1} and T_2^{-1} measurements and magnetic neutron scattering experiments have shown the existence of short–range AF correlations in the superconducting cuprates.

Next, we present results on the one–electron spectral weight $A(\mathbf{p},\omega) = -\frac{1}{\pi}\text{Im}\, G(\mathbf{p},\omega)$, where $G(\mathbf{p},\omega)$ is the one–electron Green's function. $A(\mathbf{p},\omega)$ is of experimental interest, since, within the sudden approximation, the photoemission intensity is proportional to $A(\mathbf{p},\omega)f(\omega)$, where $f(\omega)$ is the fermi factor. In Figure 2, $A(\mathbf{p},\omega)$ versus ω is plotted for three representative points of the Brillouin zone [3,4]. Because of the many–body effects, the quasiparticle peak in $A(\mathbf{p},\omega)$ is damped, and the quasiparticle bandwidth is reduced. We also observe that $A(\mathbf{p},\omega)$ has significant dependence on temperature in this temperature regime.

In this article, we have presented a brief review of the QMC data on the magnetic and one–electron excitations of the 2D Hubbard model near half–filling. In this metallic regime, there exists low–frequency short–range antiferromagnetic fluctuations, and the one–electron excitations are heavily damped and strongly renormalized. Similar electronic features have been observed experimentally for the layered cuprates.

Acknowledgments

The author would like to thank D.J. Scalapino for many helpful discussions. This review is based on work done in collaboration with D.J. Scalapino and S.R. White. The author also gratefully acknowledges support from the National Science Foundation under Grant No. DMR92–25027. The numerical calculations reported in this paper were performed at the San Diego Supercomputer Center.

References

1. S.R. White, D.J. Scalapino, R.L. Sugar, E.Y. Loh, J.E. Gubernatis and R.T. Scalettar, *Phys. Rev.* B **40**, 506 (1989).
2. S.R. White, *Phys. Rev.* B **44**, 4670 (1991).
3. N. Bulut, D.J. Scalapino and S.R. White, *Phys. Rev. Lett.* **72**, 705 (1994).
4. N. Bulut, D.J. Scalapino and S.R. White, *Phys. Rev.* B **50**, 7215 (1994).

Magnetic Excitations in High-T_c Cuprates: the 41meV Problem

Hiroshi Kohno, Bruce Normand[1] and Hidetoshi Fukuyama
*Department of Physics, University of Tokyo,
7-3-1 Hongo, Bunkyo-ku, Tokyo 113, Japan.*
[1] *Theoretische Physik, ETH-Hönggerberg,
CH-8093 Zürich, Switzerland.*

We examine the 'η-pairing' scenario for the resonance-like spin-flip excitation observed in YBa$_2$Cu$_3$O$_7$, by taking proper account of the realistic Fermi surface shape and the effect of the superconducting gap on the η-pairs, for single-layer and bilayer t-J models. The results show that inclusion of η-pair processes modifies the RPA spectral weight only slightly, and does not give rise to a desired peak.

Through intensive studies of spin excitations in high-T_c cuprates, the '41meV peak' problem has emerged as one of the mysterious features in optimally doped YBa$_2$Cu$_3$O$_7$. It is a spin triplet excitation strongly peaked around the wavevector $\vec{q} = \vec{Q} \equiv (\pi, \pi)$ and energy $\omega = 41$meV, which has been observed by inelastic neutron scattering in the superconducting state.[1,2,3] While a mean-field theory of the t-J model seems successful in describing spin excitations in LSCO, spin gap phenomena seen in NMR-T_1[4] and certain phonon anomalies,[5] the 41meV peak still awaits a definitive explanation.

Recently an appealing proposal was made that the 41meV peak is due to 'η-pairing,' a finite-momentum collective mode in the particle-particle (p-p) channel.[6] In this scenario, (i) there is a well-defined collective mode branch in the p-p channel around $\vec{q} = \vec{Q}$, and (ii) this mode becomes experimentally accessible in the superconducting state by mixing with particle-hole (p-h) channels. In this initial study, the superconducting order parameter, Δ, was treated perturbatively (*i.e.*, only as a matrix element mixing p-p and p-h channels), and only a nearest-neighbor (n.n.) transfer integral was assumed. However, more distant transfer integrals (t', t'') are known to be essential to describe both the Fermi surface and the commensurate nature of the magnetic peak[4,7] observed in optimally-doped YBCO.

In this paper, we extend the analysis of Ref.6 to include both the realistic Fermi surface shape and Δ to all orders, and calculate explicitly the dynamical spin susceptibility $\chi(\vec{q}, \omega)$. Our analysis is based on the t-J model with extended transfer integrals, in the slave-boson-based mean-field approximation with uniform and singlet RVB order parameters, the latter having $d_{x^2-y^2}$ symmetry.[4,5] To calculate $\chi(\vec{q}, \omega) = \langle S_{\vec{q}}^-; S_{\vec{q}}^- \rangle$, we may write it as $\chi = \chi_1/(1 + J_{\vec{q}} \chi_1)$, in terms of a susceptibility χ_1 irreducible with respect to the exchange interaction $J_{\vec{q}} = J(\cos q_x + \cos q_y)$. Here $\langle A; B \rangle \equiv i \int_0^\infty dt e^{i(\omega+i0)t} \langle [A(t), B^\dagger] \rangle$, $S_{\vec{q}}^- = \sum_{\vec{k}} f_{\vec{k}-\downarrow}^\dagger f_{\vec{k}+\uparrow}$ is the spin-lowering operator, $f_{\vec{k}\sigma}$ is a spinon operator and $\vec{k}\pm \equiv \vec{k} \pm \vec{q}/2$. The previous RPA treatment[4] corresponds to taking $\chi_1 = \chi_0 \equiv \langle S_{\vec{q}}^-; S_{\vec{q}}^- \rangle_0$, where the subscript 0 denotes the lowest-order estimate (Lindhard-type function). The η-pair process is taken into account, following Ref.6, by the 'T-matrix approximation' to the J-term (Fig. 1). Then $S_{\vec{q}}^-$ at $\vec{q} = \vec{Q}$ mixes with three other channels, represented by $O_1 = \sum_{\vec{k}} (\sin k_x + \sin k_y) f_{\vec{k}-\downarrow}^\dagger f_{\vec{k}+\uparrow}$ (p-h), $O_2 = \sum_{\vec{k}} (\sin k_x - \sin k_y) f_{\vec{k}+\uparrow} f_{-(\vec{k}-)\uparrow}$ (p-p) and $O_3 = \sum_{\vec{k}} (\sin k_x - \sin k_y) f_{\vec{k}-\downarrow}^\dagger f_{-(\vec{k}+)\downarrow}^\dagger$ (h-h), the latter two corresponding to η-pairs. After summing the ladder, one obtains

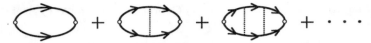

Figure 1: Irreducible susceptibility χ_1 in the Nambu representation. The dashed line represents $J/8$.

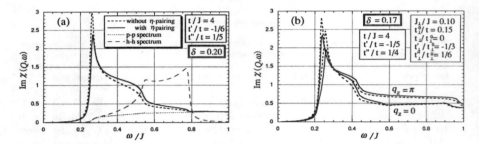

Figure 2: $\text{Im}\,\chi(\vec{Q},\omega)$ for (a) single-layer and (b) bilayer[8] t-J models with YBCO-type Fermi surface.

$\chi_1 = \chi_0 - \sum_{i,j=1}^{3} X_i \left(\hat{J}^{-1} + \hat{D}\right)_{ij}^{-1} X_j$, where $D_{ij} = \langle O_i ; O_j \rangle_0$, $X_i = \langle S_q^- ; O_i \rangle_0$ and $\hat{J} = \text{diag}(J/4, J/8, J/8)$.

Numerical results for $\text{Im}\,\chi(\vec{Q},\omega)$ as a function of ω are shown in Fig. 2. The effect of the η-pair process is to increase spectral weight around the 'shoulder' above the strongest peak, but it does not itself give a peak. In Fig.2(a) are also shown the p-p and h-h spectra, $\text{Im}\,[D_{ii}/(1 + JD_{ii}/8)]/2$ ($i = 2, 3$), both of which have continuum due to t' and t''. There is a peak in the h-h spectrum at the top of its continuum, which may correspond to the η-pair excitation, but not an isolated peak. Its contribution to $\chi(\vec{Q},\omega)$ is small since the mixing matrix element is small. We have further examined the η-pair contribution by varying several parameters such as the doping rate (δ) and the ratio t/J, which may change quantitative aspects. However, the general feature is that the correction remains small, and the overall spectral shape is determined by the lowest-order RPA results. Therefore the η-pair process does not seem to be relevent to the '41meV peak' observed in $YBa_2Cu_3O_7$.

References

1. J.M. Rossat-Mignod et al., Physica **C185-189**, 86 (1991).
2. H.A. Mook et al., Phys. Rev. Lett. **70**, 3490 (1993).
3. H.F. Fong et al., Phys. Rev. Lett. **75**, 316 (1995).
4. T. Tanamoto, H. Kohno and H. Fukuyama, J. Phys. Soc. Jpn. **62**, 1455 (1993); **63**, 2741 (1994).
5. B. Normand, H. Kohno and H. Fukuyama, Phys. Rev. B **53**, 856 (1996).
6. E. Demler and S.C. Zhang, Phys. Rev. Lett. **75**, 4126 (1995).
7. Q. Si, Y. Zha, K. Levin, J.P. Lu and J.H. Kim, Phys. Rev. B **47**, 9055 (1993).
8. B. Normand, H. Kohno and H. Fukuyama, J. Phys. Soc. Jpn. **64**, 3903 (1995).

Anomalous Charge Excitation Spectra in the t-J model

T.K. LEE
Dept. of Physics, Virginia Tech, Blacksburg, VA 24060, USA

R. EDER
Dept. of Applied and Solid State Physics, Univ. of Groningen, NL

Y.C. CHEN
Dept. of Physics, Tung-Hai Univ., Taichung, Taiwan

H.Q. LIN
Dept. of Physics, Chinese Univ. of Hong Kong, Hong Kong

Y. OHTA
Dept. of Physics, Chiba Univ., Chiba, Japan

C.T. SHIH
Dept. of Physics, National Tsing Hua Univ., Hsinchu, Taiwan

By using the exact diagonalization method to study small clusters, we show that the lowest spin and charge excitation energy dispsersion of the $t-J$ model are very similar in one and two dimensions. While the lowest spin excitation occurs at wave vector $2k_F$ the lowest charge excitation is at a different wave vector which seems to coincide with the $2k_F$ of the holes. This particular excited state is more suitably described as a collective excitation instead of a state with excited particle-hole pairs.

Many anomalous properties observed in the normal state of high temperature superconductors have inspired the debate about the Fermi liquid state vs. the non-Fermi liquid state [1,2]. Recently, there is an increasing number of numerical evidences indicating anomalous behavior in the $t-J$ model. Based on the spectral density of the dynamical correlation functions Tohyama et al [3] reported the strong suppression of $2k_F$-backscattering in the charge channel but not in the spin channel. This leads to the question whether this is a state similar to a one-dimensional Tomonaga-Luttinger liquid (TLL) with separation of spin and charge degrees of freedom. In this paper we examine the low-energy excitation spectra in one and two dimensions and find strong similarity. Only preliminary result will be given here, most of the result will be published elsewhere.

The charge and spin excitations (CE and SE) are obtained by using the Lanczos algorithm to compute the dynamical spin and charge or density correlation functions. Here we shall only report the result of a two-dimensional 4×4

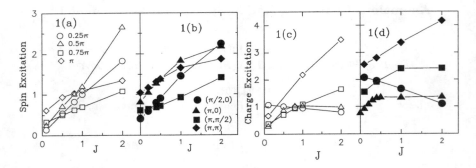

Figure 1: Lowest spin excitation energies in (a) 1D and (b) 2D; and charge excitation energies in (c) 1D and (d) 2D, as a function of J. In 1D, (a) and (c), the wave vectors are $q_0 = 0.25\pi$, $2k_F = 0.75\pi$, $4k_F = 0.5\pi$ and π. In 2D, (b) and (d), the wave vectors are $\mathbf{q}_0 = (\pi/2, 0)$, $2\mathbf{k}_F = (\pi, \pi/2)$, $\mathbf{Q} = (\pi, 0)$ and (π, π).

cluster with 10 electrons. To make a direct comparison we also examined a small one-dimensional ring of 8 sites with 6 electrons. The energies of the lowest excited state with finite spectral weight are plotted as function of J ($t = 1$) for several wave vectors in Fig. 1. The similarity between 1D and 2D, is very striking. The SE for both systems behave as what is expected from an ideal Fermi gas. Depending on J/t, either $2\mathbf{k}_F$ or the smallest wave vector $\mathbf{q}_0$ has the lowest energy. On the other hand, the CE dispersion is quite anomalous. In 1D the lowest CE at very small J/t is at $4k_F$ in accordance with the theory of TLL, that charges are described as non-interacting spinless Fermions. Here we have seen that even in 2D the lowest CE is not at $2\mathbf{k}_F$ but at $\mathbf{Q} = (\pi, 0)$. If we treat all particles or holes as spinless Fermions in this 2D system, then $\mathbf{Q}$ is indeed the corresponding $2\mathbf{k}_F$ of the spinless Fermions, i.e. $\mathbf{Q} = 2\mathbf{k}_F^{SF}$. Another consistency with 1D at small J/t is that at $\mathbf{q}_0$ the CE energy is much larger than SE energy. Those energies in 1D have the physical meaning of charge and spin velocities in the TLL. The property of this excited state at $\mathbf{Q}$ is unlike other states that can be described as particle-hole excitations from the ground state. Details will be published elsewhere.

References

1. P.A. Lee, P. 96 in *High Temperature Superconductivity: Proceedings*, K.S. Bedell et al., (Addison Wesley, Reading, MA 1990).
2. P.W. Anderson, Phys. Rev. Lett. **64**, 1839 (1990).
3. T. Tohyama et al., Phys. Rev. Lett. **74**, 980 (1995).

EXACT DIAGONALIZATION STUDY OF THE SINGLE-HOLE t-J MODEL ON A 32-SITE LATTICE

P. W. Leung
Department of Physics, Hong Kong University of Science and Technology
Clear Water Bay, Hong Kong

R. J. Gooding
Department of Physics, Queen's University
Kingston, Ontario, Canada K7L 3N6

In this paper we present results from a numerical exact-diagonalization study of the well-known t-J model:[1]

$$\mathcal{H} = -t \sum_{\langle ij \rangle \sigma} (\tilde{c}_{i\sigma}^\dagger \tilde{c}_{j\sigma} + \text{H.c.}) + J \sum_{\langle ij \rangle} (\mathbf{S}_i \cdot \mathbf{S}_j - \frac{1}{4} n_i n_j), \qquad (1)$$

where $\langle ij \rangle$ denotes nearest-neighbor sites, and $\tilde{c}_{i\sigma}^\dagger$, $\tilde{c}_{j\sigma}$ are the no-double-occupancy constrained operators. Throughout this paper we use $t = 1$ and $J = 0.3$. It is well-known that this many fermion model is difficult to study numerically. Although exact results can be obtained by diagonalization, only small clusters can be dealt with even using modern supercomputers. In a recent paper[2] we reported the study of the single-hole t-J model on a 32-site square lattice — this is the largest system that has been studied using this approach. We presented results for the spectral function and the dispersion relation of the single-hole t-J model, and determined that the self-consistent Born approximation (SCBA) led to a superb analytical representation of the dispersion relation. In this paper we are mostly concerned with the ground state properties. The ground state is found to have momentum $(\pi/2, \pi/2)$, in agreement with various studies.[1] Even using all possible symmetry operations, we still have to work in a Hilbert space with 150 million basis states in calculating the ground state wavefunction.

If no hole is present, the t-J model is just the 2D Heisenberg model, which is known to possess long-range antiferromagnetic (AFM) order. An interesting question is how will a mobile hole affect the spin background around it? Semiclassical theory[3] predicts that the spin distortions introduced by a mobile hole have dipolar symmetry — this is supported by smaller size diagonalization[4] and SCBA[5] studies. The AFM correlations of nearest-neighbor spins can be displayed by calculating $\langle \mathbf{S}_i \cdot \mathbf{S}_j \rangle$, and Fig. 1(a) shows this quantity evaluated at the ground state momentum $\mathbf{k} = (\pi/2, \pi/2)$. In the Heisenberg AFM, $\langle \mathbf{S}_i \cdot \mathbf{S}_j \rangle_{\text{HAF}} = -0.34009$. We find that the deviations from $\langle \mathbf{S}_i \cdot \mathbf{S}_j \rangle_{\text{HAF}}$ are quite small near the edge of our cluster. A related quantity is $\langle \mathbf{S}_i \times \mathbf{S}_j \rangle$. This term is the backflow of the spin current in the semiclassical theory which causes the spin distortion. Fig. 1(b) shows the z component of $\langle \mathbf{S}_i \times \mathbf{S}_j \rangle$ evaluated at the ground state momentum $\mathbf{k} = (\pi/2, \pi/2)$. Note that the configuration displays dipolar symmetry.

Finally, we present some preliminary results on the inclusion of farther than nearest-neighbor hopping terms to the t-J model. In Ref. 2 we have shown that some aspects of the

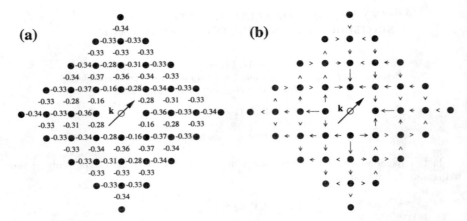

Figure 1: Nearest neighbor spin correlation functions in the 32-site t-J model at $\mathbf{k} = (\pi/2, \pi/2)$. The hole, denoted as an empty circle, is at the center of the lattice. In (a), the numbers are $\langle \mathbf{S}_i \cdot \mathbf{S}_j \rangle$. In (b), the length of the arrows are proportional to the magnitude of $\langle (\mathbf{S}_i \times \mathbf{S}_j) \cdot \hat{\mathbf{z}} \rangle$.

t-J model do not agree with experimental results on the CuO_2 plane. In particular, while the quasiparticle bandwidth is close to the experimental result, the dispersion relation is different except along the direction from $(0,0)$ to (π,π). It is known that a t' (next-nearest-neighbor hopping) is required to approximate the hole motion using a single-band model in a CuO_2 plane. However, t' reduces the bandwidth, making it less than the experimental result. Recently it has been suggested [6] that inclusion of the t'' (third-neighbor hopping) makes the dispersion relation closer to the experimental result. This is indeed found to be the case on our 32-site cluster. The bandwidths of the t-J, t-t'-J, and t-t'-t''-J models at $t' = -0.2$, and $t'' = 0.2$, are found to be $2.00J$, $1.55J$, and $2.24J$ respectively, while the experimental result is roughly $2.2J$. Hence, the t-t'-t''-J model is the most promising candidate among the three to explain the experimental data. Initial comparison of its spectral function with experiment has produced encouraging results.

This work was supported by Hong Kong RGC grant HKUST619/95P (PWL) and the NSERC of Canada (RJG). Numerical diagonalizations of the 32-site system were performed on the Intel Paragon at HKUST.

References

1. For a recent review, see E. Dagotto, *Rev. Mod. Phys.* **66**, 763 (1994).
2. P. W. Leung and R. J. Gooding, *Phys. Rev.* B **52**, 15711 (1995).
3. B. I. Shraiman and E. D. Siggia, *Phys. Rev. Lett.* **60**, 740 (1988).
4. V. Elser, D. A. Huse, B. I. Shraiman, and E. D. Siggia, *Phys. Rev.* B **41**, 6715 (1990).
5. A. Ramšak and P. Horsch, *Phys. Rev.* B **48**, 10559 (1993).
6. B. Kyung and R. Ferrell (preprint).

ENTROPY OF 2D STRONGLY CORRELATED ELECTRONS

W. O. PUTIKKA

Physics Department, University of Cincinnati, Cincinnati, OH 45221-0011

Abstract

I discuss the entropy of the 2D t-J model calculated by high temperature series for parameters corresponding to optimally doped high temperature superconductors. The entropy is found to saturate at a high temperature value of $S \sim k_B$ at $n = 0.8$ and to fall more slowly than for the non-interacting tight binding model as the temperature is reduced. At low temperatures the t-J entropy is much larger than that for the tight binding model, indicating a large number of low energy excitations.

1 Introduction

The low energy behavior of 2D strongly correlated electrons is of great interest for understanding the properties of high temperature superconductors. Below I present results for the entropy S_{tJ} of the 2D t-J model calculated by high temperature series. The data for S_{tJ}, varying the ratio of the coupling constants J/t at a fixed electron density $n = 0.8$, show a strong increase in the low temperature value [1] of S_{tJ} as J/t is reduced. This is consistent with the number of low energy states increasing as J/t becomes smaller. Together with the observed behavior of the magnetic spin susceptibility $\chi(\mathbf{q})$ this supports the hypothesis that the t-J model near half filling and $J/t \sim 1/3$ is magnetically frustrated [2].

2 Entropy

The series for the entropy is calculated by $S = -\partial F/\partial T$ from the series for the free energy [3]. The high temperature limit of S_{tJ} per site is given by $S_\infty = -n \ln n - (1-n)\ln(1-n) + n \ln 2$. For $n = 0.8$ we have $S_\infty = 1.05 k_B$, a value in reasonable agreement with the estimate $S \sim k_B$ for the high temperature entropy per Cu site of $YBa_2Cu_3O_{6.97}$ by Loram et al. [4].

The entropy of the 2D t-J model as a function of temperature for $n = 0.8$ and a range of J/t values is shown in Fig. 1, along with the entropy of the non-interacting tight binding (TB) model S_{TB} at the same density. We see that while S_{tJ} is less than S_{TB} at high temperatures (due to the removal of doubly occupied sites in the t-J model), S_{tJ} decreases much more slowly as T is reduced. This leads to S_{tJ} being much larger than S_{TB} at low temperatures [1]. The temperature scale where S_{tJ} starts to decrease is approximately $T \sim J$ consistent with the experiments on $YBa_2Cu_3O_{6.97}$, but S_{tJ} is also smoothly varying and does not have any sharp changes.

If we assume that the 2D t-J model does not have a very small energy scale and that the low temperature behavior of S_{tJ} is linear we can make an estimate of γ from $S_{tJ} = \gamma T$ at low temperatures. For the data shown in Fig. 1 this estimate gives $\gamma \sim a/J$, where $a \approx 1.4$. Having $\gamma \propto 1/J$ suggests that the low energy excitations are dominated by the spin degrees of freedom. This is also consistent with $\sim J$ being the temperature scale for the decrease in S_{tJ}.

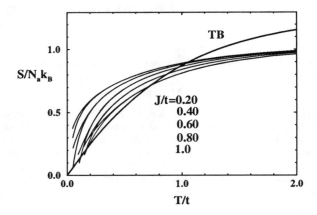

Figure 1: Entropy per sit of the 2D t-J model as a function of temperature for $n = 0.8$ and various values of J/t. Shown are two integral approximant extrapolations of the series at each value of J/t. Also shown as the heavy line is the entropy of the non-interacting tight binding model at $n = 0.8$.

3 Discussion

The entropy of the 2D t-J model gives a picture of the low energy excitations of 2D strongly correlated electrons being dominated by the spin degrees of freedom. Also, as J/t is decreased, the number of low energy states is increasing. For the same parameter range the magnetic spin susceptibility is observed to be enhanced over the non-interacting case, but also to be relatively flat as a function of wavevector in the Brillouin zone.[2] This is also consistent with recent neutron scattering measurements which show the antiferromagnetic correlation length to be about one lattice spacing for $La_{1.86}Sr_{0.14}CuO_4$.[5] As J/t is decreased the nature of the low temperature magnetic behavior changes from AF to FM. All of these observations are consistent with the spin degrees of freedom dominating the low energy excitations near optimal doping and being frustrated between FM and AF tendencies.

Acknowledgements

This work was supported by NSF grant DMR-9357199.

1. J. Jaklič and P. Prelovšek, *Phys. Rev. Lett.* **75**, 1340 (1995).
2. W. O. Putikka, *J. Phys. Chem Solids* **56**, 1747 (1995).
3. W. O. Putikka, M. U. Luchini and T. M. Rice, *Phys. Rev. Lett.* **68**, 538 (1992).
4. J. W. Loram, K. A. Mirza, J. R. Cooper and W. Y. Liang, *Phys. Rev. Lett.* **71**, 1740 (1993).
5. S. M. Hayden, *et al.*, *Phys. Rev. Lett.* **76**, 1344 (1996).

SPIN DIFFUSION AS A TEST FOR SPIN-CHARGE SEPARATION

Qimiao Si

Department of Physics, Rice University, Houston, TX 77251-1892, USA

We propose that the temperature dependence of the electron spin diffusion constant, when combined with the T−linear resistivity, can be used to test spin-charge separation in the metallic cuprates.

One of the most basic questions in the normal state of the high T_c cuprates is the coupling, or lack thereof, between the underlying spin and charge excitations. Consider the well known T−linear electrical resistivity. For electron-electron scatterings to give rise to such a temperature dependence, instead of the Fermi liquid T^2 behavior, the usual restriction of the scattering phase space imposed by the Pauli principle has to be violated. One possibility involves a minimal modification to a Fermi liquid, via low energy scales. Here, the current carriers are still quasiparticles; the scattering rate is enhanced due to the scattering of these quasiparticles from soft spin and/or charge fluctuations. A more dramatic possibility is that, the elementary excitations do not resemble quasiparticles at all; instead, spin and charge excitations are separated.

Existent experimental data do indicate contrasting behaviors between spin and charge dynamics, as can be seen from comparing the temperature dependences of the NMR $(\frac{1}{T_1})_{Cu}$ and the electrical resistivity (ρ)[1] plotted in Fig. 1. While striking, this contrast unfortunately involves two probes of quite different nature: one measures the local (i.e. **q**−averaged) dynamical spin susceptibility, the other the **q**=0 charge transport.

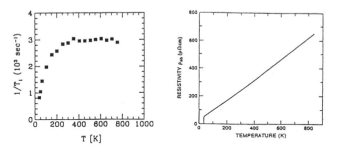

Figure 1: The NMR relaxation rate and electrical resistivity in $La_{1.85}Sr_{0.15}CuO_4$ (from Refs. 1).

Recently, we proposed that the temperature dependence of the yet to be measured electron spin diffusion can be used as a test for spin-charge separation in the normal state of the optimally doped cuprates. In analogy to the electrical conductivity, the spin conductivity σ_s is the transport coefficient relating spin current and magnetic field gradient: $j_s = \sigma_s \nabla(-H)$. The Einstein relation states that, $\sigma_s = \chi_s \times D_s$, where χ_s is the spin susceptibility and D_s the spin diffusion constant. Hence, given χ_s, measuring the spin diffusion constant is equivalent to measuring σ_s, or equivalently, the "spin resistivity", $\rho_{\text{spin}} \equiv \frac{1}{\sigma_s} = \frac{1}{\chi_s D_s}$. It is possible to measure the spin diffusion constant via the spin-injection[2] and related techniques.

We consider three examples representative of different kinds of couplings between the underlying spin and charge excitations: 1) Fermi liquids with low energy scales, describing quasiparticles coupled to soft fluctuations – no spin-charge separation; 2) Luttinger liquid in 1D – a complete separation of spin and charge excitations; 3) gauge theory of the 2D t-J model describing "holons" and "spinons" coupled by a massless gauge field – an intermediate situation. The electrical resistivities have been studied by various groups for (1), by Giamarchi for (2), and by Nagaosa and Lee and by Ioffe et al. for (3), respectively.

<u>Fermi Liquids with low energy scales:</u> Within a general effective model describing quasiparticles coupled to soft spin and/or charge fluctuations, we found that, $\rho \sim T \iff \rho_{\text{spin}} \sim T$, provided the following three conditions are met: a) no significant ferromagnetic spin fluctuations; b) a large Fermi surface for the quasiparticles, so that enough electron-electron scatterings are Umklapp in nature; c) direct transport by the soft spin/charge fluctuations negligible compared to the single particle transport.

The condition (a) is well satisfied in the cuprates. (b) is satisfied at least for the optimally doped cuprates. Finally, for the optimally doped cuprates, it is reasonable to assume the validity of (c) given the simple thermodynamic behavior and the short spin correlation length (as seen in the neutron scattering experiments). The situation with underdoped cuprates is more complex, as (b), and to a less extent, (c), might be violated.

<u>Luttinger Liquid:</u> The electric current is carried entirely by the charge excitations. It is dissipated through the Umklapp interaction (g_3), leading to $\rho \sim \rho^o (N_o g_3)^2 (N_o T)^{4K_c - 3}$, for $T \gg \Delta^* \sim \delta/N_o$. Here, K_c is the characteristic charge coupling constant, N_o is the density of states, and δ measures the deviation from the half-filling. The spin current, on the other hand, is carried entirely by the spin excitations. It is dissipated via the backscattering interaction (g_1): $\rho_{\text{spin}} \sim \rho^o_{\text{spin}} (N_o g_1)^2 (N_o T)^{4K_\sigma - 3}$, for arbitrary temperatures. Here, K_σ is the characteristic spin coupling constant.

<u>Gauge Theory of the t-J Model:</u> The electrical resistivity has contributions from both "holons" and "spinons", as a result of backflow induced by the gauge coupling: $\rho = \rho_{\text{holon}} + \rho_{\text{spinon}} \sim \rho^o \frac{1}{\delta} \frac{T}{J}$. The spin resistivity has contributions from "spinons" only: $\rho_{\text{spin}} \sim (\rho_{\text{spin}})_{\text{spinon}} \sim \rho^o_{\text{spin}} (\frac{T}{J})^{\frac{4}{3}}$.

To summarize, if the T–linear resistivity of the optimally doped cuprates originates from a Fermi-liquid-like state with low energy scales, ρ_{spin} should also be T–linear in the same temperature range. On the other hand, if the T–linear resistivity occurs as a result of spin-charge separation, ρ_{spin} is in general not T–linear. An experimental measurement of the electron spin diffusion constant can hence provide a test for spin-charge separation.

The author is an A. P. Sloan Research Fellow.

References

1. T. Imai et al., Phys. Rev. Lett.70, 1002 (1993); B. Batlogg et al., *Electronic Properties of High-T_c Superconductors*, eds. H. Kuzmany et al. (Springer-Verlag, 1993).
2. N. Hass et al., Physica C235-240, 1905 (1994) and references therein.

c-AXIS ELECTRONIC STRUCTURE OF COPPER OXIDES

J. M. WHEATLEY and J. R. COOPER
*IRC in Superconductivity, Cambridge University,
Cambridge CB3 OHE, U.K.*

A 3D extended t-J model for Copper-Oxides is introduced. It is shown that the strong $k_\parallel$ dependence of $t_\perp(k_\parallel)$ found in bandstructure calculations is also a feature of the t-J model. Possible experimental evidence for this effect, and it's crucial importance for normal and superconducting state c-axis physics is described.

1 Three Dimensional t-J model

LDA bandstructure calculations reveal a strongly $k_\parallel$ dependent c-axis hopping matrix element, which nearly vanishes along $(0,0)$-(π,π) (Γ-M) for flat planes in both simple tetragonal and body-centered-tetragonal Copper-Oxides. The full 3D electronic structure is well described by nearest neighbor overlap of Cu $d_{x^2-y^2}$ and Cu "4s" with O $p_{x(y)}$ orbitals[1]. Bonding combinations of p orbitals, given by Wannier states $\alpha_k = \frac{2}{\omega_k}(p_x \cos\frac{k_x}{2} + p_y \cos\frac{k_y}{2})$, have matrix elements $-t_{pd}\omega_k$ and $-t_{ps}\mu_k$ for α-d and α-s hopping respectively, where $\omega_k = 2\sqrt{\cos^2(\frac{k_x}{2}) + \cos^2\frac{k_y}{2}}$ and $\mu_k = \frac{4}{\omega_k}(\cos k_x - \cos k_y)$. Elimination of high-lying 4s states leads to effective O-O hopping which therefore contains a factor μ_k^2, which vanishes along Γ-M. This effect produces hole-like Fermi surfaces in the ab-plane where it acts in addition to α-d hybridisation. However, it provides the *dominant* mechanism for c-axis band formation, which is via s-s and α-s. If the Copper site interaction U is neglected, O orbitals can also be eliminated to give a single band model, Fig. (a). In reality, perturbative elimination of O orbitals is invalid as $U > \Delta = \epsilon_p - \epsilon_d$; the correct approach [2] is to solve the correlation problem within the unit cell, then derive degeneracy lifting intercell terms. Upon hole doping the resulting charge-transfer insulating state at $\frac{1}{2}$-filling, electrons (doublets, denoted $|1\sigma\rangle$) are mobile with hopping integrals $t_{ij} = \sum_k e^{ikR_{ij}} \epsilon_k$ where,

$$\epsilon_k \simeq -2t\omega_k - t'\mu_k^2 - 2t_\perp \cos(k_z)\mu_k^2.$$

The renormalized hopping integrals are derived from 4-band model parameters and U dependent *intra-cell* matrix elements $\langle 2|d(\alpha)^\dagger_{-\sigma}|1\sigma\rangle$ where $|2\rangle$ denotes a Zhang-Rice singlet. These are $t = t_{pd}\langle 2|d^\dagger_{-\sigma}|1\sigma\rangle\langle 2|\alpha^\dagger_{-\sigma}|1\sigma\rangle$, $t' = t_{ps}^2 \epsilon_s^{-1} \langle 2|\alpha^\dagger_{-\sigma}|1\sigma\rangle^2$ and $t^\perp = t_{ss}\epsilon_s^{-1}t'$.

2 Implications for c-axis physics

At sufficiently high doping levels, ϵ_k contours may approximate t-J model Fermi surfaces. When this is so, Fig. (c) indicates only modest changes in the resistivity anisotropy $\frac{\rho_c}{\rho_{ab}}$ compared to bandstructure values. However, underdoped Copper-Oxides show unexpectedly large $\frac{\rho_c}{\rho_{ab}}$. A possible explanation may be the inability of coherent spectral weight known to be concentrated near $(\frac{\pi}{2}, \frac{\pi}{2})$ in the lightly doped t-J model to contribute to c-axis transport due to small $t_\perp(\frac{\pi}{2}, \frac{\pi}{2})$. There are further implications in a d-wave superconducting state where gap nodes lie on Γ-M. While nodal lines give an ab-plane penetration depth $\lambda_{ab}(T) \sim T$, there is a much weaker T dependence of $\lambda_c(T)^3$. Recently, c-axis magneto-transport has been used to probe ab-plane electronic anisotropy in overdoped $Tl2201^4$. Zener-Jones expansion of Bloch-Boltzmann theory to order B^4 gives, $\sigma^{zz}(B, \phi) \sim \sigma_2 B^2 + \sigma_4(1 + \lambda \cos(4\phi))B^4 + O(B^6)$. Assuming a circularly symmetric ab-plane Fermi surface and $t_\perp(\theta) = t_\perp^0 \cos^2(2\theta)$ we find $\lambda = \frac{1}{6}$, which is of the same order as the observed values[4].

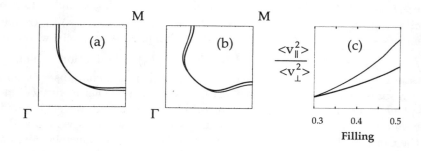

Figure 1: (a) $U = 0$ bandstructure Fermi surface for hole doping 0.1 at $k_z = 0$ and $k_z = \pi$ (thin line). (b) Corresponding contours of constant ϵ_k. (c) Mean square "velocity" ratios $\langle v_{\parallel k}^2 \rangle_{FS} / \langle v_{\perp k}^2 \rangle_{FS}$ for ϵ_k and for bandstructure (thin line) as a function of filling. Parameters (in eV): $\Delta = 3.5$, $t_{pd} = 1.6$, $\epsilon_s - \epsilon_p = 8$, $t_{ps} = 2.2$, $t_{ss} = 0.3$, $U = 9$.

References

1. O. K. Andersen et al., Phys. Rev. B **49**, 4145 (1995).
2. D. C. Mattis, Phys. Rev. Lett. **74**, 3676 (1995).
3. J. M. Wheatley and T. Xiang, to be published.
4. N. E. Hussey et al., Phys. Rev. Lett. **76**, 122 (1996); these proceedings.

III. THEORY

Order Parameter Symmetry

Impurity scattering and localization in d-wave superconductors

M. Franz, C. Kallin and A. J. Berlinsky

*Department of Physics and Astronomy and Brockhouse Institute for Materials Research,
McMaster University, Hamilton, Ontario, L8S 4M1 Canada*

We present strong evidence for the localization of low energy quasiparticle states in disordered d-wave superconductors. Within the framework of the Bogoliubov-de Gennes (BdG) theory applied to the extended Hubbard model with a finite concentration of non-magnetic impurities we carry out a fully self-consistent numerical diagonalization of the BdG equations on finite clusters containing up to 50 × 50 sites. We identify localized states by probing their sensitivity to the boundary conditions and by analyzing the finite size dependence of inverse participation ratios.

In conventional s-wave superconductors nonmagnetic impurities have little effect on the transition temperature, which may be understood from Anderson's theorem. [1] The situation is dramatically different in the high temperature cuprate superconductors where nonmagnetic impurities exhibit a strong pair breaking effect. This is interpreted as a manifestation of unconventional, most likely d-wave pairing symmetry. Understanding the role that impurities play in such superconductors is crucial for the interpretation of experimental data. A good example of this is the interpretation of experiments measuring the temperature dependent penatration depth [2,3] in YBa$_2$Cu$_3$O$_{7-x}$. The behavior of various quantities in the presence of disorder also serves as an important test for theoretical models of microscopic pairing interaction.

In the present paper we concentrate on the possibility of localization of the low-energy quasiparticle states in d-wave superconductors with line nodes in the gap on the fermi surface. We wish to address the question of whether these low lying states are strongly localized with a relatively short localization length ξ_L or whether they are essentially extended. This is a question of significant importance for many experiments since the transport and thermodynamic properties are largely determined by these low energy excitations.

The problem of localization in d-wave superconductors was first considered by Lee [4] who, appealing to arguments involving the scaling theory of localization, found that in the limit of unitary (i.e. strong) scatterers the quasiparticle states are strongly localized below the mobility gap γ_0, even if the impurity concentration is sufficiently small so that the normal state wavefunctions are essentially extended. For moderate concentrations of strong scatterers, γ_0 is a reasonable fraction of the maximum gap which allows for the possibility of experimental confirmation. One consequence of such a scenario is the prediction of the universal limit of conductivity $\sigma(\omega \to 0) \simeq (e^2/2\pi\hbar)\xi_0/a$, independent of the scattering rate τ^{-1} (ξ_0 is a coherence length and a is the lattice constant).

More recently Balatsky and Salkola [5] presented results that contradict this scenario. Their argument is based on a physical picture of a single impurity wavefunction which, according to an earlier calculation within the self-consistent T-matrix approximation,[6] is highly anisotropic with slowly decaying ($\sim 1/r$) tails along the [11] diagonals. Overlaps between these tails lead to strong interactions between the quasiparticle states on distant impurities, which then form a network of extended impurity states capable of carrying current. This network percolates across the entire system and inhibits the localization by disorder at the lowest energies.

Both theories described above resort to approximations when treating the scattering from impurities and they ignore the gap relaxation near the impurity sites resulting from their pair breaking property. Since the arguments for and against the localization are quite subtle it may well be that the above mentioned details are important, in which case perhaps the best way to address the problem is through numerical calculations. Hatsugai and Lee [7] studied localization numerically in a simpler but related model of Dirac fermions on the lattice, whose excitation spectrum is similar to that of a d-wave superconductor. By examining the sensitivity of the wavefunction to boundary conditions they concluded that the low energy states are indeed strongly localized. This would seem to be in qualitative agreement with the theory of Lee,[4] however since the system possessed no off-diagonal long range order, the quantitative predictions of this theory could not be verified. While this work was strongly suggestive, without treating a system with superconducting order one cannot establish the existence of localization in a d-wave superconductor. A convenient framework for studying the effect of impurities within a simple lattice model of d-wave superconductivity was outlined by Xiang and Wheatley,[8] however they did not address the question of localization.

In the present paper we solve a simple tight binding model with an on-site repulsion and nearest neighbor attraction which give rise to superconductivity in the d-wave channel, in the presence of point disorder, on finite clusters using the selfconsistent Bogoliubov-de Gennes (BdG) technique. The main advantage of such a model is that the impurity scattering is treated *exactly* and that it is *directly* relevant to d-wave superconductors. Our principal result is that quasiparticle states are localized at low energies below a mobility gap γ_0. The value of γ_0 as well as the localization length ξ_L are in a good agreement with the theory of Lee. [4]

The Hamiltonian we consider has been used previously to study the vortex structure [9,10] and impurities [8] in a d-wave superconductor:

$$H = -t \sum_{\langle ij \rangle \sigma} c_{i\sigma}^\dagger c_{j\sigma} - \mu \sum_{i\sigma} n_{i\sigma} + V_0 \sum_i n_{i\uparrow} n_{i\downarrow} + \frac{V_1}{2} \sum_{\langle ij \rangle \sigma \sigma'} n_{i\sigma} n_{j\sigma'} + \sum_{i\sigma} V_i^{imp} n_{i\sigma}. \quad (1)$$

Here $\langle ij \rangle$ stands for nearest neighbor pairs, and the notation is otherwise standard. The last term models strongly repulsive impurities: $V_i^{imp} = V^{imp} > 0$ at randomly chosen sites with density n_{imp} and $V_i^{imp} = 0$ on all other sites. If one takes $V_0 > 0$ and $V_1 < 0$ this model gives rise to pairing in the d-wave channel. In mean field theory (1) can be solved by defining the pairing amplitudes

$$\Delta_0(\mathbf{r}_i) = V_0 \langle c_{i\uparrow} c_{i\downarrow} \rangle, \quad \Delta_\delta(\mathbf{r}_i) = V_1 \langle c_{i+\delta \uparrow} c_{i\downarrow} \rangle, \quad (2)$$

where $\delta = \pm \hat{x}, \pm \hat{y}$ are nearest neighbor vectors for a square lattice. The resulting mean field Hamiltonian is then diagonalized using the Bogoliubov transformation[11]

$$c_{i\uparrow} = \sum_n [\gamma_{n\uparrow} u_n(\mathbf{r}_i) - \gamma_{n\downarrow}^\dagger v_n^*(\mathbf{r}_i)], \quad c_{i\downarrow} = \sum_n [\gamma_{n\downarrow} u_n(\mathbf{r}_i) + \gamma_{n\uparrow}^\dagger v_n^*(\mathbf{r}_i)].$$

to the quasiparticle operators $\gamma_{n\sigma}$. Within mean field theory the Hamitonian (1) is diagonalized when u_n and v_n satisfy the BdG equations [11,9]

$$-t \sum_\delta u_n(\mathbf{r}_i + \delta) + (V_i^{imp} - \mu) u_n(\mathbf{r}_i) + \Delta_0(\mathbf{r}_i) v_n(\mathbf{r}_i) + \sum_\delta \Delta_\delta(\mathbf{r}_i) v_n(\mathbf{r}_i + \delta) = E_n u_n(\mathbf{r}_i),$$

$$t \sum_\delta v_n(\mathbf{r}_i + \delta) - (V_i^{imp} - \mu) v_n(\mathbf{r}_i) + \Delta_0^*(\mathbf{r}_i) u_n(\mathbf{r}_i) + \sum_\delta \Delta_\delta^*(\mathbf{r}_i) u_n(\mathbf{r}_i + \delta) = E_n v_n(\mathbf{r}_i), \quad (3)$$

for all $\mathbf{r}_i$, subject to the constraints of self-consistency

$$\Delta_0(\mathbf{r}) = V_0 \sum_n u_n(\mathbf{r}) v_n^*(\mathbf{r}) \tanh(E_n / 2k_B T),$$

$$\Delta_\delta(\mathbf{r}) = \frac{V_1}{2} \sum_n [u_n(\mathbf{r} + \delta) v_n^*(\mathbf{r}) + u_n(\mathbf{r}) v_n^*(\mathbf{r} + \delta)] \tanh(E_n / 2k_B T), \quad (4)$$

where the summation is over positive eigenvalues E_n only.

For a system of linear size L, solving the system of equations (3) requires diagonalizing a $2L^2 \times 2L^2$ matrix. For a suitably chosen initial order parameter distribution we solve the system (3) using the standard LAPACK diagonalization routine. We than compute new gap functions from Eqs. (4) and iterate until the desired convergence is achieved. Typically 7-8 iterations are necessary to establish five-digit convergence in the free energy and the average gap. Periodically we perform longer runs of 25-30 iterations to confirm the accuracy of the solutions. No significant deviations are found between short and longer runs.

We probe for localization in two distinct ways. First, for a given system size and disorder configuration we evalute a generalized inverse participation ratio,[1,2]

$$a_n = \frac{\langle |u_n|^4 \rangle + \langle |v_n|^4 \rangle}{(\langle |u_n|^2 \rangle + \langle |v_n|^2 \rangle)^2}, \qquad (5)$$

where $\langle \ldots \rangle$ stands for the spatial average. As a function of increasing system size L this quantity decreases as $\sim 1/L^2$ for extended states and approaches a constant value $\sim (a/\xi_L)^2$ for a state localized within the characteristic length ξ_L. Alternatively one can study the sensitivity of states to boundary conditions by applying a uniform phase twist χ across the length of the system. We consider wavefunctions with the "twisted" boundary condition along the x-direction, $\Psi(x+L, y) = e^{i\chi}\Psi(x,y)$, and periodic along y, and we compute the stiffness[1,2,7] of the n-th eigenvalue

$$\kappa_n = \frac{1}{2}\left(\frac{\partial^2 E_n}{\partial \chi^2}\right)_{\chi=0}. \qquad (6)$$

Extended states will be sensitive to twist, and thus κ_n is expected to be large. Localized states with $\xi_L \ll L$ will be insensitive to twist and κ_n is expected to be small.

We have solved the BdG equations (3) for the following set of model parameters: $\mu = -t$, corresponding to the band filling factor $n \simeq 0.68$ ($n = 1$ is a half-filled band); $V_0 = 3t$, and $V_1 = -4.5t$. In the absence of disorder Eqs. (3) are easily solved analytically by appealing to translational invariance. One obtains the usual BCS type excitation spectrum $E_{\mathbf{k}} = (\epsilon_{\mathbf{k}}^2 + |\Delta_{\mathbf{k}}|^2)^{1/2}$ with $\epsilon_{\mathbf{k}} = -2t(\cos k_x + \cos k_y) - \mu$ and $\Delta_{\mathbf{k}} = 2\Delta_d(\cos k_x - \cos k_y)$. For the above parameters we obtain $\Delta_d = 0.67t$. With respect to real materials this value is exaggerated, however our choice of parameters was motivated by the desire to have the large mobility gap γ_0 (which according to Lee[4] scales with Δ_d), necessary to study localization numerically.

We have studied systems with $n_i = 0.015$ and $n_i = 0.06$ of strong repulsive impurities with $V^{imp} = 100t$. In Fig. 1a we display typical results for the inverse participation ratio a_n and stiffness κ_n as a function of energy for $n_i = 0.06$. There is a pronounced qualitative difference in a_n between the low energy states below $E/t \simeq 1$ and the high energy states. This is strongly suggestive of a mobility gap with a magnitude close to that predicted by Lee,[4] $\gamma_0 \approx 0.84t$ for the parameters listed above. While not so pronounced, this mobility gap can also be observed in the behavior of κ_n. For lower density of impurities, $n_i = 0.015$, we obtain similar results with the apparent mobility gap reduced by roughly a factor of two. This is in agreement with Lee's prediction that $\gamma_0 \sim n_i^{1/2}$.

Even stronger evidence for the existence of the mobility gap can be obtained by analyzing the system size dependence of a_n. To this end we have carried out calculations for $L = 20, 24, 28, 32, 36, 40, 50$ and six independent impurity configurations for each size. In Fig. 1b we display two quantities, $a^< \equiv \langle\langle a_n \rangle\rangle_{E<\gamma_0}$ and $a^> \equiv \langle\langle a_n \rangle\rangle_{E>\gamma_0}$, defined as averages of the inverse participation ratios a_n taken over the indicated range of energy, as a function of $1/L^2$. Data points for $a^>$ lie on a straight line which extrapolates to zero for $L \to \infty$, just as one would expect for the extended states. Values of $a^<$ appear to extrapolate to a finite value of $a^<_\infty \approx 0.003$, which implies that these states are localized with $\xi_L/a \approx 20$. We note that this value of ξ_L is in a reasonable agreement with the rough estimate given in Ref. [4], which gives $\xi_L/a \approx 38$.

We have inspected visually the amplitudes of wavefunctions $u_n(\mathbf{r})$ and $v_n(\mathbf{r})$ for signs of localization. We have found that, in agreement with the above analysis, these states are spatially localized

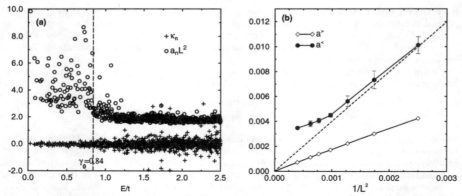

Figure 1: a) Inverse participation ratio a_n and stiffness κ_n plotted as a function of energy for $L = 40$ and $n_i = 0.06$. Dashed line marks the mobility gap $\gamma_0 = 0.84$ estimated from the theory of Lee.[4] b) Finite size scaling of the inverse participation ratio averaged over the energies below ($a^<$) and above ($a^>$) the mobility gap γ_0. The error bars reflect the scatter of the data from 6 different impurity configurations. Solid and dashed lines are guides to the eye only.

below the mobility gap with the characteristic length scale of about $20 - 30a$, and appear to be extended at higher energies (for lack of space we do not present these plots here).

To conclude, we have presented clear evidence for strong localization by impurities in d-wave superconductors at low energies. Compared to earlier analytical work[4,5] our approach treats impurity scattering exactly within BdG theory and accounts for order parameter relaxation near the impurity sites. Another advantage of the present model is that it can be easily extended to finite temperatures and frequencies, thus opening possibilities for the analysis of the temperature dependent penetration depth, infrared conductivity and other physically interesting quantities in the presence of disorder.

The authors are grateful to M. I. Salkola for numerous helpful suggestions, and to A. V. Balatsky, V. J. Emery and R. J. Gooding for valuable discussions. This work has been partially supported by the Natural Sciences and Engineering Research Council of Canada and by the Ontario Centre for Materials Research.

1. P. W. Anderson, *Phys. Rev. Lett.* **3**, 325 (1959).
2. A. T. Fiory et al., *Phys. Rev. Lett.* **61**, 1419 (1988); J. F. Annett, N. D. Goldenfeld and S. R. Renn, *Phys. Rev. B* **43**, 2778 (1991); W. N. Hardy et al., *Phys. Rev. Lett.* **70**, 3939 (1993).
3. P. J. Hirschfeld and N. D. Goldenfeld, *Phys. Rev. B* **48**, 4219 (1993).
4. P. A. Lee, *Phys. Rev. Lett.* **71**, 1887 (1993).
5. A. V. Balatsky and M. I. Salkola, *Phys. Rev. Lett.* , (in press, 1996).
6. A. V. Balatsky, M. I. Salkola and A. Rosengren, *Phys. Rev. B* **51**, 15547 (1995).
7. Y. Hatsugai and P. A. Lee, *Phys. Rev. B* **48**, 4204 (1993).
8. T. Xiang and J. M. Wheatley, *Phys. Rev. B* **51**, 11721 (1995).
9. P. I. Soininen, C. Kallin and A. J. Berlinsky, *Phys. Rev. B* **50**, 13883 (1994).
10. Y. Wang and A. H. MacDonald, *Phys. Rev. B* **52**, R3876 (1995).
11. P. G. de Gennes, *Superconductivity of Metals and Alloys*, (Addison-Wesley, Readings, MA, 1989).
12. D. J. Thouless, *Phys. Rep.* **13**, 93 (1974).

ASPECTS OF D-WAVE SUPERCONDUCTIVITY IN HIGH T_c CUPRATES

K. MAKI, Y. SUN
Department of Physics and Astronomy, University of Southern California,
Los Angeles, California 90089-0484, USA

H. WON
Department of Physics, Hallym University,
Chuchon 200-702, South Korea

We review here our recent works on impurity scattering and on vortex state of d-wave superconductivity. The resonance scattering on the Fermi surface due to impurity has a number of consequences on the low temperature properties of superconductivity. In the vortex state fourfold symmetry inherent to d-wave superconductivity is revealed in a surprising clarity, which is readily accessible experimentally.

1 Introduction

Now the evidences supporting d-wave superconductivity in high T_c cuprates are simply overwhelming[1] with possible exception of NCCO.[2] In the past few years we have been trying to establish the validity of d-wave model by confronting the theoretical results with existing experimental observations. Our first success[3] in interpreting the inelastic neutron scattering data from monocrystals of LSCO[4] and YBCO[5] in terms of d-wave superconductivity convinced us the correctness of our approach. In the present review we discuss two aspects of d-wave superconductivity.[6]

2 Impurity Scattering

It is now well documented that substitution of Zn in the Cu-O_2 plane of high T_c cuprates has dramatic effects; a rapid suppression of superconducting transition temperature T_c and a rapid rise of the residual density of states $N(0)$ (i.e. the electronic density of states on the Fermi surface). The former behavior is the same as s-wave superconductor with magnetic impurities[7] while the latter is completely different.[8,9,10,11] In Fig. 1 we show T_c/T_{c0} versus $N(0)/N_0$ and $\rho_n(\Gamma, 0) = 1 - \rho_s(\Gamma, 0)$ together with data points obtained from low temperature specific heat data of Zn-substituted LSCO by Momono and Ido.[12] In particular the agreement between the theory and the experiment is spectacular for $x = 0.22$ (over doped region), while for $x = 0.10$ and 0.18 the data points are clearly inside the theoretical curve obtained for impurities in the unitarity limit[11] (i.e. the resonance on the Fermi surface). Note that $N(0)/N_0$ is accessible both from the Knight shift at $T = 0$ K and low temperature specific heat. Therefore the NMR data from Zn-substituted YBCO and LSCO[13] can be presented in the same figure. In Fig. 1 $\rho_s(\Gamma, 0)$ is the superfluid density at $T = 0$ K, which can be accessed by muon spin rotation for example. The T_c/T_{c0} dependencies of $N(0)/N_0$ and $\rho_n(\Gamma, 0)$ are very similar but not exactly the same.[11] In the same model we show that the electronic thermal conductivity is linear in T at low temperature ($T \leq 0.1T_c$) and the Wiedemann-Franz law is established between the real part of the electric conductivity. Also the thermal conductivity increases rapidly with impurity scattering due to rapid generation of quasi-particles.

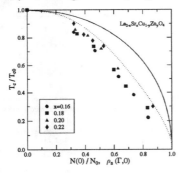

Fig. 1: T_c/T_{c0} vs $N(0)/N_0$ (dashed line) and $\rho_n(\Gamma,0)$ ($\equiv 1-\rho_s(\Gamma,0)$) (solid line) are shown here with $N(0)/N_0$ data obtained from the specific heat experiment [12] of $La_{2-x}Sr_xCu_{1-y}Zn_yO_4$. Also $\rho_n(\Gamma,0)$ is very similar to $N(0)/N_0 = \rho_{spin,n}(\Gamma,0)$ but not the same [11].

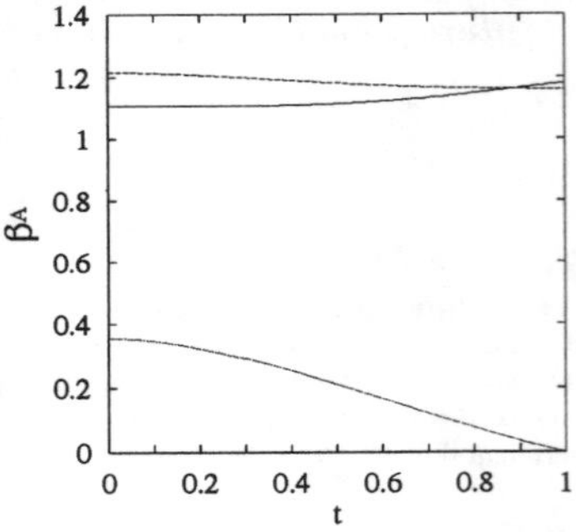

Fig. 2: The Abrikosov parameter $\beta_A^\square(\frac{\pi}{4})$ (solid line), $\Delta\beta_A(t)$ (dotted line) and $\beta_A^\triangle$ (dashed line) are shown as function of t.

3 Vortex State

The fourfold symmetry inherent to d-wave superconductivity is most directly seen in the vortex state.

3.1 B ∥ c

When a magnetic field normal to the a-b plane is applied we found that the quasi-particle spectrum around a vortex exhibits a clear fourfold symmetry [14,15], which has not been seen in a recent STM experiment from a YBCO monocrystal.[16] Recently we have succeeded [17] in implementing the quantization of the angular momentum with a discrete set of the bound states.[18] Then identifying $E_{1/2} \simeq \Delta/4$ (which implies [3,19] $\xi p_F \simeq 4/\pi$), we can describe the Geneva data.[16] In high T_c cuprates the residual density of states appears to arise entirely outside of vortex cores in sharp contrast to the one in ordinary s-wave superconductors.

Perhaps the most striking result is that a square vortex lattice tilted by 45° from the a axis is the most stable for $T \leq 0.88T_c$ in contrast to the usual triangular lattice common to the s-wave superconductor.[20] Indeed the Abrikosov parameter for a square lattice tilted by an angle θ from the a axis is given by

$$\beta_A^\square(\theta,t) = \beta_A^\square\left(\frac{\pi}{4},t\right) + \Delta\beta_A(t)\cos^2(2\theta) \qquad (1)$$

In Fig. 2 we show the temperature dependence of $\beta_A^\square(\pi/4,t)$, $\Delta\beta_A(t)$ and $\beta_A^\triangle(t)$ as function of the reduced temperature $t = T/T_c$ where $\beta_A^\triangle(t)$ is the one for the triangular lattice. As is readily seen from Fig. 2 the square lattice is more stable than the triangular lattice for $T \leq 0.88 T_c$. Further the presence of $\Delta\beta_A(t)$ implies that it will cost a large energy to rotate the square lattice. Incidently a tilted square lattice or rather rhombus has been already observed in YBCO monocrystals though elongated in the b direction by a small angle neutron scattering [21] and by STM imaging.[16] We believe this distortion of the square lattice comes from the orthorhombicity of YBCO [22] but further study is clearly desirable. Also there are alternative calculations based on the $d+s$ mixing, which predict an oblique vortex lattice. [23,24]

3.2 B $\parallel$ a-b

When a magnetic field is applied within the a-b plane, the upper critical field $H_{c2}(\theta,t)$ exhibits the fourfold symmetry [25]

$$H_{c2}(\theta,t) = H_{c2}\left(\frac{\pi}{4},t\right) + \Delta H_{c2}(t)\cos^2(2\theta) \tag{2}$$

where now θ is the angle **B** makes from the a axis. Indeed such a fourfold symmetry in $H_{c2}(\theta,t)$ has been already observed in an untwinned LSCO monocrystals [26] sometime ago by magnetoresistance. The torque in the vortex state is then approximately given by [27]

$$\tau(\theta) = \frac{B}{4\pi\kappa_2^2\beta_A}\ln\left(\frac{H_{c2}(\theta)}{B}\right)H_{c2}\left(\frac{\pi}{4},t\right)(-a_2\sin(2\theta) - 2a_4\sin(4\theta)) \tag{3}$$

where $t = T/T_c$ and we neglected the θ-dependence of $\kappa_2(\theta,t)$ and $\beta_A(\theta,t)$ for simplicity. Also we include the twofold symmetric term a_2 arising from the orthorhombicity of high T_c cuprates like YBCO and $a_4 = \Delta H_{c2}(t)/H_{c2}(\pi/4,t)$. Here $a_4 \simeq -0.170\ln t$ in the vicinity of $T = T_c$. Indeed a recent sensitive torque measurement [28] from a monocrystal YBCO indicates the presence of both $a_2 \simeq -0.0826$ and $a_4 \simeq 10^{-2}$, which are not inconsistent with the theoretical prediction. We obtain $H_{c2}(0,0) \simeq 800$ Tesla if we take $-\partial H_{c2}(0,t)/\partial T = 12$ T/K, which is far larger than the observed value of 350 Tesla, though there is only one experimental data for YBCO available. [29] In this analysis we have neglected the Pauli paramagnetism. Inclusion of the Pauli paramagnetism not only suppress the upper critical field but also indicates the appearance of a sine-wave like superconducting state predicted a long time ago. [30,31]

Acknowledgments

We would like to thank M. T. Béal-Monod and Nils Schopohl for useful discussion and collaboration. One of us (HW) acknowledges the support from the Hallym Academy of Science. The present work is supported by National Science Foundation under grant number DMR92-18371.

References

1. See for instance, B. Goss Levi, *Physics Today* **49**, (1996) 19 and D. Clery, *Science* **271**, (1996) 288.
2. B. Stadlober *et al*, *Phys. Rev. Lett.* **74**, (1995) 4911.
3. K. Maki and H. Won, *Phys. Rev. Lett.* **72**, (1994) 1758; H. Won and K. Maki, *Phys. Rev.* B **49**, (1994) 15305.
4. T. E. Mason *et al*, *Phys. Rev. Lett.* **68**, (1992) 1414; T. E. Mason *et al*, *Phys. Rev. Lett.* **71**, (1993) 919.
5. J. Rossat-Mignod *et al* in *Selected Topics in Superconductivity*, eds. L. C. Gupta and M. S. Multani (World Scientific, Singapore, 1993).
6. A brief and preliminary report in the same spirit will be published in K. Maki, Y. Sun, and H. Won, *Chinese Journal of Physics*.
7. A. A. Abrikosov and L. P. Gor'kov, *Soviet Phys. JETP* **12**, (1961) 1243.
8. T. Hotta, *J. Phys. Soc. Jpn* **62**, (1993) 274.
9. P. Hirschfeld and N. Goldenfeld, *Phys. Rev.* B **48**, (1993) 4219; P. Hirschfeld, W. O. Putikka, and D. J. Scalapino, *Phys. Rev. Lett.* **71**, (1993) 3705.
10. H. Kim, G. Preost, and P. Muzikar, *Phys. Rev.* B **49**, (1994) 3544; G. Preost, H. Kim, and P. Muzikar, *Phys. Rev.* B **50**, (1994) 1259.
11. Y. Sun and K. Maki, *Phys. Rev.* B **51**, (1994) 6059; *Europhys. Lett.* **32**, (1995) 355.
12. N. Momono *et al*, *Physica C* **233**, (1994) 395; N. Momono and M. Ido, preprint.
13. K. Ishida *et al*, *J. Phys. Soc. Jpn* **62**, (1993) 2803.
14. K. Maki, N. Schopohl, and H. Won, *Physica B* **204**, (1995) 214.
15. N. Schopohl and K. Maki, *Phys. Rev.* B **52**, (1995) 490.
16. I. Maggio-Aprile *et al*, *Phys. Rev. Lett.* **75**, (1995) 275.
17. N. Schopohl and K. Maki, preprint.
18. C. Caroli, P. G. de Gennes, and J. Matricon, *Phys. Lett.* **9**, (1964) 307; C. Caroli and J. Matricon, *Phys. Kondense. Mater.* **3**, (1965) 380.
19. M. Lavagna and G. Stemmann, *Phys. Rev.* B **49**, (1994) 4235.
20. H. Won and K. Maki, *Europhys. Lett.* **30**, (1995) 421; *Phys. Rev.* B **53**, (March 1996) 1.
21. B. Keimer *et al*, *Phys. Rev. Lett.* **73**, (1994) 3459.
22. K. Maki and M. T. Béal-Monod, *Phys. Lett. A* **202**, (1995) 313; M. T. Béal-Monod, *Phys. Rev.* B **53**, (Feb. 1996) 1.
23. A. J. Berlinsky *et al*, *Phys. Rev. Lett.* **75**, (1995) 2200.
24. J. H. Xu, Y. Ren, and C. S. Ting, *Phys. Rev.* B **53**, (1996) 2991.
25. H. Won and K. Maki, *Physica B* **199-200**, (1994) 353; H. Won and K. Maki, preprint.
26. T. Hanaguri *et al*, *Physica B* **165&166**, (1990) 1449.
27. H. Won and K. Maki, preprint.
28. T. Ishida *et al*, preprint.
29. J. D. Goettee *et al*, *Physica C* **235-240**, (1994) 2090.
30. P. Fulde and R. A. Ferrell, *Phys. Rev.* **135**, (1964) A550.
31. A. I. Larkin and Y. N. Ovchinikov, *Sov. Phys. JETP* **20**, (1965) 762.

$s+d$ PAIRING SYMMETRY AND VORTEX STRUCTURES IN YBCO

J.H. XU, Y. REN and C.S. TING
Texas Center for Superconductivity, University of Houston, Houston,
Texas 77204, USA

YBCO exhibits a large anisotropy between the a (or y) and b (or x) axes in the CuO_2 plane. This anisotropy can be modeled by introducing an anisotropic mass parameter $\lambda = m_x/m_y \simeq 2$. Assuming a d-wave pairing interaction together with a repulsive on site Coulomb interaction, we developed a Ginzburg-Landau theory for superconducting YBCO. We show that the order parameter always has s+d symmetry. The strength of the s-wave component may reach 10% of that of the d-wave component. The vortex structures have been numerically studied. The single vortex has an elliptic shape, and the vortex lattice has an oblique structure. All these results are in good agreements with experimental measurements.

While many experiments, which probe directly the phase of the pairing state in YBCO, have provided strong evidence for a sign change of the order parameter [1,2,3], consistent with a predominantly d-wave pairing symmetry, the tunneling experiment [4] on a Josephson junction made of c-oriented YBCO and Nb superconductors yields a finite tunneling current. This latter result shows that the pairing symmetry in YBCO has an s-wave character. It was argued [5] recently that these conflicting experimental results can be reconciled by assuming that there exist two order parameters with $s+d$ symmetry and both of the s and d pairings have the same transition temperature. In view of YBCO exhibits large anisotropy between a and b directions in the measurements of penetration depth [6] because of the presence of linear CuO chains, we consider a simple model based upon the anisotropic effective mass within a single CuO_2 plane. The a-b mass anisotropy of YBCO is taken into account by a single parameter $\lambda = m_x/m_y \simeq 2$ which has been fitted to the measured penetration depth anisotropy [6]. For orthorhombic YBCO the pairing interaction between the charge carriers in the CuO_2 plane is expected to have the following form

$$V(k,k') = V_s - V_d(\eta^2 \hat{k}_x^2 - \hat{k}_y^2)(\eta^2 \hat{k}_x'^2 - \hat{k}_y'^2), \tag{1}$$

here $V_s > 0$ corresponds to the on site repulsive Coulomb interaction and $-V_d(V_d > 0)$ is the attractive pairing interaction in the "distorted d-wave" channel. In the above equation $\eta = b/a \simeq 1.015$, b and a are respectively the lattice constants along x(b)- and y(a)- directions for orthorhombic YBCO. With the above pairing interaction, the order parameter can be shown to have the following form

$$\Delta(R,k) = \Delta_s(R) + \Delta_d(R)(\eta^2 \hat{k}_x^2 - \hat{k}_y^2). \tag{2}$$

In order to simplify the calculation we first approximate $\eta = 1$, and only study the mass anisotropy effect which is primarily due to the presence of Cu-O chains in YBCO. In the discussion given below, we shall set $\eta = 1$ so that the pairing interaction in Eq.(1) is purely d-wave like. Based upon the procedure presented in Ref.[7], we obtain the following Ginzburg-Landau free energy density divided by the density of state at the Fermi surface $N(0)$ in the presence of a vector potential **A**

$$F = \left[1 - \gamma_d \ln\left(\frac{2e^\gamma \omega_D}{\pi T}\right) \frac{2(1+\lambda)}{(1+\sqrt{\lambda})^2}\right] |\Delta_d|^2 + \alpha_s |\Delta_s|^2 - \frac{\lambda-1}{\lambda+1}(\Delta_s^* \Delta_d + \Delta_d^* \Delta_s)$$

$$+ 2\gamma_d \alpha \mu \left[\frac{|\Pi_x \Delta_s|^2}{2m_x} + \frac{|\Pi_y \Delta_s|^2}{2m_y} + \frac{1-\sqrt{\lambda}+3\lambda+\lambda\sqrt{\lambda}}{(1+\sqrt{\lambda})^3} \frac{|\Pi_x \Delta_d|^2}{2m_x} \right.$$

$$+ \frac{1+3\sqrt{\lambda}-\lambda+\lambda\sqrt{\lambda}}{(1+\sqrt{\lambda})^3} \frac{|\Pi_y \Delta_d|^2}{2m_y}$$

$$+ \left(\frac{\lambda-2\sqrt{\lambda}-1}{(1+\sqrt{\lambda})^2} \frac{\Pi_x \Delta_s \Pi_x^* \Delta_d^*}{2m_x} - \frac{2\sqrt{\lambda}+1-\lambda}{(1+\sqrt{\lambda})^2} \frac{\Pi_y \Delta_s \Pi_y^* \Delta_d^*}{2m_y} + \text{h.c.} \right) \Big]$$

$$+ 2\gamma_d \alpha \left[\frac{(\lambda+1)(1+\sqrt{\lambda}+\lambda)}{(1+\sqrt{\lambda})^4} |\Delta_d|^4 + \frac{1}{2}|\Delta_s|^4 + \frac{2(1+\lambda)}{(1+\sqrt{\lambda})^2}|\Delta_d|^2 |\Delta_s|^2 \right.$$

$$+ \frac{\lambda+1}{2(1+\sqrt{\lambda})^2}(\Delta_s^{*2}\Delta_d^2 + \Delta_s^2 \Delta_d^{*2}) + \frac{\sqrt{\lambda}-1}{\sqrt{\lambda}+1}|\Delta_s|^2(\Delta_s^* \Delta_d + \Delta_s \Delta_d^*)$$

$$+ \frac{\lambda\sqrt{\lambda}-1}{(1+\sqrt{\lambda})^3}|\Delta_d|^2(\Delta_s^* \Delta_d + \Delta_s \Delta_d^*) \Big]. \tag{3}$$

where $\alpha = 7\zeta(3)/8(\pi T)^2$, μ is the chemical potential, $\gamma_d = N(0)V_d/2$, $\Pi = i\nabla_R - 2e\mathbf{A}$, and

$$\alpha_s = \frac{(1+\sqrt{\lambda})^2}{1+\lambda}\left(1 + \frac{(1+\sqrt{\lambda})^2}{1+\lambda}\frac{V_s}{V_d}\right). \tag{4}$$

From Eq.(3) and when $\mathbf{A} = 0$, the superconducting transition temperature T_c is determined by

$$\ln \frac{T_c}{T_{c0}} = \frac{1}{\gamma_d}\left\{1 - \frac{(1+\sqrt{\lambda})^2}{2(1+\lambda)}\left[1 - \frac{1}{\alpha_s}\left(\frac{\lambda-1}{\lambda+1}\right)^2\right]\right\}, \tag{5}$$

where $T_{c0} = T_c(\lambda = 1)$ satisfies the relation $\ln(2e^\gamma \omega_D/\pi T_{c0}) = 1/\gamma_d$ with γ as the Euler constant. This result implies that T_c increases as λ deviates from $\lambda = 1$. Assuming $\Delta_s = |\Delta_s|e^{i\theta_s}$ and $\Delta_d = |\Delta_d|e^{i\theta_d}$, the phase difference $\theta = \theta_s - \theta_d$ between Δ_s and Δ_d can be determined through $\partial F/\partial \theta = 0$ and $\partial^2 F/\partial \theta^2 > 0$. We find that the only stable solution is for $\theta = 0$, which means that the pairing state in the bulk superconductor has s+d mixed symmetry and both of the s- and d-wave order parameters have the same transition temperature. For $\lambda = 2$, it is easy to show that $\Delta_s \simeq 0.076\Delta_d$ with $V_s/V_d \simeq 1$. For $V_s/V_d > 1$, the value of Δ_s will be slightly depressed. At this stage it should be interesting to examine the case when $\eta = b/a > 1$ in YBCO. Under this condition, Eq.(2) can be rewriten as

$$\Delta(R, k) = [\Delta_s(R) + \frac{1}{2}\delta\Delta_d(R)] + (1 - \frac{1}{2}\delta)\Delta_d(R)(\hat{k}_x^2 - \hat{k}_y^2). \tag{6}$$

with $\delta = \eta^2 - 1 = 0.032$. Therefore the effect of $\eta > 1$ is to enhance the strength of the s-wave component by approximately 3%. Together with the anisotropic mass effect, the s-wave component may reach 10% of the d-wave component. Such an s+d state is exactly what has been expected [5] to explain various experimental measurements in YBCO. In the presence of a magnetic field, we numerically studied the single vortex structure according to Eq.(3). The contour plots of Δ_s and Δ_d near a vortex core for $T/T_c = 0.5$ are respectively presented in Figs.1 and 2, with $\xi_b = \sqrt{\mu\alpha/2m_x(1-T/T_c)}$ as the coherent length along

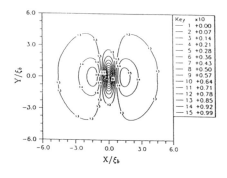

Figure 1: Contour plot of Δ_s.

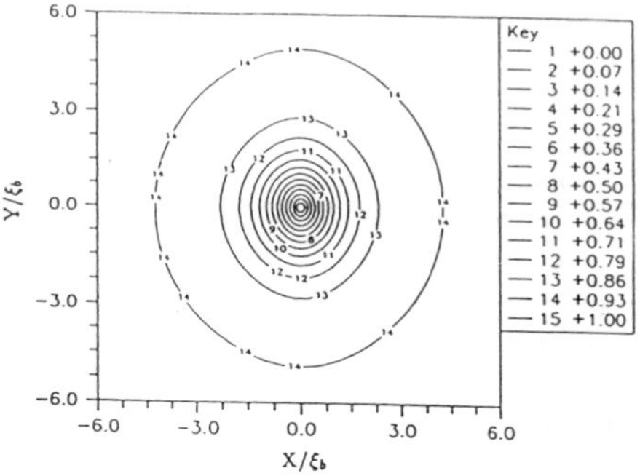

Figure 2: Contour plot of Δ_d.

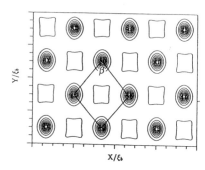

Figure 3: Contour plot of the local magnetic field distribution in the vortex lattice.

b-direction. Here both the s- and d-wave components show a two fold symmetry which is very different from the isotropic d-wave case [8].

We also numerically studied the structure of a vortex lattice at $T/T_c = 0.5$ for a magnetic field strength $H \simeq H_c$, the thermodynamic critical field. The profile of the local d-wave order parameter or the magnetic field is shown in Fig.3. The vortex lattice is found to be oblique with $\beta = 75^0$ and local magnetic field distribution in a single vortex has an elliptic shape with $\xi_a/\xi_b = 1.414$. This result is in good agreement with the STM measurements on YBCO by Maggio-Aprile et al[9].

Finally it is important to point out that for $\lambda = 2$, the shape of the vortex lattice is always oblique for $T < T_c$. For isotropic mass with $\lambda = 1$, the vortex lattice still could be oblique [7]. However the triangular lattice may be stable as the overlapping between the s-wave components of the order parameters on the nearest neighboring vortices becomes extremely small. This will happen [8] whenever the applied field is not too strong, such that the vortex lattice constant is much larger than the coherent length, or $T \to T_c$, or $V_s/V_d \gg 1$.

Acknowledgments

This work has been supported by a grant from the Robert A. Welch Foundation and the state of Texas from the Texas Center for Superconductivity at the University of Houston.

References

1. D.A. Wollman, D.J. Van Harlingen, W.C. Lee, D.M. Ginzburg, and A.J. Leggett, *Phys. Rev. Lett.* **71**, 2134 (1993).
2. C.C. Tsuei, J.R. Kirtley, C.C. Chi, L.S. Yu-Jahnes, A. Gupta, T. Shaw, J.Z. Sun, and M.B. Ketchen, PRL **73**, 593 (1994).
3. J.H. Miller, Jr., Q. Y. Ying, Z.G. Zou, N.Q. Fan, J.H. Xu, M.F. Davis, and J.C. Wolfe, *Phys. Rev. Lett.* **74**, 2347 (1995).
4. A.G. Sun, D.A. Gajewski, M.B. Maple, and R.C. Dynes, *Phys. Rev. Lett.* **72**, 2267 (1994).
5. K.A. Müller, *Nature* **377**, 133 (1995).
6. K. Zhang, D.A. Bonn, S. Kamal, R. Liang, D.J. Baar, W.N. Hardy, D. Basov, and T. Timusk, *Phys. Rev. Lett.* **73**, 2484 (1994).
7. Y. Ren, J.H. Xu, and C.S. Ting, *Phys. Rev. Lett.* **74**, 3680 (1995); J. H. Xu, Y. Ren, and C.S. Ting, *Phys. Rev.* B **52**, 7663 (1995).
8. J.H. Xu, Y. Ren, and C.S. Ting, *Phys. Rev.* B **53**, 2991 (1996).
9. I. Maggio-Aprile, C. Renner, A. Erb, E. Walker, and Ø. Fisher, *Phys. Rev. Lett.* **75**, 2754 (1995).

Boundary Effects and the Order Parameter Symmetry of High-T_c Superconductors

Safi R. Bahcall
Department of Physics, University of California, Berkeley CA 94720

Apparently conflicting phase-sensitive measurements of the order parameter symmetry in the high-T_c cuprate superconductors may be explained by regions near surfaces in which the order parameter symmetry is different than in the bulk. These surface states can lead to interesting and testable effects.

Phase-sensitive measurements on the high temperature superconductor $YBa_2Cu_3O_{7-\delta}$ have yielded two potentially conflicting sets of results for the symmetry of the superconducting order parameter [1]. Measurements involving currents flowing in the CuO_2 planes, such as the corner-junction SQUID experiments [2,3,4], the corner-junction flux modulation experiments [5,6], and the tricrystal ring [7] and grain-boundary [8] experiments indicate an order parameter with primarily $d_{x^2-y^2}$ symmetry under rotations in the plane ($\Delta(\mathbf{k}) \sim \cos k_x - \cos k_y$). The presence, however, of Josephson tunneling perpendicular to the CuO_2 planes between heavily-twinned YBCO and a conventional s-wave superconductor [9,10,11] suggests an order parameter with a significant s-wave component [12,13].

A bulk order parameter of mixed s and $d_{x^2-y^2}$ symmetry could explain both sets of experiments. An order parameter with this mixed symmetry, for a material which is otherwise macroscopically symmetric under 90° rotations (heavily twinned YBCO), requires either a first order transition or two separate bulk phase transitions. So far, there has been no convincing evidence for either of these. Here we therfore assume that the order parameter in the bulk superconductor transforms as one irreducible representation of the rotation group D_{4h}, either s or $d_{x^2-y^2}$.

Using a Ginzburg-Landau model in which both s and $d_{x^2-y^2}$ order parameter symmetries are allowed, but only one is favored in the bulk, we find that there are two possibilities consistent with both the CuO_2 plane and c-axis tunneling experiments.

The first possibility is that the order parameter is s-wave in the bulk and a d-wave component is mixed in at faces normal to the CuO_2 planes. This does not require any special choice of parameters; there is an instability to mixing near these faces. The symmetry being tested is rotation in the CuO_2 plane and placing an edge in that plane breaks the symmetry explicitly. This always causes mixing. The amount of mixing depends on the energetics: if the d-wave component is strongly disfavored (as might be expected in a conventional superconductor), the mixing is small. If there is a close competition, the mixing may be large. In addition, we find that for this case the mixing can explain the CuO_2 plane experiments only if the order parameter breaks time reversal invariance at the surface: it must have the form $s + id$ there.

The second possibility is that the order parameter is d-wave in the bulk and a surface state forms which mixes in an s component on the face perpendicular to c-axis. This occurs only under certain conditions. The two components must inhibit each other, in the sense that the presence of one makes the other energetically less favorable. In addition, the effect of the c-axis boundary must be such that the magnitude of the d-wave component decreases significantly from its bulk value near the edge. In that case, the s-wave component is less suppressed near the surface and a localized region of mixed symmetry can develop.

The presence of the surface state normal to the c-axis is sensitive to the boundary conditions. This may explain the difficulty in achieving c-axis junctions, as well as the variability among samples of angle-resolved photoemission spectroscopy studies of the gap magnitude[14]. The photoemission studies see the topmost CuO_2 layer. Variations in surface properties affect the boundary conditions, which in turn affect whether the order parameter has the form $d+s$, $d+is$ or pure d at the surface, each of which has a different momentum dependence.

The c-axis surface state may also lead to "π-junction" behavior. In a SQUID loop between YBCO and a conventional superconductor, with junctions normal to the c-axis, the configuration with opposite relative phases on the two junctions will lead to a net phase difference of π in the absence of an applied magnetic field.

Acknowledgments

This work was stimulated by many interesting discussions with S. Kivelson, V. Emery, D. S. Rokhsar, D.-H. Lee, and J. Clarke, and was supported by a Miller Research Fellowship from the Miller Institute for Basic Research in Science.

References

1. Reviewed in D. J. Van Harlingen, Rev. Mod. Phys. **67**, 515 (1995).
2. D. A. Wollman *et al.*, Phys. Rev. Lett. **71**, 2134 (1993)
3. D. A. Brawner and H. R. Ott, Phys. Rev. B **50**, 6530 (1994).
4. A. Mathai *et al.*, Phys. Rev. Lett. **74**, 4523 (1995).
5. D. A. Wollman *et al.*, Phys. Rev. Lett. **74**, 797 (1995).
6. J. H. Miller *et al.*, Phys. Rev. Lett. **74**, 2347 (1995).
7. C. C. Tsuei *et al.*, Phys. Rev. Lett. **73**, 593 (1994); J. R. Kirtley *et al.*, Nature **373**, 225 (1995).
8. J. R. Kirtley *et al.*, Phys. Rev. Lett. **76**, 1336 (1996).
9. A. G. Sun, D. A. Gajewski, M. B. Maple, and R. C. Dynes, Phys. Rev. Lett. **72**, 2267 (1994).
10. I. Iguchi and Z. Wen, Phys. Rev. B **49**, 12388 (1994).
11. R. Kleiner *et al.*, Phys. Rev. Lett., to appear.
12. We do not consider the abundant evidence for nodes in the gap function, which is consistent with either symmetry when variations with $|\mathbf{k}|$ are allowed.
13. The relevance of the order parameter symmetry to the microscopic pairing mechanism is reviewed in D. J. Scalapino, Phys. Rep. **250**, 329 (1995).
14. Reviewed in Z. X. Shen and D. S. Dessau, Phys. Rep. **253**, 1 (1995). (See Fig. 5.18.)

ANISOTROPIC PAIRING CAUSED BY UNSCREENED LONG RANGE INTERACTIONS

V. HIZHNYAKOV
Physics Dept., Tartu University, Tähe 4, Tartu, Estonia

A. BILL and E. SIGMUND
Lawrence Berkeley National Lab., 1 cyclotron road, MS 62-203, Berkeley, CA 94720, USA
Institut für Theoretische Physik, BTU Cottbus, P.O. Box 101344, 03013 Cottbus, Germany

The anisotropy of the order parameter in high-T_c superconductors is explained by unscreened interaction of charge carriers with long-wave optical phonons. Screening is absent due to the low frequency of long-wave plasmons in layered structures.

According to recent investigations the order parameter $\Delta(\vec{k})$ (i.e. the gap) in high-T_c superconductors is strongly anisotropic (see references in [1,2]). The anisotropy is observed along the ab plane, where the maximal values of $\Delta(\vec{k})$ are in a and b directions and the minima in between. A remarkable fact concerning this observation is that the density of states (DOS) at the Fermi surface (i.e. the inverse of the Fermi velocity) has the same anisotropy. The experimentally observed anisotropy of $\Delta(\vec{k})$ is often considered as a support of d-wave pairing models, especially those in which the pairing interaction is caused by spin excitations (see e.g. Refs. in [1,2]), as high-T_c superconductors exhibit strong antiferromagnetic (AF) fluctuations. Nevertheless, at the present stage of experimental investigations the assignment of the symmetry of the superconducting order parameter remains controversial. Several experiments suggest indeed a d-wave type gap while others observe an s-wave anisotropy.[1,2]

In our considerations (see also [1]) we take into account that a strongly anisotropic s-wave type gap may result from a pairing interaction which has a long-range part. Indeed, the superconducting gap results from the mixing of free-particle states near the Fermi surface through interactions. In particular, long-range interactions mix states with close $\vec{k}$ vectors and may therefore lead to a $\vec{k}$-dependent gap. In this case the anisotropy of $\Delta(\vec{k})$ is of the same type as that of the DOS (ρ) on the Fermi surface; Δ increases with ρ. As already mentioned above, this is exactly the situation found in high-T_c superconductors.

Long-range pairing interaction in high-T_c superconductors is associated with the non-screened coupling of charge carriers with long-wave optical phonons. In ordinary superconductors this interaction is very weak due to its almost perfect screening by free charge carriers. However, in high-T_c superconductors this screening is strongly reduced. The reason for this lies in the totally different spectrum of plasmons of high-T_c materials compared to usual metals. Due to the layered structure of the cuprates the minimal frequency of long-range plasmons (in the CuO_2 plane) is very small.[3] As a result the long-range plasmons cannot follow the motion of the ions and screen the Coulomb field which appears when the ions shift from their equilibrium position.

To find the functional dependence of the long-range interaction potential U on the phonon wave vector $\vec{q}$ at small $q_\parallel$ (component parallel to the CuO_2 plane we use the effective charge carrier interaction potential [4] (for a complete derivation of Eq. 1 see [1]). Taking into account that the charge carriers in cuprate superconductors have very small bandwidth along c ($\delta\varepsilon_z < \hbar\omega_\nu$), and averaging the interaction V_{eff} over $q_\perp$ one finds for small $q_\parallel$

$$U(q) = 2\pi U/(\kappa^2 + q_x^2 + q_y^2) , \qquad (1)$$

where $U \sim 0.03$eV, $\kappa \sim 0.3$eV.[1] We stress that this interaction takes also into account the Coulomb repulsion.

The energy spectrum of the free charge carriers (holes) in CuO_2 planes in high-T_c superconductors can be described by

$$\varepsilon_{\vec{k}} = -2t(\cos k_x + \cos k_y) - 2\tau(\cos 2k_x + \cos 2k_y) + 4t'\cos k_x \cos k_y + 4(t + \tau + \tau') - \varepsilon_F \quad (2)$$

(k_x and k_y correspond to the wave vector components in a and b directions). The parameters used are: $t = 0.14$ eV, $\tau = 0.25t$, $t' = 0.2t$.[5] This energy spectrum gives a good description of the experimentally observed "extended Van-Hove singularities" of the DOS. Using Eq. 1 and 2

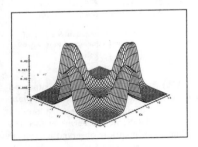 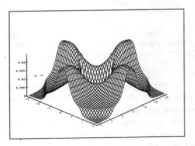

Figure 1: $\Delta(\vec{k})$ for $U_0 = 0.03$eV, $\kappa = 0.3$, $\varepsilon_F = 0.34$ eV (left) and with account of retarding effects $U_0 = 0.025$ eV, $\kappa = 0.3$, $\varepsilon_F = 0.34$ eV, $\alpha = 1$, $\epsilon = 0.015, \omega_\nu = 0.05$ eV (right). $\Delta_{max} \simeq 25$meV in both cases.

we solved numerically the BCS gap equation for different ε_F, U and κ. We found that Δ_{max} reaches its highest value for optimal doping (i.e. $\varepsilon_F \approx 0.35$ eV). For this (and close) ε_F the gap Δ_φ (as well as ρ_φ) is strongly anisotropic with maxima at $\varphi \approx n\pi/2$, $n = 0,1,2,3$, which is in agreement with experiment (see Fig. 1, left). The above given long-range interaction potential $U(q)$ is attractive for all wave vectors and does therefore not account for retarding effects. These effects are small if $\omega_{\vec{k},\vec{q}} = \varepsilon_{\vec{k}+\vec{q}} - \varepsilon_{\vec{k}} \ll \omega_\nu$, $q \sim \kappa$, which is true when $q \sim \kappa \ll \hbar\omega_\nu\rho$ (ω_ν is a characteristic phonon energy). For small values of $|\varphi - n\pi/2|$ ($n = 0,1,2,3$) the DOS is large and the condition is well fullfilled. For small $|\varphi - n\pi/4|$, however, ρ is relatively small and the condition is not well fulfilled and the account of the retarding effects may be important. This can be done in the model by replacing Eq. 1 by the sign alternating form $U_1(\vec{k},\vec{q}) = U(q)R(\vec{k},\vec{q})$ with $R(\vec{k},\vec{q}) = (1+\alpha)(\omega_{\vec{k},\vec{q}}^2 + \epsilon^2)/(\omega_\nu^2 - (\omega_{\vec{k},\vec{q}} + i\epsilon)^2)$ (α and ϵ are parameters). Our calculations show that this type of interaction leads to oscillations of Δ along $\varphi \approx n\pi/4$ (see Fig. 1, right).

We conclude that the nontotally screened long-range electron-phonon interactions in high-T_c superconductors, give an anisotropic contribution to the superconducting order parameter (gap) of the type and magnitude observed in experiment. The anisotropy implied by this interaction is also in agreement with calculations of the phonon renormalization below T_c taking into account the $\vec{k}$-dependency of the gap.[2]

We thank D. Nevedrov who did the numerical calculations.

1. V. Hizhnyakov, *Proc. Estonian Acad. Sci.,Phys. Math.* **43**, 485 (1994);
   V. Hizhnyakov and E. Sigmund, *Phys. Rev.* B , 1. march (1996).
2. A. Bill *et al*, *Phys. Rev.* B **52**, 7637 (1995) and *this volume*.
3. V. Kresin and H. Morawitz, *Phys. Rev.* B **37**, 7854 (1988); *solid state commun.***74**, 1203(1990);
4. D. Pines and P. Nozières, *The Theory of Quantum Liquids, Normal Fermi Liquids* (Addison-Wesley publishing Inc., 1989)
5. K. Gofron *et al*, *Phys. Rev. Lett.* **73**, 3302 (1994).

ZERO-BIAS CONDUCTANCE PEAK AS A RESULT OF MIDGAP INTERFACE STATES — A MODEL STUDY

CHIA-REN HU

Department of Physics, Texas A&M University,
College Station, TX 77843-4242
and Texas Center for Superconductivity, University of Houston,
Houston, TX 77204-5932

A model calculation is presented which shows that c-axis tunneling into a d-wave superconductor containing interfaces separating grains of different ab-axes orientations clearly exhibits a zero-bias conductance peak (ZBCP) at low temperatures. If the tunneling probe is an s-wave superconductor with a small energy gap, a dip in the middle of the (broadened) ZBCP appears. These and other characteristics are in good agreement with those of the ZBCPs observed ubiquitously in high-T_c superconductors.

I report here a model calculation of c-axis NIS_d and S_sIS_d tunneling characteristics into a d-wave superconductor (DWSC) which contains two slab-shaped domains and *two* interfaces (due to periodic boundary conditions assumed). The two domains have their c-axes aligned parallel to each other and to the interfaces, and are characterized by ϕ_ℓ and ϕ_r, which are the rotation angles of the respective a-axes of their $d_{x_a^2-x_b^2}$-order parameters from the interface normal. ($\phi_\ell = \pi/4$ and $\phi_r = 0$ used). A tunneling Hamiltonian in real space is employed in order to allow for spatially inhomogeneous situations. Tunneling is assumed to occur at a single point **R**. The result has been averaged with respect to **R** over the whole DWSC, so that the tunneling characteristics can depend only on the eigenspectra or density of states of the two electrodes, and not on the quasi-particle wave functions. (Actual tunneling is likely more localized, and can have larger contributions from states whose local amplitudes are larger.) The Bogoliubov equations are solved in order to obtain the eigenspectrum. Assuming $\mathbf{k}_F \equiv (k_{Fx}, k_{Fy}) = k_F(\cos\phi, \sin\phi)$ for the Fermi wave-vector on a cylindrical Fermi surface, with the x-axis perpendicular to the interfaces, and the two domains occupy $-L_\ell < x < 0$ and $0 < x < L_r$, respectively, the pair-potential order parameters of the two domains are then: $\Delta_\ell(\mathbf{k}_F) = \Delta_0 \cos(2\phi - 2\phi_\ell)$, and $\Delta_r(\mathbf{k}_F) = \Delta_0 \cos(2\phi - 2\phi_r)$. The eigenspectrum for a given ϕ is found to show clearly a "midgap interface state" (MIS) per interface at the center of the ϕ-dependent energy gap, for $-\pi/4 < \phi < 0$ and $\pi/4 < \phi < \pi/2$, but not for the remaining regions in $-\pi/2 < \phi < \pi/2$. (The eigenspectrum is the same for ϕ and $\phi+\pi$.) This is consistent with the condition $\Delta_\ell(\mathbf{k}_F)\Delta_r(\mathbf{k}_F) < 0$ for the occurance of a MIS per interface for a given ϕ, if no energy barriers exist at the interfaces.[1] The total density of states of this DWSC is found to be V-shaped near the Fermi energy, plus a sizable peak right at the Fermi energy due to the MISs. (We have used $L_\ell = 10\xi_0$ and $L_r = 20\xi_0$ which are not large, where $\xi_0 \equiv \hbar v_F/\Delta_0$ is essentially the ab-plane coherence length of the DWSC. In order to obtain a continuous density of states above the gap for a given ϕ that is similar to the bulk behavior, we have to introduce a finite width $\Gamma = 0.05\Delta_0$ for each energy level. The zero-energy peak therefore also acquires such a width. Even so, the density of states still exhibits some size-quantization oscillations.)

Using the so-obtained total density of states, we have then calculated the (**R**-averaged) (quasi-particle) tunneling conductance for c-axis NIS_d tunneling into such a DWSC. At low temperatures ($k_BT = 0.01\Delta_0$), we clearly obtained a sizable ZBCP. Raising k_BT to $0.05\Delta_0$, the ZBCP is found reduced but not widened, and it totally disappears at $k_BT = 0.10\Delta_0$. We have also calculated the c-axis S_sIS_d tunneling conductance into such a DWSC, assuming that the s-wave superconducting counter-electrode has a gap $\Delta_s = 0.10\Delta_0$. The ZBCP is found to be taller (than the $\Delta_s = 0$ result at the same temperature), but with a center dip due to the gap Δ_s. Even at $k_BT = 0.10\Delta_0$, when the NIS_d-tunneling already shows no ZBCP, the S_sIS_d-tunneling still shows a small ZBCP with a small center dip, and the width of the ZBCP is almost unchanged from its value at $k_B T = 0.01\Delta_0$. All of these features are in good qualitative agreement with those of the ZBCPs observed ubiquitously in high-T_c superconductors, including the enhanced ZBCP with a center dip if a low-T_c superconducting probe is used, and the subthermal T-dependence of its width.[2]

Acknowledgments

This work is partially supported by the Texas Center for Superconductivity.

References

1. C.-R. Hu, *Bull. Amer. Phys. Soc.* **40**, 789 (1995), which extended my earlier works in C.-R. Hu, *Phys. Rev. Lett.* **72**, 1526 (1994); and J. Yang and C.-R. Hu, *Phys. Rev.* B **50**, 16766 (1994); from surface to interface states.
2. J. Geerk *et al.*, *Z. Phys.* B **73**, 329 (1988); M. Lee *et al.*, *Phys. Rev.* B **39**, 801 (1989); I. Iguchi and Z. Wen, *Physica* C **178**, 1 (1991); T. Walsh *et al.*, *Phys. Rev. Lett.* **66**, 516 (1991); T. Walsh, *Int. J. Mod. Phys.* B **6**, 126 (1992); J. Lesueur *et al.*, *Physica* C **191**, 325 (1992); Th. Becherer *et al.*, *Phys. Rev.* B **47**, 14 650 (1993).

Ginzburg-Landau equations for a d-wave superconductor with nonmagnetic impurities

Wang Xu, Yong Ren and C. S. Ting
Department of Physic and
Texas Center for Superconductivity, University of Houston, Houston, TX 77204

Abstract

The Ginzburg-Landau equations for a d-wave superconductor with nonmagnetic impurities in both the weak scattering Born limit and the strong scattering unitary limit are derived in the framework of the Gorkov's weak-coupling theory of superconductivity. The effect of impurities on the magnetic field penetration depth and the induced s-wave order parameter are also studied.

Recently, the Ginzburg-Landau equations for the d-wave superconductor have been derived microscopically by Ren et al.[1]. Using these equations the single and lattice vortex structures as well as the surface problem for the two-dimensional d-wave superconductor have been studied[1,2]. However, their studies are confined to pure superconductors and the impurity effect has not been taken into account. In this work, we will present the Ginzburg-Landau equations for the d-wave superconductor with nonmagnetic impurities in both the Born and the unitary limits by employing the impurity configuration averaging technique [3,4]. The impurity effects on the magnetic field penetration depth and the induced s-wave order parameter are also discussed.

Assuming the pairing interation $V(\mathbf{k},\mathbf{k}') = V_s - V_d(\hat{k}_x^2 - \hat{k}_y^2)(\hat{k}_x'^2 - \hat{k}_y'^2)$ with $V_s > 0$ as the repulsive on site coulomb interaction and $-V_d < 0$ is the attractive interaction in d-wave channel, the order parameter can be shown to have following form $\Delta^*(\mathbf{R},\mathbf{k}) = \Delta_s^* + \Delta_d^*(\hat{k}_x^2 - \hat{k}_y^2)$. Based upon the method described in Refs.[3] and [4], the Ginzburg-Landau free energy functional for a 2-dimensional d-wave superconductor in the magnetic field with the vector potential **A** in the presence of the nonmagnetic impurities can be shown to have the following form

$$f = \bar{\alpha}_s N_0(0)|\Delta_s|^2 - \frac{N_0(0)}{2}\left[\ln\frac{T_{cp}}{T} + \psi(\frac{1}{2}) - \psi(\frac{1}{2} + \frac{c\alpha}{\pi T})\right]|\Delta_d|^2$$
$$+ \frac{N_0(0)\delta}{2}\left[|\Delta_s|^4 + \frac{3}{8}\chi_3(\frac{c\alpha}{\pi T_c})|\Delta_d|^4 + (\chi_1(\frac{c\alpha}{\pi T_c}) + \chi_2(\frac{c\alpha}{\pi T_c}))|\Delta_s|^2|\Delta_d|^2\right.$$
$$\left.+ \frac{1}{2}\chi_1(\frac{c\alpha}{\pi T_c})(\Delta_s^{*2}\Delta_d^2 + \Delta_d^{*2}\Delta_s^2)\right] + \frac{N_0(0)\delta v_F^2}{8}\left[2\chi_1(\frac{c\alpha}{\pi T_c})|\Pi\Delta_s^*|^2 + \chi_3(\frac{c\alpha}{\pi T_c})|\Pi\Delta_d^*|^2\right.$$
$$\left.+ \chi_2(\frac{c\alpha}{\pi T_c})(\Pi_x^*\Delta_s\Pi_x\Delta_d^* - \Pi_y^*\Delta_s\Pi_y\Delta_d^* + H.c.)\right] + \frac{1}{8\pi}(\nabla \times \mathbf{A})^2, \quad (1)$$

where T_{cp} is the transition temperature of the pure superconductor, $\alpha = \pi U^2 N_0(0)$ and $\alpha = \frac{1}{\pi N_0(0)}$ are for the Born limit and the unitary limit respectively, and c is the concentration of impurities. U is the impurity scattering potential and $N_0(0)$ is the density of states at the Fermi level. Here we define $\Pi = -i\nabla_\mathbf{R} - 2e\mathbf{A}(\mathbf{R})$, $\bar{\alpha}_s = \frac{1}{\lambda_d}\left[1 - \lambda_d \ln\frac{T_c}{T_{cp}} + \frac{2V_s}{V_d}(1 - \lambda_d \ln\frac{T_c}{T_{cp}})^2\right]$ with $\lambda_d = \frac{N_0(0)V_d}{2}$ and $\delta = \frac{7\zeta(3)}{8}(\frac{1}{\pi T_c})^2$. $\psi(z)$ is the digamma function and $\chi_i(\beta) = \left[\frac{7\zeta(3)}{8}\right]^{-1}\sum_{n=0}^\infty \frac{1}{(2n+1+\beta)^i(2n+1)^{3-i}}$. A straightforward variational differentiations of the free energy with respect to Δ_s, Δ_d and $\mathbf{A}$ lead to the Ginzburg-Landau equations and the expression of the supercurrent. One can also see from Eq.(1) that the critical temperature T_c determined by $\ln\frac{T_{cp}}{T_c} + \psi(\frac{1}{2}) - \psi(\frac{1}{2} + \frac{c\alpha}{\pi T_c}) = 0$ decreases with the increasing impurity concentration.

The magnetic field penetration depth λ_L in the Ginzburg-Landau region $T_c - T \ll T_c$ can be obtained by using the above derived equations. For the extreme type-II superconductor and in the London limit, we find

$$1/\lambda_L^2 = (8\pi e^2 n/3m)(T_c/T_{cp})(1 - T/T_c). \tag{2}$$

for $c\alpha/\pi T_c \ll 1$, and

$$1/\lambda_L^2 = (\pi e^2 n/9m)(\pi T_c/c\alpha)^2(1 - (T/T_c)^2) \tag{3}$$

for $c\alpha/\pi T_c \gg 1$. Therefore the penetration depth will change from linear T behavior to the T^2 behavior as the concentration or the scattering strength of the impurities increases. This result gives an explanation to the experimental measurements in the high-T_c superconductors and is consistent with the result obtained in Ref.[5]. We can also see from solving the vortex structure [1] that the ratio of the induced s-wave magnitude to the d-wave value decreases as impurity strength increases. This might explain why in some samples, the characters of a d-wave vortex lattice are not observed.

Acknowledgments

We would like to thank Dr. J. H. Xu for discusssions. This work is supported by the Robert A. Welch Foundation and the Texas Center for Superconductivity at the University of Houston.

References

1. Y. Ren, J. H. Xu, and C. S. Ting, Phys. Rev. Lett. **74**, 3680(1995), and Phys. Rev. B**52**, 7663(1995).
2. J. H. Xu, Y. Ren, and C. S. Ting, Phys. Rev. B**53**, R2991(1996).
3. A. A. Abrikosov and L. P. Gorkov, Zh. Exp. Theor. Fiz. **39**, 1781(1960) (Soviet Phy. JETP, **12**, 1243(1961)).
4. Wang Xu, Yong Ren and C. S. Ting, Phys. Rev. B. May 1st(1996).
5. J. Hirschfeld and N. Goldenfeld, Phys. Rev. B**48**, 4219(1993).

III. THEORY

Flux Dynamics

PANCAKE VORTICES IN HIGH-TEMPERATURE SUPERCONDUCTING THIN FILMS

JOHN R. CLEM, MAAMAR BENKRAOUDA, AND THOMAS PE
Ames Laboratory and Department of Physics and Astronomy,
Iowa State University, Ames, IA 50011, USA

A tilted stack of 2D pancake vortices in an infinite set of Josephson-decoupled superconducting layers has been found to be unstable when the angle of tilt is greater than 52°. We describe how to calculate the magnetic-field and current-density distributions generated by a single pancake vortex in an arbitrary layer of a multilayer structure of finite thickness, and we discuss the stability of a tilted stack of pancake vortices, where the tilt is maintained by application of transport currents in the top and bottom superconducting layers. There is a critical decoupling current, at which pancake vortices are driven off the top and bottom of the stack.

1 Introduction

Many of the high-temperature superconductors are highly anisotropic, with strong superconductivity parallel to the CuO_2 layers and weak superconductivity perpendicular to the layers. To model the vortex structure in the most anisotropic high-temperature superconductors, it has been useful to consider a model in which there is no Josephson coupling between superconducting layers[1,2,3]. The magnetic-field and current-density distribution generated by an arbitrary vortex distribution then can be calculated by superposition.

The name that was given to the basic building block that generates one of these contributions is the *two-dimensional (2D) pancake vortex*[1]. The magnetic-field and current-density distribution associated with one pancake vortex can be readily computed by placing a vortex in only one superconducting layer at $z = 0$ (let us call this the central layer) and then accounting for the screening effect of all the other layers. For a vortex that carries magnetic flux up through the layers, the screening current in the central layer flows with a large magnitude in the counterclockwise direction when viewed from above. The screening currents in all the other layers flow with a much smaller magnitude in the clockwise direction. The highly concentrated current distribution, localized chiefly in the central layer, is reminiscent of a pancake, and this is the origin of the descriptive name.[1]

A vortex line perpendicular to the layers thus can be represented as a straight stack of 2D pancake vortices, one on top of the next. For a stack centered on the z axis, the magnetic field, calculated by summing the individual pancake contributions, has only a z component,[1]

$$b_z(\rho) = (\phi_0/2\pi\lambda_\|^2) K_0(\rho/\lambda_\|), \qquad (1)$$

where $\phi_0 = hc/2e$ is the superconducting flux quantum, ρ is the radial coordinate, and $\lambda_\|$ is the effective penetration depth for the decay of screening currents parallel to the layers (often called λ_{ab} in the high-temperature superconductors). Since the pancake vortices also must have cores of size approximately $\xi_\|$ (or ξ_{ab}), in computing the energy per unit length of the vortex ϵ, one may cut off the kinetic energy integrals at $\rho = \xi_\|$ to obtain

$$\epsilon(\theta = 0) = (\phi_0/4\pi\lambda_\|)^2 K_0(\xi_\|/\lambda_\|), \qquad (2)$$

where the argument $\theta = 0$ indicates that the vortex line is not tilted with respect to the z axis.

For a straight stack of pancake vortices tilted at an angle θ relative to the z axis, the energy per unit length of the vortex line is, to logarithmic accuracy when $\lambda_{\|} \gg \xi_{\|}$,[1]

$$\epsilon(\theta) = (\phi_0/4\pi\lambda_{\|})^2 \ln[(\lambda_{\|}/\xi_{\|})(1+\cos\theta)/2\cos\theta]\cos\theta. \tag{3}$$

In this paper we discuss some recent results concerning the stability of tilted stacks of magnetically coupled 2D pancake vortices in Josephson-decoupled multilayers. In Sec. 2 we point out that the form of Eq. (3) implies that a tilted straight stack in an infinite set of multilayers becomes unstable for angles larger than 52°. In Sec. 3 we briefly discuss some recent results on the behavior of tilted stacks of pancake vortices in a finite set of multilayers, and in Sec. 4 we present some preliminary conclusions.

2 Infinite Sample

The line tension $P(\theta)$ of a tilted stack can be computed from the relation[5,6]

$$P(\theta) = \epsilon(\theta) + d^2\epsilon(\theta)/d\theta^2, \tag{4}$$

where the second term on the right-hand side arises from the anisotropy of the system. From Eqs. (3) and (4) we obtain

$$P(\theta) = (\frac{\phi_0}{4\pi\lambda_{\|}})^2 \frac{\cos\theta - \sin^2\theta}{\cos\theta(1+\cos\theta)}. \tag{5}$$

From this result we have shown[4] that the line tension and the tilt modulus are positive for small values of θ but change sign at $\theta = 52°$ and remain negative for angles $52° < \theta < 90°$.

This indicates that a tilted stack of magnetically coupled pancake vortices is unstable and that other arrangements of the pancake vortices are energetically preferred, even if the pancake vortices at the ends of the vortex line are fixed in position. We have shown,[4] for example, that for sufficiently large values of θ a kinked structure, in which the pancake vortices are aligned perpendicular to the layers but kinked in the middle, is energetically favored over a uniformly tilted stack.

We next discuss vortex configurations in a sample of finite thickness.

3 Sample of Finite Thickness

Consider a multilayer structure consisting of N superconducting layers centered on the planes $z_n = ns$, where $n = 0, 1, 2, ..., N-1$. We are ultimately interested in determining the configuration of a stack of N pancake vortices, one in each layer. To calculate the interaction of the pancake vortices, one first needs to derive a general expression for the current generated in layer n in response to a 2D pancake vortex in layer m, using a method[7] that extends our earlier approaches.[1,8,9,10] Consider, for example, a pancake vortex centered on the z axis at $z = ms$. In cylindrical coordinates (ρ, ϕ, z), the vector potential a has only a ϕ component, a_ϕ. Using the gauge in which the phase γ of the order parameter is $\gamma = -\phi$ in layer m and is constant in all the other layers, we may write

$$a_\phi(\rho, z; z_m) = \int_0^\infty dq A(q; z_m) J_1(q\rho) Z(q, z; z_m), \qquad (6)$$

where $J_1(q\rho)$ is a Bessel function of order 1, and $\rho = (x^2 + y^2)^{1/2}$. The function $Z(q, z; z_m)$ is determined by the requirement that Maxwell's equations must be satisfied everywhere and that the London equation must be satisfied within each of the superconducting layers.

We assume that the thickness d of each layer is much less than the bulk penetration depth λ_s within a layer. The sheet current density $\mathbf{K}$ (the integral of the current density $\mathbf{j}$ over the thickness d) also has only a ϕ component,

$$K_\phi(\rho, z_m; z_m) = -(c/2\pi\Lambda)[a_\phi(\rho, z_m; z_m) - \phi_0/2\pi\rho] \qquad (7)$$

in the superconducting layer $n = m$ containing the pancake vortex, where $\Lambda = 2\lambda_s^2/d$, and

$$K_\phi(\rho, z_n; z_m) = -(c/2\pi\Lambda) a_\phi(\rho, z_n; z_m) \qquad (8)$$

in all the other superconducting layers, $n \neq m$. It is convenient to take the limit as d and λ_s both vanish while $\Lambda = 2\lambda_s^2/d$ is held constant. Then $\Lambda = 2\lambda_\|^2/s$, and K_ϕ in superconducting layer n is given by the discontinuity of the radial component b_ρ of the local magnetic field $\mathbf{b} = \nabla \times \mathbf{a}$ across the layer:

$$K_\phi(\rho, z_n; z_m) = (c/4\pi)[b_\rho(\rho, z_n^+; z_m) - b_\rho(\rho, z_n^-; z_m)]. \qquad (9)$$

The function $A(q, z_m)$ is determined by applying Eqs. (7)-(9) and requiring continuity of a_ϕ and b_z at $z = ns$ for all n.

The Lorentz force the above vortex exerts on a second vortex at (ρ, z_n) in another layer $n \neq m$ has only a radial component,[1]

$$F_\rho(\rho, z_n; z_m) = K_\phi(\rho, z_n; z_m) \phi_0/c. \qquad (10)$$

It can be shown that this force is always negative, corresponding to a restoring force tending to align the two pancake vortices so that one is directly above the other. For example, when $\rho \ll \lambda_\|$ and $0 < |n - m|s \ll \lambda_\|$, we obtain, to good approximation,[7]

$$F_\rho(\rho, z_n; z_m) = -\left(\frac{\phi_0}{2\pi\Lambda}\right)^2 \frac{(\rho^2 + |z_n - z_m|^2 s^2)^{1/2} - |z_n - z_m|s}{\rho}. \qquad (11)$$

For the case that there is one pancake vortex in each of the N superconducting layers and no net current flows along any of the layers, the minimum-energy configuration (force-balance condition) occurs when the vortices form a straight line parallel to the z axis. A more interesting case is that for which the pancake vortex in the top layer ($n = N - 1$) is displaced to the right (in the x direction) and the pancake vortex in the bottom layer ($n = 0$) is displaced by the same distance to the left. This could be accomplished by applying a sheet current K in the y direction in the top layer and a sheet current of the same magnitude but opposite direction in the bottom layer. We have solved the force-balance equations for this case and have found that the equilibrium configuration of pancake vortices is one in which the other $N-2$ vortices (in layers $n = 1, 2, ..., N-2$) remain close to the z axis and align themselves in a nearly straight stack approximately perpendicular to the layers, while the top and bottom pancakes have large displacements to the right and left. This configuration is very different from that of N balls connected together by nearest-neighbor springs. This difference arises because the lateral restoring force between magnetically coupled pancake vortices has a long range, set by the length scale $\lambda_\parallel$, as seen in Eq. (11).

Another interesting result is that the magnitude of the restoring force on the top and bottom displaced vortices reaches a maximum when their displacement is of order $\lambda_\parallel$. This means that there is a critical value K_d of the sheet current K, above which the top and bottom 2D pancake vortices are decoupled from the other pancakes in the stack. This is an effect similar to that studied theoretically[8,9] and experimentally[11,12] in Giaever dc transformers consisting of two superconducting films separated by an insulating layer thick enough to prevent Josephson coupling.

4 Conclusions

We have found that tilted stacks of 2D pancake vortices are unstable when subjected to large tilts. If the pancake vortices at the top and bottom of a finite stack are displaced in opposite directions, they slip off the stack, leaving the other pancakes aligned very nearly parallel to a line perpendicular to the layers.

5 Acknowledgments

We thank V. G. Kogan for helpful discussions. Ames Laboratory is operated for the U.S. Department of Energy by Iowa State University under Contract No. W-7405-Eng-82. This work was supported by the Director for Energy Research, Office of Basic Energy Sciences.

References

1. J. R. Clem, *Phys. Rev.* B **43**, 7837 (1991).
2. S. N. Artemenko and A. N. Kruglov, *Phys. Lett.* A **143**, 485 (1990).
3. A. Buzdin and D. Feinberg, *J. Phys. (Paris)* **51**, 1971 (1990).
4. M. Benkraouda and J. R. Clem, *Phys. Rev.* B **53**, 438 (1996).
5. E. H. Brandt and U. Essmann, *Phys. Status Solidi* **144**, 13 (1987).
6. A. Sudbo and E. H. Brandt, *Phys. Rev. Lett.* **66**, 1781 (1991).
7. T. Pe, M. Benkraouda, and J. R. Clem, to be published.
8. J. R. Clem, *Phys. Rev.* B **9**, 898 (1974).
9. J. R. Clem, *Phys. Rev.* B **12**, 1753 (1975).
10. J. R. Clem, *Physica* C **235-240**, 2607 (1994).
11. J. W. Ekin, B. Serin, and J. R. Clem, *Phys. Rev.* B **9**, 912 (1974).
12. J. W. Ekin and J. R. Clem, *Phys. Rev.* B **12**, 1753 (1975).

NEW ASPECTS OF VORTEX DYNAMICS

Anne van Otterlo, Vadim Geshkenbein, and Gianni Blatter
Theoretische Physik, ETH-Hönggerberg, CH-8093 Zürich, Switzerland

We discuss the dynamics of vortices in the mixed state of Type II superconductors in terms of their equation of motion, which is derived from the microscopic BCS theory. A coherent view on vortex dynamics is obtained, in which both hydrodynamic flow around the vortex and the quasi-particles in the vortex core contribute to the forces on a vortex. The competition between these two provides an interpretation of the observed sign change in the Hall effect in superconductors with mean free path of the order of the coherence length in terms of broken particle-hole symmetry, which relates to a charge density variation in the vortex core. Also the damping force and the several contributions to the vortex mass are discussed, as well as the experimental observability of the vortex line charge.

1 Introduction

Recent years showed a revival of interest in the dynamical properties of vortices in superconductors, mainly due to the advent of the High Temperature Superconductors (HTSC), see Ref.[1] for a review. The first theories for vortex dynamics were formulated by Bardeen and Stephen[2], and Nozieres and Vinen[3]. These can, however, not explain the observed sign change in the Hall angle in both conventional and high temperature superconductors[4] and are limited to the low frequency domain. Many explanations for the Hall-anomaly were put forward[5], but none of these seems to explain the experimental data. More recently the Hall-anomaly has been interpreted in terms of particle-hole asymmetry of the band structure[6,7,8]. A second issue is the dynamical vortex mass, which has been discussed first by Suhl[9]. Since then different points of view were put forward[10,11], and the matter seems not to be settled at the moment. Finally, a third fascinating subject is the vortex line charge, that is discussed in Refs.[12,13] and is related to the Hall-anomaly[6,7].

It is the purpose of this paper to give a discussion of the various forces (Lorentz, Hall, and dissipative) that enter the vortex equation of motion in the dirty ($l < \xi$), clean ($\xi < l$), and superclean ($\epsilon_F \xi/\Delta < l$) limits. We review our work from Ref.[7], comment on the vortex mass, and discuss the vortex line charge that is intimately connected to the Hall-anomaly.

2 The Vortex Equation of Motion

In general, the dynamics of vortices is described by a vortex equation of motion, which is often written in the form $M\dot{\mathbf{v}}_L + \eta\mathbf{v}_L = \mathbf{j} \times \mathbf{\Phi}_0 - \gamma\mathbf{v}_L \times \hat{\mathbf{z}} + \mathbf{F}_{\text{extr}}$. Here the vortex mass M, the damping coefficient η, and the Hall force coefficient γ have been introduced, together with the vortex velocity $\mathbf{v}_L$, the applied current $\mathbf{j}$, and the flux quantum $\Phi_0 = h/2e = \pi/e$ (we use units in which $\hbar = k_B = c = 1$). Pinning and vortex-vortex interactions contribute to the extrinsic forces $\mathbf{F}_{\text{extr}}$ that we will not discuss here. In a flux flow state the Hall angle α_H is given by $\tan\theta_H = \gamma/\eta$, and therefore the interesting point is the derivation of the coefficients γ and η.

The vortex equation of motion can be derived from a model Hamiltonian H that includes a short range attractive BCS interaction with coupling constant Λ, as well as a long range repulsive Coulomb interaction (see Ref.[7] for more details). In the context of an imaginary time path integral over the electronic fields ψ and the gauge field A_α ($\alpha = \tau, x, y, z$), we consider the Euclidean action

$$S = \int d^3\mathbf{r} \int_0^\beta d\tau \left(\bar\psi_\sigma [\partial_\tau - ieA_0 + \xi(\nabla - ie\mathbf{A})]\psi_\sigma - \Lambda \bar\psi_\uparrow \bar\psi_\downarrow \psi_\downarrow \psi_\uparrow + ieA_0 n_i + \frac{\mathbf{E}^2 + \mathbf{B}^2}{8\pi} \right).$$

Here $\xi(\nabla) \equiv -\nabla^2/2m - \mu$ describes a band at chemical potential μ, and en_i denotes the background charge density of the ions. By integrating out the electronic degrees of freedom, an effective action $S = S_H + S_C$ for the vortex coordinate $\mathbf{R}$ is constructed, which reads[7]

$$S_H = \int dx \left(\frac{\mathbf{E}_v^2 + \mathbf{B}_v^2}{8\pi} + \frac{\lambda_L^2}{8\pi}(-\partial_\tau \mathbf{E}_v + \nabla \times \mathbf{B}_v)^2 + \frac{\lambda_{TF}^2}{8\pi}(\nabla \cdot \mathbf{E}_v)^2 + 2\pi \lambda_{TF}^2 n_\Delta^2 + \frac{i}{2} n_\Delta \dot\varphi_v \right)$$

$$S_C = \frac{1}{2} \int_0^\beta d\tau \int_0^\beta d\tau' \left[K_C^+(\tau - \tau') \mathbf{R}(\tau) \cdot \mathbf{R}(\tau') + i K_C^-(\tau - \tau') \mathbf{z} \cdot (\mathbf{R}(\tau) \times \mathbf{R}(\tau')) \right]. \quad (1)$$

We made a separation between the hydrodynamic and the vortex core contributions. The *hydrodynamic* part S_H of the action is due to the supercurrent flow around the vortex at scales larger than ξ. It is a function of the vortex coordinate $\mathbf{R}$ through the electric and magnetic fields $\mathbf{E}_v$ and $\mathbf{B}_v$ generated by the moving vortex. The London penetration depth λ_L and the Thomas-Fermi length λ_{TF} describe transverse and longitudinal screening, respectively. The action S_H gives rise to a small vortex mass due to the energy stored in the electric field $\mathbf{E} = (1/2e)\nabla(\mathbf{v}_L \cdot \nabla \varphi_v) - \mathbf{v}_L \times \mathbf{B}$. The first contribution to the electric field gives a contribution to the mass from scales $\sim \xi$ and is of order $M_H \sim \Phi_0^2/\xi^2$. The second contribution yields an even smaller mass $M_H' \sim \Phi_0^2/\lambda_L^2$ from the magnetic field on the scale λ_L, which is a relativistic effect. The hydrodynamic masses are never important, as the core mass (see below) is always larger by at least a factor $(\xi/\lambda_{TF})^2$.

More important is the hydrodynamic contribution γ_H to the Hall coefficient. It arises from the topological[14] term $in_\Delta \dot\varphi_v/2$ in the action and is due to the particle-hole asymmetry in the band structure. The parameter $n_\Delta = \partial_\mu N(\mu) \Delta^2/(\Lambda N(\mu))$ arises, since the electronic density in a superconductor depends on both the electro-chemical potential $\mu + e\phi$ and the energy gap, i.e. $n(\mu + e\varphi, \Delta) \approx 2eN(\mu)\varphi + n_\Delta$ in the Thomas-Fermi approximation, with the density of states per spin $N(\mu)$. Within BCS theory n_Δ can be expressed via the chemical potential dependence of the transition temperature as $n_\Delta = N(\mu)\Delta^2(\partial \ln T_c/\partial \mu)$, which is experimentally accessible. The resulting hydrodynamic Hall coefficient is $\gamma_H = \pi n_\Delta$.

The *vortex core* part S_C of the action is due to the localized Caroli–de Gennes–Matricon quasi-particle states in the core. The kernels $K_C^\pm$ that determine the mass and damping (K^+), and the Hall force (K^-), describe the effect of transitions between quasi-particle levels in the core on the vortex motion. In Fourier components they read $K_C^\pm(\omega_n) = (\omega_0 k_F^2)/4[i\omega_n/(i\bar\omega_n - \omega_0) \pm i\omega_n/(i\bar\omega_n + \omega_0)]$, where $\bar\omega_n = \omega_n + \tau_r^{-1}\text{sign}(\omega_n)$ in the relaxation time approximation and $\omega_0 = \Delta^2/\epsilon_F$ is the level spacing in the core. After analytic continuation to real time, the kernels $K_C^\pm$ can be interpreted as frequency dependent coefficients η_C and γ_C. They are[7,11]

$$\eta_C = \frac{\pi n \omega_0 (\tau^{-1} + i\omega)}{(\tau^{-1} + i\omega)^2 + \omega_0^2}, \quad \gamma_C = \frac{\pi n \omega_0^2}{(\tau^{-1} + i\omega)^2 + \omega_0^2}. \quad (2)$$

Table 1: The coefficients in the vortex equation of motion in the dirty, clean, and superclean limit.

Coefficient	dirty, $\tau \lesssim \Delta^{-1}$	clean, $\Delta^{-1} \lesssim \tau \lesssim \omega_0^{-1}$	superclean, $\omega_0^{-1} \lesssim \tau$	Galilei, $\tau = \infty$
Hall	$\sim \pi n_\Delta + \pi n(\Delta/\epsilon_F)^2$	$\pi n_\Delta + \pi n(\omega_0\tau)^2/(1+(\omega_0\tau)^2)$		πn
Damping	$\sim \pi n(\Delta/\epsilon_F)$	$\pi n(\omega_0\tau)/(1+(\omega_0\tau)^2)$		0
Mass	$\sim k_F m$	$k_F m(\epsilon_F/\Delta)^2(\omega_0\tau)^2/(1+(\omega_0\tau)^2)$		$k_F m(\epsilon_F/\omega_0)$

The frequency dependence of η_C can be interpreted as a large mass term. The vortex equation of motion can be obtained from Eq. (1) as the corresponding Euler-Lagrange equation. One aspect is, however, still missing, as can be seen from the following argument. In the Galilei invariant limit without scattering ($\tau \to \infty$) and at zero temperature we have $n_s = n$, i.e. the superfluid density equals the total electronic density. In this limit the Hall and Lorentz forces combine into one Magnus force [14]. The vortex equation of motion should be Galilei invariant as well, and only the difference $\mathbf{v}_L - \mathbf{v}_s$ can appear in the vortex equation of motion $M(\omega)[\dot{\mathbf{v}}_L - \dot{\mathbf{v}}_s] = \mathbf{j} \times \mathbf{\Phi}_0 - \pi n \mathbf{v}_L \times \hat{\mathbf{z}} = \pi n_s [\mathbf{v}_s - \mathbf{v}_L] \times \hat{\mathbf{z}}$ for $\tau \to \infty$, where the applied current density has been written in terms of the superfluid velocity as $\mathbf{j} = e n_s \mathbf{v}_s$. We see that the superfluid velocity should have a massive term as well, with the same frequency dispersion as the vortex mass. The acceleration of the superfluid can not be accounted for in our effective action formalism, and we conjecture the natural generalization to the non Galilei invariant case to be $M_s = (n_s/n)[\eta_C(\omega) - \eta_C(0)]$. Finally, we find

$$\eta_C(\omega)\mathbf{v}_L + M_s(\omega)\dot{\mathbf{v}}_s = \kappa \mathbf{v}_s \times \hat{\mathbf{z}} - [\gamma_C(\omega) + \gamma_H]\mathbf{v}_L \times \hat{\mathbf{z}} + \mathbf{F}_{\text{extr}} , \qquad (3)$$

with $\kappa = \pi n_s$. The coefficients in the equation of motion are listed for frequencies $\omega \ll \omega_0$ for the various limits in Table 1. The dirty limit results can be obtained from the expressions above by taking the dirty limit expression for H_{c2} in $\omega_0 \equiv \omega_{c2} = eH_{c2}/m$ and considering the limit $\omega_0\tau \ll 1$. Or in other words, we replace $\omega_0\tau$ by $\pi\Delta/(4\epsilon_F)$.

The resulting Hall angle is determined by the sum of γ_C and γ_H and a sign change may occur if $n_\Delta \sim (\partial \ln T_c/\partial \mu) < 0$. The different temperature dependence of γ_C and γ_H can also account for a double sign change as explained in Ref. [7].

3 Vortex Electrostatics

Apart from its relation to the Hall-anomaly, the particle-hole asymmetry also causes vortices to carry a line charge-density [12,13]. The *bare* charge density is given by $-e[n(\mu, \Delta(R)) - n(\mu, \Delta(\infty))]$, where R is the distance from the vortex phase singularity. The real charge density is reduced considerably by screening, but may still be observable [13]. The conclusion of Ref. [13] is that outside the superconductor, a single vortex produces the field of a surface dipole of atomic size $\mathbf{d} \sim e a_B$, with Bohr radius $a_B = 1/(me^2)$.

The field of a vortex can in principle be observed by atomic force microscopy (AFM) or with other techniques like SET-electrometry. The vortex dipole induces an image dipole in the AFM-tip, which leads to a force $F \sim 10^{-17}$N. If the AFM is biased at voltage V against the superconductor a much larger force $F \sim 10^{-13}$N can be obtained for typical system parameters. The observation of a force of this size is on the border of present day resolution. A voltage biased AFM-tip can also distinguish the sign of the vortex charge which is correlated to the presence of a Hall-anomaly.

The sign of the vortex core charge is opposite to the sign of the hydrodynamic contribution γ_H to the Hall conductivity. In an electron-like superconductor, electrons are expelled from (attracted to) the vortex core if $\partial_\mu N(\mu) > 0 (< 0)$. Thus, an electron-like superconductor with Hall anomaly will have a negatively charged core with a positively charged screening cloud around it, and in a hole-like superconductor with Hall-anomaly the situation will be reversed.

Acknowledgments

We thank Mikhail Feigel'man and Eduard Sonin for discussion. Furthermore, the support of the Swiss National Foundation is gratefully acknowledged.

References

1. G. Blatter et al., Rev. Mod. Phys. **66**, 1125 (1994).
2. J. Bardeen and M.J. Stephen, Phys. Rev. **140**, A1197 (1965).
3. P. Nozieres and W.F. Vinen, Philos. Mag. **14**, 667 (1966).
4. S.J. Hagen *et al.*, Phys. Rev. B **47**, 1064 (1993); J.M. Harris et al., Phys. Rev. Lett. **71**, 1455 (1993), and Phys. Rev. Lett. **73**, 1711 (1994); and many more.
5. Z. D. Wang and C.S. Ting, Phys. Rev. Lett. **67**, 3618 (1991); R.A. Ferrell, Phys. Rev. Lett. **68**, 2524 (1992); A. Freimuth et al., Phys. Rev. B **44**, 10396 (1991); Z. D. Wang et al., Phys. Rev. Lett. **72**, 3875 (1994); and many more.
6. M.V. Feigel'man et al., Physica C **235-240**, 3127 (1994); Pis'ma Zh. Eksp. Teor. Fiz. **62**, 811 (1995) [JETP Lett. **62**, 834 (1995)].
7. A. van Otterlo et al., Phys. Rev. Lett. **75**, 3736 (1995); and unpublished.
8. A.T. Dorsey, Phys. Rev. B **46**, 8376 (1992); N.B. Kopnin et al., J. Low Temp. Phys. **90**, 1 (1993); A.I. Larkin and Yu.N. Ovchinnikov, Phys. Rev. B **51**, 5965 (1995).
9. H. Suhl, Phys. Rev. Lett. **14**, 226 (1965).
10. J-M. Duan and A.J. Leggett, Phys. Rev. Lett. **68**, 1216 (1992), (E) **69**, 1148 (1992); J-M. Duan, Phys. Rev. B **48**, 333 (1993)
11. N.B. Kopnin and A. Lopatin, Phys. Rev. B **51**, 15291 (1995); N.B. Kopnin and M.M. Salomaa, Phys. Rev. B **44**, 9667 (1991); E. Simanek, Phys. Lett. A **194**, 323 (1994).
12. D.I. Khomskii and A. Freimuth, Phys. Rev. Lett. **75**, 1384 (1995)
13. G. Blatter et al., unpublished
14. P. Ao and D.J. Thouless, Phys. Rev. Lett. **70**, 2158 (1993)

NUMERICAL STUDIES ON THE VORTEX MOTION IN HIGH-T_c SUPERCONDUCTORS

Z. D. WANG

Department of Physics, University of Hong Kong, Pokfulam Road, Hong Kong

In the framework of the Bardeen-Stephen and Noziéres-Vinen approach and taking into account the backflow current due to the pinning and some other interactions, we derive a general phenomenological equation for the vortex motion in the presence of thermal fluctuations. Based on this equation, a simple analytical analysis on the dc vortex motion, particularly on the Hall effect, is introduced first. Then numerical simulation results are presented. Finally, we simulate numerically the interesting washboard effect of the moving vortex lattice. Our theoretical results are also compared with the relevant experimental observations for high-T_c superconductors.

Since the discovery of high-temperature superconductors (HTSC), there have been extensive experimental and theoretical studies on the vortex motion. A variety of interesting phenomena in connection with the vortex motion, such as giant flux creep,[1] anomalous Hall effect,[2,3] and washboard effect[4] etc., are exhibited in the mixed state of HTSC. So far, several theoretical models for the vortex motion have been proposed.[5,6,7] More particularly, a microscopic effective action analysis on the vortex motion was reported very recently.[8] Nevertheless, such kind of microscopic calculation can hardly take into account directly the backflow current effect on the vortex motion arising from the pinning as well as some other interactions; while this backflow is believed to exist[5] and plays an important role in the vortex motion in HTSC.[6] In this brief report, we present a general phenomenological theory for the vortex motion, in which both the backflow current and the thermal fluctuation are included. We then report our numerical analyses on the vortex motion based on this theory.

Let us consider a vortex carrying a quantum flux $\Phi_0 = hc/2e$ in the $\hat{z}$-direction. Then an en equation of motion for the charge fluid inside the core of the vortex can be established,[5,6]

$$\mathbf{F}_{nc} + \mathbf{F}_T^{(in)} + \mathbf{F}_b^{(in)} = (Nm/\tau)\pi a^2 \mathbf{v}_{nc}, \quad (1)$$

where N is the charge carrier density and τ is the momentum relaxation time. The term $(Nm/\tau)\pi a^2 \mathbf{v}_{nc}$ denotes the momentum dissipated inside the normal core with $\mathbf{v}_{nc}$ as the drift velocity of carriers. $\mathbf{F}_T^{in}$ is the force due to thermal fluctuations. $\mathbf{F}_b^{in}$ and $\mathbf{F}_{nc}$ are the effective backflow and the external driving forces acting on the charge fluid inside the normal core. Using the approach similar to that adopted in Ref.[6], we arrive at an effective equation of vortex motion

$$\eta \mathbf{v}_\phi = \mathbf{F}_L + \mathbf{F}_p + \mathbf{F}_T - \beta_0(1 - \bar{\gamma})\mathbf{F}_L \times \hat{\mathbf{n}} - \beta_0(1 + \bar{\gamma})\mathbf{F}_p \times \hat{\mathbf{n}}, \quad (2)$$

where $\mathbf{F}_L = \mathbf{J} \times \mathbf{\Phi}_0$ is the Lorentz force, $\mathbf{F}_T$ and $\mathbf{F}_b = \mathbf{F}_p + \mathbf{F}_{other}$ are, respectively, the thermal noise and effective backflow forces on the vortex, $\mathbf{v}_\phi$ is the velocity of the vortex line, $\beta_0 = \tau e H_{c2}/m = \omega_{c2}\tau \ll 1$ with H_{c2} the usual upper critical field, $\eta = Ne\Phi_0\beta_0$ is the usual viscous coefficient, $\bar{\gamma} = \gamma(1 - \overline{H}/H_{c2})$ with $\overline{H}$ as the average magnetic field over the core and γ as the parameter describing contact force on the surface of the core, which depends on T in the following way:[6] $\gamma \sim 0$ (NV limit) for $\xi/l \ll 1$ and $\gamma \sim 1$ (BS limit) for $\xi/l \geq 1$ with l as the mean free path of the carrier. Here, we note that the effective backflow consists of two parts: the pinning force $\mathbf{F}_P$ and $\mathbf{F}_{other}$ arising from the other effective interactions[9]. Equation (2) is an effective phenomenological equation to describe the over-damped vortex motion in the presence of backflow current and thermal fluctuations.

We now first present a brief qualitatively analysis. Taking time average, we obtain

$$\eta <\mathbf{v}_\phi> = \mathbf{F}_L + <\mathbf{F}_b> -\beta_0(1-\bar{\gamma})\mathbf{F}_L \times \hat{n} - \beta_0(1+\bar{\gamma})<\mathbf{F}_b> \times \hat{n}. \tag{3}$$

Considering that $\beta_0 \ll 1$ and $\langle F_b \rangle$ should approximately be antiparallel to $\langle v \rangle_\phi = v_L$, i.e., $\langle F_b \rangle \approx -\Gamma(v_L)\mathbf{v}_L = -[\Gamma_P + \Gamma_{other}]\mathbf{v}_L$ with Γ as a positive scale function, we obtain the Hall resistivity or Hall conductivity as

$$\rho_{xy} \approx \frac{\beta_0 \rho_{xx}^2}{\Phi_0 B}[\eta(1-\bar{\gamma}) - 2\bar{\gamma}(\Gamma_p + \Gamma_{other})], \tag{4}$$

or

$$\sigma_{xy} \approx \frac{\beta_0}{\Phi_0 B}[\eta(1-\bar{\gamma}) - 2\bar{\gamma}(\Gamma_p + \Gamma_{other})]. \tag{5}$$

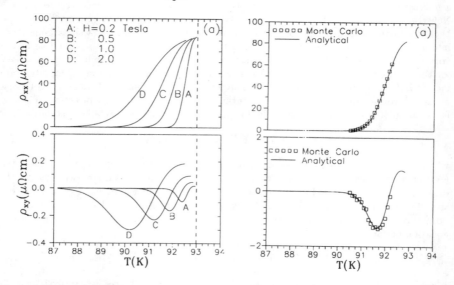

Figure 1: (a) Analytical resistivities as a function of temperature; (b) Monte Carlo simulation (open squares) and analytical resistivities. (For the details of parameters, see Ref. [10].)

From the above equation, we are able to qualitatively explain the experimentally observed sign reversal and the scaling behavior of Hall effect. More interestingly, at a fixed field and above certain temperature T^*, the irradiation could not change the value of $\Gamma = \Gamma_p + \Gamma_{other}$ significantly if the vortex motion in the region of 'free' flux flow, so that the Hall conductivity will scarcely exhibit eminent difference between the values before and after irradiation for either $T < T_0(\xi(T_0) \approx l)$ and $T > T^*$, which is also roughly consistent with some recent experimental data.[10] For some specific kinds of pinning potentials and by neglecting Γ_{other}, a self-consistent solution of Eq. (2) as well as analytical results can be found.[11] In Fig. 1, we plot these analytically calculated data and some comparisons with direct Monte Carlo simulations. We also notice that, a scaling behavior with the exponent $\beta \sim 1.7$ near the onset of negative ρ_{xy} was reported for the YBCO system.[12] To elucidate this feature, we have performed direct Monte Carlo simulation for the vortex motion described by Eq. (2), by using random pinning potentials and parameters of YBCO materials. Numerical results are consistent with experimental observation within reasonable errors (see Fig. 2).

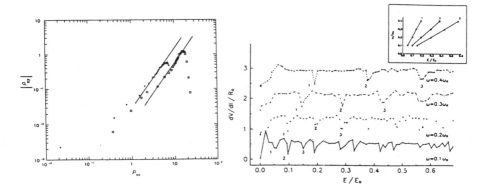

Figure 2: Log-log plot of $|\rho_{xy}|$ against ρ_{xx} at two magnetic fields: 1.5 T (*) and 4t (□).

Figure 3: dV/dI against E at the frequencies ω=0.1, 0.2, 0.3 and $0.4\omega_0$ (the curves are displaced vertically for clarity).

Finally, we turn to address numerically a very interesting ac phenomenon associated with the vortex lattice motion in random pinning potentials —washboard effect, which was reported to have been observed in YBCO system very recently.[4] As we know, the washboard effect is an interference between the applied ac current and intrinsic oscillation, the velocity of the vortex lattice acquiring a weak ac component in the frequency $\omega_{int} = 2\pi\langle v\rangle/a_0$ with a_0 the vortex lattice constant. For simplicity, here, we neglect the transverse terms and consider only a one-dimensional vortex motion with N_{eff} effective vortices. Thus, Eq. (2) may be rewritten as

$$\eta\dot{x} = F_L + F_T + F_p \qquad (6)$$

with

$$F_p = \frac{1}{N_{eff}} \sum_{s=1}^{N_{eff}} F_p^{(s)}, \qquad (7)$$

where $\dot{x} = v_\phi(s)$ $(s = 1, \cdots, N_{eff})$. Although the present model is quite simple, it captures the essential physics of the washboard effect. We solve Eq. (6) for the YBCO materials. The typical result is shown in Fig 3. We can see clearly that, (i) significant negative peaks appear in dV/dI. Whenever $p = \omega_{int}/\omega = 1, 2, 3, \ldots$ and the position of the fixed p increases linearly with ω, which agrees well with the experimental results reported in Ref.[4] It is worth to point out that, as the external ac frequency is relatively high $(\omega/\omega_0 \gg \eta/m$ with m as the effective mass of a vortex), the mass term $(m\ddot{x})$ may need to be taken into account. Interestingly, the peaks at sub-harmonic frequencies: $1/2, 3/2$, appear, which is also in good agreement with the experiment measurement.

Acknowledgments

The author greatly acknowledges many valuable discussions with Profs. C. S. Ting and J. Dong, and Dr. Q. H. Wang. The work was supported by the RGC grant of Hong KOng under No.: HKU262/95p and a CRCG grant at the University of Hong Kong.

References

1. K. A. Müller, M. Takashige, and J. G. Bednorz, Phys. Rev. Lett. **58**, 1143 (1987).
2. M. Galffy and E. Zirngiebl, Solid State Commun. **68**, 929 (1988); S. J. Hagen, C. J. Lobb, R. L. Greene, M. G. Forrester, and J. H. Kang, Phys. Rev. B **41**, 11 630 (1990); T. R. Chien, T. W. Jing, N. P. Ong, and Z. Z. Wang, Phys. Rev. Lett. **66**, 3075 (1991); M. D. Lan, J. Z. Liu, Y. X. Jia, and R. N. Shelton, J. Phys. Chem. Solids **55**, 803 (1994).
3. A. V. Samoilov, Phys. Rev. Lett. **71**, 617 (1993); P. J. M. Wöltgens, C. Dekker, and H. W. de Wijn, Phys. Rev. Lett. **71**, 3858 (1993).
4. J. M. Harris, N. P. Ong, R. Gagnon, L. Tallefer, Phys. Rev. Lett. **74**, 3684 (1995).
5. J. Bardeen and M. J. Stephen, Phys. Rev. **140**, A1197 (1965); P. Noziéres and W. F. Vinen, Philos. Mag. **14**, 667 (1966).
6. Z. D. Wang and C. S. Ting, Phys. Rev. Lett. **67**, 3618 (1991); Z. D. Wang, Jinming Dong, and C. S. Ting, Phys. Rev. Lett. **72**, 3875 (1994).
7. N. B. Kopnin, B. I. Ivlev, and V. A. Kalatsky, JETP Lett. **55**, 750 (1992); V. M. Vinokur, V. B. Geshkenbein, M. V. Feigel'man, and G. Blatter, Phys. Rev. Lett. **71**, 1242 (1993).
8. A. van Otterlo, M. Feigel'man, V. Geshkenbein, and G. Blatter, Phys. Rev. Lett. **75**, 3736 (1995).
9. R. S. Thompson and C.-R. Hu, Phys. Rev. Lett. **27**, 1352(1971).
10. A. V. Samoilov *et al.*, Phys. Rev. Lett. **74**, 2351 (1995).
11. Q. Wang, Z. D. Wang, and X. Yao, Physica C **254**, 285 (1995).
12. J. Lo *et al*, Phys. Rev. Lett. **68**, 690(1992).

Hall Anomaly in the Mixed State of Type II Superconductors

P. Ao

Department of Physics, Box 351560, University of Washington, Seattle, WA 98195, USA; and Department of Theoretical Physics, Umeå University, 901 87, Umeå, SWEDEN

A theory based on the competition between the vortex many-body correlation and pinning is constructed to explain the Hall anomaly.

A vacancy or an interstitial moving in a vortex lattice feels a Magnus force

$$\mathbf{F}_M^d = \mp q \frac{\rho_s}{2} h d \, (\mathbf{v}_s - \dot{\mathbf{r}}_0) \times \hat{z} \, , \qquad (1)$$

with '−' for a vacancy and '+' for an interstitial. Here $q = \pm$ is the vorticity, ρ_s the superfluid electron density, d the superconductor thickness, $\mathbf{v}_s$ the superfluid velocity, and $\dot{\mathbf{r}}_0$ the defect velocity. This is similar to the dynamics of a hole or a particle in a semiconductor in the presence of a magnetic field, with a pinned perfect vortex lattice as a filled valence band. It is clear that in the present situation the formation of point defects, vacancies and interstitials, is a result of the many-body correlation: a property of a vortex lattice. The pinning down of the lattice maximizes the defect contributions to the resistivities. Other manifestations of the competition between the many-body correlation and pinning, the source of the rich physics in the mixed state, have been amply studied [1].

Equation (1) shows that both vacancies and and interstitials will be carried along the externally applies current. This implies that in the case of a vacancy motion the vortex effectively moves against the current. A schematic illustration of this dynamics for a vacancy is given in the figure, where the preferring motion of the vacancy to B over A is caused by the Magnus force on vortices. Quantitatively, vacancies and interstitials may be considered as independent particles moving in the periodic potential formed by the vortex lattice and a random potential due to the residue effect of pinnings. Since the vacancy formation energy can be lower than that of interstitial [2], its motion will dominate the defect contributions to the resistivities. This leads to the conclusion that at low enough temperatures the sign of the Hall resistivity is different from its sign in the normal state because of this dominance. This explains the Hall anomaly, the sign change of the Hall effect in the mixed state [3], with the total transverse force [4] felt by an individual vortex.

Within this vacancy motion theory, several quantitative results have been found. First, I have obtained the vacancy formation energy, the energy scale in the problem. Second, a scaling relation between the Hall and longitudinal resistivities at low temperatures have been found. Both of them agree with experimental observations. My results are also consistent with the Nernst effect measurement. [5]

A particular nice expression which I have found is for the Hall conductivity near the superconducting transition temperature T_{c0}, where the superfluid density $\rho_s = \rho_{s0}(1 - T/T_{c0})$, and the vacancy formation energy $\epsilon_v = \epsilon_{v0}(1 - T/T_{c0})$. If I take the vortex density

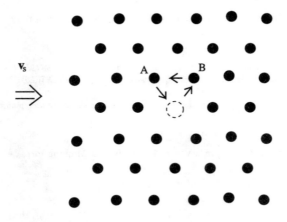

Figure: The motion of a vacancy in a vortex lattice.

● vortex ; ⊙ vacancy

$n_0 = B/\Phi_0$, with $\Phi_0 = hc/2e$ the flux quantum, I find the Hall conductivity

$$\sigma_{xy} = \alpha_1 \frac{(1 - T/T_{c0})^2}{B}, \qquad (2)$$

with $\alpha_1 = -q(2e^2/h)\rho_{s0}\Phi_0\gamma\epsilon_{v0}/4k_B T_{c0}$. The temperature and magnetic field dependences in Equation (2) have been observed experimentally [6]. Taking $\rho_{s0} = 10^{21}/cm^3$, $\gamma = 1$, and $\epsilon_{v0}/k_B T_{c0} = 50$, I find $|\alpha_1| \sim 20\, T\mu\Omega^{-1}cm^{-1}$, which is close to the experimental values.

Acknowledgments This work was supported in part by Swedish Natural Science Research Council and by US NSF grant no. DMR-9220733.

1. See, for example, E.H. Brandt, Rep. Prog. Phys. **58**, 1465 (1995), for a recent review.
2. E.H. Brandt, Phys. Stat. Sol. **36**, 371 (1969);
   D.C. Hill, D.D. Morrison, and R.M. Rose, J. Appl. Phys. **40**, 5160 (1969).
3. See, for example, S.J. Hagen et al., Phys. Rev. **B47**, 1064 (1993), for a review.
4. P. Ao and D.J. Thouless, Phys. Rev. Lett. **70**, 2158 (1993);
   P. Ao, Q. Niu, and D.J. Thouless, Physica **B194-196**, 1453 (1994).
5. P. Ao, J. Supercond. **8**, 503 (1995).
6. D.M. Ginsberg and J.T. Manson, Phys. Rev. **B51**, 515 (1995);
   C.C. Almasan et al., Phys. Rev. **B51**, 3981 (1995);
   X.G. Qiu et al., Phys. Rev. **B52**, 12994 (1995).

NONLINEAR DYNAMICS IN THE MIXED STATE OF HIGH TEMPERATURE SUPERCONDUCTORS

M. W. COFFEY
*Department of Chemistry, Regis University, Denver,
CO 80221, USA*

A new formulation of the (1+1)-dimensional coupled nonlinear electrodynamics of high-temperature superconductors in the mixed state is presented. Special coordinates are used to combine the governing Maxwell and London partial differential equations. A novel nonlinear evolution equation for the specific area A is derived for a type-II superconductor, including the effects of dissipation, nonlocal vortex interaction, and vortex inertia.

1 Introduction

Recent theoretical studies have shown that soliton propagation could result from a type-II superconductor in the mixed state in the ultraclean regime [1]. In this paper the assumption of ultracleanliness is dropped by including a viscous drag force in the vortex equation of motion. A new nonlinear evolution equation for the specific area A is then derived based upon London theory. This fourth order equation for A completely encompasses the vortex dynamics in (1+1) dimensions in the absence of pinning.

2 Derivation of nonlinear evolution equation

Here an isotropic superconductor and frequencies well below the gap frequency are assumed. Furthermore, for simplicity, a normal current density contribution is neglected. Assuming a superconductor geometry with magnetic induction $B = B_z$, vortex velocity $v = v_x$, and supercurrent density $J = J_y$, depending only on the x spatial coordinate, the governing equations are [1]

$$\frac{\partial n}{\partial t} + \frac{\partial(nv)}{\partial x} = 0, \quad \mu \frac{dv}{dt} + \eta_v v = -\frac{\phi_0}{\mu_0}\frac{\partial B}{\partial x}, \quad n\phi_0 = B - \lambda_L^2 \frac{\partial^2 B}{\partial x^2}, \quad (1)$$

where $d/dt = \partial/\partial t + v\partial/\partial x$ is the convective derivative.

The London equation in (1), with penetration depth λ_L and flux quantum ϕ_0, has been written in terms of the vortex areal density $n(x,t)$ [1]. On the left-hand side of the vortex equation of motion in (1), μ is the mass per unit length [2], η_v is the drag coefficient [3], and the right-hand side is the Lorentz force $\phi_0 J_y B_z$. The first Eq. in (1) expresses conservation of flux lines [1].

By perturbing Eqs. (1) about constant vortex density $n_0 = B_0/\phi_0$ and magnetic induction B_0 and zero vortex velocity, with the dependence $\exp i(kx - \omega t)$, it is seen that the dispersion relation for the linear damped propagation problem is

$$\omega(-i\mu\omega + \eta_v) = (-i\phi_0 B_0/\mu_0)k^2(1 + \lambda_L^2 k^2)^{-1}. \tag{2}$$

Now the Eqs. (1) can be scaled with $t' = \omega_0 t$, $x' = x/\lambda_L$, $B' = B/B_0$, $n' = n/n_0$, $v' = v/\omega_0\lambda_L$, where $\omega_0 = \sqrt{\phi_0 B_0/\mu_0\mu}/\lambda_L$. For notational ease the primes are dropped in this paper, and η_v below is written for $\eta_v \lambda_L/\sqrt{\mu_0\mu\phi_0 B_0}$.

In order to establish a novel single scalar evolution equation containing the whole system (1), a new number per length variable η is introduced,

$$\eta(x,t) = \int^x n(x',t)dx'. \tag{3}$$

The inverse length η and time t are taken as new independent variables and an equation for the specific area $A(\eta,t) = 1/n(x(\eta,t),t)$ is sought. The transformation back to real space is accomplished by $x(\eta,t) = \int^\eta A(\eta',t)d\eta'$. From the relations $\partial/\partial x = n\partial/\partial \eta$, $d/dt = \partial/\partial t$, and Eq. (1) follows

$$\frac{\partial v}{\partial t} + \eta_v v = -n\frac{\partial B}{\partial \eta}. \tag{4}$$

So far the vortex continuity equation has been used, in order to define η, together with the equation of motion. Next the London Eq. (1) is converted to $\ddot{A} = 1 - BA$ where $\ddot{} \equiv \partial_t^2$. Finally, eliminating the magnetic induction B yields

$$\ddot{A} - \eta_v \dot{A} + \frac{\partial}{\partial \eta}\frac{1}{A}\frac{\partial}{\partial \eta}\left[\frac{1}{A}\left(1 - \ddot{A} + \eta_v \dot{A}\right)\right] = 0. \tag{5}$$

Equation (5) is a new formulation of the vortex dynamical system in terms of the specific area A. Small amplitude for A has not been assumed; the equation for $A = A(\eta,t)$ is exact. Equation (5) accounts for vortex drag and therefore energy loss. It ignores pinning, which breaks translational symmetry of the superconductor.

References

1. M. W. Coffey, Phys. Rev. B **52**, R13122 (1995); Phys. Rev. B **53**, 63 (1996); Phys. Lett. A **211**, 53 (1996).
2. H. Suhl, Phys. Rev. Lett. **14**, 226 (1965).
3. J. Bardeen and M. J. Stephen, Phys. Rev. A **140**, 1197 (1965).

Vortex Dynamics in Superfluids: Cyclotron Type Motion

E. Demircan[1], P. Ao[2], and Q. Niu[1]
[1] Department of Physics The University of Texas at Austin
Austin, TX 78712
[2] Department of Theoretical Physics
Umeå University, S-901 87, Umeå, SWEDEN

April 8, 1996

Abstract

We have investigated the dyanmics of vortices in superfluid systems within the framework of the nonlinear Schrödinger equation. We have obtained the fundemantal dynamical equations that govern the motion of the vortex and the condensate. The natural motion of the vortex is found to be of cyclotron type, whose frequency is on the order of phonon velocity divided by the coherence length, and may be heavily damped due to phonon radiation.

Vortex dynamics have been an active research area for some time [1]. In this work we have tried to give a simple understanding of the physical picture [2] that underlies the dyanmics of vortices. We have considered a single vortex line in an infinite neutral superfluid at zero temperature. The Lagrangian that governs the dyanmics of this system within the framework of the nonlinear Scrödinger theory is given by

$$L = \int d^2 r \left[i\hbar \psi^* \frac{\partial}{\partial t} \psi - \frac{\hbar^2}{2m} |\nabla \psi|^2 - \frac{1}{2} V \left[|\psi|^2 - \bar{\rho} \right]^2 \right] + \frac{1}{2} M_e \dot{\mathbf{r}}_0^2, \qquad (1)$$

where m is the mass of the superfluid particle and, V represents a repulsive interaction. We have also included a term to account for the kinetic energy of trapped particles in the core which also serves as a parameter to control the time scale of the vortex motion. The bare vortex case corresponds to $M_e = 0$. The phase and density of the condensate wavefunction corresponding to a static vortex is given by $S_0 = \tan^{-1} \frac{y-y_0}{x-x_0}$ and $\rho_0 \approx \frac{2|\mathbf{r}-\mathbf{r}_0|^2}{1+2|\mathbf{r}-\mathbf{r}_0|^2}$ in scaled units with $\xi = \hbar/(mV\bar{\rho})^{1/2}$ and $\tau = \hbar/(V\bar{\rho})$.

When the vortex is in motion, the fields of density and phase of the condensate gets corrections over the static forms: $S = S_0(\mathbf{r}-\mathbf{r}_0(t)) + S_1(\mathbf{r},t)$, $\rho = \rho_0(\mathbf{r}-\mathbf{r}_0(t)) + \rho_1(\mathbf{r},t)$, We have found that the vortex coordinate and the density and pahse corrections satisfy the following equations

$$-2\pi \dot{\mathbf{r}}_0 \times \hat{\mathbf{z}} + \int \left[\dot{S}_1 \nabla \rho_0 - \dot{\rho}_1 \nabla S_0 \right] d^2 r - M_e \ddot{\mathbf{r}}_0 = 0, \qquad (2)$$

$$\dot{S}_1 + \nabla S_0 \cdot \nabla S_1 + \frac{1}{4\rho_0^2}\nabla\rho_0 \cdot \nabla\rho_1 - \frac{1}{4\rho_0}\nabla^2\rho_1 - \frac{|\nabla\rho_0|^2}{4\rho_0^3}\rho_1 + \frac{\nabla^2\rho_0}{4\rho_0^2}\rho_1 + \rho_1 = \dot{\mathbf{r}}_0 \cdot \nabla S_0, \quad (3)$$

$$\dot{\rho}_1 + \nabla\rho_0 \cdot \nabla S_1 + \rho_0 \nabla^2 S_1 + \nabla S_0 \cdot \nabla\rho_1 = \dot{\mathbf{r}}_0 \cdot \nabla\rho_0. \quad (4)$$

These equations are the analogue of the Maxwell equations of charged particles in vortex dynamics.

Upon analysis of these equations we showed that the natural motion of the vortex is of cyclotron type, in a general sense, the analogue of the massive branch of the helical vortex waves in classical fluids: $\mathbf{r}_0 = \mathrm{Re}\left[be^{-i\omega t}(\hat{\mathbf{x}} + i\hat{\mathbf{y}})\right]$. The cyclotron motion is a consequence of the Magnus force acting on the vortex and sets a new length scale which separates the two qualitatively different regimes: the adiabatic regime ($\xi < r < \lambda$) where the condensate follows the motion of the vortex adiabatically (here λ is the phonon wavelength at the cyclotron frequency), and the radiation regime ($r > \lambda$) where the collective excitations radiated by the motion of the vortex are superimposed on the static vortex. We have performed some numerical calculations, and found that this wavelength is roughly given by $\lambda \approx \xi(1 + M_e/M_c)$. When the mass of the trapped particles is large compared to the core mass, the cyclotron frequency is low, and there is a logarithmic correction to the vortex mass due to the adiabatic following of the density and phase fluctuations in the large region of adiabatic regime [3]. For a bare vortex on the other hand, the cyclotron period is on the order of the time that a phonon travels a coherence length, and the vortex mass is on the order of the mass of the fluid that can occupy the core. The adiabatic regime is essentially empty, and the logarithmic correction is absent. However, radiation damping is heavy and may completely overshadow the cyclotron motion. For large external mass, the cyclotron frequency is $\propto M_e^{-1}$, whereas the damping is $\propto M_e^{-2}$ and, it may be possible to overcome the radiaton damping.

Similar calculations can also be performed for charged superfluids. The physical picture will remain the same, yet become more complicated due to the existance of another length scale set by the London penetration length. For a superconducting thin film the London pentration length is large and the previous analyses essentially remain the same. Then a bare vortex in a superconductor should have a cyclotron frequency $\omega = \frac{\hbar}{m_e \xi^2}$, which is on the order of $\omega = 45$ MHz for Al thin films.

We gratefully acknowledge communications with D. J. Thouless, A. L. Fetter, M. Stone, D. P. Arovas, J. A. Freire, G. A. Georgakis, M. C. Chang, and A. Barr. This work is supported by the Welch Foundation.

References

[1] Q. Niu, P. Ao, D. J. Thouless, Phys. Rev. Lett. **72**, 1706 (1994); **75**, 975 (1995), P. Ao and D. J. Thouless, Phys. Rev. Lett. **70**, 2158 (1993).

[2] E. Demircan, P. Ao, Q. Niu, cond-matt/9604010

[3] J. M. Duan Phys. Rev. B **48**, 333 (1993); J. M. Duan and A. J. Leggett, Phys. Rev. Lett. **68**, 1216 (1992).

NUMERICAL STUDY OF WASHBOARD EFFECT IN HIGH T_C SUPERCONDUCTORS

Z. D. WANG and K. M. HO
Department of Physics, University of Hong Kong, Pokfulam Road, Hong Kong

Using a simple model of the moving vortex lattice, the washboard effect in the mixed state of high-T_c superconductors, which is due to the interference between the intrinsic oscillation of the vortex lattice and the applied ac driving current, is investigated by numerical simulation. We observe that interference peaks appear in the dV/dI-V curve whenever the intrinsic and external frequencies are harmonically or sub-harmonically related. The obtained numerical results are in good agreement with recent experimental measurements on $YBCO$ crystal.

Even in the present of random pining certers, the velocity of the vortex lattice acquires a weak ac component at a frequency ω_{int}. Interference between intrinsic oscillations and the applied ac current $I_{ac}e^{-\omega t}$ occurs whenever ω_{int} and ω are harmonically related. As a result, many steps exist in the current-voltage characteristics, which is referred as the "Washboard Effect" and was first measured by Fiory in conventional superconductors [1]. The relevant analytical theory which neglects the thermal fluctuations was proposed quite long time ago [2]. Very recently, Harris *et al* reported the detection of the washboard oscillations in a $YBa_2Cu_3O_{6.93}$ crystal [3]. In this paper, the washboard effect in high-T_c superconductors is numerically exhibited.

In the mixed state of a type-II superconductor, the periodic vortex structure could be considered to be rigid one approximately. For simplicity and to capture the essential feature of the washboard effect, we focus our attention on the simulation in the framework of one-dimensional vortex motion model. The equation of motion along the longitudinal direction can then be simply written as [4]

$$\sum_s (m\ddot{x}_{(s)} + \eta\dot{x}_{(s)}) = \sum_s (F_L^{(s)} + F_{th}^{(s)} + F_p^{(s)}), \tag{1}$$

where m is the effective inertial mass of a vortex, η the viscosity coefficient, $x_{(s)}$ the displacement of the s-th vortex in the lattice, $F_L^{(s)} = J\Phi_o/c$ the external ac driving force, $F_{th}^{(s)}$ is the thermal noise force, and $F_p^{(s)}$ is the pinning force on the s-th vortex, which is the interaction with a number of random pinning centers. As usual, we first neglects the effective mass term in Eq.(1) Here, two cases of pinning density, namely, 100 pinning sites in the systems of length 100 ξ_{ab} and length $800\xi_{ab}$ with periodic boundary conditions, are considered. In the calculation, 10^5 units of the time $(1/\omega)$ are used to average the velocity with the warming-up time $\sim 10^4$ units. Parameters are chosen to be more suitable for Y-Ba-Cu-O materials and close to the experiment [3]. Typical curves of dV/dI against the electric field E are shown in Fig.1 (a) and (b) respectively. From the figures, we can find that, in both cases, significant negative peaks in dV/dI appear whenever $p = 1, 2, 3 \cdots$. Such a behavior is actually in agreement with the experimental results reported in Ref.3, except that weak peaks at $p = 1/2, 3/2, \cdots$ are also observed there. To elucidate this sub-harmonic peaks, we also take into account the effect of the inertial mass term in Eq.(1). The plots of dV/dI against E with and without the mass term under the same conditions are shown in Fig.2 (a) and (b), respectively. From these two precise figures, it is easy to rule out the noise signal and identify the peaks at sub-harmonic frequencies: 1/2, 3/2.

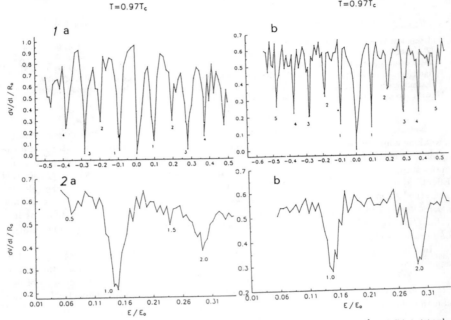

Figure 1: dV/dI against E at the frequency $\omega = 0.1\omega_o$ at the pinning densities (a) ξ_{ab}^{-1} and (b) $(1/8)\xi_{ab}^{-1}$.

Figure 2: dV/dI against E at $\omega = 0.3\omega_o$ and the density of $(1/8)\xi_{ab}^{-1}$: (a)$m = 2\eta/\omega_o$ and (b)$m = 0$.

Acknowledgments

This work was supported by the RGC Research Grant of Hong Kong under Grant No. HKU262/95P, CRCG grant at the University of Hong Kong.

References

1. A. T. Fiory, Phys. Rev. Lett. **27**, 501(1971); Phys. Rev. B. **7**, 1881(1973).
2. A. I. Larkin and Yu. N. Ovchinnikov, Sov. Phys. JETP **38**, 854(1974)].
3. J. M. Harris, N. P. Ong, R. Gagnon, and L. Taillefer, Phys. Rev. Lett. **74**, 3684(1995).
4. Z. D. Wang, J. M. Dong and C. S. Ting, Phys. Rev. Lett. **72**, 3875(1994).

TIME-WINDOW EXTENSION FOR MAGNETIC RELAXATION FROM MAGNETIC HYSTERESIS LOOP MEASUREMENTS

Qianghua Wang[1,2], Xixian Yao[2], Z. D. Wang[1] and Jian-Xin Zhu[1]
[1] *Department of Physics, University of Hong Kong, Pokfulam, Hong Kong*

[2] *Physics Department and National Laboratory of Solid State Microstructure, Nanjing University, Nanjing 210093, China*

A one to one map is established between the time t in a relaxation measurement and the field sweeping rate $\dot{H}$ in a magnetic-hysteresis-loop (MHL) measurement. Utilizing this map, we are able to extend the time-window of the relaxation curve from the MHL measurements. The advantage over previous extension schemes lies in the fact that in the present scheme no microscopic uncertainties are involved. The present scheme can be applied conveniently to extend the time window of relaxation, to connect multiple relaxation measurements (e.g., starting from different stages of magnetization) performed under the same external conditions, and thereby to investigate the activation energy for pinned flux motion in a wider range of currents.

The early stages of the magnetic relaxation (MR) are usually difficult to monitor due to the giant flux creep effect. In this paper, we suggest that the field sweeping rate $\dot{H}$ dependence of the magnetization $M(\dot{H})$ in magnetic hysterisis loop (MHL) measurements can be mapped to an MR curve, and thus extend the time-window on the short time side. This scheme is complementary to an earlier scheme [1], and has marked advantages to be discussed.

The general flux-creep equation can be written as [1,2], $-\frac{dM}{dt} = \dot{H} - \omega_0 H_e e^{-U(M)/k_B T}$ where M is the magnetization (being proportional to the current in view of the Bean model applied in the subcritical states), $\dot{H}$ the field sweeping rate, H_e the external field and ω_0 a sample-dependent constant. From this equation, we have approximately $U(M) - U(M^*) = k_B T \ln\left(\dot{H}^*/\dot{H}\right)$ in MHL measurements, and $U(M) - U(M^*) = k_B T \ln(1 + \frac{t-t^*}{\tau})$ in MR measurements, respectively. Here $\tau = \tau^* <|dM/dU|>/|dM/dU|^*$, with τ^* a reference-point dependent quantity to be fitted, $<|dM/dU|>$ is an average over the range from M^* to M. (All superscripts '*' refer to a reference point in both systems of measurements). Thus an effective time t_{eff} can be associated with $\dot{H}$:

$$t_{eff} = t^* + (\dot{H}^*/\dot{H} - 1)\tau. \tag{1}$$

Of course this is only approximate. Notice that τ^* depends on the choice of the reference point t^* (so that $M^* \equiv M(t^*)$ in MR, and $M^* = M(\dot{H}^*)$ in MHL). However, the final relaxation curve mapped out from MHL measurements should not depend on t^* at all. No microscopic uncertainties are invoked in this scheme.

The MHL and MR data from Jirsa *et al* [1] are replotted in Fig. 1. Using the map algorithm above we are able to extend the time window up to three decades on the short time side. The extended curve and the measured data in Fig. 1 follow the same master curve, of equal quality of the counterpart given in Ref. 1. However, the present algorithm is obviously more practical.

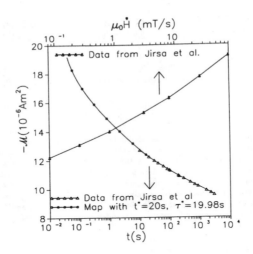

Figure 1: MHL and MR data from Jirsa *et al.*, and the mapped relaxation curve.

By utilizing the present algorithm, it is possible and convenient to extend the time window of relaxation, to connect multiple relaxation measurements (e.g., starting from different stages of magnetization) performed under the same external conditions, and thereby to investigate the $U(J)$ behavior in a wider range of currents.

Acknowledgments

We acknowledge the support from the National Center For Research and Development on Superconductivity of China under the Contract Number J-A-4102 and the RGC grant of Hong Kong.

References

1. M. Jirsa, L. Pust and H. G. Schnack and R. Griessen, Physica C**207**, 85(1993). See also M. Jirsa, L. Pust and J. Kadlecova, J. Magn. Mater. **101**, 105(1991) for experimental details.
2. Q. H. Wang and X. X. Yao, Physica C**249**, 69(1995).

III. THEORY

Transport and Spectroscopy

I. THEORY

Transport and Spectroscopy

Quasilocalized states as an explanation of some properties of cuprates

M. Cyrot

National High Magnetic Field Laboratory
Tallahassee Florida 32306
and laboratoire Louis Neel, CNRS, 38042 Grenoble, cedex France.

In particular cases, quasi localized states can exist in the metallic phase close to a Mott transition. We show that such an hypothesis can explain some experimental results in inelastic neutron scattering as the resonance frequency near $q=(\pi,\pi)$ in the cuprates. Photoemission spectroscopy and optical conductivity can also find a reasonable explanation both in the transfer of weight and in the mid infrared behavior.

We consider a doped Mott insulator whose doping concentration is large enough to destroy long range antiferromagnetic order and to obtain metallic behavior. We also suppose that this concentration is not too large in order that strong short range antiferromagnetic fluctuations persist. We first show that the electrons at the Fermi level are weakly scattered by the magnetic moments on each site as in a periodic Kondo model or s,d model. Then we analyse some consequences. The photoemission spectrum presents a three peaks structure. The inelastic neutron scattering experiments close to (π,π) have a simple explanation as the resonance is concerned. Moreover in this framework, it could exist charge excitations which are quasi localized and whose existence would appear in anomalous optical absorption in the mid infrared region.

In order to study such a situation, we start from the Hubbard model and we suppose that U is of order of, or greater than the bandwidth. The magnetic fluctuations are on a time scale large compared to the inverse of the bandwidth. Thus in order to study the magnetic moment on a site, we first use the functional integral technic in the static approximation[1,2]. This introduces a potential on each site

$$-\sigma \mu_i (n_{i\sigma} - n_{i-\sigma})$$

We can write the Green function as a function of the potential on site i

$$G_{lm}(\mu_i) = G_{lm}(\mu_i=0) + G_{li}(\mu_i=0)\, T_i\, G_{im}(\mu_i=0)$$

where T_i is the T-matrix corresponding to the localized potential on site i. We assume that the magnetic moment on site i is close to one and is well localized on site i, i.e. we have a well defined virtual bound state due to the localized perturbation. This virtual bound state is below the Fermi level for one spin direction and above it for the other spin direction.

We introduce the phase shift due to the perturbation. The phase shift at the Fermi level is given by the Friedel sum rule[3].

$$\delta_\sigma(E_f) + \delta_{-\sigma}(E_f) = 0$$

and the magnetic moment is

$$\pi\mu = \delta_\sigma(E_f) - \delta_{-\sigma}(E_f)$$

If the magnetic moment is close to one, we have

$$\delta_\sigma(E_f) = \pi/2 - \alpha$$
$$\delta_{-\sigma}(E_f) = -\pi/2 + \alpha$$

with α small. We now consider the coupling between electrons at E_f and the magnetic moment on site i. If we flip the moment on site i, the two phase shifts permute and by considering the change of the polarization induced by the moment, one can map the result on an s.d model

$$J_{eff}\, s.S$$

with an effective coupling

$$J_{eff} = -4\alpha / \pi\rho(E_f)$$

Thus if α is small, the Fermi electrons are only weakly coupled to the magnetic moment on each site. Here we have only treated the z component of the spin. However the result is the same as that of the effective exchange Hamiltonian derived by Schrieffer and Wolff[4] from the Anderson Hamiltonian. Thus many results obtained in the periodic Kondo model can be applied for the electrons close to the Fermi level. We quote Schrieffer and Kampf[5] analysis of ARPES which predicts novel effects in the energy and momentum distributions. The photoemission spectrum would present a three peaks structure. The Fermi energy lies in the middle peak close to its Hartree-Fock value as the Fermi electrons see the average of the weak perturbation due the magnetic moments.

For a discussion of the inelastic neutron scattering experiments[6] and the behavior of $S(q,\omega)$ close to (π,π), we have to calculate the dynamical susceptibility $\chi(q,\omega)$. If the magnetic moment is well defined on a site, as we just show, the model in this case can be mapped on a periodic Kondo model with fixed magnetic moments. Thus to calculate the dynamical susceptibility, we can use a two site model where we introduce U only on two sites. Of course we loose the translational invariance and have to calculate $\chi(q, q^*, \omega)$. This can be done in a generalized random phase approximation. For $q=q^* \sim (\pi,\pi)$ the frequency dependence of χ stems from the zero of the denominator i.e. one obtains a resonance frequency. If one introduces U on site 0 and site R and the Hartree-Fock susceptibility χ^0 (R_i, R_j, ω), the resonance frequency is given by the equation

$$[1- \chi^0(0,0,\omega)] [1-\chi^0(R,R,\omega)] - U^2 \chi^0(0,R,\omega) \chi^0(R,0,\omega) = 0$$

If on each site there is a magnetic virtual bound state, we have for small value of ω[7]

$$\text{Re}[1- \chi^0(0,0,\omega)] = \gamma\omega$$

The second term of the equation which gives the resonance frequency, is the square of the field induced by one site on the other site. It means that one magnetic moment resonates in the field induced by the other one. However this resonance has a width due to the itinerant electrons. The width is proportional to the square of the density of states at the Fermi level. We propose that in the cuprates the spins on the copper sites resonate in the field created by the short range antiferromagnetic order. However this would be difficult to observe in the metallic state due to the important width. In the superconducting state, this width decreases and the resonance becomes sharper. This fact could explain the experiments. The resonance field varies with α^2. Thus the resonance frequency should also increase with doping as it is observed[6].

We now consider a possible consequence of the assumption of a well defined localized moment on each site on the charge excitation spectrum. If by a photoemission experiment or by an optical absorption experiment, a hole is created on a site, during its life time, does the hole move or is it self trapped? During the life time of the hole, the magnetic fluctuations are frozen and one can rely on a static approximation. To answer the question, one can consider the hole as a particle creating an attractive potential for the electrons. The possibility of self trapping of a particle which can move in a Fermi liquid, has been considered by Yamada et al[8]. Self trapping requires that the phase shift introduces by the particle on the electrons at the Fermi level is much larger than $\pi/2$. This is usually not possible due to the Friedel sum rule. Thus we have to consider the phase shift created by the localized hole in the strongly correlated Fermi liquid. As each site is magnetic with nearly one electron of a given spin, creating a hole destroy magnetism and creates an empty virtual bound state for both spin directions. The phase shift just above the empty bound state is nearly equal to π. In order to fulfill Friedel sum rule, in an optical experiment where the total number of electrons is constant, the phase shift at the Fermi level for both spin direction has to be close to π. Applying Yamada et al results[8], it would mean that the hole can be self trapped at least during a time smaller than that of magnetic fluctuations. The experimental verification of the self trapped behavior of the hole would be in the energy of excitation. Indeed, if the hole is localized, its attractive potential would strongly decrease the Hartree-Fock energy needed to create it. Numerical computations[9] show that in the best cases, the energy of excitation can be divided by a factor of one hundred compared to the Hartree-Fock value. This is due to the fact that the energy of relaxation is the integral of the phase shift and that this phase shift is large of order of π. In an optical experiment, this excitation would exist in the mid infrared region. It is one of the possible explanation of the anomalous mid infrared absorption in the cuprates.

In conclusion, the assumption of well defined magnetic moment on the copper sites of the cuprates due to the quasi two dimensional behavior of these materials, would permit to understand some anomalous properties obtained in inelastic neutron scattering and in the optical absorption spectrum. Our main result is that the electrons at the Fermi level are only weakly coupled to the underlying magnetic moments.

Acknowledgements

The author wants to thank Professor J.R. Schrieffer for a stay in Tallahassee and for interesting discussions.

References

1. J.R. Schrieffer, Phys. Rev. Lett. 23, 92 (1969).
2. M. Cyrot, Phys. Rev. Lett. 25, 871 (1970).
3. J. Friedel, Phil. Mag. 63, 153 (1952).
4. J.R. Schrieffer and P.A. Wolff, Phys. Rev. 149, 491 (1966).
5. J.R. Schrieffer and A.P. Kampf, J. Phys. Chem. Solids, 56, 1673 (1995).
6. P. Bourges, L.P. Regnault, J.Y. Henry, C. Vetier, Y. Sidis and P. Burlet, Physica B (1995).
7. D.L. Mills and P. Lederer, Phys. Rev 160, 590 (1967).
8. K. Yamada, A. Sakurai and M. Takeshige, Progress Theor. Phys. 70, 73 (1983).
9. U. Muschelknautz and M. Cyrot, Phys. Rev. B, 53, (1996).

NEUTRON SCATTERING AND GAP ANISOTROPY IN HIGH-T_c SUPERCONDUCTORS

A. BILL

Lawrence Berkeley National Lab., 1 cyclotron road, MS 62-203, Berkeley, CA 94720, USA

V. HIZHNYAKOV

Physics Dept., Tartu University, Tähe 4, Tartu, Estonia

E. SIGMUND

Institut für Theoretische Physik, BTU Cottbus, P.O. Box 101344, 03013 Cottbus, Germany

The influence of the gap anisotropy on the phonon dispersion of high-T_c materials is studied. Calculations of the superconductivity induced energy shift along [100] and [110] for s- and d-wave symmetries indicate that this quantity is strongly influenced by the anisotropy of the gap. In this context recent measurements done on the 340 cm^{-1} mode of an YBCO single crystal are discussed. We show that the data can be well described in terms of an anisotropic s-wave gap obtained from the unscreened interaction of charge carriers with long range optical phonons.

One of the major issues in the field of high-T_c superconductivity is presently the characterisation of the gap anisotropy. Most of the phase sensitive experiments performed to determine its symmetry consist in the investigation of superconducting currents through oriented grain boundaries. As pointed out in Ref.[1] the obtained data depend on the boundary conditions that varies for different experiments and may therefore lead to different conclusions. It is thus of interest to use other methods that are also sensitive to the phase of the gap. In this context we have shown that one can use $\mathbf{q}$ dependent neutron scattering measurements of the superconductivity induced phonon renormalization $\delta\omega_{\nu\mathbf{q}}$ to characterise the anisotropy of the gap.[2] At $\mathbf{q} = 0$ this renormalization depends only on the absolute value of the gap, whereas for $\mathbf{q} \neq 0$ it is also sensitive to its phase.

Because of the difficulty to grow large untwinned single crystal samples only few experiments measuring the phonon dispersion below T_c were performed. Recently, the renormalization of the 340 cm^{-1} mode of a YBa$_2$Cu$_3$O$_7$ single crystal has been observed along [100] and [110].[3] In the following we present the calculation of the renormalization along these directions for two different cases. First, we consider the s-wave gap resulting from the unscreened interaction of charge carriers to long wave optical phonons (LR).[4] In the second case, we take the d-wave gap resulting from the spin-fluctuation model (SF).[5] We restrict our analysis to the energy shift of the 340 cm^{-1} mode. Results for the linewidth as well as for other phonons or gap symmetries are presented elswhere.[2]

We use the expression derived in [6] for the superconductivity induced renormalization of the phonons. It is of second order in the electron-phonon coupling function $g_\nu(\mathbf{k}, \mathbf{q})$. For the B_{1g} mode considered here it has been shown[7] that the coupling function can be written in first approximation as $g(\mathbf{k}, \mathbf{q}) = (\cos k_x - \cos k_y) \cdot \bar{g}(\mathbf{q})$. In the figures we normalize the energy shift of the phonon to $|\bar{g}(\mathbf{q})|^2$. We consider one electronic band with nearest ($t = 0.14$eV), second nearest ($t_1/t = 0.2$) and third nearest ($t_2/t = 0.25$) neighbour hopping.[4] The results presented below are obtained near optimal doping ($\varepsilon_F = 0.42$eV [4]). For the phonon dispersion we take results of lattice dynamical calculations for YBCO.[2] Finally, the BCS gap equation is solved for the following pairing potentials:

$$V_\mathbf{q} = 2\pi V / \left[(q-Q)^2 + \kappa^2\right] \quad , \tag{1}$$

where $\mathbf{q} = \mathbf{k} - \mathbf{k}'$ is a two dimensional vector parallel to the CuO$_2$ plane, κ^{-1} is the screening length of the interaction. Depending on the physical origine of the pairing one has different values of $\mathbf{Q}$, V and κ. We chose the value of V so that $\Delta_{\max}$ is slightly above 340 cm^{-1}.[6] For the long range model, $V \simeq 0.087$eV, $\mathbf{Q} = 0$ and $\kappa = 0.3$. This leads to an anisotropic gap with s-wave symmetry. For the spin fluctuation model, $V \simeq -0.083$, $\mathbf{Q} = (\pm\pi, \pm\pi)$ and $\kappa = 0.3$.[2] In this last case, the gap has d-wave symmetry. It should be emphasized that the gap function resulting from these two models *cannot* be written as a sum or difference of two cosine.[4]

Fig. 1 show the shift in energy of the 340 cm^{-1} mode for both models along [100] (left) and [110]

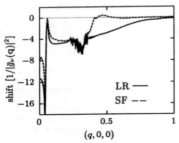 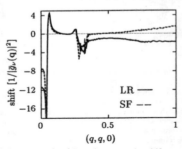

Figure 1: Energy shift for s (LR) and d-wave (SF) gaps along [100] (left) and [110] (right). A negative shift correspond to a softening.

(right) respectively. A preliminary remark concerns the two "peak like" structures at small q and at $q \sim 0.3$. These are related to the band structure [8] and will be discarded here since it turns out that they do not depend significantly on the gap symmetry. Two features distinguish the s and d-wave scenarios. In Fig. 1 (left) the shift for the d-wave case remains approximatively constant up to $(\pi/2, 0, 0)$ and then drops abruptly to zero. In the s-wave situation, instead, the shift decreases almost linearly between $(\pi/2, 0, 0)$ and $(\pi, 0, 0)$. On the other hand, along [110] (Fig. 1, right) the shift vanishes rapidly with increasing q both for the s and the d-wave cases. Thus, the main feature that distinguishes the two symmetries is seen at intermediate q along [100]. The experiment performed in [3] shows a behaviour that is in agreement with the above s-wave case. Indeed, they observe an almost constant shift below $(\pi/2, 0, 0)$ and a linear decrease above this point whereas it vanishes rapidly along [110].

The question in how far the present calculations depend on the chosen model has been discussed thouroughly.[2] It turns out that the electronic dispersion does not influence the result as long as it describes the saddle points at $(\pm\pi, 0)$ and $(0, \pm\pi)$. The main effect on the phonons is due to the k dependency of the gap. However, it is not only the symmetry of the gap but also its specific structure as a function of k that is important. Interpreting experiments solely in terms of the gap symmetry (e.g. with a phenomenological model) is thus generally insufficient.

A.B. is grateful to the organizing committee and to the Swiss National Science Foundation for financial support.

References

1. C.M. Varma, *preprint*.
2. A. Bill, V. Hizhnyakov, and E. Sigmund, to appear in *Journ. of Supercond.*.
3. D. Reznik *et al*, *Phys. Rev. Lett.* **75**, 2396 (1995).
4. V. Hizhnyakov and E. Sigmund, *Phys. Rev. B* , 1 march (1996) and *this volume*.
5. A.J. Millis *et al*, *Phys. Rev. B* **42**, 167 (1990).
6. A. Bill *et al*, *Phys. Rev. B* **52**, 7637 (1995).
7. T.P. Devereaux *et al*, *Phys. Rev. B* **51**, 505 (1995).
8. F. Marsiglio, *Phys. Rev. B* **47**, 5419 (1993).

IMPURITY VERTEX CORRECTION IN THE NMR COHERENCE PEAK OF CONVENTIONAL SUPERCONDUCTORS

HAN-YONG CHOI
Sung Kyun Kwan University, Department of Physics, Suwon 440-746, Korea

The vertex correction due to nonmagnetic impurities is considered for the NMR coherence peak. The coherence peak remains unrenormalized under the impurity vertex correction in the clean limit in agreement with the previous work. It, however, increases by about a factor of 2 as one goes from the clean to dirty limit. This result agrees well with the experimental observations on the Al- and In-based alloys.

It is well known that the impurity vertex correction (IVC) is very important in the transport properties of metals.[1] The scattering lifetime τ is changed to the transport lifetime τ_{tr} due to IVC, which appears in, for example, $\sigma(0) = ne^2\tau_{tr}/m$. Because T_1^{-1} and $\lim_{\omega\to 0}\sigma(\omega)$ have the same "plus coherence factor",[2] we may expect that the IVC may also be important for NMR relaxation rate. Maki and Fulde (MF)[3] studied this problem some 30 years ago, and found that the impurity vertex function $\hat{\Gamma}(\omega, \vec{q}) \to 1$ in the limit of the momentum transfer $\vec{q} \to \infty$ and $\omega \to 0$. It was concluded that T_1^{-1}, consequently, remains unrenormalized under IVC. This conclusion, however, is not rigorous because T_1^{-1} is given by an integral over the momentum transfer $\vec{q}$, not by the limit $\vec{q} \to \infty$. This observation calls for a more careful analysis of the IVC effects on NMR relaxation rate than the $q \to \infty$ of MF. We find that T_1^{-1} remains unrenormalized under IVC for $\ell/\xi \gg 1$, where ℓ is the electron mean free path and ξ is the coherence length. As ℓ/ξ is reduced, the NMR coherence peak is found to increase due to IVC by a factor of about 2.

The nuclear spin-lattice relaxation rate T_1^{-1} is given by

$$\frac{1}{T_1} = \lim_{\omega \to 0} \frac{1}{1 - e^{-\beta\omega}} \sum_{\vec{q}} Im\left[\chi_{+-}(\omega + i\delta, \vec{q})\right], \qquad (1)$$

where $\chi_{+-}(\omega, \vec{q})$ is a spin-spin correlation function, obtained through analytic continuation from the Matsubara function $\chi_{+-}(i\omega, \vec{q})$. It should be calculated with nonmagnetic impurities included fully self-consistently. We do this by renormalizing the quasi-particle Green's function due to impurity scattering in the t-matrix approximation and also by including the impurity vertex correction in ladder approximation, in accordance with the Ward identity.[1] This procedure results in the Bethe-Salpeter equation as shown below, which is to be solved for the vertex function, $\hat{\Gamma}$.

$$\hat{\Gamma}(ip_n, ip_m, \vec{q}) = \tau_0 + \sum_{\vec{k}'} \hat{T}(ip_n)\hat{G}(ip_n, \vec{k}')\hat{\Gamma}(ip_n, ip_m, \vec{q})\hat{G}(ip_m, \vec{k}' + \vec{q})\hat{T}(ip_m). \qquad (2)$$

Then T_1^{-1} can be calculated from Eq. (1) and

$$\chi_{+-}(i\omega) = \frac{1}{\beta}\sum_{ip}\sum_{\vec{k}',\vec{q}} Tr\left[\hat{G}(ip + i\omega, \vec{k}' + \vec{q})\hat{G}(ip, \vec{k}')\hat{\Gamma}(ip, ip + i\omega, \vec{q})\right]. \qquad (3)$$

We find with an approximation about angular average[4]

$$\frac{1}{T_1 T} \propto \int_\Delta^\infty d\epsilon \frac{\partial f}{\partial \epsilon} \left\{ \frac{\epsilon^2}{\epsilon^2 - \Delta^2} + \frac{\Delta^2}{\epsilon^2 - \Delta^2} \left[1 + \frac{2}{1 + (4\tau)^2(\epsilon^2 - \Delta^2)} \right] \right\}. \tag{4}$$

In the limit $\tau \to \infty$, Eq. (4) reduces to the standard expression.

$$\frac{1}{T_1 T} \to \int_\Delta^\infty d\epsilon \frac{\partial f}{\partial \epsilon} \left\{ \frac{\epsilon^2 + \Delta^2}{\epsilon^2 - \Delta^2} \right\}. \tag{5}$$

This agrees with the previous result of MF that $1/T_1$ remains unrenormalized under the IVC. In the limit $\tau \to 0$, on the other hand,

$$\frac{1}{T_1 T} \to \int_\Delta^\infty d\epsilon \frac{\partial f}{\partial \epsilon} \left\{ \frac{\epsilon^2 + 3\Delta^2}{\epsilon^2 - \Delta^2} \right\}. \tag{6}$$

The dominant contribution to $1/(T_1 T)$ comes from the region $\epsilon \approx \Delta$. Consequently, comparing Eqs. (5) and (6), we can easily see that the ratio of coherence peak to normal state value is bigger in dirty limit than in clean limit by a factor of about 2. This is in good agreement with the experimental observations on Al- and In-based alloys.[5] This observation raises an interesting possibility that the experimental observation of the NMR coherence peak increase may be understood in terms of impurity vertex correction rather than the gap anisotropy smearing. Experimental distinction between the two effects is highly desirable in this regard.

To summarize, we considered the effects of impurity vertex correction on nuclear spin-lattice relaxation rate within the Eliashberg formalism. We found $1/(T_1 T)$ remains unrenormalized under the impurity vertex correction in the clean limit. As the scattering lifetime is decreased, on the other hand, the coherence peak in $1/(T_1 T)$ was found to increase due to the impurity vertex correction. The nuclear spin-lattice relaxation data measured on conventional superconductors as impurity scattering rates are varied may be understood in terms of impurity vertex correction rather than the gap anisotropy smearing effects induced by impurity scatterings.

Acknowledgments

The author was supported by Korea Science and Engineering Foundation through Grant No. 951-0209-035-2 and through Center for Theoretical Physics, Seoul National University, and by the Ministry of Education through Grant No. BSRI-95-2428.

References

1. G. D. Mahan, *Many-Particle Physics*, 2nd ed., (Plenum, New York, 1990), chap. 7.
2. K. Maki, in *Superconductivity*, ed. R. D. Parks, (Dekker, New York, 1967), p. 1035.
3. K. Maki and P. Fulde, Phys. Rev **140**, A1586 (1965).
4. H.-Y. Choi, Phys. Rev. B **52**, 7549 (1995).
5. D. E. McLaughlin, in *Solid State Physics* Vol. 31, ed. H. Ehrenreich *et al.*, (Academic Press, New York, 1976), p. 1.

CURRENT INSTABILITIES IN REENTRANT SUPERCONDUCTORS

DAVID M. FRENKEL, JEFFREY A. CLAYHOLD
Texas Center for Superconductivity, University of Houston, Houston,
TX 77204-5932, USA

Reentrant superconductors have a portion of their temperature-dependent resisitivity curve where $d\rho(T)/dT < 0$. We consider a thin film of such a material on top of thermally conducting substrate. We show that when large enough current bias is applied to the film, an interplay between Joule heating and heat flow in the substrate leads to an instability towards an inhomogenous current and temperature distribution in the film. The resulting nonuniform current pattern has a 'hot channel' where most of the current is concentrated. The nonuniformity sets in with increasing current in a hysteretic manner for wide samples and via a bifurcation for narrower ones.

The temperature-dependent resistivity curve of a reentrant superconductor has a portion (right below T_{c_2}) in which $d\rho(T)/dT$ is negative (see, e.g., Ref. 1). An elementary argument suggests possible thermal instability in this region. Indeed, the total Joule heat dissipated in the film is $P = V^2/R$, and as the sample heats, $R(T)$ *decreases* , further increasing the dissipated power. To fully explore the instability issue, the counteracting effects of the substrate have to be treated. The model we choose is that of a thin film of a reentrant superconductor material on top of a substrate whose bottom is kept at a fixed temperature. The current is applied in the y-direction, and the (possible) temperature and current density variations are in the x-direction. Periodic boundary conditions are used in the x-direction. The relevant equations are

$$C_V \frac{\partial T}{\partial t} - \kappa \nabla^2 T = 0 \tag{1}$$
$$\kappa \frac{\partial T(x, z = 0)}{\partial z} = \sigma(T) E^2 d_{film},$$

where the first equation is the time-dependent heat diffusion in the substrate, and the second is the *nonlinear* boundary condition stating that the local Joule heat produced in the film is equal to the thermal current into the substrate at that point.

A solution with a uniform current distribution is

$$T^{(0)}(z) = T_{top} + \frac{T_{top} - T_{bottom}}{L_z} z, \tag{2}$$

where $z = 0$ corresponds to the top of the substrate and L_z is its thickness. T_{top} is found from the condition that the heat production in the film be equal to the heat outflow through the substrate.

$$\kappa \frac{T_{top} - T_{bottom}}{L_z} = \sigma(T_{top}) E^2 d_{film}. \tag{3}$$

For a constant current source, we replace $\sigma(T_{top})E^2 \to \rho(T_{top})j^2$ in the right-hand side.

For constant voltage source, we find a critical voltage above which there is a runaway thermal instability. For constant current source, the uniform current solution appears

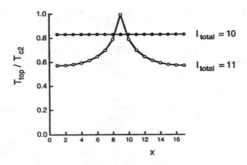

Figure 1: Temperature distribution in the film below and above the instability.

stable against uniform temperature fluctuations. However, adding a *spatially modulated* perturbation leads an instability. In a linearized analysis, one studies solutions of the form $T^{(0)}(z) + \varepsilon e^{\pm at}\cos(k_x x)f(z)$. For currents above a critical current, one finds solutions exponentially growing in time.

To find the resulting nonuniform current distributions, we numerically integrate the time-dependent heat diffusion equation with the nonlinear boundary condition. As expected from linear stability analysis, the uniform state is unstable for $I > I_{crit}$. The unexpected feature is that for wide enough samples, even as $I \to I_{crit}^+$, the instability is towards a current distribution 'far away' from the uniform one. If I is then cycled below I_{crit}, the nonuniform state persists for a range of $I < I_{crit}$, for which the uniform solutions are stable as well. Thus we have *bistability* and *hysteresis*. The nonuniform current state has a sharply localied current distribution — a *hot channel*. For narrower samples, the bistability disappears, and the transition is 2nd order-like. The Figure contains the results of integration on a 16×5 cite lattice with the conductivity function $\sigma(T) = \frac{1}{1-T/T_{c_2}}$ for two different total current values. Note the proximity of the highest temperature in the nonuniform distribution to T_{c_2}. Its attendant high conductivity results in the lion's share of the current being concentrated in a narrow strip.

Finally, by examining the Frobenius integrability condition for the flow induced by our equations in the space of temperature distributions, we have found that for a special model, with the conductivity function $\sigma(T) = ae^{bT}$, a variational principle exists. For that model, we have found an analytical explanation of the bistability and hysteresis, and their disappearance in favor of a bifurcation in narrower samples.

Acknowledgments

This work was supported by TCSUH.

References

1. W. A. Fertig, et. al, Phys. Rev. Lett. **38**, 988 (1977).

IV. APPLICATIONS

IV. APPLICATIONS

Large Current

PROGRESS AND ISSUES IN HTS POWER CABLES

A. BOLZA, P. METRA
Pirelli Cavi Spa, Milano, Italy
M. M. RAHMAN
Pirelli Cable Corporation, Lexington, SC

The discovery of HTS in 1986 has revitalized a great interest in SC cables and transmission systems, after the SC technology in LHe had proved to be technically valid but not economically competitive with conventional systems. The availability of a less costly solution, cooled by LN, is the driving factor for various possible applications of SC in the electric power system. The practical actions originating from this opportunity have started and are continuing to grow in the parallel directions of both a tailored improvement of materials and technology and an increasingly detailed technical and economical analysis of SC applications for power transmission in the grid. The present advances, such as the manufacture of a machine stranded 50 m, 1800 Ampere HTS conductor, and the possible future evolution of HTS cable systems, some of which are already in the development stage while others are being systematically analyzed for feasibility and for the associated technical and economic opportunities, are described.

1. Historical Background

Bearing in mind the fact that the adoption of LHe superconducting cable systems was prevented by their excessive cost, mostly related to the cryogenics, it is clear why the discovery ten years ago of High Temperature Superconductors (HTS) has revitalized a great interest for SC cables.

In this context, the most attractive characteristic of HTS materials is obviously their critical temperature, in the range 90 to 120 K, which enables them to operate in the superconducting state in liquid nitrogen (LN).
Furthermore, among the various possible applications of HTS, power cables have attracted particular attention from the beginning a for a number of reasons:

- a significant previous experience with LTS cables was available as a positive technical background,
- the low magnetic field application in cables was expected to give rise to a relatively easy development, even considering the intrinsically limited field tolerance of HTS materials at LN temperatures,
- HTS technology did appear as an option likely to effectively address the need for more compact power transmission systems and provide additional system design flexibility
- favorable economics were expected for the cryogenic aspects even at relatively moderate power ratings

Starting in 1988, preliminary analyses of the power cable application were performed by Pirelli and by a number of industries and experts all around the world, to assess their potential advantages in practical use. From this activity three detailed studies analyzing HTS power transmission cable systems are worthy of being noted.

In Europe, Pirelli and other major cable manufacturers, with the support of the EC, carried out from 1990 to 1992 a detailed techno-economic study of HTS cable systems, focused on both a high power (1 to 3 GVA) solution and a medium power (500 MVA) solution. [1]

In USA, EPRI funded since 1989 a feasibility study for HTS cables, which rather soon focused on a retrofit application to upgrade existing pipe type cable routes, as the easiest step to take advantage of the new technology to solve one of the emerging needs of the system. This cable concept was designed in detail by Pirelli in 1992, for the case of a 115 kV-400 MVA circuit, based on a three-phase geometry and on a room temperature dielectric. [2]

In Japan, utilities and cablemakers started since 1988 to evaluate the technical performances and the economic perspectives for HTS cables, mostly considering the replacement of conventional 275 kV circuits in congested urban environments, with the target of increasing transmissible power and saving a transformation step (600 to 1000 MVA at 66 kV). [3]

2. Development of HTS Materials and Wires

Also as a result of the interest generated by the previously mentioned studies, the development of materials technology for HTS wires has rapidly progressed.

As is well known, the most advanced technology for long, flexible HTS tapes, as needed for cables, is at present the powder-in-tube process with BSCCO 2223. Pirelli is a key player in this field through its joint development with American Superconductor Co..
A few remarks on the state of the art of this HTS cable wire development, which is extensively analyzed in another contribution [4], are worthy of being summarized here.

Long lengths of wire (in excess of 100 m) are currently available with critical current in excess of 30 A/tape. The mechanical performance of such tapes is of some concern because their flexibility does not allow them to be handled as easily as copper or aluminum wires. However, the development of suitable cabling technologies has already allowed to assemble them into long conductors capable of withstanding actual manufacturing and handling events in an industrial setting.

The perspectives for the continuing HTS wire development are expected to follow three main directions:
- an R&D effort towards the better understanding of materials and processes and the achievement of top intrinsic performances (mostly in terms of Jc, strain tolerance and losses)

- the scaling to long practical lengths of the best technology, with the least deterioration of the performance
- the industrialization of the process for the achievement of suitable cost targets

3. Development of HTS Cable and System

The prototype of a pipe retrofit HTS system for 115 kV is presently being developed by Pirelli under a EPRI/DOE contract (1994-1998), which has the goal to develop, manufacture and test the conductor, the cable and the accessories.

In the initial phase of this project, a fundamental milestone to demonstrate the conductor assembly technology has been recently achieved, by manufacturing with industrial equipment a continuous length of multilayer conductor. A 50 m long section of this conductor has shown a d.c critical current of 1800 A at 1 µV/cm in LN, after

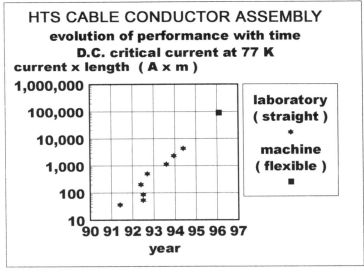

Fig 1 HTS multilayer conductors: evolution of the current and length performance achieved with laboratory prototypes and with factory assembled long and flexible HTS conductors

repeated rewinding operations. The conductor, designed by Pirelli according to the goals of the EPRI/DOE contract, is a consistent evolution of the previous work on conductor models (Fig.1). The HTS tapes, for a total of 6000 m, were supplied by American Superconductor Co. in 100 m lengths, according to the specifications jointly developed. This program is now actively progressing towards the next phases and targets.

While developing the technology, system studies for different types of application in the grid are being carried out by Pirelli with particularly interested users.

A contract is running with Florida Power & Light to cover the technical and economic aspects of designing and demonstrating a pilot HTS high voltage cable transmission system for a specific application in their grid.
Also, detailed design studies are being carried out with several European utilities in connection with the technical and economic aspects of using HTS cable systems for various types of application and the possible impact on the future structure of the electric grid.

4. Conclusions

Ten years since the discovery of the HTS materials, the indications emerging from the above mentioned studies, and the practical experience from the ongoing developments are progressively generating a more justified confidence on the opportunities and a more precise understanding of the issues related to the HTS power transmission technology:

- HTS technology has the potential to offer unique opportunities to solve, in the typical time frame of a few years, system problems like the upgrading of existing underground links, the matching of the rating of overhead line sections to be undergrounded, the saving of a transformation step to feed urban areas, the increase of transmission efficiency and stability of high power routes.

- for some applications ratings in the range of 1000 MVA and even as low as 200 to 400 MVA have been found potentially competitive with conventional underground cable systems

- different designs (room temperature vs. cryogenic dielectric, coaxial vs. three phase conductor configuration) are feasible to optimize the system for different applications

- demonstration programs to manufacture and test HTS cable and system prototypes have entered the phase of experimental activity, with very positive results, and will be followed by pilot installations to establish the relevant operational procedures prior to the introduction of the new technology into commercial use .

References

1. P.Metra et al. ,London, Nov 1993, IEE Public. N° 382, p. 248-252
2. D.Von Dollen et al. ,Chicago, Apr 1993 ,Am. Power Conf.p.1206-1211
3. T.Hara et al. IEEE Trans. Power Delivery, Vol.7, n° 4, Oct 1992 p.1745-1753
4. A. Malozemoff , 10th Anniversary HTS Workshop, Houston, March 1996

IMAGING OF VORTICES IN SUPERCONDUCTORS WITH A MAGNETIC FORCE MICROSCOPE

Chun-Che Chen, Qingyou Lu, Caiwen Yuan, and Alex de Lozanne
Department of Physics, University of Texas; Austin, TX 78712-1081

James N. Eckstein
Varian Associates, Inc., E. L. Ginzton Research Center, Palo Alto, CA 94304-1025

Marco Tortonese
Park Scientific Instruments, 1171 Borregas Ave., Sunnyvale, CA 94089

We report the observation of flux lines in BSCCO(2212) thin films with a home-made Magnetic Force Microscope (MFM). Vortices are seen as round features with a diameter of about one micrometer at 77 K. The vortex to background contrast decreases and the size increases as the temperature approaches T_{c0}. We also observe vortices in the superconducting transition temperature range from BSCCO film which indicates that vortices can exist in this range.

The importance of vortex dynamics of high T_c superconductors has driven the developments of several techniques in recent years.[1-3] The Magnetic Force Microscope (MFM) is a new technique which directly measures the force between the magnetic particle on the tip and a vortex in a superconductor. We have recently developed a home-made low temperature MFM for the study of superconductors.[4] Here we present data obtained with this instrument on BSCCO(2212) films.

The MFM tips were coated with 20-50 nm of iron. The $T_{c\,onset}$ and T_{c0} of the BSCCO film were 91.5K and 81K, respectively. The experimental details can be found elsewhere.[5] In brief, we first cooled down the samples to 77K in a 10-100G magnetic field. We then measured the film topography by scanning the MFM tip in contact with the sample surface (we will call it "DC" image) while the vortex images were taken by oscillating the cantilever at a distance of 100-500 nm away from the surface (we will call it "AC" images).

Fig. 1 shows some selected AC images of BSCCO film at temperatures from 78.7K to 86.2K taken from a sequence of images spanning 78K to 94K. The shifts of the vortices from upper-right to lower-left are due to the thermal drift of the scanning piezotube, i.e. not due to the vortices moving. Comparing the circles marked in Fig. 1(a) and (b), we can clearly see that there is a vortex missing in Fig. 1(b). We believe that the vortex has been moved out of the image by the tip-vortex magnetic interaction because it has a weaker pinning center on that site.

The dimension of the vortex is directly determined by the penetration depth, which is a function of temperature. Therefore the vortex size is expected to increases as the temperature increases. We took the cross sections from one fixed vortex as marked by a vertical bar in Fig. 1. The results are shown in Fig. 2. As we expect, the vortex size increases as the temperature rises. On the other hand, the vortex to background contrast is related to the force gradient which is proportional to the second derivative of the magnetic induction **B**. Qualitatively speaking, as the temperature increases the vortex to background contrast will be reduced. This is also seen in Fig. 2.

We gratefully acknowledge support from the Texas Advanced Research Project (Grant 3658-172).

References
1. Akira Tonomura, *Physica* **C235-240**, 33 (1994).
2. H.J. Hug, A. Moser, I. Parashikov, B. Stiefel, O. Fritz, H.-J. Güntherodt and H. Thomas, Physica C **235-240**, 2695 (1994); A. Moser, H.J. Hug, I. Parashikov, B. Stiefel, O. Fritz, H. Thomas, A. Baratoff, and H.-J. Güntherodt, Phys. Rev. Lett. **74**, 1847 (1995).
3. J. R. Kirtley, M. B. Ketchen, K. G.Stawiasz, J. Z. Sun, W. J. Gallagher, S. H. Blanton, and S. J. Wind, *Appl. Phys. Lett.* **66**, 1138 (1995)
4. C.W. Yuan, E. Batalla, A. deLozanne, M. Kirk, & M. Tortonese, *Appl. Phys. Lett.* **65**, 1308 (1994).
5. C.W. Yuan, Z. Zheng, A.L. de Lozanne, M. Tortonese, D. A. Rudman, J. N. Eckstein, *J. Vac. Sci. Technol. B* **14**(2) 1210 (1996)

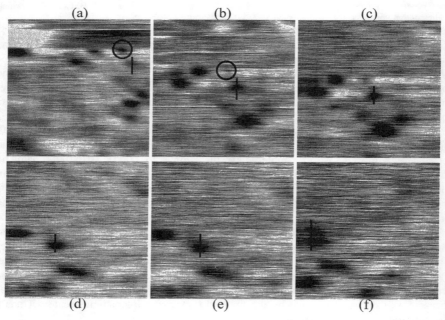

Figure 1(a)-1(f): AC images of BSCCO film . (a) 78.7K, 6.1x6.3µm^2 ,(b) 79.5K, 6.2x6.5µm^2 , (c) 80.2K, 6.3x6.5µm^2 , (d) 83.2K, 6.7x6.9µm^2 , (e) 83.5K, 6.7x7.0µm^2 , and (f) 86.2K, 7.0x7.3µm^2 .

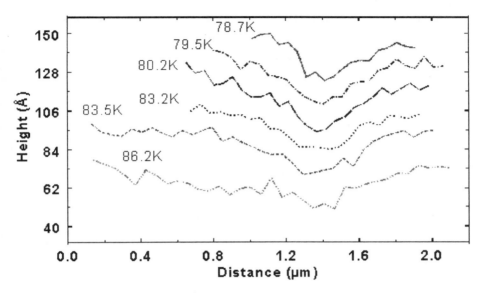

Figure 2: The cross sections of a vortex at different temperatures.

SUPERCONDUCTING MAGNETIC BEARING AND ITS APPLICATIONS IN FLYWHEEL KINETIC ENERGY STORAGE

Z. XIA, K.B. MA, R. COOLEY, P. FOWLER AND W.K. CHU
Texas Center for Superconductivity at the University of Houston
3201 Cullen Blvd.
Houston, Texas 77204

ABSTRACT

The feasibility of superconducting magnetic bearings (SMB) for industrial applications has been investigated since the discovery of high temperature superconducting materials in the late 80's. For example, TcSUH, Nuclear Research Center of Germany, ISTEC of Japan, DM & DP of Italy, Argonne National Lab.& Commonwealth Research Corp., and others, all have flywheel activities using HTS-magnet levitation bearings. At TcSUH, we have constructed a flywheel prototype using hybrid superconducting magnetic bearing (HSMB). The hybrid bearing design uses magnetic forces from permanent magnets for levitation and HTS in between the magnets for stabilization. Currently a *42* lb. flywheel has been rotated up to *6,000* RPM with a kinetic energy of *8* Wh. The recent rotor spin-down experiment lasting six days indicates an average frictional energy loss < *1%* per hour in a vacuum of 5×10^{-6} torr.

Introduction

Flywheels, as one of many energy storage devices, were intensively studied in the 70's and 80's[1,2]. An integrated flywheel system using advanced technology is mainly composed of magnetic bearings, a rotor made of composite materials and a non-contact, electronically commuted motor/generator.[1] With the advent of high temperature superconductivity, a new avenue to circumvent the inherent instability with magnetic systems so as to build a totally passive magnetic levitation becomes more readily accessible.[3] The interaction between magnets and superconductors gives rise to two distinct effects which can be exploited for stable magnetic levitation.[4] One is the Meissner effect; the other, the flux pinning effect. In the Meissner effect, the superconductor behaves as a perfect diamagnet and excludes magnets fields completely from within its volume. As a result, a repulsive force develops between a superconductor and a magnet. The Meissner effect is dominant at lower magnetic fields, and hence yields lower forces in general. Under higher magnetic fields, the superconductor allows partial or complete penetration of the external applied magnetic field into its interior. Thus, the force that develops between the superconductor and magnet can be attractive or repulsive depending on their relative position and orientation. In the "field-cooled" condition, in which a high temperature superconductor (HTS) is cooled below the critical temperature in the presence of a magnetic field, the flux lines are trapped within the HTS, then any attempt for subsequent field change in the vicinity of the superconductor will be resisted.[5] If a proper size gap between the magnet and HTS is arranged, the magnet will be maintained in its position statically as well as dynamically in all directions, by the force and stiffness from both the HTS and the magnets. Furthermore, if the magnetic field has axial symmetry, a rotor consisting of the magnet can rotate freely about the symmetry axis with almost no energy dissipation. Such superconducting bearings are expected to have a friction loss per hour of ~ *0.1%* of the total energy.[6]

Activities of Flywheel Designs Using Superconducting Magnetic Bearings (SMB)

The feasibility of superconducting magnetic bearings for industrial applications has been investigated since the discovery of HTS materials in the late 80's.[7,8,9] The research efforts have been concentrated on building superconducting bearings of industry-scale capability with totally passive levitation and extremely low frictional loss. One can find some preliminary results in the past few years with emphasize on the different research interests. This paper will just show some of the results published within the past 1~2 years.

1. A flywheel system with superconducting magnetic bearings was built and tested at Nuclear Research Center of Germany.[10] The bearing consists of six melt-textured YBCO pellets and a disk of 2.3 kg was rotated at a speed of 15,000 RPM, attaining a maximum energy capacity of 3.8 Wh and a maximum power of 1.5 kW. A very recent result from Dr. Bornemann indicates that a flywheel of 10 kg can be rotated up to 56,000 RPM, at which an energy of 300 Wh is stored, with frictional loss of 3% per hour under a vacuum of 10^{-5} torr.

2. In Argonne National Lab., a research program in collaborating with Commonwealth Research Corp. aims to build an energy storage facility composed of flywheel units with energy capacities of 5 M Wh.[6] Current work includes constructing a flywheel prototype, in which a 25 lb. rotor will be run up to over 30,000 RPM with 1~2 kWh energy stored. Preliminary studies reported that a coefficient of friction of 3×10^{-7} was obtained with an 0.32 kg rotor. More recent studies show primary results based on the analysis of different factors related to the energy loss in a magnet/superconductor system, including the effect of size on levitation force, amplitude and frequency dependence of hysteresis loss and reduced hysteresis loss, etc.[11]

3. In Japan, an early description about SMB design using bulk YBCO and its performance could be found in 1991.[12] In this design a rotor of 2.4 kg is levitated by a pair of YBCO rings and rotated at 30,000 RPM. Several SMB research programs have been ongoing based on different applications and interests. A recent data from Tokyo Electric Power Company shows a comparison of energy loss between superconducting bearings and mechanical bearings in a flywheel generator, in which a 140 kg, 1-meter diameter rotor runs at 2,000 RPM. The system can store a kinetic energy of 50 Wh and release a power at 2 kW.

3. In Italy, a design group headed by Robert Albanese has constructed a flywheel storage device using a hybrid bearing layout, in which a ceramic ball bearing is installed on the top and a SMB on the bottom to supply the levitation for a rotor of 3.3 kg. The flywheel has been successfully run up to 18,000 RPM with an energy of 14 Wh stored.[13]

Flywheel Designs Using Hybrid Superconducting Magnetic Bearing (HSMB)n at TcSUH

It has been commonly recognized that insufficient levitating force from the HTS alone in the SMB is one of the main obstacles for practical industry-scale designs. The typical lifting pressure observed is only 70~140 kPa as the gap varies from 2~0 mm. At TcSUH, we use a hybrid design concept as shown in Fig. 1 to achieve a high levitating force. In this design, the rotor weight is levitated by forces from the magnets, rather than from the HTSs. The HTSs are placed in between the magnets just for stabilization in both axial and radial directions. Such a configuration offers a

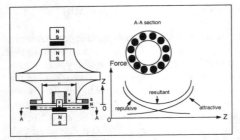

Fig. 1 Schematic representation of magnet and superconductor placements.

much higher levitating force than that from the direct interaction between magnets and HTSs. Besides, with the "field-cooled" procedure, the system is less susceptible to flux creep because the differences between the external and internal fields are small, due to flux pinning. The system is composed of a 42 lb. rotor levitated by 2 HSMBs and 1 SMB in which the HTSs are in disk or ring shapes enclosed in three cold-stages. the flywheel will have an energy of 8 Wh stored at a speed of 6,000 RPM. A detailed description about the system assembly and operation could be found from the previous publications.[3,4,14,15]

Recent Rotor Spin-down Experiment

In order to have a more comprehensive understanding the friction loss mechanism and to find the dominant factors among the variables for further improvement, we have been conducting experiments by changing the system parameters or the designs independently. Fig. 2 shows 3 rotor spin-down curves under different conditions.

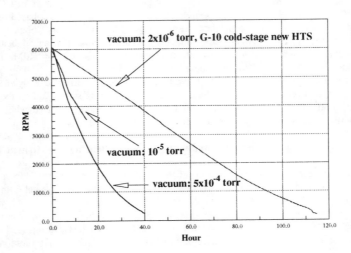

Fig. 2 Flywheel spin-down in different conditions

In Fig. 2 the 40-hour and 15-hour spin-down experiments were conducted under different vacuum conditions to determine the frictional effects coming from air drag alone as discussed in the previous publications. After additional efforts focused on reducing the system's frictional energy loss due to eddy current effect, we recently rotated our flywheel up to *6,000* RPM via a unique non-contact eddy current clutch. The flywheel had been continuously running freely for *143.5 hours*, almost six days, with an average frictional energy loss as low as *0.7%* per hour, as shown in the figure. This significant advance on spin-down performance is accomplished by improvement in HTS materials, in system alignment, in cold-stage design, in rotor balancing and in design of engaging mechanism.

Conclusion

The advance in bulk HTS materials brings an opportunity to build passive magnetic levitation systems for kinetic energy system designs. At TcSUH we have designed and tested a hybrid superconducting bearing by constructing a flywheel prototype. The *6-day* spin-down test indicates the frictional loss in HSMB can be extremely low in comparison with that in conventional and active magnetic bearings. We believe that kinetic energy storage system is a feasible application for hybrid superconducting magnetic bearings.

Acknowledgment

We acknowledge the technical support and HTS material from Ruling Meng, Paul C.W. Chu and Kamel Salama. We would like to thank Dr. Quark Chen, Dr. Nan-jui Zheng and Mark Lamb for many a stimulating discussion. We are also indebted to the crew at the machine shop of the Physics Department of UH, for their expertise in precision machining and technical discussion. We acknowledge ARPA MDA 972-90-S-1001, US DOE Grant DE-FC48-95R810542 and the State of Texas through the Texas Center for Superconductivity for support on this work.

References

1. J.A. Kirk, *Int. J. Mech. Sci.*, vol. 19, pp 223-245, 1977
2. C.R. Keckler, et.al, *NASA Conference Publication 2290*, Proceedings of a workshop held at NASA Goddard Space Flight Center, Greenbelt, Maryland, Aug.2-3, 1983.
3. Q.Y. Chen, et.al, *Applied Superconductivity*, Vol.2, No.7/8, July/Aug. 1994, pp.457-464.
4. K.B. Ma, et.al, submitted to *the 3rd International Symposium on Magnetic Suspension Technology*, Dec.1995.
5. K.B. Ma, et.al, *Proceedings of the 1992 TCSUH Workshop on HTS Materials, Bulk Processing, and Bulk Applications*, 1992.
6. J.R. Hull, et.al, *Applied Superconductivity*, Vol.2, No.7/8, July/Aug. 1994, pp.449-455.
7. B.R. Weinberger, et.al, *Supercond. Sci. Technol.* 3(1990), pp.381-388.
8. R. Takahata, et.al, *IEEE Transactions on Magnetics*, Vol. 27, No.2, March 1991, pp. 2423-2426.
9. K.B. Ma, et.al, *J. Appl. Phys.* 70(7), Oct.1991, pp.3961-3963.
10. H.J. Bornemann, et.al, *Applied Superconductivity*, Vol.2, No.7/8, July/Aug. 1994, pp.439-448.
11. J.R. Hull, et.al, submitted to *J. Appl. Supercond.*
12. H.Takaichi, et.al, *Proceedings of the Third International Symposium on Magnetic Bearings*, pp.307-316.
13. R. Albanese, et.al, *Proceedings of the Fourth International Symposium on Magnetic Bearings*.
14. Z.Xia, et.al, *Proceedings of MAG'95*, pp.321-329.
15. Z. Xia, et.al, *IEEE Transc. on Appli. Supercond.* Vol.5, No.2, pp.623-625.

Superconducting Homopolar Motor

Donald U. Gubser
Naval Research Laboratory
Washington, DC 20375-5343, U.S.A.

ABSTRACT

The U.S. Navy has been pursuing the development of superconducting homopolar motors since 1969. Successful demonstration of the motor, using NbTi wire for the magnet, was achieved in the early 1980's. Recently this same motor was used as a test bed to demonstrate progress in high temperature superconducting magnet technology. The conductor used in the magnet was bismuth-cuprate in the so called 2223 phase. In the Fall of 1995 the motor achieved a performance of 167 hp operating at a temperature of 4.2K and 122 hp while operating at 28K. Future tests are scheduled in the Spring of 1996 using new magnets with conductors of both the 2223 and the 2212 bismuth-cuprate phases.

1. Introduction

The U.S. Navy has been developing dc homopolar motors since 1969. The primary reasons for choosing dc homopolar motors is that these motors develop high torques and are easy to control at low speeds. Homopolar motors are also electrically and acoustically quiet since there is no alternating electromagnet field during operation. Finally, the magnet in the motor is a simple dc solenoid type which sees no ac fields and develops no reaction torque during operation. This makes the incorporation of superconducting technology relatively easy in the motor design.

The homopolar motor is the only strictly dc machine. In the motor a radial dc magnetic field intersects a conducting cylinder through which an axial dc current is passed. The product of the current times the magnetic field produces a radial torque which causes the cylinder to rotate. The speed of the rotation is easily controlled by adjusting the current flow in the cylinder.

These compact homopolar machines are smaller in size and lighter in weight than conventional ac machines. They also have greater efficiency. The biggest advantage of electric motor drive for ship propulsion, however, is that direct coupling from a high speed generator through a gear box to the propeller can be eliminated. Instead, the generator can be remotely located to the motor and connected by flexible current leads. This concept can lead to revolutionary new ship design concepts, leading to quieter, smaller, more fuel efficient ships.

In the early 1980's a superconducting homopolar motor and a superconducting generator were both installed in a small Navy craft which operated in the Chesapeake Bay in Maryland. This demonstration clearly showed that the engineering associated with a completely superconducting propulsion technology could be achieved.

One disadvantage of dc homopolar motors is that they are inherently a low voltage, high current motor. High currents mean large and often cumbersome current leads; but the main technical problem with high currents is with the slip ring current contacts to the rotating disk. Conventional metal brush sliding contacts have too much contact resistance to produce compact, homopolar motors. Only by going to superconducting magnets which produce fields to 5 Tesla or greater, thus generating higher voltages and lower the current requirements, and by using liquid metal sliding contacts, which have much lower contact resistance than the brushes, can compact homopolar dc machines be built.

For military applications, the two biggest drawbacks to superconducting homopolar motors are the logistic concerns (availability, storage, and transport) of liquid helium, and the volatility and stability of liquid metal NaK, which is used as the current collectors. Overall, these technical risks, coupled with the costs of developing new ship concepts, have outweighed the perceived operational benefits of superconducting motors. The U.S. Navy has thus not implemented superconducting drive in any full scale ship.

Recent advances in refrigeration systems have eliminated the need to operate the superconducting magnet in a liquid helium bath. Instead, closed cycle refrigeration systems coupled to the magnet by conduction cooling has been used to cool relatively large magnets to temperatures slightly below 5K. These systems still use NbTi superconducting wire, but completely eliminate the helium logistics problem, since all that is needed to operate the superconducting system is an electrical outlet to supply power to the compressor of the refrigerator. Design of these closed systems is very exacting and expensive but can be done.

Use of high temperature superconducting (HTS) leads and the higher operating temperature of Nb_3Sn superconducting wires, can lead to a system with a much greater margin against thermal instabilities, with less stringent demands on the cryogenic design, with a reduced power requirement on the refrigerator. These systems could operate at temperatures up to 10K, a factor of 2 higher than systems using NbTi superconducting wire. Still greater reduction in power consumption, significantly reduced complexity in engineering design, and much greater overall system reliability can be achieved by using HTS conductors for the magnet system. Present HTS conductors can be wound into magnets suitable for homopolar motors which permit operation up to 20K and hopefully up to 40K in the not too distant future.

The present demonstration program, sponsored by the Navy and ARPA, is to assess the status of HTS conductors and magnets for potential use in homopolar motors.[1] Successful demonstration of these magnets will have greater military utility than just for homopolar motors. Studies are being conducted for using a HTS magnet system in mine sweeping applications and for shipboard uses of magnetic energy storage systems.

2. HTS Homopolar Motor Demonstration

The demonstration motor is shown in figure 1. The radial field is produced in the center of the magnet assembly by operating the split magnet system in series opposition. This motor has been operated with a rating of 400 hp using NbTi magnets operating in liquid helium at 4.2K. In the demonstration, the NbTi magnets were replaced with HTS magnets having the same physical dimensions. No other operating or system aspects were changed.

Since the motor performance depends critically on the current density of the HTS conductor, a major goal of the program was to chart the progress of conductor development in industry. Figure 3 shows a plot of the overall conductor current density for approximately 30 meters of commercial wire, which was wound into 6-inch coils. It is clear that steady progress has been made over the past four years. Presently, the wires tested in this program are operating near 2×10^4 amperes/cm^2 range (1µv criteria) which is sufficient for our demonstration. To realize a sustaining technology for widespread military motor applications, one would need a current density of 2×10^5 amperes/cm^2. If progress continues over the next four years as it has in the past, industry should achieve this level of performance by the year 2000.

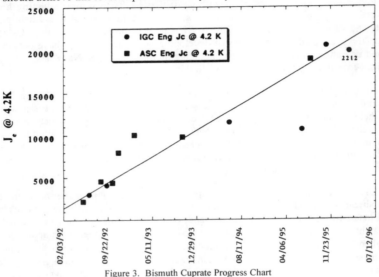

Figure 3. Bismuth Cuprate Progress Chart

Four of the initially delivered motor magnets (1/2 the total requirement) were assembled into the motor in the Fall of 1995, and the motor was operated. The superconducting motor was connected through a gear box to an induction motor, which in this case functioned as a generator and fed power back into the utility power system. The maximum power generated by the motor operating at 4.2K was 167 hp. A subsequent test was performed where the magnet system was immersed in liquid neon at 28K. In this run the motor achieved a performance level of 22 hp.

Future demonstration runs for the homopolar motor will occur in the early summer of FY96 when a set of 8 improved motor magnets will be installed in the motor. It is hoped that a 400 hp performance level will then be achieved.

3. References

1. D. U. Gubser, *Superconducting Coil Development and Motor Demonstration: Overview*, Journal of Electronic Materials **24** (1995) pp. 1843-1850.

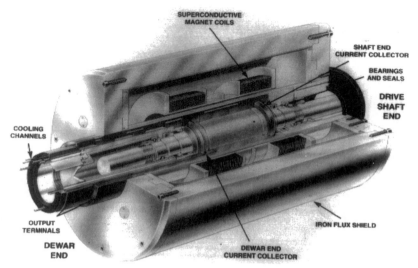

Figure 1. HTS Homopolar Motor

To simplify the construction of the HTS magnets, each magnet of the split coil arrangement was further sub-divided into 4 segments (referred to as motor magnets) and connected in series. Each motor magnet is 9.625 inches (24.448 cm) in outer diameter with a 1 inch square (6.45 cm square) cross section. The HTS motor magnets are designed to operate at 130 amperes. The ultimate performance of the motor depends on the current density in the motor magnets. Figure 2 shows the calculated performance rating for the motor as a function of the magnet current density.

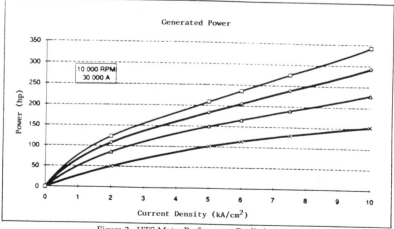

Figure 2. HTS Motor Performance Prediction

HIGH TEMPERATURE SUPERCONDUCTING FAULT CURRENT LIMITER

EDDIE M. LEUNG
Lockheed Martin Corporation
Advanced Development Operations
15250 Avenue of Science
San Diego, CA 92129
USA

ABSTRACT

One of the most near term High Temperature Superconductor (HTS) applications is a Fault Current Limiter (FCL). It is a device that can provide significant cost savings and operational efficiency increase for the power utility industry. This is especially important amid a deregulating environment. The Lockheed Martin Team, which also includes Southern California Edison (SCE), American Superconductor Corporation (ASC), and Los Alamos National Laboratory, has been developing a 2.4 kV, 2.2 kA HTS FCL since October, 1993, under the auspices of the Department of Energy (DoE) Superconductivity Partnership Initiative (SPI) program. This two-year Phase I program was successfully completed in October 1995, following six weeks of extensive testing at the Center Test Substation of SCE in the summer. This paper represents a description of the development and test results will be presented. A Phase II, for the development of a 15 kV, 10.6 kA FCL unit is scheduled to commence in March, 1996. Planning effort and schedule for this new phase will be given. A brief description of the underlying principle of a FCL and how it can benefit the power utility industry will also be given.

The Need For A Fault Current Limiter

The capacity of an electric power system to supply high short circuit current is an essential characteristic of a robust and stable system. This capability is necessary for the system to meet large or new load demands without excessive voltage or frequency drops. However, high current surges caused by short circuits, if not handled by appropriately rated equipment can result in severe damage to utility and user installations with serious economic and safety consequences.

Fault currents are usually momentary events and are originated by a variety of sources such as: equipment failures, lightning strokes, metallic balloons, tree branches, kites, and animals shorting out the circuits. In all cases the resulting current surges can easily damage transmission line and substation components, and under certain conditions the faults can even destabilize a regional power grid which extends over several states. Clearly there is a need for a current limiter that reduces current only during abnormal or fault conditions, and does not interfere with normal power flows, resulting in a safe but yet stiff system.

Traditionally, large resistors or series reactors have been used to limit short circuit currents to safe levels. Unfortunately these brute-force fixes limit current not only during disturbances, but also under normal operating conditions, thus weakening the system. To date, utilities have been lacking an affordable fault current limiter which can not only automatically detect and limit fault currents, but also disconnect itself anytime current flows do not exceed those required for normal operation of the loads. By reducing fault current to a desired level lower than those reached by uncontrolled current surges, a utility can potentially achieve: significant savings due to deferred or avoided equipment upgrades; increased reliability as expensive transmission and distribution will be less stressed; and increased power quality. These benefits are directly in line with the utility industry's goal of enhancing their competitive position under a deregulated market in the immediate future.

Why High Temperature Superconductor?

Having a FCL for the power utility industry is not a new concept. The earliest paper on such a device appeared some 40 years ago. In the mid-seventies, when various Low Temperature Superconducting (LTS) applications were considered due to the sudden availability of NbTi, ANL and EPRI (reference 1) have conducted an extensive study of applying LTS technology to a superconducting FCL. The conclusion was that it was too expensive and the heat loss (refrigeration cost) made it impractical. Indeed, an ideal FCL is one that draws little or no current and therefore losses (notice that superconductor is not lossless with an a.c. current) during normal operation and that it can be switched into the power grid automatically and instantaneously upon the detection of a fault. High Temperature Superconductor (HTS) has the advantage of operating at a higher temperature. The Carnot and therefore practical efficiency of the refrigerator/liquefier required to keep the device cold during operation increases by a factor of 40. The HTS conductor is also more stable because of the higher thermal capacity of material at higher temperatures. However, a HTS conductor with a high resistivity matrix that is designed for a.c. application does not exist and that its current performance in terms of current density at high field is still miniscule compared to that of LTS conductor. Nevertheless, the existing Bi-2223 conductor has been proven to be a useful conductor for the construction of our prototype HTS FCL. The concept we have selected requires a only a modest current density at a low field (<0.1 T).

Description Of The Phase I Project

With the advances in power electronics and HTS technology, we submitted a HTS FCL proposal to the Department of Energy (DoE) Superconductivity Partnership Initiative (SPI) program solicitation in 1993. Our vertically integrated team, which consists of Lockheed Martin (LMC), American Superconductor Corporation (ASC), Southern California Edison (SCE) and Los Alamos national Laboratory (LANL) was awarded a $4.2M Phase I contract in October 1993. This innovative government/industry team approach that consists of a system integrator, a HTS supplier, a utility and a national lab represents a new concept initiated by DoE to pursue HTS power utility applications. The availability of a real life utility that helps the team define the requirements turned out to be very important. In September, 1994, a 12 V, 32A tabletop HTS FCL unit, recording a 46% fault current reduction was on display at the first annual review of the project. DoE was so encouraged that the LMC team was requested to upgrade the final goal for the two-year program from a 240 V, 3 kA FCL unit to a 2.4 kV, 2.2 kA HTS FCL prototype. After another year and the advancement of an additional $600k by LMC into the project, we are glad to announce that we have successfully completed the design, fabrication and testing of a 2.4 kV, 2.2 kA. The performance of the unit actually superseded the design goals.

Test Results Of The 2.4 kV, 2.2 kA FCL Unit

The operating principle of our HTS FCL consists of the use of fast power electronics and a HTS coil. Upon detection of a fault, the power electronics rapidly introduces a HTS coil with pre-determined inductance into the power grid, hence lowering the fault current almost immediately. For more details of the working principle and potential applications of the unit, please consult an earlier paper by the author (reference 2).

The 2.4 kV, 2.2 kA unit was completed in August, 1995 in LMC facility in San Diego, CA. It was then shipped to the SCE Center Test Substation in Norwalk, CA for six weeks of intensive testing and evaluation. Figure 1 below summarizes the performance of our 2.4kV, 2.2kA HTS FCL.

Performance Characteristics	Design Goal	Actual
Maximum Operating Voltage (kV)	2.40	2.38
Maximum Interrupted Fault Current (kA)	2.20	3.03
Maximum Limited Fault Current (kA)	1.10	1.79
Multiple faults Interrupting Capability	2 within 15s	2 within 1s
Automatic Detection Capability	Expected	Achieved
Automatic Recovery Capability	Expected	Achieved
Intercepting At Different Portion of Cycle	Expected	Achieved
Ability To Work As A Circuit Breaker	Bonus	8ms Breaker

Fig. 1 The Performance Of Our 2.4 kV, 2.2 kA FCL Exceeded Our Expectations

Phase II Plan

The success demonstrated in Phase I represents true milestones in terms of an HTS coil's ability to intercept multiple high magnitude fault currents within a short time period, and to offer the flexibility to dial-in a desired level of fault reduction. As a result of these accomplishments, the Department of Energy (DoE) has selected the Lockheed Martin team for negotiation to start a $7.4 M, Phase II SPI project. In this second phase, which may start as early as March, 1996, Intermagnetics General Corporation (IGC) will replace ASC as the HTS supplier of the team. Southern California Edison whose utility expertise proved to be a significant factor for our Phase I success, will again be our utility partner. Los Alamos will supply its HTS and power electronics expertise as part of the government collaboration. Phase II will build upon Phase I successes and a 15 kV Class, 10.6 kA alpha unit is scheduled for field testing in late 1997. Commercial units will soon follow after this milestone is reached. Concurrently, a 69 kV Class unit will be under design.

Concluding Remarks

A revolutionary device such as a HTS FCL is expected to be a valuable tool for power system operators amid the present environment of electric utility deregulation. With further advances in the HTS, cryocooler and power electronics technologies and the corresponding decrease in the component costs, the HTS FCL can conceivably become a near term commercial application.

References

1. K. E. Gray and E. E. Fowler, *Superconductive Fault Current Limiter*, EPRI EL-329 (Dec., 1976)
2. E. M. W. Leung, G. W. Albert et al, *High Temperature Superconducting Fault Current Limiter For Utility Applications*, presented at the 1995 CEC/ICMC meeting in Albuquerque, NM.

ELECTRONIC EYES BASED ON DYE/ SUPERCONDUCTOR ASSEMBLIES

J. T. McDEVITT*, D. C. JURBERGS, S. M. SAVOY, S. J. EAMES AND J. ZHAO
Department of Chemistry and Biochemistry,
The University of Texas at Austin, Austin, TX, 78712

ABSTRACT

Methods for the deposition of molecular dyes onto cuprate thin film structures have been developed. These hybrid systems are found to be sensitive to the influence of light making them suitable for optical sensor applications. Here, the dye structures serve as light harvesting antenna layers which funnel efficiently and rapidly the light energy into the superconductor. Importantly, the molecular antenna layer and superconductor sensor structures can be tailored independently, making these systems suitable for a variety of different optical sensor applications.

Thermally-based sensors or bolometers have been studied for over a hundred years following the pioneering work of S.P. Langley.[1,2] Initial bolometric sensors were prepared using simple metal structures. Large improvements in sensitivity of these structures were achieved when superconductor materials, with their high temperature coefficient of resistance near T_c, were utilized as transition-edge bolometers.[1,3] Moreover, composite bolometers have been developed in which a separate light-absorbing layer is combined with a superconducting thermal element.[1,4-7] These composite detectors exhibit improved responsivity when compared to similar single-element structures. Figure 1 follows the important developments which have occurred in the field of bolometric sensor research.

Currently, high-T_c superconductors fabricated from $YBa_2Cu_3O_{7-\delta}$ thin films have been demonstrated to possess high sensitivities, broad spectral range, and reasonably fast response times. Several high-T_c superconductor detectors approach the theoretical limit of sensitivity for thermally based detector systems. Furthermore, high-T_c superconductor detector designs have been utilized to sense electromagnetic radiation well into the far-infrared and millimeter regions. Here, photon-based detectors do not function well because of the lack of materials with ultra-low band-gaps and problems due to thermal excitation of carriers which occurs in the absence of light absorption.

Antenna-coupled bolometers have considerable advantages over simple bolometers as Richards and others have previously discussed.[1,7-9] In simple single element bolometers, the sensor is limited by the ability of the material both to absorb effectively the incident radiation as well as convert the thermal energy derived therefrom into an electrical signal. Typically, those materials which absorb light efficiently are not the most efficient material for temperature sensing. On the other hand, composite bolometers utilize separate materials for the two functions, each of which can be optimized for their separate roles. Consequently, the absorbing element converts electromagnetic radiation into thermal energy, and the bolometric sensing element acts as a thermal energy to electrical signal transducer.

Here, we report the fabrication and characterization of a wavelength-selective high-T_c bolometric sensor. This composite detector utilizes a molecular organic chromophore material which possesses wavelength-dependent absorbing properties. The light-harvesting antenna layer is shown to absorb effectively specific wavelengths of light and transfer this optical energy to the superconductor-based detector element (*vida infra*). Furthermore, we demonstrate that through careful selection of the light-harvesting layer, the maximum sensitivity of the detector can be tuned to various wavelengths of light in the visible and near-infrared spectral regions.

In order to fabricate the hybrid organic chromophore/high-T_c superconductor detector structures, several processing steps are utilized as described previously.[10-14] First, the superconducting detector is formed using a $YBa_2Cu_3O_{7-\delta}$ thin film (1000 Å to 1500 Å thick) which is deposited using the method of pulsed laser deposition (PLD). The superconductor film is then patterned by a laser etching method to form a microbridge

(~6 mm long x 150 μm wide). Finally, the organic dye such as rhodamine 6G (Rh6G) is deposited by thermal evaporation over the previously patterned region of the high-temperature superconductor thin film.

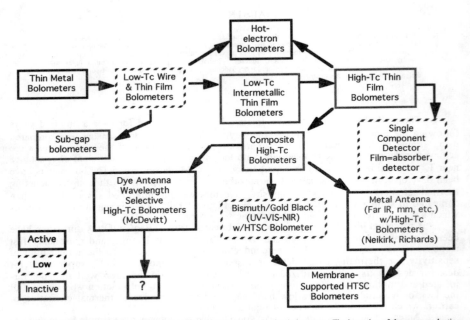

Figure 1. Diagram showing the evolution and classification of superconducting bolometers. The invention of the superconducting bolometer originated with the study of simple metal structures which were operated at low temperatures. Advances in superconducting materials led to a large number of types of superconducting bolometers which exhibit increased sensitivities and increased response rates. Recent work has resulted in the preparation of optically tunable wavelength selective optical sensors based on hybrid dye/superconductor structures.

In order to investigate the ability of the light-harvesting dye layer to affect the optical response of the $YBa_2Cu_3O_{7-\delta}$ detector, microbridge junctions were prepared in which a portion of the microbridge was coated with the dye material and another portion of the microbridge was left uncoated to serve as a reference. Optical measurements were completed using a mechanically chopped monochromatic light beam which originated from a 400-watt xenon arc lamp prior to its passage through the monochromator. Using a dc bias current, the in-phase voltage which developed across the junction was measured with a lock-in amplifier. Measurements were made over the entire length of the microbridge prior to deposition of the molecular material to verify that the optical response of the microbridge was homogenous across its length. An additional measurement was made on the uncoated region of the bridge that involved first passing the incident radiation through a thin film of Rh6G supported on a quartz slide which was not in physical contact with the optical sensor. In Figure 2, the optical responsivity of a microbridge and a region of the same microbridge which was partially coated with a 5000 Å layer of Rh6G is shown. For comparison purposes, the responsivity of the bare microbridge region and that collected through a Rh6G coated glass slide are also provided.

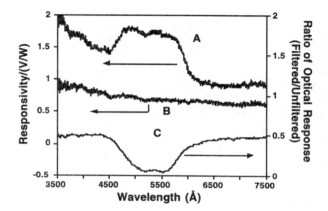

Figure 2. Optical response curves acquired as a function of wavelength for three detector structures. (A) Rhodamine 6G coated $YBa_2Cu_3O_{7-\delta}$ microbridge, (B) uncoated $YBa_2Cu_3O_{7-\delta}$ microbridge and (C) similar optical detector in front of which is placed a thin film of rhodamine 6G which is deposited onto a quartz slide.

As expected, the responsivity versus wavelength for the uncoated region of the $YBa_2Cu_3O_{7-\delta}$ microbridge is found to be relatively flat and featureless (Curve B). Interestingly, the wavelength-dependent responsivity of the Rh6G-coated regions is not flat, but instead consists of a peak feature which is similar to the absorptance spectrum of thermally evaporated thin films of Rh6G. The modification of the measured optical response characteristics of the dye-coated superconductor structure is significant. Clear increases in the responsivity at the wavelength absorbed by the dye are noted. This enhancement has been noted for every dye we have tested to date in which appropriate thickness values of 200 to 10,000 Å have been utilized. Thus, the spectral sensitization of superconductor junctions with molecular dye layers appears to be a general phenomenon. Importantly, the enhancement in response at the absorbed wavelength demonstrates that the dye functions as a light-harvesting antenna layer rather than a light filtering agent. It should be emphasized that this important difference in the role of the dye relative to that currently used for color filter/semiconductor structures may provide future technological opportunities for the described composite dye/superconductor optical sensors, where more precise color recognition is anticipated along with an expansion of the available spectral region. Furthermore, this work documents the first example of energy transfer between molecular dyes and superconductor surfaces. Interestingly, the energy transfer occurs in an efficient manner over macroscopically large distances. Although the exact mechanism of energy transfer between the molecular and superconducting elements has yet to be established, experiments are currently underway to elucidate the nature of this process. Curve C from Figure 2 demonstrates the spectral response properties acquired when the same dye material is separated spatially from the optical sensor. Here, a filtering effect is noted as a decrease in sensor response is obtained at the wavelengths absorbed by the dye, the exact opposite behavior noted for the composite dye/superconductor structure.

Further support for the light harvesting antenna role of the organic dye layer is found in superconducting assemblies that were prepared using an overlayer of H_2-OEP. This dye exhibits four distinct absorbance bands in the visible light region as is evident from UV-visible absorbance spectroscopy.[14] The responsivity of the hybrid detector formed from a 5000 Å thick film of H_2-OEP deposited onto a $YBa_2Cu_3O_{7-\delta}$ microbridge also displays similar spectral features. In both the Rh6G and H_2-OEP cases, the overall responsivity of the device increases after deposition of the dye layer, although the degree of increase depends both on the nature of the dye layer deposited and its thickness. Furthermore, both structures exhibit peak features which match closely the visible-light absorption properties of the dye thin film overlayer. This behavior strongly suggests that the phenomenon is dependent on the properties of the dye layer and that the enhancement is due to an interaction between the dye layer and the superconducting element. Thus, molecular dye assemblies can be exploited to sensitize superconductor junctions to enhance both the responsivity and the wavelength discrimination capabilities of the device.

In summary, methods for the creation of sensitive and wavelength selective bolometric sensors have been described. Here, it has been shown that a variety of molecular dye layers can be used to alter in a significant manner the spectral properties of superconductor optical sensors. Importantly, efficient energy transfer from dye molecules to superconductor structures over macroscopic distances has been illustrated.

Acknowledgments

This work was supported by the National Science Foundation as well as by the Electric Power Research Institute and the Texas Advanced Research Program.

References

1. E.L. Dereniak; D.G. Crowe in *Optical Radiation Detectors*; (John Wiley & Sons: New York, 1984)
2. W. Budde in *Optical Radiation Measurements: Physical Detectors of Optical Radiation*; (Academic Press: New York, 1983)
3. D.H. Andrews; B.W. F.; W.T. Ziegler; E.R. Blanchard *Rev. Sci. Inst.* **13**, 281-292 (1942).
4. J.C. Brasunas; S.H. Moseley; B. Lakew; R.H. Ono; D.G. McDonald; J.A. Beall; J.E. Sauvageau *J. Appl. Phys.* **66**, 4551-4554 (1989).
5. S. Verghese; P.L. Richards; K. Char; S.A. Sachtjen *IEEE Trans. Magn.* **27**, 3077-80 (1991).
6. J.C. Brasunas; B. Lakew *Appl. Phys. Lett.* **64**, 777-778 (1994).
7. P.L. Richards *J. Appl. Phys.* **76**, 1-24 (1994).
8. J. Clarke; G.I. Hoffer; P.L. Richards; N.H. Yeh *J. Appl. Phys.* **48**, 4865-4879 (1977).
9. J. Clarke; G.I. Hoffer; P.L. Richards *Revue de Physique Appliquée* **9**, 69-71 (1974).
10. D. Jurbergs; J. Zhao; S.G. Haupt; J.T. McDevitt *Mater. Res. Soc. Symp. Proc* **328**, 751-6 (1994).
11. D. Jurbergs; R.-K. Lo; J. Zhao; J.T. McDevitt in *SPIE Symposium Series*; SPIE: 1994; pp 138-149.
12. D. Jurbergs; S.G. Haupt; R.-K. Lo; J. Zhao; J.T. McDevitt *Mol. Cryst. Liq. Cryst.* **256**, 577-582 (1994).
13. D.C. Jurbergs; S.G. Haupt; R.-K. Lo; C.T. Jones; J. Zhao; J.T. McDevitt *Electrochem. Acta* **40**, 1319-1329 (1995).
14. J. Zhao; D. Jurbergs; B. Yamazi; J.T. McDevitt *J. Am. Chem. Soc.* **114**, 2737-2738 (1992).

PROGRESS OF HTS BISMUTH-BASED TAPE APPLICATION

KEN-ICHI SATO
Osaka Research Lab., Sumitomo Electric Ind., Ltd.
1-1-3, Shimaya, Konohana-ku, Osaka, 554 Japan

ABSTRACT

A great deal of progress on HTS bismuth-based tape application has been achieved. Especially, actual application using bismuth-based HTS tapes for large current conductors and magnets was demonstrated.

1. Introduction

Towards actual application of high-Tc materials, bismuth-based high-Tc tapes are being developed for large current conductors and magnets. Typical application prototypes using silver-sheathed BSCCO tapes are as follows: (1) current leads [1,2], (2) busbars [3], (3) power cables [4,5], and (4) magnets [6-9]. Among these prototypes, 2,000 A-class current leads for NbTi compact synchrotron radiation magnets have been operated for over 2 years, stably. Also, various type of magnets were constructed and evaluated. For future application, high-Tc power cables are promising from the evaluation results of 66 kV/1,000 Arms/3-phase prototype.

2. High Performance Tape Development

It was widely known that silver-sheathed BSCCO multi-filamentary tapes have many advantages because of their flexibility. Typical properties of these tapes are as follows: $Jc(77K)=42,500$ A/cm^2 for rolled short sample, $Jc(77K) > 20,000$ A/cm^2 for 500 m length and $Jc(77K) > 10,000$ A/cm^2 for 1,200 m length.

To improve critical current properties, it is important to consider the crystal alignment and inter-grain connectivity. Critical current properties were well explained with magnetic field normal to wide surface of tapes, which was caused by mis-aligned angle between BSCCO grains. A sample with $Jc=33,000$ A/cm^2 still has a mis-alignment angle of 8 degree. It is expected to improve Jc properties with better alignment of grains. From the results of inter-grain connectivity, it is also expected that Jc properties can be improved with grain boundary improvement.

For large magnet application, high amperage conductor and high strength conductor are needed because of magnet protection and high electromagnetic stress. Figs. 1 and 2 show the magnetic field dependence of critical current of high amperage conductor. These conductor have critical current of 320 ~ 377 A at 77 K and self field. The conductor showed the Ic of 1,380 A at 4.2 K and 15 T, 480 A at 27 K and 15 T, and

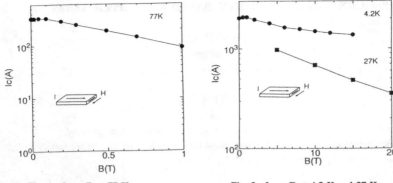

Fig. 1   Ic vs B at 77 K.   Fig. 2   Ic vs B at 4.2 K and 27 K.

91 A at 77 K and 1 T. Such conductor will be used for large scale magnet application. To get high strength conductor, better tensile strain behavior with silver-alloy sheath can be achieved. Antimony is considered to be effective alloying element to silver just neighboring to BSCCO, and manganese is effective to external region of silver. With this technique, tensile strain properties were improved to 4 times (up to 200 MPa without Jc degradation) compared with pure silver-sheathed tapes.

3. Application Prototypes

Table I shows the present status of application prototypes of Ag/BSCCO tapes. Typical results will be discussed below.

APPLICATION		PRESENT STATUS	
HIGH QUALITY TAPE		1,200m LONG HIGH AMPERAGE	$Jc > 10,000 A/cm^2$ $Ic = \sim 1,400A$ (4.2K, 15T) $Ic = \sim 100A$ (77K, 1T)
CURRENT LEADS		2.5kA, 0.4W/kA 0.5kA, 0.1W/kA	IMPLEMENTATION (SR) Ag-Au ALLOY SHEATH
BUSBAR		10kA GO & RETURN	TARGET: 50kA, 200m
CABLE		3kA(5m), 12kA(1m) 66kV/1kArms/7m	TARGET: 1GVA CABLE (66kV, 8.8kArms, 500m)
MAGNET	STAND-ALONE		
	60mm ID DP	0.66T (77K) 3T (21K) 4T (4.2K)	TARGET: ~ 10T (CRYOCOOLER)
	LAYER-W	1.4T (4.2K)	
	HYBRID (4.2K)		
	40mm ID DP	24.0T (Bex=22.54T)	23.5T (1GHz/NMR)
	LAYER-W	21.8T (Bex =21T)	

Table I

Summary of Application Prototypes.

3.1 Current Leads

Current leads of Ag/BSCCO can reduce helium consumption down to 1/4 of copper. From December 1993, 4-pairs of 2,000 A-class Ag/BSCCO leads were operated in synchrotron magnets, and for over 2 years, they worked very stably. They experienced over 100 heat cycles between room temperature and 4.2 K, and over 1,500 times current ramping up to 1,800 A.

3.2 Cables

A prototype of 66 kV/1,000 Arms/3-phase cable (7 m length) was built. From the test results, ac losses were well evaluated with 4-probe method and calorie metric method. Also, high-Tc shielding structure proved to work well judging from the strayed field measurement and inductance measurement. Main features of high-Tc cables are compact and large power capacity. For the future power demand, it is considered to be technically and economically difficult to construct new ducts or tunnels for another copper cable, but it is feasible to replace high-Tc cables for copper cables.

3.3 Magnets

High-Tc magnet application is expected at a wide temperature range between 4.2 K and 77 K. Among these application, ac coils at 77 K, 20 K operation with cryocooler cooled magnets, and super-high field magnets are deemed to be good targets, because only high-Tc can play at these temperature and field conditions. For an example, a react & wind processed magnet was built. A winding bore of this magnet was 60 mm. This magnet was evaluated at 4.2 K, 77 K and was cooled down to 21 K with GM cryocooler. As shown in Table II, the magnet generated 4 T at 4.2 K, and 0.66 T at 77 K, continuously. When cooled with GM cryocooler, the magnet

Operating Temperature (K)	4.2K Liq.He	21K Refrigerator		77.3K Liq.N_2
Max. field (T)	4.0	3.0	2.5*	0.66
Central field (T)	3.9	2.9	2.4	0.65
Operating current (A)	200	150	123	33
KA×turn	520	390	320	66
Coil Jop (A/cm^2)	6,700	5,000	4,100	860
Jop of the wire (A/cm^2)	27,600	20,700	17,000	3,530

Table II

Summary of 60 mm Winding ID Magnet.

*Continuous operation over 150 hours.

generated 3 T, and 2.5 T for 150 hours, continuously. A next target for actual use following current leads is considered to be magnets. High-Tc magnets are able to cover many application fields, such as ac use, high field and large bore magnets, and 1GHz NMR insert coil. In Table I, it is shown that solenoidal coil generated 0.81 T at 21 T backup field [9]. Because of field homogeneity and persistent mode operation, the high performance results of solenoidal coil is very expecting.

4. Acknowledgements

The author would like to express many thanks to Drs. T. Hara and H. Ishii of Tokyo Electric Power Company for joint development of cables, Drs. T. Ando, T. Isono and H. Tsuji of Japan Atomic Energy Research Institute for joint development of busbars, Drs. T. Kishida, Y. Imai and A. Ryoman of Kansai Electric Power Company for joint developments of current leads and SMES coils, Drs. T. Kiyoshi, K. Inoue, H. Kumakura, K. Togano and H. Maeda of National Research Institute for Metals for development of 1GHz NMR insert coils, Professor K. Watanabe of Tohoku University for high field evaluation at 4.2 K, and Professor Y. Iwasa of M.I.T. for high field evaluation at 4.2 K and 27 K. Part of this study was performed as " Cooperative Technology Development", being consigned by Research Development Corporation of Japan.

5. References

1. T. Kato, K. Sato, T. Masuda, T. Shibata, Y. Hosoda, S. Isojima, S. Terai, K. Kishida and E. Haraguchi: *Adv. in Superconductivity* VI (Springer, Tokyo, 1994) 1273.
2. T. Masuda, C. Suzawa, T. Shibata, S. Isojima, T. Kato and K. Sato: *Adv. in Superconductivity* VII (Springer, Tokyo, 1995)1235.
3. T. Kato, K. Sato, T. Ando, T. Isono and H. Tsuji: *Adv. in Superconductivity* VII (Springer, Tokyo, 1995)1215.
4. J. Fujikami, N. Shibuta, K. Sato, H. Ishii and T. Hara: *Adv. in Superconductivity* VII (Springer, Tokyo, 1995)1195.
5. T. Shibata, J. Fujikami, S. Isojima, K. Sato, H. Ishii and T. Hara: To be published in *Adv. in Superconductivity* VIII (Springer, Tokyo, 1996)
6. K. Sato: *JOM* **47**, No.8(1995)65.
7. K. Ohkura, K. Sato, M. Ueyama, J. Fujikami and Y. Iwasa: *Appl. Phys. Lett.* **67**(1995)1923.
8. K. Sato, K. Ohkura, K. Hayashi, M. Ueyama, J. Fujikami and T. Kato: *PHYSICA B* **216**(1996)258.
9. K. Inoue: *4th meeting of Superconductivity Application*, Cryogenic Association of Japan (1996).

DEVELOPMENT of HIGH T_c SUPERCONDUCTING WIRES for APPLICATIONS at 20 K

T. HAUGAN, F. WONG, J. YE, S. PATEL, D. T. SHAW
New York State Institute on Superconductivity and Department of Electrical and Computer Engineering, 330 Bonner Hall, State University of New York at Buffalo, Amherst, NY 14260 U.S.A.

and
L. MOTOWIDLO
Intermagnetics General Corporation Advanced Superconductors, 1875 Thomaston Ave., Waterbury CT 06704 U.S.A.

ABSTRACT

Two issues related to the development of technologically useful high T_c superconducting wires at 20 K are presented: (i) the effect of mechanical deformation parameters (% step size reduction, sheath material, or multifilament) on core thickness variations ('saugaging'), and (ii) a comparison of isothermal melt processing (IMP) and partial-melt growth (PMG) methods for processing oxide powder-in-tube (OPIT) $Bi_2Sr_2CaCu_2O_{8+x}$ (Bi-2212) Ag-sheathed tapes.

1. Introduction

The development of technologically useful superconducting wires at 20 K has progressed steadily since the discovery of high T_c superconductors in 1986 - 1990.[1] Bi-based compounds show great promise for applications at 20 K because of their reduced weak-link behavior, and their ability to be processed in long lengths (>100 meters) with technologically useful critical current densities.[1]

While much work has been achieved in high T_c wire development, additional studies will assist in improving wire properties and increasing the cost-effectiveness of manufacturing processes. This paper describes efforts with Bi-2212 conductors for improving wire properties and long length manufacturing capability. The work in these studies, however, should be applicable to the development of other high T_c phase conductors.

2. Experimental

Ag-sheathed $Bi_2Sr_2CaCu_2O_{8+x}$ (Bi-2212) tapes were prepared by the oxide powder-in-tube (OPIT) method,[2-4] and processed by partial-melt growth (PMG)[2-4] or isothermal melt process (IMP)[5] methods. For single filamentary tapes, the starting Ag tube OD was typically 0.635 cm, and the powder-packed tube was rolled to a tape with typical dimensions (0.0060 cm thickness, 0.9 cm width) using a 0.01 cm rolling step size reduction to 0.1 cm tape thickness, and a 10% reduction rate at lower tape thickness, unless specified otherwise. The rolling speed was 150 cm/minute, and twin stainless steel rollers (diameter 6.5 cm each) were used. A slight amount of oil lubrication (SAE-30 oil) was used, and the front and reverse tension was minimal. Multifilament tapes were prepared by Intermagnetics General Corporation using Ag or Ag-Zr alloys. Temperatures reported in this paper were measured using R-type thermocouples (±2°C accuracy).

To quantify core thickness variations ('sausaging'), optical microscope images of polished longitudinal cross-sections were collected with a CCD camera, digitized (512 x 512 pixels) using imaging software, and binarized using edge detection methods. Core thickness variations ('sausaging') were defined in this paper as $\delta = \sigma_{xn-1}/t_{avg}$, where t_{avg} is the mean core thickness and σ_{xn-1} is the standard deviation of the core thickness.

3. Mechanical Deformation

It is well known that the oxide powder-in-tube (OPIT) method produces core thickness variations ('sausaging') in the wire. However, there is little experimental information available about the causes of 'sausaging', or systematic studies quantifying 'sausaging'.

In the following, experimental studies on 'sausaging' are described.

3.A Core Thickness Variations

Figure 1 shows core thickness variations in monofilament 2212/Ag tape for a 10% reduction rate. While the surface roughness of the the core layers in Figure 1 remained fairly constant (σ_{xn-1} = 3 to 4 µm), Figure 2 shows that 'sausaging' (δ) increased from 4 to 20% as the average core thickness was reduced to ~0.0019 cm thickness.

Figure 2 shows that 'sausaging' in these tapes increased significantly as the step size reduction was increased from 10 to 20%. For a 20% reduction rate, the surface roughness (σ_{xn-1}) was ~6 to 7 µm, and 'sausaging' increased as high as 32% for 0.0019 core thickness. Multifilament wires (37 and 84 filaments) produced by Intermagnetics General Corporation showed marked improvement in 'sausaging', maintaining ~10% sausaging for average core thickness as small as 0.0010 cm.

The effect of 'sausaging' on J_c was investigated in single filament tapes for this work, and was found to be compare well with the theoretical drop expected from core thicknesses that vary periodically with a sinusoidal shape.[6] This theory suggests that 'sausaging' should be maintained to less than 10% to minimize J_c reductions <10%.[6] This was achieved in the multifilament tapes, as shown in Figure 2.

3.B Sheath Material

To increase the rolling pressure and final packing density after rolling, Ag tubes were clad with stainless steel 304 tubing prior to rolling. Tapes mechanically processed by this method, however, showed large sausaging (>30%) when the core thickness was quite large (~0.02 to 0.04 cm). The 'sausaging' increased as the tape thickness was reduced further, causing the silver and powder layers to break through the stainless steel cladding.

To determine if increased sheath strength would reduce 'sausaging', stainless-steel 304 PIT tapes were mechanically processed similar to Ag PIT tapes. Down to ~0.03 cm tape thickness, the stainless-steel PIT tapes showed much improvement in interface roughness (~2 µm) compared to Ag PIT tapes (3 - 4 µm). Similar results at the steel-powder interface were also observed for stainless-steel rod in Ag PIT tapes.

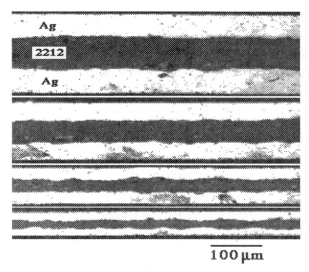

Fig. 1. Longitudinal cross-sections of Bi-2212 single filament tapes achieved with 10% rolling rate reduction below 0.1 cm.

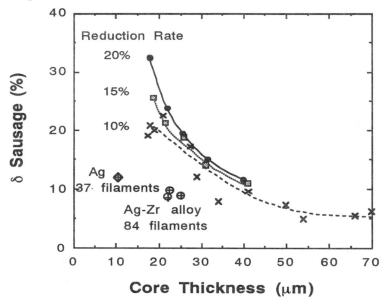

Fig. 2. Core thickness variations for Bi-2212 single and multifilament tapes.

4. Isothermal Melt Processing vs. Partial-Melt Growth

There is increasing evidence that isothermal melt processing reduces the melting temperature and desensitizes the J_c to process temperature variations.[5,7] Figure 3 shows how the $J_c(T)$ sensitivity is decreased from ±2°C to ±4.5°C when comparing IMP to PMG processing. Reduced $J_c(T)$ sensitivity is expected to be beneficial for scaled-up processing.

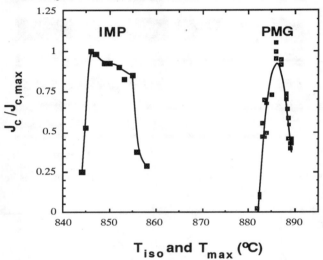

Fig. 3. The effect of temperature on J_c(20 K) for partial-melt growth (PMG)[2] and isothermal melt processing (IMP)[5] of single filament Bi-2212 tapes.

References

1. G. B. Lubkin, *Physics Today* 49 (1996) 48.
2. T. Haugan, S. Patel, M. Pitsakis, F. Wong, S. J. Chen and D. T. Shaw, *J. Elec. Mater.* 24 (1995) 1811.
3. T. Haugan, M. Pitsakis, J. Ye, F. Wong, S. Patel and D. T. Shaw, *Appl. Supercond.* 3 (1995) 85.
4. T. Haugan, *Partial-Melt Growth of $Bi_2Sr_2CaCu_2O_{8+x}$/Ag Superconducting Tapes*, Ph.D. Dissertation, State University of New York at Buffalo (UMI Dissertation Services, Ann Arbor MI U.S.A., 1995).
5. J. Ye, S. Patel and D. T. Shaw, Proc. of 7th Annual U.S.- Japan Workshop on High T_c Superconductors (Tsukuba Japan, Oct. 1995).
6. S. Patel, T. Haugan, F. Wong, S. Chen, S. S. Li, J. Ye and D. T. Shaw, *Cryogenics* 35 (1995) 249.
7. T. G. Holesinger, J. M. Johnson, J. Y. Coulter, H. Safar, D. S. Philips, J. F. Bingert, B. L. Bingham, M. P. Maley, J. L. Smith, D. E. Peterson, *Physica C* 253 (1995) 182.

VERY HIGH TRAPPED FIELDS: CRACKING, CREEP, AND PINNING CENTERS

R. WEINSTEIN, J. LIU, Y. REN, R-P SAWH, D. PARKS,
C. FOSTER* and V. OBOT**

Physics Department, University of Houston Houston, TX 77204 USA
*Indiana University Cyclotron Facility, Bloomington, IN 47405 USA
**Math Department, Texas Southern University, Houston, TX 77004 USA

ABSTRACT

The achievement of very high trapped fields, B_t, in HTS is limited by cracking under magnetic pressure, by quenches, and by creep. Cracking is avoided by keeping the applied field, B_A, as low as possible during the field cooling (FC) activation process. Large dB_A/dt during FC induces voltages which generate quenches. Therefore, the rate at which $B_A \rightarrow 0$ is decreased, but this results in reduced B_t due to creep. This problem has been solved by use of additional cooling, during activation, to slow creep.

Using the above methods, a trapped field of 10.1 Tesla at 42K has been achieved. The mini-magnet used was fabricated from four disks of proton-irradiated, melt-textured Y123, each 2 cm diameter x 0.8 cm thick. This is the highest trapped field ever achieved in an ingot of any material at any temperature.

A newer radiation process, which approximately doubles J_c compared to proton irradiation, has been used to introduce short homogeneous isotropic columnar pinning centers into Y123. In a mini-magnet of this material 3.1 Tesla was trapped at 77K. This is a record field for an ingot of any material at 77K.

1. Introduction

Superconducting materials can sustain current vortices indefinitely, and act like permanent magnets. Such trapped field magnets (TFMs) were developed in the early 1970's, using LTS materials.[1] A trapped field of 2.3 Tesla at 4K was achieved.

HTS materials make TFMs more attractive. Higher fields are available, bulky cryostats can be simplified and, for $T \geq 77K$, liquid nitrogen can be used as a coolant. Our group made the first HTS TFMs from sintered and melt textured $YBa_2Cu_3O_{7-\delta}$ (Y123).[2] An approximate guideline to achieving high B_t is given by: $B_t \propto J_c d$, where d is the diameter of the current carrying entity. In sintered materials, although d was large, intergrain J_c was small, and TFMs trapped $\sim 3 \times 10^{-5}$ Tesla.[2] Improvements in chemicals, reduction of weak links by melt texturing, and aligning B_t parallel to the $\hat{c}$ axis resulted in $B_t \sim 0.1280$ Tesla.[2] The use of excess Y, to reduce the size of the Y211 deposits further increased B_t.[3] Pinning centers were introduced by light ion bombardment at high energy, and increased J_c by a factor of 3-4, resulting in $B_t \sim 1.4$ Tesla.[4] Once B_t in excess of 1 Tesla was achieved several world groups became involved in TFMs, and in their use in levitating bearings, flywheels, motors, generators, levitated trains, and spacecraft bumper/tether.

An investigation[5] of the T and B dependence of J_c yielded an empirical law for J_c below 65K: $J_c(T) = J_c(65)[(T_c-T)/(T_c-65)]^2$, valid for $20 \leq T \leq 65K$, and $B \leq 6T$. A mini-magnet made of seven small grains of Y123 + 0.3Y, which had been irradiated by 200MeV light ions, trapped 4.0 Tesla at 64.5K, and eclipsed the record field of Ref. 1.[6] Similar irradiated magnets later trapped fields of up to 7 Tesla at 55K.[7,8]

The use of Sm seeds and Pt admixture further improved both J_c and d [7] and, permitted production of *reproducible* disks. Disks of 2 cm diameter, 0.8 cm thick, with $J_c \sim$ 10kA/cm^2 and $B_t \sim$ 0.45 Tesla, prior to irradiation, could be produced in any desired quantity with only 10% variation in J_c and B_t. Mini-magnets composed of 4 such disks, proton irradiated, trapped fields of 8.3 Tesla at 54K.[9]

As B_t increased three additional problems were addressed: flux jump, creep, and cracking under magnetic pressure, $J_c \times B_t$. We will consider these below.

2. Flux Jump

The applied field, B_A, needed to achieve trapped field, B_t, by the Field Cool method, must exceed B_t. As B_A is decreased to zero during the activation, an induced voltage appears in the superconductor: $V_{induced} = -A\, dB_A/dt$, where A is the area of the TFM. We have observed that flux jumps occur in a TFM for $V_{ind} \geq 1\mu V$. To avoid this, B_A must be reduced to zero slowly; e.g., for a 2 cm diameter TFM, with $B_{A,max} = 11$ Tesla, B_A should be reduced to zero in no less than 1 hour. Except for quenches due to induced V, we have not observed any flux jumps.

3. Creep

Creep is fairly accurately represented by the law: $B_t(t_2) = B_t(t_1)[1-\beta \log(t_2/t_1)]$. At 77K, $\beta \sim 4.5 \pm 0.5\%$ and $5.5 \pm 0.5\%$ per decade of time, for unirradiated and radiated material respectively. β initially increases as temperature decreases.

A procedure was developed to reduce creep to practically zero, after activation.[9] The procedure utilizes post-activation cooling, PAC, in which T is decreased by an amount ΔT *after* B_A has reached zero. (See Fig. 1.) For a TFM activated to 4.3 Tesla at 65K, for $\Delta T = 2, 4, 6K$, β is reduced by factors of 6.25, 197, and ≥ 1000 respectively.

4. Cracking

An object with current density J, and a self-magnetic field B_t, undergoes an outward pressure $J \times B_t$ due to the Lorentz force. As J, and hence B_t, increase, this magnetic pressure will eventually crack the object. A theory of TFM cracking in the literature[10] treated the static case of a previously activated TFM, with established current J_c and trapped field B_t. TFMs, however, undergo the greatest stress due to magnetic pressure *during* the activation process. We generalized the cracking theory to apply to the dynamic conditions of activation.[11] During activation the conditions of cracking depend on a varying $J_c = J_c(r)$, varying $B_t(r)$, and varying total field due to B_t and B_A. An experiment was performed on cracking.[11] Theory and experiment agree well: a) cracks initiate near the center of the TFM; b) cracks propagate nearly along radii; and c) the experimental fracture strength[11] is in agreement with cracking strengths measured by standard methods of mechanical engineering.[12]

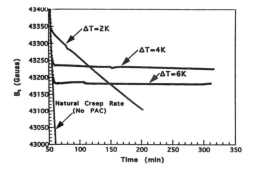 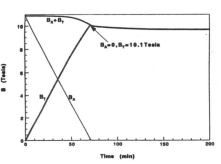

Fig. 1. Post activation cooling (PAC). Creep rates are shown for a 4.3T magnet FC activated at 65K and further cooled, after $B_A=0$, by ΔT as shown.

Fig. 2. Activation of a 4 grain mini-magnet, 2 cm. diam x 3.3 cm. B_A is reduced from 10.7T to 0T in ~ 70 min., during which IAC lowers T to 42K. 10.1 Tesla is trapped.

5. Trapped Field of 10.1 Tesla

A mini-magnet was fabricated from four proton-irradiated single grain TFMs, each 2 cm diameter and 0.8 cm long. T was chosen to achieve $B_t \geq 10T$, using Ref 5. An applied field $B_A = 10.7T$ was used to activate the magnet, guided by the need to use values of $B_A \sim B_t$, in order to avoid cracking.[11] In order to avoid flux jump, B_A was linearly decreased in t = 70 min. This kept $V_{ind} < 1\mu V$. In order to avoid a significant loss of B_t due to creep, during activation, the method of post activation cooling[9] was applied, *during* the activation, to the portion of $B_t(t)$ already trapped (i.e., intra-activation cooling, IAC, was used). With these precautions and procedures a trapped field of 10.1 Tesla was achieved at 42K. (See Fig. 2.)

This is the highest field ever trapped in an ingot of any material at any temperature.

The cracking theory[11] indicates that the material used can trap somewhat over 12 Tesla. We estimate that an ultimate trapped field of over 20 Tesla is possible with evident developments of present materials and techniques.

6. Improvement in B_t at 77K via Pinning Centers

Fields of 2.2T were achieved at 77K [8], using the same proton irradiated material used to achieve 10.1T at 42K. In order to achieve higher B_t at 77K, either J_c or d must be increased. We are exerting efforts to increase d while maintaining high, uniform J_c. However, potential increases also remain in J_c, and we report on this here.

Columnar pinning centers are being studied by several groups.[13] We have studied short, isotropic, homogeneous, columnar centers.[8] These centers increase J_c at 77 from

~40 kA/cm^2, resulting from proton bombardment, to ~85 kA/cm^2. This newer material trapped 3.1 Tesla in a fabricated mini-magnet of 4 tiles, each 2 cm diameter x 0.8 cm long, at 77K. This is the highest field achieved using HTS at 77K.

7. Acknowledgments

We thank the staffs of the National High Field Magnet Lab, and the Indiana University Cyclotron Facility for their assistance with activation and irradiation experiments, and the staff of the Texas Center for Superconductivity at UH, and the Texas A&M Reactor for their assistance with microstructure and irradiation experiments. This work is supported by ARO, NSF, NASA, and the State of Texas via TCSUH.

8. References

1. M. Rabinowitz, H. Arrowsmith and S. Dahlgren, Appl. Phys. Lett., **30**, 607 (1977)
2. R. Weinstein, I.G. Chen, J. Liu, D. Parks, V. Selvamanickam, and K. Salama, *Persistent Magnetic Fields Trapped in High T_C Superconductor*, Appl. Phys. Lett., **56**, 1475 (1990)
3. R. Weinstein and I.G. Chen, Invited Paper, *Persistent Field: Techniques for Improvement, and Applications*, Proc. of TCSUH Internat. Workshop on HTS Materials, Bulk Processing, and Applications, pg. 478, World Scientific (1992)
4. R. Weinstein, Invited Paper, *Permanent Superconducting Magnets*, Proc. World Congress on Superconductivity III, Munich, pg. 1145, Pergammon Press (1992)
5. I.G. Chen, Y. Ren, R. Weinstein, et al, *Quasi Permanent Magnets of High T_C Superconductor: Temperature Dependence*, Appl. Phys. Lett., **62**, 3366 (1993)
6. Y. Ren, R. Weinstein, et al, *Quasi Permanent Superconducting Magnet of Very High Field*, Jour. Appl. Phys., **74**, 718 (1993)
7. R. Weinstein, Y. Ren, J. Liu, I.G. Chen, R. Sawh, C. Foster, and V. Obot, Invited Paper, *Progress in J_C, Pinning, and Grain Size for Trapped Field Magnets*, Proc. Internat. Sympos. on Superconductivity, Hiroshima, pg. 855, Springer-Verlag (1993)
8. R. Weinstein, et al, Invited Paper, *Effects of High Energy Irradiation of MT Y123 on J_C, Trapped Field, Creep, and the Irreversibility Line*, Proc. International Workshop on Superconductivity, Kyoto, Japan (1994)
9. J. Liu, R. Weinstein, et al, *Very High Field Quasi Permanent HTS Magnet with Low Creep*, Proc. of the 1995 International Workshop on Superconductivity, pg. 353, Maui (1995)
10. M. D. Sumption and E. W. Collings, Proc. of Fourth World Congress on Superconductivity, Orlando, Florida (1994)
11. Y. Ren, R. Weinstein, J. Liu, and R. Sawh, *Damage Caused by Magnetic Pressure at High Trapped Field in Quasi-Permanent Magnets Composed of Melt-Textured YBaCuO Superconductor*, Physica C, **251**, 15 (1995)
12. D. Lee and K. Salama, Jpn. J. Appl. Phys., **29** (1990)
13. L. Civale, et al, Phys. Rev. Lett., **67**, 648 (1991)

MAGNETIC LEVITATION TRANSPORTATION SYSTEM BY TOP-SEEDED MELT-TEXTURED YBCO SUPERCONDUCTOR

IN-GANN CHEN, JEN-CHOU HSU, GWO JAMN
Department of Materials Science and Engineering,
National Cheng Kung University,
Tainan, TAIWAN, R.O.C.

ABSTRACT

With the Top-Seeding and Melt-Texturing (TSMT) methods, we have produced large single grain Y-Ba-Cu-O (YBCO) samples over 4 cm in diameter. A magnetic levitation (MagLev) transportation system prototype with TSMT-YBCO high temperature superconducting (HTS) sample has been built to validate the concept of HTS-MagLev system based on Meissner effect. This HTS-MagLev is an inherent stable levitation system, unlike traditional MagLev systems that require sensors and feed-back circuits to dynamically adjust their unstable levitation position. A total weight over 2 Kg (loading capability over 1 Kg) has been tested on this HTS-MagLev prototype.

1. Introduction

The HTS-MagLev with significant loading capability was not demonstrated until recently, due to the relative low levitation force of sintered HTS materials. This low levitation force has been improved by materials processing techniques, such as Top-Seeding and Melt-Texturing methods [1-3]. Centimeter sized single grain YBCO samples can be produced. Each sample is capable of levitating over 1 Kg of weight. This single grain YBCO bulk material with strong pinning force and strong links have shown their superior superconducting properties of strong magnetic levitation force, strong magnetic flux shielding or trapping capabilities[4,5]. The enhancement in superconductivity can be attributed to the elimination of grain boundary weak-links and introduction of strong pinning centers.

Recently, HTS-MagLev prototypes have been built by using the single grain YBCO bulk materials [6,7]. Unlike classical MagLev systems which were based on dynamically controlled electromagnetic suspension system. Various types of electromagnetic coil and complicate electric control device and sensor were required to levitate or suspend a permanent magnet in standing still status. Large electrical power consumption and sophisticated on-line circuit control were required to maintain a stable levitation position. The HTS MagLev system can by-pass all these problems. The HTS MagLev is based on the Meissner effect, which can stand still without consuming any powder and requires no electromagnetic coil or feed-back circuit. HTS MagLev only requires is a simple refrigeration system to maintain HTS materials below its superconducting temperatures, e.g., 77 K.

2. Single Grain YBCO Superconductor

Two processing techniques are involved in the growth of single grain YBCO materials, e.g., melt-texturing (MT) processing and seeding technique. A typical MT process can be found in Ref.[1]. This MT process only enhances the crystal growth condition, i.e., to enlarge the grain size of YBCO materials. MT-YBCO material is still a polycrystalline sample with randomly distributed grain orientations and sizes. The YBCO crystal nucleation can be controlled by introducing a Sm-Ba-Cu-O seed [2,3] to grow into a single grain YBCO materials with pre-determined orientation. Since the matrix phase is a continuous superconducting phase, no grain boundary acts as weak link to block the superconducting current. Therefore, a single grain YBCO sample can be classified as a quasi single crystal materials, although there is second phase precipitation within the single crystalline $YBa_2Cu_3O_7$ matrix. Fig. 1 shows a single grain YBCO sample produced at NCKU with over 4 cm diameter by using the TSMT technique.

3. HTS MagLev Prototype

Considering the Earnshaw's theory, the conventional levitation system between permanent magnets (or electromagnets) has an unstablized force in at least one dimension. Therefore, a dynamic controlling device is needed to maintain the unstable levitation position. Due to the mixed vortex and superconducting current in a type II superconductor, magnetic field can be shielded or trapped within the superconductor. An inherent stable levitation (or suspension) between a magnet and a superconductor can be maintained. We have built a HTS MagLev system at NCKU (Fig. 2) by TSMT YBCO samples which can move near frictionless over an elliptical magnetic rail. This HTS MagLev prototype has shown a capability of levitating over 2 Kg of weight (with a loading capability over 1 Kg).

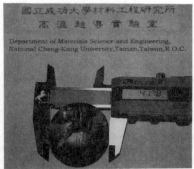 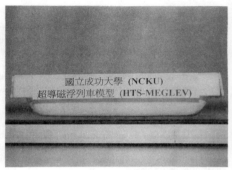

Fig. 1. A single grain YBCO sample, processed by TSMT techniques, grows to a grain size larger than 4 cm diameter. A clear square facet parallel to the a-b crystal orientation can be observed on top of the single grain YBCO sample.

Fig. 2 shows the HTS MagLev built at NCKU with an air gap of about 0.5 cm. The TSMS-YBCO samples were attached to the bottom within the carrier filled with liquid nitrogen.

5. Summary

TSMS YBCO superconductor exhibits attractive magnetic levitation and suspension properties, i.e., Meissner effect. A HTS MagLev transportation prototype with TSMS-YBCO material has been constructed to validate the concept of HTS-MagLev system based on Meissner effect. This HTS-MagLev is a stable levitation system, unlike traditional MagLev systems that require sensors and feedback circuits to dynamically adjust their unstable levitation position. In the future, research on the effects of different permanent magnet configurations, as well as their correlation with TSMT-YBCO sample's superconducting properties will be studied.

Acknowledgments

This work was supported by the National Science Council, Taiwan, R.O.C., under contract No. NSC84-2112-M006-020. We would like to thank Prof. M.K.Wu, Prof. T.S. Chin, and Prof. C.E. Lin. for many stimulating discussions.

References

1. K. Salama, V. Selvamnickam, L.Gao, and K. Sun, Appl. Phys. Lett., 54, 2352 (1989).
2. M. Morita, S. Takkebayashi, M. Tanaka, K. Kimura, K. Miyamoto, and K. Sawano, Adv. in Super., 3, 733 (1991).
3. R.L. Meng, L. Gao, P. Gautier-Picard, D. Ramirez, Y.Y. Sun, C.W. Chu, Physica C, 232, 337(1994).
4. I-G. Chen, J. Liu, Y. Ren, R. Weinstein, G. Kozlowski, and C.E. Oberly, Appl. Phys. Lett., 62, 3366(1993).
5. L. Gao, Y.Y. Xue, R.L. Meng, C.W. Chu, S. Hannahs, V. Thang, and L. Rubin, Appl. Phys. Lett., (1994).
6. H. Weh, H. Pahl, H. Hupe, A. Steingrover, H. May, Proceeding of 14th Intern. Conf. on Magnetically Levitated Systems (MAGLEV '95), Bremen, Germany. Nov. 26 - 29, 1995, p.217(1995).
7. H. Ohsaki, H. Kitahara, E. Masada, ibid.,Ref. 10. p. 203(1995).

HIGH-T_c CERAMIC SUPERCONDUCTORS FOR ROTATING ELECTRICAL MACHINES : FROM FABRICATION TO APPLICATION

A.G. Mamalis
Department of Mechanical Engineering, National Technical University of Athens, Greece.
I. Kotsis
Department of Silicate Chemistry and Technology, Veszprem University, Hungary
I. Vajda
Department of Electrical Machines and Drives, Technical University of Budapest, Hungary
A. Szalay
METALLTECH Ltd., Budapest, Hungary
G. Pantazopoulos
Department of Mechanical Engineering, National Technical University of Athens, Greece.

ABSTRACT

The fabrication of silver-sheathed superconducting $Y(Ba,K)_2Cu_3O_7$ rods by axisymmetric explosive compaction and subsequent multiple-pass warm extrusion at 470°C was employed for the production of electromagnetic components for rotating electrical machines.

1. Introduction

Axisymmetric explosive compaction and subsequent plastic forming using multiple-pass warm extrusion, were employed to fabricate silver/$Y(Ba,K)_2Cu_3O_7$ composite rods [1,2]. Applications of these superconducting components can be found in the electrical and electronic industry, for example in construction of electrical switches for measuring equipment and as HTSC conductors in high power rotating electrical machines.

2. Experimental

Dynamic powder compaction was employed for the fabrication of silver-sheathed superconductive material, with a diameter up to 12 mm, using "Paxit" high explosive with 3800 m.s^{-1} detonation velocity. The experimental details related with powder compaction are reported in previous works, see Refs. [1-2]. Heat treatment at 920°C and for 5 hours, in flowing oxygen, is realized after explosive compaction.

Multiple-pass warm extrusion was performed at 470°C on a SMG 1000 MN hydraulic press, using a punch speed up to 7 mm.s^{-1}. The final diameter of the extruded rod of the first pass was 8 mm (r_{T1}=55%), of the second pass 5 mm (r_{T2}=83%) and of the third pass 3 mm (r_{T3}=93%).

3. Results and Discussion

Mechanical damage during post-compaction deformation processing, due to brittle behaviour of the ceramic core, was further limited employing extrusion at higher temperatures, increasing, therefore, the plastic flow and formability of the working material. Punch load - punch travel diagram (P-δ) is presented in Fig. 1(a); the three curves (1, 2 and 3) correspond to the three extrusion passes. These curves are associated to the mechanical behaviour of the composite billet, namely the stress-strain curve during compression testing at the same temperature and strain rate, see Fig. 1(b).

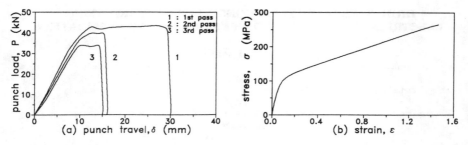

Fig. 1 (a) P - δ diagram and (b) stress-strain curve of the compacted billet

X-ray diffraction indicates a small decay of the 123 phase ($YBa_2Cu_3O_y$) accompanied also with an almost slight increase of the second phases content (Y_2BaCuO_5 and $BaCuO_2$) during deformation processing (Fig. 2(a)). This observation is consistent to the further decrease of transition temperature from 95 to 90K. The deterioration of superconductivity is expected during hot working due to oxygen loss [2]; herein this degradation has been minimized optimizing the extrusion temperature at 470°C and balancing the oxygen diffusion in and out of the sample.

An attempt was made to design and construct a high-T_c superconducting rotating electrical generator with permanent magnet excitation, see Fig. 2(b). Details about the design and construction of this machine will be presented in a future paper.

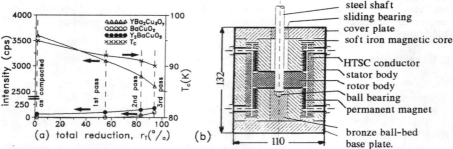

Fig. 2 (a) Evolution of phases content and T_c as a function of total reduction (%) and (b) cross-sectional view of the model machine.

4. Conclusion

Silver-sheathed superconducting rods, with final diameter up to 3 mm, were fabricated by explosive compaction and subsequent multiple-pass warm extrusion at 470°C, possessing superconducting transition at 90K. The 5 mm intermediate component was directly placed as a HTSC conductor in a high power rotating electrical machine.

5. References

1. A.G. Mamalis, A. Szalay, D. Pantelis, G. Pantazopoulos, I. Kotsis and M. Enisz, *J. Mat. Proc. Tech.* **57** (1996) 155.
2. A.G. Mamalis, A. Szalay, D.I. Pantelis and G. Pantazopoulos, *Proc. EXPLOMET '95*, 6-10 August 1995, El-Paso Texas (USA).

PROPERTIES OF JOINTED BPSCCO COMPOSITES

G. YANG AND C. VIPULANANDAN
Materials Engineering Laboratory
Texas Center for Superconductivity (TCSUH)
University of Houston
Houston, TX 77204-4791, USA

ABSTRACT

In this study a method to join bulk BPSCCO composite was investigated by using a combination of pressure and temperature. Electro-mechanical properties of the BPSCCO composites with Ag fibers was also investigated and the fibers were used to improve mechanical properties without degrading superconducting properties. After processing XRD and SEM studies were performed on the specimens to evaluate the purity of 2223 phase and fracture surface respectively. Using the mechanical testing system BPSCCO composites were tested in bending at 77K while the critical current was measured in situ. Stress-critical current-strain relationships have been developed for the composite bulk materials. For the joined specimens critical current was measured both across the joint and in the bulk. The current capacity of the joints varied from 85 to 95% of the bulk material.

1. Introduction

For bulk BPSCCO materials to be used in magnets and other applications, it is important to produce high current materials with long length by joining. Several methods have been investigated to improve the joining of BPSCCO composites without significant current loss at the joint. These methods used either uniaxial or isostatic pressure with heat treatment. to join the BPSCCO bulk. In this study butt joining of BPSCCO components was investigated. Also the electro-mechanical properties of BPSCCO with Ag fiber was also investigated.

2. Testing Program

Commercially available Bi-2223 powder was uniaxially cold pressed into rectangular beam with dimension of 1.5" x 0.4" x 0.15" and initially sintered at 830°C for 20 hours to get enough mechanical strength for the handling. These beams were placed in membrane, vacuumed, and CIPed at around 50 ksi The influence of CIPing on critical current was also studied by CIPing and sintering the specimens many times over. After each CIPing process the specimens were sintered to produce superconductivity. All the sintering was done in a reduced oxygen atmosphere at 830°C for 70 hours. The final dimension of the specimens were around 1.3" x 0.4" x 0.13".

Specimens were tested in three-point bending at 77 K. The broken specimens were then butt joined together using a thermo-mechanical method. Four-probe dc method and four-point bending test were conducted on the jointed specimens to evaluate the superconducting and mechanical properties of the joined specimens. Critical current (Ic) measurement were done across the joint and in the bulk material.

3. Results and Discussion

3.1. Density

Density is often considered as one of the main factors influencing the critical current of high-Tc superconductors. Most of the specimens manufactured during this study had an average density of 75% of the theoretical density.

3.2. Stress - Strain - Critical Current Relationship

Electro-mechanical properties of BPSSCO with and without Ag fibers was studied. The BPSSCO monolithic showed a linear stress-strain behavior and the critical current did not change up to peak stress and then dropped to zero. For the BPSCCO/fiber composites the critical current did not vary with strain during the initial linear behavior. Beyond peak stress, not only the critical current, but also the stress was decreased to 22% and 500 psi respectively. With further loading critical current decreased to 10% with strain of 0.9%, while the stress did not change.

3.3. Microstructure

X-ray diffraction patterns show pure Bi-2223 phase in the bulk and joint of the sintered BPSCCO. SEM studies showed that the sintering samples had randomly distributed grains. After the thermo-mechanical joining process the grains at the joint were oriented with the ab-plane resulting in good current capacity.

3.4. Transport Critical Current

The zero-field values of critical current were measured by the four-contact method. Ic in the Bulk depended strongly on the specimen density, increased with density. The critical current (I_c) of the monolithic specimens varied between 160 to 200 A. The efficiency factor, which is defined as the ratio of Ic across the joint and I_c in the bulk, varied from 85 to 95%.

4. Conclusion

Critical current depends on the density for BPSCCO bulk specimens and CIPing was used to increase the density. The maximum critical current achieved for the monolithic BPSSCO was about 200 A. The thermo-mechanical method adopted for joining the bulk material produced a joint current efficiency of 85-95% of the bulk material. The Ag fibers were effective in preventing the total collapse of the bulk material beyond peak stress.

4. Acknowledgments

This work was supported by the Texas Center for Superconductivity at the University of Houston under Prime Grant MDA 972-88-G-0002 from the Defense Advanced Research Agency and the State of Texas.

5. References

1. H. G. Lee, G. W. Hong, J. J. Kim and M.Y. Song, *Physica C*, **242** (1995) 81.

2. N. Murayama, Y. Kodama, F. Wakai, S. Sakaguchi and Y. Torii, *Japanese Journal of Applied Physics*, **28** (1989) 10, L 1740.

3. Y. Mutoh, M. Inoue and T. Komatsu, *Japanese Journal of Applied Physics*, **29** (1989) 8, L 1432.

STRAIN TOLERANCE OF SUPERCONDUCTING PROPERTIES AND CRYOGENIC MECHANICAL BEHAVIOR OF BULK MgO-WHISKER-REINFORCED HTS BPSCCO COMPOSITE

G.Z. ZHANG, M.S. WONG, S.S. WANG
Texas Center for Superconductivity and Mechanical Engineering Department
University of Houston, Houston, TX 77204, USA

High-temperature superconductors (HTS) are recognized to possess excellent superconducting properties with inherently weak mechanical properties. One of the critical concerns is the limited strain tolerance of superconducting properties. In this study, monolithic HTS BPSCCO is reinforced by strong MgO whiskers, critical strain tolerance of superconducting properties and thermomechanical behavior of $(MgO)_w$/BPSCCO composite at 77K are investigated. Results obtained show that the strain tolerance of $(MgO)_w$/BPSCCO composite is significantly improved.

1 Introduction

In earlier papers[1,2], mechanical properties of monolithic HTS materials are improved by additions of continuous fibers and MgO whiskers. Among the HTS composites developed, $(MgO)_w$/BPSCCO composite has been shown to possess combined excellent superconducting and room-temperature mechanical properties[3,4]. However, no information is currently available on the mechanical behavior and strain tolerance of $(MgO)_w$/BPSCCO composite at cryogenic temperatures. In this study, a unique electromechanical experimental system is developed to study the critical electromechanical behavior of $(MgO)_w$/BPSCCO composite at 77K. Important relationships between Jc and strain tolerance of the HTS composite are established.

2 Experiments

A $(MgO)_w$/BPSCCO HTS composite was fabricated by a solid-state processing method[2]. A schematic of the unique electromechanical experimental system developed in this study is shown in Fig. 1. This system can effectively eliminate undesirable thermal stress generated in the gripping area, and also ensure good alignment of the specimen. Electromechanical experiments were conducted under a displacement-control loading mode. Simultaneous critical current and density measurements were performed using the standard four-probe technique at 77K.

3 Results

3.1 Mechanical Behavior of $(MgO)_w$/BPSCCO Composite

Linear stress-strain relationships were obtained for both monolithic and composite HTS materials up to final failure at 77K. Owing to the presence of high-modulus/high-strength MgO whiskers, Young's modulus and the ultimate strength of the HTS composite were found to increase with the volume fraction of the MgO whiskers, as shown in Figs. 2 and 3. Strengthening effects by the MgO whiskers were clearly demonstrated.

3.2 Strain Tolerance of $(MgO)_w$/BPSCCO Composite

As shown in Fig. 4, the strain tolerance of superconducting properties of the $(MgO)_w$/BPSCCO composite was improved up to 20%, as compared with that of the monolithic BPSCCO. Strengthening and toughening mechanisms in the composite may include whisker bridging, whisker pull-out, and crack deflection, which prevent or inhibit further crack growth and coalescence. In general, current-carrying capability of bulk HTS material was retained up to the final failure, where critical cracks propagated through the specimen. No J_C degradation before complete failure was detected as strains increased.

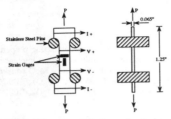

Fig.1 Schematic of electromechanical testing setup & specimen geometry

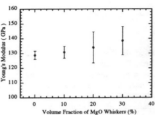

Fig. 2 Young's modulus vs. volume fraction of MgO whiskers at 77K

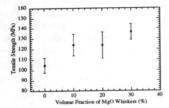

Fig. 3 Tensile strength vs. volume fraction of MgO whiskers at 77K

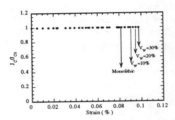

Fig. 4 Critical current density vs. tensile strain for monolithic and composite at 77K

4 Conclusions

Based results obtained, the following conclusions can be drawn:

(1) High-strength/high-modulus MgO whiskers are good reinforcement materials for HTS BPSCCO material in terms of thermodynamic and thermomechanical compatibilities. At 77K, both monolithic BPSCCO and $(MgO)_W$/BPSCCO composite exhibit almost linear elastic behavior up to failure.

(2) Current-carrying capability of monolithic BPSCCO and the $(MgO)_W$/BPSCCO composite is virtually unaffected by tensile strain up to the final failure. Strain tolerance of superconducting properties of $(MgO)_w$/BPSCCO composite can be improved by the MgO whiskers, resulted from beneficial whisker bridging, crack deflection, etc.

References

1. A. Miyase, Y.S. Yuan, M.S. Wong, J. Schon and S.S. Wang, *Supercon. Sci. Technol.*, **8**, (1995) 626-637.
2. Y.S. Yuan, M.S. Wong and S.S. Wang, *J. Mater. Res.*, **11**[1] (1996) 8-17.
3. Y.S. Yuan, M.S. Wong and S.S. Wang, *J. Mater. Res.*, **11**[1] (1996) 18-27.
4. Y.S. Yuan, M.S. Wong and S.S. Wang, *J. Mater. Res.* (1996), in pres.

IV. APPLICATIONS

Small Current

MAGNETOCARDIOGRAPHY IN A MAGNETICALLY NOISY ENVIRONMENT USING HIGH-T_C SQUIDS

J.H. MILLER, JR., N. TRALSHAWALA, J.R. CLAYCOMB, J.H. XU

Texas Center for Superconductivity, University of Houston, 4800 Calhoun Road, Houston, Texas, 77204-5932 USA

K. NESTERUK

Department of Magnetic Materials, Institute of Physics, Polish Academy of Sciences, 00 662 Warsaw, Poland

We have combined several approaches towards the goal of reducing ambient magnetic field noise, including localized high-T_c superconducting magnetic shielding, active noise compensation and digital signal processing. Thus far we have successfully measured the QRS complex and T-wave of a magnetocardiogram in a magnetically noisy environment, such as that encountered in a typical hospital catheterization laboratory.

Background and Motivation

The goal of our research is to develop a magnetocardiography (MCG) system that is capable of measuring magnetocardiograms in an unshielded environment. In order to carry out such a measurement, one has to contend with various ambient noise sources. These include power line and RF interference, microphonics pickup, fluctuations in the earth's magnetic field, and electrostatic pickup. Earlier solutions devised to overcome these problems have entailed the use of a second-order gradiometer inside a deep mine or inside a magnetically shielded enclosure [1]. Apart from these problems, which are common to both low-T_c and high-T_c SQUIDs, high-T_c SQUIDs exhibit additional $1/f$ noise [2]. This noise is the result of hopping of flux vortices that are trapped in the body of the field cooled high-T_c SQUID, and can only be eliminated by zero-field cooling the SQUID.

Methodology

Active noise compensation [3] techniques have been utilized to eliminate some of the noise. Figure 1 depicts a schematic of one of our systems. A fluxgate magnetometer is used to track earth's field and its low-frequency (LF) fluctuations as well as powerline noise. This also enables us to zero-field cool (to within 1% accuracy) our high-T_c SQUIDs [4] and passive shields. We have incorporated small, localized high-T_c superconducting shields that selectively shield out uniform noise fields without affecting the dipole source fields of interest [5]. The entire setup is supported by a non-metallic gantry to eliminate eddy current noise.

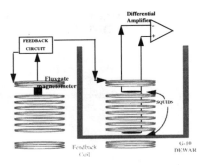

Figure 1. Schematic of our MCG instrument. In this particular example, a room temperature fluxgate magnetometer is used as a reference sensor to primarily cancel earth's field and LF fluctuations.

We have implemented a finite-impulse-response (FIR) comb filter to remove power line interference. This filter has stop bands at 0, ±60, ±120, ...etc., with stop-bandwidths of 1.4 Hz each. In addition, we have also implemented a nonlinear filter based on mathematical morphology operations to remove baseline wander. Namely, set-theoretical operations such as erosion and dilation are utilized (see [6] and references therein). Mathematical morphology has been extensively used in image processing, shape analysis and pattern recognition. Specifically, for our application, we can suppress baseline wander by choosing proper structuring elements. Fluctuations with widths less than that of the structuring element are suppressed. In an MCG signal, the P-QRS-T pulses are between about 75 msec to 250 msec in duration, so a simple flat structuring element is sufficient for our purposes.

These filters have been implemented on a dedicated DSP chip from Texas Instruments (TMS320C50). We have also implemented FIR filters and nonlinear median filters on a Windows NT workstation using the LabVIEW package. We have electronically compensated for the differences in the sensitivities between individual high-T_c SQUIDs. The individual SQUID outputs are then appropriately combined to make electronic gradiometers.

Results

We have succeeded in obtaining an adult human MCG in an unshielded, clinical environment of a catheterization laboratory at Texas Children's Hospital. This data (Figure 2) was obtained using a 3 cm baseline planar gradiometer and taking 100 ECG-gated averages.

With an improved setup we have succeeded in obtaining real time, unshielded MCG data in our laboratory, which is next to a construction site and is located on the third floor. This data (Figure 3) was obtained using a 9 cm baseline first-order axial gradiometer.

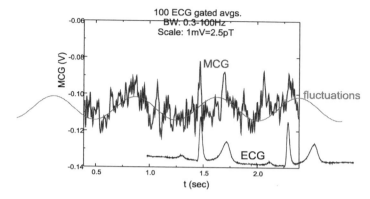

Figure 2. ECG-gated MCG data obtained at Texas Children's Hospital. The graph also depicts LF fluctuations. Due to variations in the heart rate over 100 averages, only the central QRS complex, which was at the chosen trigger point, is clearly visible, while one to the right is somewhat distorted (for comparison, the ECG record from the same patient is also shown).

Discussion

In summary, we have utilized a modular approach to designing our system. We have started out in a noisy environment and systematically removed the various contributions to ambient noise. At this stage, we have two major sources of noise which need to be eliminated. Namely, efforts are underway to de-couple the entire setup from building vibrations (fundamental frequencies of about 1 Hz and 10 Hz) and to design a better RF shielding (one of the buildings next to ours has an 88.9 MHz FM transmitter). More details about our passive HTS shields and DSP techniques are presented in a separate article in these proceedings [7].

Acknowledgments

We gratefully acknowledge the assistance of Y. Bai, M. Boyd, A.K. Guzeldere, and L.M. Xie. This project was supported by the State of Texas through the Texas Center for Superconductivity and the Advanced Technology Program, and by the Robert A. Welch Foundation.

References

1. Zimmerman, J. E. SQUID instruments and shielding for low-level magnetic measurements, Jnl. of Appl. Phys., 1977, 48: 702-710.
2. Miklich, A.H., Koelle, D., Shaw, T.J., Ludwig, F., Nemeth, D.T., Dantsker, E., Clarke, J., Alford, N.McN., Button, T.W., and Colclough, M.S. Low-frequency excess noise in YBCO dc

superconducting devices cooled in static magnetic fields, Appl. Phys. Lett., 1994, 64: 3494-3496.
3. Aarnink, W.A.M., van den Bosch, P.J., Roelofs, T.-M., Verbiesen, M., Holland, H.J., ter Brake, H.J.M., and Rogalla, H. Active noise compensation for multichannel magnetocardiography in an unshielded environment, IEEE Trans. on Applied Superconductivity, 1995, 5: 2470-2473.
4. Commercial high-T_c "iMAG" SQUID system made by Conductus, Inc., Sunnyvale, CA.
5. Tralshawala, N., Miller, J. H., Jr., and Jackson, D. R. High-Tc superconducting image surface magnetometers, IEEE Trans. on Applied Superconductivity, 1995, 5: 2354-2357.
6. Ying, Q., Tralshawala, N., Fan, N., Miller, J.H., Jr., and Jackson, D.R. Digital signal processing of magnetocardiography signals, Proceedings of the 1995 DSP[x] Technical Program, 1995, pp. 520-527.
7. Tralshawala, N., Claycomb, J.R., Xu, J.H., and Miller, J.H., Jr. Noise reduction techniques for operating high-T_c SQUIDs in an unshielded environment, (elsewhere in these proceedings).

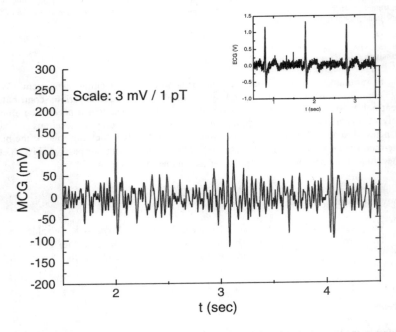

Figure 3. Real time MCG obtained in our unshielded laboratory. The QRS complex can be seen in the MCG. The simultaneously measured ECG signal is shown in the inset at the top right corner.

NEAR-TERM COMMERCIALIZATION OF HTS TECHNOLOGY AT CONDUCTUS

RANDY W. SIMON
Conductus, Inc., 969 West Maude Avenue, Sunnyvale, CA 94086, USA

ABSTRACT

Ten years after the initial discovery of high-temperature superconductors, Conductus has introduced or is poised to introduce commercial products based on thin-film HTS technology for three major markets: communications, healthcare and instrumentation. In communications, filter subsystems for wireless base stations are being field tested and evaluated by the telecommunications industry and could enter the commercial market in 1996. They have the potential to offer major economic and performance benefits to that industry. In the healthcare market, superconducting receivers for magnetic resonance instruments are reaching commercialization for NMR spectrometers and are under development for use in MRI medical scanners. These receivers offer dramatic performance enhancements to magnetic resonance instruments. The first instrumentation products, based on HTS SQUIDs, were introduced to the laboratory market in 1992. The use of HTS SQUIDs for applications in geophysics, non-destructive evaluation and medicine is just beginning. In all these markets, superconducting components are the enabling elements of high-performance subsystems. After nearly a decade of intense development, HTS technology is ready to become a significant contributor to electronic systems.

1. Introduction

The discovery of superconducting materials with critical temperatures above the boiling point of liquid nitrogen stimulated intense worldwide interest in applications development because of the economic and practical consequences of their simplified cooling requirements. Among the consequences of this expanded interest was the nearly immediate formation of several companies dedicated to the commercialization of superconductive technology. In this paper, the commercialization activities of one such company, Conductus, are discussed at a point some nine years after the discovery of YBCO, the first liquid-nitrogen-temperature superconductor. It is only fair to point out, however, that a new industry is taking shape based on HTS technology and that there are a number of companies apart from Conductus that are developing applications of the technology. The ensuing discussion pertains only to the electronic applications of HTS at Conductus; the development of the technology for power, energy and other applications is ongoing as well but is outside the scope of this paper.

The HTS discovery was greeted with tremendous enthusiasm in part because there had already been numerous applications envisioned for superconductivity long before the advent of the new materials. For this reason, even early on, there was tremendous optimism that mainstream insertion of the technology could not be far off. In the intervening years, however, challenges in materials development coupled with a slow process of identifying the real market opportunities for the technology have dominated the evolution of the industry. It is only in the past few years that specific and substantial opportunities for HTS electronics technology have been identified. Those under development at Conductus are discussed below.

2. Communications

Microwave filters made using superconductor technology can provide extraordinary frequency selectivity while maintaining excellent efficiency because of their inherent low insertion losses. Thin-film superconducting filters based on microstrip technology have the added benefit that they are significantly more compact than conventional filter designs. Such superconducting filters in conjunction with cryogenically-cooled amplifiers have the potential to offer solutions to several operating problems in cellular base stations as well as to provide solutions to anticipated problems in forthcoming personal communications services (PCS) base stations.

The performance of cellular base stations and PCS base stations could be improved significantly by using HTS components in both their receivers and their transmitters. On the receive side, the use of superconducting filters along with cryogenically-cooled low-noise amplifiers can lower the noise figure of the receiver and thereby increase its sensitivity. Increased sensitivity translates into increased range for the base station which could be utilized to improve the performance of existing stations as well as to allow for fewer base stations in a given area in new installations. HTS narrow-band notch filters and bandpass filters could also significantly reduce signal interference in base station receivers with minimal signal loss.

From the perspective of HTS technology itself, the key issues for communications applications are the availability of large-area (5-7.5 cm diameter) double-sided superconducting films with good electrical properties. Over the years, the defining characteristics of "good films" has gradually evolved past the point of merely including such features as high T_c, high J_c and low R_s; it now includes such requirements as minimal nonlinearities so as to meet requirements for low intermodulation distortion. At this point, the best films being made are good enough for the proposed applications and the key issues are being able to make them routinely and economically to meet the needs of a developing commercial market.

3. Healthcare

A relatively recent application for superconductivity is in the receivers of magnetic resonance instruments. Instruments based on NMR physics are routinely used in the diagnostic imaging of organs and structures within the body, in three-dimensional imaging of biological specimens and in chemical analysis. The existing rf receivers in these instruments use composite metal coils in tuned circuits that detect the resonant frequencies generated by the object being studied. Replacing normal-metal coils with superconducting receiver coils increases their quality factor and thereby increases the signal-to-noise ratio (SNR) of the receiver provided that the receiver noise is the dominant noise source in the system. Under appropriate dc field and/or sample size conditions, this situation is realized in practice. Higher SNR leads to either increased image quality or faster imaging time in imaging machines or to improved chemical sensitivity or faster chemical measurements in spectroscopic machines.

The first application to market is receiver probes for NMR spectrometers. NMR spectroscopy is a well-established technique for chemical analysis that reveals the structures of even very complex proteins and large molecules such as those used in the development of pharmaceuticals. The sensitivity of the technique, however, is limited, requiring large or highly concentrated samples and/or long data collection times. The use of HTS receiver coils in spectroscopy probes enables the study of smaller or more dilute samples or could greatly increase the throughput of the machines by reducing data collection times. The first commercial HTS probes provide an SNR enhancement in excess of 4 over conventional probes; data collection time is inversely proportional to the square of the SNR.

Magnetic resonance imaging (MRI) provides detailed images of internal organs and structures within the body without invasive procedures or exposure to harmful radiation. Important applications include imaging soft tissues, diagnosing certain brain and spinal cord disorders and joint and muscle injuries. The cost of MRI machines is significant, with prices often in the $1,000,000 to $2,000,000 range. More recently, so called "low-field" MRI machines have been developed whose magnetic field strength is less than that of conventional research quality machines and whose price is considerably lower as well. However, such machines do not have the image quality of high-field machines. The use of HTS receiver coils could significantly improve the SNR in these machines and potentially allow them to produce an image similar in quality to that of the more expensive machines. The development of higher performance low-field machines may lead to the use of MRI in a wider range of diagnostic imaging applications, such as routine screening for breast cancer.

4. Instrumentation

SQUIDs, which are the most sensitive magnetometers known, are used in various types of specialized laboratory equipment, non-destructive test equipment, geophysical surveying instruments and advanced medical instruments. For the most part, these applications have utilized LTS SQUIDs; HTS SQUIDs have only achieved performance levels suitable for many applications in the last couple and years and have only recently become commercially available. The advent of HTS SQUID technology has made a number of SQUID applications much more attractive.

Geophysical surveying is one application whose practicality is greatly enhanced by HTS SQUID technology. Magnetic sensors are already being used in some circumstances to locate oil and other mineral deposits through a technique known as cross-hole electromagnetic borehole logging. A second geophysical magnetic surveying technique is known as magnetotellurics, a surface technique that makes use of multiple magnetic sensors to map underlying strata over large areas of the earth's surface. HTS SQUID technology offers the advantage of portability by virtue of the small size of the sensors along with the simplified handling associated with liquid nitrogen.

SQUIDs can also be used in a variety of instruments to inspect and analyze materials and structures on the basis of their magnetic signatures that cannot be effectively measured using competing technologies. These instruments can measure materials that are hidden or submerged, such as mines or storage drums; analyze metallic objects in structures, such as buildings, nuclear power plants or airplanes, without physically invading or contacting the structure; and measure electrical currents in, for example, semiconductor integrated circuits.

Biomedical applications of SQUID technology are arguably the most significant opportunity for the technology. Whereas LTS SQUID technology is already being applied to the detection of signals within the brain (magnetoencephalography), SQUID sensors can also detect electrical impulses in the heart, without electrodes or other contacts to the patient required by conventional EKGs. The technique is called magnetocardiography ("MCG") and is an attractive opportunity for HTS SQUID technology. Under conditions in which attaching electrodes is impractical or impossible such as at accident sites, on burn patients or for fetal heart diagnostics, the non-contact MCG approach could have advantages over conventional EKG. If accepted by the medical community, MCG used to map electrical currents in the heart could become a significant diagnostic tool for pointing the sources of cardiac arrhythmias (i.e., irregular heart beats).

5. The Nature of the Business

For the foreseeable future, the commercial opportunities for HTS electronic technology will be in the form of complete cryoelectronic subsystems that incorporate HTS components, various forms of interface electronics and whatever refrigeration technology is required. At this point, superconductor manufacturers are the only ones with the technical skills and capabilities required to utilize HTS technology. Therefore, the technology must be provided in the form of complete subsystems whose interfaces with systems and the external world are entirely conventional in nature. As a result, the early HTS products will be such things as NMR probe subsystems and wireless filter subsystems whose superconducting elements are well-hidden within the confines of the product. Perhaps at a later time there will be sufficient utilization of HTS technology throughout the electronics industry that conventional manufacturers will develop capabilities for using superconductors at the component level. Until such time, there is both a challenge and an opportunity for companies in the superconductor industry.

MUTUAL HIGH-FREQUENCY INTERACTION OF HIGH T_C JOSEPHSON JUNCTIONS

M. DARULA[a], G. KUNKEL[b], S. BEUVEN

Institute of Thin Film and Ion Technology, Forschungszentrum Jülich, D-52425 Jülich, Germany

We report about recent development in a study of mutual high-frequency interaction between high-T_C (HTS) Josephson junctions integrated in two types of arrays: (a) multi-junction superconducting loops (MSL) and series arrays biased in parallel (parallel biasing scheme -PBS). In both cases the phase locking between junctions has been observed and the radiation emitted from arrays measured.

1 Introduction

Josephson junctions are natural voltage-tunable high-frequency sources which, in the case of HTS junctions, can emit radiation up to frequencies of several THz. When integrating N nearly identical junctions into an array, the emitted power scales with N, and linewidth with 1/N, provided that all single oscillators are mutually phase-locked at the same frequency. We present the results of a study of mutual high-frequency interaction (phase locking) between high-T_C (HTS) Josephson junctions integrated in two types of arrays: (a) multi-junction superconducting loops (MSL) integrated together with the on-chip detector junction, and (b) parallel biased series arrays (PBS) synchronized by auxiliary low-Q coplanar feed line resonances and integrated into bow-tie antenna structures. Both circuits are schematically shown in Fig. 1.

2 Experimental

In the course of investigation of phase locking in the high-T_C Josephson junction arrays we used different substrates ($SrTiO_3$, $LaAlO_3$) as well as different type of junctions (bicrystal, step-edge). In all cases junctions were fabricated from $YBa_2Cu_3O_7$ films. The typical thickness of the films was 100 - 200 nm. Details about fabrication can be found elsewhere[1]. In the case of MSL a layout was designed which provides access to each junction [1]. The dc voltages of all junctions can be measured simultaneously and the phase locking can be infered. On the same chip the detector junction has been integrated and coupled to the array using a radial stub resonator structure. This allowed us to measure radiation emitted from the array. The direct off-chip radiation measurements from PBS have been done using a Dicke radiometer system in W-band [2].

We report here on results of two experiments. Detection of radiation from MSL using a detector junction and a direct radiation measurement from PBS. Let discuss first the case of MSL. The MSL containing four junctions was biased within the locking-interval, so that the dc voltages of all junctions measured simultaneously were the same. At three different bias points within the locking interval the current voltage characteristic of the detector junction

---
[a]permanent adress: Institute of Electrical Engineering, SAS, SK-83249 Bratislava, Slovakia
[b]present adress: University of Erlangen, Erwin-Rommel-Str. 1, D-91058 Erlangen, Germany

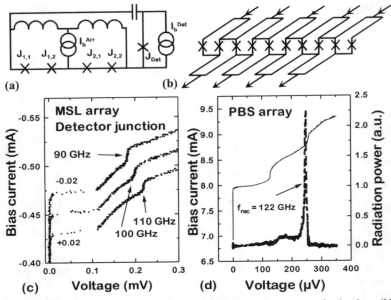

Figure 1: Josephson junction arrays investigated: (a) Multi-junction superconducting loop (MSL) with integrated detector junction, (b) Parallel biased series arrays (PBS). (c) Current-voltage characteristic of detector junction in MSL (bicrystal junctions on $SrTiO_3$). Three curves were recorded for different biasing points within the locking interval of MSL. (d) Current voltage characteristic of PBS (step-edge junctions on $LaAl_2O_3$) and direct radiation measurement at 122 GHz (dotted line).

was recorded as shown in Fig. 1(c). We see clear Shapiro steps due to the radiation from MSL. The position of Shapiro steps changes according to the changing bias point of array and, therefore, changing the frequency of emitted radiation. From the height of the step we estimate the power delivered from array to the detector junction to be approx. 0.2 nW. When biasing the array outside of the locking interval no radiation was detected. In Fig. 1(d) the current-voltage characteristic of PBS together with a radiation curve detected off-chip by radiometer at 122 GHz is shown. On the current-voltage characteristic clear steps are visible due to coplanar feed line resonances. Due to these resonances relatively narrow radiation peak is observed as show in the Fig. 1(d)(dotted curve). Experiments with circuits similar to those described in this paper (using different substrates and also PBS with on-chip detector) show phase locking up to 1 THz and up to 50 K.

References

1. S. Beuven *et al*, IEEE Trans. Appl. Supercond. **5**, 3288 (1995)
2. G. Kunkel *et al*, Supercond. Sci. Technol. , accepted. (1995)

DIRECTLY COUPLED DC-SQUIDS OF YBCO STEP-EDGE JUNCTIONS FABRICATED BY A CHEMICAL ETCHING PROCESS

JUNHO GOHNG, C. Y. DOSQUET*, J. P. HONG, E.-H. LEE, J.-W. LEE

Materials and Device Research Center, Samsung Advanced Institute of Technology, Suwon, Korea
* Ecole Nationale Superieure de Chimie et de Physique de Bordeaux, France

High T_c directly coupled dc-SQUIDs have been successfully fabricated on chemically etched MgO substrate steps. The chemical etching was performed in a mixed acid solution of H_3PO_4 and H_2SO_4 for the best control of etched surface and roughness. Characteristics of the directly coupled dc-SQUID have been studied following the patterning and fabrication of the device. The chemically etched steps show sharper edges at the bottom of the step as well as the top unlike those made by ion milling. AFM and Raman Spectroscopy studies on the YBCO thin film deposited and patterned on chemically etched areas show no sign of appreciable degradation.

Following the discovery of high T_c superconductors many types of high T_c Josephson junctions have been made [1]. Among them, step-edge junctions are one of the interesting category due to its intrinsic simplicity and relative ease in fabrication. Especially for those who are interested in the application of Josephson junctions as SQUID (Superconducting Quantum Interference Device) magnetometers, these are one of the most commonly used type of junctions along with the bicrystal junctions. Despite of its simplicity and relatively good junction properties, reproducibility and homogeneity of these junction are not guaranteed, since the junction properties differ greatly from small changes in step-edge shapes. For the ion-milling process, which is most commonly used for the fabrication of step-edge junctions, it is of no exception [2,3]. Especially, the shape of the lower edge varies greatly for typical ion-milled edges. There have been reports of chemically etched steps in single acid solutions, but the effects of chemical etching processes regarding etchant solutions, step angle and surface morphology etc. are not well defined [4,5]. The reduction of T_c for the YBCO deposited on the chemically etched portion of substrate is yet another common problem. It was not until very recently that the key elements of the chemical etching process with respect to the etchant solutions, step angle, and surface morphology of the step, etc. were studied in detail. In this earlier work we had developed and optimized an alternative process of step-edge fabrication by chemical etching in a mixed solution of H_3PO_4 and H_2SO_4 and fabricated a dc-SQUID that operated at 67 K. In this paper, application of this chemical etching process for the fabrication of directly coupled dc-SQUIDs of YBCO step-edge junctions is discussed. Most of problems observed in conventional chemical etching processes are greatly reduced. Comparisons are made between step-edges made by ion milling and chemical etching in mixed acids. Micro-Raman studies on the YBCO thin films deposited on the chemically etched area in mixed acids have been carried out, and was compared with the results from non-etched areas.

To fabricate directly coupled dc-SQUIDs, bare MgO substrates cut in $10 \times 10 \times 1$ mm^2 pieces were cleaned and then put into an annealing process at 1000 °C for two hours under a flowing oxygen environment. This process is essential for the removal of any impurities on the MgO substrate surface including $Mg(OH)_2$. Patterning for the step-edge junction (SEJ) was done by standard lithography method on a thin layer photoresist of 280 nm thickness. The patterned MgO substrate was then put into a chemical etching process using a solution of phosphoric acid (H_3PO_4) and sulfuric acid (H_2SO_4), mixed in 20 to 1 volume ratio. The advantage of using mixed acids over single acid had been studied comparatively and discussed in detail in our prior report [6]. For this experiment a step angle of 33° ~ 35° has been routinely obtained, and etching rate was about 63 nm/min. For the deposition of high quality thin YBCO films on MgO, the Pulsed Laser Deposition (PLD) method was used. It is well known that the quality of YBCO film is also greatly affected by the deposition temperature and oxygen temperature, and in our case they are optimized at 800 °C and 100 mTorr, respectively. Following the deposition of YBCO film, directly coupled dc-SQUID was patterned and fabricated. The measurements of the SQUID showed the average critical current density of 0.9 mA, which corresponded to a critical current density of 10^5 A/cm^2, with a normal resistance of 1.2 Ω. The maximum peak to peak modulation voltage reached 160 µV, and the noise power value was measured as 100 µΦ_0/rtHz at 10 Hz. At liquid nitrogen temperature of 77 K, The peak to peak modulation voltage was 6 µV, and the corresponding noise level was 450 µΦ_0/rtHz. The Raman spectroscopy studies on both chemically etched and unetched areas show no degradation of YBCO thin films deposited on chemically etched region, and exhibited good epitaxial growths.

References
[1] K. Char, M. S. Colclough, S. M. Garrison, N. Newman and G. Zaharchuk, Appl. Phys. Lett. 59 (1990) 727.
[2] S. Tanaka, H. Kado, T. Matsuura and H. Itozaki, IEEE Trans. Appl. Supercond. 3 (1993) 2365.
[3] F. Schmidl, L. Alff, R. Gross, K. D. Husemann, H. Schneidewind and P. Seidel, IEEE Trans. Appl. Supercond. 3 (1993) 2349.
[4] S. Kuriki, T. Kamiyama, D. Suzuki, and M. Matsuda, IEEE Trans. Appl. Supercond. 3 (1993) 2461.
[5] M. Sugiura, K. Kato, Ka. Asada and T. Sugano, Jpn. J. Appl. Phys. 31 (1992) 1595.
[6] C. H. Park, J. P. Hong, I. H. Song, E. -H. Lee, C. W. Moon, J. -W. Lee, Physica C 234 (1994) 85.

NOISE REDUCTION TECHNIQUES FOR OPERATING HIGH-T_c SQUIDS IN AN UNSHIELDED ENVIRONMENT

N. TRALSHAWALA, J.R. CLAYCOMB, J.H. XU, J.H. MILLER, JR.

Texas Center for Superconductivity, University of Houston, Houston, Texas, USA

ABSTRACT

We have presented results of our investigations on the ambient noise reduction to enable magnetocardiography in an unshielded environment in another article in these proceedings[1]. Here we elaborate on some of the noise reduction techniques that were utilized. Specifically, we address localized high-T_c (HTS) passive magnetic shields employed in our setup and discuss the adaptive noise cancellation (ANC) technique.

Localized Passive Shielding

We have carried out extensive finite element calculations to determine optimum shapes and sizes of bulk or thick film HTS materials that act as partial noise shields without affecting signals of interest. These calculations were done for uniform noise fields and for model magnetic and current dipole fields, such as the ones encountered in magnetocardiography (MCG) and non-destructive evaluation (NDE) of materials. Specifically, using an HTS H-shield[2] or a simple tube, with optimized dimensions, one can obtain 20 dB noise reduction without affecting the MCG or NDE signals. We have experimentally verified our calculations. The noise spectrum is shown in Figure 1. In such a configuration, we obtain signal-to-noise-improvement ratio (SNIR) of about five for dipoles within one diameter from the mouth of the HTS tube.

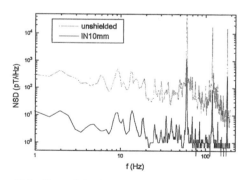

Figure 1.   Noise Spectral density measured by an high-T_c SQUID sensor in an unshielded environment and when it is centrally placed 1 cm inside an HTS tube (BSCCO, 6 cm diameter and 10 cm long), with its axis parallel to the tube axis.

Thus, not only is the SNIR improved, but also the ambient DC field is reduced 20 dB and, consequently, intrinsic $1/f$ noise of the high-T_c SQUIDs is also reduced. Also, in a clinically noisy environment where various instruments are operating simultaneously, a 20 dB reduction in ambient noise could mean a difference between an unlocked SQUID loop and a locked one (e.g., in our case, once the ambient powerline noise exceeds about 200 nT, it becomes difficult to keep the FLLs locked). In addition, this is a simple passive shield and one doesn't have to worry about electronics.

A major disadvantage, though, is that the inductance of such an HTS tube is so small (about 20-80 nH) that even some hard floor vibrations (e.g. due to movement of heavy equipment) in the earth's field induce enough screening current to drive the tube normal and trap flux. One solution is to use an H-shield instead[2]. However, we find that our tubes have high enough critical current density to stay superconducting in upto 4 mT DC field (roughly 100 times the earth's field).

Adaptive Noise Cancellation

Conventionally, outputs from two (or three) SQUIDs are electronically subtracted to make a first-order (or second-order) gradiometer. The next step is to mechanically balance the electronic gradiometer. This is a tractable problem for a single channel system but, as the number of channels increase, it becomes rather inefficient to mechanically balance many gradiometers. Adaptive noise cancellation techniques[3] can be effectively used in such a case to compensate for the gradiometer imbalance (see Figure 2). Moreover, one can utilize just two to three reference SQUIDs which are well removed from the signal sources to balance any number of nearby signal channels.

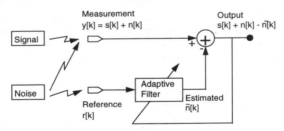

Figure 2. Schematic of ANC. Reference channels pickup only the ambient noise while signal channels pickup noise (correlated with reference channels but not identical) and the signal.

An adaptive filter automatically adjusts its impulse response through an algorithm that responds to an error signal proportional to its output (Figure 2). This objective is generally accomplished by minimizing the total output power of the filter. This corresponds to the maximization of the output signal-to-noise ratio[3].

Acknowledgments

This work was supported by the State of Texas through TcSUH and the ATP and by the R. A. Welch Foundation.

References

1. J.H. Miller, Jr. et al. (elsewhere in these proceedings).
2. N. Tralshawala et al, *IEEE Trans. on Applied Superconductivity* **5**, 2354 (1995).
3. B. Widrow et al, *Proc. of IEEE*, **65**, 1692 (1975).

ENHANCED-RESOLUTION MAGNETIC RESONANCE IMAGING USING HIGH-T_C SUPERCONDUCTING RF RECEIVER COILS

J. WOSIK, K. NESTERUK,[1] L. M. XIE, X. P. ZHANG, P. GIERLOWSKI, C. JIAO, and J. H. MILLER, Jr.

Texas Center for Superconductivity at the University of Houston, Houston, TX 77204; [1]*Institute of Physics, Polish Academy of Sciences, Warsaw, Poland*

A low-loss high temperature superconducting (HTS) *rf* receiver surface probe developed primarily for high-resolution magnetic resonance imaging (MRI) of spinal cord injuries in small animals is described. This probe is used *in lieu* of an implanted copper coil being currently used in research on spinal cord injuries. The SNR improvement due to the use of HTS thin films was theoretically analyzed for the case of MRI of a rat spine. It was found that, for each coil size, there is a position away from the body which yields the highest SNR (while imaging the voxel deep in the body). This indicates that separation of the HTS probe from the body, resulting from the cryostat implementation, does not decrease the SNR improvement. The SNR improvement due to the use of a HTS rf receiver coil, for the analyzed configuration used for MRI of a rat spine, was found to be at least five-fold. Even for the high field MRI at 84.4 MHz this improvement is about a factor of two.

1. Introduction

In the case of small-volume, high-resolution imaging, the noise of the probe coil and/or preamplifier set the system noise floor and hence the MRI performance; body noise no longer dominates the SNR of the system.[1] Thus, it is desirable to reduce the thermal coil noise to improve the image resolution and reduce the image acquisition time. High-temperature superconductors are extremely attractive for such applications due to their very low losses. Indeed, preliminary studies have shown that for selected applications, HTS MRI receiver coils perform significantly better than comparable copper coils.[1,2,3] Even in high-field MRI, when the coil and the corresponding region of interest (ROI) are reduced in size, there is a crossover point beyond which the receiver noise becomes the dominant source of noise. Magnetic resonance imaging has become one of the most widely used diagnostic tools in medicine. It is also a promising technique for studing spinal cord injuries in clinical (humans) and experimental (animals) cases. Recently it has been demonstrated[4] that a substantial improvement in the image quality can be achieved by **implanting** a receiver coil near the structure of interest, such as a heart, kidney, or spinal cord. However, despite these advantages, there are a number of disadvantages with the use of such implanted coils and what is even more important, this approach clearly can not be used in the case of human subjects.

We present another approach to improve SNR ratios for MRI of the spinal cord of small animals using HTS superconducting receiver coils.

2. Probe Design

We have designed and fabricated a two-coil probe consisting of a pair of balanced (with a virtual ground plane) planar HTS surface coils, consisting of patterned YBCO films on 2" x 2" LaAlO$_3$ substrates, suitable for imaging of vertebrae in the spine. In principle, this design follows that shown by R. Withers et. al. [1] Measurements of the Q of a copper two-coil probe at room temperature and 77 K showed $Q = 200$ and 640, respectively. By contrast, the Q of the YBCO two-coil probe was found to be more than an order of magnitude higher.

3. Signal-to Noise Ratio Calculations

The design of surface coils optimized for maximum SNR requires knowledge of the basic physical effects determining the signal and noise generated in a coil by the body after the rf excitation. To achieve this goal the body was modeled as a 60 mm long dielectric cylinder of given conductivity σ. A rectangular 15 by 30 mm coil is facing the cylinder of radius a (Fig. 1). A DV volume voxel was assumed to be 5 mm deep in the body. This configuration is an adequate representation of the actual situation for spinal cord imaging in rats using the 84.4 MHz MRI scanner (Fig. 2). The body noise was simulated by losses in the coil due to the existence of the body and defined as a equivalent body resistance R$_{body}$ for the coil. Thus the rms thermal noise voltage of the coil is assumed to be caused by both the resistance of the coil R$_{coil}$ and the body resistance R$_{body}$. Signal and noise voltage calculations are carried out using the reciprocity principle and line integrals along the coil contour.

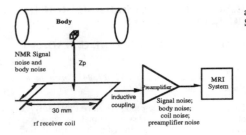

Fig. 1. The SNR was calculated for a voxel placed inside of the cylindrical lossy body in z_p distance from the rectangular receiver coil.

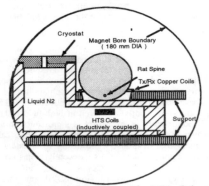

Fig. 2. Experimental arrangment for MRI of the rat spine. HTS pick-up probe is placed under the rat's body. Transmittal coil works also as the coupling coil. A superconducting 2 Tesla magnet is used.

The calculations were done for four cases: when the coil is made out of (1) copper and used at room temperature, (2) copper cooled to 77 Kelvin, (3) high temperature superconductor with R_s equal to 69 mΩ, and (4) an idealized material with R_s equal to zero. Surface resistance value for the HTS thin film was

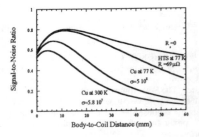

Fig. 3 shows the variations of SNR with the respect to the coil position at 84.4 MHz for 15 by 30 mm rectangular coil for a tissue of conductivity equal to 1 $(\Omega m)^{-1}$.

chosen very conservatively. This is rather a high R_s value for superconductors at this frequency. The calculations were performed for two frequencies: 10 MHz and 84.4 MHz, which the latter case is shown in Fig. 3. It can be seen from this figure that there is a position away from the body which yields the highest SNR while imaging a voxel deep in the body. This is extremely important feature for the HTS application, where the cryostat implementation introduces a gap between the body and the coil. It proves that carefully a designed probe still can be placed at z_p equal to 10-15 mm away from the body at the maximum SNR point. The SNR of the HTS probe is about 5 times larger than that of the warm copper coil at 10 MHz. Such an improvement is smaller at 84.4 MHz, however still it is equal to about 2.

4. Acknowledgments

We are grateful to Dr. Ponnada Narayana for stimulating discussions and for making his MRI scanner available for measurements. This work was supported by the Robert A. Welch Foundation, by the State of Texas via Texas Center for Superconductivity at University of Houston, and by the Texas Higher Education Coordinating Board (ARP grant).

5. References

1. R. S. Withers, B. F. Cole, M. E. Johansson, G.-C. Liang, and G. Zaharczuk, *SPIE Proceedings Series, Vol. 2156, High-T_c Microwave Superconductors and Applications*, 24-27 January 1994, Los Angeles, CA, R. B. Hammond and R. S. Withers, editors, pp. 27-35.
2. R. D. Black, T. A. Early, P. B. Roemer, O. M. Mueller, A. Mogro-Campero, L. G. Tuner, and G. A. Johnson, *Science* 259, 793 (1993).
3. J. G. van Heteren, T. W. James, and L. B. Bourne, *Magn. Reson. Med.* 32, 396 (1994).
4. J. C. Ford, D. B. Hackney, P. M. Joseph, M. Phelan, D. C. Alsop, S. L. Tabor, C. M. Hand, R. S. Markowitz, and P. Black, *Magn. Reson. Med.* 31, 218 (1994).

INTERFACE ROUGHNESS EFFECT ON DIFFERENTIAL CONDUCTANCE OF HIGH-T_c SUPERCONDUCTOR JUNCTIONS

JIAN-XIN ZHU, Z. D. WANG
Department of Physics, University of Hong Kong, Pokfulam Road, Hong Kong

D. Y. XING, Z. C. DONG
National Laboratory of Solid State Microstructures, Nanjing University, Nanjing 210093, People's Republic of China

Within the framework of Blonder-Tinkham-Klapwijk model, we find that, due to the interface roughness, the tunneling spectrum of normal metal–insulator–d-wave superconductor junctions differs significantly from that for normal metal–s-wave counterpartr in the absence of the insulating layer.

A $d_{x_a^2-x_b^2}$ symmetry of the order parameter in high-T_c superconductors has been proposed recently.[1] The d-wave pairing symmetry produces an anisotropic energy gap, which drops to zero on some nodal lines of an essentially cylindrical Fermi surface, giving rise to novel thermodynamic and transport properties at low temperatures.[2] Here we investigate the interface roughness effect on the differential conductance for a normal metal–insulator–d-wave superconductor junction.

To relate our study to high-T_c superconductors, we assume the system under consideration is two-dimensional, and choose the x-axis to be perpendicular to the interface of the junction, while the y-axis to be parallel to the interface. The interface is located at $x = 0$, and $x > 0$ is occupied by a superconductor. The Hamiltonian modelling the scattering of electrons by impurities at the rough interface $x = 0$ can be written as $H_I = V_1 g(y)\delta(x)$. As usual, the random roughness $g(y)$ is assumed to be a Gaussian-distributed white noise. Using the perturbation theory of the Green function in the Morn approximation, we find that the elastic scattering of electrons in the rough interface only leads to a finite life time of electrons states. The scattering potential at the interface can be modelled as $V(x) = H\delta(x)$ with $H = V_0 - i\Gamma$. Moreover, the pair potential is taken to be $\Delta(\mathbf{k}, x) = \Delta(\hat{\mathbf{k}})\Theta(x)$.

By solving the Bogoliubov-de Gennes equations for anisotropic superconductors,[5] in the WKBJ approximation, subject to the matching conditions at the interface, $\Psi_S|_{x=0^+} = \Psi_N|_{x=0^-}$, and $\frac{\partial \Psi_S}{\partial x}|_{x=0^+} = \frac{\partial \Psi_N}{\partial x}|_{x=0^-} + \frac{2m}{\hbar^2}(V_0\hat{1} - i\Gamma\hat{\tau}_3)\Psi_N|_{x=0^-}$ with $\hat{1}$ and $\hat{\tau}_3$ the 2×2 unit matrix and the z-component of the Pauli matrix, the differential conductance at zero temperature can be calculated according to the formula,[3] $G_{NS}(E) = G_0 \cos\theta[1 + A(E) - B(E)]$, where $G_0 = 2N(0)e^2 v_F \mathcal{A}$ with $N(0)$ the one-spin density of states at the Fermi energy, v_F the Fermi velocity, and $\mathcal{A}$ an effective cross-sectional area. Here $A(E)$ and $B(E)$ are the Andreev reflection and the normal reflection coefficients, respectively.

The effective pair potentials of $d_{x_a^2-x_b^2}$-wave symmetry, $\Delta(\theta_+)$ and $\Delta(\theta_-)$ felt by electron-like and hole-like excitations in the superconducting region, can be expressed as[2] $\Delta(\theta_+) = \Delta_0 \cos(2\theta - 2\alpha)$ and $\Delta(\theta_-) = \Delta_0 \cos(2\theta + 2\alpha)$, where θ and α is the incident angle and the angle between the normal direction of the interface and the crystalline a-axis of the superconductor along which the magnitude of the pair potential is arranged to reach a maximum.

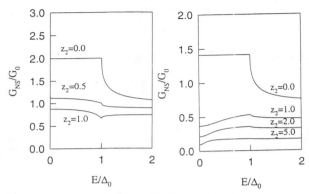

Figure 1: The normalized conductance G_{NS}/G_0 as a function of E/Δ_0 for various values z_2 representing the interface roughness at $z_1 = 0$ and $T = 0$ K. (a) $\theta = \alpha = 0$; (b) $\theta = \alpha = \pi/4$.

If θ, α, and the dimensionless complex barrier-strength $Z = z_1 - iz_2$ with $z_1 = mV_0/\hbar^2 k_F$, $z_2 = m\Gamma/\hbar^2 k_F$, are fixed, the differential conductance is a function of the bias voltage E/Δ_0. For our purpose, we concentrate exclusively on the interface roughness effect on the conductance by setting $z_1 = 0$. As a result, we plot in Fig. 1 the conductane against the bias voltage by varying z_2 for two typical cases: (a) $\theta = \alpha = 0$ and (b) $\theta = \alpha = \pi/4$. It can be seen in Fig. 1(a) that a dip is induced at Δ_0 in the behavior of differential conductance when the interface roughness is enhanced. This dip has been observed but not yet well understood in point-contact junctions formed on a conventional superconductor.[4] While for case (b), as shown in Fig. 1(b), there exists no dip in the behavior of tunneling conductance, which, as compared with Fig. 1, is essentially different from that in the case (a). This interesting feature might be used as some signal of the $d_{x_a^2 - x_b^2}$ pairing symmetry in high-T_c superconductors.

Acknowledgments

This work was supported by the RGC research grant of Hong Kong under Grant No. HKU262/95P and the CRCG grant at the University of Hong Kong.

References

1. B. G. Levi, Phys. Today **49**, 19 (Jan. 1996).
2. Y. Tanaka and S. Kashiwaya, Phys. Rev. Lett. **74**, 3451 (1995).
3. G. E. Blonder, M. Tinkham, and T. M. Klapwijk, Phys. Rev. B **25**, 4515 (1982).
4. H. Srikanth and A. K. Raychaudhuri, Phys. Rev. B **45**, 383 (1992).
5. C. Bruder, Phys. Rev. B **41**, 4017 (1990).

WORKSHOP PROGRAM

MONDAY, MARCH 11

2 p.m.-on	**Hotel Check-In**
03:00-06:00, 07:00-09:30	**Early Registration, Check-In & Manuscript Submission** *La Salle Pre-Function Foyer*
04:00-08:00	**Exhibitor Registration & Set-Up & Poster Sessions 1A & 1B Set-Up** Poster Set-Up will continue on Tuesday, March 12, 7:30 a.m. - 12 noon Speaker Preparation Room, *Travis I*, and Poster Preparation Room, *Travis II*, open

TUESDAY, MARCH 12

07:30 a.m.	**Press Room Open** *Fannin*
08:00	**EXHIBITS Open** *de Zavala*
08:00-08:30	**Continental Breakfast** *Granger Foyer*
	OPENING CEREMONIES *La Salle Ballroom A & B*
08:30-08:50	**Welcome to Participants and Workshop Overview** **C. W. Chu (TCSUH) & K. A. Müller (IBM Zürich)**
08:50-09:00	**Greetings from State of Texas, City of Houston & University of Houston** **Dr. John M. Ivancevich**, Senior Vice President for Academic Affairs & Provost, UH
09:00-09:05	**Introduction of Speaker** **C. W. Chu (TCSUH)**
9:05-9:40	**Opening Address** **The Honorable Tom DeLay**, House Majority Whip, U. S. House of Representatives, Congressional District 22 Followed by 15-minute Q & A with Press & Workshop Participants
09:40-09:50	**Technical Announcement - Pirelli/EPRI**
09:50-10:20	**Break** *Refreshments served in La Salle Pre-Function Foyer & Granger Foyer*
10:20-11:40	**PLENARY SESSION 1** *La Salle Ballroom A & B* **Chairs:** J. R. Schrieffer (NHMFL, Florida State U), K. Kitazawa (U Tokyo) **PL1.1** K. A. Müller (IBM Zürich), "The Development of Superconductivity Research in Oxides" **PL1.2** C. W. Chu (TCSUH), "Superconductivity Above 90 K and Beyond"
11:40-01:20	**Box Lunch Pick-Up** *Nautile Room, 1st Floor* Seating available in lobby bar area & outside tables to left of bar area, 1st Floor **EXHIBITS**, *de Zavala*, & Poster Sessions 1A & 1B, *Granger A & B*, open for viewing
12:00-12:30	**Pirelli/EPRI Press Conference** *La Salle Ballroom A & B*

01:20-02:40 **PLENARY SESSION 2** *La Salle Ballroom A & B*
 Chairs: M. L. Cohen (UC Berkeley), Z. X. Zhao (Chinese Academy of Sciences)
 PL2.1 **J. R. Schrieffer (NHMFL, Florida State U)**, "The Role of Vertex Corrections to the Pairing Interaction in Spin Fluctuation Superconductors"
 PL2.2 **S. Tanaka (ISTEC)**, "Superconductors and Semiconductors"

02:40-03:00 **Break** *Refreshments served in La Salle Pre-Function Foyer & Granger Foyer*

03:00-03:40 **PLENARY SESSION 3** *La Salle Ballroom A & B*
 Chair: M.-K. Wu (National Tsing Hua U)
 PL3.1 **J. M. Rowell**, "Superconducting Electronics – The Next Decade"

04:00-06:00 **POSTER SESSIONS 1A & 1B: HTS Theory, Experiment, Material, & Properties** *Granger A & B*
 Posters will be removed from Granger A & B by 8:30 p.m.

06:00-07:30 **Welcome Reception** *La Salle Ballroom A & B*
 Hosted by Dr. and Mrs. Glenn A. Goerke, UH Office of the President, for Workshop Participants and members, Houston Consular Corps

WEDNESDAY, MARCH 13

08:00 a.m. **EXHIBITS Open** *de Zavala*

08:00-08:30 **Continental Breakfast** *Granger Foyer*

08:30-08:40 **General Announcements** *La Salle Ballroom A & B*

08:40-10:00 **PLENARY SESSION 4** *La Salle Ballroom A & B*
 Chairs: M. Tachiki (Tohoku U, Sendai), P. M. Grant (EPRI)
 PL4.1 **A. J. Leggett (U Illinois, Urbana-Champaign)**, "The Symmetry of the Order Parameter in the Cuprate Superconductors"
 PL4.2 **C. Rosner (IGC)**, "HTS Technology: The Road from Research to Applications"

10:00-10:20 **Break** *Refreshments served in La Salle Pre-Function Foyer & Granger Foyer*

10:20-12:40 **CONCURRENT SESSIONS A & B**

SESSION A: HTS Properties
La Salle Ballroom A & B
Chairs: M. V. Klein (U Illinois, Urbana-Champaign),
R. L. Greene (U Maryland)

A.1 **Y. Ando (Lucent Technologies)**, "Normal State Resistivity of Superconducting LaSrCuO in the Zero-Temperature Limit"

A.2 **M. Osofsky (NRL)**, "Transport Properties of Overdoped HTS at High Fields and Low Temperatures"

A.3 **S. Uchida (U Tokyo)**, "'Spin-Gap' Effects on the Charge Dynamics of HTSC"

A.4 **M. Cardona (Max Planck Inst.)**, "Raman Spectroscopy of High T_c Superconductors"

A.5 **D. C. Johnston (Iowa State U)**, "Magnetic and Structural Properties and Phase Diagrams of Sr_2CuO_2Cl and Lightly-Doped $La_{2-x}Sr_xCuO_{4+\delta}$"

A.6 **G. Aeppli (NEC)**, "Enhanced Magnetic Response in the Superconducting State of $La_{1.86}Sr_{0.14}CuO_4$"

A.7 **Y. Uemura (Columbia U)**, "Correlations Between T_c and n_s/m^* in High-T_c Cuprates"

SESSION B: Bulk Applications *Granger A & B*
Chairs: R. Eaton III (DOE), R. A. Hawsey (Oak Ridge)

B.1 **A. P. Malozemoff (ASC)**, "Bulk Applications of HTS Wire"

B.2 **A. F. Bolza (Pirelli)**, "Progress and Issues in HTS Power Cables"

B.3 **D. Driscoll (Reliance Electric/Rockwell)**, "125 hp HTS Synchronous Motor Design and Preliminary Test Results"

B.4 **D. Gubser (NRL)**, "Superconductor Homopolar Motor Demonstration"

B.5 **W. K. Chu (TCSUH)**, "Superconducting Magnetic Bearing and Its Application in Flywheel Kinetic Energy Storage"

B.6 **E. M. W. Leung (Lockheed Martin)**, "High Temperature Superconducting Fault Current Limiter"

B.7 **D. T. Shaw (SUNY Buffalo)**, "Development of Technologically-Useful Superconducting Wires at 20K"

12:40-01:00 **Box Lunch Pick-Up** *Nautile Room, 1st Floor*
Seating available in lobby bar area & outside tables to left of bar area, 1st Floor; may also bring lunch into press briefing
EXHIBITS open for viewing, *de Zavala*

01:00-01:50 **Press Briefing: "Achievements in HTS Science and Technology"** *La Salle Ballroom A & B*
Press Moderator: P. M. Grant (EPRI)
Panel:
    Science: J. R. Schrieffer (NHMFL, Florida State U)
    Materials: C. W. Chu (TCSUH)
    Film Applications: D. Gubser (NRL)
    Bulk Applications: W. K. Chu (TCSUH)

01:50-01:55 **Introduction of Keynote Speaker**
C. W. Chu (TCSUH) *La Salle Ballroom A & B*

01:55-02:40 **Keynote Address** *La Salle Ballroom A & B*
Dr. Neal Lane, Director, NSF, "Public Support for Science – Public Roles for Scientists"
Followed by 20 minute Q & A with Workshop Participants & Press
Individual Interviews can be scheduled in Press Room, Fannin

02:40-06:20 **CONCURRENT SESSIONS C & D**

SESSION C: Symmetry *La Salle Ballroom A & B*
Chairs: A. L. de Lozanne (UT Austin), D. G. Naugle (Texas A&M U)

C.1 R. C. Dynes (UC San Diego), "Josephson Tunneling Between a Conventional Superconductor and $YBa_2Cu_3O_{7-\delta}$"
C.2 C. C. Tsuei (IBM Yorktown Heights), "Half-Integer Flux Quantum Effect and Pairing Symmetry in Cuprate Superconductors"
C.3 K. Tanabe (ISTEC), "Intrinsic SIS Josephson Junction in Bi-2212 and Symmetry of Cooper Pair"
C.4 G. E. Blumberg (U Illinois, Urbana-Champaign), "Magnetic Raman Scattering in Cuprate Antiferromagnetic Insulators and Doped Superconductors"
C.5 Ø. Fischer (U Geneva), "STM Observation of the Vortex Lattice in $YBa_2Cu_3O_7$"
C.6 A. M. Goldman (U Minnesota), "Determination of the Pairing State of High-T_c Superconductors Through Measurements of the Transverse Magnetic Moment"
C.7 Z.-X. Shen (Stanford U), "Excitation Gap in the Normal and Superconducting State of Underdoped Cuprate Superconductors"
C.8 J. Mannhart (IBM Zürich), "Single Grain Boundary Josephson Junction Devices - New Insights from Basic Experiments"
C.9 K. Levin (U Chicago), "What Does d-Wave Symmetry Tell Us About the Pairing Mechanism?"
C.10 M. Franz (McMaster U), "Impurity States in a d-Wave Superconductor"

SESSION D: HTS Materials *Granger A & B*
Chairs: A. J. Jacobson (TCSUH), J. Crow (NHMFL, Florida State U)

D.1 Y. Y. Xue (TCSUH), "Phase Stability, Defects, and Formation of $HgBa_2Ca_{n-1}Cu_nO_{2n+2+\delta}$"
D.2 B. Raveau (U Caen), "Stabilization of New HTcS Cuprates and Oxycarbonates: From the Bulk Materials to the Thin Films"
D.3 M. Takano (Kyoto U), "Quantum Spin Ladder Oxides"
D.4 A. M. Hermann (U Colorado, Boulder), "Normal State Transport Properties of Tl-2201 Single Crystals"
D.5 Y. Maeno (Hiroshima U), "Superconductive Sr_2RuO_4 and Related Materials"
D.6 P. H. Hor (TCSUH), "Electronic Phase Separations in $La_2CuO_{4+\delta}$"
D.7 R. J. Cava (Lucent Technologies), "Borocarbide and Other Unusual Intermetallic Compounds"
D.8 J.-I. Shimoyama (U Tokyo), "Strong Flux Pinning, Anisotropy and Microstructure of $(Hg,Re)M_2Ca_{n-1}Cu_nO_y$ (M=Ba,Sr)"
D.10 K. Kitazawa (U Tokyo), "A View on HTS Performances Under Magnetic Fields from the Key Term 'Anisotropy Factor'"

08:00 **Exhibits Close**
Exhibits will be dismantled by 10 p.m.

THURSDAY, MARCH 14

8-8:30 a.m. **Continental Breakfast** *La Salle Pre-Function Foyer*

08:30-09:10 **PLENARY SESSION 5** *La Salle Ballroom A & B*
Chair: T. H. Geballe (Stanford U)
PL5.1 D. C. Larbalestier (U Wisconsin, Madison), "Critical Current Limiting Mechanisms in Polycrystalline High Temperature Superconductor Conductor Forms"

09:10-09:20 **Break** *Refreshments served in La Salle Pre-Function Foyer & Granger Foyer*

09:20-01:00 **CONCURRENT SESSIONS E & F**

SESSION E: HTS Theory I *La Salle Ballroom A & B*
Chairs: R. C. Dynes (UC San Diego), C. S. Ting (TCSUH)

E.1 D. J. Scalapino (UC Santa Barbara), "Numerical Results and the High T_c Problem: From Ladders to Planes"
E.2 D. Pines (U Illinois, Urbana-Champaign), "Spin Fluctuations, Magnetotransport and $d_{x^2-y^2}$ Pairing in the Cuprate Superconductors"
E.3 R. B. Laughlin (Stanford U), "Quantitative Explanation of 10 Peculiar Behaviors of High-T_c Superconductors Using 'Anyon' Approach to the t-J Hamiltonian"
E.4 T. M. Rice (ETH Hönggerberg), "Cuprate Ladder Compounds"
E.5 V. J. Emery (Brookhaven), "Charge Inhomogeneity and High Temperature Superconductivity"
E.6 Z. Y. Weng (TCSUH), "Phase String and Superconductivity in the t-J Model"
E.7 N. Nagaosa (U Tokyo), "Gauge Theory of the Normal State Properties of High-T_c Cuprates"
E.8 E. R. Dagotto (NHMFL, U Florida), "Superconductivity in the Cuprates as a Consequence of Antiferromagnetism and Large Hole Density of States"
E.9 A. H. Luther (Nordita), "Superconductivity and the Square Fermi Surface"
E.10 M. Weger (Hebrew U), "A Possible Primary Role of the Oxygen Polarizability in High Temperature Superconductivity"
E.11 J. D. Fan (Southern U), "A Complete Pairing Mechanism in Superconductivity"

SESSION F: HTS Processing *Granger A & B*
Chairs: D. T. Shaw (SUNY Buffalo), U. Balachandran (Argonne)

F.1 K.-I. Sato (Sumitomo), "Progress of HTS Bismuth-Based Tape Application"
F.2 N. Chikumoto (ISTEC), "Strong Flux Pinning in RE123 Grown by Oxygen-Controlled Melt Process"
F.3 D. E. Peterson (Los Alamos), "IBAD Deposition of Thick Film Superconductors"
F.4 R. L. Meng (TCSUH), "Fabrication of $HgBa_2Cu_3O_{8+\delta}$ Tape"
F.5 Y. Shiohara (ISTEC), "New Process to Control Critical Currents of $Nd_{1+x}Ba_{2-x}Cu_3O_{7-\delta}$"
F.6 H. W. Weber (Atomic Institute of the Austrian Universities), "Improvement of Flux Pinning in High Temperature Superconductors by Artificial Defects"
F.7 V. Selvamanickam (IGC), "HTS Conductors: Challenges and Progress"
F.8 K. Salama (TCSUH), "A Novel Approach to High Rate Melt-Texturing in 123 Superconductors"
F.9 U. Balachandran (Argonne), "Processing and Properties of Ag-Clad BSCCO Superconductors"
F.10 D. M. Kroeger (Oak Ridge), "Grain Boundary Misorientation Distributions and Percolation in HTS Conductors"

01:00-02:00 **Box Lunch Pick-Up** *Nautile Room, 1st Floor*
For those not taking an afternoon tour, seating available in lobby bar area & outside tables to left of bar area, 1st Floor

Optional Afternoon Activities

06:30-08:30 **Reception & Banquet** *La Salle Ballroom A & B*
Speaker: R. Smalley (Rice U), "Nanoscience and Technology"
Reception, *La Salle Pre-Function Foyer*, 6:30-7 p.m.; Dinner, 7-8:30 p.m. (Talk begins at 8 p.m.)

09:00-10:00 **Poster Sessions 1B-cont, 2 & 3 Set-Up** *Granger A & B & de Zavala* (Set-up will continue Friday, March 15, 7 a.m. - 12 noon)

FRIDAY, MARCH 15

8-8:30 a.m. **Continental Breakfast** *La Salle Pre-Function Foyer*

08:30-09:10 **PLENARY SESSION 6** *La Salle Ballroom B*
Chair: B. Batlogg (Lucent Technologies)
PL6.1 S. A. Wolf (NRL), "Applications of HTS to Electronics-A DoD Perspective"

09:10-9:30 **Break** *Refreshments served in La Salle Pre-Function Foyer & Granger Foyer*

09:30-12:30 **CONCURRENT SESSIONS G & H**

SESSION G: HTS Theory II *La Salle Ballroom B*
Chair: K. Levin (U Chicago)

G.1 **C. M. Varma (Lucent Technologies)**, "Current Fluctuations in Copper-Oxide Metals"
G.2 **K. Maki (USC)**, "Aspects of D-Wave Superconductivity"
G.3 **C. S. Ting (TCSUH)**, "Ginzburg-Landau Theory of Superconductors with $d_{x^2-y^2}$ Symmetry"
G.4 **A. J. Freeman (Northwestern U)**, "Inside HT$_c$ Superconductors: An Electronic Structure View"
G.5 **V. Z. Kresin (Lawrence Berkely Lab)**, "Energy Spectrum and 'Intrinsic' T_c of the Cuprates: Effects of Pairing-Breaking, Pressure and Non-Adiabaticity"
G.6 **M. Cyrot (CNRS)**, "Localized States as an Explanation of Some Properties of Cuprates"
G.7 **J.-S. Zhou (UT Austin)**, "Thermopower of the Cuprates Under High Pressure"
G.8 **A. Ferraz (U Brasilia)**, "Quasiparticles, Strong-Coupling Regime and Fermi Liquid Theory Breakdown"
G.9 **L. P. Gor'kov (NHMFL, Florida State U)**, "Hole Spectrum in Three Band Model"

SESSION H: Film & Applications *La Salle Ballroom A*
Chairs: A. Lauder (DuPont), J. C. Wolfe (TCSUH)

H.1 **I. Bozovik (Varian)**, "Synthesis of Novel High-T_c Superconductors by Atomic-Layer Epitaxy"
H.2 **A. Ignatiev (TCSUH)**, "Very High Growth Rates of High Quality YBCO Under Photo-Assisted MOCVD"
H.3 **K. K. Likharev (SUNY Stony Brook)**, "Ultrafast Superconductor Digital Electronics"
H.4 **S. Sridhar (Northeastern U)**, "In-Plane and C-Axis Linear and Nonlinear Microwave Response of Cuprate Superconductors"
H.5 **Q. Xiong (U Arkansas, Fayetteville)**, "Fabrication of Highly Textured Superconducting Thin Films on Polycrystalline Substrates Using Ion Beam Assisted Deposition"
H.6 **H. C. Freyhardt (U Göttingen)**, "Biaxially Textured YBaCuO Thick Films on Technical Substrates"
H.7 **J. Z. Wu (U Kansas)**, "Fabrication of High Quality Hg-1223 Thin Films Using Rapid Thermal Hg-Vapor Annealing"
H.8 **R. Simon (Conductus)**, "Near-Term Commercialization of HTS Technology at Conductus"

12:30-1:30 **Box Lunch Pick-Up** *Nautile Room, 1st Floor*
Seating available in lobby bar area & outside tables to left of bar area, 1st Floor
Poster Sessions 1B-cont, 2 & 3 open for viewing, *Granger A & B & de Zavala*

01:40-05:40   **CONCURRENT SESSIONS I & J**

SESSION I: Properties & Theory *La Salle Ballroom A*
Chairs: W.-Y. Liang (U Cambridge), H. H. Wickman (NSF)

I.1   **M. P. Maley (Los Alamos)**, "Enhancements of HTS Conductor Critical Currents by Splayed Columnar Tracks from Fission Fragments"
I.2   **A. van Otterlo (ETH Hönggerberg)**, "New Aspects of Vortex Dynamics"
I.3   **Z. D. Wang (U Hong Kong)**, "Numerical Studies on the Vortex Motion in High-T_c Superconductors"
I.4   **M. Suenaga (Brookhaven)**, "Effects of 5.8 GeV Pb Irradiation on Magnetic Vortex Dynamics in Bi (2212 and 2223) Tapes"
I.5   **R. P. Sharma (U Maryland)**, "Phonon Anomaly in High Temperature Superconducting $YBa_2Cu_3O_{7-\delta}$ Crystals"
I.6   **J. A. Clayhold (TCSUH)**, "Unusual Magnetic Field Dependence of the Electrothermal Conductivity in Cuprate Superconductors"
I.7   **R. J. Birgeneau (MIT)**, "Structures and Excitations in Monolayer Copper Oxides"
I.8   **H. A. Mook (Oak Ridge)**, "Recent Neutron Scattering Results on $YBa_2Cu_3O_{7-\delta}$"
I.9   **D. van der Marel (U Groningen)**, "Electrodynamical Properties of High-T_c Superconductors Studied with Angle Dependent Infrared Spectroscopy"
I.10 **T. Timusk (McMaster U)**, "Transport in the ab-Plane of HTSC, New Results"
I.11 **Z. Fisk (NHMFL, U Florida)**, "Properties of Li-Doped La_2CuO_4"

SESSION J: Materials & Properties
La Salle Ballroom B
Chair: H. C. Ku (National Tsing Hua U)

J.1   **J. C. Campuzano (U Illinois, Chicago)**, "The Electronic Structure of the High Temperature Superconductors as Seen by Angle-Resolved Photoemission"
J.2   **W. N. Hardy (U British Columbia)**, "Microwave Measurements of the Penetration Depth in High T_c Single Crystals"
J.3   **K. A. Moler (Princeton U)**, "Specific Heat of $YBa_2Cu_3O_{7-\delta}$"
J.4   **A. R. Junod (U Geneva)**, "Specific Heat of HTS in High Magnetic Fields"
J.5   **J. W. Loram (U Cambridge)**, "Thermodynamic Evidence on the Superconducting and Normal State Energy Gaps in $La_{2-x}Sr_xCuO_4$ and $Y_{0.8}Ca_{0.2}Ba_2Cu_3O_{6+x}$"
J.6   **I. Terasaki (ISTEC)**, "Scattering Time: A Unique Property of High-T_c Cuprates"
J.7   **C. Uher (U Michigan)**, "Heat Transport in High-T_c Perovskites - Effect of Magnetic Field"
J.8   **J. L. Tallon (New Zealand Inst. for Industrial Research)**, "Thermoelectric Power: A Simple, Highly Instructive Probe of High-T_c Superconductors"
J.9   **F. Chen (TCSUH)**, "High Pressure Study on Hg-Based Cuprates"
J.10 **J. D. Jorgensen (Argonne)**, "Structural Control of Transition Temperature and Flux Pinning in High-T_c Superconductors"
J.11 **M. Marezio (MASPEC-CNR)**, "The Effect of Pressure on Superconducting Copper Mixed Oxides"

05:40-07:30   **POSTER SESSION 1B-cont: HTS Theory, Experiment, Material, & Properties**
POSTER SESSION 2: HTS Bulk Processing, Characterization & Application
POSTER SESSION 3: HTS Film Processing, Characterization & Application
Granger A & B & de Zavala
Posters will be removed from Granger A & B & de Zavala by 9 p.m.

SATURDAY, MARCH 16

8-8:30 a.m.   **Continental Breakfast** *La Salle Pre-Function Foyer*

08:30-08:40   **General Announcements** *La Salle Ballroom B*

08:40-09:20   **PLENARY SESSION 7** *La Salle Ballroom B*
Chair: C. Y. Huang (National Taiwan U)
PL7.1   **G. W. Crabtree (Argonne)**, "Collective Behavior of Vortices in Superconductors"

09:20-09:40   **Break** *Refreshments served in La Salle Pre-Function Foyer & Granger Foyer*

09:40-01:00   **CONCURRENT SESSIONS K & L**

SESSION K: Vortex, etc. *Granger A & B*
Chairs: D. C. Larbalestier (U Wisconsin, Madison),
               M. B. Maple (UC San Diego)

K.1   **E. Zeldov (Weizmann Inst. of Science)**, "Thermodynamic Vortex-Lattice Phase Transitions in BSCCO"
K.2   **J. R. Clem (Iowa State U)**, "Pancake Vortices in High-Temperature Superconducting Thin Films"
K.3   **P. L. Gammel (Lucent Technologies)**, "Correlation Lengths in the Flux Line Lattice of Type-II Superconductors"
K.4   **M. B. Maple (UC San Diego)**, "Magnetoresistivity of Thin Films of the Electron-Doped High T_c Superconductor $Nd_{1.85}Ce_{0.15}CuO_{4\pm\delta}$"
K.5   **F. de la Cruz (Centro Atómico Bariloche)**, "Comparitive Study of Vortex Correlation in Twinned and Untwinned $YBa_2Cu_3O_{7-\delta}$ Single Crystals"
K.6   **I.-S. Yang (Ewha Women's U)**, "High-Pressure Raman Study of the Mercury-Based Superconductors and the Related Compounds"
K.7   **R. Weinstein (TCSUH)**, "Very High Trapped Fields: Cracking, Creep and Pinning Centers"
K.8   **D. K. Christen (Oak Ridge)**, "Local Texture, Current Flow, and Superconductive Transport Properties of Tl1223 Deposits on Practical Substrates"
K.9   **J. Budnick (U Connecticut)**, "Muon Spin Rotation Studies of Magnetism in $La_{2-x}Sr_xCuO_4$ and $Y_{1-x}Ca_xBa_2Cu_3O_5$"
K.10 **S. Moss (TCSUH)**, "X-Ray Search for CDW in Single Crystal $YBa_2Cu_3O_{7-\delta}$"

SESSION L: Film & Properties *de Zavala*
Chairs: H. Weinstock (USAFOSR),
               T. Claeson (Chalmers U of Technology)

L.1   **J. Clarke (UC Berkeley)**, "HTS SQUIDs and Their Applications"
L.2   **J. R. Kirtley (IBM Yorktown Heights)**, "SQUID Imaging"
L.3   **C. C. Chen (UT Austin)**, "Imaging of Superconducting Vortices with a Magnetic Force Microscope"
L.4   **J. T. McDevitt (UT Austin)**, "Electronic Eyes Based on Dye/Superconductor Assemblies"
L.5   **A. Kapitulnik (Stanford U)**, "The Search for Broken-Time-Reversal-Symmetry in High-T_c Superconductors: Status Report"
L.6   **J. G. Lin (National Taiwan U)**, "Pressure Effect on the Superconducting and Normal-State Properties for the YBCO/PBCO Superlattice"
L.7   **J. H. Miller, Jr. (TCSUH)**, "Magnetocardiography in a Magnetically Noisy Environment Using High-T_c SQUIDs"
L.8   **C. Rossel (IBM Zürich)**, "Pinning and Anisotropy Properties of High-T_c Microcrystals by Miniaturized Torquemeter"
L.9   **D. W. Face (DuPont)**, "HTS Materials for High Power rf and Microwave Applications"
L.10 **A. Schilling (ETH Hönggerberg)**, "High-Accuracy Specific-Heat Study on $YBa_2Cu_3O_7$ and $Bi_2Sr_2CaCu_2O_8$ Around T_c in External Magnetic Fields"

01:00-02:30   **Luncheon & Closing Ceremony** *La Salle Ballroom A & B*
Session Summaries by B. Batlogg (Lucent Technologies), D. Gubser (NRL), W. K. Chu (TCSUH), M. B. Maple (UC San Diego)
Closing Remarks
Adjournment

Optional Afternoon & Evening Activities

Poster Sessions

TUESDAY, MARCH 12
4-6 p.m., Granger A & B

Poster Sessions 1A & 1B: HTS Theory, Experiment, Material & Properties

Poster Session 1B continued: Friday, March 15, 5:40-7:30 p.m., Granger A & B and de Zavala

POSTER SESSION 1A

- **P1A.1** T. K. Lee (Virginia Tech), "Anomalous Charge-Excitation Spectra in t-J and Hubbard Models"
- **P1A.2** P. W. Leung (Hong Kong U of Science & Technology), "Exact Diagonalization Study of the Single Hole t-J Model on a 32-Site Lattice"
- **P1A.3** N. Bulut (UC Santa Barbara), "Electronic Properties of the Layered Cuprates"
- **P1A.4** Q. M. Si (Rice U), "Electron-Spin Diffusion Constant as Diagnostics for Spin-Charge Separation in the Metallic Cuprates"
- **P1A.5** W. Putikka (U Cincinnati), "Magnetic Frustration and Spin-Charge Separation in 2D Strongly Correlated Electron Systems"
- **P1A.6** H. Kohno (U Tokyo), "Magnetic Excitation in High-T_c Cuprates"
- **P1A.9** A. Bill (Lawrence Berkeley Lab), "Neutron Scattering: A Signature of the Gap Symmetry in High-T_c Superconductors"
- **P1A.10** A. Bill (Lawrence Berkeley Lab), "Anisotropy of the Gap Induced by Unscreened Long Range Interactions"
- **P1A.11** S. R. Bahcall (UC Berkeley), "Boundary Effects and the Order Parameter Symmetry of HTC Superconductors"
- **P1A.12** W. Xu (TCSUH), "The Ginzburg-Landau Equations for d-Wave Superconductors with Nonmagnetic Impurities"
- **P1A.13** J. Tahir-Kheli (Caltech), "Inter-Band Pairing: Resolution of Observed d-Wave and s-Wave Tunneling with Isotropic s-Wave Pairing"
- **P1A.14** H.-Y. Choi (Sung Kyun Kwan U), "Effects of Impurity Vertex Correction on NMR Coherence Peak of Conventional Superconductors"
- **P1A.15** G. Reiter (TCSUH), "Comparison of Three-Band and t-t'-J Model Calculations of the One-Hole Spectral Function in an Antiferromagnet with Photoemission Experiments"
- **P1A.16** J. M. Wheatley (U Cambridge), "c-Axis Electronic Structure and Transport in Copper-Oxides"
- **P1A.17** A. Kallio (U Oulu), "Paramagnetic Meissner Effect and Time Reversal Non-Invariance from Spin Polarization"
- **P1A.18** A. Kallio (U Oulu), "Hall-Effect Scaling and Chemical Equilibrium in Normal States of High-T_c Superconductors"
- **P1A.19** M. Eremin (Kazan State U), "Spin-Susceptibility of Strong Correlated Bands in Fast Fluctuating Regime"
- **P1A.20** Y. M. Malozovsky (Southern U), "Pairing Instability and Anomalous Response in an Interacting Fermi Gas"
- **P1A.21** D. Zhou (General Superconductor), "The Connections of the Experimental Results of Universal Stress Experiments and of Thermal Expansion Measurements and the Mechanisms of Microscopic Dynamics Process on CuO_2 Planes"
- **P1A.22** J. D. Dow (Arizona State U.), "Phenomenology: What the Data Say"
- **P1A.23** D. Frenkel (TCSUH), "Current Instabilities in Reentrant Superconductors"
- **P1A.24** J.-X. Zhu (U Hong Kong), "Time-Window Extension for Magnetic Relaxation from Magnetic-Hysteresis-Loop Measurements"
- **P1A.25** Z. D. Wang (U Hong Kong), "Numerical Study of Washboard Effect in High-T_c Superconductors"
- **P1A.26** P. Ao (U Washington), "Vortex Vacancy Motion as the Origin of the Hall Anomaly"
- **P1A.27** E. Demircan (UT Austin), "Vortex Dynamics in Superfluids: Cyclotronic Motion"
- **P1A.28** C.-R. Hu (Texas A&M U), "Zero-Bias (Tunneling-)Conductance Peak (ZBCP) as a Result of Midgap Interface States (MISs) – Model Calculations"

POSTER SESSION 1B

- **P1B.1** D. J. Derro (UT Austin), "Low Temperature Scanning Tunneling Microscopy and Spectroscopy of the CuO Chains in $YBa_2Cu_3O_{7-x}$"
- **P1B.2** H. Ding (U Illinois, Chicago), "Momentum Dependence of the Superconducting Gap of $Bi_2Sr_2CaCu_2O_8$"
- **P1B.3** J. T. Lin (TCSUH), "Use of Tricrystal Microbridges to Probe the Pairing State Symmetries of Cuprate Superconductors"
- **P1B.4** G. S. Boebinger (AT&T Bell Labs), "Ground State of Superconducting LaSrCuO in 61-Tesla Magnetic Fields"
- **P1B.5** A. B. Agafonov (B. Verkin Inst. for Low Temperature Physics and Engineering), "Behavior of ScN and ScS Contacts Under Microwave Irradiation"

P1B.6   T. Zhou (Max Planck Inst.), "Electronic Raman Scattering in $YBa_2Cu_4O_8$ at High Pressure"

P1B.7   X. J. Zhou (Max Planck Inst.), "Raman Scattering on $HgBa_2Ca_{n-1}Cu_nO_{2n+2+\delta}$ (n=1,2,3,4,5) Superconductors"

P1B.8   P. F. Henning (Florida State U), "Anisotropy of Thermal Conductivity of YBCO and Selectively Doped YBCO Single Crystals"

P1B.9   M. Houssa (U Liège), "Thermal Conductivity of High-T_c Superconductors"

P1B.10   M. Sugahara (Yokahama National U), "Dielectric Anomaly of $La_{2-x}Sr_xCuO_4$ Film at x=1/4n"

P1B.11   N. E. Hussey (U Cambridge), "Angular Dependence of the c-Axis Normal State Magnetoresistance in Single Crystal $Tl_2Ba_2CuO_6$"

P1B.12   F. F. Balakirev (Rutgers U), "Hall Effect and Magnetoresistance in Normal State in LaSrCuO"

P1B.13   R. J. Gooding (Queen's U), "The Effect of Sr Impurity Disorder on the Magnetic and Transport Properties of $La_{2-x}Sr_xCuO_4$, $0.02 \leq x \leq 0.05$"

P1B.14   D. Sánchez (Centro Brasileiro de Pesquisas Físicas), "Mössbauer Studies of $Re_{1.85}Sr_{0.15}CuO_4$ T' Phase"

P1B.15   D. Sánchez (Centro Brasileiro de Pesquisas Físicas), "Observation of a Pair-Breaking Field in $RENi_2B_2C$ Compounds"

P1B.16   Y. Cao (TCSUH), "Pressure Effects on T_c of $HgBa_2Ca_{n-1}Cu_nO_{2n+2+\delta}$ with $n \geq 4$"

P1B.18   Y. Eckstein (Technion), "Strong Overdoping, Similar Depression of T_c by Zn and Ni Substitution and Departure from the Universal Thermopower Behaviour in (Ca, La)-1:2:3"

P1B.19   M. W. Coffey (Regis U), "Nonlinear Dynamics in the Mixed State of High Temperature Superconductors"

P1B.20   W. Y. Guan (Tamkang U), "The Drude Model of Transport Properties in Pure, Pr- and Ca-Doped $RBa_2Cu_3O_{7-\delta}$ Systems"

P1B.21   H.-C. Ku (National Tsing Hua U), "Anomalous Pr Ordering and Filamentary Superconductivity for the Pr2212 Cuprates"

P1B.22   J. A. Aguiar (U Federal de Pernambuco), "Superconductivity and Structural Aspects of $Y_{1-x}Ca_xBa_2Cu_3O_{7-\delta}$ with Variable Oxygen Content"

P1B.23   J. A. Aguiar (U Federal de Pernambuco), "Superconducting Properties of Nb Thin Films"

P1B.24   J. A. Aguiar (U Federal de Pernambuco), "On the Thickness Dependence of Irreversibility Line in $YBa_2Cu_2O_{7-x}$ Thin Films"

P1B.25   S. W. Chan (Columbia U), "Twin Structures in Large Grains of $YB_2Cu_3O_{7-\delta}$ as Affected by the Dispersion and Volume Fractions of Y_2BaCuO_5"

P1B.26   M. B. Field (U Wisconsin, Madison), "Electromagnetic Coupling of Melt-Textured $YBa_2Cu_3O_{6+x}$ Bicrystals"

P1B.26A   Y. W. Park (Seoul National U), "Thermoelectric Power of Superconducting Alloys YNi_2B_2C and $LuNi_2B_2C$"

FRIDAY, MARCH 15
5:40-7:30 p.m., Granger A & B & de Zavala

Poster Session 1B-cont: HTS Theory, Experiment, Material & Properties

Poster Session 2: HTS Bulk Processing, Characterization & Application

Poster Session 3: HTS Film Processing, Characterization & Application

---

POSTER SESSION 1B-cont

P1B.27   C.-L. Lin (National Tsing Hua U), "Evaluation of Overdoping Effect in $Y_{1-x}Ca_xBa_2Cu_3O_{7-\delta}$ Films"

P1B.31   E. Baggio-Saitovitch (Centro Brasileiro de Pesquisas Físicas), "Electronic Structure and Magnetic Properties of $RENi_2B_2C$ (RE=Pr, Nd, Sm, Gd, Tb, Dy, Ho, Tm, Er)"

P1B.32   E. Baggio-Saitovitch (Centro Brasileiro de Pesquisas Físicas), "The Magnetic Properties of the Quaternary Intermetallics RNiBC (R=Er, Ho, Tb, Gd, Y)"

P1B.33   E. Baggio-Saitovitch (Centro Brasileiro de Pesquisas Físicas), "Related Y-Ba-Cu-O Superconducting Oxides Containing Oxyanions"

P1B.34   J. T. Markert (UT Austin), "Physical Properties of Infinite-Layer and T'-Phase Copper Oxides"

P1B.37   Z. L. Du (TCSUH), "Superconducting Phases in the Sr-Cu-O System"

P1B.38   D.-C. Ling (National Tsing Hua U), "Superconductivity in $Sr_2YRu_{1-x}Cu_xO_6$ System"

P1B.40   S.-I. Lee (Pohang U of Science Technology), "Magnetic Properties of $HgBa_2Ca_{0.86}Sr_{0.14}Cu_2O_{6-\delta}$"

P1B.41   T. Mertelj (U Ljubljana), "Photoexcited Carrier Relaxation in Metallic $YBa_2C_3O_{7-\delta}$: A Probe of Electronic Structure"

P1B.42   V. Eremenko (Inst. for Low Temperature Physics & Engineering), "Superstructure of Potential Created by Impurity Oxygen Ions and its Effect on Resistive and Spectral Characteristics of 1-2-3 HTSC"

P1B.43   V. Eremenko (Inst. for Low Temperature Physics & Engineering), "Photostimulation of Critical Temperature, Critical Current and Normal State Conductivity in Epitaxial Y – Ba – Cu – O Films Nonsaturated by Oxygen"

P1B.45   E. Silva (U "La Sapienza"), "Resistive Transitions of HTS Under Magnetic Fields: Influence of Fluctuations and Viscous Vortex Motion"

P1B.46   B. Brown (Oregon State U), "Vortex Phase Transition Critical Parameters in Single Crystals of YBCO – Apparent Translational Order Dependence"

P1B.47   D. Lopez (Centro Atómico Bariloche), "Discontinuous Onset of the c-Axis Vortex Correlation at the Melting Transition in $YBa_2Cu_3O_{7-\delta}$"

P1B.48   C. A. Balseiro (Centro Atómico Bariloche), "Longitudinal Superconductivity and Percolation Transition of the Vortex Lattice"

P1B.49   H. Ikuta (U Tokyo), "Vortex Phase Transition in $Bi_2Sr_2CaCu_2O_y$ Single Crystals and the Doping Level Dependence"

P1B.50   T. W. Li (Leiden U), "Oxygen Dependence of First-Order Melting Transition Lines in $Bi_2Sr_2Ca_2Cu_2O_{8+x}$ Single Crystals"

P1B.51   T. W. Li (Leiden U), "Flux Pinning by Ti Doping in $Bi_2Sr_2Ca_1Cu_2O_{8+x}$ Single Crystals"

P1B.52   V. K. Vlasko-Vlasov (Inst. for Solid State Physics), "Meissner Holes in Remagnetized Superconductors"

P1B.53   B. Hickey (TCSUH), "Formation of $HgBa_2Ca_2Cu_3O_{8+\delta}$ with Additions Under Ambient Conditions"

P1B.54   J. Karpinski (ETH Hönggerberg), "Single Crystals of $HgBa_2Ca_{n-1}Cu_nO_{2n+2+\delta}$ Compounds: Growth at 10 kbar Gas Pressure and Properties"

P1B.55   H. Schwer (ETH Hönggerberg), "X-Ray Single Crystal Structure Analysis of $HgBa_2Ca_{n-1}Cu_nO_{2n+2+\delta}$ Compounds"

P1B.60   B. O. Wells (MIT), "Intercalation and Staging Behavior in Superoxygenated $La_2CuO_{4+\delta}$"

P1B.61   X. Z. Xiong (TCSUH), "In-Plane Ordering in Phase-Separated and Staged Single Crystal $La_2CuO_{4+\delta}$"

P1B.62   I. Rusakova (TCSUH), "Microstructural Changes of YBCO Induced by Lanthanide Doping"

P1B.64   M. Ausloos (U Liège), "Fractal Grain Boundaries in Composite $YBa_2Cu_3O_{7-\delta}/Y_2O_3$ Resulting from a Competition Between Growing Grains: Experiments and Simulations"

P1B.65   M. Mironova (TCSUH), "TEM Study of Low-Angle Grain Boundaries in Polycrstalline YBCO"

P1B.66   C. Y. Huang (National Taiwan U), "Magnetic Properties of Some High-T_c Superconductors"

P1B.67   N. T. Cherpak (National Academy of Sciences), "Temperature Dependence of Radiofrequency Absorption and Critical Current Density in High-T_c YBaCuO Thin Samples"

POSTER SESSION 2

P2.1   J. K. Meen (TCSUH), "Phase Relations in the Bi_2O_3-CaO-CuO and Bi_2O_3-CaO-SrO Systems at 750° to 1000° in Pure Oxygen at 1 atm"

P2.2   J. K. Meen (TCSUH), "Subsolidus Pressure-Temperature Phase Relations in CaO-CuO to 30 kbar"

P2.3   M.-L. Carvalho and K. L. Senes (TCSUH), "Liquidus Phase Relations in the Bismuth-Rich Portion of the Bi_2O_3-SrO-CaO-CuO System at 1 Atm in Pure Oxygen"

P2.4   J. Geny (TCSUH), "Phase Equilibria in the La_2O_3-SrO-CuO System at 950°C and 10-30 kbar"

P2.5   S. Sinha (U Illinois, Chicago), "Y123 Superconductor via in-situ Reaction/Deoxidation of a Submicrometer Precursor Containing $BaCuO_{2.5}$"

P2.6   Q. M. Lin (TCSUH), "Stability Study of $Hg_{1-x}Ba_2Ca_{n-1}Cu_nO_{2n+\delta}$ With $n \leq 6$"

P2.9   Y. Shiohara (ISTEC), "Control of 211 Particle Dispersion And J_c Property In Melt-Textured YBCO Superconductor"

P2.10   S. S. Wang (TCSUH), "Partial-Melt Processing of Bulk MgO-Whisker Reinforced $(Bi,Pb)_2Sr_2Ca_2Cu_3O_{10-x}$ Superconductor"

P2.11   G. Z. Zhang (TCSUH), "Strain Tolerance of Superconducting Properties of Bulk MgO-Whisker-Reinforced HTS BPSCCO Composite"

P2.12   G. Yang (TCSUH), "Electro-Mechanical Properties of Jointed BPSCCO Composites"

P2.13   I.-G. Chen (National Cheng Kung U), "Magnetic Levitation Transporation System by Top-Seeded Melt-Textured YBCO Superconductor"

P2.14   A. G. Mamalis (National Technical U of Athens), "High-T_c Ceramic Superconductors for Rotating Electrical Machines: From Fabrication to Application"

POSTER SESSION 3

P3.1   J. Ritchie (UT Austin), "Molecular Level Control of the Interfacial Properties of High-T_c Superconductor Structures and Devices"

P3.2   J. T. McDevitt (UT Austin), "Crystal Engineering of Chemically Stabilized, Cation Substituted $YBa_2Cu_3O_{7-\delta}$ Thin Film Structures"

P3.5   J. Gohng (Samsung Advanced Inst. of Technology), "Directly Coupled DC-SQUIDs of YBCO Step-Edge Junctions Fabricated by a Chemical Etching Process in Mixed Acids"

P3.6   S. S. Shoup (Oak Ridge), Growth of Epitaxial $LaAlO_3$ and CeO_2 Films Using Sol-Gel Precursors"

P3.7   M.-Y. Li (National Tsing Hua U), "Study of the In-Plane Epitaxy of Bi-Epitaxial Superconducting Grain Boundary Junctions"

P3.8   K. E. Myers (DuPont), "*in situ* Deposition of Thallium Cuprate and Other Thallium-Containing Oxides"

P3.9   Z. F. Ren (SUNY Buffalo), "Growth of Superconducting Epitaxial $Tl_2Ba_2CuO_{6+\delta}$ Thin Films with Tetragonal Lattice and Continuously Adjustable Critical Temperature"

P3.10   **L. Schmirgeld-Mignot (C. E. A. Saclay)**, "Transport Critical Currents of Bi-2212 Tapes Prepared by Sequential Electrolytic Deposition"

P3.11   **S. Salib (TCSUH)**, "Role of Constituents on the Behavior of Composite BPSCCO Tapes"

P3.12   **X. T. Cui (TCSUH)**, "Ion Channeling Studies in YBCO Thin Film at Low Temperature"

P3.13   **R. D. Dittmann (Inst. of Thin Film and Ion Technology)**, "Current Transport Across YBCO-Au Interfaces"

P3.15   **X.-M. Zhu (Umeå U)**, "Direct Measurement of the Magnus Force in YBCO Films"

P3.16   **J.-X. Zhu (U Hong Kong)**, "Interface Roughness Effect on Differential Conductance of High-T_c Superconductor Junctions"

P3.17   **V. I. Nikitenko (Inst. for Solid State Physics)**, "Modern Magneto-Optical Techniques for Superconductors"

P3.18   **A. A. Polyanskii (U Wisconsin, Madison)**, "Magneto-Optical Study of Flux Penetration and Critical Current Densities in [001] Tilt $YBa_2Cu_3O_{7-\delta}$ Thin Film Bicrystals"

P3.19   **Y. K. Tao (Teco Electric & Machinery)**, "Monolithic Terminations for Multifilamentary BSCCO Wires"

P3.21   **M. Darula (Inst. of Thin Film and Ion Technology)**, "Mutual High-Frequency Interaction of High T_c Josephson Junctions"

P3.22   **N. Tralshawala (TCSUH)**, "Magnetocardiography in an Unshielded Clinical Environment Using High-T_c SQUIDs"

P3.23   **J. Wosik (TCSUH)**, "High-T_c Superconducting rf Receiver Coils for Magnetic Resonance Imaging"

P3.24   **J. Wosik (TCSUH)**, "Investigation of the Microwave Power Handling Capability of High-T_c Superconducting Thin Films"

P3.25   **S. Sridhar (Northeastern U)**, "Linear and Nonlinear Microwave Flux Dynamics in Thin Films: Relationship and Critical State Flux Penetration"

Optional Activities

Optional activities for the afternoon of Thursday, March 14 and the afternoon and evening of Saturday, March 16 are listed below. Contact Workshop Staff members to purchase tickets or to get additional information about any of these or other activities.

Tour Options for the afternoon of Thursday, March 14:

Tour of NASA/Space Center Houston
- 1:15-5 p.m. (arrive back at hotel, 5:45 p.m.)
- Luxury Coach transportion provided
- includes astronaut briefing, IMAX, and tours of Mission Control
- box lunches & drinks provided on the bus
- $15 advance ticket required

VIP Tour of Texas Medical Center facilities, hosted by TMC, Inc.
- 1:30-4 p.m. (arrive back at hotel, 4:30 p.m.)
- Luxury Coach transportation provided
- includes personal tours of facilities, e.g. Telemedicine – Texas Children's Hospital, Cyclotron/Positron facility – Hermann Hospital/UT
- box lunches & drinks will be provided on the bus
- $6 advance ticket required

Other activities for the afternoon of Thursday, March 14 to be scheduled through the Hotel Consierge:

Metropolitan Racquet Club
- special fees for conference members

Downtown YMCA

Galleria Shopping District
- check registration desk for shuttle times

Downtown Trolley to shopping areas

Other activities for the afternoon of Thursday, March 14 to be scheduled through the Workshop Staff (some fees may apply, depending upon tour operator):

Rounds of Golf at various Houston clubs

TCSUH Lab tours
- advance sign-up required

Escorted Shopping Excursion to Galleria District or Designer Warehouses

Galveston Island tour
- Moody Gardens, historic Strand District and shopping

Optional activities for the afternoon and evening of Saturday, March 16:

American Institute of Architects Walking Tour of Downtown Houston

TCSUH Lab tours
- 3-5 p.m.
- advance sign-up required

Galleria Shopping District
- check registration desk for shuttle times

Rounds of Golf at various Houston clubs
- must be arranged in advance of Workshop

1994 & 1995 World Champion Houston Rockets vs. Miami Heat at the Summit
- 7:30 p.m. tip-off
- 30 seats available

Romeo & Juliet performed by the Houston Ballet at the Wortham Centre
- curtain at 7:30 p.m.
- reservations required in advance

Richmond Avenue Nightclub tour
- arrange through Workshop registration desk

Dinner and entertainment at the Magic Island Restaurant
- reservations with restaurant required

WORKSHOP PARTICIPANTS

Mr. Roger Adams
Pirelli Cable Corporation
700 Industrial Drive
Lexington SC 29072
P: 803 951 4992
F: 803 951 4991

Dr. Gabriel Aeppli
NEC Research Institute, Inc.
4 Independence Way
Princeton NJ 08540
P: 609 451 2658
F: 609 951 2482
e-mail: gabe@
research.nj.nec.com

Mr. Alexei Agafonov
Department of
Superconductivity
B. Verkin Institute for Low
Temperature Physics and
Engineering
47 Lenin Avenue
Kharkov 31064
UKRAINE
P: 380 572 308 530
F: 380 572 322 370
e-mail: agafonov@
ilt.kharkov.ua

Dr. J. Albino Aguiar
Departmento de Física
UFPE
Recife-PE 50670-901
BRASIL
P: 005581 2718450
F: 005581 2710359
e-mail: albino@npd.ufpe.br

Dr. Yoichi Ando
Resident Visitor
Bell Labs
Lucent Technologies
700 Mountain Avenue, Room
1D-208
Murray Hill NJ 07974-0636
P: 908 582 3293
F: 908 582 3260
e-mail: ando@physics.att.com

Dr. David Andrews
Vice President of Marketing
Oxford Instruments
600 Milik Street
Carteret NJ 07008
P: 908 541 1300
F. 908 541 7789
e-mail: Andrewsde@aol.com

Dr. Ping Ao
Department of Physics
University of Washington
Box 351560
Seattle WA 98195
P: 206 543 3901
F: 206 685 0635
e-mail: ao@
dirac.phys.washington.edu

Mr. Srinath P. Athur
Texas Center for
Superconductivity
Department of Mechanical
Engineering
University of Houston
Houston TX 77204-4792
P: 713 743 4543
F: 713 743 4513
e-mail: Srinath@uh.edu

Dr. Marcel Ausloos
Institut de Physique B5 3/54
Université de Liège, Sart
Tilman
B-4000 Liège
BELGIUM
P: ++ 32 41 66 37 52
F: ++ 32 42 66 29 90
e-mail: ausloos@
gw.unipc.ulg.ec.be

Prof. Elisa Baggio-Saitovitch
Rua Xavier Siguad
Centro Brasileiro de Pesquisas
Fisicas
EEPT 22290-180
Rio de Janeiro
BRASIL
P: 55 21 541-0337 ext. 182
F: 55 21 541-2047
e-mail: elisa@
novell.cat.cbpf.br

Dr. Safi Bahcall
Department of Physics
University of California at
Berkeley
Berkeley CA 94720
P: 510 642 0588
F: 510 643 8497
e-mail: bahcall@
physics.berkeley.edu

Dr. U. (Balu) Balachandran
Manager, Ceramics Section
Director, HTSC Technology
Program
Argonne National Laboratory
9700 South Cass Avenue
Argonne IL 60439-4838
P: 708 252 4250
F: 708 252 3604
e-mail: u_balachandran@
qmgate.anl.gov

Dr. Fedor F. Balakirev
Department of Physics &
Astronomy
Rutgers University
P. O. Box 849
Piscataway NJ 08855
P: 905 445 5011
F: 905 445 4343
e-mail: balakir@
physics.rutgers.edu

Prof. Carlos A. Balseiro
Centro Atómico Bariloche
Comisión Nacional de Energía
Atómica
Ar. Bastillo KM 9
S. C. de Bariloche RN/8400
ARGENTINA
P: 0054 944 45163
F: 0054 944 61001

Dr. Bertram Batlogg
Bell Labs
Lucent Technologies
700 Mountain Avenue
Room 1D-369
Murray Hill NJ 07974
P: 908 582 6663
F: 908 582 3260
e-mail: batlogg@
bell-labs.com

Mr. Sai Bhavaraju
Texas Center for
Superconductivity
Department of Chemistry
University of Houston
Houston TX 77204-5641
P: 713 743 2785
F: 713 743 2787

Dr. Andreas Bill
Postdoctoral Fellow
Lawrence Berkeley Laboratory
1 Cyclotron Road, MS 62-203
Berkeley CA 94720
P: 510 486 6691
F: 510 486 4995
e-mail: abill@lbl.gov

Dr. Robert J. Birgeneau
Dean of Science
Massachusetts Institute of
Technology
School of Science, 6-123 MIT
77 Massachusetts Avenue
Cambridge MA 02319
P: 617 253 8900
F: 617 253 8901
e-mail: robertJB@mit.edu

Mr. Aaron Bitterman
Editor
Superconductor Week
P. O. Box 411433
San Francisco CA 94141
P: 415 864 3122
F: 415 864 1423
e-mail: abitterman@aol.com

Dr. Girsh E. Blumberg
Department of Physics
University of Illinois at Urbana-Champaign
1110 W. Green St.
Urbana IL 61801
P: 217 244 8038
F: 217 244 8544
e-mail: blumberg@uiuc.edu

Dr. Greg S. Boebinger
Bell Labs
Lucent Technologies
700 Mountain Avenue, Room 1D-208
Murray Hill NJ 07974
P: 908 582 7573
F: 908 582 3260
e-mail: bo@physics.att.com

Dr. Aldo F. Bolza
Director, Energy & Materials R&D
Pirelli Cavi SpA
viale Sarea 202
20126 Milano
ITALY
P: 011 39 2 6442 3098
F: 011 39 2 6442 2205

Dr. Ivan Bozovic
Senior Research Scientist
Varian Research Center
3075 Hansen Way
K-114
Palo Alto CA 94304-1075
P: 415 424 6358
F: 415 424 6988
e-mail: ivan.bozovic@vrc.varian.com

Mr. Brandon Brown
Department of Physics
Oregon State University
301 Weniger Hall
Corvallis OR 97331
P: 541 737 1721
F: 541 737 1683
e-mail: browbran@ucs.orst.edu

Prof. Joseph I. Budnick
Department of Physics
University of Connecticut
2152 Hillside Road
Storrs CT 06289
P: 203 486 5541 or 4915
F: 203 486 3346

Nejat Bulut
Department of Physics
University of California at Santa Barbara
Santa Barbara CA 93106
P: 805 893 2246
F: 805 893 8838
e-mail: nejat@spock.physics.ucsb.edu

Mr. Calvin Burnham
World Congress on Superconductivity
c/o Houston Lighting and Power Company
P. O. Box 1700
Houston TX 77002
P: 713 207 3423
F: 713 623 3102
e-mail: calvin-burnham@hlp.com

Dr. Juan Carlos Campuzano
Department of Physics
Argonne National Laboratory
University of Illinois at Chicago
4700 South Cass Ave.
Chicago IL 60439
P: 708 252 5018
F: 708 252 7777
e-mail: jcc@uic.edu

Mr. Yong Cao
Texas Center for Superconductivity
University of Houston
4800 Calhoun
Houston TX 77204-5932
P: 713 743 8304
F: 713 743 8221
e-mail: Phys8x@menudo.uh.edu

Dr. Manuel Cardona
Max-Planck-Institut für Festkörperforschung
Heisenbergstraße 1
70569 Stuttgart
GERMANY
P: 49 711 689 1710
F: 49 711 689 1712
e-mail: cardona@cardix.mpi.stuttgart.mpg.de

Ms. Marie-Laure Carvalho
Texas Center for Superconductivity
Department of Chemistry
University of Houston
4800 Calhoun
Houston TX 77204-5641
P: 713 743 8288
F: 713 743 8281
e-mail: Marie@tem.tcs.uh.edu

Dr. Robert J. Cava
Bell Labs
Lucent Technologies
700 Mountain Avenue, Room 1T-304
Murray Hill NJ 07074
P: 908 582 2180
F: 908 582 2521
e-mail: cava@allwise.mh.att.com

Prof. Siu-Wai Chan
Department of Material Science & Metallurgy
Columbia University
1136 Mudd Building
New York NY 10027
P: 212 854 8519
F: 212 854 3054
e-mail: sc174@columbia.edu

Mr. Chun Che Chen
Department of Physics
University of Texas at Austin
RLM, Room 5.208
26th and Speedway Streets
Austin TX 78712-1081
P: 512 471 5544
F: 512 471 9637
e-mail: ccc@UTPAPA.UTEXAS.EDU

Mr. Feng Chen
Texas Center for Superconductivity
University of Houston
Houston TX 77204-5932
P: 713 743 8307
F: 713 743 8201
e-mail: fchen@uh.edu

Dr. In-Gann Chen
Department of Materials Science & Engineering
National Cheng Kung University
1-Ta-Hsueh Road
Tainan
Taiwan
REPUBLIC OF CHINA
P: 886 6 2757575, ext. 62947
F: 886 6 234 6290
e-mail: ingann@mail.ncku.edu.tw

Dr. Quark Chen
Texas Center for Superconductivity
University of Houston
Houston TX 77204-5932
P: 713 743 8253
F: 713 743 8201
e-mail: Qchen@uh.edu

Prof. S. S. Chern
Mathematics Institute
University of California at Berkeley
1000 Centennial Drive
Berkeley CA 84720
P: (510) 642-8204

Prof. C. C. Chi
Materials Science Center
National Tsing Hua University
Kuang Fu Road, Station 1
Hsinchu, Taiwan
REPUBLIC OF CHINA
P: 886 35 723695
F: 886 35 713113
e-mail: cchi@phys.nthu.edu.tw

Dr. Noriko Chikumoto
Superconductivity Research Laboratory
ISTEC-SRL
16-25, Shibaura 1-chome
Minato-ku
Tokyo 105
JAPAN
P: 81 3 3454 9284
F: 81 3 3454 9287
e-mail: 65124651@people.or-jp

Dr. Hsiao-Mei Cho
Department of Physics
National Taiwan University
1 Sec. 4, Roosevelt Road
Taipei 10764
Taiwan
REPUBLIC OF CHINA
P: 886 2 3626937, ext. 101
F: 886 2 3639984
e-mail: hcyang2@
phys.ntu.edu.tw

Prof. Han-Yong Choi
Department of Physics
Sung Kyun Kwan University
Suwon 440-746
KOREA
P: 82 331 290 5902
F: 82 331 290 5370
e-mail: hychoi@
phys1.skku.ac.kr or
hychoi@yurim.skku.ac.kr

Dr. Pen-Chu Chou
Texas Center for
Superconductivity
Space Vacuum Epitaxy Center
University of Houston
Houston TX 77204-5507
P: 713 743 3621
F: 713 747 7724
e-mail: penchou@
space.svec.uh.edu

Dr. David K. Christen
Oak Ridge National Laboratory
P. O. Box 2008, MS 6061, Bldg 3115
Oak Ridge TN 37831-6061
P: 423 574 6269
F: 423 574 6263
e-mail: dkc@ornl.gov

Dr. Roy Christoffersen
Texas Center for
Superconductivity
University of Houston
Houston TX 77204-5641
P: 713 743-8273
F: 713 743-2787

Mr. Kam-Hong Chu
Texas Center for
Superconductivity
University of Houston
Houston TX 77204-5932
P: 713 743 8259
F: 713 743 8201

Dr. Paul C. W. Chu
Texas Center for
Superconductivity
University of Houston
Houston TX 77204-5932
P: 713 743 8222
F: 713 743 8201
e-mail: cwchu@uh.edu

Dr. Wei-Kan Chu
Texas Center for
Superconductivity
University of Houston
Houston TX 77204-5932
P: 713 743 8250
F: 713 743 8201
e-mail: WKChu@uh.edu

Mr. Paul L. Cinquemani
Pirelli Cables North America
Manager, Engineering Services
and Electrical Lab
710 Industrial Drive
Lexington SC 29072
P: 803 951 4000
F: 803 957 8654

Prof. Tord Claeson
Physics Department
Chalmers University of
Technology
Gothenburg S-412 96
SWEDEN
P: +46 31 772 3304
F: +46 31 772 3371
e-mail: F4ATC@
FY.CHALMERS.SE

Dr. John Clarke
Professor of Physics
Department of Physics
University of California at
Berkeley
366 LeConte Hall
Berkeley CA 94720-7300
P: 510 642 3069
F: 510 642 1304
e-mail: jclarke@
physics.berkeley.edu

Mr. James Claycomb
Texas Center for
Superconductivity
University of Houston
Houston TX 77204-5932
P: 713 743 8272
F: 713 743 8201

Dr. Jeffrey A. Clayhold
Texas Center for
Superconductivity
University of Houston
Houston TX 77204-5932
P: 713 743 8316
F: 713 743 8201
e-mail: jclayhold@uh.edu

Dr. John R. Clem
Distinguished Professor of
Physics
Iowa State University and Ames
Laboratory
Department of Physics - A517
Ames IA 50011-3020
P: 515 294 4223
F: 515 294 0689
e-mail: clem@ameslab.gov

Prof. Mark W. Coffey
Department of Chemistry
Regis University
3333 Regis Boulevard
Denver CO 80221
P: 303 458-4027
F: 303 964 5480
e-mail: chem@regis.edu or
coffe@cmg.eeel.nist.gov

Prof. Marvin L. Cohen
Department of Physics
University of California at
Berkeley
539 Birge Hall, PY-01
Berkeley CA 94720
P: 510 642 4753
F: 510 643 9473
e-mail: cohen@
jungle.berkeley.edu

Ms. K. C. Cole
Science Writer
The Los Angeles Times
3366 Deronda Drive
Los Angeles, CA 90068
P: 213 237 3534
F: 213 237 4712
e-mail: kc.cole@latimes.com

Mr. Rodger Cooley
TCSUH Levitation Laboratory
Texas Center for
Superconductivity
University of Houston
Houston Texas 77204-5932
P: 713 743 8218
F: 713 843 8201
e-mail: rcooley@uh.edu

Mr. Willard Cooper
610 Society Hill
Cherry Hill NJ 08003
P: 609 751 1411

Dr. George W. Crabtree
Senior Scientist
Argonne National Laboratory
9700 South Cass Avenue
Building 223
Argonne IL 60439
P: 708 252 5509
F: 708 252 7777
e-mail: crabtree@anl.gov

Prof. Jack Crow
National High Magnetic Field
Laboratory
Florida State University
1800 E. Paul Dirac Drive
Tallahassee FL 32306-3016
P: 904 644 0850
F: 904 644 9462
e-mail: crow@magnet.fsu.edu

Mr. Xingtian Cui
Texas Center for
Superconductivity
University of Houston
Houston TX 77204-5932
P: 713 743 8256
F: 713 743 8201
e-mail: Phys61@jetson.uh.edu

Dr. M. Cyrot
Laboratoire Louis Néel
CNRS
25, Avenue des Martyrs
38042 Grenoble BP166
FRANCE
P: 33 76 88 10 91
F: 33 76 88 11 91
e-mail: cyrot@
labs.polycnrs-gre.fr

Dr. Elbio R. Dagotto
National High Magnetic Field
Laboratory
Florida State University
Magnet Lab, Innovation Park
Tallahassee FL 32306
P: 904 644 1726
F: 904 644 5038
e-mail: dagotto@
theory.nhmfl.fsu.edu

Dr. Weiming Dai
Oxford Instruments
600 Milik Street
Carteret NJ 07008
P: 908 541 1300
F: 908 541 7769

Dr. Marian Darula
Forschonszentrum Jülich-
GmbH (KFA)
Institute of Thin Film and Ion
Technology
D-52425 Jülich
GERMANY
P: 49 246 161 2346 or 2961
F: 49 246 161 2940 or 2333
e-mail: darula@
isitel1.isi.kfa-juelich.de

Dr. Francisco de la Cruz
Centro Atómico Bariloche
8400 Bariloche RN
Rio Negro
ARGENTINA
P: 54 944 61001
F: 54 944 61001
e-mail: delacruz@
cab.cnea.edu.ar

Dr. Alex de Lozanne
Department of Physics
University of Texas at Austin
Austin TX 78712-1081
P: 512 471 6108
F: 512 471 6518 or 9637
e-mail: lozanne@
physics.utexas.edu

The Honorable Tom DeLay
Majority Whip and Member of
U.S. Congress - District 22
United States House of
Representatives
203 Cannon House Office
Building
Washington DC 20515
P: (202) 225-5951
F: (202) 225-5241

Mr. Erturgrul Demircan
Department of Physics
University of Texas at Austin
RLM 5.208
Austin TX 78741
P: 512 471 4023
e-mail: demircan@
physics.utexas.edu or
demircan@utpapa.ph.utexas.edu

Mr. David Derro
Department of Physics
University of Texas at Austin
Austin TX 78712-1081
P: 512 471 5544
F: 512 471 9637 or 6518
e-mail: derro@
utpapa.utexas.edu

Mr. Luiz Marcos Dezaneti
Texas Center for
Superconductivity
University of Houston
Houston TX 77204-5932
P: 713 743 8304
F: 713 743 8201
e-mail: Marcos@
bambam.tcs.uh.edu

Hong Ding
Department of Physics
University of Illinois at Chicago
845 W. Taylor Street
Chicago IL 60607
P: 312 996 9236
F: 312 996 9016
e-mail: hongding@uic.edu

Dr. Regina Dömel Dittmann
Institute für Schicht- und
Ionentechnik, KFA-Jülich
Jülich D-52425
GERMANY
P: 49 2461 61 2357
F: 49 2461 61 2940
e-mail: rdittmann@
kfa.juelich.de

Mr. Sasa Djordjevic
Space Vacuum Epitaxy Center
University of Houston
4800 Calhoun
Houston TX 77204-5507
P: 713 743 3621
F: 713 747 7724

Prof. John D. Dow
Arizona State University
6031 East Cholla Lane
Scottsdale AZ 85253
P: 602 423 8540
F: 602 423 8540

Dr. David Driscoll
Reliance Electric/Rockwell
Automation
24800 Tungsten
Cleveland OH 44117
P: 216 266 6002
F: 216 266 1040
e-mail: DID@
RRCORPMHS.COMPUSERVE.COM

Mr. Guoping Du
Texas Center for
Superconductivity
Department of Mechanical
Engineering
University of Houston
Houston TX 77204-4792
P: 713 743 4543
F: 713 743 4513
e-mail: gxd25601@
jetson.uh.edu

Mr. Zhong Lian Du
Visiting Researcher
Texas Center for
Superconductivity
University of Houston
Houston TX 77204-5932
P: 713 743 8315
F: 713 743 8221

Dr. Robert Dynes
Sr. Vice Chancellor for
Academic Affairs
University of California at San
Diego
9500 Gilman Drive
La Jolla CA 92093-0001
P: 619 534 3130
F: 619 534 5355
e-mail: dynes@ucsd.edu

Mr. Russell Eaton III
Director
Advanced Utility Concepts
Division
Department of Energy
1000 Independence Avenue, SW
Washington DC 20585
P: 202 586 0205
F: 202 586 0784

Dr. Yakov Eckstein
Department of Physics
Kiryat Technion
Haifa 32000
ISRAEL
P: 972 4 822 7401
F: 972 422 221514
e-mail: phrecks@
physics.technion.ac.il

Dr. Demetre Economou
Texas Center for
Superconductivity
Department of Chemical
Engineering
University of Houston
Houston TX 77204-4792
P: 713 743 4320
F: 713 743 4323
e-mail: economou@uh.edu

Prof. Don Elthon
Texas Center for
Superconductivity
Department of Chemistry
University of Houston
Houston TX 77204-4792
P: 713 743 8282
F: 713 743 8281
e-mail: Elthon@uh.edu

Dr. Victor J. Emery
Senior Scientist
Brookhaven National
Laboratory
Department of Physics
20 Pennsylvania St.
Upton NY 11973-5000
P: 516 282 3765
F: 516 282 2918
e-mail: emery@bnl.gov

Dr. Dean W. Face
DuPont Superconductivity
Experimental Station E304/
C118
Wilmington DE 19880-0304
P: 302 695 9227
F: 302 695 2721
e-mail: faced@
esvax.dnet.dupont.com

Mr. Robert Fairhurst
Texas Center for
Superconductivity
Department of Chemistry
University of Houston
4800 Calhoun
Houston TX 77204-5641
P: 713 743 8288
F: 713 743 8281
e-mail: Fairhurst@
elthon.tcs.uh.edu

Dr. J. D. Fan
Department of Physics
Southern University and A&M
P. O. Box 10554
128 W. James Hall
Baton Rouge LA 70813
P: 504 771 3926
F: 504 771 3926
e-mail: phjola@
lsuvax.sncc.lsu.edu or
fan@newton.engr.subr.edu

Dr. Nongqiang Fan
Texas Center for
Superconductivity
University of Houston
Houston TX 77204-5932
P: 713 743 8267
F: 713 743 8201
e-mail: Fan@miller.tcs.uh.edu

Dr. Hung-Hsu Feng
Texas Center for
Superconductivity
University of Houston
Houston TX 77204-5932
P: 713 743 8300
F: 713 743 8301

Dr. Alvaro Ferraz
International Center of
Condensed Matter Physics-
ICCMP
Universidade de Brasilia,
Campus Universitario
Caixa Postal 04667
70919-970 DF
BRASIL
P: 55 61 348 2572
F: 55 61 273 3884
e-mail: ferraz@iccmp.br or
ferraz@solids.iccmp.br

Dr. Michael B. Field
Applied Superconductivity
Center
University of Wisconsin-
Madison
1500 Engineering Drive
Madison WI 53706
P: 608 263 4462
F: 608 263 1087
e-mail: field@cae.wisc.edu

Prof. Øystein Fischer
Section de Physique
Université de Genève
D. P. M. C.
24, quai Ernest Ansermet
CH-1211 Genève 4
SWITZERLAND
P: 41 22 702 6270
F: 41 22 781 2192
e-mail: fischer@sc2a.unige.ch

Mr. Brandon Fisher
Engineering Assistant
Argonne National Laboratory
9700 S. Cass Ave.
Argonne IL 60439
P: 708 252 5029
F: 708 252 3604
e-mail: b45313@
achilles.ctd.anl.gov

Dr. Zachary Fisk
National High Magnetic Field
Laboratory
Florida State University
1800 E. Paul Dirac Drive
Tallahassee FL 32306-3016
P: 904 644 2922
F: 904 644 5038
e-mail: fisk@magnet.fsu.edu

Dr. Marcel Franz
Department of Physics &
Astronomy
McMaster University
1280 Main Street W
Hamilton, Ontario L85 4M1
CANADA
P: 905 525 9140, ext. 23182
F: 905 521 2773
e-mail: franz@
physun.mcmaster.ca

Dr. Arthur J. Freeman
Physics Department
Northwestern University
Evanston IL 60208
P: 708 491 3343
F: 708 491 5082
e-mail: art@
freeman.phys.nwu.edu

Dr. David Frenkel
Texas Center for
Superconductivity
University of Houston
Houston TX 77204-5932
P: 713 743 8276
F: 713 743 8201
e-mail: frenkel@ox.tcs.uh.edu

Prof. Herbert C. Freyhardt
LLM/KL Institut für
Metallphysik
University of Göttingen
Windausweg 2
37073 Göttingen
GERMANY
P: 49 551 394492
F: 49 551 60717 50
e-mail: freyzfw@
umpsun1.gwdg.de

Dr. Yoshito Fukumoto
Brookhaven National
Laboratory
Upton NY 11973
P: 516 344 3954
F: 516 344 4071

Prof. Marc Gabay
Universite Paris-Sue, Bât 510
Laboratoire de Physique de
Solides
Orsay 91405
FRANCE
P: 33 1 69 41 69 26
F: 33 1 69 41 60 86
e-mail: arc@solrt.lps.u-psud.fr

Mr. J. R. Gaines, Jr.
Vice President
Superconducting Components,
Inc.
1145 Chesapeake Avenue
Columbus OH 43212
P: 614 486 0261
F: 614 486 0912
e-mail: JRGSCI@AOL.COM

Dr. Peter L. Gammel
Bell Labs
Lucent Technologies
Room ID-221
700 Mountain Avenue
Murray Hill NJ 07974
P: 908 582 2611
F: 908 582 3260
e-mail: plg@physics.att.com

Dr. Li Gao
Texas Center for
Superconductivity
University of Houston
Houston TX 77204-5932
P: 713 743 8309
F: 713 743 8201
e-mail: lgao@uh.edu

Prof. Theodore H. Geballe
Professor Emeritus
Department of Applied Physics
Stanford University
Ginzton Laboratory
via Palou
Stanford CA 94305-4085
P: 415 723 0215
F: 415 725 2189
e-mail: geballe@
loki.stanford.edu

Mr. Joel Geny
Texas Center for
Superconductivity
Department of Chemistry
University of Houston
Houston TX 77204-5932
P: 713 743 8288
F: 713 743 8281
e-mail: Geny@tem.tcs.uh.edu

Dr. Piotr Gierlowski
Texas Center for
Superconductivity
University of Houston
4800 Calhoun
Houston TX 77204-5932
P: 713 743 8239
F: 713 743 8201
e-mail: Piotr@
menudo.uh.edu

Dr. Junho Gohng
New Materials Laboratory
Samsung Advanced Institute of Technology
P. O. Box 111
Suwon 440-600
KOREA
P: 82 331 280 9341
F: 82 331 280 9349
e-mail: gohng@
saitgw.sait.samsung.co.kr

Ms. Oya Gokcen
Texas Center for
Superconductivity
Department of Chemisty
University of Houston
4800 Calhoun
Houston TX 77204-5641
P: 713 743 8288
F: 713 743 8281
e-mail: ogokcen@
bayou.uh.edu

Dr. Allen M. Goldman
School of Physics and
Astronomy
University of Minnesota
116 Church Street, SE
Minneapolis MN 55455
P: 612 624 6525
F: 612 626 8029
e-mail: goldman@
physics.spa.umn.edu

Prof. Robert J. Gooding
Department of Physics
Queen's University
1 Queen's Cres.
Kingston, Ontario K7L 3N6
CANADA
P: 613 545 2696
F: 613 545 6463
e-mail: bob@
cezanne.phy.queensu.ca

Prof. Lev P. Gor'kov
National High Magnetic Field
Laboratory
Florida State University
1800 E. Paul Dirac Drive
Tallahassee FL 32306
P: 904 644 4187
F: 904 644 5038
e-mail: gorkov@
magnet.fsu.edu

Dr. Paul M. Grant
Electric Power Research
Institute
3412 Hillview Avenue
P. O. Box 10412
Palo Alto CA 94303
P: 415 855 2234
F: 415 855 2287
e-mail: pgrant@
eprinet.epri.com

Dr. Richard L. Greene
Department of Physics
University of Maryland
College Park MD 20742
P: 301 405 6128
F: 301 314 3779
e-mail: rgreene@
umdd.umd.edu

Prof. R. Gross
Lehrstuhl Experimentalphysik II
University of Tübingen
Morgenstelle 14
D-72076 Tübingen
GERMANY
P: +49 7071 296318
F: +49 7071 296322

Dr. W. Y. Guan
Department of Physics
Tamkang University
Tamsui
Taiwan
REPUBLIC OF CHINA
P: 886 35 710461
F: 886 35 723052

Dr. Donald U. Gubser
Superintendent, Materials
Science and Technology
Division
Naval Research Laboratory
4555 Overlook Drive, SW
Code 6300
Washington DC 20375-5343
P: 202 767 2926
F: 202 404 8009
e-mail: gubser@
anvil.nrl.navy.mil

Dr. Arnold Guloy
Texas Center for
Superconductivity
Department of Chemistry
University of Houston
4800 Calhoun
Houston TX 77204-5641
P: 713 743 2792
F: 713 743 2787
e-mail: aguloy@uh.edu

Dr. John E. Hack
Director, Research & Product
Development
Midwest Superconductivity, Inc.
1315 Wakarusa Drive
Lawrence KN 66049
P: 913 749 3613
F: 913 749 0738
e-mail: JEHACK@aol.com

Dr. Jeff Hahn
NREL-Golden Field Office
1617 Cole Boulevard
Golden CO 80401
P: 303 275 4775
F: 303 275 4753
e-mail: Hahnj@
TCPLINK.NERL.GOV

Dr. Alejandro Hamed
Texas Center for
Superconductivity
University of Houston
Houston TX 77204-5932
P: 713 743 8300
F: 713 743 8301

Prof. Walter N. Hardy
Department of Physics
University of British Columbia
6224 Agricultural Road
Vancouver, BC V6T 171
CANADA
P: 604 822 6341
F: 604 822 5324
e-mail: walter@
cryos.physics.ubc.ca

Dr. Robert A. Hawsey
Manager, Superconductivity
Program for Electric Power
Systems
Superconducting Technology
Program
Oak Ridge National Laboratory
P. O. Box 2008
Oak Ridge TN 37831-6040
P: 423 574 8057
F: 423 574 6073
e-mail: hawseyra@ornl.gov or
RAV@ORNL.gov

Ms. Xiaopei (Sherry) He
Texas Center for
Superconductivity
Department of Electrical
Engineering
University of Houston
4800 Calhoun
Houston TX 77204-4792
P: 713 743 4400
F: 713 743 4444
e-mail: HXL0631@
jetson.uh.edu

Dr. Patrick F. Henning
Department of Physics
Florida State University
Keen Building, 1800 E. Paul
Dirac Dr.
Tallahassee FL 32306
P: 904 644 6913
F: 904 644 9462
e-mail: henning@
magnet.fsu.edu

Prof. Allen M. Hermann
Superconductivity Laboratories
Department of Physics
Room E-032 Duane
University of Colorado at
Boulder
Campus Box 390
Boulder CO 90309-0390
P: 303 492 0744
F: 303 492 2998
e-mail: allen.hermann@
colorado.edu

Mr. Brian Hickey
Texas Center for
Superconductivity
University of Houston
Houston TX 77204-5932
P: 713 743 8305
F: 713 743 8201
e-mail: hickey@
pebbles.tcs.uh.edu

Mr. James Hoehn
Intermagnetics General
Corporation
450 Old Niskayuna Road
Latham NY 12110-0461
P: 518 782 1122
F: 518 783 2615

Dr. Pei-Herng Hor
Texas Center for
Superconductivity
University of Houston
Houston TX 77204-5932
P: 713 743 8300
F: 713 743 8301
e-mail: PHor@uh.edu

Mr. Michel Houssa
Institut de Physique B5
Université de Liège
B-4000 Liège
BELGIUM
P: 32 41 66 37 03
F: 32 41 66 29 90
e-mail: houssa@
gw.unipc.ulg.ac.be

Prof. Chia-Ren Hu
Department of Physics
Texas A&M University
College Station TX 77843-4242
P: 409 845 3531
F: 409 845 2590
e-mail: hu@phys.tamu.edu

Prof. Chao-Yuan Huang
Director
Center for Condensed Matter
Sciences
National Taiwan University
1, Roosevelt Road, Section 4
Taipei 10764
Taiwan
REPUBLIC OF CHINA
P: 011 886 2 365 5403
F: 011 886 2 363 9984
e-mail: cyhuang@
ccms.ntu.edu.tw or
cyhuang@phys.ntu.edu.tw

Dr. Zhijun Huang
Texas Center for
Superconductivity
University of Houston
Houston TX 77204-5932
P: 713 743 8314
F: 713 743 8201
e-mail: Peter@
bambam.tcs.uh.edu

Dr. Nigel E. Hussey
IRC in Superconductivity
University of Cambridge
Madingley Road
Cabridge CB3 OHE
UNITED KINGDOM
P: 0 1223 337443
F: 0 1223 337074
e-mail: neh1000@
hermes.cam.ac.uk

Dr. Alex Ignatiev
Texas Center for
Superconductivity
Space Vacuum Epitaxy Center
University of Houston
4800 Calhoun
Houston TX 77204-5507
P: 713 743 3621
F: 713 747 7724
e-mail: Ignatiev@uh.edu

Dr. Hiroshi Ikuta
Department of Applied
Chemistry
University of Tokyo
Hongo 7-3-1
Bunkyo-ku, Tokyo 113
JAPAN
P: 81 3 3812 2111, ext. 7782
F: 81 3 5689 0574
e-mail: h-ikuta@
tansei.cc.u-tokyo.ac.jp

Prof. Allan J. Jacobson
Texas Center for
Superconductivity
Department of Chemistry
University of Houston
4800 Calhoun
Houston TX 77204-5641
P: 713 743 2785
F: 713 743 2787
e-mail: ajjacob@uh.edu

Mr. Zhiyong (John) Jiang
Texas Center for
Superconductivity
Department of Electrical
Engineering
University of Houston
4800 Calhoun
Houston TX 77204-4792
P: 713 743 4400
F: 713 743 4444
e-mail: ZYJiang@uh.edu

Mr. Cheng Jiao
Texas Center for
Superconductivity
University of Houston
Houston TX 77204-5932
P: 713 743 8239
F: 713 743 8201

Mr. Peng Jin
Texas Center for
Superconductivity
Department of Physics
University of Houston
Houston TX 77204-5932
P: 713 743 8307
F: 713 743 8221
e-mail: PXJ41610@
jetson.uh.edu

Dr. David C. Johnston
Department of Physics and
Astronomy
Iowa State Univesity
Ames IA 50011
P: 515 294 5435
F: 515 294 0689
e-mail: johnston@
ameslab.gov

Dr. James D. Jorgensen
Materials Science Division and
Science and Technology Center
for Superconductivity
Argonne National Laboratory
Building 223, Room D-221
Argonne IL 60439
P: 708 252 5513
F: 708 252 7777
e-mail: jim_jorgensen@
qmgate.anl.gov

Dr. Alain R. Junod
D. P. M. C., Institute de
Physique
University of Genève
24 Quai Ernest-Ansermet
CH-1211 Genève 4
SWITZERLAND
P: 41 22 702 62 04, 702 6308
F: 41 22 702 6869
e-mail: JUNOD@
SC2A.UNIGE.CH

Prof. Alpo J. Kallio
Department of Physical
Sciences
University of Oulu
Linnanmaa, P.O. Box 400
Oulu, FIN-90571
FINLAND
P: 358 81 553 1882
F: 358 81 553 1884
e-mail: toi@jussi.oulu.fi

Dr. Aharon Kapitulnik
Department of Applied Physics
Edward L. Ginzton Laboratory
Stanford University
Stanford CA 94305-4085
P: 415 723 3847
F: 415 725 2189
e-mail: AK@
LOKI.STANFORD.EDU

Dr. Janusz Karpinski
Solid State Physics Laboratory
ETH Hönggerberg
Zürich 8093
SWITZERLAND
P: 41 01 633 22 54
F: 41 01 633 10 72
e-mail: karpinski@
solid.phys.ethz.ch

Mr. Chris Kinalidis
Texas Center for
Superconductivity
University of Houston
Houston, Texas 77204-5932
P: 713 743 8308
F: 713 743 8201

Prof. Tony King
Texas Center for
Superconductivity
Department of Electrical &
Computer Engineering
University of Houston
4800 Calhoun
Houston TX 77204-4793
P: 713 743 4423
F: 713 743 4444
e-mail: TKing@uh.edu

Dr. John R. Kirtley
IBM T. J. Watson Research
Center
P. O. Box 218
Yorktown Heights NY 10598
P: 914 945 2043
F: 914 945 2536
e-mail: kirtley@
watson.ibm.com

Prof. Koichi Kitazawa
Department of
Superconductivity
University of Tokyo
7-3-1 Hongo, Tokyo 113
JAPAN
P: 81 3 3812 2111, ext. 7201
F: 81 3 3815 5632
e-mail: supercom@
tansei.cc.u-tokyo.ac.jp

Dr. Miles Klein
NSF Science & Technology
Center for Superconductivity
University of Illinois
1110 West Green Street
Urbana IL 61501
P: 217 333 1744
F: 217 244 2278
e-mail: miles_klein@
stcs.mrl.uiuc.edu

Dr. Hiroshi Kohno
Department of Physics
University of Tokyo
7-3-1 Hongo, Bunkyo-ku
Tokyo 113
JAPAN
P: 81 3 3812 2111, ext. 4185
F: 81 3 5800 6791
e-mail: skohno@
hongo.ecc.u-tokyo.ac.jp

Prof. Tamotsu Koyano
Cryogenics Center
University of Tsukuba
Tsukuba
Ibaraki 305
JAPAN
e-mail: Koyano@
bukko.bk.tsukuba.ac.jp

Dr. Vladimir Z. Kresin
Lawrence Berkeley National
Laboratory
M/S 62-203
1 Cyclotron Road
Berkeley CA 94720
P: 510 486 6991
F: 510 486 5401
e-mail: vzkresin@lbl.gov

Dr. Kumar Krishen
Chief Technologist
NASA/Johnson Space Center
2101 NASA Road 1
Houston TX 77058
P: 713 483 0695
F: 713 244 8452
e-mail: KKrishen@
JP101.JSC.NASA.GOV

Dr. Don M. Kroeger
Metal and Ceramics Division
Oak Ridge National Laboratory
Bldg. 4500 South-MS6116
P. O. Box 2008
Oak Ridge TN 37831-6116
P: 423 574 5155
F: 423 574 7659
e-mail: okg@ornl.gov

Dr. H. C. Ku
Department of Physics
National Tsing Hua University
Hsinchu 30042
Taiwan
REPUBLIC OF CHINA
P: 011 886 35 71 59 28
F: 011 886 35 72 30 52 or 72 72 94

Dr. Joseph Kulik
Materials Characterization
Facility
Texas Center for
Superconductivity
University of Houston
Houston TX 77204-5932
P: 713 743 8283
F: 713 743 8281
e-mail: JKulik@uh.edu

Dr. Pradeep Kumar
National Science Foundation
4201 Wilson Boulevard
Arlington, VA 22230
P: 703 306 1997
F: 703 306 0902
e-mail: pkumar@nsf.gov

Mr. Mark Lamb
Texas Center for
Superconductivity
University of Houston
Houston, Texas 77204-5932
P: 713 743 8258
F: 713 743 8201

Dr. Michael T. Lanagan
Argonne National Laboratory
9700 S. Cass Avenue
Argonne IL 60439
P: 708 252 4251
F: 708 252 3604
e-mail: M_Lanagan@
qmgate.anl.gov

Dr. Neal Lane
Director
National Science Foundation
4201 Wilson Boulevard, Suite 1245
Arlington VA 22230
P: 703 306 1070
F: 703 306 0159

Prof. David C. Larbalestier
University of Wisconsin-Madison
1500 Engineering Drive, 909 Eng. Research Building
Madison WI 53706
P: 608 263 2194
F: 608 263 1087
e-mail: larbales@engr.wisc.edu

Mr. Michael Larkin
Columbia University
Department of Physics
538 W. 120th
New York NY 10027
P: 212 854 5675
F: 212 854 5888
e-mail: mlarkin@phys.columbia.edu

Dr. Alan Lauder
General Manager
DuPont Superconductivity
Experimental Station E304/114
P. O. Box 80304, Route 141
Wilmington DE 19880-0304
P: 302 695 9230
F: 302 695 2721
e-mail: lauder@a1.esvax.umc.dupont.com

Dr. Robert B. Laughlin
Department of Physics
Stanford University
Stanford CA 94305
P: 415 723 4563
F: 415 725 6544
e-mail: rbl@large.stanford.edu

Ms. Jeong Soon Lee
Samsung Advanced Institute of Technology
P. O. Box 111
Suwon 440-600
KOREA
P: 82 331 9178
F: 82 331 9158
e-mail: leejs@saitgw.samsung.co.kr

Prof. Sung-Ik Lee
Department of Physics
Pohang University of Science & Technology
Pohang 790-784
SOUTH KOREA
P: 82 562 279 2073
F: 82 562 279 6988
e-mail: silee@vision.postech.ac.kr

Prof. T. K. Lee
Department of Physics
Virginia Tech
Blacksburg VA 24060
P: 540 231 8998
F: 540 231 751
e-mail: tklee@tklee1.phys.vt.edu

Prof. Anthony J. Leggett
Department of Physics
University of Illinois at Urbana-Champaign
307 Loomis Laboratory, 1100 West Green St.
Urbana IL 61801
P: 217 333 2077
F: 217 333 9819
e-mail: tony@cromwell.physics.uiuc.edu

Dr. Eddie Leung
Lockheed Martin Corporation
15250 Avenue of Science
San Diego CA 92128
P: 619 673 6584
F: 619 673 6500
e-mail: leung@mmcado.com

Prof. Pakwo Leung
Department of Physics
Hong Kong University of Science & Technology
Clear Water Bay
HONG KONG
P: 852 23587438
F: 852 23581652
e-mail: phleung@usthk.ust.hk

Prof. Kathryn Levin
The James Franck Institute
The University of Chicago
5640 South Ellis Ave.
Chicago IL 60637
P: 312 702 7186
F: 312 702 5863
e-mail: levin@control.uchicago.edu

Dr. Ting-wei Li
Kammerlingh Onnes Laboratory
Leiden University
Nieuwsteeg 18
P. O. Box 9506
2300 RA Leiden
THE NETHERLANDS
P: 31 71 5275626
F: 31 71 5275404
e-mail: twli@rulkol.leidenuniv.nl

Prof. Wei-Yao Liang
Director
Interdisciplinary Centre in Superconductivity
University of Cambridge
Madingley Road
Cambridge, CB3 OHE
UNITED KINGDOM
P: 011 44 1223- 33 70 77
F: 011 44 1223- 33 70 74
e-mail: wyl1@hermes.cam.ac.uk

Prof. Konstantin K. Likharev
Department of Physics
SUNY at Stony Brook
Stony Brook NY 11794-3800
P: 516 632 8159
F: 516 632 8774
e-mail: klikharev@ccmail.sunysb.edu

Dr. C. L. Lin
Materials Science Center
National Tsing Hua University
Hsinchu 30042
Taiwan
REPUBLIC OF CHINA
P: 001 886 35 715131, ext. 5381
F: 001 886 35 713113
e-mail: cll@phys.nthu.edu.tw

Dr. Jauyn Grace Lin
Department of Physics
National Taiwan University
1, Roosevelt Road, Section 4
Taipei, Taiwan 10764
REPUBLIC OF CHINA
P: 1-886-2-3630231, ext. 2871-140
F: 1-886-2-3655404
e-mail: JGLin@phys.ntu.edu.tw

Mr. Jiangtao Lin
Texas Center for Superconductivity
University of Houston
Houston TX 77204-5932
P: 713 743 8272
F: 713 743 8281
e-mail: Phys7s@menudo.uh.edu

Mr. Qiuming Lin
Texas Center for Superconductivity
University of Houston
Houston TX 77204-5932
P: 713 743 8307
F: 713 743 8221
e-mail: TCSUHZ@menudo.uh.edu

Dr. Dah-Chin Ling
National Tsing Hua University
Materials Science Center
Hsinchu 30042
Taiwan
REPUBLIC OF CHINA
P: 886 35 715131, ext. 3815
F: 886 35 713113
e-mail: dcling@msc.nthu.edu.tw

Dr. Jia Liu
Texas Center for
Superconductivity
Institute for Beam Particle
Dynamics
University of Houston
Houston TX 77204-5506
P: 713 743 3600
F: 713 747 4526

Dr. Jia-Rui Liu
Texas Center for
Superconductivity
University of Houston
Houston TX 77204-5932
P: 713 743 8255
F: 713 743 8201
e-mail: JRLiu@uh.edu

Mr. Ingo Loa
Technical University of Berlin
Seler. PN 5-4, Hardenbergstr. 36
D-10623 Berlin
GERMANY
P: 49 30 314 23468
F: 49 30 314 27705
e-mail: INGOR@
MAIL.PHYSIK.TU-BERLIN.DE

Dr. Charles Lockerby
American Superconductor
Corporation
Manager, Government
Programs
2 Technology Drive
Westborough MA 01581
P: 508 836 4200
F: 508 836 4242
e-mail: CLockerb@
asc.MHScompuserve.com

Dr. Daniel Lopez
Centro Atómico Bariloche and
Instituto Balseiro
Av. Bustillo KM 9
S. C. de Bariloche, RN 8400
ARGENTINA
P: 00 944 61002
F: 00 54 944 61006
e-mail: lopez@cab.cnea.edu

Dr. John Loram
Interdisciplinary Research
Center in Superconductivity
University of Cambridge
Madingley Road
Cambridge CB3 OHE
ENGLAND
P: 44 1223 337 445
F: 44 1223 337 074
e-mail: jwl15@
hermes.cam.ac.uk

Dr. Bernd Lorenz
Research Scientist
Texas Center for
Superconductivity
University of Houston
Houston TX 77204-5932
P: 713 743 8300
F: 713 743 8301

Mr. Qingyou Lu
Department of Physics
University of Texas at Austin
Austin TX 78712
P: 512 471 5544
F: 512 471 9637
e-mail: U235@
utpapa.utexas.edu

Prof. Alan Luther
Nordita
Blegdamsvej 17
DK-2100 Copenhagen
DENMARK
P: 45 35 325 221
F: 45 31 389 389 157
e-mail: luther@nordita.dk

Dr. Ki Ma
Texas Center for
Superconductivity
University of Houston
Houston TX 77204-5932
P: 713 743 8254
F: 713 743 8201
e-mail: kma@uh.edu

Dr. Yoshiteru Maeno
Department of Physics, Faculty
of Science
Hiroshima University
1-3-1 Kaga Miyama
Higashi-Hiroshima 739
JAPAN
P: 81 824 24 7367
F: 81 824 24 0716
e-mail: maeno@
butsuri.sci.hiorsha-u.ac.jp

Dr. Kazumi Maki
Department of Physics and
Astronomy
University of Southern
California
Los Angeles CA 90089-0484
P: 213 740 8405
F: 213 740 6653
e-mail: kmaki@usc.edu

Dr. Martin P. Maley
Team Leader
Los Alamos National
Laboratory/Superconductivity
Technology Center
P. O. Box 1663, MS K763
Los Alamos NM 87545
P: 505 665 0189
F: 505 665 3164
e-mail: marty.maley@lanl.gov

Dr. Alexis P. Malozemoff
Chief Technical Officer
American Superconductor
Corporation
2 Technology Drive
Westborough MA 01581-1727
P: 508 836 4200
F: 508 836 4248
e-mail: amalozem@
asc.mhs.compuserve.com

Dr. Y. M. Malozovsky
Department of Physics
Southern University and A&M
College
128 W. James Hall
Baton Rouge LA 70813
P: 504 771 3926
F: 504 771 3926
e-mail: yuriy@
newton.engr.subr.edu

Prof. A. G. Mamalis
Department of Mechanical
Engineering
National Technical University
of Athens
42, 28th October Avenue
10682 Athens
GREECE
P: 301 3811 988
F: 301 3813 897
e-mail: mamalis@
naxos.esd.ece.ntua.gr

Dr. Jochen Mannhart
IBM Research Division
Zürich Research Laboratory
Saumerstrasse 4
CH-8803 Rüschlikon
SWITZERLAND
P: 41 1 724 8391
F: 41 1 724 0084
e-mail: MHA@
zurich.IBM.COM

Dr. M. Brian Maple
Bernd T. Matthias Professor of
Physics
University of California at San
Diego
Department of Physics, 0319
9500 Gilman Drive
La Jolla CA 92093-0319
P: 619 534 3969
F: 619 534 1241
e-mail: mbmaple@ucsd.edu

Mr. F. Marciano
Pirelli Cables North America
Market Development Manager,
Superconductivity
710 Industrial Drive
Lexington SC 29072
P: 803 951 4000
F: 803 957 8654

Dr. Massimo Marezio
Senior Scientist
MASPEC-CMR
via Chiavari 18/A
Parma 43100
ITALY
P: 39 521 2691
F: 39 521 269206

Prof. John T. Markert
Department of Physics
University of Texas at Austin
Austin TX 78712
P: 512 471 1039
F: 512 471 9637
e-mail: markert@
physics.utexas.edu

Dr. Ian McCallum
Department of Energy
(Energetics)
7164 Gateway Drive
Columbia MD 20146
P: 410 290 0370
F: 410 290 0377
e-mail: ian_mccallum@
energetics.com

Dr. J. T. McDevitt
Department of Chemistry and
Biochemistry
University of Texas at Austin
Austin TX 78712
P: 512 471 0046
F: 512 471 8696
e-mail: mcdevitt@
huckel.cm.utexas.edu

Ms. Julia McNair
Naval Research Laboratory
4555 Overlook Avnue, SW
Washington DC 20375 5322
P: 202 767 2541
F: 202 767 6991
e-mail: mcnair@
ccf.nrl.navy.mil

Prof. James Meen
Texas Center for
Superconductivity
University of Houston
Houston TX 77204-5932
P: 713 743 8289
F: 713 743 8281
e-mail: JMeen@uh.edu

Ms. Ruling Meng
Texas Center for
Superconductivity
University of Houston
Houston TX 77204-5932
P: 713 743 8306
F: 713 743 8221
e-mail: ruruling@uh.edu

Dr. Tomaz Mertelj
Faculty of Mathematics and
Physics
University of Ljubljana
Jaamova 39/P. O. Box 100
61111 Ljubljana
SLOVENIA
P: 386 61 17 73 900
F: 386 61 21 93 85
e-mail: tomaz.mertelj@eijs.si

Dr. John H. Miller, Jr.
Texas Center for
Superconductivity
University of Houston
Houston TX 77204-5932
P: 713 743 8257
F: 713 743 8201
e-mail: JHMiller@uh.edu

Dr. Maria Mironova
Texas Center for
Superconductivity
University of Houston
Houston TX 77204-4792
P: 713 743 8201
F: 713 743 8221
e-mail: mkm3797@
jetson.uh.edu

Prof. Akira Miyase
Texas Center for
Superconductivity
Department of Mechanical
Engineering
University of Houston
Houston TX 77204-4792
P: 713 743 4537
F: 713 743 4503

Dr. Kathryn A. Moler
Department of Physics
Princeton University
Jadwin Hall
Princeton NJ 08544-0708
P: 609 258 5928
F: 609 258 6360
e-mail: kam@
pupgg.princeton.edu

Dr. Herbert A. Mook, Jr.
Solid State Division
Oak Ridge National Laboratory
P. O. Box 2008
Oak Ridge TN 37831-6393
P: 423 574 5242
F: 423 574 6268
e-mail: ham@ornl.gov

Prof. Simon Moss
Department of Physics
Texas Center for
Superconductivity
University of Houston
Houston TX 77204-5506
P: 713 743 3539
F: 713 743 3550
e-mail: SMoss@uh.edu

Dr. Joel Muehlhauser
University of Tennessee Space
Institute
B. H. Goethert Parkway
Tullahoma TN 37388
P: 615 393 7286
F: 615 455 7266
e-mail: jmuehlha@utsi.edu

Prof. Karl Alex Müller
IBM Research Division
Zürich Research Laboratory
Säumerstrasse 4
CH-8803 Rüschlikon
SWITZERLAND
P: 41 1 724 8111 or 8238
F: 41 1 724 0724
e-mail: man@
zurich.ibm.com

Dr. Kirsten E. Myers
DuPont Superconductivity
Experimental Station E304/
C118
Wilmington DE 19880-0304
P: 302 695 3357
F: 302 695 2721
e-mail: myerske@
esvax.dnet.dupont.com

Prof. Naoto Nagaosa
Department of Applied Physics
University of Tokyo
7-3-1 Hongo, Bunkyo-ku
Tokyo 113
JAPAN
P: 81 3 3812 2111, ext. 6811
F: 81 3 3816 7805
e-mail: nagaosa@
tansei.cc.u-tokyo.ac.jp

Mr. Ben Nashumi
Columbia University
538 W. 100th Street
New York NY 10027
P: 212 854 5675
F: 212 854 5888
e-mail: scylla@
susol.phys.columbia.edu

Prof. Don Naugle
Department of Physics
Texas A&M University
College Station TX 77843-4242
P: 409 845 4429
F: 409 845 2590
e-mail: naugle@
phys.tamu.edu

Dr. Gladys L. Nieva
Centro Atómico Bariloche
8400 Bariloche RN
Rio Negro
ARGENTINA
P: 54 944 45171
F: 54 944 61006
e-mail: gnieva@
cab.cnea.edu.ar

Prof. V. I. Nikitenko
Laboratory Head
Institute for Solid State Physics
Russian Academy of Sciences
142432 Chernogolovka
Moscow District
RUSSIA
P: 7 095 913 2324
F: 7 095 576 4111
e-mail: nikiten@issp.ac.ru

Dr. Rick O'Neal
American Superconductor Corporation
2 Technology Drive
Westborough MA 01581
P: 508 836 4200
F: 508 836 4242

Dr. Michael Osofsky
Naval Research Laboratory
Code 6344
4555 Overlook Avenue, SW
Washington DC 20375-5343
P: 202 767 6149
F: 202 767 1697
e-mail: osofsky@anvil.nrl.navy.mil

Mr. Apurva Parikh
TCSUH HTS Manufacturing Division
Texas Center for Superconductivity
University of Houston
Houston TX 77204-5932
P: 713 743 4547
F: 713 743 4513
e-mail: apurva@uh.edu

Dr. Y. W. Park
Department of Physics
Seoul National University
Seoul 151-472
KOREA
P: 82 2 880 6607
F: 82 2 2873 7037
e-mail: ywpark@alliant.snu.ac.kr

Dr. Peter Penfold
Oxford Instruments, North America
130A Baker Avenue
Concord MA 01742
P: 508 369 9933
F: 508 369 6616

Dr. Dean E. Peterson
Superconducting Technology Center
Los Alamos National Laboratory
P. O. Box 1663, MS-K763
Los Alamos NM 87545
P: 505 665 3030
F: 505 665 3164
e-mail: dpeterson@lanl.gov

Mr. Duc Pham
Texas Center for Superconductivity
University of Houston
Houston, Texas 77204-5932
P: 713 743 8308
F: 713 743 8201

Prof. David Pines
Research Professor of Physics
Department of Physics
University of Illinois at Urbana-Champaign
1110 West Green Street
Urbana IL 61801-3080
P: 217 333 0115
F: 217 244 7559
e-mail: maryo@uiuc.edu

Dr. A. A. Polyanskii
Institute of Solid State Physics
Russian Academy of Sciences
Chernogolovka, Moscow 53706
RUSSIA
e-mail: polyansk@coefac.engr.wisc.edu

Dr. Deva Ponnusamy
Texas Center for Superconductivity
Department of Mechanical Engineering
University of Houston
Houston TX 77204-4792
P: 713 743 4548
F: 713 743 4513
e-mail: DPonnusamy@jetson.uh.edu

Dr. William Putikka
Department of Physics
University of Cincinnati
ML-0011
Cincinnati OH 45221-0011
P: 513 556 0639
F: 513 556 3425
e-mail: putikka@physunc.phy.uc.edu

Mr. Philip Putman
Texas Center for Superconductivity
Department of Mechanical Engineering
University of Houston
Houston TX 77204-4792
P: 713 743 4547
F: 713 743 4513
e-mail: ST7LP@jetson.uh.edu

Mr. Xiaodong Qiu
Texas Center for Superconductivity
University of Houston
Houston TX 77204-5932
P: 713 743 8305
F: 713 743 8221
e-mail: XQiu@uh.edu

Mr. Jhijun (James) Qu
Texas Center for Superconductivity
University of Houston
Houston TX 77204-5932
P: 713 743 8259
F: 713 743 8201
e-mail: TCSUHY@jetson.uh.edu

Mr. M. M. Rahman
Pirelli Cables North America
VP Chief Engineer
710 Industrial Drive
Lexington SC 29072
P: 803 951 4000
F: 803 957 8654

Mr. Diego Ramirez
Texas Center for Superconductivity
University of Houston
Houston, Texas 77204-5932
P: 713 743 8305
F: 713 743 8201

Dr. Udaya Rao
Department of Energy
Project Manager
P. O. Box 10940
Pittsburgh PA 15236
P: 412 892 4743
F: 412 892 4604
e-mail: RAO@PETC.DOE.GOV

Prof. Bernard Raveau
Laboratory of Crystallography and Material Sciences
University of Caen
ISMRA 6
Boulevard du Marechal Juin
Caen Cedex 104050
FRANCE
P: 33 31 95 26 17
F: 33 31 95 16 00
e-mail: raveau@crismat.ismra.fr

Prof. George Reiter
Texas Center for Superconductivity
Department of Physics
University of Houston
4800 Calhoun
Houston TX 77204-5506
P: 713 743 3527
F: 713 743 3589
e-mail: GReiter@uh.edu

Dr. Yanru Ren
Texas Center for Superconductivity
Institute for Beam Particle Dynamics
University of Houston
4800 Calhoun
Houston TX 77204-5506
P: 713 743 3600
F: 713 747 4526

Dr. Yong Ren
Texas Center for Superconductivity
University of Houston
Houston TX 77204-5932
P: 713 743 8234
F: 713 743 8201
e-mail: tcsuh11@menudo.uh.edu

Prof. Zhifeng Ren
Superconductive Materials
Laboratory
SUNY Buffalo
Box 835, NS&M Complex
Buffalo NY 14260-3000
P: 716 645 6800, ext. 2241
F: 716 645 6949
e-mail: zren@
ubvms.cc.buffalo.edu

Prof. T. Maurice Rice
Institute of Theoretical Physics
ETH-Hönggerberg
CH-8093 Zürich
SWITZERLAND
P: 41 1 633 2581
F: 41 1 633 1115
e-mail: rice@itp.phys.ethz.ch

Mr. Jason Ritchie
Department of Chemistry
University of Texas at Austin
Welch Hall, 26th & Speedway
Streets
Austin TX 78712-1167
P: 512 471 0042
F: 512 471 8696
e-mail: jaritchie@
mail.utexas.edu

Mr. Dwight Ritums
Texas Center for
Superconductivity
Space Vacuum Epitaxy Center
University of Houston
4800 Calhoun
Houston TX 77204-5507
P: 713 743 3621
F: 713 747 7724
e-mail: dritums@uh.edu

Dr. Carl Rosner
President and CEO
Intermagnetics General
Corporation
450 Old Niskayuna Road
P. O. Box 461
Latham NY 12110-0461
P: 518 782 1122
F: 518 783 2610

Dr. Kent Ross
Materials Characterization
Facility
Texas Center for
Superconductivity
University of Houston
Houston TX 77204-5932
P: 713 743 8284
F: 713 743 8281
e-mail: DKRoss@uh.edu

Mr. Michael Ross
Staff Communications
Specialist
IBM-Almaden Research Center
650 Harry Road
San Jose CA 95120
P: 408 927 2923
F: 408 927 3204
e-mail: ross@
almaden.ibm.com or
donna@almaden.ibm.com

Dr. Christophe Rossel
IBM Research Division
Zürich Research Laboratory
Saumerstrasse 4
CH-8803 Rüschlikon
SWITZERLAND
P: 41 1 724 8522
F: 41 1 724 0084
e-mail: rsl@zurich.ibm.com

Dr. John Rowell
John Rowell, Inc
102 Exeter
Berkeley Heights NJ 07422
P: 908 464 6994
F: 908 665 9589
e-mail: jmrberkhts@aol.com

Dr. Irene Rusakova
Texas Center for
Superconductivity
University of Houston
Houston TX 77204-5932
P: 713 743 8286
F: 713 743 8201
e-mail: Irene@
bambam.tcs.uh.edu

Dr. Kamel Salama
Texas Center for
Superconductivity
Department of Mechanical
Engineering
University of Houston
4800 Calhoun
Houston TX 77204-4792
P: 713 743 4514
F: 713 743 4513

Dr. Sherif Salib
Texas Center for
Superconductivity
University of Houston
4800 Calhoun
Houston TX 77204-4791
P: 713 743 4291
F: 713 743 4260

**Mr. Dalber Ruben
Sánchez Candela**
Centro Brasilero de Pesquisa
Fisicas
Rua Xavier Sigaud 150
22290-180 Rio de Janeiro CEP
BRASIL
P: 55 21 5410337, ext. 131
F: 55 21 5412047
e-mail: dalber@
novell.cat.cbpf.br

**Mr. Srivatsan
Sathyamurthy**
Texas Center for
Superconductivity
Department of Mechanical
Engineering
University of Houston
4800 Calhoun
Houston TX 77204-4792
P: 713 743 4547
F: 713 743 4513
e-mail: mece2hz@
menudo.uh.edu

Dr. Ken-ichi Sato
Manager
Osaka Research Laboratory
Sumitomo Electric Industries,
Ltd.
1-1-3, Shimaya 1-chome
Konohana-ku
Osaka, 554
JAPAN
P: 81 6 466 5633
F: 81 6 466 5704
e-mail: sato@
okk.sumiden.co.jp

Mr. Ravi-Presad Sawh
Texas Center for
Superconductivity
Institute for Beam Particle
Dynamics
University of Houston
4800 Calhoun
Houston TX 77204-5506
P: 713 743 3600
F: 713 747 4526

Prof. Douglas Scalapino
Department of Physics
University of California at
Santa Barbara
Santa Barbara CA 93106-9530
P: 805 893 2871
F: 805 893 8838
e-mail: leslie@
spock.physics.edu

Mr. Darren Scarfe
Texas Center for
Superconductivity
Department of Chemistry
University of Houston
4800 Calhoun
Houston TX 77204-5641
P: 713 743 2785
F: 713 743 2787

Dr. Andreas Schilling
c/o N. E. Phillips
Department of Chemistry
University of California
Berkeley CA 94720-1460
P: 510 642 2971
F: 510 642 2835
e-mail: 102757.505@
compuserve.com

Dr. Lelia Schmirgeld-
Mignot
SRMP/DECM
C.E.A. Saclay
F91191 Gif-Sur-Yvette Cedex
FRANCE
P: 33 1 69 08 20 68
F: 33 1 69 08 68 67
e-mail: lelia@
srmp04.saclay.cea.fr

Dr. J. Robert Schrieffer
Chief Scientist
National High Magnetic Field
Lab
Florida State University
1800 E. Paul Dirac Drive
Tallahassee FL 32306-3016
P: 904 644 3203
F: 904 644 5038
e-mail: schrieff@
magnet.fsu.edu

Dr. Hansjörg Schwer
Laboratory of Solid State
Physics
ETH Hönggerberg
Zürich 8093
SWITZERLAND
P: 41 1633 2256
F: 41 1633 1072
e-mail: schwer@
solid.phys.ethe.ch

Dr. Venkat
Selvamanickam
High Temperature
Superconductors
Intermagnetics General
Corporation (IGC)
450 Old Niskayuna Road
P. O. Box 461
Latham NY 12110-0461
P: 518 782 1122
F: 518 782 2615

Ms. Karine Senes
Texas Center for
Superconductivity
Department of Chemisty
University of Houston
4800 Calhoun
Houston TX 77204-5641
P: 713 743 8288
F: 713 743 8280
e-mail: Karine@
tem.tcs.uh.edu

Dr. Suvankou Sengupta
Superconductive Components
Inc.
1145 Chesapeake Avenue
Columbus OH 43212
P: 614 486 0261
F: 614 486 0912
e-mail: ssengupta@aol.com

Dr. Robert F. Service
Science Magazine
1333 H Street, NW
Washington DC 20005
P: 202 326 7013
F: 202 371 9227
e-mail: rservice@aaas.org

Dr. Rajeshwar P. Sharma
University of Maryland
Center for Superconductivity
Research
Department of Physics
College Park MD 20742
P: 301 405 7674
F: 301 314 9541
e-mail: rps@squid.umd.edu

Prof. David T. Shaw
Director
New York State Institute on
Superconductivity
SUNY at Buffalo
330 Bonner Hall
Electrical and Computer
Engineering
Buffalo NY 14260
P: 716 645 3112
F: 716 645 3349
e-mail: DSHAW@
ENGR.BUFFALO.EDU

Prof. Zhi-xun Shen
C.M.R.
McCullough Building 232
Stanford University
Stanford CA 94305-4055
P: 415 725 8254
F: 415 725 5457
e-mail: shen@ee.stanford.edu

Mr. Jun-ichi Shimoyama
Faculty of Engineering
Department of Applied
Chemistry
University of Tokyo
Hongo 7-3-1, Bunkyo-ku
Tokyo 113
JAPAN
P: 81 3 3812 2111, ext. 7777
F: 81 3 5689 0574
e-mail: jras@
tansei.cc.u-tokyo.ac.jp

Dr. Yuh Shiohara
Director, Division 4 (Materials
Processing Division)
Superconductivity Research
Laboratory
ISTEC
1-10-13 Shinonome, Koto-ku
Tokyo 135
JAPAN
P: 81 8 3536 5710
F: 81 8 3536 5717

Dr. Shara S. Shoup
Chemical and Analytical
Sciences
Oak Ridge National Laboratory
Box 2008, MS-6110
Oak Ridge TN 37931-6110
P: 423 574 4418
F: 423 574 4961
e-mail: shoupss@ornl.gov

Prof. Qimiao Si
Department of Physics
Rice University
MS-61, 6100 Main Street
Houston TX 77251-1892
P: 713 285 5204
F: 713 527 9033
e-mail: qusi@kaitum.rice.edu

Dr. Enrico Silva
Dipartimento di Fisica-G20
Universita di Roma Tre
Via Della Vasca navale 84
00146 Roma
ITALY
P: 35 6 49913434
F: 35 6 4463158
e-mail: silva@roma1.infn.it

Dr. Randy Simon
Vice President
Conductus, Inc.
969 West Maude Ave.
Sunnyvale CA 94086
P: 408 523 9473
e-mail: Simon@
conductus.com

Dr. Shoma Sinha
CME Department - M/C 246
University of Chicago at Illinois
842 W. Taylor
Chicago IL 60607
P: 312 996 3430
F: 312 996 2426
- and -
President
Adtech Neph, Inc.
220 North Lombard Avenue
Oak Park IL 60302 2504
P: 708 386 9657

Dr. Richard E. Smalley
Rice University
Department of Chemistry
6100 Main Street, MS-100
P. O. Box 1892
Houston TX 77005
P: 713 527 4845
F: 713 285 5320

Dr. Srinivas Sridhar
Physics Department
Northeastern University
360 Huntington Avenue
Boston MA 02115
P: 617 373 2930
F: 617 373 2943
e-mail: Srinivas@neu.edu

Mr. Vance Jason Styve
Texas Center for
Superconductivity
Department of Chemisty
University of Houston
4800 Calhoun
Houston TX 77204-5641
P: 713 743 8288
F: 713 743 8281
e-mail: Fairhurst@
Elthon.tcs.uh.edu

Dr. Wu-Pei Su
Texas Center for
Superconductivity
Department of Physics
University of Houston
Houston TX 77204-5506
P: 713 743 8280
F: 713 743 8201
e-mail: WPSu@uh.edu

Dr. Masaki Suenaga
Department of Applied Physics
Brookhaven National
Laboratory
P. O. Box 5000
Building 480
76 Cornell
Upton NY 11973-5000
P: 516 344 3518
F: 516 344 4071
e-mail: mas@bnlux1.bnl.gov

Prof. Masanori Sugahara
Faculty of Engineering
Yokohama National University
156 Tokiwadai, Hodogaya
Yokohama 240
JAPAN
P: 81 45 335 1451
F: 81 45 338 1157
e-mail: sugahara@
dnj.ynu.ac.jp

Mrs. Yan Yi Sun
Texas Center for
Superconductivity
University of Houston
Houston TX 77204-5932
P: 713 743 8311
F: 713 743 8201
e-mail: YSun@uh.edu

Prof. Masashi Tachiki
Professor Emeritus
Institute for Materials Research
Tohoku University
2-1-1, Katahira, Aoba-ku
Sendai 980-77
JAPAN
P: 011 81 22 215 2005
F: 011 81 22 215 2006
e-mail: tachiki@
tacsun.imr.tohoku.ac.jp

Dr. Jamil Tahir-Kheli
Beckman Institute
Caltech 139-74
Pasadena CA 91125
P: 213 933 4232
F: 213 930 0430
e-mail: djamil@
wag.caltech.edu

Dr. Mikio Takano
Institute for Chemical Research
Kyoto University
Gokasho, Uji
Kyoto-ku 611
JAPAN
P: 81 774 32 3419
F: 81 774 32 3419
e-mail: takano@
scl.kyoto-U.ac.jp

Dr. Jeffrey Tallon
Program Manager
Materials Physics
New Zealand Institute for
Industrial Research
P. O. Box 31310
Lower Hutt
NEW ZEALAND
P: 64 4 569 0293
F: 64 4 569 0117
e-mail: j.tallon@IRL.CRI.NZ

Dr. Keiichi Tanabe
Director, Division II
Superconductivity Research
Laboratory
ISTEC
1-10-13 Shinonome-1, Koto-ku
Tokyo 135
JAPAN
P: 81 3 3536 0618
F: 81 3 3536 5714

Dr. Shoji Tanaka
Director of Superconductivity
Research Laboratory (SRL)
Vice President of International
Superconductivity Technology
Center (ISTEC)
1-10-13 Shinonome 1-chome,
Koto-ku
Tokyo 135
JAPAN
P: 81 3 3536 5700; 5703
F: 81 3 3536 5714; 5717

Dr. Y. K. Tao
Teco Electric & Machinery Co.
Ltd.
11 An-Tung Road
Chung-Li City
Taiwan
REPUBLIC OF CHINA
P: 011 886 3 452 5101
F: 011 886 3 452 5113
e-mail: yktao@
msc.nthu.edu.tw

Mr. Sahoko Tashiro
Superconductivity Research
Laboratory
ISTEC
1-10-13 Shinonome, l-chome,
Koto-ku
Tokyo 135
JAPAN
P: 011 81 3 3536 5700 or 5703
F: 011 81 3 3536 5714 or 5717

Ms. Lori Telson
EPRI
Senior Public Information
Officer
3412 Hillview Avenue
Palo Alto CA 94303
P: 415-855-2272
F: 415-855-2900
e-mail: LTelson@
msm.epri.com

Dr. Ichiro Terasaki
Superconductivity Research
Laboratory
ISTEC
1-10-13 Shinonome, 1-chome
Koto-ku, Tokyo 135
JAPAN
P: 81 3 3536 0618
F: 81 3 3536 5714

Oleg Teshernysheu
Columbia University
538 W. 100th Street
New York NY 10027
P: 212 854 5675
F: 212 854 5888
e-mail: scylla@
susol.phys.columbia.edu

Prof. Christian Thomsen
TU Berlin
Hardenbergstr. 36
Berlin 10623
GERMANY
P: 49 30 314 23187
F: 49 30 314 27705
e-mail: thomsen@
mail.Physik.tu-berlin.de

Dr. Thomas Timusk
Department of Physics and
Astronomy
McMaster University
Hamilton, Ontario L8S 4M1
CANADA
P: 905 525 9140, ext. 24290
F: 905 546 1252
e-mail: timusk@mcmaster.ca

Dr. C. S. Ting
Texas Center for
Superconductivity
University of Houston
Houston TX 77204-5932
P: 713 743 8275
F: 713 743 8201
e-mail: Ting@uh.edu

Dr. Shaw-Tsong Ting
Texas Center for
Superconductivity
University of Houston
Houston TX 77204-5932
P: 713 743 8300
F: 713 743 8301
e-mail: Ting@
bambam.tcs.uh.edu

Mr. Srinivas Tirumala
Texas Center for
Superconductivity
Department of Mechanical
Engineering
University of Houston
4800 Calhoun
Houston TX 77204-4792
P: 713 743 4547
F: 713 743 4513
e-mail: TPS@uh.edu

Dr. Nilesh Tralshawala
Texas Center for
Superconductivity
University of Houston
Houston TX 77204-5392
P: 713 743 8272
F: 713 743 8201
e-mail: Nilesh@
miller.tcs.uh.edu

Dr. C. C. Tsuei
IBM T. J. Watson Research
Center
P. O. Box 218, Rt. 134
Yorktown Heights NY 10598
P: 914 945 2799
F: 914 945 4407
e-mail: tsuei@
watson.ibm.com

Prof. Shin-ichi Uchida
Department of
Superconductivity
University of Tokyo
Hongo 7-3-1, Bunkyo-ku
Tokyo 113
JAPAN
P: 81 3 3812 2111
F: 81 3 5689 0574
e-mail: s-uchida@
tansei.cc.u-tokyo.ac.jp

Prof. Yasutomo J. Uemura
Columbia University
Department of Physics, Room
1310
538 West 120th Street
New York NY 10027
P: 212 854 8370
F: 212 854 5888
e-mail: tomo@
cusol.phys.columbia.edu

Dr. Ctirad Uher
University of Michigan
Department of Physics
505 E. University
2071 Randall Lab, 1120
Ann Arbor MI 48109
P: 313 936 0657
F: 313 763 9694
e-mail: cuher@umich.edu

Prof. D. van der Marel
Solid State Physics Lab
University of Groningen
Nijenborgh 4
9747 AG Groningen
THE NETHERLANDS
P: 31 503 637229
F: 31 503 634825
e-mail: marel@phys.rug.nl

Dr. Anne van Otterlo
ETH-Hönggerberg
Theoretische Physik
CH-8093 Zürich
SWITZERLAND
P: 41 1 633 25 73
F: 41 1 633 11 15
e-mail: avo@itp.phys.ethz.ch

Dr. Chandra M. Varma
Bell Labs
Lucent Technologies
700 Mountain Avenue
Murray Hill NJ 07974
P: 908 582 2358
F: 908 582 4702
e-mail: cmv@physics.att.com

Dr. Cumaraswamy
Vipulanandan
Texas Center for
Superconductivity
Department of Civil and
Environmental Engineering
University of Houston
Houston TX 77204-4791
P: 713 743 4278
F: 713 743 4260
e-mail: CVipulanandan@
uh.edu

Dr. V. K. Vlasko-Vlasov
Institute for Solid State Physics
Argonne National Laboratory
MSD-223, Rm. C125
9700 S. Cass Avenue
Argonne IL 60439
P: 708 252 5512
F: 708 252 7777
e-mail: vitalii_vlaso@
qmgate.anl.gov

Dr. Don Von Dollen
Electric Power Research
Institute
3412 Hillview Avenue
Palo Alto CA 94303
P: 415 855 2272
F: 415 855 2900

Mrs. Qi Wang
Texas Center for
Superconductivity
Department of Mechanical
Engineering
University of Houston
Houston TX 77204-4792
P: 713 743 8243
F: 713 743 8201

Mr. Yaqi Wang
Texas Center for
Superconductivity
University of Houston
Houston TX 77204-5932
P: 713 743 8315
F: 713 743 8221
e-mail: Wang@
dino.tcs.uh.edu

Prof. Z. D. Wang
Department of Physics
University of Hong Kong
Pokfulam Road
HONG KONG
P: 852 2859 1961
F: 852 2559 9152
e-mail: zwang@hkucc.hku.hk

Prof. Harald W. Weber
Atomic Institute of the Austrian
Universities
Schüttelstrasse 115
A-1020 Vienna
AUSTRIA
P: 43 1 72701 240, 263
F: 43 1 7289220
e-mail: weber@ati.ac.at

Prof. Meir Weger
The Hebrew University
Dacinger-B
Jerusalem 91904
ISRAEL
P: 972 2 714 512
F: 972 2 658 4437
e-mail: weger@vms.huji.ac.il

Dr. Roy Weinstein
Texas Center for
Superconductivity
Institute for Beam Particle
Dynamics
University of Houston
4800 Calhoun
Houston TX 77204-5506
P: 713 743 3600
F: 713 747 4526
e-mail: Weinstein@uh.edu

Dr. Harold Weinstock
Air Force Office of Scientific
Research
Electronic and Materials
Sciences
AFOSR/NE
110 Duncan Avenue Ste. B115
Bolling AFB DC 20332-0001
P: 202 767 4933
F: 202 767 0486
e-mail: harold.weinstock@
afosr.af.mil

Dr. Barry O. Wells
Department of Physics
Massachusetts Institute of
Technology
77 Massachusetts Avenue
Cambridge MA 02139
P: 617 253 6800
F: 617 258 6883
e-mail: bwells@x-ray.mit.edu

Dr. Zheng Yu Weng
Texas Center for
Superconductivity
University of Houston
Houston TX 77204-5932
P: 713 743 8279
F: 713 743 8201
e-mail: zyweng@uh.edu

Dr. Joseph M. Wheatley
Interdisciplinary Research
Centre in Superconductivity
University of Cambridge
Madingley Road
CB3 OHE
UNITED KINGDOM
P: 44 1223 337071
F: 44 1223 337074
e-mail: jmw19@
cus.cam.ac.uk

Dr. H. Hollis Wickman
National Science Foundation
Division of Materials Research
4201 Wilson Boulevard
Arlington VA 22230
P: 703 306 1818
F: 703 306 0902
e-mail: hwickman@nsf.gov

Mr. Philip W. Winkler
Air Products and Chemicals,
Inc.
Government Systems
7201 Hamilton Boulevard
Allentown PA 18195-1501
P: 610 481 4284
F: 610 481 2576

Dr. Stuart Wolf
Head
Materials Physics Branch
U. S. Naval Research
Laboratory
Code 6340
4555 Overlook Avenue, SW
Washington DC 20375-5345
P: 202 767 4163
F: 202 767-1697
e-mail: swolf@arpa.mil

Prof. John C. Wolfe
Texas Center for
Superconductivity
Department of Electrical and
Computer Engineering
University of Houston
Houston TX 77204-4793
P: 713 743 4449
F: 713 743 4444
e-mail: wolfe@jetson.uh.edu

Dr. Ming-Shih Wong
Texas Center for
Superconductivity
Department of Mechanical
Engineering
University of Houston
Houston TX 77204-4792
P: 713 743 8242
F: 713 743 8201
e-mail: MWong@uh.edu

Dr. Jaroslaw Wosik
Texas Center for
Superconductivity
University of Houston
Houston TX 77204-5932
P: 713 743 8237
F: 713 743 8201
e-mail: Jarek@uh.edu

Dr. Judy Z. Wu
University of Kansas
Department of Physics and
Astronomy
1082 Malott Hall
Lawrence KS 66045
P: 913 864 3240
F: 913 864 5262
e-mail: jwu@
kuphsx.phsx.ukans.edu

Prof. Maw-Kuen Wu
National Tsing Hua University
Department of Physics
Number 101, Section 2, Kuang
Fu Road
Hsinchu 30043
Taiwan
REPUBLIC OF CHINA
P: 011 886 35 717470
F: 011 886 35 729115
e-mail: mkwu@
phys.nthu.edu.tw

Dr. Nai-Juan Wu
Texas Center for
Superconductivity
Space Vacuum Epitaxy Center
University of Houston
4800 Calhoun
Houston TX 77204-5507
P: 713 743 3621
F: 713 747 7724
e-mail: naijwu@uh.edu

Dr. Harold Xia
TCSUH Levitation Laboratory
Texas Center for
Superconductivity
University of Houston
Houston TX 77201-5932
P: 713 743 8254
F: 713 843 8201
e-mail: harold@
menudo.uh.edu

Dr. Lei Ming Xie
Texas Center for
Superconductivity
University of Houston
Houston Science Center
Houston TX 77204-5932
P: 713 743 8236
F: 713 743 8201
e-mail: patxie@uh.edu

Dr. Q. Xiong
Physics Department
University of Arkansas
Fayetteville AR 72701
P: 501 575 4313
F: 501 575 4580

Mr. Xiaozhong Xiong
Texas Center for
Superconductivity
Department of Physics
University of Houston
3201 Cullen Boulevard
Houston TX 77204-5506
P: 713 743 3541
F: 714 743 3589
e-mail: xiong@
xray.phys.uh.edu

Dr. Wang Xu
Texas Center for
Superconductivity
University of Houston
Houston Science Center
Houston TX 77204-5932
P: 713 743 8313
F: 713 743 8201
e-mail: wxu@ox.tcs.uh.edu

Dr. Y. Y. Xue
Texas Center for
Superconductivity
University of Houston
Houston TX 77204-5932
P: 713 743 8310
F: 713 743 8201
e-mail: yxue@uh.edu

Mrs. Guangping Yang
Texas Center for
Superconductivity
Department of Mechanical
Engineering
University of Houston
4800 Calhoun
Houston TX 77204-4792
P: 713 743 8291
F: 713 743 8201
e-mail: gxy25603@
bayou.uh.edu

Prof. In-Sang Yang
Ewha Woman's University
Seoul 120-750
KOREA
P: 82 2 360 2332
F: 82 2 312 2367 or 82 2 360 2372
e-mail: yang@mm.ewha.ac.kr

Mr. Feng Yu
Texas Center for
Superconductivity
Department of Mechanical
Engineering
University of Houston
4800 Calhoun
Houston TX 77204-4792
P: 713 743 4547
F: 713 743 4513
e-mail: FXY5571@
jetson.uh.edu

Dr. Eli Zeldov
Department of Condensed
Matter Physics
Weizmann Institute of Science
Rehovot 71600
ISRAEL
P: 972 8 342 892
F: 972 8 344 106
e-mail: fnzeldov@
weizmann.weizmann.ac.il

Mr. Guozheng Zhang
Texas Center for
Superconductivity
Department of Mechanical
Engineering
University of Houston
Houston TX 77204-4792
P: 713 743 8243
F: 713 743 8201

Mr. Xiaoping Zhang
Research Assistant
Texas Center for
Superconductivity
University of Houston
Houston TX 77204-5932
P: 713 743 8235
F: 713 743 8201
e-mail: XPZhang@
jetson.uh.edu

Dr. Zu-Hua Zhang
Texas Center for
Superconductivity
University of Houston
4800 Calhoun
Houston TX 77204-5932
P: 713 743 8256
F: 713 743 8201
e-mail: TCSUH5@
jetson.uh.edu

Prof. Zhong-Xian Zhao
National Laboratory for
Superconductivity
Institute of Physics
Chinese Academy of Sciences
P. O. Box 603
Beijing 100 080
PEOPLE'S REPUBLIC OF
CHINA
P: 011 86 10 256 9220
F: 011 86 10 256 8834 or 2605
e-mail: zhaozx@
sun.ihep.ac.cn

Dr. Qun Zhong
Texas Center for
Superconductivity
Space Vacuum Epitaxy Center
University of Houston
4800 Calhoun
Houston TX 77204-5507
P: 713 743 3621
F: 713 747 7724
e-mail: Qun@
space.svec.uh.edu

Dr. Dawei Zhou
R&D Department
General Superconductor, Inc.
1663 Technology Avenue
Alachua FL 32615
P: 904 375 2020
F: 904 462 1414

Dr. Jianshi Zhou
Department of Materials
Science & Engineering
University of Texas at Austin
ETH-8.174
Austin TX 78712
P: 512 471 3588
F: 512 471 7681
e-mail: jszhou@
mail.utexas.edu

Mr. Tao Zhou
Max-Planck-Institute für
Festkorperforschung
Heisenbergstrasse 1
D-70509 Stuttgart
GERMANY
P: 0049 711 689 1404
F: 0049 711 689 1010
e-mail: zhou@
servix.mpi-stuttgart.mpg.de

Mr. Xingjiang Zhou
Max-Planck-Institute für
Festkörperforschung
Heisenbergstrasse 1
70569 Stuttgart
GERMANY
P: 0049 711 689 1704
F: 0049 711 689 1010
e-mail: zhou@
cardix.mpi-stuttgart.mpg.de

Mr. Jian-Xin Zhu
University of Hong Kong
Pokfulam Road
HONG KONG
P: 852 2859 2373
F: 852 2559 9152
e-mail: h9390017@
hkusua.hku.hk

Dr. Wen-Jie Zhu
Texas Center for
Superconductivity
University of Houston
Houston TX 77204-5932
P: 713 743 8300
F: 713 743 8301

Dr. X.-M. Zhu
Experimental Physics
Umeå University
S-90187 Umeå
SWEDEN
P: 46 90 167487
F: 46 90 166673

WORKSHOP PHOTOS

Marvin Cohen (UC Berkeley), Paul C. W. Chu (TCSUH), Shoji Tanaka (ISTEC), Karl Alex Müller (IBM Zürich), Bertram Batlogg (Lucent Technologies), and Arthur J. Freeman (Northwestern U)

Neal Lane (NSF)

The Hon. Tom DeLay, House Majority Whip and member of the U.S. House of Representatives, Congressional District 22

Paul M. Grant and Don Von Dollen (EPRI), Russell Eaton III (DOE), Alexis P. Malozemoff (ASC), and Mujla M. Rahman (Pirelli)

Stuart Wolf (NRL)

John Rowell (John Rowell, Inc.)

Workshop participants

Francisco de la Cruz and Daniel Lopez (Centro Atómico Bariloche) and David C. Larbalestier (U Wisconsin, Madison)

Robert J. Birgeneau (MIT)

Jianshi Zhou (UT Austin)

Ichiro Terasaki (ISTEC)

Paul C. W. Chu (TCSUH), Shoji Tanaka (ISTEC), and Karl Alex Müller (IBM Zürich) with students from St. John's School in Houston

Carl Rosner (IGC)

Herbert A Mook, Jr. (ORNL)

Manuel Cardona (Max Planck Inst.)

Robert Schrieffer (NHMFL), Paul C. W. Chu (TCSUH), Donald U. Gubser (NRL), and Wei-Kan Chu (TCSUH)

Robert Dynes (UC San Diego)

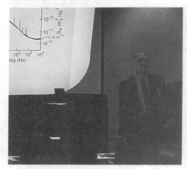

John Clarke (UC Berkeley)

Hsiao-Mei Cho and Jauyn Grace Lin (National Taiwan U) and Quan Xiong (U Arkansas)

H. C. Ku, Maw-Kuen Wu, and Dah-Chin Ling (National Tsing Hua U)

David Driscoll (Reliance Electric/Rockwell)

John R. Kirtley (IBM T. J. Watson)

Miles Klein (U Illinois), Richard L. Greene (U Maryland), and Manuel Cardona (Max Planck Inst.)

Keiichi Tanabe (ISTEC)

John T. Markert (UT Austin)

Eddie M. W. Leung (Lockheed Martin)

Jeffrey Tallon (New Zealand Inst. for Industrial Research)

Juan Carlos Campuzano (ANL)

Masashi Tachiki (Tohoku U)

Dawei Zhou (General Superconductor) and Alex Ignatiev and Qun Zhong (TCSUH)

J. T. McDevitt (UT Austin) and Harold Weinstock (AFOSR)

David C. Johnston (Iowa State U)

Wei-Yao Liang (U Cambridge) and Masaki Suenaga (BNL)

George W. Crabtree (ANL)

Richard E. Smalley (Rice U)

Alvaro Ferraz (U Brasilia)

C. C. Tsuei (IBM T. J. Watson)

Zhijun Huang (TCSUH), Maw-Kuen Wu (National Tsing Hua U), Zidan Wang (U Hong Kong), and Chao-Yuan Huang (National Taiwan U)

Aldo F. Bolza (Pirelli)

David C. Larbalestier
(U Wisconsin, Madison)

Zachary Fisk (NHMFL)

Bernard Raveau (U Caen)

Zhi-xun Shen (Stanford U)

Anthony J. Leggett (U Illinois, Urbana-Champaign)

Øystein Fischer (U Genève)

Chandra M. Varma (Lucent Technologies)

Simon Moss (TCSUH)

In-Sang Yang (Ewha Woman's U)

M. Cyrot (CNRS)

Yoichi Ando (Lucent Technologies)

Lev P. Gor'kov (NHMFL)

Walter N. Hardy (U British Columbia)

Marcel Franz (McMaster U)

Massimo Marezio (MASPEC-CMR)

Girsh E. Blumberg (U Illinois, Urbana-Champaign)

Janusz Karpinski (ETH Hönggerberg) and Brian Hickey and Ruling Meng (TCSUH)   Gabriel Aeppli (NEC)

James D. Jorgensen (ANL)   Vladimir Z. Kresin (LBNL)

Shin-ichi Uchida (U Tokyo)

Ctirad Uher (U Michigan)

Thomas Timusk (McMaster U) and D. van der Marel (U Groningen)

Kathryn Levin (U Chicago)

Judy Z. Wu (U Kansas)

John R. Clem (Iowa State U)

Michael Osofsky (NRL)

Allen M. Hermann (U Colorado, Boulder) at the TCSUH High Pressure & Low Temperature Physics Lab exhibit

Department of Energy exhibit

Intermagnetics General Corporation exhibit

J. R. Gaines, Jr. (SCI) and Paul M. Grant (EPRI) at the Superconducting Components, Inc., exhibit; background: Rodger Cooley (TCSUH) at the TCSUH Levitation Laboratory exhibit

Julia McNair (NRL) at the Naval Research Laboratory exhibit

American Superconductor Corporation exhibit

Oxford Instruments exhibit

Suvankou Sengupta (SCI) at the Superconductive Components, Inc., exhibit

Brandon Fisher (ANL) at the Argonne National Laboratory exhibit

Apurva Parikh, Sharif Salib, and Deva Ponnusamy (TCSUH) at the TCSUH HTS Manufacturing Processes Division exhibit

AUTHOR INDEX

Aeppli, G.	345	Chen, B.	213	de Lozanne, A. L.	433, 599
Afonso, S.	167	Chen, C. C.	599	de Melo, M. A. C.	389
Agafonov, A. B.	314	Chen, F.	263	Deluca, J. A.	193
Aguiar, J. A.	171, 403, 429	Chen, F. Z.	129	Demircan, E.	573
Ando, Y.	259, 320	Chen, I.-G.	629	Derro, D. J.	433
Ang, S.	167	Chen, K.	181	Dikin, D. A.	314
Annett, J. F.	63	Chen, K. Y.	167	Dilley, N. R.	280
Ao, P.	569, 573	Chen, Q.	365	Dmitriev, V. M.	314
Askenazy, S.	284	Chen, T. Y.	129	Dogan, F.	345
Athanassopoulou, N.	341	Chen, Y. C.	523	Dominguez, A. B.	385
Ausloos, M.	173, 328	Cherpak, N. T.	316	Dong, Z. C.	655
Awana, V. P. S.	171	Chi, C. C.	248	Dosquet, C. Y.	649
Bäckström, G.	417	Chikumoto, N.	139	Dow, John D.	485
Baggio-Saitovitch, E. M.	381, 383, 385, 389, 391	Cho, J. H.	373	Du, G.	179
		Choi, H.-Y.	587	Du, Z. L.	123
Bahcall, S. R.	547	Choi, Y. S.	332	Dunn, B. C.	437
Balachandran, U.	189	Chopra, M.	175	Dynes, R. C.	213
Balakirev, F. F.	318	Chou, F. C.	322, 373, 421	Dzick, J.	143
Baran, M.	97	Chou, P. C.	147	Eames, S. J.	613
Barr, A.	433	Christen, D. K.	193	Eckart, D.	89
Basov, D. N.	223	Chu, C. W.	17, 81, 93, 99, 123, 175, 183, 263, 267, 369, 387, 431	Eckstein, J. N.	103, 267, 284, 599
Bauer, P.	377			Eckstein, Y.	435
Beach, D. B.	134			Eder, R.	523
Benkraouda, M.	557	Chu, R.	353	Edwards, H.	433
Berlinsky, A. J.	535	Chu, W. K.	147, 353, 365, 602	El Massalami, M.	383
Beuven, S.	647			Elliott, A. V.	437
Bhattacharya, A.	219	Chudzik, M.	189	Ellis, D. E.	381
Bill, A.	549, 585	Clarke, J.	213	Elthon, D.	125, 131
Birgeneau, R. J.	322, 421	Claycomb, J. R.	639, 651	Emery, V. J.	451
Blackstead, H. A.	485	Clayhold, J. A.	267, 589	Endo, A.	185
Blatter, G.	561	Clem, J. R.	557	Endoh, Y.	421
Boebinger, G. S.	259, 320	Cloots, R.	173	Eremin, I. M.	517
Boffa, V.	413	Coffe, G.	284	Eremin, M. V.	517
Bolza, A.	595	Coffey, M. W.	571	Fan, J. D.	455, 489
Bonn, D. A.	223, 232	Cohn, J.	284	Fastampa, R.	413
Borsa, F.	373	Conder, K.	97, 439	Feenstra, R.	193
Boyko, V. S.	175	Cooksey, J.	167	Fendrich, J. A.	411
Bozovic, I.	103, 267, 284	Cooley, R.	602	Ferraz, A.	497
Brändström, E.	417	Cooper, J. R.	330, 341, 531	Ferreira, J. M.	429
Broto, J. M.	284	Corti, M.	373	Field, M. B.	405
Brown, W. D.	167	Coulter, J. Y.	203	Fisher, I. R.	330
Buan, J.	219	Crabtree, G. W.	411, 415	Fisk, Z.	107
Bud'ko, S. L.	391	Crow, J. E.	326	Florence, G.	167
Bulut, N.	519	Cui, X. T.	365	Fontes, M. B.	391
Cao, G.	326	Cyrot, M.	581	Foster, C.	625
Cao, Y.	123, 263, 431	da Silva, E. F., Jr.	429	Fowler, P.	602
Cardona, M.	72, 367, 369	Dagotto, E.	447	Franz, M.	535
Carretta, P.	373	Dai, P.	345	Freeman, A. J.	459
Cava, R. J.	259	Dantsker, E.	213	Frenkel, R. M.	589
Chaillout, C.	113	Darula, M.	647	Freyhardt, H. C.	143
Chan, F. T.	167	de Albuquerque, J. C. C.	429	Fukuyama, H.	521
Chan, S.-W.	175			Gajewski, D. A.	213
Char, K.	213	de la Cruz, F.	395, 411	Gao, L.	81, 123, 263

Gapud, A.	89	
Garcia-Moreno, F.	143	
Gautier-Picard, P.	183	
Genoud, J.-Y.	228	
Geny, J.	125	
Gerber, C.	232, 236	
Geshkenbein, V.	561	
Giapintzakis, J.	425	
Gierlowski, P.	337, 653	
Ginsburg, D. M.	425	
Giura, M.	413	
Gohng, J.	649	
Goldenfeld, N. D.	63	
Goldman, A. M.	219	
Goldschmidt, D.	435	
Golubnichaya, G. V.	316	
Goodenough, J. B.	310	
Gooding, R. J.	322, 373, 525	
Gor'kov, L. P.	501	
Goyal, A.	193	
Greene, R.	284	
Grupp, D.	219	
Gu, C.	213	
Gu, G. D.	288	
Guan, W. Y.	324	
Gubser, D. U.	606	
Guenzburger, D.	381	
Guha, S.	318	
Gunter, D.	415	
Gupta, A.	248	
Ha, Y. S.	332	
Haldar, P.	189	
Hammel, P. C.	107	
Han, S. H.	213, 280	
Han, X.-Y.	335	
Haneji, N.	335	
Hardy, W. N.	223, 232	
Harris, J. M.	240	
Haugan, T.	621	
He, Q.	193	
Heinemann, K.	143	
Henning, P. F.	326	
Hermann, A. M.	271	
Herrmann, J.	280	
Hervieu, M.	117	
Hickey, B. R.	81, 93	
Hidaka, Y.	244	
Hilgenkamp, H.	232, 236	
Hirayama, T.	159	
Hizhnyakov, V.	549, 585	
Ho, K. M.	575	
Hofer, J.	377	
Hoffmann, J.	143	
Homes, C. C.	223	
Hong, J. P.	649	
Honkala, K.	252, 487	
Hor, P. H.	387	
Houssa, M.	328	
Hsu, D. Y.	127	
Hsu, J.-C.	629	
Hsu, Y. Y.	127	
Hu, C.-R.	551	
Huang, C. C.	219	
Huang, C. Y.	276, 387	
Huang, Z. J.	276	
Hur, N. H.	361, 409	
Hussey, N. E.	330	
Ichikawa, N.	320	
Ignatiev, A.	147	
Ikuhara, Y.	159	
Ikuta, H.	151, 407	
Isaacs, E.	425	
Israeloff, N.	219	
Iyer, A. N.	189	
Izhyk, E. V.	316	
Jacobs, T.	288	
Jammy, R.	189	
Jamn, G.	629	
Jeandupeux, O.	349	
Jiang, X.	284	
Jiao, C.	653	
Johnston, D. C.	373	
Jones, C. E.	177	
Junod, A.	228	
Jurbergs, D. C.	613	
Kabanov, V.	415	
Kaldis, E.	367	
Kallin, C.	535	
Kallio, A.	252, 487	
Kamal, S.	223	
Kaneda, H.	335	
Kang, B. W.	89	
Kang, W. N.	89	
Kao, C. H.	205	
Karimoto, S.	244	
Karpinski, J.	97, 367, 377, 439	
Kastner, M. A.	421	
Katz, A. S.	213	
Keller, H.	377	
Kes, P. H.	197, 399	
Khaykovich, B.	399	
Kiehl, W.	271	
Kim, M.-S.	409	
Kimura, T.	259, 320	
Kirichenko, A. Y.	316	
Kirtley, J. R.	232, 236, 248	
Kishio, K.	85, 151, 259, 320, 407	
Kitazawa, K.	85, 151, 407	
Kivelson, S. A.	451	
Klabunde, C. E.	193	
Kleiner, R.	213	
Kohno, H.	521	
Konczykowski, M.	399	
Koshizuka, N.	288	
Kotsis, I.	631	
Koyano, T.	433	
Kresin, V. Z.	463	
Kroeger, D. M.	193	
Ku, H. C.	127	
Kumar, P.	501	
Kunkel, G.	647	
Kwok, W. K.	411	
Larbalestier, D. C.	41, 405	
Lascialfari, A.	373	
Lee, E.-H.	649	
Lee, H.-G.	361	
Lee, J. H.	332	
Lee, J.-W.	649	
Lee, S.-I.	361, 409	
Lee, T. K.	523	
Lee, Y. S.	421	
Legendre, F.	203	
Leggett, A. J.	63	
Leung, E. M.	610	
Leung, P. W.	525	
Levin, K.	467	
Li, D.	337	
Li, Q.	288	
Li, T. W.	197, 399	
Liang, R. X.	223, 232	
Liang, W. Y.	341	
Lin, H. Q.	523	
Lin, J.	254	
Lin, J. G.	276, 387	
Lin, M. L.	276	
Lin, Q. M.	93, 99, 369, 431	
Lin, S. R.	127	
Lindenfeld, P.	318	
Ling, D. C.	129	
List, F. A.	193	
Litterst, F. J.	389	
Liu, D. Z.	467	
Liu, J.	625	
Liu, J. R.	147, 353, 365	
Lo, R.-K.	181	
Loeser, A. G.	240	
Long, S. A.	337	
López, A.	389	